HALF-LINEAR DIFFERENTIAL EQUATIONS

NORTH-HOLLAND MATHEMATICS STUDIES 202

ELSEVIER

2005

Amsterdam – Boston – Heidelberg – London – New York – Oxford
Paris – San Diego – San Francisco – Singapore – Sydney – Tokyo

HALF-LINEAR DIFFERENTIAL EQUATIONS

ONDŘEJ DOŠLÝ
Department of Mathematics
Masaryk University
Brno, Czech Republic

PAVEL ŘEHÁK
Mathematical Institute
Academy of Sciences of the Czech Republic
Brno, Czech Republic

ELSEVIER

2005

Amsterdam – Boston – Heidelberg – London – New York – Oxford
Paris – San Diego – San Francisco – Singapore – Sydney – Tokyo

ELSEVIER B.V.
Radarweg 29
P.O. Box 211, 1000 AE Amsterdam
The Netherlands

ELSEVIER Inc.
525 B Street
Suite 1900, San Diego
CA 92101-4495, USA

ELSEVIER Ltd.
The Boulevard
Langford Lane, Kidlington,
Oxford OX5 1GB, UK

ELSEVIER Ltd.
84 Theobalds Road
London WC1X 8RR
UK

First edition 2005

Library of Congress Cataloging in Publication Data
A catalog record is available from the Library of Congress.

British Library Cataloguing in Publication Data
A catalogue record is available from the British Library.

ISBN: 0-444-52039-2
ISSN: 0304-0208 (Series)

∞ The paper used in this publication meets the requirements of ANSI/NISO Z39.48-1992 (Permanence of Paper).
Printed in The Netherlands.

PREFACE

The principal concern of this book is the qualitative theory of the half-linear second order differential equation

$$\text{(HL)} \qquad (r(t)\Phi(x'))' + c(t)\Phi(x) = 0, \quad \Phi(x) := |x|^{p-1} \operatorname{sgn} x, \quad p > 1.$$

The investigation of (HL) has attracted considerable attention in the last two decades. Among others, it was shown that solutions of this equation behave in many aspects like those of the Sturm-Liouville differential equation

$$\text{(SL)} \qquad (r(t)x')' + c(t)x = 0,$$

which is the special case of (HL) when $p = 2$. The aim of this book is to present the substantial results of this investigation. The main attention is focused to the oscillation theory and asymptotic theory of (HL), and to boundary value problem associated with this equation. Note that the term *half-linear* equations is motivated by the fact that the solution space of (HL) has just one half of the properties which characterize linearity, namely homogeneity (but not additivity).

The investigation of qualitative properties of nonlinear second order differential equations has a long history. Recall here only the papers of Emden [151], Fowler [166], Thomas [349], and the book of Sansone [336] containing the survey of the results achieved in the first half of the last century. In the fifties and the later decades, the number of papers devoted to nonlinear second order differential equations increased rapidly, so we mention here only treatments directly associated with (HL). Even if some ideas concerning the properties of solutions of (HL) can already be found in the papers of Bihari [36, 38], Elbert and Mirzov with their papers [139, 290] are the ones usually regarded as pioneers of the qualitative theory of (HL).

In later years, in particular in the nineties, the striking similarity between properties of solutions of (HL) and of (SL) was revealed. On the other hand, various problems associated with (HL) have been indicated, where the situation is completely different comparing the linear and half-linear case, and where the absence of the additivity of the solution space of (HL) brings completely new phenomena.

One of the reasons for the research in the field of half-linear equations is that variety of physical, biological, and chemical phenomena (let us mention at least non-Newtonian fluid theory or some models in glaceology) are described by the partial differential equations with the so-called p-Laplacian, and this PDE's can be reduced under some assumptions to the ODE's of the form (HL). Another reason can be "purely mathematical", it is natural to ask what results of the deeply developed qualitative theory of (SL) can be extended to (HL).

The book is divided into 9 chapters. The first one deals with classical topics like the existence, uniqueness, and the Sturmian theory for (HL). In particular, it is shown that the linear Sturmian theory extends verbatim to (HL). An attention is also focused to some elementary half-linear differential equations. The next two chapters are devoted to the oscillation theory of (HL). Chapter 2 presents basic methods of the half-linear oscillation theory which are essentially the same as in the linear case (variational principle, Riccati technique and comparison theorems). Chapter 3 then delas with the particular oscillation and nonoscillation criteria. These criteria are sorted according to the method used in their proofs and according to the types of conditions that are involved.

Chapter 4 presents a very recent material concerning nonoscillatory solutions of (HL). In particular, this chapter deals with the asymptotic analysis of nonoscillatory solutions of (HL) and with properties of the so-called principal solution of (HL). The largest chapter of this book is Chapter 5, where we collected various statements related to half-linear oscillation theory. Among them, let us mention at least the half-linear Sturm-Liouville problem and the theory of half-linear equations with almost periodic coefficients. Chapters 6 and Chapter 7 contain mostly the results concerning boundary value problems associated with (HL) and with the partial differential equations with p-Laplacian. The main attention is focused to the Fredholm alternative for the the so-called resonant BVP's. Note that this is an example of the problem where linear and half-linear cases are completely different. Chapter 8 presents basic facts of the recently established discrete half-linear oscillation theory and the results presented there can be understood as discrete counterparts of some statements presented in previous chapters. The last chapter collects miscellaneous material related to half-linear equations, like more general differential equations, functional differential equations and various inequalities related to (HL).

The origin of this book lies in Chapter 3 of the Handbook of Differential Equations [56] written by the first author. Here, comparing with that chapter in the Handbook, all material is elaborated in more details and the whole treatment is approximately three times larger. Also, comparing our book with Chapter 3 of the recent monograph [6] (this book is devoted to the oscillation theory of various differential equations), there are some common points, but the most part of our presentation differs from that of [6]. We present here a systematic approach to the qualitative theory of half-linear differential equations, while [6] is directed only to oscillation problems.

The book is addressed to a wide audience of mathematicians interested in differential equations and related topics. It can be used as a textbook at the graduate level and as the reference book in the field of half-linear differential equations.

The preparation of this book was supported by the grants KJB1019407 and A1163401/04 of the Grant Agency of the Czech Academy of Sciences, by the Research Project MSM0021622409 of the Czech Government, by the Academy of Sciences of the Czech Republic, Institutional Research Plan AV0Z10190503, and by the grant G201/04/0580 of the Czech Grant Agency.

Finally, we would like to express our thanks to Mariella Cecchi, Zuzana Došlá, Jaroslav Jaroš, Mauro Marini, Robert Mařík, and Karel Šrot for their comments, suggestions and help in various periods of the preparation of this book.

April 2005 Ondřej Došlý & Pavel Řehák

Contents

CHAPTER 1

BASIC THEORY

In this first chapter we present basic properties of solutions of the half-linear second order differential equation

$$(1.1.1) \qquad (r(t)\Phi(x'))' + c(t)\Phi(x) = 0, \quad \Phi(x) := |x|^{p-1} \operatorname{sgn} x, \; p > 1,$$

which is the main concern of this book. Recall that the terminology *half-linear* differential equation (systematically used by Bihari and Elbert for the first time) reflects the fact that the solution space of (1.1.1) is homogeneous, but not additive.

We suppose that the functions r, c are continuous and $r(t) > 0$ in the interval under consideration. Note that the most of our results can be formulated under weaker assumptions that the functions $1/r$, c are locally integrable. However, since we are interested in solutions of (1.1.1) in the classical sense (i.e., a solution x of (1.1.1) is a C^1 function such that $r\Phi(x') \in C^1$ and satisfies (1.1.1) in an interval under consideration), the continuity assumption is appropriate for this setting.

As we have already mentioned in the Preface, half-linear equations are closely related to the partial differential equations with p-Laplacian. In fact, (1.1.1) is sometimes called the *differential equation with the one-dimensional p-Laplacian*. Recall that the p-Laplacian is a partial differential operator of the form

$$\Delta_p u := \operatorname{div}\left(\|\nabla u\|^{p-2}\nabla u\right),$$

where (for $u = u(x) = u(x_1, \dots, x_N)$) $\nabla u = \left(\frac{\partial}{\partial x_1}, \dots, \frac{\partial}{\partial x_N}\right)$ is the Hamilton nabla operator and (for $v(x) = (v_1(x), \dots, v_N(x))$ $\operatorname{div} v(x) = \sum_{i=1}^{N} \frac{\partial v}{\partial x_i}(x)$ is the usual divergence operator. If u is a radially symmetric function, i.e., $u(x) = y(t)$, $t = \|x\|$, $\|\cdot\|$ being the Euclidean norm in $\mathbb{R}^N$, the (partial) differential operator Δ_p can be reduced to the ordinary differential operator

$$\Delta_p u(x) = t^{1-N}\left(t^{N-1}\Phi(y'(t))\right)', \quad ' = \frac{d}{dt}.$$

In the first section of this chapter we deal with the existence and the unique solvability of (1.1.1). The principal tool is the half-linear Prüfer transformation, we also offer an alternative approach to the existence and uniqueness problem. In Section 1.2 we present the basic oscillatory properties of (1.1.1), in particular, we show that the linear Sturmian oscillation theory extends almost verbatim to half-linear equations, and that basic concepts of the linear oscillation theory as Picone's identity, Riccati equation and other concepts have a natural half-linear extension. In Section 1.3 we show basic differences between second order linear and half-linear equations. The last section of this chapter deals with some special half-linear equations. In particular, we present basic properties of solution of half-linear equations with constant coefficients and of Euler-type half-linear equation.

1.1 Existence and uniqueness

The first section of this chapter is devoted to the existence and uniqueness theory for (1.1.1). We introduce the half-linear trigonometric functions and the half-linear Prüfer transformation which are the basic tools used in the proof of the existence and uniqueness statement for (1.1.1). We also deal with the relationship of half-linear equation (1.1.1) to various similar equations and systems. An alternative approach to the uniqueness theory (based on a Gronwall type lemma) is discussed as well.

1.1.1 First order half-linear system and other forms of half-linear equations

Consider the Sturm-Liouville *linear* differential equation

$$(r(t)x')' + c(t)x = 0, \tag{1.1.2}$$

which is a special case $p = 2$ in (1.1.1). Then, given $t_0, x_0, x_1 \in \mathbb{R}$, there exists the unique solution of (1.1.2) satisfying the initial conditions $x(t_0) = x_0$, $x'(t_0) = x_1$, which is extensible over the whole interval where the functions r, c are continuous and $r(t) > 0$. This follows e.g. from the fact that (1.1.2) can be written as the 2-dimensional first order linear system

$$x' = \frac{1}{r(t)}u, \quad u' = -c(t)x,$$

and the linearity (hence the Lipschitz property) of this system implies the unique solvability of (1.1.2). On the other hand, if we rewrite (1.1.1) into the first order system (substituting $u = r\Phi(x')$), we get the system

$$x' = r^{1-q}(t)\Phi^{-1}(u), \quad u' = -c(t)\Phi(x), \tag{1.1.3}$$

where q is the conjugate number of p, i.e., $1/p + 1/q = 1$, and Φ^{-1} is the inverse function of Φ. The right hand-side of (1.1.3) is no longer Lipschitzian in x, u, hence the standard existence and uniqueness theorems do not apply directly to this system. Moreover, it is known that the Emden-Fowler type differential equation

$$(r(t)|x'|^{\alpha-1}\operatorname{sgn} x')' + p(t)|x|^{\beta-1}\operatorname{sgn} x = 0, \quad \alpha,\beta > 1, \tag{1.1.4}$$

(which looks similarly to (1.1.1), in a certain sense) may admit the so-called singular solutions (see Section 9.1 and e.g. the books [202, 292]) i.e., solutions which violate uniqueness and continuability of solutions of (1.1.4).

Half-linear differential equations occur in various forms in the literature. Elbert (and some of his collaborators) mostly considered equations of the form

$$(r(t)(x')^{n*})' + c(t)x^{n*} = 0, \tag{1.1.5}$$

where $u^{n*} := |u|^n \operatorname{sgn} u$ with $n > 0$. The alternative notation for $n*$ is n^* or $\overset{*}{n}$. Clearly, (1.1.5) is equivalent to (1.1.1) when $n = p - 1$, and so there is, in fact, no difference between (1.1.1) and (1.1.5). Note that the equations of these forms appear in the absolute majority of the quoted papers dealing with half-linear differential equations. Another form (considered in particular in [70, 197, 262], but also elsewhere) is

$$u'' + a(t)|u|^{\alpha}|u'|^{1-\alpha}\operatorname{sgn} u = 0, \tag{1.1.6}$$

where $\alpha \in (0, 1]$ and $a(t)$ is a real function. When taking so-called proper solutions, see the beginning of Section 3.3, this equation is equivalent to (1.1.1) (i.e., their solution spaces are the same) provided $p = \alpha + 1$, $p \in (1, 2]$, $r(t) \equiv 1$ and $(p-1)c(t) = a(t)$, in view of the identity $(\Phi(x'))' = (p-1)x''|x'|^{p-2}$. Observe also that the form (1.1.1) enables to consider the case $\alpha = p - 1 > 1$, and many of the results for (1.1.6) can be extended in this direction without difficulties. Sometimes, slightly more general forms than (1.1.1) are considered, e.g.,

$$(R(t)x')' + Q(t)f(x, R(t)x') = 0, \tag{1.1.7}$$

where R, Q are continuous real functions and $R(t) > 0$. At least two sets of restrictions, which are imposed on $f(x, y)$, appear in the literature:

(i) Bihari [36] required $f(x, y)$ such that it is defined on $\mathbb{R}^2$ and is Lipschitzian on every bounded domain in $\mathbb{R}^2$, $xf(x, y) > 0$ if $x \neq 0$ (consequently, $f(0, y) = 0$ for all $y \in \mathbb{R}$), $f(\lambda x, \lambda y) = \lambda f(x, y)$ for all $\lambda \in \mathbb{R}$, $(x, y) \in \mathbb{R}^2$.

(ii) In [143, 145] it is assumed that $f(x, y)$ is defined and continuous on $\mathbb{R} \times \mathbb{R}_0$, $\mathbb{R}_0 = [0, \infty)$, $xf(x, y) > 0$ if $xy \neq 0$, $f(\lambda x, \lambda y) = \lambda f(x, y)$ for $\lambda \in \mathbb{R}^+$, $(x, y) \in \mathbb{R} \times \mathbb{R}_0$, the functions $F_+(t), F_-(t)$ defined by $F_+(t) = tf(t, 1)$, $F_-(t) = -tf(-t, -1)$ satisfy the relations

$$\int_{-\infty}^{\infty} \frac{dt}{1 + F_+(t)} < \infty, \quad \int_{-\infty}^{\infty} \frac{dt}{1 + F_-(t)} < \infty, \quad \lim_{|t|\to\infty} F_{\pm}(t) = \infty,$$

the functions $\log(1 + F_+(t))$ and $\log(1 + F_-(t))$ are uniformly Lipschitzian on $\mathbb{R}$. Some strange effects may occur for (1.1.7) satisfying the latter conditions: For example, it may have an eventually positive solution, but it has no eventually negative solution, or the zeros of two (linearly independent) solutions do not necessarily separate each other (observe that the latter effect may occur also for equation (1.1.6), see the beginning of Section 3.3). However, all this "deviations" disappear if $f(-x,-y) = -f(x,y)$, which is sometimes assumed there, see also Remark 1.3.1 below. A nontrivial example when the last relation holds is $f(x,y) = \Phi(x)|y|^{2-p}$. Then (1.1.7) reads as $(R(t)x')' + Q(t)\Phi(x)|R(t)x'|^{2-p} = 0$ or, equivalently, $|R(t)x'|^{p-2}(R(t)x')' + Q(t)\Phi(x) = 0$ or $(R^{p-1}(t)\Phi(x'))' + (p-1)Q(t)\Phi(x) = 0$, which is an equation of the form (1.1.1). Another approach to examining second order half-linear equations is to consider a (more general) first order system (studied by Mirzov, see [290, 291, 292] and also Subsection 1.2.9) of the form

$$u_1' = a_1(t)|u_2|^{\lambda_1} \operatorname{sgn} u_2, \quad u_2' = -a_2(t)|u_1|^{\lambda_2} \operatorname{sgn} u_1, \tag{1.1.8}$$

where $\lambda_i > 0$ $(i = 1, 2)$ with $\lambda_1\lambda_2 = 1$ are constants and a_1, a_2 are real functions. Its relation to equation (1.1.1) becomes apparent looking at system (1.1.3). Finally note that the half-linear system

$$y' = b_1(t)y + b_2(t)z^{(1/n)*}, \quad z' = -b_3(t)y^{n*} + b_4(t)z$$

appeared in [141, 142], and it is closely related to the equation with a damping term

$$(A(t)\Phi(x'))' + B(t)\Phi(x') + C(t)\Phi(x) = 0,$$

which is also considered sometimes.

1.1.2 Half-linear trigonometric functions

In proving the existence and uniqueness result for (1.1.1), the fundamental role is played by the generalized Prüfer transformation introduced in [139]. Consider a special half-linear equation of the form (1.1.1)

$$(\Phi(x'))' + (p-1)\Phi(x) = 0 \tag{1.1.9}$$

and denote by $S = S(t)$ its solution given by the initial conditions $S(0) = 0$, $S'(0) = 1$. We will show that the behavior of this solution is very similar to that of the classical sine function. Multiplying (1.1.9) (with x replaced by S) by S' and using the fact that $(\Phi(S'))' = (p-1)|S'|^{p-2}S''$, we get the identity $[|S'|^p + |S|^p]' = 0$. Substituting $t = 0$ and using the initial condition for S, we have the generalized Pythagorian identity

$$|S(t)|^p + |S'(t)|^p \equiv 1. \tag{1.1.10}$$

The function S is positive in some right neighborhood of $t = 0$ and using (1.1.10), $S' = \sqrt[p]{1 - S^p}$, i.e., $\frac{dS}{\sqrt[p]{1-S^p}} = dt$ in this neighborhood, hence

$$t = \int_0^{S(t)} (1 - s^p)^{-\frac{1}{p}} \, ds. \tag{1.1.11}$$

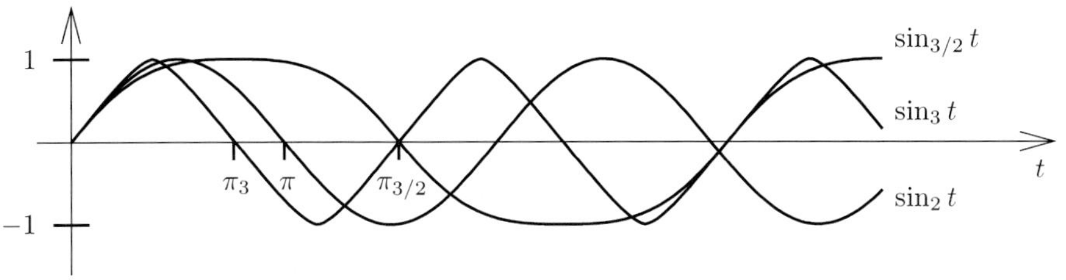

Figure 1.1.1: Generalized sine functions for $p = 3/2$, $p = 2$, and $p = 3$

Following the analogy with the case $p = 2$, we denote

$$\begin{aligned}\frac{\pi_p}{2} &= \int_0^1 (1-s^p)^{-\frac{1}{p}}\,ds = \frac{1}{p}\int_0^1 (1-u)^{-\frac{1}{p}}u^{-\frac{1}{q}}\,du \\ &= \frac{1}{p}B\left(\frac{1}{p},\frac{1}{q}\right),\end{aligned}$$

where

$$B(x,y) = \int_0^1 t^{x-1}(1-t)^{y-1}\,dt$$

is the Euler beta function. Using the formulas

$$B(x,y) = \frac{\Gamma(x)\Gamma(y)}{\Gamma(x+y)}, \quad \Gamma(x)\Gamma(1-x) = \frac{\pi}{\sin \pi x}$$

with the Euler gamma function $\Gamma(x) = \int_0^\infty t^{x-1}\,e^{-t}\,dt$, we have

$$\pi_p = \frac{2\pi}{p\sin\frac{\pi}{p}}. \tag{1.1.12}$$

The formula (1.1.11) defines uniquely the function $S = S(t)$ on $[0, \pi_p/2]$ with $S(\pi_p/2) = 1$ and hence by (1.1.10) $S'(\pi_p/2) = 0$. Now we define the *generalized sine function* $\sin_p$ on the whole real line as the odd $2\pi_p$ periodic continuation of the function

$$\begin{cases} S(t), & 0 \le t \le \frac{\pi_p}{2}, \\ S(\pi_p - t), & \frac{\pi_p}{2} \le t \le \pi_p. \end{cases}$$

The *generalized cosine function* $\cos_p$ is defined as $\cos_p t = (\sin_p t)'$. The functions $\sin_p$ and $\cos_p$ reduce to the classical functions sin and cos, respectively, in case $p = 2$. The generalized Pythagorian identity

$$|\sin_p t|^p + |\cos_p t|^p \equiv 1 \tag{1.1.13}$$

holds.

It is not difficult to see that $\sin_p$ is a bijective mapping of $[-\pi_p/2, \pi_p/2]$ onto $[-1,1]$. Hence there exists its inverse function; we denote it by $\arcsin_p$. Similarly, via $\cos_p$, we introduce the half-linear extension of arccosine function, denoted by arccos_p.

In addition, we introduce the half-linear tangent and cotangent functions $\tan_p$ and $\cot_p$ by

$$\tan_p t = \frac{\sin_p t}{\cos_p t}, \quad \cot_p t = \frac{\cos_p t}{\sin_p t}.$$

The function $\tan_p$ is periodic with the period π_p and has discontinuities at $\pi_p/2 + k\pi_p$, $k \in \mathbb{Z}$. The function $\cot_p$ is also π_p periodic, with discontinuities at $t = k\pi_p$, $k \in \mathbb{Z}$. By (1.1.9) and (1.1.13) we have

$$(\tan_p t)' = \frac{1}{|\cos_p t|^p} = 1 + |\tan_p t|^p, \quad (\cot_p t)' = -|\cot_p t|^{2-p}(1 + |\cot_p t|^p). \tag{1.1.14}$$

Hence $(\tan_p t)' > 0$, $(\cot_p t)' < 0$ on their definition domains, and there exist the functions $\arctan_p$, arccot_p that are defined as inverse functions of $\tan_p$ and $\cot_p$ in the domains $(-\pi_p/2, \pi_p/2)$ and $(0, \pi_p)$, respectively. From (1.1.14) we have

$$(\arctan_p t)' = \frac{1}{1 + |t|^p}.$$

Remark 1.1.1. (i) The definition of the "half-linear" π_p just introduced comes from the original paper of Elbert [139]. One can find a slightly different definition in the literature, namely,

$$\pi_p := 2(p-1)^{1/p} \int_0^1 \frac{ds}{(1-s)^{1/p}}, \tag{1.1.15}$$

see e.g. [89]. This difference is caused by the fact that instead of (1.1.9), the initial value problem

$$(\Phi(x'))' + \Phi(x) = 0, \quad x(0) = 0, \ x'(0) = 1$$

is taken as defining for the half-linear sine and cosine functions. Note also that if x solves the last initial value problem, then we have the identity $|x'|^p + |x|^p/(p-1) \equiv 1$ instead of (1.1.13) in this modified setting. However, the definition of π_p by (1.1.12) is now more common, even if sometimes the definition (1.1.15) is still used in some works, see e.g. [331, 332].

(ii) Interesting results concerning an extension of the sine and cosine functions can be found in [259] (see also [260]). For $m, k > 0$, the functions $S = S_{m/k}(\xi)$ and $C = C_{m/k}(\xi)$ are defined (implicitly) as the inverses of the Abelian integrals

$$\xi = \int_0^S \frac{dx}{(1-x^k)^{m/k}} \quad \text{and} \quad \xi = -\int_1^C \frac{dx}{(1-x^k)^{m/k}}.$$

Among others, the following two identities are proved

$$S_{m/k}^k(\xi) + C_{n/l}^l(\eta) = 1 \ \text{provided} \ \frac{m}{k} + \frac{1}{l} = 1, \ \frac{n}{l} + \frac{1}{k} = 1, \ k\xi = l\eta, \tag{1.1.16}$$

and

$$\left(1 + S_{2/k}^{k/2}(\xi)\right)\left(1 + C_{2/l}^{l/2}(\eta)\right) = 2 \text{ provided } \frac{1}{k} + \frac{1}{l} = \frac{1}{2},\ l^{4^l\eta} = k^{4^k\xi}.$$

Clearly, a special choice of k, l, m, n in (1.1.16) yields the generalized Pythagorian identity (1.1.13).

1.1.3 Half-linear Prüfer transformation

Using the above defined generalized trigonometric functions and their inverse functions, we can introduce the generalized Prüfer transformation as follows. First note that this transformation will be very useful in proving many qualitative results for (1.1.1), including the existence and uniqueness of the initial value problem. Let x be a nontrivial solution of (1.1.1). Put

$$\rho(t) = \sqrt[p]{|x(t)|^p + r^q(t)|x'(t)|^p}$$

and let φ be a continuous function defined at all points where $x(t) \neq 0$ by the formula

$$\varphi(t) = \operatorname{arccot}_p \frac{r^{q-1}(t)x'(t)}{x(t)},$$

where q is the conjugate number of p, i.e., $1/p + 1/q = 1$. Hence

$$x(t) = \rho(t)\sin_p \varphi(t), \quad r^{q-1}(t)x'(t) = \rho(t)\cos_p \varphi(t). \tag{1.1.17}$$

Differentiating the first equality in (1.1.17) and comparing it with the second one we get

$$r^{1-q}(t)\rho(t)\cos_p \varphi(t) = \rho'(t)\sin_p \varphi(t) + \rho(t)\cos_p(\varphi(t))\varphi'(t). \tag{1.1.18}$$

Similarly, applying the function Φ to both sides of the second equation in (1.1.17), differentiating the obtained identity and substituting from (1.1.1) we get

$$\begin{aligned} &-c(t)\rho^{p-1}(t)\Phi(\sin_p \varphi(t)) \\ &\quad = (p-1)\left[\rho^{p-2}(t)\rho'(t)\Phi(\cos_p \varphi(t)) - \rho^{p-1}(t)\Phi(\sin_p \varphi(t))\varphi'(t)\right]. \end{aligned} \tag{1.1.19}$$

Now, multiplying (1.1.18) by $\Phi(\cos_p \varphi)/\rho$, (1.1.19) by $\sin_p \varphi/\rho^{p-1}$ and combining the obtained equations we get the first order system for φ and ρ

$$\begin{aligned} \varphi' &= \frac{c(t)}{p-1}\left|\sin_p \varphi\right|^p + r^{1-q}(t)\left|\cos_p \varphi\right|^p, \\ \rho' &= \Phi(\sin_p \varphi(t))\cos_p \varphi(t)\left[r^{1-q}(t) - \frac{c(t)}{p-1}\right]\rho. \end{aligned} \tag{1.1.20}$$

Remark 1.1.2. Similarly to the definition of π_p, also the half-linear Prüfer transformation appears in the literature in various modifications. The definition presented here has been established by Elbert [139]. Some of these modifications will appear later in this book, see e.g. Subsection 5.6.1 and the section devoted to half-linear Sturm-Liouville problems. For some other modifications we refer to [54, 268, 331] and the references given therein.

1.1.4 Half-linear Riccati transformation

In the alternative proof of the uniqueness result the following transformation will find the application. Later we will see its extreme usefulness in the oscillation theory of (1.1.1). Let x be a solution of (1.1.1) such that $x(t) \neq 0$ in an interval I. Then $w(t) = r(t)\Phi(x'(t))/\Phi(x(t))$ is a solution of the Riccati type differential equation (or the generalized Riccati differential equation)

$$w' + c(t) + (p-1)r^{1-q}(t)|w|^q = 0, \tag{1.1.21}$$

where q is the conjugate number of p, i.e., $q = \frac{p}{p-1}$. Indeed, in view of (1.1.1) we have

$$\begin{aligned} w' &= \frac{(r\Phi(x'))'\Phi(x) - (p-1)r\Phi(x')|x|^{p-2}x'}{\Phi^2(x)} = -c - (p-1)\frac{r|x'|^p}{|x|^p} \\ &= -c - (p-1)r^{1-q}|w|^q. \end{aligned}$$

Remark 1.1.3. Using the above Riccati equation (1.1.21) one can derive the first equation in (1.1.20) as follows. From (1.1.17) we have

$$w = \frac{r(t)\Phi(x'(t))}{\Phi(x(t))} = \frac{\Phi(\cos_p \varphi(t))}{\Phi(\sin_p \varphi(t))} =: v(\varphi(t)).$$

The function $v(t) = \Phi(\cos_p t)/\Phi(\sin_p t)$ satisfies the Riccati equation corresponding to (1.1.9). This implies

$$\begin{aligned} [v(\varphi(t))]' &= v'(\varphi(t))\varphi'(t) = \left[-(p-1) - (p-1)\left|\frac{\Phi(\cos_p \varphi(t))}{\Phi(\sin_p \varphi(t))}\right|^q\right]\varphi'(t) \\ &= -(p-1)\left[1 + \left|\frac{\cos_p \varphi(t)}{\sin_p \varphi(t)}\right|^p\right]\varphi'(t), \end{aligned}$$

Substituting from (1.1.21)

$$w'(t) = -c(t) - (p-1)r^{1-q}(t)|w(t)|^q = -c(t) - (p-1)r^{1-q}(t)\left|\frac{\cos_p \varphi}{\sin_p \varphi}\right|^p,$$

and hence

$$c(t) + (p-1)r^{1-q}(t)\left|\frac{\cos_p \varphi(t)}{\sin_p \varphi(t)}\right|^p = (p-1)\left[1 + \left|\frac{\cos_p \varphi(t)}{\sin_p \varphi(t)}\right|^p\right]\varphi'(t).$$

Multiplying this equation by $|\sin_p \varphi(t)|^p$ and using (1.1.13) we get really the first equation in (1.1.20).

1.1.5 Existence and uniqueness theorem

Since the right-hand side of system (1.1.20) is Lipschitzian in both ρ and φ, the initial value problem for this system is uniquely solvable and its solution exists on the whole interval where r, c are continuous and $r(t) > 0$. Hence, the same holds for (1.1.1). This statement is summarized in the next theorem.

Theorem 1.1.1. *Suppose that the functions r, c are continuous in an interval $I \subseteq \mathbb{R}$ and $r(t) > 0$ for $t \in I$. Given $t_0 \in I$ and $A, B \in \mathbb{R}$, there exists a unique solution of* (1.1.1) *satisfying $x(t_0) = A$, $x'(t_0) = B$ which is extensible over the whole interval I. This solution depends continuously on the initial values A, B.*

Remark 1.1.4. Note that the uniqueness of the IVP for half-linear equation (1.1.1) is a subtle problem as pointed out in [331]. In that paper, it is shown that the uniqueness is not generally preserved if we add the forcing term in (1.1.1), i.e., we consider the equation

$$(r(t)\Phi(x'))' + c(t)\Phi(x) = f(t),$$

where f is a continuous function. More precisely, in [331] it is shown using the concept of the subsolution of the initial value problem

$$(t^\alpha \Phi(x'))' + g(t,x) = 0, \; x(a) = x_0, \; x'(a) = x_1, \quad x_0, x_1 \in \mathbb{R}, \; a \geq 0,$$

that the initial value problem

$$(\Phi(x'))' - \Phi(x) = -1, \quad x(0) = 1, \; x'(0) = 0 \tag{1.1.22}$$

with $p > 2$ has in addition to the "obvious" solution $x(t) \equiv 1$ also a solution for which $x(t) > 1$ in a right neighborhood of $t = 0$. More precisely, the function $\tilde{x}(t) = 1 + t^{1+\varepsilon}$, $\varepsilon > 2/(p-2)$, is a subsolution of (1.1.22), i.e., it satisfies the inequality $(\Phi(x'))' - \Phi(x) \leq -1$, and the general theory developed in [331, 356] states that there is a solution of (1.1.4) which is greater than this subsolution for small $t > 0$, i.e., it is a solution different from $x(t) \equiv 1$. We refer to the above mentioned paper [331] for details.

1.1.6 An alternative approach to the existence theory

In this subsection we offer an alternative approach (comparing with the Prüfer transformation) to the global existence and uniqueness theorem for the IVP

$$(1.1.1), \; x(t_0) = A, \; x'(t_0) = B. \tag{1.1.23}$$

This statement can be inferred from the following five lemmas and Remark 1.1.5. The classical results, like the Peano theorem and the Gronwall inequality, play important roles.

Throughout this subsection, similarly as before, we suppose that r, c are continuous functions in $I = [a, \infty)$ and $r(t) > 0$ in this interval.

Lemma 1.1.1 (Global Existence). *The initial value problem* (1.1.23) *with $t_0 = a$ has at least one solution defined on the whole interval $[a, \infty)$.*

Proof. The local existence follows from the Peano theorem. We will show that this solution can be extended over the entire interval $[a, \infty)$. Integrating system (1.1.3) over the interval $[a, b]$, $b \in [a, \infty)$, we obtain

$$\begin{aligned} x(t) &= x(a) + \int_a^t \Phi^{-1}\left(\frac{u(s)}{r(s)}\right) ds, \\ u(t) &= u(a) - \int_a^t c(s)\Phi(x(s))\, ds \end{aligned}$$

for $t \in [a,b]$. Using Hölder's inequality, we have

$$|x(t) - x(a)|^p \le \left(\int_a^t (1/r(s))^{q/(p-1)}\, ds \right)^{p/q} \int_a^t |u(s)|^q\, ds$$

and

$$|u(t) - u(a)|^q \le \left(\int_a^t |c(s)|^p\, ds \right)^{q/p} \int_a^t |x(s)|^p\, ds.$$

Taking into account that

$$|\lambda + \nu|^p \le 2^{p-1} \left(|\lambda|^p + |\nu|^p \right)$$

for $\lambda, \nu \in \mathbb{R}$, we get

$$\begin{aligned} |x(t)|^p + |u(t)|^q &\le 2^{p-1}|x(a)|^p + 2^{q-1}|u(a)|^q \\ &\quad + 2^{p-1} \left(\int_a^t (1/r(s))^{q/p-1)}\, ds \right)^{p/q} \int_a^t |u(s)|^q\, ds \\ &\quad + 2^{q-1} \left(\int_a^t |c(s)|^p\, ds \right)^{q/p} \int_a^t |x(s)|^p\, ds \\ &\le K + H(b) \int_a^t \left(|x(s)|^p + |u(s)|^q \right) ds \end{aligned}$$

for $t \in [a,b]$, where

$$K = 2^{p-1}|x(a)|^p + 2^{q-1}|u(a)|^q \tag{1.1.24}$$

and

$$H(b) = \max_{t\in[a,b]} \left\{ 2^{p-1} \left(\int_a^t |c(s)|^p\, ds \right)^{q/p},\ 2^p \left(\int_a^t (1/r(s))^{q/(p-1)}\, ds \right)^{p/q} \right\}.$$

Using the Gronwall inequality, we have from the above estimate

$$|x(t)|^p + |u(t)|^q \le K \exp[H(b)(t-a)]$$

for $t \in [a,b]$, and thus the solution can be extended over the whole interval $[a,b]$. Since $b \in [a,\infty)$ is arbitrary, the assertion of the lemma follows. □

Lemma 1.1.2. *The initial value problem* (1.1.23) *with* $A = B = 0$, $t_0 \in [a,\infty)$, *possesses only the trivial solution on* $[a,\infty)$.

Proof. The statement follows from the proof of the previous lemma since the zero initial conditions imply $K = 0$ in (1.1.24), and so $|x(t)|^p + |u(t)|^q \le 0$ for all $t \in [a,\infty)$. □

Lemma 1.1.3. *Suppose that either* (i) $A \ne 0, B \ne 0$, *or* (ii) $A = 0, B \ne 0$ *and* $p \ge 2$, *or* (iii) $A \ne 0, B = 0$ *and* $p \le 2$. *Then the IVP* (1.1.23) *with* $t_0 = a$ *has a unique solution in a right neighborhood of the point* a.

Proof. It is easy to see that the right-hand side of system (1.1.3) satisfies locally a Lipschitz condition on the set $\mathcal{D}$ defined in one of the following ways (with respect to the cases (i) – (iii)):

$$\begin{aligned}\mathcal{D} &= [a,\infty)\times(\mathbb{R}\setminus\{0\})\times(\mathbb{R}\setminus\{0\}) \text{ for } p>1 \text{ arbitrary},\\ \mathcal{D} &= [a,\infty)\times\mathbb{R}\times(\mathbb{R}\setminus\{0\}) \text{ for } p\geq 2,\\ \mathcal{D} &= [a,\infty)\times(\mathbb{R}\setminus\{0\})\times\mathbb{R} \text{ for } p\leq 2.\end{aligned}$$

Hence, the right-hand side of (1.1.3) is Lipschitz in some right neighborhood of a and the statement follows. □

Lemma 1.1.4. *Suppose that $A=0$, $B\neq 0$ and $p<2$. Then the IVP (1.1.23) with $t_0=a$ has a unique solution in a right neighborhood of the point a.*

Proof. Let y_1 and y_2 be two (local) solutions of (1.1.23) with with $t_0=a$, $A=0$ and $B\neq 0$. Integrating (1.1.1) with $y=y_i$ twice from a to $t\in \operatorname{dom}(y_1)\cap\operatorname{dom}(y_2)\cap [a,\infty)$ we get

$$y_i(t)=\int_a^t r^{1-q}(s)\Phi^{-1}\left[\tilde{B}-\int_a^s c(\tau)\Phi(y_i(\tau))\,d\tau\right]ds,\quad i\in\{1,2\},$$

where $\tilde{B}=r(a)\Phi(B)$. Hence

$$y_1(t)-y_2(t)=\int_a^t r^{1-q}(s)\left[\Phi^{-1}(\tilde{B}-I_1(s))-\Phi^{-1}(\tilde{B}-I_2(s))\right]ds,$$

where

$$I_i(t)=\int_a^t c(s)\Phi(y_i(s))\,ds,\quad i\in\{1,2\},$$

and, by the Mean Value Theorem,

$$y_1(t)-y_2(t)=(q-1)\int_a^t r^{1-q}(s)|\eta(s)|^{q-2}(I_2(s)-I_1(s))\,ds,$$

where $\eta(t)$ lies between $\tilde{B}-I_1(t)$ and $\tilde{B}-I_2(t)$. Since $\tilde{B}-I_i(t)\to\tilde{B}$ as $t\to a$, $i\in\{1,2\}$, one can find $\delta>0$ such that $|\eta(t)|\leq 2|\tilde{B}|$ for $t\in[a,a+\delta]$. Noting that $q-2>0$ and using integration by parts, we obtain

$$\begin{aligned}\frac{|y_1(t)-y_2(t)|}{(q-1)(2|\tilde{B}|)^{q-2}} &\leq \int_a^t r^{1-q}(s)|I_1(s)-I_2(s)|\,ds\\ &\leq \int_a^t r^{1-q}(s)\int_a^s |c(\tau)||\Phi(y_1(\tau))-\Phi(y_2(\tau))|\,d\tau\,ds\\ &= \left[\int_a^s |c(\tau)||\Phi(y_1(\tau))-\Phi(y_2(\tau))|\,d\tau\int_a^s r^{1-q}(\tau)\,d\tau\right]_a^t\\ &\quad -\int_a^t\left(\int_a^s r^{1-q}(\tau)\,d\tau\right)|c(s)||\Phi(y_1(s))-\Phi(y_2(s))|\,ds\\ &= \int_a^t\left(\int_s^t r^{1-q}(\tau)\,d\tau\right)|c(s)||\Phi(y_1(s))-\Phi(y_2(s))|\,ds\end{aligned}$$

for $t \in [a, a+\delta]$. Define the continuous function

$$\tilde{y}_i(t) = \begin{cases} y_i(t)\left(\int_a^t r^{1-q}(s)ds\right)^{-1} & \text{for } t \in (a, a+\delta], \\ Br^{q-1}(a) & \text{for } t = a, \end{cases}$$

$i \in \{1,2\}$. Then, using the above estimates,

$$|\tilde{y}_1(t) - \tilde{y}_2(t)| \le (q-1)(2|\tilde{B}|)^{q-2} \int_a^t \left(\int_a^s r^{1-q}(\tau)\,d\tau\right)^{p-1} \{|c||\Phi(\tilde{y}_1) - \Phi(\tilde{y}_2)|\}(s)\,ds$$

for $t \in [a, a+\delta]$. By the Mean Value Theorem,

$$|\Phi(\tilde{y}_1(t)) - \Phi(\tilde{y}_2(t))| \le (p-1)|\xi(t)|^{p-2}|\tilde{y}_1(t) - \tilde{y}_2(t)|$$

for $t \in [a, a+\delta]$, where $\xi(t)$ lies between $\tilde{y}_1(t)$ and $\tilde{y}_2(t)$. Since $\tilde{y}_i(t) \to Br^{q-1}(a)$ as $t \to a$, $i \in \{1,2\}$, $\xi(t)$ can be made to satisfy $|Br^{q-1}(a)|/2 \le |\xi(t)|$, if $t \in [a, \infty)$ is taken sufficiently close to a. Hence there exists $0 < \omega < \delta$ such that

$$|\Phi(\tilde{y}_1(t)) - \Phi(\tilde{y}_2(t))| \le (p-1)\left(|Br^{q-1}(a)|/2\right)^{p-2}|\tilde{y}_1(t) - \tilde{y}_2(t)|$$

for $t \in [a, a+\omega]$. Now, using the above inequality and the fact that

$$(q-1)(p-1)(2|\tilde{B}|)^{q-2}\left(|Br^{q-1}(a)|/2\right)^{p-2} = 2^{q-p}(r(a))^{(q-1)(p+q-4)},$$

we find that

$$\begin{aligned} |\tilde{y}_1(t) - \tilde{y}_2(t)| \le\ & 2^{q-p}(r(a))^{(q-1)(p+q-4)} \\ & \times \int_a^t \left(\int_a^s r^{1-q}(\tau)\,d\tau\right)^{p-1} |c(s)||\tilde{y}_1(s) - \tilde{y}_2(s)|\,ds \end{aligned}$$

for $t \in [a, a+\omega]$. Applying the Gronwall inequality we conclude that $\tilde{y}_1(t) \equiv \tilde{y}_2(t)$ on $[a, a+\omega]$, which implies that the solution of the IVP is unique in a right neighborhood of the point a. □

Lemma 1.1.5. *Suppose that $A \ne 0$, $B = 0$ and $p > 2$. Then the IVP* (1.1.23) *with $t_0 = a$ has a unique solution in a right neighborhood of the point a.*

Proof. If $c(a) \ne 0$, the statement can be proved in a similar way as the previous lemma. Another possibility in this case is to use the method based on the reciprocal equation (1.2.11) (see the below Subsection 1.2.8). Indeed, if y is a solution of (1.1.1) satisfying $y(a) \ne 0$, $y'(a) = 0$, then $u = r\Phi(y')$ solves (1.2.11) with $u(a) = 0$ and $u'(a) \ne 0$. Moreover, $q < 2$ (since $p > 2$) and so the previous lemma can be applied.

We have to show that the statement is valid also for the case where $c(a) = 0$ (in fact, the following approach applies for any value of $c(a)$). Let y_1 and y_2 be two (local) solutions of (1.1.23) with with $t_0 = a$, $A \ne 0$ and $B = 0$. Then there exists $\delta > 0$ such that $y_i(t) \ne 0$ on $[a, a+\delta]$, $i \in \{1,2\}$. Then $w_i = r(t)\Phi(y_i'(t)/y_i(t))$,

$i \in \{1, 2\}$, solve the generalized Riccati equation (1.1.21) on $[a, a + \delta]$. Further, $w_1(a) = 0 = w_2(a)$. The initial value problem (1.1.21), $w(a) = w_0$ with $w_0 \in \mathbb{R}$, which corresponds to a nonzero solution y of (1.1.23) on $[a, a+\delta]$, has exactly one solution on $[a, a + \delta]$ as the function $-c(t) - (p - 1)r^{1-q}(t)|w|^q$ is Lipschitzian in w. Hence $w_1(t) = w_2(t)$ and thus $y_1'(t)/y_1(t) = y_2'(t)/y_2(t)$, which implies $y_1(t) = Ky_2(t)$ for $t \in [a, a + \delta]$, where K is a real constant. Since $y_1(a) = A = y_2(a)$, we get $K = 1$, and the statement follows. □

Remark 1.1.5. It is not difficult to see that similar statements as above can be proved also in the case, where we examine the problem of the existence and u-niqueness of IVP (1.1.23) with $t_0 \in (a, \infty)$ in a left neighborhood of t_0.

1.2 Sturmian theory

In this section we establish the basic oscillatory properties of half-linear equation (1.1.1). In particular, we show that the fundamental methods of the half-linear oscillation theory are similar to those of the oscillation theory of Sturm-Liouville linear equations (1.1.2), and that the Sturmian theory extends verbatim to (1.1.1).

1.2.1 Picone's identity

The original Picone's identity [315] for the linear second order differential equation (1.1.2) was established in 1910. Since that time, this identity has been extended in various directions and the half-linear version of this identity reads as follows.

Theorem 1.2.1. *Consider a pair of half-linear differential operators*

$$\mathcal{L}[x] = \mathcal{L}_{r,c}[x] = (r(t)\Phi(x'))' + c(t)\Phi(x), \quad \mathcal{L}_{R,C}[y] = (R(t)\Phi(y'))' + C(t)\Phi(y)$$

and let x, y be continuously differentiable functions such that $r\Phi(x'), R\Phi(y')$ are also continuously differentiable and $y(t) \neq 0$ in an interval $I \subset \mathbb{R}$. Then in this interval

$$\text{(1.2.1)} \quad \left\{ \frac{x}{\Phi(y)} [\Phi(y)r\Phi(x') - \Phi(x)R\Phi(y')] \right\}' = (r - R)|x'|^p + (C - c)|x|^p$$
$$+ pR^{1-q}P\left(R^{q-1}x', R\Phi(xy'/y)\right) + \frac{x}{\Phi(y)} [\Phi(y)\mathcal{L}_{r,c}[x] - \Phi(x)\mathcal{L}_{R,C}[y]],$$

where

$$\text{(1.2.2)} \qquad P(u, v) := \frac{|u|^p}{p} - uv + \frac{|v|^q}{q} \geq 0$$

with equality if and only if $v = \Phi(u)$.

Proof. The identity (1.2.1) can be verified by a direct computation, inequality (1.2.2) is the classical Young inequality, see e.g. [173]. □

In the particular case when $r = R$, $C = c$, y is a nonzero solution of the equation $\mathcal{L}_{r,c}[y] = 0$ and $w = r\Phi(y')/\Phi(y)$, identity (1.2.1) reduces to

$$(1.2.3)\qquad r(t)|x'|^p - c(t)|x|^p = (w(t)|x|^p)' + pr^{1-q}(t)P(r^{q-1}(t)x', \Phi(x)w(t)).$$

This reduced Picone's identity will be used frequently in the sequel.

1.2.2 Energy functional

The p-degree functional

$$(1.2.4)\qquad \mathcal{F}(y; a, b) = \int_a^b [r(t)|y'|^p - c(t)|y|^p]\, dt$$

considered over the Sobolev space $W_0^{1,p}(a,b)$ is usually called the *energy functional* of (1.1.1). Recall that the Sobolev space $W_0^{1,p}(a,b)$ consists of absolutely continuous functions x whose first derivative is in $L^p(a,b)$ and $x(a) = 0 = x(b)$, with the norm $\|x\| = \left(\int_a^b [|x'|^p + |x|^p]\, dt\right)^{1/p}$ or with the equivalent norm $\|x\| = \left(\int_a^b |x'|^p\, dt\right)^{1/p}$. This set is sometimes (in the variational context) called the class of *admissible functions*. Equation (1.1.1) is the Euler-Lagrange equation of the functional $\mathcal{F}$. Moreover, if x is a solution of (1.1.1) satisfying $x(a) = 0 = x(b)$, then using integration by parts we have

$$(1.2.5)\quad \mathcal{F}(x; a, b) = [r(t)x(t)\Phi(x'(t))]_a^b - \int_a^b x(t)[(r(t)\Phi(x'))' + c(t)\Phi(x)]\, dt = 0.$$

1.2.3 Roundabout theorem

This theorem relates Riccati equation (1.1.21), the energy functional (1.2.4) and the basic oscillatory properties of solutions of (1.1.1). The terminology Roundabout theorem (or Reid type Roundabout Theorem) is due to Reid [320] (it concerns the linear case), and it is motivated by the fact that the proof of this theorem consists of the "roundabout" proof of several equivalent statements.

Definition 1.2.1. Equation (1.1.1) is said to be *disconjugate* on the closed interval $[a, b]$ if the solution x given by the initial condition $x(a) = 0$, $r(a)\Phi(x'(a)) = 1$ has no zero in $(a, b]$ (by a zero of a solution x we mean such a t_0 that $x(t_0) = 0$). In the opposite case (1.1.1) is said to be *conjugate* on $[a, b]$.

Theorem 1.2.2. *The following statements are equivalent.*

(i) *Equation* (1.1.1) *is disconjugate on the interval* $I = [a, b]$.

(ii) *There exists a solution of* (1.1.1) *having no zero in* $[a, b]$.

(iii) *There exists a solution* w *of the generalized Riccati equation* (1.1.21) *which is defined on the whole interval* $[a, b]$.

(iv) The energy functional $\mathcal{F}(y;a,b)$ is positive for every $0 \not\equiv y \in W_0^{1,p}(a,b)$.

Proof. (i) $\Rightarrow$ (ii): Consider the solution $\tilde{x}$ of (1.1.1) given by the initial condition $\tilde{x}(a) = \varepsilon$, $r(a)\Phi(\tilde{x}'(a)) = 1$, where $\varepsilon > 0$. Then, according to the continuous dependence of solutions of (1.1.1) on the initial conditions, disconjugacy of (1.1.1) on $[a,b]$ implies that this solution is positive on this interval if ε is sufficiently small.

(ii) $\Rightarrow$ (iii): This implication is the immediate consequence of the Riccati substitution from Subsection 1.1.4.

(iii) $\Rightarrow$ (iv): If there exists a solution w of (1.1.21) defined in the whole interval $[a,b]$, then by integrating the reduced Picone identity (1.2.3) with $x \in W_0^{1,p}(a,b)$ we get

$$\mathcal{F}(x;a,b) = p\int_a^b r^{1-q}(t)P(r^{q-1}(t)x', \Phi(x)w(t))\,dt \geq 0$$

with equality if and only if $\Phi(r^{q-1}(t)x') = \Phi(x)w(t)$, i.e., $x' = \Phi^{-1}(w(t)/r(t))x$ in $[a,b]$, thus

$$x(t) = x(a)\exp\left\{\int_a^t \Phi^{-1}\left(\frac{w(s)}{r(s)}\right)ds\right\} \equiv 0$$

since $x(a) = 0$. This means that $\mathcal{F}(x;a,b) \geq 0$ over $W_0^{1,p}(a,b)$ with equality only if $x(t) \equiv 0$.

(iv) $\Rightarrow$ (i): Suppose that $\mathcal{F} > 0$ for nontrivial $y \in W_0^{1,p}(a,b)$ and (1.1.1) is not disconjugate in $[a,b]$, i.e., the solution x of (1.1.1) given by the initial condition $x(a) = 0$, $r(a)\Phi(x'(a)) = 1$ has a zero $d \in [a,b]$. Define the function $y \in W_0^{1,p}(a,b)$ as follows

$$y(t) = \begin{cases} x(t) & t \in [a,d], \\ 0 & t \in [d,b]. \end{cases}$$

Then by (1.2.5)

$$\mathcal{F}(y;a,b) = \mathcal{F}(y;a,d) = \mathcal{F}(x;a,d) = 0$$

which contradicts the positivity of $\mathcal{F}$. □

Remark 1.2.1. (i) It is not difficult to see that the Picone identity could be also used to show that the existence of a nonzero solution of (1.1.1) implies positive definiteness of $\mathcal{F}$ (the implication (ii)$\Rightarrow$(iv)). But we include the generalized Riccati equation into the Roundabout theorem since such an equivalence is important for applications of our theory.

(ii) The following statement was proved in [245]: If there exists a solution y of (1.1.1) such that $y(t) \neq 0$ on (a,b), then for every $u \in W_0^{1,p}(a,b)$ one has $\mathcal{F}(u;a,b) \geq 0$, with equality if y and u are proportional. This is essentially the same statement as the implication (i) $\Rightarrow$ (iv) of Theorem 1.2.2, its proof is based on the Hardy type inequalities (see formula (9.5.4) given in the last chapter).

(iii) Later, in Section 5.8, we give a variant of the Roundabout theorem where an argument of $\mathcal{F}(\cdot;a,b)$ satisfies a different type of boundary conditions; recall that here we assume the zero boundary conditions.

The following lemma shows that disconjugacy allows an alternative definition in terms of zeros of solutions.

Lemma 1.2.1. *Equation* (1.1.1) *is disconjugate on* $[a, b]$ *if and only if every its nontrivial solution has at most one zero in* $[a, b]$.

Proof. The proof of the "if" part is trivial. The "only if" part will be proved by a contradiction. Thus suppose that (1.1.1) is disconjugate on $[a, b]$ and that there is a solution y of (1.1.1) such that $y(c_1) = 0 = y(c_2)$, $a \le c_1 < c_2 \le b$. Since the solution space is homogeneous, there is $\lambda \in \mathbb{R}$ such that $z = \lambda y$ satisfies (1.1.1) with $z(c_1) = 0 = z(c_2)$ and $z'(c_1) = \Phi^{-1}(1/r(c_1))$. Then there is no solution of (1.1.1) without zeros on $[c_1, c_2]$ by Theorem 1.2.2 (the implication (ii)$\Rightarrow$(i)) and hence there is no solution of (1.1.1) without zeros on $[a, b]$. The implication (i)$\Rightarrow$(ii) of Theorem 1.2.2 then says that the solution x satisfying $x(a) = 0$, $x'(a) = \Phi^{-1}(1/r(a))$ has a zero in $(a, b]$, and so (1.1.1) is not disconjugate on $[a, b]$, a contradiction. □

Remark 1.2.2. Similarly to the linear case, two points $t_1, t_2 \in \mathbb{R}$ are said to be *conjugate relative to* (1.1.1) if there exists a nontrivial solution x of this equation such that $x(t_1) = 0 = x(t_2)$. Due to Lemma 1.2.1, disconjugacy and conjugacy of (1.1.1) on a compact interval $I \subset \mathbb{R}$ can be equivalently defined as follows. Equation (1.1.1) is said to be *disconjugate* on an interval I if this interval contains no pair of points conjugate relative to (1.1.1) (i.e., every nontrivial solution has at most one zero in I). In the opposite case (1.1.1) is said to be *conjugate* on I (i.e., there exists a nontrivial solution with at least two zeros in I). In Subsection 1.2.6 we will discus the disconjugacy on various types of intervals. In Section 4.2 we will show that using the concept of the principal solution of (1.1.1), this equivalent definition applies also to unbounded intervals or to intervals whose end points are singular points of (1.1.1).

1.2.4 Sturmian separation and comparison theorems

The interlacing property of zeros of linearly independent solutions of linear equations is one of the most characteristic properties, which among others justifies the definition of oscillation/nonoscillation of equation. The next Sturm type separation theorem claims that this property extends to (1.1.1). The proof is based on the Riccati transformation. In the subsequent subsection we offer also some alternative methods.

Theorem 1.2.3. *Let* $t_1 < t_2$ *be two consecutive zeros of a nontrivial solution* x *of* (1.1.1). *Then any other solution of this equation which is not proportional to* x *has exactly one zero on* (t_1, t_2).

Proof. Let $w = r\Phi(x')/\Phi(x)$, then w is a solution of (1.1.21) which is defined on (t_1, t_2) and satisfies $w(t_1+) = \infty$, $w(t_2-) = -\infty$. Suppose that there exists a solution $\tilde{x}$ of (1.1.1), linearly independent of x, which has no zero in (t_1, t_2) and let $\tilde{w} = r\Phi(\tilde{x}'/\tilde{x})$. See also Figure 1.2.1. Since $\tilde{x}(t_1) \ne 0$, $\tilde{x}(t_2) \ne 0$ (otherwise $\tilde{x}$ would be a multiple of x), we have $\tilde{w}(t_1) < \infty$, $\tilde{w}(t_2) > -\infty$. Hence the graph of $\tilde{w}$ has to intersect the graph of w at some point in (t_1, t_2), but this contradicts the unique solvability of (1.1.21) (which follows from the fact that the function $c + (p-1)r^{1-q}|w|^q$ is Lipschitzian with respect to w). □

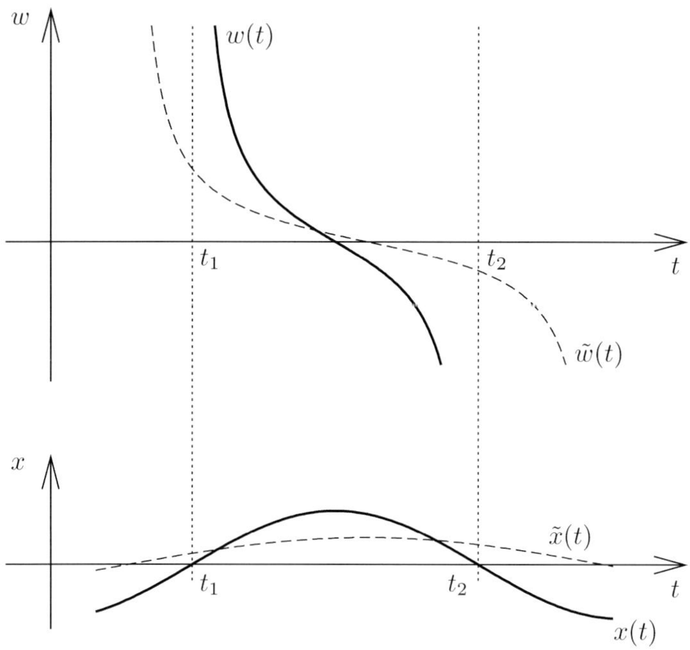

Figure 1.2.1: Proof of the separation theorem based on the Riccati technique

Now we introduce the definition of oscillation and nonoscillation of (1.1.1), which can be the same as in the linear case.

Definition 1.2.2. *Equation* (1.1.1) *is said to be* *nonoscillatory* (more precisely, nonoscillatory at ∞), if there exists $T_0 \in \mathbb{R}$ such that (1.1.1) is disconjugate on $[T_0, T_1]$ for every $T_1 > T_0$. In the opposite case, (1.1.1) is said to be *oscillatory.*

According to Theorem 1.2.3, the above definition is correct, in the sense that one solution of (1.1.1) is oscillatory if and only if any other solution of (1.1.1) is oscillatory. Oscillation of a (nontrivial) solution of (1.1.1) means the existence of a sequence of zeros of this solution tending to ∞. That is why we can speak about *(non)oscillation of equation* (1.1.1). Note also that similarly to the linear case, if the functions r, c are continuous and $r(t) > 0$ in an interval $[T, \infty)$, then according to the unique solvability of the initial value problem associated with (1.1.1), the sequence of zeros of any nontrivial solution of (1.1.1) cannot a have a finite cluster point.

Along with (1.1.1) consider another equation of the same form

$$(R(t)\Phi(y'))' + C(t)\Phi(y) = 0, \tag{1.2.6}$$

where the functions R, C satisfy the same assumptions as r, c, respectively, in (1.1.1).

The next statement is an extension of well-known Sturm comparison theorem (sometimes called Sturm-Picone comparison theorem, when r and R are different) to (1.1.1).

Theorem 1.2.4. *Let $t_1 < t_2$ be consecutive zeros of a nontrivial solution x of* (1.1.1) *and suppose that*

$$C(t) \geq c(t), \quad r(t) \geq R(t) > 0 \tag{1.2.7}$$

for $t \in [t_1, t_2]$. Then any solution of (1.2.6) *has a zero in (t_1, t_2) or it is a multiple of the solution x. The last possibility is excluded if one of the inequalities in* (1.2.7) *is strict on a set of positive measure.*

Proof. Let x be a solution of (1.1.1) having zeros at $t = t_1$ and $t = t_2$. Then by (1.2.5) we have $\mathcal{F}(x; t_1, t_2) = 0$ and according to (1.2.7)

$$\mathcal{F}_{RC}(x; t_1, t_2) := \int_{t_1}^{t_2} [R(t)|x'|^p - C(t)|x|^p]\, dt \leq 0. \tag{1.2.8}$$

This implies, by Theorem 1.2.2, that the solution y of (1.2.6) given by the initial condition $y(t_1) = 0$, $y'(t_1) > 0$ has a zero in $(t_1, t_2]$ and by Theorem 1.2.3 any linearly independent solution of (1.2.6) has a zero in (t_1, t_2). Finally, suppose that the first zero of y on the right of t_1 is just t_2, i.e., $y(t_1) = 0 = y(t_2)$. Let $v = R\Phi(y')/\Phi(y)$ be the solution of the Riccati equation associated with (1.2.6). Since $\lim_{t \to t_1+}[x(t)/y(t)] = \lim_{t \to t_1+}[x'(t)/y'(t)] = x'(t_1)/y'(t_1)$ exists finite, we have

$$\lim_{t \to t_1+} v(t)\,|x(t)|^p = \lim_{t \to t_1+} R(t)x(t)\Phi(y'(t))\frac{\Phi(x(t))}{\Phi(y(t))} = 0.$$

Similarly $\lim_{t \to t_2-} v(t)\,|x(t)|^p = 0$. This implies $\mathcal{F}_{RC}(x; t_1, t_2) \geq 0$ (in view of (1.2.3) with R, C, v instead of r, c, w respectively), and hence $\mathcal{F}_{RC}(x; t_1, t_2) = 0$ by (1.2.8). This implies, by the same argument as in the part (iii) $\Rightarrow$ (iv) of the proof of Theorem 1.2.2, that $v = R\Phi(x'/x)$, i.e., x and y are proportional. However, this is impossible if one of inequalities in (1.2.7) is strict on an interval of positive length. □

We will employ the same terminology as in the linear case. If (1.2.7) are satisfied in a given interval I, then (1.2.6) is said to be the *majorant* equation of (1.1.1) (or Sturmian majorant) on I and (1.1.1) is said to be the *minorant* equation of (1.2.6) (or Sturmian minorant) on I.

1.2.5 More on the proof of the separation theorem

We have already seen that Theorem 1.2.3 can be proved by means of the Riccati technique (i.e., the equivalence (i)$\Leftrightarrow$(iii) of Theorem 1.2.2). Here we offer another four ways of its proof in order to show a wide variety of different approaches which are possible in spite of the fact that the additivity of the solution space of (1.1.1) is lost.

Before we present the proofs, let us give some auxiliary results. First note that in Subsection 1.3.1 below it is proved that the Wronskian identity applies only when $p = 2$. This means, among others, that there is no extension of the reduction of order formula (see also Subsection 1.3.1), which is a basis for one of the proofs of the separation theorem in the linear case. On the other hand, the concept of Wronskian can be utilized in characterization of linear (in)dependence of solutions (Lemma 1.3.1), which will be helpful here. The following statement clearly follows from Lemma 1.3.1. Nevertheless, we offer also an alternative proof.

Lemma 1.2.2. *Two nontrivial solutions x and y of* (1.1.1) *which are not proportional cannot have a common zero.*

Proof. Suppose, by contradiction, that x and y are linearly independent solutions with $x(t_0) = 0 = y(t_0)$. Then $x'(t_0) = A \neq 0$ and $y'(t_0) = B \neq 0$. Consider the solution z of (1.1.1) satisfying $z(t_0) = 0$, $z(t_0) = 1$. Then Az and Bz are the solutions of (1.1.1) satisfying $Az(t_0) = 0 = Bz(t_0)$, $Az'(t_0) = A$, $Bz'(t_0) = B$, in view of the homogeneity property. Owing to the uniqueness we have $x = Az$ and $y = Bz$, which implies $x = (A/B)y$. Consequently, x and y are proportional, a contradiction. □

Now we are ready to give several alternative proofs of the separation theorem (Theorem 1.2.3).

1. *Proof based on the Riccati technique:* See Subsection 1.2.4.

2. *Variational proof:* This proof is based on the combination of the implication (i)⇒(ii) of Theorem 1.2.2 with Lemma 1.2.1. In fact, there is nothing to prove, in view of Lemma 1.2.1, since the Roundabout theorem then says that the coexistence of two solutions of (1.1.1), where one solution has at least two zeros in a given interval while another one has no zero, is impossible. This can be seen also from Sturmian comparison theorem where $c(t) \equiv C(t)$ and $r(t) \equiv R(t)$. We call this proof variational since an important role is played by Picone's identity involving the p-degree functional $\mathcal{F}$.

3. *Proof based on Prüfer's transformation:* Without loss of generality we assume $x(t) > 0$, $t \in (t_1, t_2)$. Hence, by (1.1.17), $\varphi(t) \in (0, \pi_p) \pmod{\pi_p}$ for $t \in (t_1, t_2)$. See also Figure 1.2.2. By (1.1.20), $\varphi'(t_i) = r^{1-q}(t_i)$, $i = 1, 2$, thus φ is increasing in some neighborhood of t_i, and so without loss of generality we may suppose that $\varphi(t_1) = 0$, $\varphi(t_2) = \pi_p$. Let us consider any other solution y different from λx, $\lambda \in \mathbb{R}$. Then $y(t_1) \neq 0$ by Lemma 1.2.2. We may suppose $y(t_1) > 0$ and by (1.1.17), $0 = \varphi(t_1) < \bar\varphi(t_1) < \pi_p$, where $\bar\varphi$ corresponds to y like φ corresponds to x. Since by (1.1.20) φ' depends only on φ and not on ϱ, there is also uniqueness for the variable φ alone. This uniqueness implies that from $\bar\varphi(t_1) > \varphi(t_1)$ it follows $\bar\varphi(t) > \varphi(t)$ for all t from the interval under consideration. Hence $\bar\varphi(t_2) > \varphi(t_2) = \pi_p$, and the continuous function $\bar\varphi(t) - \pi_p$ changes its sign in $[t_1, t_2]$. So there is a point $\bar t \in (t_1, t_2)$ such that $\bar\varphi(\bar t) = \pi_p$. By (1.1.17) with y and $\bar\varphi$ instead of x and φ, respectively, $y(\bar t) = 0$.

4. *Proof based on the Wronskian:* Let y be a solution of (1.1.1) independent on x. Then
$$-y(t_i)x'(t_i) = (xy' - yx')(t_i) =: W(x, y)(t_i) \neq 0,$$
since x, y are linearly independent. Hence $x'(t_i) \neq 0 \neq y(t_i)$ (actually, this follows also from the uniqueness and Lemma 1.2.2). Clearly, $x'(t_1)x'(t_2) < 0$ and

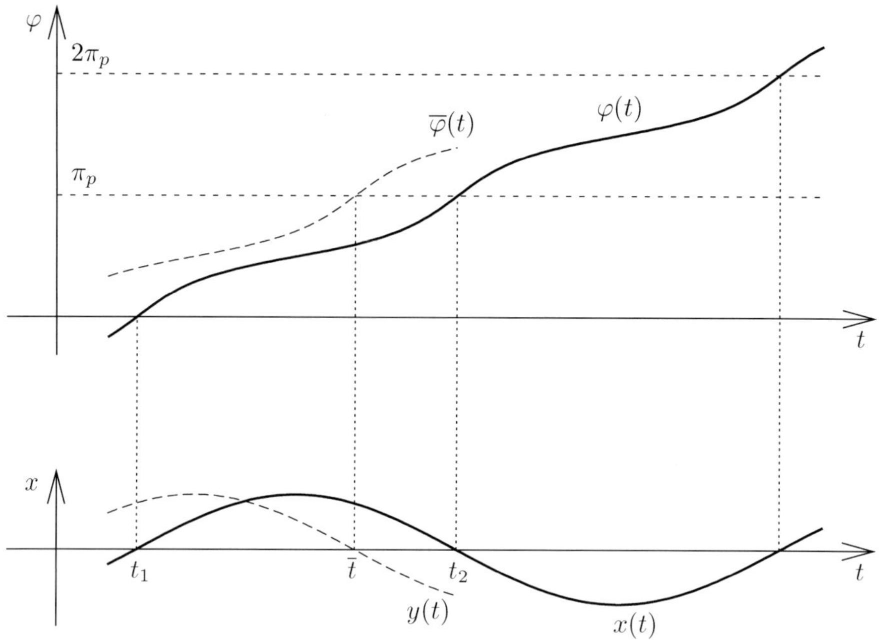

Figure 1.2.2: Proof of the separation theorem based on Prüfer's transformation

$W(x, y)(t) > 0$ [or < 0] for all t by below given Lemma 1.3.1. Consequently, $y(t_1)y(t_2) < 0$ and therefore y has to vanish somewhere between t_1 and t_2.

5. *Proof based on the uniqueness of the IVP:* Without loss of generality we assume $x(t) > 0$, $t \in (t_1, t_2)$, and we prove that another solution, linearly independent of x, has at least one zero in (t_1, t_2). Thus assume, by a contradiction, that there is a solution y such that $y(t) > 0$ for $t \in [t_1, t_2]$. Define the set $\Omega = \{\lambda > 0 : \lambda x(t) < y(t) \text{ for } t \in (t_1, t_2)\}$. Clearly, Ω is nonempty and bounded. Therefore there exists $\bar{\lambda} = \sup \Omega$. Now it is not difficult to see that there exists $\bar{t} \in (t_1, t_2)$ such that $\bar{\lambda}x(\bar{t}) = y(\bar{t})$. Indeed, if not, then we come to a contradiction with the definition of $\bar{\lambda}$. By a contradiction with this definition it can be also shown that $\bar{\lambda}x'(\bar{t}) = y'(\bar{t})$. The uniqueness now yields $\bar{\lambda}x(t) \equiv y(t)$. Consequently, x and y are linearly dependent, a contradiction.

1.2.6 Disconjugacy on various types of intervals

Following Lemma 1.2.1, we now extend the concept of disconjugacy to the general interval I, where it does not matter whether I is closed or open or half-open.

Definition 1.2.3. Equation (1.1.1) is *disconjugate on an interval* I if every its nontrivial solution has at most one zero in I.

However, hereafter we show that the change of the form of I may affect some of the properties; in particular, the existence of a solution of (1.1.1) without zeros is no more necessary for disconjugacy in case of general I.

Theorem 1.2.5. *Equation* (1.1.1) *is disconjugate on* I *if it has a solution without zeros on* I. *For a compact or an open interval this condition is also necessary.*

Proof. The sufficiency follows from the Sturmian separation theorem (see Theorem 1.2.3). The necessity for a compact interval is in fact the implication (i) $\Rightarrow$ (ii) of Theorem 1.2.2, in view of Lemma 1.2.1. Suppose next that $I = (a, b)$ is open. Let d be any point in I and choose $\{a_n\}, \{b_n\}$ so that $a_n < d < b_n$, $a_n \searrow a$ and $b_n \nearrow b$ as $n \to \infty$. Then there exists a solution $y_n(t)$ which is positive for $a_n \le t \le b_n$. Letting $n \to \infty$ through a suitable subsequence we obtain $y_n(d) \to A$, $y_n'(d) \to B$, where $A^2 + B^2 > 0$. Then $y_n(t) \to y(t)$, $y_n'(t) \to y'(t)$ for all t in I, where $y(t)$ is the solution which satisfies the initial conditions $y(d) = A$, $y'(d) = B$. Therefore $y(t) \ge 0$ for all t in I. If $y(t_0) = 0$ for some $t_0 \in I$, then also $y'(t_0) = 0$, and hence $y(t) \equiv 0$, a contradiction. □

The existence of a solution without zeros is not necessary for disconjugacy on a *half-open* interval. For example, every nontrivial solution $y(t) = A \sin_p(t - \delta)$, $A > 0$, $0 \le \delta < 2\pi$, of equation (1.1.1), with $r(t) \equiv 1$ and $c(t) \equiv (p - 1)$ (see Theorem 1.4.1), vanishes exactly once on the interval $I = [0, \pi_p)$.

Theorem 1.2.6. *Equation* (1.1.1) *is disconjugate on the half-open interval* $I = [a, b)$ *if it is disconjugate on its interior* (a, b).

Proof. We only need to show that if $y(t)$ is the solution satisfying the initial conditions $y(a) = 0$, $y'(a) = 1$, then $y(t) \ne 0$ for $a < t < b$. Suppose on the contrary that $y(d) = 0$, where $a < d < b$. Since $y'(d) \ne 0$, $y(t)$ assumes negative values in a neighborhood of d. Therefore, if $\varepsilon > 0$ is sufficiently small, the solution $\tilde{y}(t)$ satisfying the initial conditions $\tilde{y}(a + \varepsilon) = 0$, $\tilde{y}'(a + \varepsilon) = 1$ has a zero near d. But this contradicts the hypothesis that (1.1.1) is disconjugate on (a, b). □

An easy modification of the proof of Theorem 1.2.2, with using the results of this subsection, yields the variant of the Roundabout theorem, which involves the disconjugacy on an open interval. Observe how the positive definiteness of the energy functional is changed to the positive semidefiniteness.

Theorem 1.2.7. *The following statements are equivalent.*

(i) Equation (1.1.1) *is disconjugate on the interval* (a, b).

(ii) There exists a solution of (1.1.1) *having no zero in* (a, b).

(iii) There exists a solution w *of the generalized Riccati equation* (1.1.21) *which is defined on the whole interval* (a, b).

(iv) The energy functional $\mathcal{F}(y; a, b)$ *is nonnegative for every* $y \in W_0^{1,p}(a, b)$.

1.2.7 Transformation of independent variable

Let us introduce new independent variable $s = \varphi(t)$ and new function $y(s) = x(t)$, where φ is a differentiable function such that $\varphi'(t) \ne 0$ in some interval I where we consider equation (1.1.1). Then $\frac{d}{dt} = \varphi'(t)\frac{d}{ds}$ and hence this transformation transforms (1.1.1) into the equation of the same form

$$\frac{d}{ds}\left[r(t)\Phi(\varphi'(t))\Phi\left(\frac{d}{ds}y\right)\right] + \frac{c(t)}{\varphi'(t)}\Phi(y) = 0, \quad t = \varphi^{-1}(s), \tag{1.2.9}$$

where φ^{-1} is the inverse function of φ. In particular, if

$$\varphi(t) = \int_T^t r^{1-q}(\tau)\, d\tau, \quad T \in I, \tag{1.2.10}$$

then the resulting equation (1.2.9) is the equation of the form (1.1.1) with $r(t) \equiv 1$.

Observe that if $\int^\infty r^{1-q}(t)\, dt = \infty$, then (1.2.10) transforms an unbounded interval $[T, \infty)$ into the interval $[0, \infty)$ which is of the same form as $[T, \infty)$. If $\int^\infty r^{1-q}(t)\, dt < \infty$ then the interval $[T, \infty)$ is transformed into the bounded interval $[0, b)$, $b = \int_T^\infty r^{1-q}(t)\, dt$. This fact, coupled with the remark about cluster points of an oscillatory solution of (1.1.1) given below Definition 1.2.2, shows why some (non)oscillation criteria and asymptotic formulas for solutions of (1.1.1) substantially depend on the divergence (convergence) of $\int^\infty r^{1-q}(t)\, dt$.

1.2.8 Reciprocity principle

Suppose that the function c in (1.1.1) does not change its sign in an interval I and let $u = r\Phi(x')$. By a simple computation one can verify that u is a solution of the so-called *reciprocal equation*

$$\left(\frac{1}{\Phi^{-1}(c(t))}\Phi^{-1}(u')\right)' + r^{1-q}(t)\Phi^{-1}(u) = 0, \tag{1.2.11}$$

where $\Phi^{-1}(s) = |s|^{q-1}\operatorname{sgn} s$, $q = p/(p-1)$, is the inverse function of Φ. The terminology reciprocal equation is motivated by the linear case $p = 2$. The reciprocal equation to (1.2.11) is again the original equation (1.1.1)

If $t_1 < t_2$ are consecutive zeros of a solution x of (1.1.1), then by the Rolle Mean Value Theorem u has at least one zero in (t_1, t_2). Conversely, if $\tilde t_1 < \tilde t_2$ are consecutive zeros of u, then $u' = -c(t)x$ and hence also x has a zero in (t_1, t_2). This means that (1.1.1) is oscillatory if and only if the reciprocal equation (1.2.11) is oscillatory. This fact will be referred to as the *reciprocity principle*.

1.2.9 Sturmian theory for Mirzov's system

As we have mentioned in Subsection 1.1.1, half-linear equation (1.1.1) is a special case of the first order system

$$u_1' = a_1(t)|u_2|^{\lambda_1}\operatorname{sgn} u_2, \quad u_2' = -a_2(t)|u_1|^{\lambda_2}\operatorname{sgn} u_1, \tag{1.2.12}$$

where it is assumed that $\lambda_1, \lambda_2 > 0$ and

$$\lambda_1\lambda_2 = 1. \tag{1.2.13}$$

A detailed treatment of (1.2.12) (in most of the parts of that book even without assuming (1.2.13)) can be found in [292], see also extended english translation of Mirzov's book [293].

Mirzov considers a different definition of (non)oscillation where both components u_1, u_2 have "the same significance", in contrast to our approach, where the oscillatory properties of (1.1.1) (and hence of system (1.1.3)) are defined via zero point of a solution x, i.e., as zeros of the component u_1 in (1.2.12). By Mirzov's definition, a solution (u_1, u_2) of (1.2.12) oscillates if *both* components u_1, u_2 oscillate (i.e., have a sequence of zero points tending to ∞). We will return to the problem of oscillation of (1.2.12) later.

In this subsection we briefly recall the main result of [292, Chap. 10], a Sturm type theorem for (1.2.12). In fact, as we will see, its proof is based on the Riccati type substitution.

Theorem 1.2.8. *Let for some $k \in \{1, 2\}$, the inequalities*

$$0 \le a_k(t) \le a_k^*(t), \quad a_{3-k}(t) \le a_{3-k}^*(t) \text{ for } t_1 \le t \le t_2 \tag{1.2.14}$$

hold and let system (1.2.12) *have a solution u_1, u_2 such that*

$$u_k(t_1) = u_k(t_2) = 0 \text{ and } u_k(t) \ne 0 \text{ for } t_1 < t < t_2, \tag{1.2.15}$$

where $0 \le t_1 < t_2 < \infty$. Then, for any solution v_1, v_2 of the system

$$v_1' = a_1^*(t)|v_2|^{\lambda_1} \operatorname{sgn} v_2, \quad v_2' = -a_2^*(t)|v_1|^{\lambda_2} \operatorname{sgn} v_1 \tag{1.2.16}$$

the component v_k has at least one zero on the interval $[t_1, t_2]$.

Proof. In view of (1.2.15), without loss of generality we can assume that

$$u_k(t_i) = 0, \quad i = 1, 2, \quad u_k(t) > 0 \text{ for } t_1 < t < t_2. \tag{1.2.17}$$

According to the unique solvability of (1.2.12) (compare the proof of Lemma 1.1.1 and Lemma 1.1.2), $u_{3-k}(t_1) \ne 0$. Let us show that

$$(-1)^{k-1} u_{3-k}(t_1) > 0. \tag{1.2.18}$$

Assume the contrary. Then there exists $\varepsilon > 0$ such that

$$(-1)^{k-1} u_{3-k}(t) < 0 \quad \text{for } t_1 \le t \le t_1 + \varepsilon.$$

Thus, by the first inequality in (1.2.14), we have

$$u_k'(t) = a_k(t)|u_{3-k}(t)|^{\lambda_k}(-1)^{k-1} \operatorname{sgn} u_{3-k}(t) = -a_k(t)|u_{3-k}(t)|^{\lambda_k} \le 0$$

for $t_1 \le t \le t_1 + \varepsilon$, which contradicts conditions (1.2.17). Thus the validity of inequality (1.2.18) is proved. Analogously we can prove that

$$(-1)^{k-1} u_{3-k}(t_2) < 0. \tag{1.2.19}$$

From (1.2.13), (1.2.17)-(1.2.19) it is obvious that the function

$$z(t) = u_{3-k}(t)/[u_k(t)]^{\lambda_{3-k}}$$

is a solution of the generalized Riccati equation

$$z' + (-1)^{k-1}\lambda_{3-k}a_k(t)|z|^{1+\lambda_k} + (-1)^{k-1}a_{3-k}(t) = 0, \tag{1.2.20}$$

satisfying the conditions

$$\lim_{t\to t_1+}(-1)^{k-1}z(t) = \infty, \quad \lim_{t\to t_2-}(-1)^{k-1}z(t) = -\infty. \tag{1.2.21}$$

Now we suppose that there exists a solution (v_1, v_2) of system (1.2.16) such that $v_k(t) \neq 0$ for $t_1 \leq t \leq t_2$. Then, as for the function z which solves (1.2.20), the function $\rho(t) = v_{3-k}(t)\operatorname{sgn} v_k(t)/|v_k(t)|^{\lambda_{3-k}}$ is a solution of the equation

$$\rho' + (-1)^{k-1}\lambda_{3-k}a_k^*(t)|\rho|^{1+\lambda_k} + (-1)^{k-1}a_{3-k}^*(t) = 0,$$

defined on the interval $[t_1, t_2]$. In view of condition (1.2.21) and the continuity of the function ρ on $[t_1, t_2]$, there exists a point $t^* \in (t_1, t_2)$ such that

$$z(t^*) = \rho(t^*), \quad (-1)^{k-1}z(t) < (-1)^{k-1}\rho(t) \tag{1.2.22}$$

for $t^* < t < t_2$. On the other hand, by conditions (1.2.14),

$$\begin{aligned}(-1)^{k-1}z' &= -\lambda_{3-k}a_k(t)|z|^{1+\lambda_k} - a_{3-k}(t)\\ &\geq -\lambda_{3-k}a_k^*(t)|z|^{1+\lambda_k} - a_{3-k}^*(t)\end{aligned}$$

for $t_1 < t < t_2$. According to the standard theorem on differential inequalities (see [174]), (1.2.22) and (1.2.23) result in

$$(-1)^{k-1}z(t) \geq (-1)^{k-1}\rho(t), \quad \text{for } t^* \leq t < t_2.$$

The last inequality contradicts (1.2.22). Therefore, our assumption that $v_k(t) \neq 0$ on $[t_1, t_2]$ is not true. □

Remark 1.2.3. One can find in [292, Chap. 10] also several other statements related to Sturmian theory for (1.2.12). These statements are mostly based on the Prüfer transformation applied to (1.2.12), but in contrast to the Subsection 1.1.3, Mirzov uses *the classical* sine and cosine functions, i.e., his Prüfer transformation is of the form

$$u_1(t) = \varrho(t)\sin\varphi(t), \quad u_2(t) = \varrho(t)\cos\varphi(t).$$

The obtained differential equations for ϱ, φ are then, of course, different from (1.1.20), but their solutions have essentially the same properties as those of equation (1.1.20). So also this kind of the Prüfer transformation can be used to derive the results which will be obtained later on in this book using the Prüfer transformation in form (1.1.17) or in its modifications as mentioned in Remark 1.1.2.

1.2.10 Leighton-Wintner oscillation criterion

In this subsection we formulate a simple oscillation criterion for (1.1.1). Even if we will devote a special chapter to oscillation criteria for (1.1.1), we prefer to formulate this criterion already here, to show an example of the application of the two basic methods of the oscillation theory of (1.1.1). In the linear case $p = 2$, this criterion was proved first by Leighton [233] under the additional assumption $c(t) \geq 0$. This restriction was later removed by Wintner, see e.g. [341].

Theorem 1.2.9. *Equation* (1.1.1) *is oscillatory provided*

$$\int^{\infty} r^{1-q}(t)\, dt = \infty \quad \text{and} \quad \int^{\infty} c(t)\, dt = \lim_{b\to\infty} \int^{b} c(t)\, dt = \infty. \tag{1.2.23}$$

Proof. According to the definition of oscillation of (1.1.1), we need to show that this equation is not disconjugate on any interval of the form $[T, \infty)$. To illustrate typical methods of the half-linear oscillation theory, we present here two different proofs.

(i) *Variational proof.* We will find for every $T \in \mathbb{R}$ a nontrivial function $y \in W_0^{1,p}(T, \infty)$ such that

$$\mathcal{F}(y; T, \infty) = \int_T^{\infty} [r(t)|y'|^p - c(t)|y|^p]\, dt \leq 0. \tag{1.2.24}$$

The function which satisfies (1.2.24) can be constructed as follows

$$y(t) = \begin{cases} 0 & T \leq t \leq t_0, \\ \int_{t_0}^{t} r^{1-q}(s)\, ds \Big(\int_{t_0}^{t_1} r^{1-q}(s)\, ds \Big)^{-1} & t_0 \leq t \leq t_1, \\ 1 & t_1 \leq t \leq t_2, \\ \int_{t}^{t_3} r^{1-q}(s)\, ds \Big(\int_{t_2}^{t_3} r^{1-q}(s)\, ds \Big)^{-1} & t_2 \leq t \leq t_3, \\ 0 & t_3 \leq t < \infty, \end{cases}$$

where $T < t_0 < t_1 < t_2 < t_3$ will be specified later. Denote

$$K := \mathcal{F}(y; t_0, t_1) = \int_{t_0}^{t_1} [r(t)|y'|^p - c(t)|y|^p]\, dt.$$

Then by a direct computation we have

$$\mathcal{F}(y; T, \infty) \;=\; K - \int_{t_1}^{t_2} c(t)\, dt + \Big(\int_{t_2}^{t_3} r^{1-q}(t)\, dt \Big)^{1-p} - \int_{t_2}^{t_3} c(t)|y|^p\, dt.$$

Since the function $g(t) = \int_t^{t_3} r^{1-q}(s)\, ds \Big(\int_{t_2}^{t_3} r^{1-q}(s)\, ds \Big)^{-1}$ is monotonically decreasing on $[t_2, t_3]$ with $g(t_2) = 1$ and $g(t_3) = 0$, by the second mean value theorem of integral calculus there exists $\xi \in (t_2, t_3)$ such that

$$\int_{t_2}^{t_3} c(t) g^p(t)\, dt = \int_{t_2}^{\xi} c(t)\, dt.$$

Using this equality, we have

$$\mathcal{F}(y; t_0, t_3) = K - \int_{t_1}^{\xi} c(t)\, dt + \left(\int_{t_2}^{t_3} r^{1-q}(t)\, dt \right)^{1-p}.$$

Let $\varepsilon > 0$ and $t_1 > t_0$ be arbitrary. The second condition in (1.2.23) implies that t_2 can be chosen in such a way that $\int_{t_1}^{\xi} c(t)\, dt > K + \varepsilon$ and the first condition in (1.2.23) implies that $t_3 > t_2$ can be taken such that $\left(\int_{t_2}^{t_3} r^{1-q}(t)\, dt \right)^{-1} < \varepsilon$. Summarizing these estimates, we have

$$\mathcal{F}(y; t_0, t_3) \le K - (K + \varepsilon) + \varepsilon \le 0,$$

hence (1.1.1) is oscillatory by Theorem 1.2.2.

(ii) *Proof by the Riccati technique.* Suppose, by contradiction, that (1.2.23) holds and (1.1.1) is nonoscillatory. Then there exists $T \in \mathbb{R}$ and a solution w of Riccati equation (1.1.21) which is defined in the whole interval $[T, \infty)$. By integrating (1.1.21) from T to t we get

$$w(t) = w(T) - \int_T^t c(s)\, ds - (p-1) \int_T^t r^{1-q}(s) |w(s)|^q\, ds.$$

The second condition in (1.2.23) implies the existence of $T_1 > T$ such that we have $w(T) - \int_T^t c(s)\, ds \le 0$ for $t > T_1$ and hence

$$w(t) \le -(p-1) \int_T^t r^{1-q}(s) |w(s)|^q\, ds \quad \text{for } t > T_1.$$

Denote $G(t) = \int_T^t r^{1-q}(s) |w(s)|^q\, ds$, then $|w| = \left[G' r^{q-1} \right]^{1/q}$ and the last inequality reads

$$\frac{G'(t)}{G^q(t)} \ge (p-1)^q r^{1-q}(t).$$

By integrating this inequality from T_1 to t we get

$$\frac{1}{q-1} G^{1-q}(T_1) > \frac{1}{q-1} \left[G^{1-q}(T_1) - G^{1-q}(t) \right] \ge (p-1)^q \int_{T_1}^t r^{1-q}(s)\, ds.$$

Letting $t \to \infty$ we have a contradiction with the first condition in (1.2.23). □

1.3 Differences between linear and half-linear equations

The basic difference between linear and half-linear equations has already been mentioned at the beginning of this chapter. In this section we point out some other differences (in fact, they are more or less consequences of the lack of the additivity of the solution space).

1.3.1 Wronskian identity

If y_1, y_2 are two solutions of the *linear* Sturm-Liouville differential equation (1.1.2), then by a direct differentiation one can verify the so-called *Wronskian identity*

$$r(t)[y_1(t)y_2'(t) - y_1'(t)y_2(t)] \equiv \omega, \tag{1.3.1}$$

where ω is a real constant. In [140], Elbert showed that there is no half-linear version of this identity. In fact, he considered equation (1.1.1) with $r(t) \equiv 1$, i.e.,

$$(\Phi(x'))' + c(t)\Phi(x) = 0, \tag{1.3.2}$$

which however is not a serious restriction. More precisely, we assume a *generalized Wronskian* as a function of four variables $W = W(y_1, y_1', y_2, y_2')$. This function is subject to the following "reasonable" restrictions (typical for Wronskian), and the subsequent theorem says that the only possibility is then, in fact, the linear case. The restrictions are:

(i) W is continuously differentiable with respect to each variable,

(ii) W is not identically constant on $\mathbb{R}^4$,

(iii) for every pair of the solutions y_1, y_2 of (1.3.2) the Wronskian W is constant, i.e., independent of t and of the coefficient c,

(iv) W has antisymmetry property $W(x_3, x_4, x_1, x_2) = -W(x_1, x_2, x_3, x_4)$.

Theorem 1.3.1. *Let y_1 and y_2 be two arbitrary solutions of* (1.3.2) *and let the function $W(y_1, y_1', y_2, y_2')$ be the generalized Wronskian for* (1.3.2) *satisfying the conditions* (i)–(iii). *Then the only possibility is $p = 2$ and $W = \Psi(y_1y_2' - y_1'y_2)$, where the function $\Psi = \Psi(u)$ is continuously differentiable. If, in addition, the condition* (iv) *is satisfied, then the function $\Psi(u)$ is odd.*

Proof. By the polar transformation (1.1.17) (with $r(t) \equiv 1$) we can look for the Wronskian also in the form $W = \Psi^0(\rho_1, \rho_2, \varphi_1, \varphi_2)$, where the functions ρ_i, φ_i are the polar coordinates of the points (y_i', y_i), the function $\Psi^0(u_1, u_2, u_3, u_4)$ is defined on $\mathcal{D} = \mathbb{R}^+ \times \mathbb{R}^+ \times \mathbb{R} \times \mathbb{R}$ and the partial derivatives $\Psi_i^0 = \partial\Psi^0/\partial u_i$ $(i = 1, 2, 3, 4)$ are continuous functions in $\mathcal{D}$. Let us compute the derivative $d\Psi^0/dt$. By (1.1.20) we have $d\Psi^0/dt = \Gamma_1 + c\Gamma_2$, where

$$\begin{aligned} \Gamma_1 &= \Psi_1^0\rho_1 S_1'\Phi(S_1) + \Psi_2^0\rho_2 S_2'\Phi(S_2) + \Psi_3^0|S_1'|^p + \Psi_4^0|S_2'|^p, \\ \Gamma_2 &= -\Psi_1^0\rho_1 S_1'\Phi(S_1) - \Psi_2^0\rho_2 S_2'\Phi(S_2) + \Psi_3^0|S_1|^p + \Psi_4^0|S_2|^p, \end{aligned} \tag{1.3.3}$$

and $S_i = \sin_p \varphi_i$, $S_i' = \cos_p \varphi_i$, $i = 1, 2$. The condition (iii) implies that

$$\Gamma_1 = \Gamma_2 = 0. \tag{1.3.4}$$

Hence $\Gamma_1 + \Gamma_2 = 0$ and taking into account (1.1.13) we get by (1.3.3) $\Psi_3^0 + \Psi_4^0 = 0$. Let ρ_1, ρ_2 be fixed and $\varphi_2 = \varphi_1 + k$, k being a constant. Then

$$\frac{d\Psi^0(\rho_1, \rho_2, \varphi_1, \varphi_1 + k)}{dt} = \Psi_3^0 + \Psi_4^0 = 0,$$

i.e., along the line $(\rho_1, \rho_2, \varphi_1, \varphi_1 + k)$, $-\infty < \varphi_1 < \infty$, the function Ψ^0 is constant. Hence it depends on ρ_1, ρ_2 and $\varphi_1 - \varphi_2 = \varphi$, i.e., $\Psi^0 = \Psi^1(\rho_1, \rho_2, \varphi_1 - \varphi_2)$. Substituting this function in (1.3.3) we have for Γ_2 by (1.3.4)

$$\Psi_1^1 \rho_1 S_1' \Phi(S_1) + \Psi_2^1 \rho_2 S_2' \Phi(S_2) = \Psi_3^1(|S_1|^\alpha - |S_2|^\alpha). \tag{1.3.5}$$

Let us choose specially $\varphi_1 = \varphi, \varphi_2 = -\varphi$, $\varphi \in \mathbb{R}$. Since the function $\sin_p \varphi$ is odd and $\cos_p \varphi$ is even, we have $(\Psi_1^1 \rho_1 - \Psi_2^1 \rho_2) \cos_p \varphi \Phi(\sin_p \varphi) = 0$. Thus for $\varphi \neq m\pi_p/2$, $m = 0, \pm 1, \pm 2, \dots$ it follows that

$$\Psi_1^1 \rho_1 - \Psi_2^1 \rho_2 = 0. \tag{1.3.6}$$

Now we fix the value of φ. Since $\rho_1 > 0, \rho_2 > 0$, we can define the function Θ by $\Theta(\log \rho_1, \log \rho_2) = \Psi^1(\rho_1, \rho_2, 2\varphi)$. Then for the function Θ we obtain $\partial\Theta/\partial\rho_i = \Theta_i/\rho_i$ and the relation (1.3.6) becomes $\Theta_1 - \Theta_2 = 0$. By a similar argumentation as above it follows that $\Theta(u_1, u_2) = \Theta^1(u_1 + u_2)$ and so $\Psi^1(\rho_1, \rho_2, 2\varphi) = \Theta^1(\log \rho_1 + \log \rho_2)$ for almost every $\varphi \in \mathbb{R}$ with the exceptions $\varphi = m\pi_p/2$ $(m = 0, \pm 1, \pm 2, \dots)$. But the continuity of Ψ^1 with respect to its third variable implies that this relation holds for these exceptional values of φ as well, hence $\Psi^1(\rho_1, \rho_2, \varphi_1 - \varphi_2) = \Psi^2(\rho_1 \rho_2, \varphi_1 - \varphi_2)$. By virtue of (1.3.5), the function Ψ^2 satisfies the relation

$$\Psi_1^2 \rho_1 \rho_2 (S_1' \Phi(S_1) + S_2' \Phi(S_2)) = \Psi_2^2(|S_1|^p - |S_2|^p). \tag{1.3.7}$$

Putting $\varphi_1 = \varphi, \varphi_2 = 0$ in (1.3.7) we have

$$\Psi_1^2 \rho_1 \rho_2 \cos_p \varphi \Phi(\sin_p \varphi) = \Psi_2^2 |\sin_p \varphi|^p. \tag{1.3.8}$$

Let $\sin_p \varphi > 0$, or say, let $0 < \varphi < \pi_p$ and let $\Theta(u_1, u_2)$ be defined by

$$\Psi^2(u, \varphi) = \Theta(\log u, \log \sin_p \varphi). \tag{1.3.9}$$

Then from (1.3.8) it follows that $(\Theta_1 - \Theta_2) \cos_p \varphi \Phi(\sin_p \varphi) = 0$, hence $\Theta = \bar{\Theta}(\log u + \log \sin_p \varphi) = \bar{\Theta}(\log(u \sin_p \varphi)) = \Psi(u \sin_p \varphi)$, and by (1.3.9)

$$\Psi^2(\rho_1 \rho_2, \varphi) = \Psi(\rho_1 \rho_2 \sin_p \varphi). \tag{1.3.10}$$

A similar argumentation provides the extension of (1.3.10) from $0 < \varphi < \pi_p$ to all $\varphi \in \mathbb{R}$. Hence we get

$$W(y_1, y_1' y_2, y_2') = \Psi(\rho_1 \rho_2 \sin_p(\varphi_1 - \varphi_2)), \tag{1.3.11}$$

where the function $\Psi = \Psi(u)$ is continuously differentiable.

In order to prove our theorem, we must show that relation (1.3.11) can be satisfied only in the case $p = 2$. Let us substitute this final form into (1.3.7). Then we have $\Psi' \rho_1 \rho_2 \Gamma(\varphi_1, \varphi_2) = 0$, where

$$\begin{aligned}\Gamma(\varphi_1, \varphi_2) = \sin_p(\varphi_1 - \varphi_2)[\cos_p \varphi_1 \Phi(\varphi_1) + \cos_p \varphi_2 \Phi(\varphi_2)] \\ - \cos_p(\varphi_1 - \varphi_2)[|\sin_p \varphi_1|^p - |\sin_p \varphi_2|^p].\end{aligned} \tag{1.3.12}$$

Since by (ii) the function W is not identically constant, there is a value u_0 for which $\Psi'(u_0) \neq 0$. By the continuity of $\Psi'(u)$ we may suppose that $u_0 \neq 0$. Hence for the values $\rho_1, \rho_2, \varphi_1, \varphi_2$ subject to $\rho_1\rho_2 \sin_p(\varphi_1 - \varphi_2) = u_0$ the equality $\Gamma(\varphi_1, \varphi_2) = 0$ holds. Owing to the freedom of choosing the values $\rho_1\rho_2 > 0$ the equality $\Gamma(\varphi_1, \varphi_2) = 0$ is valid for all φ_1, φ_2 subject to $u_0 \sin_p(\varphi_1 - \varphi_2) > 0$. Let us choose the value of ε small and let $u_0 \sin_p \varepsilon > 0$. Then $\Gamma(2\varepsilon, \varepsilon) = 0$, hence

$$\lim_{\varepsilon \to 0} \frac{\Psi(2\varepsilon, \varepsilon)}{|\varepsilon|^p} = 0. \tag{1.3.13}$$

On the other hand, L'Hospital's rule yields from (1.3.12)

$$\lim_{\varepsilon \to 0} \frac{\Psi(2\varepsilon, \varepsilon)}{|\varepsilon|^p} = 2^{p-1} + 1 - (2^p - 1) = 2 - 2^{p-1}.$$

Combining this with (1.3.13) we obtain the only possible value $p = 2$ in accordance with the first part of the theorem, since $S_1(\varphi) = \sin \varphi$.

Applying the restriction (iv) on the Wronskian given by (1.3.11) we have

$$\Psi(\rho_1\rho_2 \sin_p(\varphi_2 - \varphi_1)) = -\Psi(\rho_1\rho_2 \sin_p(\varphi_1 - \varphi_2)).$$

Since the function $\sin_p \varphi$ is odd, we find $\Psi(-u) = -\Psi(u)$, i.e., the function $\Psi(u)$ is odd, which completes the proof. □

The absence of a Wronskian type identity implies that we have also no analogue of the linear reduction of order formula: given a solution $\tilde{x}$ of (1.1.2) with $\tilde{x}(t) \neq 0$ in an interval I, then

$$x(t) = \tilde{x}(t) \int^t \frac{ds}{r(s)\tilde{x}^2(s)}$$

is another solution of (1.1.2).

On the other hand, the concept of Wronskian can be utilized in the characterization of linear (in)dependence of solutions of (1.1.1). If x, y are two continuously differentiable functions, then the Wronskian is defined as the function

$$W(x, y)(t) = x(t)y'(t) - x'(t)y(t).$$

The next statement shows that similarly to the linear case, the Wronskian of two solutions of (1.1.1) is either identically zero or always nonzero.

Lemma 1.3.1. *Let x and y be two nontrivial solutions of* (1.1.1) *defined on I. Then either $W(x, y) \equiv 0$, in this case x and y are linearly dependent (i.e., proportional) on I, or $W(x, y) \neq 0$ on I and then x and y are linearly independent on I*

Proof. We offer two different approaches. Let us suppose that there is $t_0 \in I$ such that $W(x, y)(t_0) = 0$. Then the system of linear equations

$$\lambda_1 x(t_0) - \lambda_2 y(t_0) = 0, \quad \lambda_1 x'(t_0) - \lambda_2 y'(t_0) = 0$$

for the unknowns λ_1, λ_2 has a nontrivial solution (λ_1, λ_2). Now we have $\lambda_1 \neq 0$, $\lambda_2 \neq 0$ because otherwise it would be either $x(t_0) = x'(t_0) = 0$ or $y(t_0) = y'(t_0) = 0$, which is impossible. Let $\lambda = \lambda_1/\lambda_2$. Then we have $y(t_0) = \lambda x(t_0)$, $y'(t_0) = \lambda x'(t_0)$, $\lambda \neq 0$. First we suppose $\lambda > 0$. Let φ_x^0, ϱ_x^0 be the polar coordinates of the point $(r^{q-1}(t_0)x'(t_0); x(t_0))$. Then φ_x^0, $\lambda\varrho_x^0$ are the corresponding polar coordinates of the point $(r^{q-1}(t_0)y'(t_0); y(t_0))$. Now let φ_x, ϱ_x and φ_y, ϱ_y be defined as in (1.1.17) and correspond to x and y, respectively. Then these functions are solutions of (1.1.20) satisfying the initial conditions $\varphi_x(t_0) = \varphi_x^0 = \varphi_y(t_0)$, $\varrho_x(t_0) = \varrho_x^0$ and $\varrho_y(t_0) = \lambda\varrho_x^0$. Owing to the uniqueness we have $\varphi_x(t) \equiv \varphi_y(t)$, $\varrho_y(t) \equiv \lambda\varrho_x(t)$ on I, consequently $y = \lambda x$ on I, i.e., the solutions x, y are linearly dependent and the Wronskian is identically zero. In the case $\lambda < 0$ we consider the function $z = -y$. Then z is also a solution of (1.1.1) and $z(t_0) = \bar{\lambda}x(t_0)$, $z'(t_0) = \bar{\lambda}x'(t_0)$ where $\bar{\lambda} = -\lambda$. Since $\bar{\lambda} > 0$, we find again $z(t) \equiv \bar{\lambda}x(t)$ on I, hence $y(t) \equiv \lambda x(t)$, i.e., the solutions x, y are linearly dependent. If the Wronskian is never vanishing on I, then there is no λ such that $y(t) \equiv \lambda x(t)$ on I, consequently the solutions x, y are linearly independent, which completes the proof.

It is worthy to mention that the proof can be also done via the Riccati equation. Actually, only the middle part of the proof is changed as follows. We have $y(t_0) = \lambda x(t_0)$, $y'(y_0) = \lambda x'(t_0)$, $\lambda \neq 0$. Here we assume that $x(t_0) \neq 0$. If not, then such a situation can be solved very easily, see Lemma 1.2.2. Put $w_x^0 = (r\Phi(x'/x))(t_0)$ and $w_y^0 = (r\Phi(y'/y))(t_0)$. The functions w_x, w_y defined by $w_x = r\Phi(x'/x)$ and $w_y = r\Phi(y'/y)$ satisfy the same Riccati equation (1.1.21) (as long as they exist) and since $w_x^0 = w_y^0$, they coincide on the interval of the existence, which implies that x and y are linearly dependent. □

Remark 1.3.1. (i) It is very easy to see that, in fact, we have two equivalences: nontrivial solutions x and y of (1.1.1) are linearly dependent [resp. independent] on I if and only if $W(x,y)(t) \equiv 0$ on I [resp. $W(x,y)(t) \neq 0$ for all $t \in I$].

(ii) If we consider a slightly more general equation, like (1.1.7), where the set (ii) of restrictions (presented after (1.1.7)) is imposed on f, then there is an example showing that the Wronskian of two solutions of (1.1.7) may be identically zero on some interval although the solutions are linearly independent, see [143]. This unpleasant phenomenon however disappears if we add the assumption $f(-x,-y) = -f(x,y)$.

1.3.2 Transformation formula

Let $h(t) \neq 0$ be a differentiable function such that rh' is also differentiable and let us introduce a new dependent variable y which is related to the original variable x by the formula $x = h(t)y$. Then we have the following (linear) identity which is the basis of the linear transformation theory (see [10, 52, 305])

$$h(t)[(r(t)x')' + c(t)x] = (r(t)h^2(t)y')' + h(t)[(r(t)h'(t))' + c(t)h(t)]y. \tag{1.3.14}$$

In particular, if x a solution of (1.1.2) then y is a solution of

$$(R(t)y')' + C(t)y = 0$$

with $R = rh^2$ and $C = h[(rh')' + ch]$. Since the function Φ is not additive, we have no half-linear analogue of this transformation identity. This has the following important consequence. Many oscillation results for linear equation (1.1.2) are based on the so-called *trigonometric transformation* which reads as follows. Let x_1, x_2 be two (linearly independent) solutions of (1.1.2) such that $r(x_1x_2' - x_1'x_2) = 1$ and let $h = \sqrt{x_1^2 + x_2^2}$. Then we have the identity (which can be verified by a direct computation)

$$h[(rh')' + ch] = \frac{1}{rh^2}.$$

This means that the transformation $x = h(t)y$ transforms (1.1.2) into the equation

$$\left(\frac{1}{q(t)}y'\right)' + q(t)y = 0, \quad q(t) = \frac{1}{r(t)h^2(t)}. \tag{1.3.15}$$

Equation (1.3.15) can be solved explicitly and

$$y_1 = \sin\left(\int^t q(s)\,ds\right), \quad y_2 = \cos\left(\int^t q(s)\,ds\right)$$

are its linearly independent solutions, in particular, (1.3.15) and hence also (1.1.2) is oscillatory if and only if

$$\int^\infty \frac{dt}{r(t)[x_1^2(t) + x_2^2(t)]} = \infty$$

for any pair of linearly independent solutions x_1, x_2 of (1.1.2). This fact is used in proofs of many oscillation results for (1.1.2), see [318, 341] and references given therein. Since half-linear version of the transformation formula (1.3.14) is missing, analogous results for half-linear equation (1.1.1) are not known.

1.3.3 Fredholm alternative

Consider the linear Dirichlet boundary value problem associated with (1.1.2)

$$\begin{aligned} &(r(t)x')' + c(t)x = f(t), \quad t \in [a, b], \\ &x(a) = 0 = x(b). \end{aligned} \tag{1.3.16}$$

It is well known that if the homogeneous problem with $f(t) \equiv 0$ has only the trivial solution, then (1.3.16) has a solution for any (sufficiently smooth) right-hand side f (the so-called *nonresonant case*). If the homogeneous problem has a solution φ_0, problem (1.3.16) has a solution if and only if

$$\int_a^b f(t)\varphi_0(t)\,dt = 0.$$

In particular, the problem

$$x'' + x = f(t), \quad x(0) = 0 = x(\pi), \tag{1.3.17}$$

has a solution if and only if $\int_0^\pi f(t)\sin t\,dt = 0$.

Now consider the half-linear version of the boundary value problem (1.3.17)

$$(1.3.18)\qquad \begin{aligned}&(\Phi(x'))' + (p-1)\Phi(x) = f(t), \quad t \in [a,b],\\ &x(0) = 0 = x(\pi_p),\end{aligned}$$

where the generalized π_p is given by (1.1.12). A natural question is whether

$$(1.3.19)\qquad \int_0^{\pi_p} f(t)\sin_p t\,dt = 0$$

is a necessary and sufficient condition for solvability of (1.3.18). This problem has attracted considerable attention in the last years, see [118] and the references given therein. It was shown that (1.3.19) is *sufficient* but *no longer necessary* for solvability of (1.3.18). We will deal with this problem in details in Chapter 6.

1.4 Some elementary half-linear equations

In this section we focus our attention to half-linear equations with constant coefficients and to Euler-type half-linear equation.

1.4.1 Equations with constant coefficients

Before passing to half-linear equations with constant coefficients, consider the equation

$$(1.4.1)\qquad (r(t)\Phi(x'))' = 0.$$

Here the situation is the same as in case of linear equations. The solution space of this equation is a two-dimensional linear space and the basis of this space is formed by the functions $x_1(t) \equiv 1$, $x_2(t) = \int^t r^{1-q}(s)\,ds$, where $q = \frac{p}{p-1}$ is the conjugate number of p.

Now, consider equation (1.1.1) with $r(t) \equiv r > 0$ and $c(t) \equiv c$. This equation can be written in the form

$$(1.4.2)\qquad (\Phi(x'))' + \frac{c}{r}\Phi(x) = 0$$

and the transformation of independent variable $t \mapsto \lambda t$ with $\lambda = \left(\frac{|c|}{r(p-1)}\right)^{1/p}$ transforms (1.4.2) into the equation

$$(\Phi(x'))' + (p-1)\operatorname{sgn c}\Phi(x) = 0$$

If $c > 0$, this equation already appeared in Subsection 1.1.3 as equation (1.1.9). Its solution given (uniquely) by the initial conditions $x(0) = 0$, $x'(0) = 1$ was denoted by $\sin_p t$. Consequently, taking into account homogeneity of the solution space of (1.1.1), we have the following statement concerning solvability of (1.1.9).

Theorem 1.4.1. *For any $t_0 \in \mathbb{R}$, $x_0, x_1 \in \mathbb{R}$, the unique solution of* (1.1.9) *satisfying $x(t_0) = x_0$, $x'(t_0) = x_1$ is of the form $x(t) = \alpha \sin_p(t - t_1)$, where α, t_1 are real constants depending on t_0, x_0, x_1.*

Now let $c < 0$, i.e., we consider the equation

$$(\Phi(x'))' - (p-1)\Phi(x) = 0. \tag{1.4.3}$$

Multiplying this equation by x' and integrating the obtained equation over $[0, t]$ we have the identity

$$|x'(t)|^p - |x(t)|^p = |x'(0)|^p - |x(0)|^p = C, \tag{1.4.4}$$

where C is a real constant. If $C = 0$ then $x' = \pm x$ thus $x_1 = e^t$ and $x_2 = e^{-t}$ are solutions of (1.4.3).

In the remaining part of this subsection we focus our attention to the generalized half-linear hyperbolic sine and cosine functions. In the linear case $p = 2$, these functions are linear combinations of e^t and e^{-t}. However, in the half-linear case the additivity of the solution space of (1.1.1) is lost and one has to use a more complicated method. Let $E = E_p(t)$ be the solution of (1.4.3) with the initial

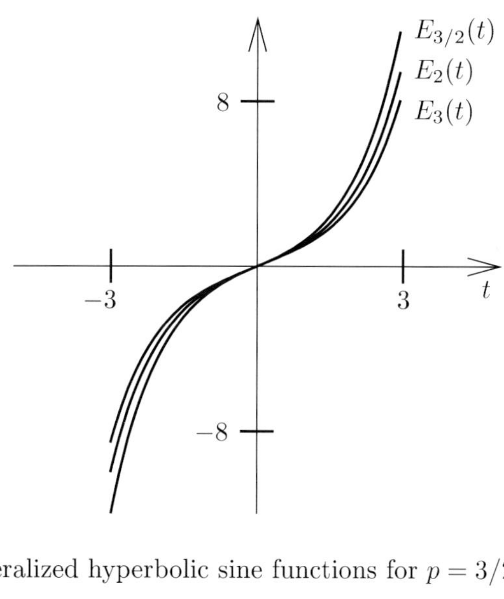

Figure 1.4.1: Generalized hyperbolic sine functions for $p = 3/2$, $p = 2$, and $p = 3$

conditions $E(0) = 0, E'(0) = 1$, and similarly let $F = F_p(t)$ be the solution given by the initial conditions $F(0) = 1, F'(0) = 0$. Let us also observe that the function E corresponds to $C = 1$ and F to $C = -1$ in (1.4.4), respectively. Moreover, for $p = 2$, i.e., if differential equation (1.4.3) is linear, we have $E_2(t) = \sinh t$ and $F_2(t) = \cosh t$.

Due to (1.4.4), the function E satisfies also the relation

$$E' = \sqrt[p]{1 + |E|^p},$$

hence $E' > 1$ for $t > 0$. Consequently

$$t = \int_0^t \frac{E'(s)\,ds}{\sqrt[p]{1+E^p(s)}} = \int_0^{E(t)} \frac{ds}{\sqrt[p]{1+s^p}}. \tag{1.4.5}$$

In order to compare (asymptotically) the function $E(t)$ with e^t, let the function $f(s)$ be defined by

$$f(s) = \begin{cases} 1 & \text{for} \quad 0 \le s \le 1, \\ 1/s & \text{for} \quad s > 1. \end{cases}$$

Then by (1.4.5) we obtain

$$t - \log E(t) = 1 + \int_0^{E(t)} \frac{ds}{\sqrt[p]{1+s^p}} - \int_0^{E(t)} f(s)ds \quad \text{for} \quad E(t) > 1.$$

Hence

$$\log \delta_p := \lim_{t\to\infty} [t - \log E(t)] = 1 - \int_0^{\infty} \left[f(s) - \frac{1}{\sqrt[p]{1+s^p}} \right] ds. \tag{1.4.6}$$

The integral in the right-hand side of (1.4.6) can be interpreted as the area of the domain in the (s, y)-plane given by the inequalities

$$\frac{1}{\sqrt[p]{1+s^p}} \le y \le f(s) \quad \text{for} \quad 0 \le s < \infty.$$

Taking y as an independent variable, we find for the integral in (1.4.6)

$$\log \delta_p = 1 - \int_0^1 \frac{1 - \sqrt[p]{1-y^p}}{y} dy = 1 - \frac{1}{p} \int_0^1 \frac{1 - u^{1/p}}{1-u} du. \tag{1.4.7}$$

Since $0 < 1 - \sqrt[p]{1-y^p} < y$ for $0 < y < 1$ we have $0 < \log \delta_p < 1$, i.e., $1 < \delta_p < e$. On the other hand, the integral in (1.4.7) can be expressed by means of the function $\Psi(z) = d \log \Gamma(z)/dz$ as

$$\Psi(z) = -\tilde{C} + \int_0^1 \frac{1 - t^{z-1}}{1-t} dt \quad \text{for} \quad \text{Re } z > 0,$$

where $\tilde{C}$ is the Euler constant and $\Gamma(z) = \int_0^\infty t^{z-1} e^{-t}\, dt$ denotes the usual Euler gamma function. Making use of this relation we obtain

$$\log \delta_p = 1 - \frac{1}{p} \left[\tilde{C} + \Psi\left(\frac{p+1}{p} \right) \right]. \tag{1.4.8}$$

Finally, the relation (1.4.6) can be rewritten as

$$\lim_{t\to\infty} \frac{e^t}{E_p(t)} = \delta_p \quad \text{where} \quad 1 < \delta_p < e. \tag{1.4.9}$$

A similar relation is expected also for the function $F_p(t)$. We will use the following auxiliary statement which we present without the proof, this proof can be found in [139].

Lemma 1.4.1. *Let $I_1(R), I_2(R)$ be integrals defined by*

$$I_1(R) = \int_0^R \frac{d\xi}{\sqrt[p]{1+\xi^p}}, \quad R > 0, \quad I_2(R) = \int_1^R \frac{d\xi}{\sqrt[p]{\xi^p - 1}}, \quad R > 1.$$

Then

$$\lim_{R\to\infty} [I_1(R) - I_2(R)] = \frac{\pi}{p} \cot \frac{\pi}{p}.$$

Now we return to the asymptotic formula for F_p. We want to obtain a relation similar to (1.4.9). Since F fulfills the differential equation

$$|F'|^p - |F|^p = -1,$$

we have

$$\frac{F'}{\sqrt[p]{F^p - 1}} = 1 \quad \text{for} \quad t > 0.$$

Integrating the last equality yields

$$\int_1^{F(t)} \frac{d\xi}{\sqrt[p]{\xi^p - 1}} = t \quad \text{for} \quad t > 0.$$

This relation implies that $\lim_{t\to\infty} F(t) = \infty$. On the other hand

$$\begin{aligned} \lim_{t\to\infty} [t - \log F(t)] &= \lim_{t\to\infty} \int_1^{F(t)} \left[\frac{1}{\sqrt[p]{\xi^p - 1}} - f(\xi)\right] d\xi \\ &= \int_1^{\infty} \left[\frac{1}{\sqrt[p]{\xi^p - 1}} - \frac{1}{\xi}\right] d\xi \end{aligned} \tag{1.4.10}$$

because the integral on the right-hand side is convergent. Let Λ_p be introduced by

$$\log \Lambda_p := \int_1^{\infty} \left[\frac{1}{\sqrt[p]{\xi^p - 1}} - \frac{1}{\xi}\right] d\xi. \tag{1.4.11}$$

It is clear that $\Lambda_p > 1$. The relation (1.4.10) can be rewritten as

$$\lim_{t\to\infty} \frac{e^t}{F_p(t)} = \Lambda_p \quad \text{with} \quad \Lambda_p > 1. \tag{1.4.12}$$

Now we want to establish a connection between Λ_p and δ_p. By (1.4.6), (1.4.10), and taking into account the definition of the function $f(\xi)$, we have

$$\begin{aligned} \log \frac{\Lambda_p}{\delta_p} &= \lim_{R\to\infty} \left[\int_1^R \left(\frac{1}{\sqrt[p]{\xi^p - 1}} - \frac{1}{\xi}\right) d\xi - 1 + \int_0^R \left(f(\xi) - \frac{1}{\sqrt[p]{1+\xi^p}}\right) d\xi\right] \\ &= \lim_{R\to\infty} [I_2(R) - I_1(R)], \end{aligned}$$

where the functions $I_1(R), I_2(R)$ were introduced in Lemma 1.4.1. Then by this lemma we get the wanted relation as

$$\log \frac{\delta_p}{\Lambda_p} = \frac{\pi}{p} \cot \frac{\pi}{p}. \tag{1.4.13}$$

We may observe here that this relation implies $\delta_2 = \Lambda_2$ in the linear case ($p = 2$). In fact, we have $\delta_2 = 2 = \Lambda_2$.

By (1.4.8) the value of δ_p can be considered to be known, consequently by the relation (1.4.13) the value of Λ_p is known as well.

Finally, there are interesting functional relations between the half-linear hyperbolic sine and cosine functions $E_p(t), F_p(t)$ as follows

(1.4.14)
$$E_p'(t) = \{F_q((p-1)t)\}^{q-1} = \Phi^{-1}(F_q((p-1)t)), \quad F_p'(t) = \Phi^{-1}\left(E_q((p-1)t)\right).$$

To prove these relations it is sufficient to show that the functions on both sides of the equalities satisfy the same differential equation and fulfill the same initial conditions, this is a matter of a direct computation (use e.g. the result of Subsection 1.2.8).

The relations (1.4.14) provide another connections between the values of δ_p and Λ_p. Indeed, by (1.4.4) and (1.4.9) we have

$$\lim_{t\to\infty} \frac{E_p'(t)}{e^t} = \lim_{t\to\infty} \frac{E_p(t)}{e^t} = \frac{1}{\delta_p}. \tag{1.4.15}$$

On the other hand, this, (1.4.12) and (1.4.14) imply (taking into account that $(p-1)(q-1) = 1$ for conjugate pair p, q)

$$\frac{1}{\delta_p^{p-1}} = \lim_{t\to\infty} \frac{(E_p'(t))^{p-1}}{e^{(p-1)t}} = \lim_{t\to\infty} \frac{F_q((p-1)t)}{e^{(p-1)t}} = \frac{1}{\Lambda_q}$$

hence

$$\Lambda_q = \delta_p^{p-1}, \tag{1.4.16}$$

and similarly

$$\delta_q = \Lambda_p^{p-1}. \tag{1.4.17}$$

We remark that the last two relations are equivalent since replacing p by q we get each from the other. By relations (1.4.13), (1.4.16) (or (1.4.17)) it is sufficient to know one of the values of $\Lambda_p, \delta_p, \Lambda_q, \delta_q$, and then all the other values can be obtained easily.

As in the linear case where the function $\sinh t$ is odd and the function $\cosh t$ is even, the functions $E_p(t), F_p(t)$ behave in a similar way:

$$E_p(-t) = -E_p(t), \quad F_p(-t) = F_p(t). \tag{1.4.18}$$

To prove this statement it is sufficient to show that the functions on the both side of the equality are solutions of differential equation (1.4.3) and satisfy the same initial conditions at $t = 0$. Then the uniqueness of the initial value problem (see Subsection 1.1.5) proves (1.4.18).

Now we know all the solutions of differential equations (1.4.3). We display them in the next theorem.

Theorem 1.4.2. *The solutions of* (1.4.3) *are:*

$$Ke^t, \quad Ke^{-t}, \quad KE_p(t+t_0), \quad KF_p(t+t_0), \tag{1.4.19}$$

where K and t_0 are real parameters. More precisely, there are two one-parameter families of solutions $x(t) = Ke^t$, $x(t) = Ke^{-t}$ and two two-parameter families satisfying the following asymptotic formula

$$\lim_{t\to\infty} \frac{x(t)}{e^t} = L,$$

where $L = Ke^{t_0}/\delta_p$ or $L = Ke^{t_0}/\Lambda_p$ with K from (1.4.19).

Proof. Since equation (1.4.3) is autonomous, it is sufficient to consider solutions whose initial values are prescribed at $t = 0$, i.e. $x(0) = x_0$, $x'(0) = x_1$. If $x_0 = 0 = x_1$, then according to the unique solvability, $x(t) \equiv 0$. If at least one of the constants x_0, x_1 is nonzero, distinguish the cases according to the value of the constant C in (1.4.4). In case $C = 0$, the only possibilities are $x(t) = Ke^t$ or $x(t) = Ke^{-t}$. More precisely, if $x_0x_1 > 0$, then $x(t) = x_0e^t$, if $x_0x_1 < 0$, then $x(t) = x_0e^{-t}$. Now, if $C > 0$, then by the definition of the function E_p we have $x(t) = KE_p(t+t_0)$. Let $K = \operatorname{sgn} x_0 \sqrt[p]{C}$. Since the function E_p is strictly increasing (observe that $E'(0) = 1$, E' is continuous and $|E'|^p = 1 + |E|^p$) there exists $t_0 \in \mathbb{R}$ such that $CE_p(t_0) = x_1$. Concerning the initial condition for the derivative x', we have

$$|x_1|^p = C + |x_0|^p = C + C|E|^p = C(1 + |E|^p) = C|E'|^p,$$

hence $x_1 = \pm KE'(t_0)$. But $E'(t_0) > 0$ and $\operatorname{sgn} K = \operatorname{sgn} x_1$, the sign $+$ is the correct one. If $C < 0$, let $K = \operatorname{sgn} x_0 \sqrt[p]{-C}$ and t_0 be the solution of $CF'(t_0) = x_1$. Then the function $KF_p(t+t_0)$ is the solution we looked for.

Finally, concerning the asymptotic formula, any solution which is not proportional to e^t or e^{-t} satisfies by (1.4.9) or (1.4.12),

$$\lim_{t\to\infty} \frac{x(t)}{e^t} = \lim_{t\to\infty} \frac{KE(t+t_0)}{e^{t+t_0}} e^{t_0} = \frac{Ke^{t_0}}{\delta_p} \quad \text{or} \quad \lim_{t\to\infty} \frac{x(t)}{e^t} = \frac{Ke^{t_0}}{\Lambda_p},$$

hence $L = Ke^{t_0}/\delta_p$ or $L = Ke^{t_0}/\Lambda_p$. □

1.4.2 Euler type half-linear differential equation

In this subsection we deal with the Euler type (or generalized Euler) differential equation

$$(\Phi(x'))' + \frac{\gamma}{t^p}\Phi(x) = 0, \tag{1.4.20}$$

where γ is a real constant. By the analogue with the linear Euler equation we look first for solutions in the form $x(t) = t^\lambda$. Substituting into (1.4.20) we get the algebraic equation for λ

$$G(\lambda) := (p-1)|\lambda|^p - (p-1)\Phi(\lambda) + \gamma = 0.$$

The function G is convex, hence the equation $G(\lambda) = 0$ has two, one or no (real) root according to the value of γ. However, even if the first possibility happens, since the additivity of the solution space of is lost in the half-linear case, we are not able to compute other solutions explicitly. To get a more detailed information about their asymptotic behavior, we use the procedure which is also typical in the linear case, namely the transformation of (1.4.20) into an equation with constant coefficients.

The change of independent variable $s = \log t$ converts (1.4.20) into the equation (where the dependent variable will be denoted again by x and $' = \frac{d}{ds}$)

$$(\Phi(x'))' - (p-1)\Phi(x') + \gamma\Phi(x) = 0. \tag{1.4.21}$$

The Riccati equations corresponding to (1.4.20) and (1.4.21) are

$$w' = -\gamma t^{-p} - (p-1)|w|^p \tag{1.4.22}$$

and

$$v' = -\gamma + (p-1)v - (p-1)|v|^q =: F(v), \tag{1.4.23}$$

respectively. The solutions w and v are related by the formula $w(t) = t^{1-p}v(\log t)$ and, moreover, we have $G(\Phi^{-1}(v)) = -F(v)$.

The function F is concave on $\mathbb{R}$ with the global maximum at $\tilde{v} = \left(\frac{p-1}{p}\right)^{p-1}$ and the value of this maximum is $\tilde{\gamma} - \gamma$, where $\tilde{\gamma} = \left(\frac{p-1}{p}\right)^p$. We distinguish the following three cases with respect to the value of the constant γ.

I) $\gamma < \tilde{\gamma}$. Then the equation $F(v) = 0$ has two real roots $v_1 < \tilde{v} < v_2$;

II) $\gamma = \tilde{\gamma}$. Then the equation $F(v) = 0$ has the double root $v = \tilde{v}$;

III) $\gamma > \tilde{\gamma}$. Then $F(v) < 0$ for every $v \in \mathbb{R}$.

Case I) The constant functions $v(s) \equiv v_1$, $v(s) \equiv v_2$ are solutions of (1.4.23). Clearly, if v is a solution of (1.4.23) such that $v(s) < v_1$, for some $s \in \mathbb{R}$, then $v'(s) < 0$, if $v(s) \in (v_1, v_2)$, then $v'(s) > 0$, and $v'(s) < 0$ for $v(s) > v_2$, a picture of the direction field of (1.4.23) helps to visualize the situation. Any solution of (1.4.23) different from $v(s) = v_{1,2}$ can be expressed (implicitly) in the form ($S \in \mathbb{R}$ being fixed)

$$\int_{v(S)}^{v(s)} \frac{dv}{F(v)} = s - S. \tag{1.4.24}$$

Observe that the integral $\int_{s_1}^{s_2} \frac{ds}{F(s)}$ is convergent whenever the integration interval does not contain zeros $v_{1,2}$ of F, in particular, for any $\varepsilon > 0$

$$\int_{-\infty}^{v_1-\varepsilon} \frac{dv}{F(v)} > -\infty, \qquad \int_{v_2+\varepsilon}^{\infty} \frac{dv}{F(v)} > -\infty.$$

Case Ia) $v(S) < v_1$. Then $v(s) < v(S)$ for $s > S$ and v is decreasing. If v were extensible up to ∞, we would have a contradiction with (1.4.24) since the right-hand side tends to ∞ while the left-hand one is bounded. Later we will show that $v(s) = v_1$ is the so-called *eventually minimal solution* of (1.4.23).

Case Ib) $v(S) \in (v_1, v_2)$. The solution v is increasing, and as $s \to \infty$, we have $v(s) \to v_2$, otherwise we have the same contradiction as in the previous case.

Next we compute the asymptotic formula for the difference $v_2 - v(s)$. We have

$$F(v) = F'(v_2)(v - v_2) + O((v - v_2)^2), \quad \text{as } v \to v_2,$$

hence

$$\begin{aligned} \frac{1}{F(v)} &= \frac{1}{F'(v_2)(v - v_2)[1 + O(v - v_2)]} = \frac{1}{F'(v_2)(v - v_2)}[1 + O(v - v_2)] \\ &= \frac{1}{F'(v_2)(v - v_2)} + O(1) \end{aligned}$$

and therefore (since $|v(s) - v(S)| < v_2 - v_1 = O(1)$), substituting into (1.4.24)

$$\int_{v(S)}^{v(s)} \frac{dv}{F(v)} = \frac{1}{F'(v_2)} \log \frac{v_2 - v(s)}{v_2 - v(S)} + O(1) = s - S,$$

i.e.,

$$v_2 - v(s) = K \exp\{F'(v_2)s\},$$

where K is a positive constant (depending on $v(S)$), and substituting $w(t) = t^{1-p}v(\log t)$ we have

$$v_2 - t^{p-1}w(t) = Kt^{F'(v_2)} \to 0 \quad \text{as} \quad t \to \infty \tag{1.4.25}$$

since $F'(v_2) < 0$. Substituting $w = \Phi(x'/x)$ in (1.4.25) we have

$$\frac{\Phi(x'(t))}{\Phi(x(t))} = t^{1-p}\left(v_2 - Kt^{F'(v_2)}\right)$$

and using the formula $(1 + \alpha)^{p-1} = 1 + (p-1)\alpha + o(\alpha)$ as $\alpha \to 0$, we obtain (with $q = p/(p-1)$)

$$\frac{x'(t)}{x(t)} = \frac{\Phi^{-1}(v_2)}{t}\left(1 - \tilde{K}t^{F'(v_2)}\right)^{q-1} \sim \frac{\Phi^{-1}(v_2)}{t}\left(1 - \tilde{K}t^{F'(v_2)}\right),$$

as $t \to \infty$, here $f \sim g$ for a pair of functions f, g means $\lim_{t\to\infty} \frac{f(t)}{g(t)} = 1$, and $\tilde{K}$ is a real constant. Thus

$$x(t) = t^{\lambda_2} \exp\{\bar{K}t^{F'(v_2)}\} \sim t^{\lambda_2} \quad \text{as} \quad t \to \infty,$$

since $F'(v_2) < 0$, where $\bar{K}$ is another real constant and $\lambda_2 = \Phi^{-1}(v_2)$ is the larger of the roots of the equation $G(\lambda) = 0$.

Case Ic) $v(S) > v_2$. Then $v'(s) < 0$ and $v(s) \in (v_2, v(S))$ for $s > S$. Using the same argument as in Ib) we have

$$t^{p-1}w(t) - v_2 = \tilde{K}t^{F'(v_2)} \to 0 \quad \text{as} \quad t \to \infty,$$

$\tilde{K}$ being a positive constant, and this implies the same asymptotic formula for the solution x of (1.4.20) which determines the solution w of (1.4.22).

Case II) $\gamma = \tilde{\gamma} = \left(\frac{p-1}{p}\right)^p$. Then the function F has the double root $\tilde{v} = \left(\frac{p-1}{p}\right)^{p-1}$. Equation (1.4.20) has a solution $x(t) = t^{\Phi^{-1}(\tilde{v})} = t^{\frac{p-1}{p}}$. This means that (1.4.20) with $\gamma = \tilde{\gamma}$ is still nonoscillatory. In the linear case $p = 2$, hence $\tilde{\gamma} = 1/4$. Thus we are able to compute a linearly independent solution using the reduction of order formula. This solution is $\tilde{x}(t) = \sqrt{t}\log t$. The reduction of order formula is missing in the half-linear case, but using essentially the same method as in Case I, we are able to show that all solutions which are not proportional to $t^{\frac{p-1}{p}}$ behave asymptotically as $t^{\frac{p-1}{p}} \log^{\frac{2}{p}} t$.

To this end, we proceed as follows. Since $F(\tilde{v}) = 0 = F'(\tilde{v})$,

$$F(v) = \frac{1}{2}F''(\tilde{v})(v - \tilde{v})^2 + O((v - \tilde{v})^3) \quad \text{as} \quad v \to \tilde{v},$$

hence, taking into account that $F''(\tilde{v}) = -\frac{1}{\tilde{v}}$,

$$\begin{aligned} \frac{1}{F(v)} &= \frac{1}{\frac{1}{2}F''(\tilde{v})(v - \tilde{v})^2[1 + O(v - \tilde{v})]} \\ &= -\frac{2\tilde{v}}{(v - \tilde{v})^2} + O((v - \tilde{v})^{-1}) \quad \text{as} \quad v \to \tilde{v}. \end{aligned}$$

On the other hand, using the same argument as in the previous part, we see from (1.4.24) that any solution v which starts with the initial value $v(S) < \tilde{v}$ fails to be extensible up to ∞ and solutions with $v(S) > \tilde{v}$ tend to $\tilde{v}$ as $t \to \infty$. Substituting for $F(v)$ in (1.4.24) we have

$$\frac{2\tilde{v}}{v - \tilde{v}} + O\left(\log|v - \tilde{v}|\right) = s - S,$$

hence

$$2\tilde{v} + (v - \tilde{v})O(\log|v - \tilde{v}|) = (v - \tilde{v})(s - S).$$

Since $\lim_{v \to \tilde{v}}(v - \tilde{v})O(\log|v - \tilde{v}|) = 0$, we have

$$\lim_{s \to \infty}(s - S)\left(v(s) - \tilde{v}\right) = \lim_{s \to \infty} s(v(s) - \tilde{v}) = 2\tilde{v}.$$

Consequently,

$$O\left(\log|v(s) - \tilde{v}|\right) = O\left(\log s^{-1}\right) = O(\log s) \quad \text{as} \quad s \to \infty,$$

and thus $(v(s)-\tilde{v})^{-1} = \frac{s}{2\tilde{v}} + O(\log s)$, which means

$$\begin{aligned} v(s)-\tilde{v} &= \frac{2\tilde{v}}{s}\cdot\frac{1}{1+O\left(\frac{\log s}{s}\right)} = \frac{2\tilde{v}}{s}\left(1+O\left(\frac{\log s}{s}\right)\right) \\ &= \frac{2\tilde{v}}{s}+O\left(\frac{\log s}{s^2}\right). \end{aligned}$$

Now, taking into account that solutions of (1.4.22) and (1.4.23) are related by $w(t) = t^{1-p}v(\log t)$, we have

$$t^{p-1}w(t)-\tilde{v} = \frac{2\tilde{v}}{\log t}+O\left(\frac{\log(\log t)}{\log^2 t}\right),$$

which means that the solution x of (1.4.20) which determines the solution w of (1.4.22) satisfies

$$\frac{x'(t)}{x(t)} \sim \frac{\Phi^{-1}(\tilde{v})}{t}\left(1+\frac{2}{\log t}\right)^{\frac{1}{p-1}} \sim \frac{p-1}{pt}+\frac{2}{pt\log t}$$

and thus

$$x(t) \sim t^{\frac{p-1}{p}}\log^{\frac{2}{p}} t.$$

Case III) The equation $F(v) = 0$ has no real root and $F(v) < 0$ for every $v \in \mathbb{R}$. Again

$$\int_{v(S)}^{v(s)} \frac{dv}{F(v)} = s - S, \quad \text{and} \quad v(s) < v(S) \quad \text{for} \quad s > S.$$

Since the left-hand side of the last equality is bounded for any value $v(s)$, while the right-hand one tends to ∞ as $s \to \infty$, no solution of (1.4.23) and hence also of (1.4.22) is extensible up to ∞, which means that (1.4.20) is oscillatory. We will show that oscillatory solutions of (1.4.21) are periodic and we will determine the value of their period.

To this end, we use the Prüfer transformation mentioned in Section 1.1 applied in a slightly modified form to (1.4.21). Any nontrivial solution of this equation can be expressed in the form

$$x(s) = \rho(s)\sin_p\varphi(s), \quad x'(s) = \rho(s)\cos_p\varphi(s),$$

where $\sin_p$ and $\cos_p$ are the generalized half-linear sine and cosine functions, respectively. The angular and radial variables φ, ρ satisfy the first order differential system

$$\varphi' = |\cos_p\varphi|^p - \sin_p\varphi\Phi(\cos_p\varphi) + \frac{\gamma}{p-1}|\sin_p\varphi|^p, \tag{1.4.26}$$

$$\rho' = \cos_p\varphi\left[\Phi(\cos_p\varphi) + \left(1-\frac{\gamma}{p-1}\right)\Phi(\sin_p\varphi)\right]\rho. \tag{1.4.27}$$

Oscillation of (1.4.21) implies that $\lim_{s\to\infty}\varphi(s)=\infty$. Denote

$$\Psi(\varphi) := |\cos_p\varphi|^p - \sin_p\varphi\Phi(\cos_p\varphi) + \frac{\gamma}{p-1}|\sin_p\varphi|^p. \tag{1.4.28}$$

Then equation (1.4.26) can be written in the form

$$\int_{\varphi(S)}^{\varphi(s)} \frac{du}{\Psi(u)} = s - S. \tag{1.4.29}$$

Here we have used the fact that $\Psi(\varphi)>0$ for $\varphi\in\mathbb{R}$ since if $\varphi=0\ (\mathrm{mod}\pi_p)$, then $\Psi(\varphi)=1$ (observe that $|\sin_p\varphi|^p+|\cos_p\varphi|^p=1$), and if $\sin_p\varphi\neq 0$, then

$$\Psi(\varphi) = |\sin_p\varphi|^p\left[|\alpha|^p - \Phi(\alpha) + \frac{\gamma}{p-1}\right] > 0, \quad \alpha := \frac{\cos_p\varphi}{\sin_p\varphi}.$$

Further, let

$$\tau = \int_0^{2\pi_p} \frac{d\varphi}{\Psi(\varphi)} = 2\int_0^{\pi_p} \frac{d\varphi}{\Psi(\varphi)}. \tag{1.4.30}$$

By (1.4.28) we have $\Psi(\varphi+\pi_p)=\Psi(\varphi)$, hence $\varphi(s+\tau)=\varphi(s)+2\pi_p$ and the substitution $t:=\sin_p\varphi/\cos_p\varphi=\tan_p\varphi$ gives

$$\tau = 2\int_{-\infty}^{\infty} \frac{dt}{\frac{\gamma}{p-1}|t|^p - t + 1}, \tag{1.4.31}$$

which is the quantity depending only on γ.

Finally, we will estimate the radial variable ρ. Denote

$$R(\varphi) := \cos_p\varphi\left[\Phi(\cos_p\varphi) + \left(1-\frac{\gamma}{p-1}\right)\Phi(\sin_p\varphi)\right].$$

By (1.4.27) and the identity $\varphi(s+\tau)=\varphi(s)+2\pi_p$ we have

$$\log\frac{\rho(s+\tau)}{\rho(s)} = \int_s^{s+\tau} R(\varphi(s))\,ds = \int_{\varphi(s)}^{\varphi(s+\tau)} \frac{R(\varphi)}{\Psi(\varphi)}d\varphi = \int_0^{2\pi_p}\frac{R(\varphi)}{\Psi(\varphi)}d\varphi.$$

Now, using the identity $\Psi' + pR = p-1$, we get

$$\log\frac{\rho(s+\tau)}{\rho(s)} = \int_0^{2\pi_p} \frac{(p-1)/p-(1/p)\Psi'(\varphi)}{\Psi(\varphi)}\,d\varphi = \frac{p-1}{p}\tau = \frac{\tau}{q}. \tag{1.4.32}$$

A consequence of (1.4.32) is that the function $\rho(s)\exp\{-\frac{s}{q}\}$ is periodic with the period τ because of

$$\frac{\rho(s+\tau)\exp\{-(s+\tau)/q\}}{\rho(s)\exp\{-s/q\}} = \frac{\rho(s+\tau)}{\rho(s)}\exp\{-\tau/q\} = 1.$$

The previous computations in Case III are summarized in the next theorem.

Theorem 1.4.3. *If $\gamma > \tilde{\gamma} = \left(\frac{p-1}{p}\right)^p$ then equation (1.4.21) is oscillatory and $x(s) = \rho(s)\exp\{-s/q\}$ is a periodic solution of (1.4.21) with the period τ given by (1.4.31).*

Remark 1.4.1. Consider the half-linear differential equation

$$(t^\alpha \Phi(x'))' + \frac{\gamma}{t^{p-\alpha}}\Phi(x) = 0. \tag{1.4.33}$$

If $\alpha \neq p-1$ and we look for a solution of this equation in the form $x(t) = t^\lambda$, then substituting into (1.4.33) we get the algebraic equation for the exponent λ

$$(p-1)|\lambda|^p - (p-1-\alpha)\Phi(\lambda) + \gamma = 0. \tag{1.4.34}$$

This equation has a real root if and only if $\gamma \leq \tilde{\gamma}_\alpha := \left(\frac{|p-1-\alpha|}{p}\right)^p$ and hence (1.4.33) with $\alpha \neq p-1$ is nonoscillatory if $\gamma \leq \tilde{\gamma}_\alpha$. If $\gamma > \tilde{\gamma}_\alpha$, using the same ideas as in case $\alpha = 0$ treated in main part of this section one can see that (1.4.33) is oscillatory.

If $\alpha = p-1$ and $\gamma > 0$, equation (1.4.34) has no real root and in this case we consider the modified Euler-type equation

$$\left(t^{p-1}\Phi(x')\right)' + \frac{\gamma}{t\log^p t}\Phi(x) = 0. \tag{1.4.35}$$

The change of independent variable $t \mapsto \log t$ transforms (1.4.35) into equation (1.4.20) and the interval $[1,\infty)$ is transformed into the interval $[e,\infty)$. The situation is summarized in the next theorem.

Theorem 1.4.4. *If $\alpha \neq p-1$, equation (1.4.33) is nonoscillatory if and only if*

$$\gamma \leq \left(\frac{|p-1-\alpha|}{p}\right)^p =: \tilde{\gamma}_\alpha.$$

Equation (1.4.35) is nonoscillatory if and only if

$$\gamma \leq \left(\frac{p-1}{p}\right)^p.$$

1.4.3 Kneser type oscillation and nonoscillation criteria

As an immediate consequence of the Sturmian comparison theorem and the above result concerning oscillation of Euler equation (1.4.20), we have the following half-linear version of the classical Kneser oscillation and nonoscillation criterion.

Theorem 1.4.5. *Suppose that*

$$\liminf_{t\to\infty} t^p c(t) > \left(\frac{p-1}{p}\right)^p =: \tilde{\gamma}. \tag{1.4.36}$$

Then the equation

$$(\Phi(x'))' + c(t)\Phi(x) = 0 \tag{1.4.37}$$

is oscillatory. If

$$\limsup_{t\to\infty} t^p c(t) < \tilde{\gamma}, \tag{1.4.38}$$

then (1.4.37) *is nonoscillatory.*

Proof. If (1.4.36) holds, then $c(t) > \tilde{\gamma} + \varepsilon/t^p$ for some $\varepsilon > 0$ and since the Euler equation (1.4.20) with $\gamma = \tilde{\gamma} + \varepsilon$ is oscillatory, (1.4.37) is also oscillatory by the Sturm comparison theorem (Theorem 1.2.4). The nonoscillatory part of theorem can be proved using the same argument. □

Remark 1.4.2. Using the results of Theorem 1.4.4 and the Sturm comparison theorem, one can prove various extensions of the previous theorem. For example, if $\alpha \neq p-1$,

$$\liminf_{t\to\infty} t^{-\alpha} r(t) > 1, \quad \limsup_{t\to\infty} t^{p-\alpha} c(t) < \tilde{\gamma}_\alpha,$$

then (1.1.1) is nonoscillatory. An oscillation counterpart of this result, as well as the criteria in the case $\alpha = p-1$ can be formulated in a similar way.

1.5 Notes and references

Some special results on qualitative theory of differential equations with a homogeneous (but not generally additive) solution space first appeared in Bihari's papers in 1957-58, see [36, 37]. Equation of the form (1.1.1) was perhaps for the first time considered by Beesack in 1961 (see [30]) in connection with an extension of Hardy inequality. He understood half-linear equations as Euler-Lagrange equations there. In the same paper, Riccati type transformation was introduced. The term *half-linear equation* was first used by Bihari in [38] in 1964, see also [39]. As pioneering works in the field of qualitative theory of half-linear differential equations are usually regarded the paper of Mirzov [290] (1976) and of Elbert [139] (1979). In Mirzov's work, the first order half-linear system was considered and a Riccati type transformation was introduced, which then serves to prove Sturm type theorem, see Subsection 1.2.9. A survey of Mirzov's results is presented in his book [292], see also its extended English translation [293]. Elbert introduced the half-linear Prüfer transformation, which was used to show the existence and uniqueness of the IVP involving half-linear differential equations (see Subsection 1.1.3) and Sturmian type theorems (in particular, see the third alternative proof of the separation theorem on p. 19). Half-linear Picone's identity is due to Jaroš and Kusano [185], which then plays an important role in the Roundabout theorem (see Subsection 1.2.1 and Subsection 1.2.3). An alternative approach was used in the work [245] of Li and Yeh . The discussion on disconjugacy on various intervals is mostly based on an extension of "linear" results of Coppel [80]. Concerning the treatment of various aspects of the Sturmian theory for half-linear equations we also refer to the papers of Binding and Drábek [42] and of Yang [374].

The absolute majority of the observations concerning Wronskian in Subsection 1.3.1 is taken from Elbert's papers [140, 143]. A note on the Fredholm alternative is based on the paper of Drábek [118], a comprehensive treatment of this problem is given in Del Pino, Drábek and Manasevich [88], see also Chapter 6. Finally, the results presented in Section 1.4 are essentially contained in Elbert's work [142]. This Elbert's paper deals with system (1.1.3) (with constant coefficients or coefficients corresponding to the generalized Euler equation), but the results can be easily reformulated to equations of the form (1.1.1).

Various phenomena in physics, chemistry etc. modeled by half-linear differential equations and equations with p-Laplacian can be found in the book of Díaz [93].

CHAPTER 2

METHODS OF OSCILLATION THEORY

In this chapter we describe main methods for the investigation of oscillatory properties of (1.1.1). As we will see, a crucial role is played by the Roundabout theorem (Theorem 1.2.2) which provides two important methods: variational principle and Riccati technique. In the first section of this chapter we formulate the variational principle, which involves p-degree functional, along with a Wirtinger type inequality and some applications. In the second section we will see that there exist more general Riccati transformations than that from the first chapter, and that for nonoscillation of (1.1.1) merely solvability of Riccati inequality is sufficient. Under additional assumptions, we will be able to consider Riccati integral equation (inequality) involving improper integrals. Some of the asymptotic properties of solutions of this equation will be then described. The applications of the results will be illustrated by various criteria. Already from Sturmian comparison theorem we have seen that knowing some oscillatory properties of a given equation, we can obtain some information about the equation whose coefficients are in certain relation with those of the original one. This idea will be extended in various directions in the last section of this chapter.

2.1 Variational principle

The variational principle is the classical method of the linear oscillation theory. In this section we show that the energy functional $\mathcal{F}$ given by (1.2.4) plays a similar role in the half-linear oscillation theory.

2.1.1 Formulation of variational principle

As an immediate consequence of the equivalence of (i) and (iv) of Theorem 1.2.2 we have the following statement which may be widely used in the proofs of oscillation and nonoscillation criteria, as will be seen later. When using such a method, we speak about using a *variational principle.*

Theorem 2.1.1. *Equation* (1.1.1) *is nonoscillatory if and only if there exists* $T \in \mathbb{R}$ *such that*

$$\mathcal{F}(y;T,\infty) := \int_T^\infty [r(t)|y'|^p - c(t)|y|^p]\, dt > 0$$

for every nontrivial $y \in W_0^{1,p}(T,\infty)$.

Note that the class $W_0^{1,p}(T,\infty)$ is called the *class of admissible functions.*

Corollary 2.1.1. *Equation* (1.1.1) *is oscillatory if and only if for any* $T \in \mathbb{R}$ *there exists a nontrivial admissible function* y *such that* $\mathcal{F}(y;T,\infty) \le 0$.

2.1.2 Wirtinger inequality

A useful tool in the variational technique is the following half-linear version of the Wirtinger inequality. See also Section 9.5 how the well-known Hardy inequality is related to this technique.

Lemma 2.1.1. *Let* M *be a positive continuously differentiable function for which* $M'(t) \neq 0$ *in* $[a,b]$ *and let* $y \in W_0^{1,p}(a,b)$. *Then*

$$\int_a^b |M'(t)||y|^p\, dt \le p^p \int_a^b \frac{M^p(t)}{|M'(t)|^{p-1}} |y'|^p\, dt. \tag{2.1.1}$$

Proof. Suppose that $M'(t) > 0$ in $[a,b]$, in the opposite case the proof is similar. Using integration by parts, the fact that $y(a) = 0 = y(b)$, and the Hölder inequality, we have

$$\begin{aligned} \int_a^b |M'(t)||y|^p\, dt &\le p \int_a^b M|y|^{p-1}|y'|\, dt \\ &\le p \left(\int_a^b |M'||y|^{(p-1)q}\, dt \right)^{\frac{1}{q}} \left(\int_a^b \frac{M^p}{|M'|^{p-1}} |y'|^p\, dt \right)^{\frac{1}{p}} \\ &= p \left(\int_a^b |M'||y|^p\, dt \right)^{\frac{1}{q}} \left(\int_a^b \frac{M^p}{|M'|^{p-1}} |y'|^p\, dt \right)^{\frac{1}{p}}, \end{aligned}$$

hence (2.1.1) holds. ∎

2.1.3 Applications

As an example of application of the above described principle (involving Wirtinger inequality) we give the next nonoscillation criterion of Hille-Nehari type.

Further applications will be given in Chapters 3 and 5.

Theorem 2.1.2. *Denote* $c_+(t) = \max\{0, c(t)\}$. *If*

$$\int^\infty r^{1-q}(t)\,dt = \infty, \qquad \int^\infty c_+(t)\,dt < \infty,$$

and

$$\limsup_{t\to\infty} \left(\int^t r^{1-q}(s)\,ds\right)^{p-1} \left(\int_t^\infty c_+(s)\,ds\right) < \frac{1}{p}\left(\frac{p-1}{p}\right)^{p-1}, \tag{2.1.2}$$

or $\int^\infty r^{1-q}(t)\,dt < \infty$ *and*

$$\limsup_{t\to\infty} \left(\int_t^\infty r^{1-q}(s)\,ds\right)^{p-1} \left(\int^t c_+(s)\,ds\right) < \frac{1}{p}\left(\frac{p-1}{p}\right)^{p-1}, \tag{2.1.3}$$

then (1.1.1) *is nonoscillatory.*

Proof. We will prove the statement in case $\int^\infty r^{1-q}(t)\,dt = \infty$. If this integral converges, the proof is analogous. Denote

$$\nu := \frac{1}{p}\left(\frac{p-1}{p}\right)^{p-1}, \qquad M(t) := \left(\int^t r^{1-q}(s)\,ds\right)^{1-p}$$

and let $T \in \mathbb{R}$ be such that the expression in (2.1.2) is less than ν for $t > T$. Using (2.1.2), the Hölder inequality and the Wirtinger inequality, we have for any nontrivial $y \in W_0^{1,p}(T,\infty)$

$$\begin{aligned}
\int_T^\infty c(t)|y|^p\,dt &\le \int_T^\infty c_+(t)|y|^p\,dt = p\int_T^\infty c_+(t)\left(\int_T^t y'\Phi(y)\,ds\right)dt \\
&\le p\int_T^\infty |y'||y|^{p-1}M(t)\frac{\int_t^\infty c_+(s)\,ds}{M(t)}\,dt < p\nu\int_T^\infty M(t)|y'||y|^{p-1}\,dt \\
&\le p\nu\left(\int_T^\infty |M'(t)||y|^p\,dt\right)^{\frac{1}{q}}\left(\int_T^\infty \frac{|M(t)|^p}{|M'(t)|^{p-1}}|y'|^p\,dt\right)^{\frac{1}{p}} \\
&\le p^p\nu\int_T^\infty \frac{|M(t)|^p}{|M'(t)|^{p-1}}|y'|^p\,dt = \int_T^\infty r(t)|y'|^p\,dt
\end{aligned}$$

since one may directly verify that

$$\frac{|M(t)|^p}{|M'(t)|^{p-1}} = (p-1)^{1-p}r(t).$$

Hence we have

$$\mathcal{F}(y;T,\infty) = \int_T^\infty [r(t)|y'|^p - c(t)|y|^p]\,dt > 0$$

for any nontrivial $y \in W_0^{1,p}(T,\infty)$. □

Remark 2.1.1. (i) Later we will show that the constant $\frac{1}{p}\left(\frac{p-1}{p}\right)^{p-1}$ in the previous nonoscillation criterion is sharp in the sense that if the lim sup in (2.1.2) (or in (2.1.3)) is greater than this constant, then (1.1.1) is oscillatory. Also, when the previous criterion is applied to Euler type differential equation (1.4.20) in the previous chapter, we reveal the constant (the so-called *critical constant*) $\tilde{\gamma} = \left(\frac{p-1}{p}\right)^p$.

(ii) In (2.1.2) and (2.1.3), the nonnegative part of the function c appeared. In the next subsection we present an improvement of the previous nonoscillation criterion, where instead of c_+ the function c directly appears. Many other related information can be found in Section 3.1. See also Subsection 9.4.2 for an extension to equations of a higher order.

2.2 Riccati technique

The technique based on the connection between the nonexistence of zeros of a solution to (1.1.1) and the solvability of generalized Riccati equation (1.1.21) has already been proved to be a very useful tool of oscillation theory. In this section we further elaborate this important technique. Introduce the generalized Riccati operator

$$\mathcal{R}[w] = w' + c(t) + (p-1)r^{1-q}(t)|w|^q. \tag{2.2.1}$$

2.2.1 Preliminaries

Consider the function

$$S(x, y, p) := (p-1)y^{1-q}|x|^q \tag{2.2.2}$$

on $D_S = \{(x, y, p) : x \in \mathbb{R}, y \in (0, \infty), p \in (1, \infty)\}$, which appears in the generalized Riccati equation. Recall that $1/p + 1/q = 1$. Often we use just the notation $S(x, y) := S(x, y, p)$ when p is fixed. The following lemma describes important properties of the function S. Later, it will find many applications when dealing with the equation $\mathcal{R}[w] = 0$. Among others, our aim is to show that S behaves essentially like its special case x^2/y, which appears in the Riccati equation corresponding to linear Sturm-Liouville equation (1.1.2).

Lemma 2.2.1. *For S the following statements hold:*

(i) The function S is nondecreasing with respect to $|x|$ on D_S.

(ii) The function S is nonincreasing with respect to $y > 0$ on D_S.

(iii) The function S is nonnegative on D_S, and $S(x, y) = 0$ if and only if $x = 0$.

(iv) If $(|x|/y)^{q-1} \le e$, where e is the basis of natural logarithm, then the function S is nondecreasing with respect to p.

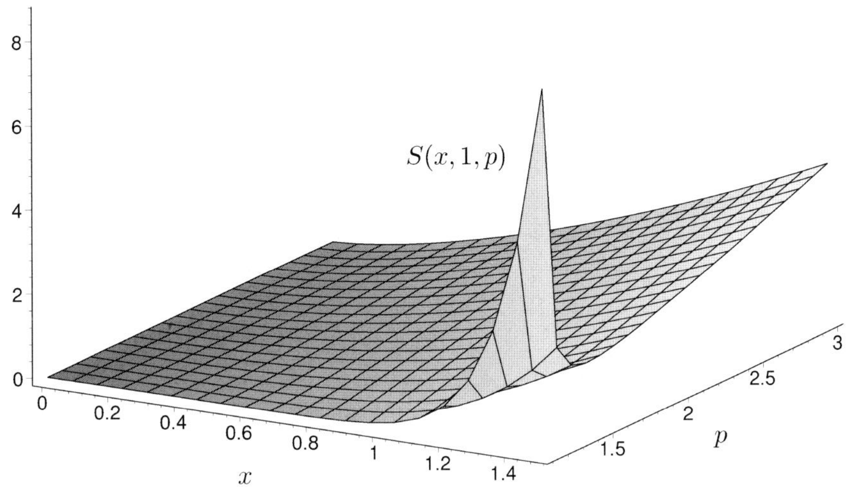

Figure 2.2.1: The function S with fixed $y = 1$; S is nondecreasing with respect to p when x is small

Proof. The proof of the parts (i)–(iii) is trivial. To prove (iv) note that S can be rewritten as $S(w, r, p) = (p-1)|x|(|x|/y)^{q-1}$, from which it is easy to compute that

$$\frac{\partial S}{\partial p} = |x| \left(\frac{|x|}{y}\right)^{q-1} \left[1 - \log\left(\frac{|x|}{y}\right)^{q-1}\right].$$

□

The next technical result will find the application in proving the existence of positive solutions of the generalized Riccati equation.

Lemma 2.2.2. *If*

$$F(a) := \liminf_{t\to\infty} \int_a^t f(s)\, ds \geq 0 \quad \textit{and} \quad F(a) \not\equiv 0 \tag{2.2.3}$$

for all large a, then there is T is such that $\int_T^t f(s)\, ds \geq 0$, $t \in [T, \infty)$.

Proof. If no such T exists, then for $T \in [a, \infty)$ fixed but arbitrary we define

$$T_1 = T_1(T) := \sup\left\{t > T : \int_T^t f(s)\, ds < 0\right\}.$$

If $T_1 = \infty$, then choosing $t_n \to \infty$ such that $\int_T^{t_n} f(s)\, ds < 0$ for all n, we get $F(T) \leq 0$. Since T was arbitrary, by the first condition in (2.2.3), it yields $F(T) \equiv 0$ for large T, a contradiction to the second condition in (2.2.3). Hence, we must have $T_1 < \infty$ which implies $\int_{T_1}^t f(s)\, ds \geq 0$, $t \in [T_1, \infty)$. □

2.2.2 More general Riccati transformation

Sometimes it is convenient to use a more general Riccati substitution

$$v(t) = \frac{f(t)r(t)\Phi(x'(t))}{\Phi(x(t))},$$

where f is a differentiable function. By a direct computation one can verify that v satisfies the first order Riccati-type equation

$$v' - \frac{f'(t)}{f(t)}v + f(t)c(t) + (p-1)r^{1-q}(t)f^{1-q}(t)|v|^q = 0. \tag{2.2.4}$$

For an application see e.g. Remark 5.1.2 and Subsection 9.2.1.

Another generalization may be grounded in considering a slightly more general half-linear equation

$$(r(t)\Phi(x'))' + b(t)\Phi(x') + c(t)\Phi(x) = 0. \tag{2.2.5}$$

The Riccati substitution $w = r\Phi(x')/\Phi(x)$ leads to the equation

$$w' + c(t) + \frac{b(t)}{r(t)}w + (p-1)r^{1-q}(t)|w|^q = 0. \tag{2.2.6}$$

Multiplying this equation by $\exp\{\int^t b(s)/r(s)\,ds\} =: g(t)$ and denoting $v = gw$, equation (2.2.6) can be written in the same form as (1.1.21), i.e.,

$$v' + c(t)g(t) + (p-1)r^{1-q}(t)g^{1-q}(t)|v|^q = 0.$$

The same effect is achieved if we multiply the original equation (2.2.5) by g since then this equation can be again written in the form (1.1.1).

2.2.3 Riccati inequality

From the Roundabout theorem (Theorem 1.2.2) it follows that nonoscillation of (1.1.1) is equivalent to solvability of the associated Riccati equation (1.1.21) (in a neighborhood of infinity). Due to the Sturm comparison theorem, nonoscillation of (1.1.1) is actually equivalent to solvability of the *generalized Riccati inequality.* This is formulated, among others, in the next statement. In its proof we also show that $y\mathcal{L}[y] = |y|^p\mathcal{R}[r\Phi(y'/y)]$ provided $y \neq 0$. Recall that the operator $\mathcal{L}$ is defined in Theorem 1.2.1 and the operator $\mathcal{R}$ by (2.2.1).

Theorem 2.2.1. *The following statements are equivalent:*

(i) *Equation* (1.1.1) *is nonoscillatory.*

(ii) *There is* $a \in \mathbb{R}$ *and a (continuously differentiable) function* $w : [a, \infty) \to \mathbb{R}$ *such that*

$$\mathcal{R}[w](t) = 0 \quad \textit{for } t \in [a, \infty).$$

(*iii*) *There is* $a \in \mathbb{R}$, *a constant* $A \in \mathbb{R}$ *and a (continuous) function* $w : [a, \infty) \to \mathbb{R}$ *such that*

$$w(t) = A - \int_a^t \{c + S[w, r]\}(s)\, ds \quad \textit{for } t \in [a, \infty).$$

(*iv*) *There is* $a \in \mathbb{R}$ *and a (continuously differentiable) function* $w : [a, \infty) \to \mathbb{R}$ *such that*

$$\mathcal{R}[w](t) \le 0 \quad \textit{for } t \in [a, \infty). \tag{2.2.7}$$

(*v*) *There is* $a \in \mathbb{R}$ *and a positive function* $y : [a, \infty) \to \mathbb{R}$ *(with* $r\Phi(y')$ *continuously differentiable) such that*

$$\mathcal{L}[y](t) \le 0 \quad \textit{for } t \in [a, \infty). \tag{2.2.8}$$

Proof. (i)⇒(ii): This implication follows from the Roundabout theorem (Theorem 1.2.2) since nonoscillation of (1.1.1) implies the existence of $a \in \mathbb{R}$ such that (1.1.1) is disconjugate on $[a, \infty)$.

(ii)⇒(iii): Trivial.

(iii)⇒(iv): Trivial.

(iv)⇒(v): Let w satisfy $\mathcal{R}[w] \le 0$ on $[a, \infty)$. The function

$$y = \exp\left\{ \int_a^t \Phi^{-1}\left(\frac{w(s)}{r(s)}\right) ds \right\}$$

is a positive solution of the initial value problem $y' = \Phi^{-1}\left(\frac{w(t)}{r(t)}\right) y$, $y(a) = 1$. We have

$$\begin{aligned} y\mathcal{L}[y] &= y(r\Phi(y'))' + yc\Phi(y) - (p-1)r|y'|^p + (p-1)r|y'|^p \\ &= |y|^p \frac{(r\Phi(y'))'\Phi(y) - (p-1)r\Phi(y')|y|^{p-2}y'}{\Phi(y)\Phi(y)} \\ &\quad + |y|^p c + |y|^p (p-1) r^{1-q} \left(\frac{r|y'|^{p-1}}{|y|^{p-1}}\right)^q \\ &= |y|^p \mathcal{R}[w] \le 0, \end{aligned}$$

hence (v) holds.

(v)⇒(i): Suppose that a function y satisfies (2.2.8) on $[a, \infty)$. Then

$$\varphi(t) := -\{y\mathcal{L}[y]\}(t)$$

is a nonnegative function on this interval. Set $\bar{c}(t) = c(t) - \varphi(t)/|y|^p$. Then $\bar{c} \ge c$ and

$$(r(t)\Phi(y'))' + \bar{c}(t)\Phi(y) = (r(t)\Phi(y'))' + \left(c(t) - \frac{\varphi(t)}{|y|^p}\right)\Phi(y) = 0.$$

Thus the equation $(r(t)\Phi(y'))' + \bar{c}(t)\Phi(y) = 0$ is disconjugate on $[a, \infty)$. Therefore, (1.1.1) is disconjugate on $[a, \infty)$ as well by the Sturmian comparison theorem (Theorem 1.2.4) and hence nonoscillatory. □

Remark 2.2.1. It is not difficult to see that (iv)⇒(i) can be proved directly. Indeed, let w be a solution of $\mathcal{R}[w] \le 0$. Denote $C(t) := -w' - (p-1)r^{1-q}(t)|w|^q$. Then w is a solution of $w' + C(t) + (p-1)r^{1-q}(t)|w|^q = 0$ which is the Riccati equation associated with a Sturmian majorant of (1.1.1) (since $C(t) \ge c(t)$). This majorant equation is nonoscillatory and hence (1.1.1) is nonoscillatory as well.

Now we give a variant of the above theorem where certain *Riccati integral inequality* appears. In the next subsections we will present another ones where improper integrals occur in Riccati type equations or inequalities.

Theorem 2.2.2. *Suppose that there is $a \in \mathbb{R}$, a constant $A \in \mathbb{R}$ and a (continuous) function $w : [a, \infty) \to \mathbb{R}$ such that either*

$$w(t) \ge A - \int_a^t \{c + S[w, r]\}(s)\, ds \ge 0$$

or

$$w(t) \le A - \int_a^t \{c + S[w, r]\}(s)\, ds \le 0$$

for $t \in [a, \infty)$, where S is defined by (2.2.2)*. Then (i)–(v) from the previous theorem hold. If, in addition, $\int_a^\infty c(t)\, dt = \infty$, then the above condition is necessary for (i)–(v).*

Proof. For sufficiency we show that the assumptions imply (iv) of Theorem 2.2.1. Let

$$v(t) = A - \int_a^t \{c + S[w, r]\}(s)\, ds.$$

Then $v' + c(t) + S[w, r](t) = 0$. We have $w \ge v \ge 0$ or $w \le v \le 0$ and hence $S[w, r] \ge S[v, r]$ by Lemma 2.2.1. Therefore we get $\mathcal{R}[v](t) \le 0$ for $t \in [a, \infty)$ and so (iv) holds. The part concerning necessity is obvious. □

2.2.4 Half-linear Hartman-Wintner theorem

The next theorem is a half-linear extension of the classical Hartman-Wintner theorem [174] which relates the square integrability of the solutions of the Riccati equation

$$w' + c(t) + w^2 = 0$$

corresponding to (1.1.2) with $r(t) \equiv 1$ to the finiteness of a certain limit involving the function c. Later we will show another variants of this important statement.

Theorem 2.2.3. *Suppose that*

$$\int^\infty r^{1-q}(t)\, dt = \infty \tag{2.2.9}$$

and (1.1.1) *is nonoscillatory. Then the following statements are equivalent.*

(i) *It holds*

$$\int^{\infty} r^{1-q}(t)|w(t)|^q \, dt < \infty \tag{2.2.10}$$

for every solution w of Riccati equation (1.1.21).

(ii) *There exists a finite limit*

$$\lim_{t\to\infty} \frac{\int^t r^{1-q}(s) \left(\int^s c(\tau)\, d\tau\right) ds}{\int^t r^{1-q}(s)\, ds}. \tag{2.2.11}$$

(iii) *For the lower limit we have*

$$\liminf_{t\to\infty} \frac{\int^t r^{1-q}(s) \left(\int^s c(\tau)\, d\tau\right) ds}{\int^t r^{1-q}(s)\, ds} > -\infty. \tag{2.2.12}$$

Proof. (i) $\Rightarrow$ (ii): Nonoscillation of (1.1.1) implies that Riccati equation (1.1.21) has a solution which is defined on an interval $[T,\infty)$. Integrating this equation from T to t and using (2.2.10) we have

$$\begin{aligned} w(t) &= w(T) - \int_T^t c(\tau)\, d\tau - (p-1)\int_T^t r^{1-q}(\tau)|w(\tau)|^q \, d\tau \\ &= w(T) - \int_T^t c(\tau)\, d\tau - (p-1)\int_T^{\infty} r^{1-q}(\tau)|w(\tau)|^q \, d\tau \\ &\quad + (p-1)\int_t^{\infty} r^{1-q}(\tau)|w(\tau)|^q \, d\tau \\ &= C - \int_T^t c(\tau)\, d\tau + (p-1)\int_t^{\infty} r^{1-q}(\tau)|w(\tau)|^q \, d\tau, \end{aligned} \tag{2.2.13}$$

where $C = w(T) - (p-1)\int_T^{\infty} r^{1-q}(\tau)|w(\tau)|^q \, d\tau$. Multiplying (2.2.13) by r^{1-q} and integrating the resulting equation from T to t, and then dividing by $\int_T^t r^{1-q}(s)\, ds$, we get

$$\begin{aligned} \frac{\int_T^t r^{1-q}(s)w(s)\, ds}{\int_T^t r^{1-q}(s)\, ds} &= C - \frac{\int_T^t r^{1-q}(s)\left(\int_T^s c(\tau)\, d\tau\right)\, ds}{\int_T^t r^{1-q}(\tau)\, d\tau} \\ &\quad + \frac{\int_T^t r^{1-q}(s)\left(\int_s^{\infty} r^{1-q}(\tau)|w(\tau)|^q\, d\tau\right) ds}{\int_T^t r^{1-q}(s)\, ds}. \end{aligned} \tag{2.2.14}$$

By the Hölder inequality, we have

$$\begin{aligned} \left|\int_T^t r^{1-q}(s)w(s)\, ds\right| &= \left|\int_T^t r^{\frac{1-q}{p}}(s) r^{\frac{1-q}{q}}(s) w(s)\, ds\right| \\ &\le \left(\int_T^t r^{1-q}(s)\, ds\right)^{\frac{1}{p}} \left(\int_T^t r^{1-q}(s)|w(s)|^q\, ds\right)^{\frac{1}{q}}, \end{aligned} \tag{2.2.15}$$

and hence, taking into account (2.2.10)

$$\left|\frac{\int_T^t r^{1-q}(s)w(s)\,ds}{\int_T^t r^{1-q}(s)\,ds}\right| \le \frac{\left(\int_T^t r^{1-q}(s)\,ds\right)^{\frac{1}{p}}\left(\int_T^t r^{1-q}(s)|w(s)|^q\right)^{\frac{1}{q}}}{\int_T^t r^{1-q}(s)\,ds}$$
$$= \left(\frac{\int_T^t r^{1-q}(s)|w(s)|^q}{\int_T^t r^{1-q}(s)\,ds}\right)^{\frac{1}{q}} \to 0, \quad t\to\infty.$$

Since also the last term in (2.2.14) tends to zero as $t \to \infty$ (again in view of (2.2.10)), we have that

$$\lim_{t\to\infty}\frac{\int_T^t r^{1-q}(s)\int_T^s c(\tau)\,d\tau}{\int_T^t r^{1-q}(s)\,ds} = C \quad \text{exists finite.}$$

(ii) $\Rightarrow$ (iii): This implication is trivial.

(iii) $\Rightarrow$ (i): Let w be any solution of (1.1.21) which exists on $[T,\infty)$. Then by (2.2.13) and using computation from the first part of this proof

$$\frac{\int_T^t r^{1-q}(s)w(s)\,ds}{\int_T^t r^{1-q}(s)\,ds} = w(T) - \frac{\int_T^t r^{1-q}(s)\left(\int_T^s c(\tau)\,d\tau\right)\,ds}{\int_T^t r^{1-q}(\tau)\,d\tau\,ds}$$
$$-(p-1)\frac{\int_T^t r^{1-q}(s)\left(\int_T^t r^{1-q}(\tau)|w(\tau)|^q\,d\tau\right)\,ds}{\int_T^t r^{1-q}(s)\,ds}.$$

Taking into account (2.2.12) and applying the Hölder inequality again, there exists a real constant K such that

$$-\frac{\left(\int_T^t r^{1-q}(s)|w(s)|^q\,ds\right)^{\frac{1}{q}}}{\left(\int_T^t r^{1-q}(s)\,ds\right)^{\frac{1}{q}}} \le K-(p-1)\frac{\int_T^t r^{1-q}(s)\left(\int_T^s r^{1-q}(\tau)|w(\tau)|^q\,d\tau\right)ds}{\int_T^t r^{1-q}(s)\,ds}.$$

Suppose that (2.2.10) fails to holds. Then by L'Hospital's rule the last term in the previous inequality tends to ∞ and hence

$$\left(\frac{\int_T^t r^{1-q}(s)|w(s)|^q\,ds}{\int_T^t r^{1-q}(s)\,ds}\right)^{\frac{1}{q}} \ge \frac{\int_T^t r^{1-q}(s)\left(\int_T^s r^{1-q}(\tau)|w(\tau)|^q\,d\tau\right)\,ds}{q\int_T^t r^{1-q}(s)\,ds}$$

for large t. Denote $M(t) = \int_T^t r^{1-q}(s)\left(\int_T^s r^{1-q}(\tau)|w(\tau)|^q\,d\tau\right)\,ds$. Then the last inequality reads

$$\left[\frac{M'(t)r^{q-1}(t)}{\int^t r^{1-q}(s)\,ds}\right]^{\frac{1}{q}} \ge \frac{M(t)}{q\int_T^t r^{1-q}(s)\,ds},$$

hence

(2.2.16)
$$\frac{M'(t)}{M^q(t)} \ge \left(\frac{1}{g}\right)^q\frac{r^{1-q}(t)}{\left(\int_T^t r^{1-q}(s)\,ds\right)^{q-1}}.$$

If $q \le 2$, we integrate (2.2.16) from T_1 to t, $T_1 > T$, and we get

$$\begin{aligned}\frac{1}{q-1}M^{1-q}(T_1) &> \frac{1}{q-1}\left[M^{1-q}(T_1) - M^{1-q}(t)\right]\\ &\ge \left(\frac{1}{q}\right)^q \begin{cases}\log\left(\int_T^t r^{1-q}(s)\,ds\right) & \text{if } q=2,\\ \frac{1}{2-q}\left(\int_T^t r^{1-q}(s)\,ds\right)^{2-q} & \text{if } q<2.\end{cases}\end{aligned}$$

Letting $t \to \infty$ we have a contradiction with the assumption that $\int^\infty r^{1-q}(t)\,dt = \infty$. If $q > 2$, we integrate (2.2.16) from t to ∞ and we obtain

$$\frac{1}{(q-1)M^{q-1}(t)} \ge \left(\frac{1}{q}\right)^q \frac{1}{(q-2)\left(\int_T^t r^{1-q}(s)\,ds\right)^{q-2}},$$

hence

$$\frac{q^q(q-2)}{q-1} \ge \left(\frac{M(t)}{\int_T^t r^{1-q}(s)\,ds}\right)^{q-1}\left(\int_T^t r^{1-q}(s)\,ds\right),$$

which is again a contradiction since $M(t)\left(\int_T^t r^{1-q}(s)\,ds\right)^{-1} \to \infty$ as $t \to \infty$. □

Based on the Hartman-Wintner theorem, the following statement says that nonoscillation of (1.1.1) can be expressed in terms of solvability of Riccati integral equation involving improper integrals.

Theorem 2.2.4. *Suppose that* (2.2.9) *holds and the integral* $\int^\infty c(t)\,dt$ *is convergent. Equation* (1.1.1) *is nonoscillatory if and only if there is* $a \in \mathbb{R}$ *and a (continuous) function* $w : [a,\infty) \to \mathbb{R}$ *satisfying the Riccati integral equation*

$$w(t) = \int_t^\infty c(s)\,ds + (p-1)\int_t^\infty r^{1-q}(s)|w(s)|^q\,ds \tag{2.2.17}$$

for $t \ge a$.

Proof. Suppose that (1.1.1) is nonoscillatory and let w be a solution of the associated Riccati equation (1.1.21) which is defined on some interval $[T_0,\infty)$. The convergence of the integral $\int^\infty c(t)\,dt$ and (2.2.9) imply that (2.2.10) holds by Theorem 2.2.3. Integrating (1.1.21) from t to T, $t \ge T_0$ and letting $T \to \infty$ we see that $\lim_{T\to\infty} w(T)$ exists and since (2.2.9) holds, this limit equals zero, i.e., w satisfies also (2.2.17). Conversely, if w is a solution of (2.2.17), then it is also a solution of (1.1.21) and hence (1.1.1) is nonoscillatory. □

A simple argument will show that for nonoscillation of (1.1.1) it is sufficient to assume a solvability of Riccati integral inequality provided an additional requirement is satisfied.

Theorem 2.2.5. *Suppose that the integral* $\int^\infty c(t)\,dt$ *is convergent and* (2.2.9) *holds. Then there is* $a \in \mathbb{R}$ *and a (continuous) function* $w : [a, \infty) \to \mathbb{R}$ *satisfying either*

$$w(t) \geq \int_t^\infty c(s)\,ds + (p-1)\int_t^\infty r^{1-q}(s)|w(s)|^q\,ds \geq 0 \tag{2.2.18}$$

or

$$w(t) \leq \int_t^\infty c(s)\,ds + (p-1)\int_t^\infty r^{1-q}(s)|w(s)|^q\,ds \leq 0 \tag{2.2.19}$$

if and only if equation (1.1.1) *is nonoscillatory.*

Proof. The proof is similar to that of Theorem 2.2.2. Indeed, let

$$v(t) = \int_t^\infty \{c + S[w, r]\}(s)\,ds,$$

where S is defined by (2.2.2). Then $v' + c(t) + S[w, r](t) = 0$. We have $w \geq v \geq 0$ or $w \leq v \leq 0$ and hence $S[w, r] \geq S[v, r]$ by Lemma 2.2.1. Therefore we get $\mathcal{R}[v](t) \leq 0$ for $t \in [a, \infty)$, where $\mathcal{R}$ is defined by (2.2.1), and so (iv) from Theorem 2.2.1 holds. The necessity follows clearly from Theorem 2.2.4. □

Remark 2.2.2. Observe that if (2.2.18) and (2.2.19) are replaced by the only condition

$$|w(t)| \geq \left| \int_t^\infty c(s)\,ds + (p-1)\int_t^\infty r^{1-q}(s)|w(s)|^q\,ds \right|, \tag{2.2.20}$$

then Theorem 2.2.5 works as well. Indeed, if $v(t) = \int_t^\infty \{c + S[w, r]\}(s)\,ds$, then $|v| \leq |w|$ and $v' + c(t) + S[w, r](t) = 0$, and so we get $\mathcal{R}[v](t) \leq 0$ by Lemma 2.2.1. Taking into account Theorem 2.2.4, altogether we get that the following statements are equivalent:

(i) Equation (1.1.1) is nonoscillatory.

(ii) Equation (2.2.17) is solvable in a neighborhood of infinity.

(iii) Inequality (2.2.20) is solvable in a neighborhood of infinity.

For related discussion see Lemma 2.2.4, the comment before it, Remark 2.2.3, Remark 2.3.1 and Section 5.5.

Compare the following result with the previous one and with Theorem 2.2.2. We omit the proof since it is similar to those of the above mentioned statements. Note only that the necessity is shown by means of Theorem 2.2.3 (see also (2.2.13)).

Theorem 2.2.6. *Suppose that there is* $a \in \mathbb{R}$, *a constant* $A \in \mathbb{R}$ *and a function* $w : [a, \infty) \to \mathbb{R}$ *such that either*

$$w(t) \geq A - \int_a^t c(s)\,ds + \int_t^\infty S[w, r](s)\,ds \geq 0$$

or

$$w(t) \le A - \int_a^t c(s)\,ds + \int_t^\infty S[w,r](s)\,ds \le 0$$

for $t \in [a,\infty)$. *Then* (1.1.1) *is nonoscillatory. If, in addition,* (2.2.9) *and* (2.2.12) *hold, then the above condition is necessary for nonoscillation of* (1.1.1).

2.2.5 Positive solution of generalized Riccati equation

In this subsection we give conditions guaranteeing the existence of positive solution w of the generalized Riccati equation. This positivity then enables to prove alternatively the results similar to those given at the end of the previous subsection, as well as to find an effective estimation for w, which will appear very handy. In spite of the fact that some of the results will require slightly stronger assumptions comparing with those in the previous subsection, we present them here in order to show a variety of different approaches. Moreover, the subsequent approach will be proved to be very useful in extending our theory to the cases where the Hartman-Wintner theorem is not applicable (see Chapter 8).

First note that if

$$\int^\infty r^{1-q}(s)\,ds = \infty$$

and $c(t) \ge 0$ for large t, then it is very easy to show that there exists an eventually positive solution of generalized Riccati equation (1.1.21) provided (1.1.1) is nonoscillatory. In this case, in fact, any solution of (1.1.21) is eventually positive provided $c(t) \not\equiv 0$ for large t. Indeed, below given Lemma 4.1.3 says that the only possibility for a nonoscillatory solution x of (1.1.1) is $x(t)x'(t) > 0$ for large t.

Next we show that the assumption of the nonnegativity of c may be somewhat relaxed. Also observe how the positivity of a solution w of (1.1.21) follows from Theorem 2.2.4, under the condition $\int_t^\infty c(s)\,ds \ge 0$. For the discussion on positivity of w under some different special conditions see the text after Corollary 3.3.5 in the next chapter.

Lemma 2.2.3. *Assume*

$$\Psi(T) := \liminf_{t\to\infty} \int_T^t c(s)\,ds \ge 0 \quad \text{and} \quad \Psi(T) \not\equiv 0 \tag{2.2.21}$$

for all large T, *and* (2.2.9) *holds. If* y *is a solution of* (1.1.1) *such that* $y(t) > 0$ *for* $t \in [T,\infty)$, *then there exists* $S \in [T,\infty)$ *such that* $y'(t) > 0$ *for* $t \in [S,\infty)$.

Proof. The proof is by contradiction. We consider two cases:

Case I. Suppose that $y'(t) < 0$ for $t \in [T,\infty)$. Then also $[\Phi(y)]'(t) < 0$ for $t \in [T,\infty)$ since

$$[\Phi(y)]'(t) = (p-1)|y(t)|^{p-2}y'(t) < 0.$$

Another argument for $[\Phi(y)]'(t) < 0$ is that if y is decreasing, then $\Phi(y)$ is decreasing as well because of the properties of the function Φ. Without loss of generality we may assume that T is such that $\int_T^t c(s)\,ds \geq 0$, $t \in [T, \infty)$, by Lemma 2.2.2. Define $Q(t,T) = \int_T^t c(s)\,ds$. Integration by parts gives

$$\begin{aligned}\int_T^t c(s)\Phi(y(s))\,ds &= \int_T^t Q'(s,T)\Phi(y(s))\,ds \\ &= Q(t,T)\Phi(y(t)) - \int_T^t Q(s,T)[\Phi(y(s))]'ds \geq 0.\end{aligned}$$

Integrating (1.1.1) we have, using the last estimate,

$$r(t)\Phi(y'(t)) - r(T)\Phi(y'(T)) = \int_T^t [r(s)\Phi(y'(s))]'ds \leq 0.$$

Hence

$$y'(t) \leq \frac{r^{q-1}(T)y'(T)}{r^{q-1}(t)} \tag{2.2.22}$$

for $t \in [T, \infty)$. Integrating (2.2.22) for $t \geq T$ we see that $y(t) \to -\infty$ by (2.2.9), a contradiction. Therefore, $y'(t) < 0$ cannot hold for all large t.

Case II. Next, if $y'(t) \not> 0$ eventually, then for every (large) T (2.2.21) holds and there exists $T_0 \in [T, \infty)$ such that $y'(T_0) \leq 0$. Since $y(t) > 0$ for $t \in [T, \infty)$, the function $w(t) = r(t)\Phi\left[y'(t)/y(t)\right]$ satisfies generalized Riccati equation (1.1.21) for $t \in [T, \infty)$. Integrating (1.1.21) from T_0 to t, $t \geq T_0$, gives

$$w(t) = w(T_0) - \int_{T_0}^t c(s)\,ds - \int_{T_0}^t S(w,r)(s)\,ds,$$

where S is defined by (2.2.2). Therefore, it follows that $\limsup_{t\to\infty} w(t) < 0$, using the facts $w(T_0) \leq 0$, $w(t)$ is eventually nontrivial, and (2.2.21) holds. Indeed, there is $M > 0$ such that $\int_{T_0}^t S(w,r)(s)\,ds \geq M$ and $\int_{T_0}^t c(s)\,ds \geq -M/2$ for all large t. Hence there exists $T_1 \in [T, \infty)$ such that $w(t) < 0$ for $t \in [T_1, \infty)$ and so $y'(t) < 0$ for $t \in [T_1, \infty)$, a contradiction to the first part. □

Example 2.2.1. Now we give an example of the function $c(t)$, which is not eventually of one sign, but (2.2.21) holds. Let $\lambda > 0$, $A \neq 0$ and $B \geq 1$ be real numbers. Put

$$c(t) = \frac{\lambda B}{t^{\lambda+1}} + \frac{\lambda \sin At}{t^{\lambda+1}} - \frac{A\cos At}{t^{\lambda}}.$$

Then we can easily see that

$$\int_t^\infty c(s)\,ds = \frac{\sin At + B}{t^\lambda} \geq 0$$

for all $t \geq 0$. Observe that in addition to (2.2.21), condition (2.2.23) is satisfied. Using Example 3.3.1 given in Subsection 3.3.1, one can easily discover a function $c(t)$, which is not eventually of one sign, and (2.2.21) holds but (2.2.23) fails to hold.

In the next lemma, a necessary condition for nonoscillation of (1.1.1) is given in terms of solvability of a generalized Riccati integral inequality involving improper integrals. Note that in Theorem 2.2.4 we obtained the integral equation, even under weaker (sign) condition on c, but for completeness we present here also an alternative (simpler) approach based on the positivity of w. As already said above, this approach will be shown to be very useful e.g. in the discrete case treated in Chapter 8.

Lemma 2.2.4. *Let the assumptions of Lemma 2.2.3 hold and assume further that*

$$\int_a^\infty c(s)\,ds = \lim_{t\to\infty}\int_a^t c(s)\,ds \quad \textit{is convergent.} \tag{2.2.23}$$

Let y be a solution of (1.1.1) *such that $y(t)>0$ for $t\in[T,\infty)$. Then there exists $T_1\in[T,\infty)$ such that*

$$w(t) \ge \int_t^\infty c(s)\,ds + \int_t^\infty S(w,r)(s)\,ds \tag{2.2.24}$$

for $t\in[T_1,\infty)$, where $w(t)=r(t)\Phi\left[y'(t)/y(t)\right]>0$.

Proof. By Lemma 2.2.3 there exists $T_1\in[T,\infty)$ such that $w(t)>0$ for $t\in[T_1,\infty)$ and w satisfies (1.1.21) for $t\in[T,\infty)$. Integrating (1.1.21) from t to s, $s\ge t\ge T_1$, gives

$$w(s)-w(t)+\int_t^s c(\xi)\,d\xi + \int_t^s S(w,r)(\xi)\,d\xi = 0. \tag{2.2.25}$$

Therefore,

$$0<w(s)=w(t)-\int_t^s c(\xi)\,d\xi - \int_t^s S(w,r)(\xi)\,d\xi,$$

and hence

$$w(t)\ge \int_t^s c(\xi)\,d\xi + \int_t^s S(w,r)(\xi)\,d\xi$$

for $s\ge t\ge T_1$. Letting $s\to\infty$ we obtain (2.2.24). □

Remark 2.2.3. Clearly, condition (2.2.24) is also sufficient for nonoscillation of (1.1.1), as shown in Theorem 2.2.5.

If we strengthen the assumptions of the previous lemma somewhat, then we may prove (differently from Theorem 2.2.4) that there exists a solution of (2.2.17).

Lemma 2.2.5. *Suppose that* (2.2.9) *and* (2.2.23) *hold with $c(t)\ge 0$ (which is eventually nontrivial). Let y be a solution of* (1.1.1) *such that $y(t)>0$ for $t\in[T,\infty)$. Then there exists $T_1\in[T,\infty)$ such that $w(t)=r(t)\Phi\left[y'(t)/y(t)\right]$ is positive, nonincreasing, tends to zero and satisfies* (2.2.17) *for $t\in[T_1,\infty)$.*

Proof. From Lemma 2.2.3 there exists $T_1 \in [T, \infty)$ such that $w(t) > 0$ for $t \in [T_1, \infty)$. Note that this can actually be proved even much easier since $c(t)$ is nonnegative here (see the beginning of this subsection). Further, w satisfies (1.1.21) on $[T, \infty)$. The fact that $w'(t) \le 0$ for $t \in [T_1, \infty)$ follows from (1.1.21). Next we show that $w(t) \to 0$ as $t \to \infty$. Since y is positive and increasing, it either converges to a positive constant L or diverges to ∞. First suppose that $y(t) \to \infty$ as $t \to \infty$. Then, since $r(t)\Phi[y'(t)]$ is nonincreasing, we have

$$w(t) = \frac{r(t)\Phi[y'(t)]}{\Phi[y(t)]} \le \frac{r(T_1)\Phi[y'(T_1)]}{\Phi[y(t)]} \to 0$$

as $t \to \infty$. Now, if $y(t) \to L$ as $t \to \infty$, then $r(t)\Phi[y'(t)] \to 0$ as $t \to \infty$ and, consequently, $w(t)$ tends to zero as $t \to \infty$. To see that $r(t)\Phi[y'(t)]$ converges to zero, note first that it converges since it is positive and nonincreasing. If, however, $r(t)\Phi[y'(t)]$ converges to a positive constant K, then we get

$$y(t) \ge y(T_1) + K^{q-1} \int_{T_1}^{t} r^{1-q}(s)\, ds \to \infty$$

as $t \to \infty$, which contradicts the boundedness of y. Finally, the fact that y satisfies (2.2.17) on $[T_1, \infty)$ follows from (2.2.25). □

Remark 2.2.4. Clearly, solvability of (2.2.17) is also a sufficient condition for nonoscillation of (1.1.1), like in Theorem 2.2.4.

Finally we show how a (positive) solution of the Riccati equation can be estimated from above in terms of r. This will be very useful in some subsequent (non)oscillatory criteria and comparison theorems.

Lemma 2.2.6. *Let the assumptions of Lemma 2.2.5 be fulfilled. Then the function w from that lemma satisfies the inequality*

$$w(t) \le \left(\int_{t_0}^{t} r^{1-q}(s)\, ds\right)^{1-p} \tag{2.2.26}$$

for $t \ge t_0$.

Proof. Since c is nonnegative, we have $w'(t) \le -(p-1)r^{1-q}(t)w^q(t)$. Hence,

$$\left(\int_{a}^{t} r^{1-q}(s)\, ds - w^{1-q}(t)\right)' = r^{1-q}(t) + (q-1)w^{-q}(t)w'(t) \le 0.$$

Integration from t_0 to t, $a \le t_0 \le t$, yields

$$\int_{a}^{t} r^{1-q}(s)\, ds - w^{1-q}(t) \le \int_{a}^{t_0} r^{1-q}(s)\, ds - w^{1-q}(t_0) \le \int_{a}^{t_0} r^{1-q}(s)\, ds$$

and so

$$w^{q-1}(t) \le \left(\int_{t_0}^{t} r^{1-q}(s)\, ds\right)^{-1}$$

from which (2.2.26) follows. □

2.2.6 Modified Riccati inequality

In the most of the above results we have assumed $\int^\infty r^{1-q}(s)\,ds = \infty$. Now we will discuss the complementary case, i.e., the convergence of this integral. It is supposed that $c(t) \geq 0$ for large t and that

$$\int^\infty r^{1-q}(t)\,dt < \infty. \tag{2.2.27}$$

We denote

$$\rho(t) = \int_t^\infty r^{1-q}(s)\,ds. \tag{2.2.28}$$

The first auxiliary statement concerns boundedness of solutions of (1.1.1) and of the associated Riccati equation.

Lemma 2.2.7. *Let x be a nonoscillatory solution of* (1.1.1) *and let $w = r\Phi(x'/x)$ be the associated solution of* (1.1.21). *Then x and the function*

$$z(t) := \rho^{p-1}w(t) \tag{2.2.29}$$

are bounded. Moreover,

$$\rho^{p-1}(t)w(t) \geq -1 \quad \textit{for large } t \tag{2.2.30}$$

and

$$\limsup_{t\to\infty} \rho^{p-1}(t)w(t) \leq 0. \tag{2.2.31}$$

Proof. Without loss of generality we can suppose that $x(t) > 0$ for $t \in [t_0, \infty)$. The function $r(t)\Phi(x')$ is nonincreasing (since its derivative equals $-c(t)\Phi(x) \leq 0$), the derivative x' is eventually of constant sign. That is, $x'(t) > 0$ for $t \geq t_0$ or there is $t_1 > t_0$ such that $x'(t) < 0$ for $t \geq t_1$, and that

$$r^{q-1}(s)x'(s) \leq r^{q-1}(t)x'(t) \quad \text{for} \quad s \geq t \geq t_0.$$

Dividing this inequality by $r^{q-1}(s)$ and integrating it over $[t, \tau]$ we obtain

$$x(\tau) \leq x(t) + r^{q-1}(t)x'(t)\int_t^\tau r^{1-q}(s)\,ds. \tag{2.2.32}$$

If $x'(t) > 0$ for $t \geq t_0$, then we have from (2.2.32)

$$x(\tau) \leq x(t) + r^{q-1}(t)x'(t)\rho(t)$$

which shows that x is bounded on $[t_0, \infty)$. If $x'(t) < 0$ for $t \geq t_1$, then x is clearly bounded and, letting $\tau \to \infty$ in (2.2.32), we have

$$0 \leq x(t) + r^{q-1}(t)x'(t)\rho(t), \quad t \geq t_0.$$

In either case we obtain

$$\rho(t)r^{q-1}(t)\frac{x'(t)}{x(t)} \geq -1,$$

which immediately implies (2.2.30).

The limit inequality (2.2.31) trivially holds if $x'(t) < 0$ for $t \geq t_1$, since in this case the function z defined by (2.2.29) is negative for $t \geq t_1$. If $x'(t) > 0$ for $t \geq t_0$, then there exist positive constants c_1, c_2 such that

$$x(t) \geq c_1 \quad \text{and} \quad r(t)\Phi(x'(t)) \leq c_2 \quad \text{for} \quad t \geq t_0,$$

which implies

$$w(t) \leq \frac{c_2}{c_1^{p-1}}, \quad t \geq t_0.$$

Since $\rho(t) \to 0$ as $t \to \infty$, we then conclude that

$$\lim_{t\to\infty} \rho^{p-1}(t)w(t) = 0.$$

This completes the proof. □

Based on the previous lemma we show that nonoscillation of (1.1.1) is equivalent to solvability of a certain modified Riccati integral inequality.

Theorem 2.2.7. *Equation* (1.1.1) *is nonoscillatory if and only if*

$$\text{(2.2.33)} \qquad \int^\infty \rho^p(t)c(t)\,dt < \infty$$

and there exists a continuous function v *such that*

$$\text{(2.2.34)} \qquad \rho^{p-1}v(t) \ \textit{is bounded,} \quad \rho^{p-1}(t)v(t) \geq -1,$$

and

$$\text{(2.2.35)} \quad \rho^p(t)v(t) \geq \int_t^\infty \rho^p(s)c(s)\,ds + p\int_t^\infty r^{1-q}(s)\rho^{p-1}(s)v(s)\,ds + (p-1)\int_t^\infty r^{1-q}(s)\rho^p(s)|v(s)|^q\,ds$$

for large t.

Proof. "⇒": Let x be a solution of (1.1.1) such that $x(t) \neq 0$ for $t \geq t_0$ and let $w = r\Phi(x')/\Phi(x)$ be the corresponding solution of Riccati equation (1.1.21). Multiplying this equation by $\rho^p(t)$ and integrating over $[t,\tau]$, $\tau \geq t \geq t_0$, we get

$$\rho^p(\tau)w(\tau) - \rho^p(t)w(t) = -p\int_t^\tau r^{1-q}(s)\rho^{p-1}(s)w(s)\,ds - \int_t^\tau \rho^p(s)c(s)\,ds - (p-1)\int_t^\tau r^{1-q}(s)\rho^p(s)|w(s)|^q\,ds. \qquad \text{(2.2.36)}$$

In view of boundedness of the function $\rho^{p-1}w(t)$ (compare with the previous lemma), we see that $\rho^p(\tau)w(\tau) = \rho(\tau)\rho^{p-1}(\tau)w(\tau) \to 0$ as $\tau \to \infty$, and

$$\left|\int_t^\infty r^{1-q}(s)\rho^{p-1}(s)w(s)\,ds\right| \le \int_t^\infty r^{1-q}(s)\left|\rho^{p-1}(s)w(s)\right|\,ds < \infty,$$

$$\int_t^\infty r^{1-q}(s)\rho^p(s)|w(s)|^q\,ds < \infty$$

for $t \ge t_0$. Therefore, letting $\tau \to \infty$ in (2.2.36), we find that $\int_t^\infty \rho^p(s)c(s)\,ds$ is convergent, i.e., (2.2.33) holds, and

$$\begin{aligned}\rho^p(t)w(t) &= \int_t^\infty \rho^p(s)c(s)\,ds + p\int_t^\infty r^{1-q}(s)\rho^{p-1}(s)w(s)\,ds \\ &\quad +(p-1)\int_t^\infty r^{1-q}(s)\rho^p(s)|w(s)|^q\,ds\end{aligned}$$

hence (2.2.35) holds as equality. The inequality $\rho^p(t)w(t) \ge -1$ follows from the previous lemma.

"$\Leftarrow$": Suppose that (2.2.33) holds and let w be a continuous function satisfying conditions of theorem. Further, let us denote by $C[t_0,\infty)$ the Fréchet space of continuous functions with the topology of uniform convergence on compact subintervals of $[t_0,\infty)$. Consider the space

$$\mathcal{V} := \{v \in C[t_0,\infty) : -1 \le v(t) \le \rho^{p-1}(t)w(t),\ t \ge t_0\}, \tag{2.2.37}$$

which is a closed convex subset of $C[t_0,\infty)$. Define the mapping $F : \mathcal{V} \to C[t_0,\infty)$ by

$$\begin{aligned}\rho(t)(Fv)(t) &= \int_t^\infty \rho^p(s)c(s)\,ds + p\int_t^\infty r^{1-q}(s)v(s)\,ds \\ &\quad +(p-1)\int_t^\infty r^{1-q}(s)|v(s)|^q\,ds.\end{aligned} \tag{2.2.38}$$

If $v \in \mathcal{V}$, then from (2.2.37), (2.2.38) and the inequality stated in theorem

$$(Fv)(t) \le F(\rho^{p-1}w)(t) \le \rho^{p-1}(t)w(t), \quad t \ge t_0,$$

and

$$\rho(t)[(Fv)(t)+1] \ge \int_t^\infty r^{1-q}(s)\left[(p-1)|v(s)|^q + pv(s) + 1\right]\,ds \ge 0,$$

where we have used also the property that the function $(p-1)|\xi|^q + p\xi$ is strictly increasing for $\xi \ge -1$, i.e.,

$$(p-1)|\xi|^q + p\xi + 1 \ge 0 \quad \text{for } \xi \ge -1. \tag{2.2.39}$$

This shows that F maps $\mathcal{V}$ into itself. It can be shown in a routine manner that F is continuous and $F(\mathcal{V})$ is relatively compact in the topology of $C[t_0,\infty)$. Therefore, by the Schauder-Tychonov fixed point theorem, there exists an element $v \in \mathcal{V}$ such that $v(t) = (Fv)(t)$. Define w by $w(t) = v(t)/\rho^{p-1}(t)$. Then, in view of (2.2.38), w satisfies the integral equation

$$\begin{aligned}\rho^p(t)w(t) &= \int_t^\infty \rho^p(t)c(s)\,ds + p\int_t^\infty r^{1-q}(s)\rho^{p-1}w(s)\,ds \\ &\quad +(p-1)\int_t^\infty r^{1-q}(s)\rho^p(s)|w(s)|^q\,ds.\end{aligned}$$

Differentiating this equality and then dividing by $\rho^p(t)$ shows that w solves Riccati equation (1.1.21) and hence (1.1.1) is nonoscillatory. □

2.2.7 Applications

Here we give some examples of (non)oscillation criteria whose proofs are based on the above described technique. Many other applications can be found in the subsequent chapters.

We start with a simple consequence of Theorem 2.2.1.

Theorem 2.2.8. *If*

$$\int_a^\infty s^{p-1}c(s)\,ds \quad \textit{is convergent,} \tag{2.2.40}$$

then

$$(\Phi(y'))' + c(t)\Phi(y) = 0 \tag{2.2.41}$$

is nonoscillatory.

Proof. Let

$$w(t) = \frac{1}{2}t^{1-p}\left\{1 + 2\int_t^\infty s^{p-1}c(s)\,ds\right\}.$$

Then

$$w'(t) = -\frac{p-1}{2}t^{-p}\left\{1 + 2\int_t^\infty s^{p-1}c(s)\,ds\right\} - c(t). \tag{2.2.42}$$

In view of (2.2.40), there exists $T \geq 0$ such that

$$0 < 1 + 2\int_t^\infty s^{p-1}c(s)\,ds \leq 2$$

for $t \geq T$, so that

$$|w(t)|^q = 2^{-q}t^{-p}\left\{1 + 2\int_t^\infty s^{p-1}c(s)\,ds\right\}^q \leq \frac{1}{2}t^{-p}\left\{1 + 2\int_t^\infty s^{p-1}c(s)\,ds\right\}.$$

Consequently, (2.2.42) implies $w'(t) + c(t) + (p-1)|w(t)|^q \leq 0$ for $t \geq T$. It follows from Theorem 2.2.1 that (2.2.41) is nonoscillatory. □

Now we give more refined criteria. The technique involving the Riccati inequality was used to prove the classical Hille-Nehari criterion in the linear case. This criterion claims that if $\int^\infty r^{-1}(t)\,dt = \infty$ and $\int^\infty c(t)\,dt$ converges, then linear Sturm-Liouville equation (1.1.2) is nonoscillatory provided

$$\limsup_{t\to\infty}\left(\int^t r^{-1}(s)\,ds\right)\left(\int_t^\infty c(s)\,ds\right) < \frac{1}{4}$$

and

$$\liminf_{t\to\infty}\left(\int^t r^{-1}(s)\,ds\right)\left(\int_t^\infty c(s)\,ds\right) > -\frac{3}{4}.$$

The next statement is a half-linear extension of this linear criterion. Compare this result with the criterion given in Subsection 2.1.3.

Theorem 2.2.9. *Suppose that* $\int^\infty r^{1-q}(t)\,dt = \infty$ *and* $\int^\infty c(t)\,dt$*converges. If*

$$\limsup_{t\to\infty}\left(\int^t r^{1-q}(s)\,ds\right)^{p-1}\left(\int_t^\infty c(s)\,ds\right) < \frac{1}{p}\left(\frac{p-1}{p}\right)^{p-1}, \tag{2.2.43}$$

$$\liminf_{t\to\infty}\left(\int^t r^{1-q}(s)\,ds\right)^{p-1}\left(\int_t^\infty c(s)\,ds\right) > -\frac{2p-1}{p}\left(\frac{p-1}{p}\right)^{p-1}, \tag{2.2.44}$$

then (1.1.1) *is nonoscillatory.*

Proof. We will find a solution of the Riccati type inequality (2.2.7), i.e., of the inequality

$$v' \le -c(t) - (p-1)r^{1-q}(t)|v|^q \tag{2.2.45}$$

which is extensible up to ∞, i.e., it exists on some interval $[T,\infty)$. To find this solution v of (2.2.45), we show that there exists an extensible up to ∞ solution of the differential inequality

$$\rho' \le (1-p)r^{1-q}(t)|\rho + C(t)|^q, \quad C(t) := \int_t^\infty c(s)\,ds \tag{2.2.46}$$

related to (2.2.45) by the substitution $\rho = v - C$. This solution ρ has the form

$$\rho(t) = \beta\left(\int^t r^{1-q}(s)\,ds\right)^{1-p}, \quad \beta := \left(\frac{p-1}{p}\right)^p.$$

Indeed, $\rho' = (1-p)\beta r^{1-q}(t)\left(\int^t r^{1-q}(s)\,ds\right)^{-p}$ and the right-hand side of inequality (2.2.46) is

$$\begin{aligned}(1-p)r^{1-q}(t)|\rho + C(t)|^q &= (1-p)r^{1-q}(t)\left|\beta\left(\int^t r^{1-q}\right)^{1-p} + C(t)\right|^q\\ &= (1-p)r^{1-q}(t)\left|\beta + \left(\int^t r^{1-q}\right)^{p-1} C(t)\right|^q \left(\int^t r^{1-q}\right)^{-p}.\end{aligned}$$

Consequently, (2.2.46) is equivalent to the inequality

$$\beta \geq \left| \beta + \left(\int^t r^{1-q} \right)^{p-1} C(t) \right|^q . \tag{2.2.47}$$

However, since (2.2.43) and (2.2.44) hold, there exists $\varepsilon > 0$ such that

$$-\frac{2p-1}{p}\left(\frac{p-1}{p}\right)^{p-1} + \varepsilon < \left(\int^t r^{1-q} \right)^{p-1} C(t) < \frac{1}{p}\left(\frac{p-1}{p}\right)^{p-1} - \varepsilon$$

for large t and by a direct computation it is not difficult to verify that (2.2.47) really holds. □

As a direct consequence of the Hartman-Wintner theorem (Theorem 2.2.3) we have the following generalized Hartman-Wintner oscillation criterion.

Theorem 2.2.10. *Suppose that $\int^\infty r^{1-q}(t)\, dt = \infty$. Then each of the following two conditions is sufficient for oscillation of* (1.1.1)*:*

$$\lim_{t\to\infty} \frac{\int^t r^{1-q}(s) \left(\int^s c(\tau)\, d\tau \right) ds}{\int^t r^{1-q}(s)\, ds} = \infty, \tag{2.2.48}$$

(2.2.49)

$$-\infty < \liminf_{t\to\infty} \frac{\int^t r^{1-q}(s) \left(\int^s c(\tau)\, d\tau \right) ds}{\int^t r^{1-q}(s)\, ds} < \limsup_{t\to\infty} \frac{\int^t r^{1-q}(s) \left(\int^s c(\tau)\, d\tau \right) ds}{\int^t r^{1-q}(s)\, ds}.$$

Proof. We will prove only sufficiency of (2.2.48); the proof of sufficiency of (2.2.49) is similar. Suppose that (1.1.1) is nonoscillatory and (2.2.48) holds. Then (2.2.12) holds and by Theorem 2.2.3 the integral in (2.2.10) converges for every solution w of (1.1.21) and hence limit (2.2.11) exists as a finite number which contradicts to (2.2.48). □

We conclude this section by the application of Theorem 2.2.7. The following criterion is its immediate consequence.

Theorem 2.2.11. *Suppose that* (2.2.27) *holds. Equation* (1.1.1) *is oscillatory if*

$$\int^\infty c(s)\rho^p(t)\, dt = \infty, \tag{2.2.50}$$

where ρ is given by (2.2.28).

This oscillation criterion opens a natural question about the oscillatory nature of (1.1.1) when the integral in (2.2.50) is convergent. This will be discussed later (see Theorem 2.3.4 and Section 3.1).

2.3 Comparison theorems

In this section we present various types of statements where oscillatory properties of two equations are compared. We usually suppose that oscillatory behavior of one equation is known and using various inequalities we deduce oscillatory nature of the second equation. In the first three subsections we give conditions on coefficients, under which new equation is (non)oscillatory provided the original one is so. Then we present the so-called telescoping principle, and finally we offer the statement where two equations with different nonlinearities are compared.

2.3.1 Hille-Wintner comparison theorems

While in the Sturm comparison theorem the coefficients are compared pointwise, the classical Hille-Wintner theorem (for the linear version see [341, Theorem 2.14]) compares the coefficients on average. Its half-linear version appeared in many works (e.g., [180, 182, 227, 247, 253, 242]), under different assumptions. Here we offer the version where the sign condition on the second coefficients is slightly relaxed comparing to the usual assumptions. Moreover, we will discuss the complementary case and various methods of the proof.

Along with equation (1.1.1) consider the equation

(2.3.1) $$(r(t)\Phi(y'))' + \tilde{c}(t)\Phi(y) = 0.$$

Theorem 2.3.1. *Assume that* (2.2.9) *holds,* $\int^\infty c(t)\,dt$ *and* $\int^\infty \tilde{c}(t)\,dt$ *converge. If*

(2.3.2) $$\left|\int_t^\infty c(s)\,ds\right| \le \int_t^\infty \tilde{c}(s)\,ds \quad \textit{for large } t,$$

and (2.3.1) *is nonoscillatory, then so is equation* (1.1.1), *or equivalently, the oscillation of* (1.1.1) *implies that of equation* (2.3.1).

Proof. If (2.3.1) is nonoscillatory, then there is a function w satisfying

$$w(t) = \int_t^\infty \tilde{c}(s)\,ds + \int_t^\infty S(w,r)(s)\,ds \ (\ge 0),$$

for large t, by Theorem 2.2.4. Hence

$$\begin{aligned} w(t) &\ge \left|\int_t^\infty c(s)\,ds\right| + \int_t^\infty S(w,r)(s)\,ds \\ &\ge \left|\int_t^\infty c(s)\,ds + \int_t^\infty S(w,r)(s)\,ds\right|, \end{aligned}$$

so (1.1.1) is nonoscillatory by Remark 2.2.2. □

Remark 2.3.1. (i) Note that without the use of Theorem 2.2.5, under the additional assumptions $c(t) \ge 0$, $\tilde{c}(t) \ge 0$ (but, in fact, $\int_t^\infty c(s)\,ds \ge 0$ suffices), some authors (e.g., [227]) use the Schauder-Tychonov fixed point theorem to prove the

Hille-Wintner theorem. A closer examination of that proof then reveals that the statement of Theorem 2.2.5 (with the additional assumptions) is actually hidden in it.

(ii) Later, in Section 5.5, we will see that there exists another possibility how to prove the Hille-Wintner theorem, namely the sequence approach. Note that in the classical work of Nehari [304], the proof of this theorem in the linear case is based on the variational technique.

In Subsection 1.4.3 we have presented the criteria of Kneser type, which are established using the Sturmian comparison theorem and the generalized Euler equation. Now we use the same idea, but with the Hille-Wintner theorem instead of the Sturm one, to obtain an improvement, namely the criteria of Hille-Nehari type. In Subsection 1.4.2 we have shown that Euler equation (1.4.20) is nonoscillatory if and only if $\gamma \leq \tilde{\gamma} = \left(\frac{p-1}{p}\right)^p$. Consider now the equation

$$(r(t)\Phi(x'))' + \frac{\gamma r^{1-q}(t)}{\left(\int^t r^{1-q}(s)\,ds\right)^p}\Phi(x) = 0 \tag{2.3.3}$$

with r satisfying (2.2.9). The transformation $t \mapsto \int^t r^{1-q}(s)\,ds$ of independent variable transforms this equation into the Euler equation (1.4.20). Hence also (2.3.3) is nonoscillatory if and only if $\gamma \leq \tilde{\gamma}$. This fact, combined with Theorem 2.3.1, leads to the following nonoscillation and oscillation criteria which are the half-linear extension of the Hille-Nehari (non)oscillation criteria, see [341, Chap. II] and also Section 3.1.

Theorem 2.3.2. *Suppose that* $\int^\infty r^{1-q}(t)\,dt = \infty$ *and the integral* $\int^\infty c(t)\,dt$ *is convergent.*

(i) If

$$0 \leq \left(\int^t r^{1-q}(s)\,ds\right)^{p-1}\left(\int_t^\infty c(s)\,ds\right) \leq \frac{1}{p}\left(\frac{p-1}{p}\right)^{p-1}$$

for large t, *then* (1.1.1) *is nonoscillatory.*

(ii) If

$$\liminf_{t\to\infty}\left(\int^t r^{1-q}(s)\,ds\right)^{p-1}\left(\int_t^\infty c(s)\,ds\right) > \frac{1}{p}\left(\frac{p-1}{p}\right)^{p-1}, \tag{2.3.4}$$

then (1.1.1) *is oscillatory.*

Proof. First of all observe that (2.3.4) implies that $\int_t^\infty c(s)\,ds > 0$ for large t. Now, since

$$\int_t^\infty \frac{\tilde{\gamma} r^{1-q}(s)}{\left(\int^s r^{1-q}(\tau)\,d\tau\right)^p}\,ds = \frac{1}{p}\left(\frac{p-1}{p}\right)^{p-1}\left(\int^t r^{1-q}(s)\,ds\right)^{1-p},$$

the statement follows from Theorem 2.3.1 with $\tilde{c}(t) = \tilde{\gamma} r^{1-q}(t)\left(\int^t r^{1-q}(s)\,ds\right)^{-p}$. □

Next we give a complement of Theorem 2.3.1 in the sense of the convergence of $\int^\infty r^{1-q}(s)\,ds$.

Theorem 2.3.3. *Let $c(t) \geq 0$, $\tilde{c}(t) \geq 0$ for large t and $\rho(t) := \int_t^\infty r^{1-q}(s)\,ds < \infty$. Assume that*

$$\int^\infty c(t)\rho^p(t)\,dt < \infty, \qquad \int^\infty \tilde{c}(t)\rho^p(t)\,dt < \infty. \tag{2.3.5}$$

If

$$\int_t^\infty c(t)\rho^p(t)\,dt \leq \int_t^\infty \tilde{c}(t)\rho^p(t)\,dt, \tag{2.3.6}$$

(2.3.1) *is nonoscillatory, then so is equation* (1.1.1), *or equivalently, the oscillation of* (1.1.1) *implies that of equation* (2.3.1).

Proof. Assume that (2.3.1) is nonoscillatory. Then, by the "only if" part of Theorem 2.2.7 there exists a continuous function w satisfying (2.2.34) and

$$\begin{aligned}\rho^p(t)w(t) \quad \geq \quad & \int_t^\infty \rho^p(s)\tilde{c}(s)\,ds + p\int_t^\infty r^{1-q}(s)\rho^p(s)w(s)\,ds \\ & +(p-1)\int_t^\infty r^{1-q}(s)\rho^p(s)|w(s)|^q\,ds.\end{aligned}$$

Using (2.3.5) and (2.3.6) we see that w satisfies integral inequality (2.2.35) and hence (1.1.1) is nonoscillatory by the "if" part of Theorem 2.2.7. □

Now we can apply the last theorem to obtain modified Hille-Nehari criteria. This will answer the question posed at the very end of Section 2.2. For related results see Subsection 2.2.6 and Theorem 2.2.11. For another information about these types of criteria see Section 3.1.

Theorem 2.3.4. *Suppose that* (2.2.27) *holds, ρ is given by* (2.2.28) *and the integral $\int^\infty \rho^p(t)c(t)\,dt$ is convergent.*

(i) Equation (1.1.1) *is oscillatory if*

$$\liminf_{t\to\infty} \rho^{-1}(t)\int_t^\infty c(s)\rho^p(s)\,ds > \left(\frac{p-1}{p}\right)^p. \tag{2.3.7}$$

(ii) Equation (1.1.1) *is nonoscillatory if*

$$\rho^{-1}(t)\int_t^\infty c(s)\rho^p(s)\,ds \leq \left(\frac{p-1}{p}\right)^p \tag{2.3.8}$$

for large t.

Proof. First we show that the equation

$$(t^{\alpha}\Phi(y'))' + g(t)\Phi(y) = 0, \tag{2.3.9}$$

where g is a continuous function and α is a constant such that $\alpha > p-1$, is oscillatory if

$$\liminf_{t\to\infty} t^{\frac{\alpha-p+1}{p-1}} \int_t^{\infty} s^{-\frac{p(\alpha-p+1)}{p-1}} g(s)\, ds > \frac{(p-1)(\alpha-p+1)^{p-1}}{p^p}, \tag{2.3.10}$$

and it is nonoscillatory if

$$t^{\frac{\alpha-p+1}{p-1}} \int_t^{\infty} s^{-\frac{p(\alpha-p+1)}{p-1}} g(s)\, ds \le \frac{(p-1)(\alpha-p+1)^{p-1}}{p^p} \tag{2.3.11}$$

for all large t. Suppose that (2.3.10) holds. Then there exist γ^* and T such that $\gamma^* > [(\alpha-p+1)/p]^p$ and

$$t^{\frac{\alpha-p+1}{p-1}} \int_t^{\infty} s^{-\frac{p(\alpha-p+1)}{p-1}} g(s)\, ds > \frac{p-1}{\alpha-p+1}\gamma^*$$

for $t \ge T$. Since

$$\frac{p-1}{\alpha-p+1}\gamma^* = t^{\frac{\alpha-p+1}{p-1}} \int_t^{\infty} s^{-\frac{p(\alpha-p+1)}{p-1}} \gamma^* s^{\alpha-p}\, ds,$$

we have

$$\int_t^{\infty} s^{-\frac{p(\alpha-p+1)}{p-1}} g(s)\, ds > \int_t^{\infty} s^{-\frac{p(\alpha-p+1)}{p-1}} \gamma^* s^{\alpha-p}\, ds,$$

$t \ge T$. Noting that the generalized Euler equation (1.4.33) with $\gamma = \gamma^*$ is oscillatory by Theorem 1.4.4 and applying Theorem 2.3.3, we conclude that (2.3.9) is oscillatory. Let $\gamma_0 = [(\alpha-p+1)/p]^p$. Then equation (1.4.33) with $\gamma = \gamma_0$ is nonoscillatory by Theorem 1.4.4. Since (2.3.11) can be written as

$$\int_t^{\infty} s^{-\frac{p(\alpha-p+1)}{p-1}} g(s)\, ds \le \frac{p-1}{\alpha-p+1}\gamma_0 t^{-\frac{\alpha-p+1}{p-1}} = \int_t^{\infty} s^{-\frac{p(\alpha-p+1)}{p-1}} \gamma_0 s^{\alpha-p}\, ds,$$

it follows from Theorem 2.3.3 that (2.3.9) is nonoscillatory provided (2.3.11) is satisfied.

The next step of the proof is the transformation of independent variable $u(s) = x(t)$, $s = s(t) = (\rho(t))^{\frac{p-1}{p-1-\alpha}}$. This transformation transforms (1.1.1) into the equation

$$(s^{\alpha}\Phi(u'))' + Q(s)\Phi(u) = 0, \tag{2.3.12}$$

where

$$Q(s) = \left(\frac{\alpha-p+1}{p-1}\right)^p r^{1-q}(t(s))\,[\rho(t(s))]^{\frac{p-1}{\alpha-p+1}}\, c(t(s)),$$

$t = t(s)$ being the inverse function of $s = s(t)$. Since $\alpha > p - 1$. Then we have $\int^\infty s^{\alpha(1-q)}\, ds < \infty$, so (2.3.12) satisfies assumption (2.2.27). The results of the first part of the proof applied to (2.3.12) show that (2.3.12) is oscillatory if

$$\liminf_{s\to\infty} s^{\frac{\alpha-p+1}{p-1}} \int_s^\infty \xi^{-\frac{p(\alpha-p+1)}{p-1}} Q(\xi)\, d\xi > \frac{(p-1)(\alpha-p+1)^{p-1}}{p^p} \tag{2.3.13}$$

and that (2.3.12) is nonoscillatory if

$$s^{\frac{\alpha-p+1}{p-1}} \int_s^\infty \xi^{-\frac{p(\alpha-p+1)}{p-1}} Q(\xi)\, d\xi \le \frac{(p-1)(\alpha-p+1)^{p-1}}{p^p} \tag{2.3.14}$$

for all large s. It is a matter of easy computation to verify that inequalities (2.3.13) and (2.3.14) transform back to (2.3.7) and (2.3.8), respectively, which are directly applicable to original equation (1.1.1). □

2.3.2 Leighton comparison theorems

First recall that (1.1.1) is said to be *disconjugate* in a given interval I if every nontrivial solution of this equation has at most one zero in I, in the opposite case, i.e., if there exists a nontrivial solution of (1.1.1) having at least two zeros in I, equation (1.1.1) is said to be *conjugate* in I. Further, two points $t_1, t_2 \in \mathbb{R}$ are said to be *conjugate relative to* (1.1.1) if there exists a nontrivial solution x of this equation such that $x(t_1) = 0 = x(t_2)$.

Consider a pair of half-linear differential equations (1.2.6) and (1.1.1). If (1.1.1) is a Sturmian minorant of (1.2.6) on $I = [a,b]$ and (1.1.1) is conjugate on this interval, then majorant equation (1.2.6) is conjugate on $[a,b]$ as well. In the next theorem we replace the pointwise comparison of coefficients by the integral one (in a different sense than in the previous subsection). In the linear case $p = 2$, this statement was proved by Leighton [234]. Here we offer its half-linear extension.

Theorem 2.3.5. *Let x be a solution of* (1.1.1) *satisfying $x(a) = 0 = x(b)$, $x(t) \neq 0$ for $t \in (a,b)$, and*

$$\mathcal{J}(x; a, b) := \int_a^b [(r(t) - R(t))|x'|^p - (c(t) - C(t))|x|^p]\, dt \ge 0. \tag{2.3.15}$$

Then every solution y of (1.2.6) *has a zero in (a,b), or equations* (1.2.6) *and* (1.1.1) *are identical, and y is a constant multiple of x. If, in addition, the strict inequality is satisfied, then y has a zero in (a,b).*

Proof. We have (with the notation introduced in the proof of Theorem 1.2.4)

$$\begin{aligned}
\mathcal{F}_{R,C}(x; a, b) &= \int_a^b [R(t)|x'|^p - C(t)|x|^p]\, dt \\
&= \int_a^b [r(t)|x'|^p - c(t)|x|^p]\, dt - \mathcal{J}(x; a, b) \\
&= [r(t) x \Phi(x')]_a^b - \int_a^b x[(r(t)\Phi(x'))' - c(t)\Phi(x)]\, dt - \mathcal{J}(x; a, b) \\
&= -\mathcal{J}(x; a, b) \le 0.
\end{aligned}$$

Recalling that $\mathcal{J}(x;a,b) = 0$ corresponds to the case where $r \equiv R$, $c \equiv C$ and $x' \equiv xy'/y$, the conclusion follows from Theorem 1.2.2. □

Remark 2.3.2. In terms of conjugacy, the last statement can be reformulated as follows: Suppose that the points a, b are conjugate relative to (1.1.1) and let x be a nontrivial solution of this equation for which $x(a) = 0 = x(b)$. If $\mathcal{J}(x;a,b) \geq 0$, then (1.2.6) is also conjugate in $[a,b]$. Clearly, the Sturm-Picone type comparison theorem is a consequence of this statement.

Now we give a variant of the Leighton type theorem, along with its dual. Their statements in terms of conjugacy are obvious.

Theorem 2.3.6. *Let R/r be continuously differentiable on (a,b) and x be a solution of (1.1.1) satisfying $x(a) = 0 = x(b)$, $x(t) \neq 0$ for $t \in (a,b)$, and*

$$\int_a^b \left\{ \left(C - \frac{R}{r} c \right) |x|^p + r \left(\frac{R}{r} \right)' x\Phi(x') \right\} (t)\, dt > 0. \tag{2.3.16}$$

Then every solution of (1.2.6) has a zero in (a,b).

Proof. Putting $r \equiv R$ on the left-hand side of (1.2.1) and rewriting (1.1.1) as

$$\left(\frac{r(t)}{R(t)} R(t)\Phi(x') \right)' + c(t)\Phi(x) = 0$$

or, equivalently,

$$(R(t)\Phi(x'))' - r(t) \left(\frac{R(t)}{r(t)} \right)' \Phi(x') + \frac{R(t)}{r(t)} c(t)\Phi(x) = 0,$$

identity (1.2.1) becomes

$$\begin{aligned} &\left\{ \frac{x}{\Phi(y)} \left[\Phi(y)R\Phi(x') - \Phi(x)R\Phi(y') \right] \right\}' \\ &= \left(C - \frac{R}{r} c \right) |x|^p + r \left(\frac{R}{r} \right)' x\Phi(x') + pR^{1-q} P\left(R^{q-1}x', R\Phi(xy'/y) \right). \end{aligned} \tag{2.3.17}$$

It is not difficult to see that the function on the left-hand side inside {} tends to zero as $t \to a+$ or $t \to b-$ (see the proof of Theorem 1.2.4), so that after integrating (2.3.17) from $a + \varepsilon$ to $b - \varepsilon$, letting $\varepsilon \to 0+$ we are led to the contradiction with (2.3.16), in view of the Young inequality (1.2.2). □

Similarly, assuming that $c \neq 0$, $C \neq 0$ and $C/c \in C^1(a,b)$ we can rewrite $(r\Phi(x'))'$ as

$$\frac{c}{C} \left(\frac{rC}{c} \Phi(x') \right)' + \left(\frac{c}{C} \right)' \frac{rC}{c} \Phi(x'),$$

so that (1.1.1) becomes

$$\left(\frac{rC}{c} \Phi(x') \right)' + \frac{C}{c} \left(\frac{c}{C} \right)' \frac{C}{c} r\Phi(x') + C\Phi(x) = 0,$$

or, equivalently,

$$\left(\frac{rC}{c}\Phi(x')\right)' - \left(\frac{C}{c}\right)' r\Phi(x') + C\Phi(x) = 0, \tag{2.3.18}$$

so that the coefficients by $\Phi(x)$ and $\Phi(y)$ in (2.3.18) and (1.2.6), respectively, are the same. From (1.2.1) with r replaced by rC/c we then have

$$\left\{\frac{x}{\Phi(y)}\left[\Phi(y)\frac{rC}{c}\Phi(x') - \Phi(x)R\Phi(y')\right]\right\}'$$
$$= \left(\frac{rC}{c} - R\right)|x'|^p + r\left(\frac{C}{c}\right)' x\Phi(x') + pR^{1-q}P\left(R^{q-1}x', R\Phi(xy'/y)\right).$$

Now we easily see the following dual comparison result to the previous theorem.

Theorem 2.3.7. *Let $c \neq 0$, $C \neq 0$ and $C/c \in C^1(a,b)$ and x be a solution of* (1.1.1) *satisfying $x(a) = 0 = x(b)$, $x(t) \neq 0$ for $t \in (a,b)$, and*

$$\int_a^b \left\{\left(\frac{rC}{c} - R\right)|x'|^p + r\left(\frac{C}{c}\right)' x\Phi(x')\right\}(t)\,dt > 0.$$

Then every solution of (1.2.6) *has a zero in (a,b).*

Remark 2.3.3. In Subsection 5.8.3 we give variants of the above theorems in a Leighton-Levin sense, which are based on the variational technique involving nonzero boundary conditions.

2.3.3 Multiplied coefficient comparison

First we mention a few background details which serve to motivate the main results of this subsection. Along with equation (1.1.1) consider the equation

$$[r(t)\Phi(y')]' + \lambda c(t)\Phi(y) = 0, \tag{2.3.19}$$

where λ is a real constant. We claim that if (1.1.1) is nonoscillatory and $0 < \lambda \leq 1$, then (2.3.19) is also nonoscillatory. If $c(t) \geq 0$, then this statement follows immediately from the Sturm comparison theorem (Theorem 1.2.4). If $c(t)$ may change sign, then dividing (2.3.19) by λ we obtain an equivalent equation which is nonoscillatory again by the Sturm theorem. This can be analogously done for oscillatory counterparts. If the constant λ is replaced by a function $a(t)$, then the situation is not so easy (when $c(t)$ may change sign; otherwise the Sturm theorem can be applied immediately). The following statements give an answer to the question: "What are the conditions which guarantee that (non)oscillation of (1.1.1) is preserved when multiplying the coefficient $c(t)$ by a function $a(t)$?". Along with equation (1.1.1) consider the equation

$$[R(t)\Phi(x')]' + a(t)C(t)\Phi(x) = 0, \tag{2.3.20}$$

where R and C satisfy the same assumptions as r and c, respectively. Related results in the linear case can be found in [152].

Theorem 2.3.8. *Assume that $a(t)$ is continuously differentiable, $r(t) \le R(t)$, $C(t) \le c(t)$, $0 < a(t) \le 1$, $a'(t) \le 0$. Further, let*

$$\Psi(T) := \liminf_{t\to\infty} \int_T^t c(s)\, ds \ge 0 \quad \textit{and} \quad \Psi(T) \not\equiv 0$$

for all large T and $\int^\infty r^{1-q}(s)\, ds = \infty$. Then nonoscillation of (1.1.1) *implies nonoscillation of* (2.3.20).

Proof. The assumptions of the theorem imply that there exists a solution y of (1.1.1) and $T \in \mathbb{R}$ such that $y(t) > 0$ and $y'(t) > 0$ on $[T, \infty)$ by Lemma 2.2.3. Therefore, the function $w(t) := r(t)\Phi(y'(t)/y(t)) > 0$ satisfies (1.1.21) on $[T, \infty)$. We have $ar^{1-q}w^q = (ra)^{1-q}(wa)^q$. Now, multiplying (1.1.21) by a, we get

$$\begin{aligned} 0 &= w'a + ca + (p-1)(ra)^{1-q}(wa)^q \ge w'a + Ca + (p-1)(ra)^{1-q}(wa)^q \\ &\ge w'a + wa' + Ca + (p-1)(ra)^{1-q}(wa)^q \\ &= (wa)' + Ca + (p-1)(ra)^{1-q}(wa)^q \end{aligned}$$

for $t \in [T, \infty)$. Hence the function $v = wa$ satisfies the generalized Riccati inequality $v' + C(t)a(t) + (p-1)(r(t)a(t))^{1-q}v^q \le 0$ for $t \in [T, \infty)$. Therefore, the equation

$$[a(t)r(t)\Phi(x')]' + a(t)C(t)\Phi(x) = 0 \tag{2.3.21}$$

is nonoscillatory by Theorem 2.2.1, and so equation (2.3.20) is nonoscillatory by Theorem 1.2.4 since $a(t)r(t) \le r(t) \le R(t)$. □

Theorem 2.3.9. *Assume that $a(t)$ is continuously differentiable, $R(t) \le r(t)$, $c(t) \le C(t)$, $a(t) \ge 1$, $a'(t) \ge 0$. Further, let*

$$\Psi_a(T) := \liminf_{t\to\infty} \int_T^t a(s)C(s)\, ds \ge 0 \quad \textit{and} \quad \Psi_a(T) \not\equiv 0 \tag{2.3.22}$$

for all large T and

$$\int^\infty R^{1-q}(s)\, ds = \infty.$$

Then oscillation of (1.1.1) *implies oscillation of* (2.3.20).

Proof. Suppose, by a contradiction, that (2.3.20) is nonoscillatory. Then there exists a solution x of (2.3.20) and $T \in \mathbb{R}$ such that $x(t) > 0$ and $x'(t) > 0$ on $[T, \infty)$ by Lemma 2.2.3. Therefore, the function $v(t) := R(t)\Phi(x'(t)/x(t)) > 0$ satisfies

$$v' + a(t)C(t) + (p-1)R^{1-q}(t)v^q = 0 \tag{2.3.23}$$

on $[T, \infty)$. We have

$$\frac{v'}{a} \ge \frac{v'a}{a^2} - \frac{va'}{a^2} = \left(\frac{v}{a}\right)'.$$

Dividing (2.3.23) by a and using the the last estimate, we get

$$\begin{aligned} 0 &= \frac{v'(t)}{a(t)} + C(t) + \frac{1}{a(t)}(p-1)R^{1-q}(t)v^q(t) \\ &\geq \left(\frac{v(t)}{a(t)}\right)' + c(t) + (p-1)[R(t)/a(t)]^{1-q}[v(t)/a(t)]^q \end{aligned}$$

for $t \in [T, \infty)$. Hence the function $w(t) = v(t)/a(t)$ satisfies the inequality $w'(t) + c(t) + (p-1)[R(t)/a(t)]^{1-q}w^q(t) \leq 0$ for $t \in [T, \infty)$. Therefore the equation

$$\left[\frac{R(t)}{a(t)}\Phi(y')\right]' + c(t)\Phi(y) = 0 \tag{2.3.24}$$

is nonoscillatory by Theorem 2.2.1. Now, since $R(t)/a(t) \leq R(t) \leq r(t)$, equation (1.1.1) is nonoscillatory by Theorem 1.2.4, a contradiction. □

Remark 2.3.4. A closer examination of the proofs shows that the last two theorems can be improved in the following way (assuming the same conditions): (a) Theorem 2.3.8: (1.1.1) is nonoscillatory implies (2.3.21) is nonoscillatory; (b) Theorem 2.3.9: (2.3.24) is oscillatory implies (2.3.20) is oscillatory. Our theorems then follow from the above by virtue of the Sturmian comparison theorem (Theorem 1.2.4).

2.3.4 Telescoping principle

Now we want to present an oscillation preserving construction – the so-called *telescoping principle*, which was first proved for linear differential equation (1.1.2) in [230] or in [231], and now it is extended for (1.1.1). The crucial role is played by the Riccati technique and a standard result concerning differential inequalities. Suppose that the coefficients of equation (1.1.1) are in C_a, where C_a denotes the set of continuous functions on $[0, a)$, $a \in \mathbb{R}^+ \cup \{\infty\}$, and assume that (1.1.1) has a solution y which is of one sign in an interval $[a', b') \subset [0, a)$. Then, similarly as in Section 1.1.4, the substitution $w = -r\Phi(y'/y)$ yields the following equivalent equations

$$w'(t) = c(t) + (p-1)r^{1-q}(t)|w(t)|^q \tag{2.3.25}$$

and

$$w(t) = w_0 + \int_{a'}^{t} c(s)\,ds + \int_{a'}^{t} (p-1)r^{1-q}(s)|w(s)|^q ds \tag{2.3.26}$$

for $t \in [a', b')$, where $w_0 = w(a')$. Clearly, the solution y has a zero at b' if and only $w(t) \to \infty$ as $t \to b'-$. Equation (1.1.1) is oscillatory if for any numbers $a' \geq 0$ and w_0, the unique solution w of (2.3.26) satisfies $w(t) \to \infty$ as $t \to b'-$ for some $b' < a$. In employing the latter equivalent condition in the proof of certain oscillation criterion we can often assume, without loss of generality, that $a' = 0$.

Let $\Omega = \bigcup_{i=1}^{n}(a_i, b_i)$ be a finite ($n < \infty$) or infinite ($n = \infty$) union of disjoint open intervals. We assume that

$$0 < a_i < b_i < a_{i+1}, \quad \text{the set } \{a_i\} \text{ has no finite accumulation points.}$$

Let

$$\tau = \tau(t) = \operatorname{mes}([0, t] \setminus \Omega), \tag{2.3.27}$$

where mes denotes the usual Lebesgue measure, and let

$$A = \tau(a), \ A_i = \tau(a_i), \quad i = 1, 2, \ldots, n. \tag{2.3.28}$$

The interval $[0, A)$ is obtained from $[0, a)$ by shrinking each interval (a_i, b_i) to its left endpoint. We construct a class of transformations $T_\Omega : C_a \to C_A$ as follows: Let $f \in C_a$. Then $F = T_\Omega(f)$ is defined by $F(\tau) = f(t)$ if $\tau = \tau(t)$, $\tau \neq A_i$, and $F(A_i) = f(a_i)$. The function F is obtained from f by collapsing each interval (a_i, b_i) to a point.

The next result is a kind of comparison theorem. We use the above notation.

Theorem 2.3.10. *Assume*

$$\int_{a_i}^{b_i} c(t)\, dt \geq 0, \quad i \in \mathbb{N}. \tag{2.3.29}$$

Let $R = T_\Omega(r), C = T_\Omega(c)$. Suppose that z is a solution of

$$(R(t)\Phi(z'))' + C(t)\Phi(z) = 0 \tag{2.3.30}$$

on $[0, A)$ such that $z(t) \neq 0$ for $t \in [0, B)$ and $z(B) = 0$ for some $B < A$. If y is a solution of (1.1.1) such that $y(0) \neq 0$, $r(0)\Phi(y'(0)/y(0)) \leq R(0)\Phi(z'(0)/z(0))$, then $y(b) = 0$ for some $b < a$. More precisely, if $B < A_i$, then there exists a point $b < a_i$ such that $y(b) = 0$, $i \in \mathbb{N}$.

Proof. The proof is by induction on n. Let $v = -R\Phi(z'/z)$ and $w = -r\Phi(y'/y)$. Then

$$v'(t) = C(t) + (p-1)R^{1-q}(t)|v(t)|^q \tag{2.3.31}$$

on $[0, B)$. If $B \leq A_1 = a_1$, then on $[0, B)$, w satisfies the same equation (2.3.31), since $c = C$ and $r = R$ on $[0, a_1) = [0, A_1)$. By hypothesis, $w(0) \geq v(0)$. Hence by [174, Theorem 4.1], $w(t) \geq v(t)$ for $t \in [0, B)$. Since $z(B) = 0$, $v(t) \to \infty$ as $t \to B-$. Therefore $w(t) \to \infty$ as $t \to b-$ for some $b \leq B$, implying that y has a zero at b. If $A_1 < B \leq A_2$, then arguing as above we obtain $w(a_1) \geq v(a_1) = v(A_1)$. Integrating (2.3.25) and using (2.3.29) we get

$$w(b_1) - w(a_1) = \int_{a_1}^{b_1} [c + (p-1)r^{1-q}|w|^q](t)\, dt \geq 0.$$

Hence

$$w(b_1) \geq w(a_1) \geq v(A_1). \tag{2.3.32}$$

Since C, R are simply the functions c and r translated to the left by $t \mapsto t - (b_1 - a_1)$, w and v satisfy the same generalized Riccati equation on the intervals $[b_1, B + (b_1 - a_1))$, $[A_1, B)$, respectively. By [174, Theorem 4.1] and (2.3.32) we conclude that whenever w and v are defined, $w(t + (b_1 - a_1)) \geq v(t)$, $A_1 \leq t < B$. As above, we see that $w(t) \to \infty$ as $t \to b-$ for some $b \leq B + (b_1 - a_1)$ implying that y has a zero at b. This completes the proof of the case $n = 1$. The proof of the inductive step from n to $n + 1$ is similar and hence omitted. □

Remark 2.3.5. In fact, the last theorem says that if the solution of the Riccati equation

$$v(\tau) = v(0) + \int_0^T T_\Omega c(t)\,dt + \int_0^T (p-1)(T_\Omega r)^{1-q}(t)|v(t)|^q dt$$

tends to ∞ at a point $B < A_i$, then the solution of the Riccati equation

$$w(t) = w(0) + \int_0^t c(s)\,ds + \int_0^t (p-1) r^{1-q}(s)|w(s)|^q ds$$

tends to ∞ at a point $b < a_i$, as long as $w(0) > v(0)$.

The previous theorem plays a crucial role in the following telescoping principle.

Theorem 2.3.11 (Telescoping principle). *Assume that the conditions of Theorem 2.3.10 hold. If* (2.3.30) *is oscillatory, then so is* (1.1.1).

Proof. Let z be a solution of (2.3.30) with $z(0) \neq 0$. Let y be a solution of (1.1.1) satisfying $y(0) \neq 0$ and $r(0)\Phi(y'(0)/y(0)) \leq R(0)\Phi(z'(0)/z(0))$. By Theorem 2.3.10, $y(b) = 0$ for some $b < \infty$ (here we set $a = \infty$). Now working with the half-line $[b, \infty)$ instead of $[0, \infty)$ and proceeding as before one shows that y must have a zero to the right of b. Continuing this process leads to the conclusion that y is oscillatory and hence all solutions of (1.1.1) oscillate. □

Remark 2.3.6. It is not difficult to derive this result also by means of the variational principle. For this approach in the linear case see [231].

This principle can be applied to get many new examples of oscillatory half-linear differential equations. We use a process that is the reverse of the construction (2.3.27)-(2.3.30) in Theorem 2.3.10. Start with any known oscillatory equation (2.3.30) and assume that $A = \infty$. Choose a sequence of $A_i \to \infty$. Cut the plane at each vertical line $t = A_i$ and pull the two half-planes apart forming a gap of arbitrary length. Now fill the gap with an arbitrary positive continuous function r and any continuous c whose integral over the length of the gap is nonnegative. Do this at each point A_i and denote the new coefficient sequences by r, c. Then equation (1.1.1) is oscillatory.

The telescoping principle is also useful in extending various known oscillation criteria. It implies that any sufficient oscillation condition need only be verified on "intervals", namely, on $\bigcup_{i=1}^{\infty}(b_i, a_{i+1})$ (only the case $n = \infty$ is of interest here), while on the complementary intervals the coefficients r and c can be arbitrary as long as $r > 0$ and c has a nonnegative integral over each such interval.

2.3.5 Comparison theorem with respect to p

The main statement of this subsection gives a kind of comparison theorem with respect to the power of Φ.

Along with (1.1.1) we consider another half-linear equation with a different power function $\Phi_\alpha(x) = |x|^{\alpha-1} \operatorname{sgn} x$, $\alpha > 1$,

$$(r(t)\Phi_\alpha(x'))' + c(t)\Phi_\alpha(x) = 0, \tag{2.3.33}$$

we denote by β the conjugate number of α, i.e., $\beta = \alpha/(\alpha-1)$ (recall also that q is the conjugate number of p, i.e., $q = p/(p-1)$).

Theorem 2.3.12. *Let $\int^\infty r^{1-\beta}(t)\, dt = \infty$ and $\int^\infty c(t)\, dt$ converge, $\Psi(t) := \int_t^\infty c(s)\, ds \geq 0$ with $\Psi(t) \not\equiv 0$ for all large t, say $t \geq T$, and*

$$\liminf_{t\to\infty} r(t) > 0. \tag{2.3.34}$$

If $\alpha \geq p$ and equation (2.3.33) is nonoscillatory, then (1.1.1) is also nonoscillatory. If, in addition, $c(t) \geq 0$ for large t, c being eventually nontrivial, then (2.3.34) may be replaced by the weaker condition

$$\frac{r^{1-q}(t)}{\int_T^t r^{1-q}(s)\, ds} \leq e \quad \text{for all large } t, \tag{2.3.35}$$

where e is the basis of natural logarithm.

Proof. If (2.3.33) is nonoscillatory, then by Theorem 2.2.4 there is a function v and $T \in \mathbb{R}$ such that

$$v(t) = \int_t^\infty c(s)\, ds + \int_t^\infty S(v, r, \alpha)(s)\, ds$$

for $t \geq T$, where the function S is defined in Subsection 2.2.1. First assume $c(t) \geq 0$. By Lemma 2.2.6 we have $0 < v(t) \leq \left(\int_T^t r^{1-q}(s)\, ds\right)^{1-p}$. Take $t_0 > T$ so large that (2.3.35) holds for $t \geq t_0$, and hence

$$\left(\frac{v(t)}{r(t)}\right)^{q-1} \leq \frac{r^{1-q}(t)}{\int_T^t r^{1-q}(s)\, ds} \leq e,$$

$t \geq t_0$. Consequently, $S(v, r, \alpha)(t) \geq S(v, r, p)(t)$, $t \geq t_0$, which yields

$$v(t) \geq \int_t^\infty c(s)\, ds + \int_t^\infty S(v, r, p)(s)\, ds,$$

and so (1.1.1) is nonoscillatory by Theorem 2.2.5. If $c(t) \geq 0$ fails to hold, but $\int_t^\infty c(s)\, ds \geq 0$ for all large t, then again we have $(v(t)/r(t))^{q-1} \leq e$, since $v(t) \to 0$ as $t \to \infty$ and (2.3.34) holds. The rest of the proof is the same as above. $\square$

2.4 Notes and references

General statements on the variational principle and the Riccati technique appeared in various works, and are usually modeled on the classical linear results. Here we have tried to treat all known approaches in as general as possible settings. An extension of the Wirtinger inequality along with the application (Theorem 2.1.2) is taken from Došlý [100]. The proof of Theorem 2.2.3 presented here is a direct extension of the proof given in Li and Yeh [244], which involves half-linear equation (1.1.1) with $r(t) = 1$. The Hartman-Wintner type theorem has already appeared in earlier Mirzov's papers (for system (1.1.8)) and it is presented e.g. in the book [292]. The results of the subsection concerning modified Riccati inequality as well as Theorem 2.2.11 are taken from Kusano and Naito [218]. Theorem 2.2.8 is due to Lie and Yeh [245], while Theorem 2.2.9 is proved in Došlý [102] . The latter one appeared also e.g. in Yang [376], but it is not quite correct. An extension of Hille-Wintner comparison theorem can be found e.g. in Lie and Yeh [242] but also in many other papers, as mentioned in the text. Modified Hille-Wintner theorem and its applications are proved in Kusano and Naito [218]. The half-linear version of Leighton type comparison theorem can be found in Jaroš and Kusano [185], related statements can also be found in the paper of Rostás [335]. The multiplied coefficient comparison result and the telescoping principle are extensions of the linear results mentioned in the text. The last comparison theorem is an improvement (in differential equations setting) of the result originally stated for (more general) half-linear dynamic equations on time scales in Řehák [328]. Finally note that a brief survey of the methods of the half-linear oscillation theory can be found in Došlý [103, 105].

CHAPTER 3

OSCILLATION AND NONOSCILLATION CRITERIA

In the previous chapters as well in some of the subsequent ones, we present oscillation and nonoscillation criteria, which immediately manifest applicability of our methods, or the criteria, which are of rather special types. The aim of this chapter is to give criteria of a "standard" type, which are mostly generalizations of classical linear criteria. As a by-product, some interesting relationships will come out. Again we will see the power of the methods, which are extensions of the "linear" ones, in particular, the variational principle and the Riccati technique; the latter one is used more frequently. We will see that if one may use both methods in proving (non)oscillation criteria, then the Riccati technique usually requires weaker assumptions (e.g. on the sign of the coefficient c, or on the constant that is involved in a criterion).

First we will discuss an extension of Hille-Nehari type criteria. In Section 3.2, we use the averaging technique to obtain generalized criteria of Coles, Kamenev and Philos type. The end of this chapter is devoted to the theory, which mainly discusses the situations where classical criteria fail.

Before we start, let us give some general observations:

(i) If $\int^\infty r^{1-q}(t)\,dt = \infty$ in (1.1.1), then this equation can be transformed into the equation $(\Phi(x'))' + \tilde{c}(t)\Phi(x) = 0$, where $\tilde{c} = cr^{q-1}$, and this transformation transforms the interval $[T, \infty)$ into an interval of the same form. For this reason, we will formulate sometimes our results for the equation

$$(\Phi(x'))' + c(t)\Phi(x) = 0 \tag{3.1.1}$$

(mainly in the situations when these results were first established for (1.1.2) with $r(t) \equiv 1$ in linear case), the extension to the equation of the form (1.1.1) with $\int^\infty r^{1-q}(t)\,dt = \infty$ is then straightforward. Recall that the generalized Riccati equation associated to (3.1.1) is

$$w' + c(t) + (p-1)|w|^q = 0. \tag{3.1.2}$$

(ii) If we suppose that $c(t) \geq 0$ for large t in (3.1.1), then the situation is considerably simpler than that in the general case when c is allowed to change its sign. Indeed, from (3.1.1) it follows that $\Phi(x')$ is nonincreasing whenever x is a positive solution. Hence x' is nonincreasing as well, and so x is concave, i.e., its graph lies below the tangent line at any point of this graph. This means that for a positive solution x, one has $x(t) \leq x(T) + x'(T)(t-T)$ for any t and T with $t \geq T$. In particular, if $x'(T) < 0$, then $x(t)$ cannot be positive eventually. Consequently, in oscillation criteria for (3.1.1) with $c(t) \geq 0$, it is sufficient to impose such conditions on the function c that for any $T \in \mathbb{R}$ the solution given by $x(T) = 0$, $x'(T) > 0$ has an eventually negative derivative $x'(t)$.

3.1 Criteria of classical type

In this section we give a generalization of Hille-Nehari type criteria along with their natural complements. In fact, some of the criteria which we want to present here occur also in some other parts of this book, either as consequences of more general criteria or as typical examples of applications of our methods. However, the principal aim of this section is to gather all information concerning these "classical" criteria (including their complements), so interesting relationships among all of them become apparent. Moreover, we will see a wide variety of different approaches to the proofs.

We introduce the notation

$$\begin{aligned}
\mathcal{A}(t) &:= \left(\int_a^t r^{1-q}(s)\,ds\right)^{p-1} \int_t^\infty c(s)\,ds,\\
\bar{\mathcal{A}}(t) &:= \left(\int_t^\infty r^{1-q}(s)\,ds\right)^{p-1} \int_a^t c(s)\,ds,\\
\mathcal{B}(t) &:= \left(\int_t^\infty r^{1-q}(s)\,ds\right)^{-1} \int_t^\infty \left(\int_s^\infty r^{1-q}(\tau)\,d\tau\right)^p c(s)\,ds,\\
\bar{\mathcal{B}}(t) &:= \left(\int_t^\infty c(s)\,ds\right)^{-1} \int_t^\infty \left(\int_s^\infty c(\tau)\,d\tau\right)^q r^{1-q}(s)\,ds,\\
\mathcal{C}(t) &:= \left(\int_a^t r^{1-q}(s)\,ds\right)^{-1} \int_a^t \left(\int_a^s r^{1-q}(\tau)\,d\tau\right)^p c(s)\,ds,\\
\bar{\mathcal{C}}(t) &:= \left(\int_a^t c(s)\,ds\right)^{-1} \int_a^t \left(\int_a^s c(\tau)\,d\tau\right)^q r^{1-q}(s)\,ds.
\end{aligned}$$

As it can be easily seen, the discussion in this section cannot be restricted to the case $r(t) \equiv 1$, since sometimes we assume $\int^\infty r^{1-q}(t)\,dt < \infty$, and one of the objectives of this section is to show the differences between the cases where this integral diverges or converges.

3.1.1 Hille-Nehari type criteria

In Subsection 1.2.10 we have proved (by means of two different methods) an extension of the classical Leighton-Wintner criterion: $\int^\infty r^{1-q}(t)\,dt = \infty = \int^\infty c(t)\,dt$ implies oscillation of (1.1.1). One of the reasoned questions which arises now is: What about the case when $\int^\infty c(t)\,dt$ converges? Note that the divergence of this integral in the sense that $\liminf_{t\to\infty} \int^t c(s)\,ds < \limsup_{t\to\infty} \int^t c(s)\,ds$ is discussed elsewhere (e.g., in Section 3.3). Thus we assume that

(3.1.3) $$\int_a^\infty c(s)\,ds := \lim_{t\to\infty} \int_a^t c(s)\,ds \text{ exists as a finite number.}$$

We start with the following criterion.

Theorem 3.1.1. *Let* (3.1.3) *hold and*

(3.1.4) $$\int_a^\infty r^{1-q}(t)\,dt = \infty.$$

If

(3.1.5) $$\liminf_{t\to\infty} \mathcal{A}(t) > \frac{1}{p}\left(\frac{p-1}{p}\right)^{p-1},$$

then equation (1.1.1) *is oscillatory.*

Proof. As said in the next remark, this statement can be viewed as a corollary of the below given more general theorem, but for convenience we give here the direct proof, which is based on the Riccati technique. Suppose, by a contradiction, that (1.1.1) is nonoscillatory. Then, by Theorem 2.2.4, there is a function w satisfying the equation $w(t) = \int_t^\infty c(s)\,ds + (p-1)\int_t^\infty r^{1-q}(s)|w(s)|^q\,ds$ in a neighborhood of ∞. Multiplying this equation by $\left(\int_a^t r^{1-q}(s)\,ds\right)^{p-1}$, using the assumptions of the theorem, and supposing that $\limsup_{t\to\infty}\left(\int_a^t r^{1-q}(s)\,ds\right)^{p-1} w(t) =: M < \infty$ (in the case $M = \infty$, to get a contradiction is even easier than for $M < \infty$) we find an $\varepsilon > 0$ such that M satisfies the inequality

$$M > \frac{1}{p}\left(\frac{p-1}{p}\right)^{p-1} + \varepsilon + |M|^q.$$

Since

$$|t|^q - t + \frac{1}{p}\left(\frac{p-1}{p}\right)^{p-1} \geq 0$$

for all $t \in \mathbb{R}$, we have the required contradiction. □

Remark 3.1.1. We have seen how the Riccati technique was directly used in the proof, but there are also another possibilities:

(i) The criterion can be seen as a consequence of the more general one (after the transformation of the independent variable, see Subsection 1.2.7): see Corollary 3.3.4. This comes from [197] and the Riccati technique is used there.

(ii) In [180], it was used the technique involving Riccati integral equation with certain weight. There is an assumption $c(t) \geq 0$ which however is not needed.

(iii) In Theorem 2.3.2, we obtained this criterion by means of the Hille-Wintner comparison theorem where general equation is compared with the generalized Euler equation (1.4.20). This approach appeared in [219]. The additional assumption $c(t) \geq 0$ from [219] may be replaced by the weaker one, $\int_t^\infty c(s)\,ds \geq 0$, which means no restriction in view of condition (3.1.5).

(iv) In Section 5.5, we use the sequence approach to prove this criterion.

(v) In [100], the variational principle was used. However, the constant on the right-hand side of (3.1.5) has to be replaced by (bigger) 1. If, in addition, $c(t) \geq 0$, then lim inf can be replaced by lim sup, as a closer examination of that proof shows. See also the next criterion.

The next theorem shows that lim inf in (3.1.5) can be replaced by lim sup provided additional conditions are satisfied.

Theorem 3.1.2. *Let* (3.1.3), (3.1.4) *hold and* $\mathcal{C}(t) \geq 0$ *for large* t. *If*

$$\limsup_{t\to\infty} \mathcal{A}(t) > 1,$$

then equation (1.1.1) *is oscillatory.*

Proof. The statement is a consequence of a stronger statement, namely below given Theorem 3.3.3, using the transformation of the independent variable. The Riccati technique plays a key role there. □

Remark 3.1.2. It is easy to see that $\mathcal{C}(t) \geq 0$ is satisfied provided e.g. $c(t) \geq 0$. Just the condition $c(t) \geq 0$ is needed in all other approaches:

(i) The proof by variational principle was mentioned in the previous remark.

(ii) In [180] and [219], the Riccati technique is used.

(iii) In Section 5.5 we will see that this criterion is a consequence of two more general ones, which are based on the sequence approach.

Now we give a nonoscillatory complement to Theorem 3.1.1.

Theorem 3.1.3. *Let* (3.1.3) *and* (3.1.4) *hold. If*

$$-\frac{2p-1}{p}\left(\frac{p-1}{p}\right)^{p-1} < \liminf_{t\to\infty} \mathcal{A}(t) \leq \limsup_{t\to\infty} \mathcal{A}(t) < \frac{1}{p}\left(\frac{p-1}{p}\right)^{p-1}, \tag{3.1.6}$$

then equation (1.1.1) *is nonoscillatory.*

Proof. This is Theorem 2.2.9, proved by the Riccati technique. □

Remark 3.1.3. Similarly as the previous criteria, also this one has frequently appeared in the literature:

(i) The same approach as in the above proof (only with $r(t) \equiv 1$, which however does not matter, in view of the transformation) was used in [197], see Theorem 3.3.6.

(ii) In [180], the sequence approach was used under the assumption $c(t) \geq 0$. However, an alternative sequence can be also used and in both cases, $c(t) \geq 0$ is relaxed to $\int_t^\infty c(s)\,ds \geq 0$, see Section 5.5. Note that (3.1.6) reads there as $\mathcal{A}(t) \leq \frac{1}{p}\left(\frac{p-1}{p}\right)^{p-1}$ for large t, the same holds in the next described approach.

(iii) In [227] and [219], the Hille-Wintner theorem is applied to compare a general equation with Euler type equation, under the assumption $c(t) \geq 0$. This is done in the proof of Theorem 2.3.2, where we see that $c(t) \geq 0$ may be relaxed to $\int_t^\infty c(s)\,ds \geq 0$.

(iv) The use of variational priciple is possible as well, see Theorem 2.1.2. However $c(t)$ in $\mathcal{A}(t)$ has to be replaced by $c_+(t) = \max\{0, c(t)\}$.

Remark that the cases where (3.1.5), (3.1.6) fail to hold are discussed in Section 3.3

3.1.2 Other criteria

In this subsection we discuss the criteria which are complementary from various points of view. We start with the question: What about the case where (3.1.4) fails to hold? Thus let us investigate the complementary case

$$\int_a^\infty r^{1-q}(t)\,dt < \infty. \tag{3.1.7}$$

First we show how the reciprocity principle simply enables to use the previous criteria under the assumption $c(t) > 0$.

Theorem 3.1.4. *Let* (3.1.7) *hold and* $c(t) > 0$ *for large* t. *If*

$$\liminf_{t\to\infty} \bar{\mathcal{A}}(t) > \frac{1}{p}\left(\frac{p-1}{p}\right)^{p-1} \quad \text{or} \quad \limsup_{t\to\infty} \bar{\mathcal{A}}(t) > 1,$$

then (1.1.1) *is oscillatory. If*

$$\limsup_{t\to\infty} \bar{\mathcal{A}}(t) < \frac{1}{p}\left(\frac{p-1}{p}\right)^{p-1},$$

then (1.1.1) *is nonoscillatory.*

Proof. From the oscillatory criteria contained in this theorem we show only the sufficiency of the first condition, since the latter one is very similar. Let us apply Theorem 3.1.1 to the reciprocal equation

$$\left(c^{1-q}(t)\Phi^{-1}(u')\right)' + r^{1-q}(t)\Phi^{-1}(u) = 0, \tag{3.1.8}$$

where $\Phi^{-1}(u) = |u|^{q-1}\operatorname{sgn} u$ is the inverse to Φ, q being the conjugate number of p. Denote $\gamma_p = \frac{1}{p}\left(\frac{p-1}{p}\right)^{p-1}$, γ_q is introduced correspondingly. First observe that the condition $\liminf_{t\to\infty} \bar{A}(t) > \gamma_p$ implies that $\int^\infty c(t)\,dt = \infty$. Taking into account that $(p-1)(q-1) = 1$, we have $\int^\infty c^{(1-q)(1-p)}(t)\,dt = \int^\infty c(t)\,dt = \infty$. Hence, equation (3.1.8) is oscillatory if

$$\liminf_{t\to\infty} \left(\int_a^t c(s)\,ds\right)^{q-1} \left(\int_t^\infty r^{1-q}(s)\,ds\right) > \gamma_q. \tag{3.1.9}$$

Now, taking the $(p-1)$-th power of both sides, we see that (3.1.9) is equivalent to

$$\liminf_{t\to\infty} \bar{A}(t) > \gamma_q^{p-1} = \gamma_p,$$

what we needed to prove.

Concerning the proof of the "nonoscillatory" part of the theorem, first consider the case $\int^\infty c(t)\,dt < \infty$. If this happens, then we use the transformation of independent variable

$$s = \int^t r^{1-q}(\tau)\,d\tau, \quad x(s) = y(t), \tag{3.1.10}$$

which transforms (1.1.1) into the equation

$$\frac{d}{ds}\left(\Phi\left(\frac{d}{ds}x\right)\right) + r^{q-1}(t(s))c(t(s))\Phi(x) = 0, \tag{3.1.11}$$

where $t = t(s)$ is the inverse function of $s = s(t)$ given by (3.1.10). The convergence of $\int^\infty r^{1-q}(t)\,dt$ implies that the new variable s runs through a bounded interval where (3.1.11) has no singularity, hence any solution of this equation has only a finite number of zeros in this interval, which means that (1.1.1) is nonoscillatory. If $\int^\infty c(t)\,dt = \infty$, we proceed in the same way as in the first part of the proof and use Theorem 3.1.3 instead of Theorem 3.1.1. □

Theorem 3.1.5. *Let* (3.1.7) *hold. If*

$$-\frac{2p-1}{p}\left(\frac{p-1}{p}\right)^{p-1} < \liminf_{t\to\infty} \bar{A}(t) \le \limsup_{t\to\infty} \bar{A}(t) < \frac{1}{p}\left(\frac{p-1}{p}\right)^{p-1},$$

then equation (1.1.1) *is nonoscillatory.*

Proof. One can show in the same way as in the proof Theorem 2.2.9 that the function

$$v(t) = -\left(\frac{p-1}{p}\right)^p \left(\int_t^\infty r^{1-q}(s)\,ds\right)^{1-p}$$

satisfies the inequality

$$v' \le (1-p)r^{1-q}(t)|v - \tilde{C}(t)|^q, \quad \tilde{C}(t) = \int^t c(s)\,ds,$$

which implies that $w = v - \tilde{C}$ satisfies the Riccati inequality (2.2.7). □

Remark 3.1.4. The variational approach in this nonoscillation criterion is shown in Theorem 2.1.2. The function c in $\bar{A}$ has to be replaced by c_+. The variational principle can be also used to show that $\liminf_{t\to\infty} \bar{A}(t) > 1$ implies oscillation of (1.1.1). If $c(t) \geq 0$, then lim sup may be replaced by lim inf. The proof is very similar to that of the corresponding case involving $\mathcal{A}(t)$.

The next criteria with the function $\mathcal{B}(t)$, where (3.1.7) is supposed, can also be understood as certain complementary cases to the criteria involving $\mathcal{A}(t)$.

Theorem 3.1.6. *Let* (3.1.7) *hold,* $\int_a^\infty \left(\int_t^\infty r^{1-q}(s)\,ds\right)^p c(t)\,dt$ *be convergent and* $c(t) \geq 0$ *for large* t. *If*

$$\liminf_{t\to\infty} \mathcal{B}(t) > \left(\frac{p-1}{p}\right)^p = q^{-p} \quad \text{or} \quad \limsup_{t\to\infty} \mathcal{B}(t) > 1,$$

then (1.1.1) *is oscillatory. If*

$$\limsup_{t\to\infty} \mathcal{B}(t) < \left(\frac{p-1}{p}\right)^p = q^{-p},$$

then (1.1.1) *is nonoscillatory.*

Proof. The criteria involving the constant q^{-p} have already been proved above, see Theorem 2.3.4. The modified Hille-Wintner comparison theorem plays an important role there. Concerning the remaining criterion, assume, by a contradiction, that $\limsup_{t\to\infty} \mathcal{B}(t) > 1$ and (1.1.1) is nonoscillatory. Let y be a solution of (1.1.1) such that $y(t) > 0$ for $t \geq t_0$. Consider the function $v = r\Phi(y'/y)$. According to Lemma 2.2.7 and Theorem 2.2.7, v satisfies (2.2.34), (2.2.35) and (2.2.31). From (2.2.35) and (2.2.39) we see that (with ρ given by (2.2.28))

$$\begin{aligned}\rho^p(t)v(t) + \rho(t) &\geq \int_t^\infty r^{1-q}(s)\left[(p-1)\rho^p(s)|v(s)|^q + p\rho^{p-1}(s)v(s) + 1\right] ds \\ &\quad + \int_t^\infty \rho^p(s)c(s)\,ds \\ &\geq \int_t^\infty \rho^p(s)c(s)\,ds,\end{aligned}$$

$t \geq t_0$, which implies

$$\rho^{-1}(t)\int_t^\infty \rho^p(s)c(s)\,ds \leq \rho^{p-1}(t) + 1,$$

$t \geq t_0$. Taking the upper limit as $t \to \infty$, in view of (2.2.31), we find

$$\limsup_{t\to\infty} \rho^{-1}(t)\int_t^\infty \rho^p(s)c(s) \leq 1,$$

which is a contradiction. □

Remark 3.1.5. (i) In [102], the variational principle was used to show that if the condition $\liminf_{t\to\infty} \mathcal{B}(t) > 1$ is satisfied, then equation (1.1.1) is oscillatory (with no sign assumption on c). If in addition $c(t) \geq 0$, then a closer examination of that proof shows that lim inf can be replaced by lim sup.

(ii) Similarly as in the previous case, we may consider "reciprocal complements" of the criteria in Theorem 3.1.6. If $c(t) > 0$ for large t, then it is easy to see that

$$\liminf_{t\to\infty} \bar{\mathcal{B}}(t) > \left(\frac{q-1}{q}\right)^q = p^{-q} \quad \text{or} \quad \limsup_{t\to\infty} \bar{\mathcal{B}}(t) > 1$$

imply oscillation of (1.1.1), while

$$\limsup_{t\to\infty} \bar{\mathcal{B}}(t) < \left(\frac{q-1}{q}\right)^q = p^{-q}$$

implies nonoscillation of (1.1.1). In Section 5.5 (Theorems 5.5.6, 5.5.7 and 5.5.11) we will obtain these criteria (more precisely, the ones involving the constant p^{-q}) by means of the sequence approach, even under the weaker assumption $\int_t^\infty c(s)\, ds \geq 0$. It is assumed there that (3.1.3) and (3.1.4) hold. Note that in [89] these two criteria were proved by means of the Riccati technique combined with the Banach fixed point theorem, under the assumptions $c(t) \geq 0$ and $r(t) \equiv 1$. See the text before Theorem 5.5.11 why we call these criteria of Willet type.

We finish this section with the criteria which are counterparts to the latter ones, namely the criteria involving the functions $\mathcal{C}(t)$ and $\bar{\mathcal{C}}(t)$. Let

$$\eta = x_0 + \frac{2p-1}{p}\left(\frac{p-1}{p}\right)^{p-1},$$

where x_0 is the least root of the equation

$$(p-1)|x|^q + px + \frac{2p-1}{p}\left(\frac{p-1}{p}\right)^{p-1} = 0.$$

It is not difficult to show that $\eta < 0$ since $\frac{2p-1}{p}\left(\frac{p-1}{p}\right)^{p-1} < 1$ for $p > 1$.

Theorem 3.1.7. *Let* (3.1.4) *hold. If*

$$\liminf_{t\to\infty} \mathcal{C}(t) > \left(\frac{p-1}{p}\right)^p = q^{-p}$$

or

$$\mathcal{A}(t) \geq 0 \text{ for large } t \quad \text{and} \quad \limsup_{t\to\infty} \mathcal{C}(t) > 1,$$

then (1.1.1) *is oscillatory. If*

$$\eta < \liminf_{t\to\infty} \mathcal{C}(t) \leq \limsup_{t\to\infty} \mathcal{C}(t) < \left(\frac{p-1}{p}\right)^p = q^{-p},$$

then (1.1.1) *is nonoscillatory.*

Proof. The criteria follow from below Corollary 3.3.4, Theorem 3.3.3 and Theorem 3.3.9, respectively, making use the transformation of the independent variable, see Subsection 1.2.7. □

Remark 3.1.6. It is easy to see how the corresponding criteria involving the function $\bar{C}(t)$ can be obtained from the previous theorem by means of the reciprocity principle, under the condition $c(t) > 0$, and so we omit this.

Table 3.1.1: Critical constants for $p = 3/2$, $p = 2$, and $p = 3$

general case	$p = \frac{3}{2}$	$p = 2$	$p = 3$
$-\frac{2p-1}{p}\left(\frac{p-1}{p}\right)^{p-1}$	$-\frac{4}{3\sqrt{3}}$	$-\frac{3}{4}$	$-\frac{20}{27}$
$\frac{1}{p}\left(\frac{p-1}{p}\right)^{p-1}$	$\frac{2}{3\sqrt{3}}$	$\frac{1}{4}$	$\frac{4}{27}$
$\left(\frac{p-1}{p}\right)^{p} = q^{-p}$	$\frac{1}{3\sqrt{3}}$	$\frac{1}{4}$	$\frac{8}{27}$
$\left(\frac{q-1}{q}\right)^{q} = p^{-q}$	$\frac{8}{27}$	$\frac{1}{4}$	$\frac{1}{3\sqrt{3}}$

3.2 Criteria by averaging technique

In the (non)oscillation criteria for (1.1.1) that have been presented so far, the integrals $\int^t c(s)\,ds$ or $\int^\infty c(s)\,ds$ appeared, sometimes the function c is multiplied by a quantity related to the function r (see the beginning of Section 3.1). In this section we present oscillation criteria involving various integral averages of the function c.

3.2.1 Coles type criteria

The results of this subsection concern the half-linear extension of the averaging technique introduced in the linear case by Coles in [79]. The statements are formulated for (3.1.1).

Let $\mathcal{J}$ be the class of nonnegative locally integrable functions f defined on $[0, \infty)$ and satisfying the condition

$$\limsup_{t\to\infty} \left(\int_0^t f(s)\,ds\right)^{q-1-\mu} [F_\mu(\infty) - F_\mu(t)] > 0 \tag{3.2.1}$$

for some $\mu \in [0, q-1)$, where

$$F_\mu(t) = \int_0^t f(s) \frac{\left(\int_0^s f(\xi)\, d\xi\right)^\mu}{\left(\int_0^s f^p(\xi)\, d\xi\right)^{q-1}}\, ds.$$

If $F_\mu(\infty) = \infty$, then $f \in \mathcal{J}$. Let $\mathcal{J}_0$ be the subclass of $\mathcal{J}$ consisting of nonnegative locally integrable functions f satisfying

$$\lim_{t\to\infty} \frac{\int_0^t f^p(s)\, ds}{\left(\int_0^t f(s)\, ds\right)^p} = 0. \tag{3.2.2}$$

Observe that if (3.2.1) or (3.2.2) holds, then

$$\int^\infty f(t)\, dt = \infty. \tag{3.2.3}$$

On the other hand, every bounded nonnegative locally integrable function satisfying (3.2.3) belongs to $\mathcal{J}_0$ and $\mathcal{J}_0 \subset \mathcal{J}$. Since all nonnegative polynomials are in $\mathcal{J}_0$, this class of functions contains also unbounded functions. Elements of $\mathcal{J}$ and $\mathcal{J}_0$ will be called *weight functions.*

For $f \in \mathcal{J}$, we define

$$A_f(s,t) := \frac{\int_s^t f(\tau) \int_s^\tau c(\mu)\, d\mu\, d\tau}{\int_s^t f(\tau)\, d\tau}.$$

The following statement reduces to the Hartman-Wintner theorem (Theorem 2.2.3) when the weight function f is $f(t) \equiv 1$.

Theorem 3.2.1. *Suppose that* (3.1.1) *is nonoscillatory.*

(i) *If there exists $f \in \mathcal{J}$ such that for some $T \in \mathbb{R}$*

$$\liminf_{t\to\infty} A_f(T,t) > -\infty \tag{3.2.4}$$

then

$$\int^\infty |w(t)|^q\, dt < \infty \tag{3.2.5}$$

for every solution w of the associated Riccati equation (3.1.2).

(ii) *Assume that* (3.2.5) *holds for some solution w of* (3.1.2). *Then for every $f \in \mathcal{J}_0$ and $T \in \mathbb{R}$ sufficiently large $\lim_{t\to\infty} A_f(T,t)$ exists finite.*

Proof. The proof of this statement copies essentially the proof of Theorem 2.2.3.

(i) Assume, by contradiction, that

$$\int^\infty |w(t)|^q\, dt = \infty \tag{3.2.6}$$

for some solution w of (3.1.2). Integrating from ξ to t, multiplying the obtained integral equation by $f(t)$ and then integrating again from ξ to t, we obtain

$$\begin{aligned}\int_\xi^t f(s)w(s)\,ds &= w(\xi)\int_\xi^t f(s)\,ds - \int_\xi^t f(s)\int_\xi^s c(\tau)\,d\tau\,ds \\ &\quad -(p-1)\int_\xi^t f(s)\int_\xi^s |w(\tau)|^q\,d\tau\,ds \\ &= w(\xi)\int_\xi^t f(s)\,ds - A_f(\xi,t)\int_\xi^t f(s)\,ds \\ &\quad -(p-1)\int_\xi^t f(s)\int_\xi^s |w(\tau)|^q\,d\tau\,ds \\ &= [w(\xi)-A_f(\xi,t)]\int_\xi^t f(s)\,ds - (p-1)\int_\xi^t f(s)\int_\xi^s |w(\tau)|^q\,d\tau\,ds,\end{aligned}$$

where $t \geq \xi \geq T$. From (3.1.2) we have

$$w(\xi) = w(T) - \int_T^\xi c(s)\,ds - (p-1)\int_T^\xi |w(s)|^q\,ds.$$

Since $f \in \mathcal{J}$, (3.2.3) holds. This implies

$$\begin{aligned}A_f(\xi,t) &= \frac{\int_T^t f(s)\,ds}{\int_\xi^t f(s)\,ds}A_f(T,t) - \int_T^\xi c(s)\,ds - \frac{\int_T^\xi f(s)\int_T^s c(\tau)\,d\tau\,ds}{\int_\xi^t f(s)\,ds} \\ &= \frac{\int_T^t f(s)\,ds}{\int_\xi^t f(s)\,ds}A_f(T,t) - \int_T^\xi c(s)\,ds + o(1) \quad \text{as } t\to\infty.\end{aligned}$$

Thus

$$w(\xi)-A_f(\xi,t) = w(T) - \frac{\int_T^t f(s)\,ds}{\int_\xi^t f(s)\,ds}A_f(T,t) - (p-1)\int_T^t |w(s)|^q\,ds + o(1) \tag{3.2.7}$$

as $t\to\infty$. Since $f \in \mathcal{J}$, there exists a positive number $\lambda > 0$ such that

$$\frac{\lambda^{1-q}}{p-1} < (q-1-\mu)\limsup_{t\to\infty}\left[\int^t f(s)\,ds\right]^{q-1-\mu}[F_\mu(\infty)-F_\mu(t)], \tag{3.2.8}$$

where μ is the same as in (3.2.1). It follows from (3.2.4), (3.2.6) and the previous computation that there exist two numbers a and b, $b \geq a \geq T$, such that

$$w(a) - A_f(a,t) \leq -\lambda \tag{3.2.9}$$

for $t \geq b$. Let $z(t) := \int_a^t f(s)w(s)\,ds$. Then the Hölder inequality implies

$$\int_a^t |w(\tau)|^q\,d\tau \geq \frac{|z(t)|^q}{\left(\int_a^t f^p(\tau)\,d\tau\right)^{q-1}}.$$

It follows from (3.2.7) and (3.2.9) that

(3.2.10)
$$z(t) \leq -\lambda \int_a^t f(s)\, ds - (p-1) \int_a^t f(s)|z(s)|^q \left(\int_a^s f^p(\tau)\, d\tau \right)^{1-q} ds =: -G(t).$$

Thus

$$G'(t) = \lambda f(t) + (p-1) f(t)|z(t)|^q \left(\int_a^t f^p(s)\, ds \right)^{1-q} \tag{3.2.11}$$

and

$$0 \leq \lambda \int_a^t f(s)\, ds \leq G(t) \leq |z(t)|. \tag{3.2.12}$$

It follows from (3.2.10), (3.2.11) and (3.2.12) that

$$\begin{aligned} G'(t) G^{\mu-q}(t) &\geq G'(t) G^{\mu}(t) |z(t)|^{-q} \\ &\geq (p-1)\lambda^{\mu} f(t) \left(\int_a^t f(s)\, ds \right)^{\mu} \left(\int_a^t f^p(s)\, ds \right)^{1-q}. \end{aligned}$$

Integrating this inequality from t $(t \geq b)$ to ∞, we get

$$\frac{1}{q-1-\mu} G^{\mu-q-1}(t) \geq (p-1)\lambda^{\mu} [F_{\mu}(\infty) - F_{\mu}(t)].$$

Inequality (3.2.12) then implies

$$\frac{\lambda^{1-q}}{p-1} \geq (q-1-\mu) \left[\int_a^t f(s)\, ds \right]^{q-1-\mu} [F_{\mu}(\infty) - F_{\mu}(t)]$$

which contradicts (3.2.8).

(ii) As in the previous part of the proof, (3.2.7) holds. This implies that

$$A_f(\xi, t) = w(\xi) - \frac{\int_\xi^t f(s) w(s)\, ds}{\int_\xi^t f(s)\, ds} - (p-1) \frac{\int_\xi^t f(s) \int_\xi^s |w(\tau)|^q\, d\tau\, ds}{\int_\xi^t f(s)\, ds}. \tag{3.2.13}$$

Since $f \in \mathcal{J}_0$, (3.2.3) holds. Thus,

$$\lim_{t \to \infty} \frac{\int_\xi^t f(s) \int_\xi^s |w(\tau)|^q d\tau\, ds}{\int_\xi^t f(s)\, ds} = \int_\xi^\infty |w(s)|^q\, ds < \infty.$$

By Hölder's inequality

$$0 \leq \lim_{t \to \infty} \frac{\left| \int_\xi^t f(s) w(s)\, ds \right|}{\int_\xi^t f(s)\, ds} \leq \lim_{t \to \infty} \frac{\left(\int_\xi^t f^p(s)\, ds \right)^{1/p} \left(\int_\xi^t |w(s)|^q\, ds \right)^{1/q}}{\int_\xi^t f(s)\, ds} = 0.$$

Hence, by (3.2.13), $\lim_{t\to\infty} A_f(\xi, t)$ exists and

$$\lim_{t\to\infty} A_f(\xi, t) = w(\xi) - (p-1)\int_\xi^\infty |w(s)|^q\, ds.$$

This completes the proof. □

As a consequence of the previous statement we have the following oscillation criterion which is the half-linear extension of the criterion of Coles [79], this statement can be also viewed as an extension of Theorem 2.2.10.

Theorem 3.2.2. *The following statements hold:*

(i) If there exists $f \in \mathcal{J}$ such that (3.2.4) *holds, then either* (3.1.1) *is oscillatory, or $\lim_{t\to\infty} A_g(\cdot, t)$ exists as a finite number for every $g \in \mathcal{J}_0$.*

(ii) If there exist two nonnegative bounded functions f, g on an interval $[T, \infty)$ satisfying $\int^\infty f(t)\, dt = \infty = \int^\infty g(t)\, dt$ such that

$$\lim_{t\to\infty} A_f(T, t) < \lim_{t\to\infty} A_g(T, t),$$

then equation (3.1.1) *is oscillatory.*

Proof. (i) Suppose that (3.1.1) is nonoscillatory. Then by Theorem 3.2.1 every solution of the associated Riccati equation (3.1.2) satisfies $\int^\infty |w(t)|^q\, dt < \infty$ and hence $\lim_{t\to\infty} A_g(\cdot, t)$ exists finite for every $g \in \mathcal{J}_0$.

(ii) Let $\alpha, \beta \in \mathbb{R}$ be such that

$$\lim_{t\to\infty} A_f(T, t) < \alpha < \beta < \lim_{t\to\infty} A_g(T, t).$$

Let $h(t) = g(t)$ for $T \le t \le t_1$, where t_1 is determined such that $A_g(T, t_1) \ge \beta$ and $\int_T^{t_1} g(s)\, ds \ge 1$. Let $h(t) = f(t)$ for $t_1 \le t_2$ where t_2 is determined such that $A_h(T, t_2) \le \alpha$ and $\int_T^{t_1} h(s)\, ds \ge 2$. This is possible because

$$\begin{aligned} A_h(T, t_2) &= \frac{\int_T^{t_2} h(s) \int_T^s c(\tau)\, d\tau\, ds}{\int_T^{t_2} h(s)\, ds} \\ &= \frac{\int_T^{t_1} [g(s) - f(s)] \int_T^s c(\tau)\, d\tau\, ds}{\int_T^{t_1} g(s)\, ds + \int_{t_1}^{t_2} f(s)\, ds} \\ &\quad + \frac{\int_T^{t_2} f(s) \int_T^s c(\tau)\, d\tau\, ds}{\int_T^{t_2} f(s)\, ds} \cdot \frac{\int_T^{t_2} f(s)\, ds}{\int_T^{t_1} g(s)\, ds + \int_{t_1}^{t_2} f(s)\, ds} \\ &= A_f(T, t_2)[1 + o(1)] + o(1), \end{aligned}$$

as $t_2 \to \infty$. Continuing in this manner, we obtain a nonnegative and bounded function h defined on $[T, \infty)$ such that

$$\limsup_{t\to\infty} A_h(T, t) \ge \beta > \alpha \ge \liminf_{t\to\infty} A_h(T, t).$$

Hence, by the part (i), equation (3.1.1) is oscillatory. □

3.2.2 Generalized Kamenev criterion

The classical Kamenev criterion (see [196]) concerns the linear equation $x'' + c(t)x = 0$ and claims that this equation is oscillatory provided there exists $\lambda > 1$ such that

$$\limsup_{t\to\infty} \frac{1}{t^\lambda} \int_0^t (t-s)^\lambda c(s)\, ds = \infty. \tag{3.2.14}$$

In what follows we offer a half-linear extension of this criterion.

Theorem 3.2.3. *Suppose that there exists $\lambda > p-1$ such that*

$$\limsup_{t\to\infty} \frac{1}{t^\lambda} \int_0^t (t-s)^{\lambda-p} \left[(t-s)^p c(s) - \left(\frac{\lambda}{p}\right)^p r(s)\right] ds = \infty. \tag{3.2.15}$$

Then (1.1.1) is oscillatory.

Proof. Suppose that (1.1.1) is nonoscillatory, i.e., there exists a solution of the associated Riccati equation (1.1.21). Multiplying this equation by $(t-s)^\lambda$ and integrating it from T to t, T sufficiently large, we get

$$\begin{aligned} -(t-T)^\lambda w(T) &+ \lambda \int_T^t (t-s)^{\lambda-1} w(s)\, ds \\ &+ (p-1)\int_T^t (t-s)^\lambda r^{1-q}(s)|w(s)|^q\, ds + \int_T^t (t-s)^\lambda c(s)\, ds = 0. \end{aligned} \tag{3.2.16}$$

Using the Young inequality (1.2.2) with

$$u = \frac{\lambda}{p}(t-s)^{\frac{\lambda-p}{p}} r^{\frac{1}{p}}, \quad v = (t-s)^{\frac{\lambda}{q}} r^{\frac{1-q}{q}} |w(s)|,$$

we obtain

$$(t-s)^\lambda r^{1-q}(s)|w(s)|^q \ge (q-1)\lambda(t-s)^{\lambda-1}|w(s)| - (q-1)\left(\frac{\lambda}{p}\right)^p (t-s)^{\lambda-p} r(s)$$

for $T \le s \le t$. This inequality and (3.2.16) imply

$$\int_T^t (t-s)^{\lambda-p}\left[(t-s)^p c(s) - \left(\frac{\lambda}{p}\right)^p r(s)\right] ds \le (t-T)^\lambda w(T).$$

Thus

$$\limsup_{t\to\infty} \frac{1}{t^\lambda}\int_T^t (t-s)^{\lambda-p}\left[(t-s)^p c(s) - \left(\frac{\lambda}{p}\right)^p r(s)\right] ds \le w(T).$$

which contradicts to (3.2.15). □

Remark 3.2.1. Clearly, if $r(t) \equiv 1$, then $\lim_{t\to\infty} \frac{1}{t^\lambda}\int_T^t (t-s)^{\lambda-p}\left(\frac{\lambda}{p}\right)^p r(s)\, ds = 0$ and hence equation (3.1.1) is oscillatory if

$$\limsup_{t\to\infty} \frac{1}{t^\lambda}\int_0^t (t-s)^\lambda c(s)\, ds = \infty \quad \text{for some } \lambda > p-1.$$

which is the half-linear extension of the classical Kamenev linear oscillation criterion.

3.2.3 Generalized H-function averaging technique – Philos type criterion

In the linear case, the method used in this subsection was introduced by Philos [314].

Theorem 3.2.4. *Let $D_0 = \{(t,s) : t > s \geq t_0\}$ and $D = \{(t,s) : t \geq s \geq t_0\}$. Assume that the function $H \in C(D; R)$ satisfies the following conditions:*

(i) *$H(t,t) = 0$ for $t \geq t_0$ and $H(t,s) > 0$ for $t > s \geq t_0$;*

(ii) *H has a continuous nonpositive partial derivative on D_0 with respect to the second variable.*

Suppose that $h : D_0 \to \mathbb{R}$ is a continuous function such that

$$-\frac{\partial H}{\partial s}(t,s) = h(t,s)[H(t,s)]^{1/q} \quad \text{for all } (t,s) \in D_0, \quad q = \frac{p}{p-1},$$

and

$$\int_{t_0}^{t} h^p(t,s)\, ds < \infty \quad \text{for all } t \geq t_0. \tag{3.2.17}$$

If

$$\limsup_{t\to\infty} \frac{1}{H(t,t_0)} \int_{t_0}^{t} \left[H(t,s)c(s) - \left(\frac{1}{p}h(t,s)\right)^p \right] ds = \infty, \tag{3.2.18}$$

then (3.1.1) is oscillatory.

Proof. Suppose that (3.1.1) is nonoscillatory and v is a solution of the associated Riccati equation (3.1.2) which exists on the interval $[T_0,\infty)$, $T_0 \geq t_0$. Since (3.2.17) holds, we have for $t \geq T \geq T_0$

$$\begin{aligned}
&\int_T^t H(t,s)c(s)\, ds \\
&= H(t,T)v(T) - \int_T^t \left(-\frac{\partial H}{\partial s}(t,s)\right) v(s)\, ds - (p-1)\int_T^t H(t,s)|v(s)|^q\, ds \\
&= H(t,T)v(T) - \int_T^t \{h(t,s)[H(t,s)]^{1/q}v(s) + (p-1)H(t,s)|v(s)|^q\} ds \\
&= H(t,T)v(T) - \int_T^t \Big\{h(t,s)[H(t,s)]^{1/q}v(s) + (p-1)H(t,s)|v(s)|^q \\
&\quad + \left(\frac{1}{p}h(t,s)\right)^p\Big\}\, ds + \int_T^t \left(\frac{1}{p}h(t,s)\right)^p ds.
\end{aligned}$$

Hence, for $t \geq T \geq T_0$, we have

$$\int_T^t \left\{ H(t,s)c(s) - \left(\frac{1}{p}h(t,s)\right)^p \right\} ds$$
$$= H(t,T)v(T) - \int_T^t \left\{ h(t,s)[H(t,s)]^{1/q}v(s) + (p-1)H(t,s)|v(s)|^q + \left(\frac{1}{p}h(t,s)\right)^p \right\} ds.$$

Since $q > 1$, by Young's inequality (1.2.2)

$$h(t,s)[H(t,s)]^{1/q}v(s) + (p-1)H(t,s)|v(s)|^q + \left(\frac{1}{p}h(t,s)\right)^p \geq 0$$

for $t \geq s \geq T_0$. This implies that for every $t \geq T_0$

$$\int_{T_0}^t \left\{ H(t,s)c(s) - \left(\frac{1}{p}h(t,s)\right)^p \right\} ds$$
$$\leq H(t,T_0)v(T_0) \leq H(t,T_0)|v(T_0)| \leq H(t,t_0)|v(T_0)|.$$

Therefore,

$$\begin{aligned}\int_{t_0}^t \left\{ H(t,s)c(s) - \left(\frac{1}{p}h(t,s)\right)^p \right\} ds &= \int_{t_0}^{T_0} \left\{ H(t,s)c(s) - \left(\frac{1}{p}h(t,s)\right)^p \right\} ds \\ &\quad + \int_{T_0}^t \left\{ H(t,s)c(s) - \left(\frac{1}{p}h(t,s)\right)^p \right\} ds \\ &\leq H(t,t_0)\int_{t_0}^{T_0} |c(s)|\,ds + H(t,t_0)|v(T_0)| \\ &= H(t,t_0)\left\{ \int_{t_0}^{T_0} |c(s)|\,ds + |v(T_0)| \right\}.\end{aligned}$$

This gives

$$\limsup_{t\to\infty} \frac{1}{H(t,t_0)} \int_{t_0}^t \left\{ H(t,s)c(s) - \left(\frac{1}{p}h(t,s)\right)^p \right\} ds \leq \int_{t_0}^{T_0} |c(s)|\,ds + |v(T_0)|,$$

which contradicts (3.2.18). □

Taking $H(t,s) = (t-s)^\lambda$, $\lambda > p-1$, the last statement reduces to the half-linear version of Kamenev's oscillation criterion presented in Subsection 3.2.2.

The next statement is presented without proof (which can be found in [243]). Similarly to the proof of the previous theorem, it follows more or less the original idea of Philos [314]. For comparison with the linear case we also refer to the papers of Yan [367, 368].

Theorem 3.2.5. *Let H and h be as in the previous theorem, and let*

$$\inf_{s\geq t_0}\left\{\liminf_{t\to\infty}\frac{H(t,s)}{H(t,t_0)}\right\}>0. \tag{3.2.19}$$

Suppose that

$$\limsup_{t\to\infty}\frac{1}{H(t,t_0)}\int_{t_0}^t h(t,s)\,ds<\infty$$

and there exists a function $A\in C[t_0,\infty)$ such that

$$\int^\infty A_+^q(s)\,ds=\infty, \tag{3.2.20}$$

where $A_+(t)=\max\{A(t),0\}$. If

$$\limsup_{t\to\infty}\frac{1}{H(t,T)}\int_T^t\left\{H(t,s)c(s)-\left(\frac{1}{p}h(t,s)\right)^p\right\}ds\geq A(T)$$

for $T\geq t_0$, then equation (3.1.1) is oscillatory.

3.3 Further extensions of Hille-Nehari type criteria

The aim of this section is to extend some of the criteria presented in Section 3.1 (in particular, those involving the expressions $\mathcal{A}(t)$ and $\mathcal{C}(t)$), and also to discuss the cases when the assumptions from those criteria fail to hold.

In [70, 197, 262], the equation

$$x''+c(t)|x|^{p-1}|x'|^{2-p}\operatorname{sgn}x=0 \tag{3.3.1}$$

with $p\in(1,2]$ is considered. The papers [197, 262] present oscillation and nonoscillation criteria for equation (3.3.1), while the paper [70] deals with conjugacy of a singular equation of this form. Equation (3.3.1) can be obtained from equation of the form (3.1.1) by using the identity

$$(\Phi(x'))'=(p-1)x''|x'|^{p-2}.$$

More precisely, every solution of the equation

$$(\Phi(x'))'+(p-1)c(t)\Phi(x)=0$$

is also a proper solution of (3.3.1) and vice versa, where a solution x of (3.3.1) is said to be *proper* if

$$\operatorname{mes}\left\{\{t\in\mathbb{R}^+:x'(t)=0\}\setminus\{t\in\mathbb{R}^+:c(t)=0\}\right\}=0.$$

Observe how some of the basic results known for (3.1.1) may fail to hold for equation (3.3.1) when non-proper solutions are not excluded. For instance, let x

be a proper solution of (3.3.1) with $p \in (1,2)$ and $c(t) \neq 0$, satisfying $x(t_0) = 1$, $x'(t_0) = 0$. Then this solution is not constant in a neighborhood of t_0, say (a,b), since it is proper, and we may construct another solution $\bar{x}$ of (3.3.1) such that $\bar{x}(t) = x(t)$ for $t \in (a, t_0]$ and $\bar{x}(t) = 1$ for $t \in (t_0, b)$. Then the different solutions x and $\bar{x}$ satisfy the same initial conditions at t_0, and so the uniqueness is violated. Also the Sturmian separation theorem may be problematic. On the one hand, it clearly applies to proper solutions of (3.3.1), on the other hand the following situation may happen: Let us consider the equation $x'' + |x|^{p-1}|x'|^{2-p} \operatorname{sgn} x = 0$ for $p \in (1,2)$. The function $\sin_p x$ is proper solution while the constant function 1 is non-proper solution of this equation. Both are linearly independent, but their zeros do not separate each other.

Thus, in general case the analogy of Sturm's separation theorem is not true for all solutions, and hence the concept of *strong nonoscillation* (different from that in Section 5.4) has to be introduced: strong nonoscillation of (3.3.1) means nonoscillation of all solutions of (3.3.1). Nonoscillatory equation (3.3.1) is defined there as the equation having at least one proper nonoscillatory solution. Otherwise it is oscillatory.

First we reformulate here the results from [197] for equation (3.1.1), where c is a continuous function on $[1, \infty)$, and we will see that they apply also to the case $p > 2$, and not only to $p \in (1,2]$. Further extension to equation (1.1.1) with r satisfying $\int^\infty r^{1-q} = \infty$ is obvious, in view of Subsection 1.2.7. The same will be done for the results of [262]; this is the content of Subsection 3.3.2.

3.3.1 Q, H type criteria

In this subsection we give the criteria in terms of the below defined functions Q_p and H_p, which can be viewed as an extension of some of the Hille-Nehari type criteria given in Section 3.1. Denote

$$c_p(t) = \frac{(p-1)}{t^{p-1}} \int_1^t s^{p-2} \int_1^s c(\tau)\, d\tau\, ds.$$

Using essentially the same idea as in the proof of Theorem 2.2.3, the following Hartman-Wintner type lemma can be proved.

Lemma 3.3.1. *Let* (3.1.1) *be nonoscillatory and* $y(t) \neq 0$ *for* $t \geq T$ *be its solution. Then* $\int^\infty |y'(s)/y(s)|^p ds < \infty$ *if and only if the finite limit* $\lim_{t\to\infty} c_p(t)$ *exists.*

As Theorem 2.2.10 follows from Theorem 2.2.3, the following statement follows from Lemma 3.3.1.

Theorem 3.3.1. *Let either* $\lim_{t\to\infty} c_p(t) = \infty$ *or*

$$-\infty < \liminf_{t\to\infty} c_p(t) < \limsup_{t\to\infty} c_p(t).$$

Then equation (3.1.1) *is oscillatory.*

Now let us examine how stands the condition from the linear case in our situation.

Corollary 3.3.1. *Let* $p \in (1,2]$ *and either* $\lim_{t\to\infty} c_2(t) = \infty$ *or*

$$-\infty < \liminf_{t\to\infty} c_2(t) < \limsup_{t\to\infty} c_2(t).$$

Then equation (3.1.1) *is oscillatory.*

Proof. Using the integration by parts, it is easy to verify that

$$c_p(t) = (p-1)c_2(t) + \frac{(2-p)(p-1)}{t^{p-1}} \int_1^t s^{p-2} c_2(s)\, ds \tag{3.3.2}$$

and

$$(p-1)c_2(t) = c_p(t) - \frac{2-p}{t} \int_1^t c_p(s)\, ds \tag{3.3.3}$$

for $t > 1$. Since $\liminf_{t\to\infty} c_2(t) > -\infty$, it follows from formula (3.3.2) that $\liminf_{t\to\infty} c_p(t) > -\infty$. Thus if the conditions of Theorem 3.3.1 are not satisfied, then the finite limit in the below formula (3.3.4) exists. In that case, the finite limit $\lim_{t\to\infty} c_2(t)$ also exists by virtue of (3.3.3), which contradicts the conditions of the corollary. □

In view of Theorem 3.3.1, in the next investigation we suppose that the following limit exists as a finite number:

$$\lim_{t\to\infty} c_p(t) =: c_p(\infty). \tag{3.3.4}$$

The next theorem shows that (3.1.1) is oscillatory if $c_p(t)$ does not tend to its limit too rapidly.

Theorem 3.3.2. *Suppose that* (3.3.4) *holds and*

$$\limsup_{t\to\infty} \frac{t^{p-1}}{\log t} (c_p(\infty) - c_p(t)) > \left(\frac{p-1}{p}\right)^p. \tag{3.3.5}$$

Then equation (3.1.1) *is oscillatory.*

Proof. Suppose, by contradiction, that (3.1.1) is nonoscillatory and $w = \Phi(y'/y)$, $y(t) \neq 0$ being a solution of (3.1.1), is a solution of the associated Riccati equation (3.1.2) for large t. Integrating this equation from a to t and using Lemma 3.3.1 we find that

$$w(t) = \tilde{c} - \int_1^t c(s)\, ds + (p-1) \int_t^\infty |w(s)|^q\, ds, \tag{3.3.6}$$

where

$$\tilde{c} = w(a) - \int_1^a c(s)\, ds - (p-1) \int_a^\infty |w(s)|^q\, ds.$$

Note that from the proof of Lemma 3.3.1 it follows that $\lim_{t\to\infty} c_p(t) = \tilde{c}$, and so equation (3.3.6) takes the form

$$w(t) = c_p(\infty) - \int_1^t c(s)\,ds + (p-1)\int_t^\infty |w(s)|^q\,ds. \tag{3.3.7}$$

Multiplying both sides of this equation by t^{p-2} and integrating from T to t, T sufficiently large, we obtain (integrating one of the terms by parts)

$$\begin{aligned} &\int_T^t s^{p-2}\left[c_p(\infty) - \int_1^s c(\tau)\,d\tau\right] ds \\ &= \int_T^t \frac{s^{p-1}w(s) - s^{p-1}|w(s)|^q}{s}\,ds - t^{p-1}\int_t^\infty |w(s)|^q\,ds + T^{p-1}\int_T^\infty |w(s)|^q\,ds. \end{aligned} \tag{3.3.8}$$

Since we have the inequality $|\lambda|^q - \lambda + \frac{1}{p}\left(\frac{p-1}{p}\right)^{p-1} \geq 0$ for every $\lambda \in \mathbb{R}$, (3.3.8) implies

$$\begin{aligned} t^{p-1}\left(c_p(\infty) - c_p(t)\right) \;\leq\; & \left(\frac{p-1}{p}\right)^p \log\frac{t}{T} + T^{p-1}c_p(\infty) \\ & + (p-1)T^{p-1}\int_T^\infty |w(s)|^q\,ds - T^{p-1}c_p(T). \end{aligned}$$

Therefore

$$\limsup_{t\to\infty} \frac{t^{p-1}}{\log t}\left(c_p(\infty) - c_p(t)\right) \leq \left(\frac{p-1}{p}\right)^p,$$

which contradicts (3.3.5). □

In many oscillation criteria in this book it is assumed that there exists a finite limit

$$\int^\infty c(s)\,ds = \lim_{t\to\infty}\int^t c(s)\,ds. \tag{3.3.9}$$

As it can be easily seen, in that case $c_p(\infty) = \int_1^\infty c(t)\,dt$. As we will see in the next example, Theorem 3.3.2 also covers the case where (3.3.9) fails to hold, see the next example; in particular the case where

$$c_* := \liminf_{t\to\infty}\int_1^t c(s)\,ds < c_p(\infty) < \limsup_{t\to\infty}\int_1^t c(s)\,ds =: c^*. \tag{3.3.10}$$

Example 3.3.1. Let $\lambda \neq 0$ and γ be real numbers,

$$g(t) = -\gamma\frac{\log t}{t^{p-1}} + \frac{\lambda}{1+\log t}(\sin\log^2 t - 1)$$

and

$$c(t) = g'(t) + \frac{1}{p-1}(tg'(t))' = \frac{p}{p-1}g'(t) + \frac{t}{p-1}g''(t)$$

for $t \geq 1$. One can easily obtain

$$\int_1^t c(s)\,ds = g(t) + \frac{1}{p-1} t g'(t) + \gamma, \quad c_p(t) = g(t) - \frac{\gamma - \lambda}{t^{p-1}} + \gamma,$$

$$\liminf_{t\to\infty} \int_1^t c(s)\,ds = \gamma - \frac{2|\lambda|}{p-1}, \quad \limsup_{t\to\infty} \int_1^t c(s)\,ds = \gamma + \frac{2|\lambda|}{p-1},$$

$$c_p(\infty) = \gamma, \quad \frac{t^{p-1}}{\log t}(c_p(\infty) - c_p(t)) = -\frac{t^{p-1}}{\log t} g(t) + \frac{\gamma - \lambda}{\log t}$$

for $t > 1$. Further,

$$\liminf_{t\to\infty} \frac{t^{p-1}}{\log t}(c_p(\infty) - c_p(t)) = -\infty, \quad \limsup_{t\to\infty} \frac{t^{p-1}}{\log t}(c_p(\infty) - c_p(t)) = \gamma$$

when $\lambda < 0$, while for $\lambda > 0$ we have

$$\liminf_{t\to\infty} \frac{t^{p-1}}{\log t}(c_p(\infty) - c_p(t)) = \gamma, \quad \limsup_{t\to\infty} \frac{t^{p-1}}{\log t}(c_p(\infty) - c_p(t)) = \infty.$$

Therefore (3.3.10) is satisfied. On the other hand, if $\lambda < 0$ or $\lambda > 0$ and $\gamma > \left(\frac{p-1}{p}\right)^p$, then equation (3.1.1) is oscillatory by Theorem 3.3.2.

To formulate the next statements, we introduce the following notation:

$$Q_p(t) := t^{p-1}\left(c_p(\infty) - \int_1^t c(s)\,ds\right), \quad H_p(t) := \frac{1}{t}\int_1^t s^p c(s)\,ds,$$

(3.3.11) $$Q_* := \liminf_{t\to\infty} Q_p(t), \quad Q^* := \limsup_{t\to\infty} Q_p(t),$$

(3.3.12) $$H_* := \liminf_{t\to\infty} H_p(t), \quad H^* := \limsup_{t\to\infty} H_p(t).$$

Observe that if $\lim_{t\to\infty} \int^t c(s)\,ds$ exists, then $Q_p(t)$ is the same as $\mathcal{A}(t)$ defined in Section 3.1, provided $r(t) \equiv 1$. Also, $H_p(t) = \mathcal{C}(t)$ for $r(t) \equiv 1$.

Corollary 3.3.2. *Suppose that* (3.3.4) *holds and* $Q_* > -\infty$. *If*

(3.3.13) $$\limsup_{t\to\infty} \frac{1}{\log t}\int_1^t s^{p-1} c(s)\,ds > \left(\frac{p-1}{p}\right)^p,$$

then equation (3.1.1) *is oscillatory.*

Proof. It is easy to verify that

$$\frac{t^{p-1}}{\log t}(c_p(\infty) - c_p(t)) = \frac{Q_p(t)}{\log t} + \frac{1}{\log t}\int_1^t s^{p-1} c(s)\,ds$$

for $t > 1$. Therefore the conditions of the previous theorem are satisfied. □

Corollary 3.3.3. *Suppose that* (3.3.4) *holds. If*

$$\liminf_{t\to\infty}[Q_p(t)+H_p(t)] > \left(\frac{p-1}{p}\right)^{p-1}, \tag{3.3.14}$$

then equation (3.1.1) *is oscillatory.*

Proof. It is not difficult to verify that

$$\frac{t^{p-1}}{\log t}(c_p(\infty)-c_p(t)) = \frac{c_p(\infty)}{\log t}+\frac{p-1}{\log t}\left[\frac{1}{t}\int_1^t Q_p(s)\,ds+\int_1^t \frac{1}{s^2}\left(\int_1^s Q_p(\tau)\,d\tau\right)ds\right] \tag{3.3.15}$$

and

$$Q_p(t)+H_p(t) = \frac{c_p(\infty)}{t}+\frac{p}{t}\int_1^t Q_p(s)\,ds \tag{3.3.16}$$

for $t>1$. Then we have

$$\liminf_{t\to\infty}\frac{1}{t}\int_1^t Q_p(s)\,ds > \frac{1}{p}\left(\frac{p-1}{p}\right)^{p-1}.$$

Hence (3.3.5) holds by (3.3.15) since

$$\liminf_{t\to\infty}\frac{p-1}{\log t}\int_1^t \frac{1}{s^2}\int_1^s Q_p(\tau)\,d\tau\,ds > \frac{p-1}{p}\left(\frac{p-1}{p}\right)^{p-1}.$$

□

Corollary 3.3.4. *Suppose that* (3.3.4) *holds. If*

$$\textit{either}\quad Q_* > \frac{1}{p}\left(\frac{p-1}{p}\right)^{p-1}\quad \textit{or}\quad H_* > \left(\frac{p-1}{p}\right)^{p},$$

then equation (3.1.1) *is oscillatory.*

Proof. Let $Q_* > \frac{1}{p}\left(\frac{p-1}{p}\right)^{p-1}$. Since

$$\frac{t^{p-1}}{\log t}(c_p(\infty)-c_p(t)) = \frac{c_p(\infty)}{\log t}+\frac{p-1}{\log t}\int_1^t \frac{Q_p(s)}{s}\,ds$$

for $t>1$, we easily find that (3.3.5) is satisfied and therefore (3.1.1) is oscillatory by Theorem 3.3.2.

Assume now that $H_* > \left(\frac{p-1}{p}\right)^{p}$. Using the fact that

$$c_p'(t) = (p-1)t^{-p}\int_1^t s^{p-1}c(s)\,ds,$$

we have

$$c_p(T) = c_p(t) + (p-1)\int_t^T \frac{\log s}{s^p}\left(\frac{1}{\log s}\int_1^s \tau^{p-1}c(\tau)\,d\tau\right)ds \tag{3.3.17}$$

for $T > t > 1$. Further we have

$$\frac{1}{\log t}\int_1^t s^{p-1}c(s)\,ds = \frac{H_p(t)}{\log t} + \frac{1}{\log t}\int_1^t \frac{H_p(s)}{s}\,ds \tag{3.3.18}$$

for $t > 1$. In view of (3.3.18) there exists $\varepsilon \in \left(0, H_* - \left(\frac{p-1}{p}\right)^p\right)$ such that

$$\liminf_{t\to\infty} \frac{1}{\log t}\int_1^t s^{p-1}c(s)\,ds > \left(\frac{p-1}{p}\right)^p + \varepsilon.$$

Hence, by (3.3.17) it holds

$$\begin{aligned} c_p(T) &> c_p(t) + (p-1)\left[\left(\frac{p-1}{p}\right)^p + \varepsilon\right] \\ &\quad \times \left(\frac{\log t}{t^{p-1}} - \frac{\log T}{T^{p-1}} + \frac{1}{(p-1)t^{p-1}} - \frac{1}{(p-1)T^{p-1}}\right) \end{aligned}$$

for $T > t > 1$. If we let $T \to \infty$, then the last estimate yields

$$\frac{t^{p-1}}{\log t}(c_p(\infty) - c_p(t)) \geq \left[\left(\frac{p-1}{p}\right)^p + \varepsilon\right]\left(1 + \frac{1}{(p-1)\log t}\right)$$

for $t > 1$. Therefore (3.3.5) is satisfied. □

Theorem 3.3.3. *Suppose that* (3.3.4) *holds. If*

$$\limsup_{t\to\infty}[Q_p(t) + H_p(t)] > 1, \tag{3.3.19}$$

then equation (3.1.1) *is oscillatory.*

Proof. Assume, by a contradiction, that there is a solution y of (3.1.1) such that $y(t) \neq 0$ for $t \geq T$. If $w = \Phi(y'/y)$, then w solves the Riccati equation (3.1.2) for $t \geq T$. Multiplying its both sides by t^p and integrating from T to t, we obtain

$$t^{p-1}w(t) = -H_p(t) + \frac{1}{t}\int_T^t [ps^{p-1}w(s) - (p-1)|s^{p-1}w(s)|^q]\,ds + \frac{T^p}{t}w(T) + \frac{T}{t}H_p(T) \tag{3.3.20}$$

for $t > T$. Using now (3.3.7) we readily find that

$$\begin{aligned} Q_p(t) + H_p(t) &= \frac{1}{t}\int_T^t [ps^{p-1}w(s) - (p-1)|s^{p-1}w(s)|^q]\,ds \\ &\quad -(p-1)t^{p-1}\int_t^\infty |w(s)|^q ds + \frac{1}{t}(T^p w(T) + TH_p(T)) \end{aligned}$$

for $t \geq T$. Hence by the inequality $px - (p-1)|x|^q \leq 1$, which holds for $x \in \mathbb{R}$, we have

$$Q_p(t) + H_p(t) \leq 1 + \frac{1}{t}(T^p w(T) + TH_p(T))$$

for $t \geq T$. This contradicts (3.3.19). □

To prove the next criterion we need the following two auxiliary statements. First we introduce the notation. Let $0 < \lambda < \left(\frac{p-1}{p}\right)^p$. Denote by $\omega_1(\lambda)$ and $\omega_2(\lambda)$ the least and the largest root, respectively, of the equation

$$(p-1)|x|^q - (p-1)x + \lambda = 0.$$

Lemma 3.3.2. *Let* (3.1.1) *be nonoscillatory and suppose that* (3.3.4) *holds. If* $0 \leq Q_* \leq \frac{1}{p}\left(\frac{p-1}{p}\right)^{p-1}$, *then for each solution* y *of* (3.1.1) *the estimate*

$$\liminf_{t\to\infty} t^{p-1}\Phi(y'(t)/y(t)) \geq \omega_1((p-1)Q_*) \tag{3.3.21}$$

holds.

Proof. Let $y(t) \neq 0$ for $t \geq T$ be some solution of (3.1.1). Then (3.3.7) holds where $w = \Phi(y'/y)$. Set $m = \liminf_{t\to\infty} t^{p-1}w(t)$. If $m = \infty$, then there is nothing to prove. Therefore assume that $m < \infty$. If $Q_* = 0$, then equality (3.3.21) is trivial by virtue of (3.3.7). So we will assume that $Q_* > 0$. For an arbitrary $\varepsilon \in (0, Q_*)$ choose $t_1 > T$ such that

$$Q_p(t) > Q_* - \varepsilon \tag{3.3.22}$$

for $t \geq t_1$. Taking (3.3.22) into account, from (3.3.7) we have $t^{p-1}w(t) \geq Q_* - \varepsilon$ for $t \geq t_1$. Hence we easily conclude that $m \geq Q_*$. Now choose $t_2 > t_1$ such that

$$t^{p-1}w(t) > m - \varepsilon \ (> 0) \tag{3.3.23}$$

for $t > t_2$. By (3.3.22) and (3.3.23), from (3.3.7) we find $t^{p-1}w(t) \geq Q_* - \varepsilon + (m - \varepsilon)^q$. Since $\varepsilon \in (0, Q_*)$ is arbitrary, we conclude that $m^q - m + Q_* \leq 0$. Therefore $m \geq \omega_1((p-1)Q_*)$. □

Lemma 3.3.3. *Let* (3.1.1) *be nonoscillatory. If* $0 \leq H_* \leq \left(\frac{p-1}{p}\right)^p$, *then for each solution* y *of* (3.1.1) *the estimate*

$$\limsup_{t\to\infty} t^{p-1}\Phi(y'(t)/y(t)) \leq \omega_2(H_*) \tag{3.3.24}$$

holds.

Proof. Let $y(t) \neq 0$ for $t \geq T$ be some solution of (3.1.1). Then $w = \Phi(y'/y)$ solves Riccati equation (3.1.2). From this we get (3.3.20). Set $M = \limsup_{t\to\infty} t^{p-1}w(t)$. If $M \leq 0$, then (3.3.24) holds trivially. Therefore we will assume that $M > 0$.

Using the inequality $px - (p-1)|x|^q \le 1$, which holds for $x \in \mathbb{R}$, from (3.3.20) we have

$$M \le 1 - H_*. \tag{3.3.25}$$

Hence, if $H_* = 0$, then (3.3.24) holds. Assume that $H_* > 0$. For arbitrary $\varepsilon \in (0, H_*)$ choose $t_1 > T$ such that

$$H_p(t) > H_* - \varepsilon, \quad t^{p-1}w(t) < M + \varepsilon < 1 \tag{3.3.26}$$

for $t \ge t_1$. Since the function $px - (p-1)|x|^q$ increases for $x \in (-\infty, 1)$, from (3.3.20), taking into account (3.3.25) and (3.3.26), we have

$$t^{p-1}w(t) \le -H_* + \varepsilon + \frac{t - t_1}{t}[p(M+\varepsilon) - (p-1)(M+\varepsilon)^q] + \frac{t_1^p}{t}w(t_1) + \frac{t_1}{t}H_p(t_1)$$

for $t \ge t_1$, and therefore $M \le -H_* + \varepsilon + p(M+\varepsilon) - (p-1)(M+\varepsilon)^q$. Now, since $\varepsilon \in (0, H_*)$ is arbitrary, we obtain $(p-1)M^q - (p-1)M + H_* \le 0$. Therefore $M \le \omega_2(H_*)$. □

Now we are ready to give two theorems which complement Corollary 3.3.4.

Theorem 3.3.4. *Let* (3.3.4) *hold. If either*

$$0 \le Q_* \le \frac{1}{p}\left(\frac{p-1}{p}\right)^{p-1} \quad \text{and} \quad H^* > 1 - \omega_1((p-1)Q_*), \tag{3.3.27}$$

or

$$0 \le H_* \le \left(\frac{p-1}{p}\right)^p \quad \text{and} \quad Q^* > \omega_2(H_*), \tag{3.3.28}$$

then equation (3.1.1) *is oscillatory.*

Proof. Suppose by a contradiction that $y(t) \ne 0$ for $t \ge T$ is some solution of equation (3.1.1). Let (3.3.27) be fulfilled. Define the function $w = \Phi(y'/y)$. Then, for such w, equation (3.3.20) is satisfied. By virtue of Lemma 3.3.2, for any $\varepsilon > 0$ there exists $t_1 > T$ such that

$$t^{p-1}w(t) > \omega_1((p-1)Q_*) - \varepsilon$$

for $t \ge t_1$. Hence (3.3.20) implies

$$H_p(t) \le -\omega_1((p-1)Q_*) + \varepsilon + 1 + \frac{1}{t}\left(t_1^p w(t_1) + t_1 H_p(t_1)\right)$$

for $t \ge t_1$. Therefore $H^* \le 1 - \omega_1((p-1)Q_*)$, which contradicts (3.3.27).

To prove the sufficiency of (3.3.28), we proceed in the same way as above. Only instead of (3.3.20) we have (3.3.7) and instead of Lemma 3.3.2 we apply Lemma 3.3.3. Then we come to

$$Q_p(t) = t^{p-1}w(t) - (p-1)t^{p-1}\int_t^\infty |w(s)|^q ds \le \omega_2(H_*) + \varepsilon,$$

which implies $Q^* \le \omega_2(H_*)$, a contradiction. □

Theorem 3.3.5. *Let* (3.3.4) *hold and*

$$0 \le Q_* \le \frac{1}{p}\left(\frac{p-1}{p}\right)^{p-1}, \quad 0 \le H_* \le \left(\frac{p-1}{p}\right)^{p}.$$

If

$$Q^* > Q_* + \omega_2(H_*) - \omega_1((p-1)Q_*) \tag{3.3.29}$$

or

$$H^* > H_* + \omega_2(H_*) - \omega_1((p-1)Q_*), \tag{3.3.30}$$

then equation (3.1.1) *is oscillatory.*

Proof. The proof is by contradiction. Let (3.3.30), resp. (3.3.29) hold and $y(t) \ne 0$ for $t \ge T$ is some solution of equation (3.1.1). Then we have (3.3.20), resp. (3.3.7), where w is the function defined by $w = \Phi(y'/y)$. We will assume that $H_* > 0$ resp. $Q_* > 0$ since for $H_* = 0$, resp. $Q_* = 0$ condition (3.3.30), resp. (3.3.29) is equivalent to condition (3.3.27), resp. (3.3.28). By virtue of Lemma 3.3.3, resp. 3.3.2, for arbitrary $\varepsilon \in (0, 1 - \omega_2(H_*))$, resp. $\varepsilon \in (0, \omega_1((p-1)Q_*))$ there exists $t_1 > T$ such that

$$t^{p-1}w(t) < \omega_2(H_*) + \varepsilon, \quad \text{resp.} \quad t^{p-1}w(t) > \omega_1((p-1)Q_*) - \varepsilon \tag{3.3.31}$$

for $t \ge t_1$. Since the function $px - (p-1)|x|^q$ increases in $(-\infty, 1)$ and $\omega_2(H_*) + \varepsilon < 1$, we obtain

$$ps^{p-1}w(s) - (p-1)|s^{p-1}w(s)|^q < p(\omega_2(H_*) + \varepsilon) - (p-1)(\omega_2(H_*) + \varepsilon)^q \tag{3.3.32}$$

for $s \ge t_1$. Taking into account (3.3.31) and (3.3.32), we find from (3.3.20), resp. (3.3.7) that

$$\begin{aligned} H_p(t) \le\ & -\omega_1((p-1)Q_*) + \varepsilon + p(\omega_2(H_*) + \varepsilon) \\ & -(p-1)(\omega_2(H_*) + \varepsilon)^q + \frac{1}{t}(t_1^p w(t_1) + t_1 H_p(t_1)), \end{aligned}$$

resp.

$$Q_p(t) \le \omega_2(H_*) + \varepsilon - (\omega_1((p-1)Q_*) - \varepsilon)^q$$

for $t \ge t_1$. Hence we easily conclude that

$$H^* \le -\omega_1((p-1)Q_*) + p\omega_2(H_*) - (p-1)(\omega_2(H_*))^q,$$

resp.

$$Q^* \le \omega_2(H_*) - (\omega_1((p-1)Q_*))^q,$$

which contradicts (3.3.30), resp. (3.3.29). □

In the second part of this subsection we give nonoscillatory criteria in terms of $Q_p(t)$ and $H_p(t)$. As it is already pointed out above, in [197] it is distinguished between nonoscillation and strong nonoscillation (in the sense of the definition given at the beginning of this section) of equation (3.3.1). It is known (see [70]) that nonoscillation of (3.3.1) is guaranteed by the existence of a solution of the associated Riccati type inequality in a neighborhood of infinity, while for strong nonoscillation we need a positivity of such a solution. However, when considering equation of the form (3.1.1), both concepts coincide. We will see that the theorems which originally deal with strong nonoscillation, usually involve somehow stronger assumptions in comparison with the criteria of the same type, in order to guarantee the positivity of the solution of the Riccati inequality. But we present them here since they, in addition to nonoscillation, discuss the existence of a solution of (3.1.1) which is eventually positive and strictly increasing. See also the text before Theorem 3.3.8.

Let $\lambda \le 0$. Denote by $\mu(\lambda)$ the greatest root of the equation $|x|^{1/q} + x + \lambda = 0$.

Theorem 3.3.6. *Let* (3.3.4) *hold. If either*

$$-\frac{2p-1}{p}\left(\frac{p-1}{p}\right)^{p-1} < Q_* \quad \text{and} \quad Q^* < \frac{1}{p}\left(\frac{p-1}{p}\right)^{p-1},$$

or

$$-\infty < Q_* \le -\frac{2p-1}{p}\left(\frac{p-1}{p}\right)^{p-1} \quad \text{and} \quad Q^* < (\mu(Q_*))^{1/q} - \mu(Q_*),$$

then equation (3.1.1) *is nonoscillatory.*

Proof. Choose $\varepsilon > 0$ and $T > 0$ such that

$$\begin{gathered} -\tfrac{2p-1}{p}\left(\tfrac{p-1}{p}\right)^{p-1} < Q_* - \varepsilon, \quad Q^* + \varepsilon < \tfrac{1}{p}\left(\tfrac{p-1}{p}\right)^{p-1}, \\ \text{resp. } Q_* + \varepsilon < (\mu(Q_* + \varepsilon))^{1/q} - \mu(Q_* + \varepsilon) \end{gathered} \tag{3.3.33}$$

and

$$Q_* - \varepsilon < Q_p(t) < Q^* + \varepsilon \tag{3.3.34}$$

for $t \ge T$. Let $\lambda = \left(\frac{p-1}{p}\right)^p$, resp. $\lambda = \mu(Q_* + \varepsilon)$ and

$$w(t) = \frac{1}{t^{p-1}}(\lambda + Q_p(t)) \tag{3.3.35}$$

for $t \ge T$. Then (3.3.33) and (3.3.34) readily imply $-\lambda - \lambda^{1/q} < Q_p(t) < \lambda^{1/q} - \lambda$ for $t \ge T$, so that $|w(t)|^q < \lambda/t^p$ for $t \ge T$. From the last inequality we readily conclude that the function w satisfies Riccati inequality (2.2.7) with $r \equiv 1$. Thus equation (3.1.1) is nonoscillatory by Theorem 2.2.1. □

The next theorem can be viewed as a nonoscillatory counterpart to Corollary 3.3.2.

Theorem 3.3.7. *Let there exist a finite limit*

$$\tilde{c} := \lim_{t\to\infty} \frac{1}{\log t} \int_1^t s^{p-1} c(s)\, ds < \left(\frac{p-1}{p}\right)^p \tag{3.3.36}$$

and

$$G^* < G_* + \omega_2(\tilde{c}) - \omega_1(\tilde{c}), \tag{3.3.37}$$

where

$$G(t) = \log t \left(\frac{1}{\log t} \int_1^t s^{p-1} c(s)\, ds - \tilde{c}\right),$$

$$G_* := \liminf_{t\to\infty} G(t) \quad \textit{and} \quad G^* := \limsup_{t\to\infty} G(t). \tag{3.3.38}$$

Then equation (3.1.1) *is nonoscillatory.*

Proof. Choose $\varepsilon > 0$ and $T > 0$ such that

$$G_* - \varepsilon < G(t) < G^* + \varepsilon, \quad G^* + 2\varepsilon < G_* + \omega_2(\tilde{c}) - \omega_1(\tilde{c}). \tag{3.3.39}$$

Put $\lambda = G_* - \varepsilon + \omega_2(\tilde{c})$ and $w(t) = (\lambda - G(t))/t^{p-1}$ for $t \geq T$. By (3.3.39) we have $\omega_1(\tilde{c}) < \lambda - G(t) < \omega_2(\tilde{c})$ for $t \geq T$. Hence $(p-1)|\lambda - G(t)|^q - (p-1)(\lambda - G(t)) + \tilde{c} < 0$ for $t \geq T$. By the latter inequality it is easy to conclude that the function w satisfies the inequality (2.2.7) with $r \equiv 1$. Therefore equation (3.1.1) is nonoscillatory by Theorem 2.2.1. □

Corollary 3.3.5. *If*

$$-\infty < \limsup_{t\to\infty} \frac{1}{p-1} \int_1^t s^{p-1} c(s)\, ds < \liminf_{t\to\infty} \frac{1}{p-1} \int_1^t s^{p-1} c(s)\, ds + 1 < \infty,$$

then equation (3.1.1) *is nonoscillatory.*

The remaining theorems in this subsection were originally dealing with strong nonoscillation of (3.3.1) in the sense of [197]. But now, rewritten for (3.1.1), they give the conditions for the existence of an eventually positive strictly increasing solution of (3.1.1), which implies nonoscillation of (3.1.1). Recall that we have *no sign condition on* c. There is a comment before Lemma 2.2.3 on positivity of a solution of the Riccati equation (implied by the existence of a solution y of (1.1.1) with $y(t)y'(t) > 0$). Here we work under different assumptions. In [70, Lemma 2.5], it was proved the following statement (for general coefficient c); we adjust it for our needs.

Lemma 3.3.4. *Let y be a nontrivial solution of* (3.1.1) *satisfying* $y(a) = 0$ *and let a function (sufficiently smooth) v be such that $v(a) \geq 0$, $v'(t) > 0$ for $t \in [a, b]$ and $\mathcal{L}_1[v](t) := (\Phi(v'))' + c(t)\Phi(v) \leq 0$ for $t \in [a, b]$. Then $y'(t) \neq 0$ for $t \in [a, b]$.*

Now, in view of the identity $v\mathcal{L}_1[v] = |v|^p \mathcal{R}_1[w]$, where $\mathcal{R}_1[u](t) := w' + c(t) + (p-1)|w|^q$ and $w = \Phi(v'/v)$ (see Theorem 2.2.1), we see that the positivity of w satisfying $\mathcal{R}_1[w] \le 0$ guarantees the existence of a solution of (3.1.1) which is positive and strictly increasing. Obviously, (3.1.1) is nonoscillatory.

Theorem 3.3.8. *Let* (3.3.4) *hold. If either*

$$-\left(\frac{p-1}{p}\right)^p < Q_* \quad \text{and} \quad Q^* < \frac{1}{p}\left(\frac{p-1}{p}\right)^{p-1} \tag{3.3.40}$$

or

$$-\infty < Q_* \le -\left(\frac{p-1}{p}\right)^p \quad \text{and} \quad Q^* < Q_* + |Q_*|^{1/q}, \tag{3.3.41}$$

then equation (3.1.1) *is nonoscillatory. Moreover, there exists an eventually positive strictly increasing solution of* (3.1.1).

Proof. The theorem can be proved similarly to Theorem 3.3.6 with the only difference that when (3.3.41) is fulfilled we should put $\lambda = Q_* - \varepsilon$ and keep in mind that under conditions (3.3.40) and (3.3.41) the function w is positive. □

For the next statement we need to introduce some notation. Let

$$\eta = x_0 + \frac{2p-1}{p}\left(\frac{p-1}{p}\right)^{p-1},$$

where x_0 is the least root of the equation

$$(p-1)|x|^q + px + \frac{2p-1}{p}\left(\frac{p-1}{p}\right)^{p-1} = 0. \tag{3.3.42}$$

It is not difficult to show that $\eta < 0$ since $\frac{2p-1}{p}\left(\frac{p-1}{p}\right)^{p-1} < 1$ for $p > 1$. Further, let $\lambda < 1$. Denote by $\psi(\lambda)$ the greatest root of the equation $(p-1)|x|^q + px + \lambda = 0$.

Theorem 3.3.9. *Let either*

$$\eta < H_* \quad \text{and} \quad H^* < \left(\frac{p-1}{p}\right)^p, \tag{3.3.43}$$

or

$$-\infty < H_* \le \eta \quad \text{and} \quad H^* < H_* + \omega_2(H_*) + \psi(H_* + \omega_2(H_*)), \tag{3.3.44}$$

where ω_2 is defined before Lemma 3.3.2. Then equation (3.1.1) *is nonoscillatory. Moreover, there exists an eventually positive strictly increasing solution of* (3.1.1).

Proof. Let (3.3.43) be satisfied. Choose $\varepsilon > 0$ and $T > 0$ such that

$$\eta < H_* - \varepsilon < H_p(t) < H^* + \varepsilon < \left(\frac{p-1}{p}\right)^{p-1} \tag{3.3.45}$$

for $t \geq T$. Now put $K = \frac{2p-1}{p}\left(\frac{p-1}{p}\right)^{p-1}$. By (3.3.45) and the definition of the number η we have

$$x_0 < H_p(t) - K < -\left(\frac{p-1}{p}\right)^{p-1} \tag{3.3.46}$$

for $t \geq T$. Since $x_1 = -\left(\frac{p-1}{p}\right)^{p-1}$ is the greatest root of the equation (3.3.42), in view of (3.3.46) we find that

$$(p-1)|H_p(t) - K|^q + p(H_p(t) - K) + K \leq 0, \quad H_p(t) - K < 0 \tag{3.3.47}$$

for $t \geq T$. Let

$$w(t) = \frac{1}{t^{p-1}}(K - H_p(t)) \tag{3.3.48}$$

for $t \geq T$. According to (3.3.47), the (positive) function w satisfies inequality (2.2.7) (with $r \equiv 1$) for $t \geq T$. Therefore (3.1.1) is nonoscillatory by Theorem 2.2.1.

Now consider the case where (3.3.44) is satisfied. Choose $\varepsilon > 0$ and $T > 1$ such that

$$H^* + \varepsilon < H_* + \omega_2(H_* - \varepsilon) - \varepsilon + \psi(H_* + \omega_2(H_* - \varepsilon)), \tag{3.3.49}$$

$$H_* - \varepsilon < H_p(t) < H^* + \varepsilon. \tag{3.3.50}$$

Let $M = H_* - \varepsilon + \omega_2(H_* - \varepsilon)$. Clearly, $-\omega_2(H_* - \varepsilon)$ satisfies the equation

$$(p-1)|x|^q + px + M = 0. \tag{3.3.51}$$

On the other hand, since $H_* - \varepsilon < 0$, we obtain $-\omega_2(H_* - \varepsilon) < -1$. Thus $-\omega_2(H_* - \varepsilon)$ is the least root of equation (3.3.51). Moreover, since $(p-1)|x|^q + px \geq -1$ for $x \in \mathbb{R}$, we have $M < -1$ and therefore $\psi(M) < 0$. Now (3.3.49) and (3.3.50) imply that $-\omega_2(H_* - \varepsilon) < H_p(t) - M < \psi(M) < 0$ for $t \geq T$. Therefore (3.3.47) is satisfied. Using (3.3.47), we readily find that the function w defined by (3.3.48) is positive and satisfies inequality (2.2.7) (with $r \equiv 1$). □

In view of Theorem 3.3.7, the proof of the next statement may be omitted.

Theorem 3.3.10. *Let the finite limit* (3.3.36) *exist and either* $\tilde{c} > 0$ *and* (3.3.37) *be fulfilled or* $\tilde{c} \leq 0$ *and* $G^* < G_* + \omega_2(\tilde{c})$, *where* G^* *and* G_* *are the numbers defined by* (3.3.38) *and* ω_2 *is defined before Lemma 3.3.2. Then equation* (3.1.1) *is nonoscillatory. Moreover, there exists an eventually positive strictly increasing solution of* (3.1.1).

Corollary 3.3.6. *Under the conditions of Corollary 3.3.5, there exists an eventually positive strictly increasing solution of* (3.1.1).

Remark 3.3.1. It is easy to see that if the finite limit (3.3.4) exists, then for the function Q_p to be bounded from below it is necessary that $c_p(\infty) = c^*$, c^* being defined in (3.3.10), while for the function Q_p to be bounded (from both sides) it is necessary that (3.3.9) holds. Since

$$c_p(t) = \int_1^t c(s)\,ds - \frac{1}{t}H_p(t) - \frac{1}{t^{p-1}}\int_1^t \frac{1}{s}H_p(s)\,ds, \quad t > 1,$$

for the function H_p to be bounded from below it is necessary that $c_p(\infty) = c_*$, while for H_p to be bounded (from both sides) it is necessary in order that (3.3.9) is fulfilled.

In view of the above observation, condition (3.3.9) is the necessary one for Theorems 3.3.5, 3.3.6, 3.3.8 and 3.3.9, while for conditions (3.3.27) and (3.3.28) of Theorem 3.3.4 to hold it is necessary that $c_p(\infty) = c^*$ and $c_p(\infty) = c_*$, respectively.

According to (3.3.17), if the finite limit (3.3.36) (not necessarily less than $\left(\frac{p-1}{p}\right)^p$) exists, then (3.3.4) holds. Therefore condition (3.3.4) is necessary for Theorems 3.3.7 and 3.3.10.

Condition (3.3.4) is not generally used in proving the oscillation criterion involving $H_* > \left(\frac{p-1}{p}\right)^{p-1}$ (see Corollary 3.3.4). However by (3.3.17) and (3.3.18), the boundedness of the function H_p from below implies that $\liminf_{t\to\infty} c_p(t) > -\infty$. Therefore, if (3.3.4) is not satisfied, then by virtue of Theorem 3.3.1 equation (3.1.1) is oscillatory. Hence for this criterion condition (3.3.4) is the necessary one, in a certain sense.

3.3.2 Hille-Nehari type weighted criteria and extensions

Now we present the results originally stated for (3.3.1) with $p \in (1, 2]$ and $c \geq 0$ in [262]. We adjust them for equation (3.1.1), and we will see that they apply for $p > 2$ as well. Thus let us consider (3.1.1) where the function c is nonnegative, as this condition was assumed in [262]. However, one could easily observed that the below results can be obtained without any further difficulties also under the relaxed condition, namely that either $\int_{t_0}^t s^\lambda c(s)\,ds > 0$ or $\int_t^\infty s^\lambda c(s)\,ds > 0$ for all large t and certain λ. Which of the two integrals is chosen depends on the value of λ; see the below given definition of $c^*(\lambda)$ and $c_*(\lambda)$. As we show in the next lemma (for a linear equation this assertion goes back to Fite and Hille), it is reasonable to assume that $\int^\infty s^\lambda c(s)\,ds$ converges for $\lambda < p - 1$.

Lemma 3.3.5. *Let for some $\lambda < p-1$ the integral $\int^\infty s^\lambda c(s)\,ds$ diverges. Then equation* (3.1.1) *is oscillatory.*

Proof. Assume for a contradiction that (3.1.1) is nonoscillatory. Then there is a positive increasing solution $y(t)$ of (3.1.1) for $t \geq T$, and $w = \Phi(y'/y) > 0$ satisfies (3.1.2). Moreover,

$$\limsup_{t\to\infty} t^{p-1}w(t) \leq 1 \tag{3.3.52}$$

(see the proof of Lemma 3.3.3). Multiplying (3.1.2) by t^λ and integrating from T to t we obtain

$$\int_T^t s^\lambda c(s)\,ds = -t^\lambda w(t) + T^\lambda w(T) + \lambda\int_T^t s^{\lambda-1}w(s)\,ds - (p-1)\int_T^t s^\lambda w^q(s)\,ds$$

for $t \geq T$. By (3.3.52) the right-hand side of the last equality has a finite limit as $t \to \infty$. Hence the integral $\int^\infty s^\lambda c(s)\, ds$ converges, a contradiction. □

Introduce the notation

$$\begin{aligned}
c_*(\lambda) &= \liminf_{t\to\infty} t^{p-1-\lambda} \int_t^\infty s^\lambda c(s)\, ds \quad \text{for} \quad \lambda < p-1, \\
c^*(\lambda) &= \limsup_{t\to\infty} t^{p-1-\lambda} \int_t^\infty s^\lambda c(s)\, ds \quad \text{for} \quad \lambda < p-1, \\
c_*(\lambda) &= \liminf_{t\to\infty} t^{p-1-\lambda} \int_1^t s^\lambda c(s)\, ds \quad \text{for} \quad \lambda > p-1, \\
c^*(\lambda) &= \limsup_{t\to\infty} t^{p-1-\lambda} \int_1^t s^\lambda c(s)\, ds \quad \text{for} \quad \lambda > p-1.
\end{aligned}$$

We give criteria in terms of the numbers $c_*(\lambda)$ and $c^*(\lambda)$. Note that $c_*(0) = Q_*$, $c^*(0) = Q^*$, $c_*(p) = H_*$ and $C^*(p) = H^*$, where Q_*, Q^*, H_*, H^* are defined in (3.3.11) and (3.3.12). Then the conditions of Corollary 3.3.4 read as $c_*(0) > \frac{1}{p}\left(\frac{p-1}{p}\right)^{p-1}$ and $c_*(p) > \left(\frac{p-1}{p}\right)^p$. Hence we assume

$$c_*(0) \leq \frac{1}{p}\left(\frac{p-1}{p}\right)^{p-1} \quad \text{and} \quad c_*(p) \leq \left(\frac{p-1}{p}\right)^p. \tag{3.3.53}$$

Also note that Theorem 3.3.5 reads as follows: If (3.3.53) holds and either $c^*(0) > B - A + c_*(0)$ or $c^*(p) > B - A + c_*(p)$, then (3.1.1) is oscillatory. Here A and B stand for the least nonnegative root of $x^q - x + c_*(0) = 0$ and for the greatest root of $(p-1)x^q - (p-1)x + c_*(p) = 0$, respectively, i.e., $A = \omega_1((p-1)Q_*)$, $B = \omega_2(H_*)$, where ω_1, ω_2 are defined before Lemma 3.3.2. The above algebraic equations are solvable provided (3.3.53) holds.

Now we are ready to give the main statement of this subsection.

Theorem 3.3.11. *Let* (3.3.53) *be satisfied. If either*

$$c^*(\lambda) > \frac{1}{p-1-\lambda}\left(\frac{\lambda}{p}\right)^p + B \tag{3.3.54}$$

for some $0 < \lambda < p-1$ *or*

$$c^*(\lambda) > \frac{1}{\lambda - p + 1}\left(\frac{\lambda}{p}\right)^p - A \tag{3.3.55}$$

for some $\lambda > p-1$, *then equation* (3.1.1) *is oscillatory.*

Proof. Assume for a contradiction that (3.1.1) has a nonoscillatory solution. Then there exists $T > 0$ such that Riccati equation (3.1.2) has a positive solution w satisfying $\liminf_{t\to\infty} t^{p-1} w(t) \geq A$ and $\limsup_{t\to\infty} t^{p-1} w(t) \leq B$ (see Lemma 3.3.2 and Lemma 3.3.3). Clearly, for any $\varepsilon > 0$ there exists $t_1 > T$ such that

$t^{p-1}w(t) < B+\varepsilon$, $t^{p-1}w(t) > A-\varepsilon$ for $t > t_1$. From (3.1.2) we easily find that

$$t^{p-1-\lambda}\int_t^\infty s^\lambda c(s)\,ds = t^{p-1}w(t) \\ + t^{p-1-\lambda}\int_t^\infty s^{\lambda-p}(sw^{q-1}(s))^{p-1}(\lambda-(p-1)sw^{q-1}(s))\,ds$$

for $t > t_1$ and $\lambda < p-1$, while

$$t^{p-1-\lambda}\int_{t_1}^t s^\lambda c(s)\,ds = -t^{p-1}w(t) + \frac{t_1^\lambda w(t_1)}{t} \\ + t^{p-1-\lambda}\int_{t_1}^t s^{\lambda-p}(sw^{q-1}(s))^{p-1}(\lambda-(p-1)sw^{q-1}(s))\,ds$$

for $t > t_1$ and $\lambda > p-1$. Since $\max\{x^{p-1}(\lambda-(p-1)x)\ :\ 0 \le x < \infty\} = (\lambda/p)^p$, from the last two inequalities we have

$$t^{p-2}\int_t^\infty s^\lambda c(s)\,ds < B+\varepsilon+\frac{1}{p-1-\lambda}\left(\frac{\lambda}{p}\right)^p$$

for $t > t_1$ and $0 < \lambda < p-1$, and

$$t^{p-2}\int_{t_1}^t s^\lambda c(s)\,ds < \frac{t_1^\lambda w(t_1)}{t}+\frac{1}{\lambda-p+1}\left(\frac{\lambda}{p}\right)^p - A+\varepsilon$$

for $t > t_1$ and $\lambda > p-1$. Consequently,

$$c^*(\lambda) \le \frac{1}{p-1-\lambda}\left(\frac{\lambda}{p}\right)^p + B \quad \text{for} \quad \lambda < p-1,$$
$$c^*(\lambda) \le \frac{1}{\lambda-p+1}\left(\frac{\lambda}{p}\right)^p - A \quad \text{for} \quad \lambda > p-1,$$

but this contradicts (3.3.54) and (3.3.55). □

Before we give consequences of the last theorem, let us prove two technical statements.

Lemma 3.3.6. *Let $c^*(\lambda) < \infty$ for $\lambda \neq p-1$. Then the mapping $\lambda \mapsto (p-1-\lambda)c^*(\lambda)$, resp. $\lambda \mapsto (p-1-\lambda)c_*(\lambda)$, does not increase, resp. does not decrease, for $\lambda < p-1$ and does not decrease, resp. does not increase, for $\lambda > p-1$.*

Proof. We prove this lemma only in the case when $\lambda < p-1$. The case $\lambda > p-1$ can be handled in a similar way. Let $\varepsilon > 0$. Choose $t_1 > 0$ so that

$$c_*(\lambda)-\varepsilon < t^{p-1-\lambda}\int_t^\infty s^\lambda c(s)\,ds < c^*(\lambda)+\varepsilon$$

for $t > t_1$. It is easy to see that whatever $\mu < p-1$ is, we have

$$t^{p-1-\mu}\int_t^\infty s^\mu c(s)\,ds = t^{p-1-\lambda}\int_t^\infty s^\lambda c(s)\,ds + (\mu-\lambda)t^{p-1-\mu}\int_t^\infty s^{\mu-\lambda-p+1}\left(\int_s^\infty \tau^\lambda c(\tau)\,d\tau\right)ds.$$

Hence if $\lambda < \mu$, then

$$(c_*(\lambda)-\varepsilon)\left(1+\frac{\mu-\lambda}{p-1-\mu}\right) < t^{p-1-\mu}\int_t^\infty s^\mu c(s)\,ds < (c^*(\lambda)+\varepsilon)\left(1+\frac{\mu-\lambda}{p-1-\mu}\right)$$

for $t > t_1$. This implies that $(p-1-\mu)c^*(\mu) \le (p-1-\lambda)c^*(\lambda)$ and $(p-1-\mu)c_*(\mu) \ge (p-1-\lambda)c_*(\lambda)$. □

Lemma 3.3.7. *Let $c^*(\lambda) < \infty$ for $\lambda \neq p-1$. Then*

$$\begin{aligned} \lim_{\lambda\to(p-1)-}(p-1-\lambda)c_*(\lambda) &= \lim_{\lambda\to(p-1)+}(\lambda-p+1)c_*(\lambda),\\ \lim_{\lambda\to(p-1)-}(p-1-\lambda)c^*(\lambda) &= \lim_{\lambda\to(p-1)+}(\lambda-p+1)c^*(\lambda). \end{aligned} \tag{3.3.56}$$

Proof. Let $\lambda < p-1$, $\mu > p-1$ and $\varepsilon > 0$. Choose $t_1 > 1$ so that

$$c_*(\lambda)-\varepsilon < t^{p-1-\lambda}\int_t^\infty s^\lambda c(s)\,ds < c^* + \varepsilon$$

and

$$c_*(\mu)-\varepsilon < t^{p-1-\mu}\int_1^t s^\mu c(s)\,ds < c^* + \varepsilon$$

for $t > t_1$. It can be easily verified that

$$\begin{aligned} &t^{p-1-\lambda}\int_t^\infty s^\lambda c(s)\,ds\\ &\quad = -t^{p-1-\mu}\int_1^t s^\mu c(s)\,ds + (\mu-\lambda)t^{p-1-\lambda}\int_t^\infty s^{\lambda-\mu-1}\left(\int_1^s \tau^\mu c(\tau)\,d\tau\right)ds \end{aligned}$$

and

$$\begin{aligned} t^{p-1-\mu}\int_1^t s^\mu c(s)\,ds &= -t^{p-1-\lambda}\int_t^\infty s^\lambda c(s)\,ds\\ &\quad + t^{p-1-\mu}\int_1^\infty s^\lambda c(s)\,ds + (\mu-\lambda)t^{p-1-\mu}\int_1^t s^{\mu-\lambda-1}\left(\int_s^\infty \tau^\lambda c(\tau)\,d\tau\right)ds. \end{aligned}$$

From these equalities we have

$$\begin{aligned} \frac{\mu-\lambda}{p-1-\lambda}(c_*(\mu)-\varepsilon) - t^{p-1-\mu}\int_1^t s^\mu c(s)\,ds &< t^{p-1-\lambda}\int_t^\infty s^\lambda c(s)\,ds\\ &< \frac{\mu-\lambda}{p-1-\lambda}(c^*(\mu)+\varepsilon) \end{aligned}$$

for $t > t_1$ and

$$-t^{p-1-\lambda}\int_t^\infty s^\lambda c(s)\,ds + \frac{\mu-\lambda}{\mu-p+1}(c_*(\lambda)-\varepsilon)(1-t^{p-1-\lambda}) < t^{p-1-\mu}\int_1^t s^\mu c(s)\,ds < \int_1^\infty s^\lambda c(s)\,ds + \frac{\mu-\lambda}{\mu-p+1}(c^*(\lambda)+\varepsilon)$$

for $t > t_1$. Hence

$$\begin{aligned}(\mu-\lambda)c_*(\mu)-(p-1-\lambda)c^*(\mu) &< (p-1-\lambda)c_*(\lambda),\\ (p-1-\lambda)c^*(\lambda) &< (\mu-\lambda)c^*(\mu),\\ -(\mu-p+1)c^*(\lambda)+(\mu-\lambda)c_*(\lambda) &< (\mu-p+1)c_*(\mu),\end{aligned}$$

and

$$(\mu-p+1)c^*(\mu) < (\mu-\lambda)c^*(\lambda)+(\mu-p+1)\int_1^\infty s^\lambda c(s)\,ds.$$

Finally by Lemma 3.3.6, we obtain

$$\begin{aligned}&\lim\nolimits_{\lambda\to(p-1)-}(p-1-\lambda)c_*(\lambda) \ge (\mu-p+1)c_*(\mu),\\ &\lim\nolimits_{\lambda\to(p-1)-}(p-1-\lambda)c^*(\lambda) \le (\mu-p+1)c^*(\mu),\\ &\lim\nolimits_{\lambda\to(p-1)+}(\mu-p+1)c_*(\mu) \ge (p-1-\lambda)c_*(\lambda),\\ &\lim\nolimits_{\lambda\to(p-1)+}(\mu-p+1)c^*(\mu) \le (p-1-\lambda)c^*(\lambda).\end{aligned}$$

From the last four inequalities we conclude that (3.3.56) holds. □

Corollary 3.3.7. *Let*

$$\lim_{\lambda\to p-1}|p-1-\lambda|c^*(\lambda) > \left(\frac{p-1}{p}\right)^p. \tag{3.3.57}$$

Then equation (3.1.1) *is oscillatory.*

Proof. We may assume that $c^*(\lambda) < \infty$, otherwise (3.1.1) is oscillatory by Theorem 3.3.11. Then by Lemma 3.3.7, the limit in (3.3.57) exists. Obviously,

$$\lim_{\lambda\to(p-1)-}\left(|p-1-\lambda|c^*(\lambda)-\left(\frac{\lambda}{p}\right)^p-(p-1-\lambda)B\right) > 0.$$

This implies that (3.3.54) is satisfied for some $\lambda < p-1$. Therefore (3.1.1) is oscillatory by Theorem 3.3.11. □

Corollary 3.3.8. *Let for some* $\lambda \ne p-1$

$$|p-1-\lambda|c_*(\lambda) > \left(\frac{p-1}{p}\right)^p. \tag{3.3.58}$$

Then equation (3.1.1) *is oscillatory.*

Proof. If for some $\lambda \neq p-1$, $c^*(\lambda) = \infty$, then (3.1.1) is oscillatory by Theorem 3.3.11. Thus we assume that $c^*(\lambda) < \infty$ for $\lambda \neq p-1$. Now, if (3.3.58) holds for some $\lambda \neq p-1$, then (3.3.57) is satisfied by Lemma 3.3.6. Hence according to Corollary 3.3.7, equation (3.1.1) is oscillatory. □

To show another consequence, we need the following lemma.

Lemma 3.3.8. *Let $c^*(\lambda) < \infty$ for some $\lambda < p-1$ and $c^*(\mu) < \infty$ for some $\mu > p-1$. Then*

$$\limsup_{t\to\infty} \frac{1}{\log t}\int_1^t s^{p-1}c(s)\,ds \le (p-1-\lambda)c^*(\lambda) \tag{3.3.59}$$

and

$$\limsup_{\lambda\to(p-1)-} (p-1-\lambda)\int_1^\infty s^\lambda c(s)\,ds \le (\mu-p+1)c^*(\mu). \tag{3.3.60}$$

Proof. Let $\varepsilon > 0$. Choose $t_1 > 1$ so that

$$t^{p-1-\lambda}\int_t^\infty s^\lambda c(s)\,ds < c^*(\lambda)+\varepsilon, \quad t^{p-1-\mu}\int_1^t s^\mu c(s)\,ds < c^*(\mu)+\varepsilon$$

for $t > t_1$. We can easily see that

$$\begin{aligned}\frac{1}{\log t}\int_1^t s^{p-1}c(s)\,ds &= -\frac{t^{p-1-\lambda}}{\log t}\int_t^\infty s^\lambda c(s)\,ds \\ &\quad +\frac{p-1-\lambda}{\log t}\int_1^t s^{p-\lambda-2}\left(\int_s^\infty \tau^\lambda c(\tau)\,d\tau\right)ds,\end{aligned}$$

and

$$\int_1^\infty s^\delta c(s)\,ds = (\mu-\delta)\int_1^\infty s^{\delta-\mu-1}\left(\int_1^s \tau^\mu c(\tau)\,d\tau\right)ds$$

for $\delta < p-1$. From these inequalities we have

$$\frac{1}{\log t}\int_1^t s^{p-1}c(s)\,ds < (p-1-\lambda)(c^*(\lambda)+\varepsilon)$$

for $t > t_1$ and

$$(p-1-\delta)\int_1^\infty s^\delta c(s)\,ds < (\mu-\delta)(c^*(\mu)+\varepsilon).$$

Hence inequalities (3.3.59) and (3.3.60) hold. □

To show the next two statements, note that according to Lemma 3.3.8, (3.3.57) follows from (3.3.61) [or (3.3.62)]. Equation (3.1.1) is then oscillatory by Corollary 3.3.7. Also compare Corollary 3.3.9 with Corollary 3.3.2.

Corollary 3.3.9. *If*

$$\limsup_{t\to\infty} \frac{1}{\log t} \int_1^t s^{p-1} c(s)\, ds > \left(\frac{p-1}{p}\right)^p, \tag{3.3.61}$$

then equation (3.1.1) *is oscillatory.*

Corollary 3.3.10. *If*

$$\limsup_{\lambda\to(p-1)-} (p-1-\lambda) \int_1^\infty s^\lambda c(s)\, ds > \left(\frac{p-1}{p}\right)^p, \tag{3.3.62}$$

then equation (3.1.1) *is oscillatory.*

Remark 3.3.2. Inequalities (3.3.57), (3.3.58), (3.3.61) and (3.3.62) cannot be weakened. Indeed, let $c(t) = \left(\frac{p-1}{p}\right)^p t^{-p}$ for $t \geq 1$. Then

$$\begin{aligned} |p-1-\lambda| c_*(\lambda) &= |p-1-\lambda| c^*(\lambda) = \frac{1}{\log t} \int_1^t s^{p-1} c(s)\, ds \\ &= (p-1-\lambda) \int_1^\infty s^\lambda c(s)\, ds = \left(\frac{p-1}{p}\right)^p, \end{aligned}$$

and equation (3.1.1) has the nonoscillatory solution $y(t) = t^{(p-1)/p}$ for $t \geq 1$.

We conclude this section with a nonoscillatory complement to Corollary 3.3.8. In [262], the following criterion (for equation (3.3.1) with $p \in (1, 2]$ and $c \geq 0$) is proved by means of this statement: If there exists a positive function v which is locally absolutely continuous together with its first derivative and satisfies the inequalities $v'(t) > 0$, $v'' + c(t) v^{p-1} (v')^{2-p} \leq 0$ for $t \geq T$ almost everywhere, then equation (3.1.1) is nonoscillatory; see also the text after Corollary 3.3.5. However, we do not need such a kind of statement since we can simply use Theorem 2.2.1 (equivalence (i)$\Leftrightarrow$(v)).

Again, as in oscillation theorems, the condition $c(t) \geq 0$ may be relaxed to either $\int_{t_0}^t s^\lambda c(s)\, ds > 0$ or $\int_t^\infty s^\lambda c(s)\, ds > 0$ for all large t and certain λ, and the results work as well. Compare the subsequent theorem with Theorem 3.3.9.

Theorem 3.3.12. *Let either for some*

$$\lambda < \frac{(p-1)^2}{p} \quad \text{or for some} \quad \lambda > p - 1 + \frac{(p-1)^p}{p[p^{p-1} - (p-1)^{p-1}]}$$

the inequality

$$|p-1-\lambda| c^*(\lambda) < \left(\frac{p-1}{p}\right)^p \tag{3.3.63}$$

be satified. Then equation (3.1.1) *is nonoscillatory.*

Proof. Introduce the notation

$$f(t) = \int_t^\infty s^\lambda c(s)\,ds \quad \text{for} \quad t > 1, \quad \lambda < \frac{(p-1)^2}{p},$$

$$f(t) = -\int_1^t s^\lambda c(s)\,ds \quad \text{for} \quad t > 1, \quad \lambda > p - 1 + \frac{(p-1)^p}{p[p^{p-1} - (p-1)^{p-1}]},$$

$$K = \left(\frac{p-1}{p}\right)^{p-1} - \frac{1}{p-1-\lambda}\left(\frac{p-1}{p}\right)^p.$$

From (3.3.63) for some $T > 1$ we have

$$0 \le K + t^{p-1-\lambda} f(t) < \left(\frac{p-1}{p}\right)^{p-1} \quad \text{for} \quad t > 1, \quad \lambda < \frac{(p-1)^2}{p} \tag{3.3.64}$$

and

(3.3.65)

$$\left(\frac{p-1}{p}\right)^{p-1} < K + t^{p-1-\lambda} f(t) \le K \quad \text{for} \quad t > 1, \quad \lambda > p - 1 + \frac{(p-1)^p}{p[p^{p-1} - (p-1)^{p-1}]}.$$

It can be easily seen that if $\lambda < \frac{(p-1)^2}{p}$ and $0 \le x \le \left(\frac{p-1}{p}\right)^{p-1}$ or

$$\lambda > p - 1 + \frac{(p-1)^p}{p[p^{p-1} - (p-1)^{p-1}]} \quad \text{and} \quad \frac{(p-1)^2}{p} < x \le K,$$

then $(p-1)x^q - \lambda x + K(\lambda - p + 1) \le 0$. Thus, according to (3.3.64) and (3.3.65) we have

$$(p-1)(K + t^{p-1-\lambda} f(t))^q - \lambda(K + t^{p-1-\lambda} f(t)) + K(\lambda - p + 1) \le 0$$

for $t > T$. The last inequality is equivalent to

$$w'(t) \le \frac{\lambda}{t}(w(t) + f(t)) - \frac{p-1}{t^{\lambda/(p-1)}}(w(t) + f(t))^q$$

for $t > T$, where $w(t) = Kt^{\lambda - p + 1}$. Then the function

$$y(t) = \exp\left(\int_T^t \left(\frac{w(s) + f(s)}{s^\lambda}\right)^{1/(p-1)} ds\right)$$

satisfies inequality $\mathcal{L}[y] \le 0$ (with $r(t) \equiv 1$) for $t > T$, $\mathcal{L}$ being defined in Theorem 1.2.1. Hence (3.1.1) is nonoscillatory by Theorem 2.2.1. □

3.4 Notes and references

The results of Subsection 3.1.1 are mostly extensions of classical linear criteria (see e.g. Swanson's book [341]). Since the detailed quotations are given within the text, here we only note that these problems have been studied, using various methods, under different conditions, in many works, see e.g. [102, 180, 197, 219, 227] by Došlý, Hoshino, Imabayashi, Kandelaki, Kusano, Lomtatidze, Naito, Ogata, Tanigawa, Ugulava and Yoshida. Concerning Subsection 3.1.2, Theorems 3.1.4 and 3.1.5 are due to Došlý [102], while Theorem 3.1.6 was proved by Kusano and Naito [218]. Theorem 3.1.7 is extracted from Kandelaki, Lomtatidze, Ugulava [197]. Besides this, Section 3.1 is supplemented by several new observations.

Coles type criteria (Subsection 3.2.1) are taken from Li and Yeh [244]. Generalized Kamenev criterion, Theorem 3.2.3, was proved also by Li and Yeh in [239]. The results concerning the generalized H-function averaging technique were established in Li and Yeh [243]. Related results are given in the papers of Manojlović, [265] Li, Zhong, Fan [254] and Yang, Cheng [370].

The results of Section 3.3 are adopted for the equation of the form (3.1.1), originally for (3.3.1), and come from Kandelaki, Lomtatidze, Ugulava [197] and Lomtatidze [262]. Only Corollary 3.3.1 is originally due to Mirzov [292].

Finally note that various oscillation and nonoscillation criteria for (1.1.1) can also be found in the papers of Kusano, Li, and Wang [228, 237].

CHAPTER 4

NONOSCILLATORY SOLUTIONS

This chapter is devoted to nonoscillatory equations of the form (1.1.1). In the first section we focus our attention to the asymptotic analysis of nonoscillatory solutions of (1.1.1). We show that these solutions can be classified into various classes according to their asymptotic behavior at ∞, and integral conditions on the functions r, c are given which guarantee that (1.1.1) does/does not have a solution in a given class. The next section contains a comprehensive treatment of the concept of the principal solution of (1.1.1). The principal solution of (1.1.2) plays the fundamental role in the oscillation theory of linear equations and we show that the half-linear extension of this concept is of the same importance. The last section is devoted to nonoscillatory equations whose solutions have a regular growth at ∞.

4.1 Asymptotic of nonoscillatory solutions

This section is devoted to the investigation of asymptotic properties of nonoscillatory solutions of equation (1.1.1) when the function c does not change its sign. In this case, it is possible to associate with (1.1.1) its reciprocal equation

$$\left(\frac{1}{\Phi^{-1}(c(t))}\Phi^{-1}(u')\right)' + r^{1-q}(t)\Phi^{-1}(u) = 0. \tag{4.1.1}$$

Recall that the so-called reciprocity principle says that (4.1.1) is nonoscillatory if and only if (1.1.1) is nonoscillatory, see Subsection 1.2.8. Note also that if $c(t) \leq 0$ for large t then (1.1.1) is nonoscillatory since the equation $(r(t)\Phi(x'))' = 0$ is its nonoscillatory majorant.

4.1.1 Integral conditions and classification of solutions

If c is different from zero for large t, then all solutions of nonoscillatory equation (1.1.1) are eventually monotone, this result is formulated in the next statement.

Lemma 4.1.1. *Let $c(t) \neq 0$ for large t and x be a solution of nonoscillatory equation* (1.1.1). *Then either $x(t)x'(t) > 0$ or $x(t)x'(t) < 0$ for large t.*

Proof. The monotonicity of x follows from the reciprocity principle which ensures that the so-called *quasiderivative* $x^{[1]} := r\Phi(x')$ does not change its sign for large t. □

Then it is possible, *a-priori*, to divide the set of solutions x of (1.1.1) into the following two classes :

$$\mathbb{M}^+ = \{x : \exists t_x \geq 0 : \ x(t)x'(t) > 0 \text{ for } t > t_x\},$$
$$\mathbb{M}^- = \{x : \exists t_x \geq 0 : \ x(t)x'(t) < 0 \text{ for } t > t_x\}.$$

Clearly, solutions in $\mathbb{M}^+$ are eventually either positive increasing or negative decreasing and solutions in $\mathbb{M}^-$ are either positive decreasing or negative increasing. The existence of solutions in these classes depends on the sign of the function c, as the following results show.

Lemma 4.1.2. *Assume $c(t) < 0$ for large t.*

(i) Equation (1.1.1) *has solutions in the class $\mathbb{M}^-$. More precisely, for every pair $(t_0, a) \in [0, \infty) \times \mathbb{R} \setminus \{0\}$ there exists a solution x of* (1.1.1) *in the class $\mathbb{M}^-$ such that $x(t_0) = a$.*

(ii) Equation (1.1.1) *has solutions in the class $\mathbb{M}^+$. More precisely, for every pair $(a_0, a_1) \in \mathbb{R}^2$, $a_0\, a_1 > 0$ and for any t_0 sufficiently large, there exists a solution x of* (1.1.1) *in the class $\mathbb{M}^+$ such that $x(t_0) = a_0$, $x'(t_0) = a_1$.*

Proof. Claim (i) follows, for instance, from [72, Theorem 1] and [292, Theorems 9.1, 9.2]. Concerning the claim (ii), let x be a solution of (1.1.1) such that $x(0)x'(0) > 0$. Since the auxiliary function

$$F_x(t) = r(t)\Phi(x'(t))x(t) \tag{4.1.2}$$

is nondecreasing, we obtain $x(t)x'(t) > 0$ for $t > 0$. The assertion follows taking into account that every solution is continuable up to ∞, see Section 1.1. □

In the opposite case, i.e., when $c(t) > 0$ for large t, the existence in the classes $\mathbb{M}^+, \mathbb{M}^-$ may be characterized by means of the convergence or divergence of the following two integrals (the notation will be used also in case $c(t) < 0$)

$$J_r = \int_0^\infty r^{1-q}(t)\, dt, \qquad J_c = \int_0^\infty |c(t)|\, dt.$$

Lemma 4.1.3. *Assume that* (1.1.1) *is nonoscillatory and $c(t) > 0$ for large t.*

(i) If $J_c = \infty$, *then* $\mathbb{M}^+ = \emptyset$.

(ii) If $J_r = \infty$, *then* $\mathbb{M}^- = \emptyset$.

Proof. (i) Let x be a solution of (1.1.1) in the class $\mathbb{M}^+$ and, without loss of generality, suppose $x(t) > 0, x'(t) > 0$ for $t \geq T \geq 0$. From (1.1.1) we obtain for $t \geq T$

$$r(t)\Phi(x'(t)) \leq r(T)\Phi(x'(T)) - \Phi(x(T)) \int_T^t c(s)ds$$

that gives a contradiction as $t \to \infty$. Claim (ii) follows by applying (i) to (4.1.1) and using the reciprocity principle. □

Lemma 4.1.4. *Assume* $c(t) > 0$ *for large* t.

(i) If (1.1.1) *is nonoscillatory and* $J_r = \infty, J_c < \infty$, *then* $\mathbb{M}^+ \neq \emptyset$.

(ii) If (1.1.1) *is nonoscillatory and* $J_r < \infty, J_c = \infty$, *then* $\mathbb{M}^- \neq \emptyset$.

(iii) If $J_r < \infty, J_c < \infty$, *then* $\mathbb{M}^+ \neq \emptyset, \mathbb{M}^- \neq \emptyset$.

Proof. Claims (i), (ii) follow from Lemma 4.1.3. The assertion (iii) follows, for instance, as a particular case from [364, Theorem 3.1, Theorem 3.3] and their proofs by observing that certain assumptions in the paper [364] (which deals with a more general equation than (1.1.1)) are not necessary in the half-linear case. □

In view of Lemma 4.1.3, if (1.1.1) is nonoscillatory, c is eventually positive and $J_r + J_c = \infty$, then all solutions of (1.1.1) belong to the same class ($\mathbb{M}^+$ or $\mathbb{M}^-$). In addition, from the same Lemma 4.1.3, the well known Leighton-Wintner type oscillation result can be obtained, see Theorem 1.2.9.

In both cases $c > 0$ and $c < 0$ eventually, the classes $\mathbb{M}^+, \mathbb{M}^-$ may be divided, *a-priori*, into the following four subclasses, which are mutually disjoint:

$$\begin{aligned}
\mathbb{M}_B^- &= \{x \in \mathbb{M}^- : \lim_{t\to\infty} x(t) = \ell_x \neq 0\}, \\
\mathbb{M}_0^- &= \{x \in \mathbb{M}^- : \lim_{t\to\infty} x(t) = 0\}, \\
\mathbb{M}_B^+ &= \{x \in \mathbb{M}^+ : \lim_{t\to\infty} x(t) = \ell_x,\ |\ell_x| < \infty\}, \\
\mathbb{M}_\infty^+ &= \{x \in \mathbb{M}^+ : \lim_{t\to\infty} |x(t)| = \infty\}.
\end{aligned}$$

In the following subsections, we consider both cases $c(t) > 0, c(t) < 0$ and we describe the existence of solutions of (1.1.1) in the above classes in terms of certain integral conditions. Similarly to the linear case, we are going to show that the convergence or divergence of the integrals

$$J_1 = \lim_{T\to\infty} \int_0^T r^{1-q}(t)\Phi^{-1}\Big(\int_0^t |c(s)|\, ds\Big)\, dt, \tag{4.1.3}$$

$$J_2 = \lim_{T\to\infty} \int_0^T r^{1-q}(t)\Phi^{-1}\Big(\int_t^T |c(s)|\, ds\Big)\, dt,$$

fully characterize the above four classes.

The next lemma describes relations between J_1, J_2, J_r, J_c.

Lemma 4.1.5. *The following statements hold.*

(a) *If* $J_1 < \infty$, *then* $J_r < \infty$.

(b) *If* $J_2 < \infty$, *then* $J_c < \infty$.

(c) *If* $J_2 = \infty$, *then* $J_r = \infty$ *or* $J_c = \infty$.

(d) *If* $J_1 = \infty$, *then* $J_r = \infty$ *or* $J_c = \infty$.

(e) $J_1 < \infty$ *and* $J_2 < \infty$ *if and only if* $J_r < \infty$ *and* $J_c < \infty$.

Proof. Claim (a): Let $t_1 \in (0, T)$. Since

$$\int_0^T r^{1-q}(s)\Phi^{-1}\Big(\int_0^s |c(t)|\, dt\Big)\, ds$$
$$> \int_0^{t_1} r^{1-q}(s)\Phi^{-1}\Big(\int_0^s |c(t)|\, dt\Big)\, ds + \Phi^{-1}\Big(\int_0^{t_1} |c(s)|\, ds\Big)\int_{t_1}^T r^{1-q}(s)\, ds,$$

the assertion follows. Claim (b) follows in a similar way. Claims (c), (d) follow from the inequalities

$$\int_0^T r^{1-q}(t)\Phi^{-1}\Big(\int_t^T |c(s)|\, ds\Big)\, dt \le \int_0^T r^{1-q}(t)\, dt\ \Phi^{-1}\Big(\int_0^T |c(s)|\, ds\Big),$$
$$\int_0^T r^{1-q}(t)\Phi^{-1}\Big(\int_0^t |c(s)|\, ds\Big)\, dt \le \int_0^T r^{1-q}(t)\, dt\ \Phi^{-1}\Big(\int_0^T |c(s)|\, ds\Big).$$

Finally, the claim (e) immediately follows from (a)–(d). □

4.1.2 The case c negative

We start by noting that if $c(t) < 0$ in the whole interval $[0, \infty)$, then for any solution $x \in \mathbb{M}^-$ we have $x(t)x'(t) < 0$ on $[0, \infty)$. This property can be proved using the auxiliary function F_x given in (4.1.2). Since, as claimed, F is a nondecreasing function and x is not eventually constant, there are only two possibilities: (a) F_x does not have zeros; (b) there exists $t_x \ge \alpha_x$ such that $F_x(t) > 0$ for all $t > t_x$. Thus the assertion follows.

The following hold.

Theorem 4.1.1. *Let* $c(t) < 0$ *for large* t.

(i) *Equation* (1.1.1) *has solutions in the class* $\mathbb{M}_B^-$ *if and only if* $J_2 < \infty$.

(ii) *Equation* (1.1.1) *has solutions in the class* $\mathbb{M}_B^+$ *if and only if* $J_1 < \infty$.

Proof. Claim (i) "⇒": Let $x \in \mathbb{M}_B^-$. Without loss of generality we can assume $x(t) > 0, x'(t) < 0$ for $t \ge T \ge 0$. Integrating (1.1.1) in (t, ∞), $t > T$, we obtain

$$-\lambda_x - r(t)\Phi(x'(t)) = \int_t^\infty |c(\tau)|\Phi(x(\tau))d\tau, \tag{4.1.4}$$

where $-\lambda_x = \lim_{t\to\infty}[r(t)\Phi(x'(t))]$. Since $x(\tau) > x(\infty) > 0$ and $\lambda_x \geq 0$, (4.1.4) implies

$$-r(t)\Phi(x'(t)) \geq \Phi(x(\infty)) \int_t^\infty |c(\tau)|d\tau.$$

Hence

$$x(t) \leq x(T) - x(\infty) \int_T^t \Phi^{-1}\Big(\frac{1}{r(s)} \int_s^\infty |c(\tau)|d\tau\Big)\, ds.$$

As $t \to \infty$ we obtain the assertion.

Claim (i) "$\Leftarrow$": Choose $t_0 \geq 0$ such that

$$\int_{t_0}^\infty \Phi^{-1}\Big(\frac{1}{r(t)} \int_t^\infty |c(\tau)|d\tau\Big)\, dt \leq \frac{1}{2}. \tag{4.1.5}$$

Denote by $C[t_0,\infty)$ the Fréchet space of all continuous functions on $[t_0,\infty)$ endowed with the topology of the uniform convergence on compact subintervals of $[t_0,\infty)$. Let Ω be the nonempty subset of $C[t_0,\infty)$ given by

$$\Omega = \Big\{u \in C[t_0,\infty) : \frac{1}{2} \leq u(t) \leq 1\Big\}. \tag{4.1.6}$$

Clearly Ω is bounded, closed and convex. Now consider the operator $T : \Omega \to C[t_0,\infty)$ which assigns to any $u \in \Omega$ the continuous function $T(u) = y_u$ given by

$$y_u(t) = T(u)(t) = \frac{1}{2} + \int_t^\infty \Phi^{-1}\Big(\frac{1}{r(s)} \int_s^\infty |c(\tau)|\Phi(u(\tau))d\tau\Big)\, ds. \tag{4.1.7}$$

We have

$$\frac{1}{2} \leq T(u)(t) \leq \frac{1}{2} + \int_t^\infty \Phi^{-1}\Big(\frac{1}{r(s)} \int_s^\infty |c(\tau)|d\tau\Big)\, ds$$

which implies, by virtue of (4.1.5), $T(\Omega) \subseteq \Omega$. In order to apply the Tychonov fixed point theorem to the operator T, it is sufficient to prove that T is continuous in $\Omega \subseteq C[t_0,\infty)$ and that $T(\Omega)$ is relatively compact in $C[t_0,\infty)$. Let $\{u_j\}, j \in \mathbb{N}$, be a sequence in Ω which is convergent to $\bar{u}$ in $C[t_0,\infty), \bar{u} \in \overline{\Omega} = \Omega$. Since for $s \geq t_0$

$$\Phi^{-1}\Big(\frac{1}{r(s)} \int_s^\infty |c(\tau)|\Phi(u_j(\tau))d\tau\Big) \leq \Phi^{-1}\left(\frac{1}{r(s)} \int_s^\infty |c(\tau)|d\tau\right) < \infty,$$

the Lebesgue Dominated Convergence theorem gives the continuity of T in Ω. It remains to prove that $T(\Omega)$ is relatively compact in $C[t_0,\infty)$, i.e., that functions in $T(\Omega)$ are equibounded and equicontinuous on every compact subinterval of $[t_0,\infty)$. The equiboundedness easily follows taking into account that Ω is a bounded subset of $C[t_0,\infty)$. In order to prove the equicontinuity, for any $u \in \Omega$ we have

$$\begin{aligned} 0 < -(T(u)(t))' &= \Phi^{-1}\left(\frac{1}{r(t)} \int_t^\infty |c(\tau)|\Phi(u(\tau))d\tau\right) \\ &\leq \Phi^{-1}\left(\frac{1}{r(t)} \int_t^\infty |c(\tau)|d\tau\right), \end{aligned} \tag{4.1.8}$$

which implies that functions in $T(\Omega)$ are equicontinuous on every compact subinterval of $[t_0, \infty)$. From the Tychonov fixed point theorem, there exists $x \in \Omega$ such that $x = T(x)$ or, from (4.1.7),

$$x(t) = \frac{1}{2} + \int_t^\infty \Phi^{-1}\left(\frac{1}{r(s)}\int_s^\infty |c(\tau)|\Phi(x(\tau))d\tau\right) ds.$$

It is easy to show that x is a positive solution of (1.1.1) in $[t_0, \infty)$ and, from (4.1.8), $x'(t) < 0$. Finally, clearly x satisfies the inequality $x(t)x'(t) < 0$ in its maximal interval of existence and the proof of claim (i) is complete.

Claim (ii) "$\Rightarrow$": Assume, by contradiction, $J_1 = \infty$. Without loss of generality let x be a solution of (1.1.1) in the class $\mathbb{M}_B^+$ such that $x(t) > 0, x'(t) > 0$ for $t \geq t_0 \geq 0$. Integrating (1.1.1) on (t_0, t) we obtain for $t > t_0$

$$x^{[1]}(t) = x^{[1]}(t_0) + \int_{t_0}^t |c(s)|\Phi(x(s))ds > \Phi(x(t_0))\int_{t_0}^t |c(s)|ds,$$

where $x^{[1]} = r\Phi(x')$. Hence

$$x'(t) > x(t_0)\Phi^{-1}\left(\frac{1}{r(t)}\int_{t_0}^t |c(s)|ds\right).$$

Integrating again over (t_0, t) we obtain a contradiction.

Claim (ii) "$\Leftarrow$": The argument is similar to that given in Claim (i) "$\Leftarrow$". It is sufficient to consider in the same set Ω, defined in (4.1.6), the operator $T : \Omega \to C[t_0, \infty)$ given by

$$y_u(t) = T(u)(t) = \frac{1}{2} + \int_{t_0}^t \Phi^{-1}\left(\frac{1}{r(s)}\int_{t_0}^s |c(\tau)|\Phi(u(\tau))d\tau\right) ds$$

and to apply the Tychonov fixed point theorem. □

Theorem 4.1.2. *Let $c(t) < 0$ for large t.*

(i) If $J_1 = \infty$ and $J_2 < \infty$, then $\mathbb{M}_0^- = \emptyset$.

(ii) If $J_1 < \infty$, then $\mathbb{M}_\infty^+ = \emptyset$.

Proof. Claim (i). Let x be a solution of (1.1.1) in the class $\mathbb{M}^-$ such that $x(t) > 0$, $x'(t) < 0$ for $t \geq T$ and $\lim_{t\to\infty} x(t) = 0$. By Lemma 4.1.5, $J_r = \infty$ and thus, by [57, Lemma 1], $\lim_{t\to\infty} r(t)\Phi(x'(t)) = 0$. Taking into account this fact and integrating (1.1.1) over (t, ∞), $t > T$, we obtain

$$\frac{x'(t)}{x(t)} > -\Phi^{-1}\left(\frac{1}{r(t)}\int_t^\infty |c(\tau)|d\tau\right).$$

Integrating over (T, t) we have

$$\log\frac{x(t)}{x(T)} > -\int_T^t \Phi^{-1}\left(\frac{1}{r(s)}\int_s^\infty |c(\tau)|d\tau\right) ds,$$

from which, as $t \to \infty$, we obtain a contradiction.

Claim (ii). Let $x \in \mathbb{M}_\infty^+$ and assume $x(t) > 0, x'(t) > 0$ for $t \geq T \geq 0$. From (1.1.21) we have (with $w = r\Phi(x')/\Phi(x)$)

$$\begin{aligned} \frac{r(t)\Phi(x'(t))}{\Phi(x(t))} &= -(p-1)\int_T^t r^{1-q}(s)|w(s)|^q\, ds + k - \int_T^t c(s)\, ds \\ &\leq k + \int_T^t |c(s)|\, ds, \end{aligned} \tag{4.1.9}$$

where $k = r(T)\Phi(x'(T))/\Phi(x(T))$. If $J_c < \infty$, then there exists a positive constant k_1 such that

$$\frac{r(t)\Phi(x'(t))}{\Phi(x(t))} \leq k_1$$

or

$$\frac{x'(t)}{x(t)} \leq \Phi^{-1}(k_1)r^{1-q}(t).$$

Integrating again over (T, t) we obtain

$$\log\frac{x(t)}{x(T)} \leq \Phi^{-1}(k_1)\int_T^t r^{1-q}(s)\, ds$$

which implies $x \in \mathbb{M}_B^+$, i.e., a contradiction. If $J_c = \infty$, choose $t_1 > T$ such that $k < \int_T^{t_1} c(s)\, ds$. Then from (4.1.9) we obtain for $t \geq t_1$

$$\frac{r(t)\Phi(x'(t))}{\Phi(x(t))} \leq 2\int_T^t |c(s)|\, ds,$$

or

$$\frac{x'(t)}{x(t)} \leq \Phi^{-1}(2)\Phi^{-1}\Big(\frac{1}{r(t)}\int_T^t |c(s)|\, ds\Big).$$

Integrating over (t_1, t) we have

$$\log\frac{x(t)}{x(t_1)} \leq \Phi^{-1}(2)\int_T^t \Phi^{-1}\Big(\frac{1}{r(s)}\int_T^s |c(\tau)|\, d\tau\Big)\, ds,$$

which gives the assertion. □

Theorem 4.1.3. *Let $c(t) < 0$ for large t. If $J_1 < \infty$ and $J_2 < \infty$, then equation* (1.1.1) *has solutions in both classes $\mathbb{M}_0^-$ and $\mathbb{M}_B^-$.*

Proof. The statement $\mathbb{M}_B^- \neq \emptyset$ follows from Theorem 4.1.1. The existence in the class $\mathbb{M}_0^-$ can be proved by using a similar argument as that given in the proof of Theorem 4.1.1. It is sufficient to consider the set

$$\Omega = \Big\{u \in C[t_0, \infty) : 0 \leq u(t) \leq \int_t^\infty r^{1-q}(s)\, ds\Big\}$$

and the operator $T : \Omega \to C[t_0, \infty)$ given by

$$y_u(t) = T(u)(t) = \int_t^{\infty} r^{1-q}(t)\Phi^{-1}\left(1 - \int_{t_0}^{s} |c(\tau)|\Phi(u(\tau))d\tau\right) ds$$

and to apply the Tychonov fixed point theorem. Details are omitted. □

Remark 4.1.1. The behavior of quasiderivatives of solutions (i.e., of expressions $x^{[1]} = r\Phi(x')$) plays an important role in the study of principal solutions, especially in their limit characterization, see the next section. Concerning the solution $x \in \mathbb{M}_B^-$, defined as a fixed point in the proof of Theorem 4.1.1, we have $\lim_{t\to\infty} x^{[1]}(t) = 0$. Concerning the solution $x \in \mathbb{M}_0^-$, defined in the proof of Theorem 4.1.3, it is easy to show that $\lim_{t\to\infty} x^{[1]}(t) = c_x < 0$. Indeed, the limit

$$\lim_{t\to\infty} x'(t)r^{q-1}(t) \tag{4.1.10}$$

exists finite and it is different from zero, because

$$\begin{aligned} -x'(t) &= r^{1-q}(t)\Phi^{-1}\left(1 - \int_{t_0}^{t} |c(\tau)|\Phi(u(\tau))d\tau\right) \\ &\geq \Phi^{-1}\left(\frac{1}{2}\right)\Phi^{-1}\left(\frac{1}{r(t)}\right) = 2^{1-q}r^{1-q}(t). \end{aligned}$$

and the function $x'r^{q-1}$ is negative increasing.

From Theorems 4.1.1, 4.1.2, 4.1.3, we can summarize the situation in the following way. Clearly, as regards the convergence or divergence of J_1, J_2, the possible cases are the following:

$$\begin{array}{lll} (\mathrm{A}_1) & J_1 = \infty, & J_2 = \infty, \\ (\mathrm{A}_2) & J_1 = \infty, & J_2 < \infty, \\ (\mathrm{A}_3) & J_1 < \infty, & J_2 = \infty, \\ (\mathrm{A}_4) & J_1 < \infty, & J_2 < \infty. \end{array}$$

Then the following result holds.

Theorem 4.1.4. *Let $c(t) < 0$ for large t.*

(i) Assume case (A_1). Then any solution of (1.1.1) in the class $\mathbb{M}^-$ tends to zero as $t \to \infty$ and any solution of (1.1.1) in the class $\mathbb{M}^+$ is unbounded.

(ii) Assume case (A_2). Then any solution of (1.1.1) in the class $\mathbb{M}^-$ tends to a nonzero limit as $t \to \infty$ and any solution of (1.1.1) in the class $\mathbb{M}^+$ is unbounded.

(iii) Assume case (A_3). Then any solution of (1.1.1) in the class $\mathbb{M}^-$ tends to zero as $t \to \infty$ and any solution of (1.1.1) in the class $\mathbb{M}^+$ is bounded.

(iv) Assume case (A_4). Then both solutions of (1.1.1) converging to zero and solutions of (1.1.1) tending to a nonzero limit (as $t \to \infty$) exist in the class $\mathbb{M}^-$. Solutions of (1.1.1) in the class $\mathbb{M}^+$ are bounded.

From Theorem 4.1.4 we obtain immediately the following statement which generalizes a well-know result stated for the linear equation in [281, Theorems 3 and 4]; see also [174, Chapters VI, XI]).

Corollary 4.1.1. *Let $c(t) < 0$ for large t.*

(a) Any solution x of (1.1.1) *in the class $\mathbb{M}^-$ tends to zero as $t \to \infty$ if and only if $J_2 = \infty$.*

(b) Any solution of (1.1.1) *is bounded if and only if $J_1 < \infty$.*

Following another classification used in [299, 347], we distinguish the next types of eventually positive solutions x of (1.1.1) (clearly a similar classification holds for eventually negative solutions):

$$
\begin{array}{ll}
\text{Type (1)} & \lim_{t\to\infty} x(t) = 0, \quad \lim_{t\to\infty} x^{[1]}(t) = 0; \\
\text{Type (2)} & \lim_{t\to\infty} x(t) = 0, \quad \lim_{t\to\infty} x^{[1]}(t) = c_1 < 0; \\
\text{Type (3)} & \lim_{t\to\infty} x(t) = c_0 > 0, \quad \lim_{t\to\infty} x^{[1]}(t) = c_1 \le 0; \\
\text{Type (4)} & \lim_{t\to\infty} x(t) = c_0 > 0, \quad \lim_{t\to\infty} x^{[1]}(t) = c_1 > 0; \\
\text{Type (5)} & \lim_{t\to\infty} x(t) = c_0 > 0, \quad \lim_{t\to\infty} x^{[1]}(t) = \infty; \\
\text{Type (6)} & \lim_{t\to\infty} x(t) = \infty, \quad \lim_{t\to\infty} x^{[1]}(t) = c_1; \\
\text{Type (7)} & \lim_{t\to\infty} x(t) = \infty, \quad \lim_{t\to\infty} x^{[1]}(t) = \infty.
\end{array}
$$

Eventually positive solutions in $\mathbb{M}^-$ are of Types (1)–(3), eventually positive solutions in $\mathbb{M}^+$ are of Types (4)–(7). From Theorem 4.1.4 and the reciprocity principle (see Subsection 1.2.8), necessary and/or sufficient conditions for their existence can be obtained. To this end observe that the integral J_r [resp. J_c] for (1.1.1) plays the same role as J_c [resp. J_r] for the reciprocal equation (4.1.1). Similarly, for the reciprocal equation (4.1.1) the integrals J_1, J_2 become

$$
R_1 = \lim_{T\to\infty} \int_0^T |c(t)|\, \Phi\Big(\int_0^t r^{1-q}(s)\, ds\Big) dt,
$$
$$
R_2 = \lim_{T\to\infty} \int_0^T |c(t)|\, \Phi\Big(\int_t^T r^{1-q}(s)\, ds\Big) dt,
$$

respectively. Then the following holds.

Theorem 4.1.5. *Let $c(t) < 0$ for large t. Then the following statements hold:*

(a) Every eventually positive solution of (1.1.1) *in $\mathbb{M}^-$ is of Type (1) if and only if $J_2 = \infty$ and $R_2 = \infty$.*

(b) Eq. (1.1.1) *has solutions of Type (2) if and only if $R_2 < \infty$.*

(c) Eq. (1.1.1) *has solutions of Type (3) if and only if $J_2 < \infty$.*

(d) Eq. (1.1.1) *has solutions of Type (4) if and only if* $J_1 < \infty$ *and* $R_1 < \infty$.

(e) Eq. (1.1.1) *has solutions of Type (5) if and only if* $J_1 < \infty$ *and* $R_1 = \infty$.

(f) Eq. (1.1.1) *has solutions of Type (6) if and only if* $J_1 = \infty$ *and* $R_1 < \infty$.

(g) Every eventually positive solution in $\mathbb{M}^+$ *is of Type (7) if and only if* $J_1 = \infty$ *and* $R_1 = \infty$.

4.1.3 Uniqueness in $\mathbb{M}^-$.

The uniqueness in the class $\mathbb{M}^-$ plays a crucial role in the study of the limit characterization of principal solutions (see Theorem 4.2.7 in the next section).

Lemma 4.1.2 states that, when c is eventually negative, the class $\mathbb{M}^-$ is nonempty. In the linear case, the assumption

$$\int_0^\infty \left(\frac{1}{r(t)} + |c(t)|\right) dt = \infty \tag{4.1.11}$$

is necessary and sufficient for uniqueness in $\mathbb{M}^-$ of such a solution; given t_0 sufficiently large and $x_0 > 0$, there exists the unique solution $x \in \mathbb{M}^-$ satisfying $x(t_0) = x_0$, see [281, Theorems 3,4]. We will show that also for (1.1.1) such a property is assured by a natural extension of condition (4.1.11).

Theorem 4.1.6. *Let* $c(t) < 0$ *for large* t. *For any* $(t_0, x_0) \in [0,\infty) \times \mathbb{R} \setminus \{0\}$, *there exists a unique solution* x *of* (1.1.1) *in the class* $\mathbb{M}^-$ *such that* $x(t_0) = x_0$ *if and only if*

$$\int_0^\infty \left(r^{1-q}(t) + |c(t)|\right) dt = \infty. \tag{4.1.12}$$

The following result can be easily proved and will be useful in the proof of Theorem 4.1.6.

Lemma 4.1.6. *Let* $c(t) < 0$ *for large* t. *If* $J_r = \infty$, *then for every solution* x *of* (1.1.1) *in the class* $\mathbb{M}^-$ *we have* $\lim_{t\to\infty} x^{[1]}(t) = 0$.

Proof of Theorem 4.1.6. *Necessity.* Assume (4.1.12) does not hold, i.e.,

$$\int_0^\infty |c(\tau)| d\tau < \infty, \qquad \int_0^\infty r^{1-q}(t)\, dt < \infty,$$

and let t_0 be so large that

$$\Phi^{-1}\left(\int_{t_0}^\infty |c(\tau)| d\tau\right) \int_{t_0}^\infty r^{1-q}(t)\, dt < \frac{1}{\Phi^{-1}(2)}. \tag{4.1.13}$$

Consider the solutions x_1, x_2 of (1.1.1) with the initial values $x_1(t_0) = x_2(t_0) = 1$ and

(4.1.14)

$$x_1'(t_0) = -\Phi^{-1}\left(\frac{c_1}{r(t_0)} \int_{t_0}^\infty |c(\tau)| d\tau\right), \quad x_2'(t_0) = -\Phi^{-1}\left(\frac{c_2}{r(t_0)} \int_{t_0}^\infty |c(\tau)| d\tau\right),$$

where c_i are positive constants such that $c_1 \neq c_2$ and

$$1 \leq c_i \leq 2. \tag{4.1.15}$$

Let us show that $x_i \in \mathbb{M}^-$, $i = 1, 2$. It is easy to show that solutions x_i are positive decreasing on $[0, t_0]$. In order to prove that $x_i \in \mathbb{M}^-$, it will be sufficient to show that $x_i(t)x_i'(t) < 0$ for any $t \geq t_0$. Clearly solutions x_i are positive decreasing in a right neighborhood of t_0. Assume there exists $t_i > t_0$ such that $x_i(t_i)x_i'(t_i) = 0, x_i(t) > 0, x_i'(t) < 0$ for $t_0 \leq t < t_i$, $i = 1, 2$. Integrating (1.1.1) on (t_0, t_i) we have

$$r(t_i)\Phi(x_i'(t_i)) - r(t_0)\Phi(x_i'(t_0)) = \int_{t_0}^{t_i} |c(\tau)|\Phi(x_i(\tau))d\tau. \tag{4.1.16}$$

If $x_i'(t_i) = 0$, from (4.1.14) and (4.1.16) we obtain

$$c_i \int_{t_0}^{\infty} |c(\tau)|d\tau = \int_{t_0}^{t_i} |c(\tau)|\Phi(x_i(\tau))d\tau \leq \Phi(x_i(t_0)) \int_{t_0}^{t_i} |c(\tau)|d\tau = \int_{t_0}^{t_i} |c(\tau)|d\tau$$

which implies

$$\int_{t_i}^{\infty} |c(\tau)|d\tau \leq 0,$$

that is a contradiction. Now suppose $x_i(t_i) = 0$. For $t \in (t_0, t_i)$ from

$$r(t)\Phi(x_i'(t)) \geq r(t_0)\Phi(x_i'(t_0)) = -c_i \int_{t_0}^{\infty} |c(\tau)|dt,$$

we obtain

$$x_i'(t) \geq -\Phi^{-1}(c_i)r^{1-q}(t)\Phi^{-1}\Big(\int_{t_0}^{\infty} |c(\tau)|d\tau\Big)$$

or

$$x_i(t_i) - x(t_0) = -1 \geq -\Phi^{-1}(c_i)\Phi^{-1}\left(\int_{t_0}^{\infty} |c(\tau)|d\tau\right) \int_{t_0}^{t_i} r^{1-q}(t)\, dt.$$

Thus, by virtue of (4.1.15),

$$1 \leq \Phi^{-1}(2)\Phi^{-1}\left(\int_{t_0}^{\infty} |c(\tau)|d\tau\right) \int_{t_0}^{\infty} r^{1-q}(t)\, dt,$$

which contradicts (4.1.13) and the necessity of (4.1.12) is proved.

Sufficiency. Let us show that for any $(t_0, x_0) \in [0, \infty) \times \mathbb{R} \setminus \{0\}$, there exists at most one solution x of (1.1.1) in the class $\mathbb{M}^-$ such that $x(t_0) = x_0$ when $J_r = \infty$ or $J_c = \infty$. Let x, y be two solutions of (1.1.1) in the class $\mathbb{M}^-$ such that $x(t_0) = y(t_0), x'(t_0) > y'(t_0)$. Consider the function d given by $d(t) = x(t) - y(t)$. Then $d(t_0) = 0, d'(t_0) > 0$. We claim that d does not have positive points of maximum greater than t_0, i.e.,

$$d(t) > 0, \quad d'(t) > 0 \ \text{ for } t > t_0.$$

Assume there exists $t_1 > t_0$ such that $d(t_1) > 0$, $d'(t_1) = 0$ and $d'(t) > 0$ in a suitable left neighborhood I of t_1. Without loss of generality suppose that $d(t) > 0$ for $t \in I$. Now consider the function G given by

$$G(t) = r(t)[\Phi(x'(t)) - \Phi(y'(t))].$$

Hence $G(t_1) = 0$. Taking into account that Φ is increasing and $d'(t) > 0$, we have $G(t) > 0$, $t \in I$. In addition, from

$$G'(t) = |c(t)|[\Phi(x(t)) - \Phi(y(t))],$$

we obtain $G'(t) > 0$, $t \in I$, which gives a contradiction, because $G(t_1) = 0$. Hence the function d is increasing.

If $J_c = \infty$ then, by Lemma 4.1.5 we have $J_2 = \infty$. Then, in view of Corollary 4.1.1-(a), we obtain $d(\infty) = 0$, that is a contradiction. If $J_r = \infty$, then taking into account that $d'(t) > 0$ for $t > t_0$, the function G satisfies $G(t) > 0$, $G'(t) > 0$ for $t > t_0$ and, by Lemma 4.1.6, $\lim_{t\to\infty} G(t) = 0$, that is a contradiction. Finally the existence of at least one solution $x \in \mathbb{M}^-$ such that $x(t_0) = x_0$ is assured by Lemma 4.1.2-(i). □

4.1.4 The case c positive

As already stated before, when c is eventually positive, equation (1.1.1) may be either oscillatory or nonoscillatory. In this subsection, similarly to Subsection 4.1.2, we provide a complete description of the asymptotic behavior of nonoscillatory solutions of (1.1.1) also when c is positive.

Lemma 4.1.7. *Let $c(t) > 0$ for large t and $J_r = \infty$.*

(i) If $J_2 = \infty$, then $\mathbb{M}_B^+ = \emptyset$.

(ii) If $J_2 < \infty$, then equation (1.1.1) is nonoscillatory and $\mathbb{M}_B^+ \neq \emptyset$.

Proof. In view of Lemma 4.1.3-(ii) any nonoscillatory solution of (1.1.1) is in the class $\mathbb{M}^+$. Then claims (i) and (ii) follow from [180, Theorem 4.2]. □

The following "uniqueness" result will be useful in the proof of the existence of unbounded solutions. It is, in some sense, an analogue to Theorem 4.1.6 and its proof can be found in [180, Theorem 4.3].

Theorem 4.1.7. *Let $c(t) > 0$ for large t. Let $\eta \neq 0$ be a given constant and assume $J_r = \infty, J_2 < \infty$. Then there exists a unique solution x of (1.1.1), $x \in \mathbb{M}^+$, such that $\lim_{t\to\infty} x(t) = \eta$.*

By using the previous result we obtain the next statement.

Theorem 4.1.8. *Let $c(t) > 0$ for large t and assume $J_r = \infty, J_2 < \infty$. Then (1.1.1) has unbounded solutions, i.e, $\mathbb{M}_\infty^+ \neq \emptyset$.*

Proof. Assume, by contradiction, that $M^+_\infty = \emptyset$. In view of Lemma 4.1.7, let u be a solution of (1.1.1) in the class $\mathbb{M}^+_B$ and let x be another solution of (1.1.1) such that $x(0) = u(0), x'(0) \neq u'(0)$. Hence $x \neq u$ and from Lemma 4.1.3-(ii) we have $x \in \mathbb{M}^+_B$. In view of Theorem 4.1.7 we have $u(\infty) \neq x(\infty)$. Now consider the solution w of (1.1.1) given by

$$w(t) = \frac{u(\infty)}{x(\infty)} x(t).$$

We have $w \in \mathbb{M}^+_B$. But $w(\infty) = u(\infty)$, that gives a contradiction. □

Lemma 4.1.8. *Let $c(t) > 0$ for large t. If $J_r < \infty$, then* (1.1.1) *has no unbounded nonoscillatory solution, i.e.,* $\mathbb{M}^+_\infty = \emptyset$.

Proof. The assertion follows, with minor changes, from [159, Lemma 2]. □

Concerning the existence in the class $\mathbb{M}^-$, we have the following statement.

Theorem 4.1.9. *Let $c(t) > 0$ for large t.*

(i) *If $J_c = \infty$, $J_1 < \infty$, then* (1.1.1) *is nonoscillatory and* $\mathbb{M}^-_B \neq \emptyset$.

(ii) *If $J_1 = \infty$, then* $\mathbb{M}^-_B = \emptyset$.

Proof. Claim (i) follows from [159, Theorem 4]. As for the claim (ii), let $x \in \mathbb{M}^-_B$ and, without loss of generality, assume $x(t) > 0, x'(t) < 0$ for $t \geq T$ and $x(\infty) = c_x > 0$. From (1.1.1) we have

$$x^{[1]}(t) = x^{[1]}(T) - \int_T^t c(s)\Phi(x(s))ds < -\Phi(c_x) \int_T^t c(s)ds$$

or

$$x'(t) < -c_x \Phi^{-1} \left(\frac{1}{r(t)} \int_T^t c(s)ds \right).$$

Integrating over (T, t) we obtain

$$x(t) - x(T) < -c_x \int_T^t \Phi^{-1} \left(\frac{1}{r(s)} \int_T^s c(\tau)d\tau \right) ds$$

that gives a contradiction as $t \to \infty$. □

Lemma 4.1.9. *Let $c(t) > 0$ for large t. If* (1.1.1) *is nonoscillatory and $J_r < \infty$, then* $\mathbb{M}^-_0 \neq \emptyset$.

Proof. If $J_c < \infty$, the assertion follows, as a particular case, from [286, Theorem 2.2]. When $J_c = \infty$ the assertion follows from [61, Lemma 2-(ii)]. □

By considering the mutual behavior of integrals J_r, J_c, J_1, J_2 it is possible to summarize the situation in a complete way. Indeed, as regards the convergence or divergence of the above integrals, in view of Lemma 4.1.5, we have the following six possible cases:

$$\begin{aligned}
&\text{(C1)} \quad J_r = J_c = J_1 = J_2 = \infty, \\
&\text{(C2)} \quad J_r = J_1 = J_2 = \infty, \ J_c < \infty, \\
&\text{(C3)} \quad J_r = J_1 = \infty, \ J_c < \infty, \ J_2 < \infty, \\
&\text{(C4)} \quad J_c = J_1 = J_2 = \infty, \ J_r < \infty, \\
&\text{(C5)} \quad J_c = J_2 = \infty, \ J_r < \infty, \ J_1 < \infty, \\
&\text{(C6)} \quad J_r < \infty, \ J_c < \infty, \ J_1 < \infty, \ J_2 < \infty.
\end{aligned}$$

If (C1) holds, then, as already claimed in Subsection 4.1.1, equation (1.1.1) is oscillatory. In the remaining cases, from the above results we obtain the following theorem, which is a natural extension of the linear case [64, Theorem 1].

Theorem 4.1.10. *Let $c(t) > 0$ for large t.*

If (C2) *holds and equation* (1.1.1) *is nonoscillatory, then* $\mathbb{M}^+_\infty \neq \emptyset$, $\mathbb{M}^+_B = \mathbb{M}^-_B = \mathbb{M}^-_0 = \emptyset$.

If (C3) *holds, then equation* (1.1.1) *is nonoscillatory and* $\mathbb{M}^+_\infty \neq \emptyset$, $\mathbb{M}^+_B \neq \emptyset$, $\mathbb{M}^-_B = \mathbb{M}^-_0 = \emptyset$.

If (C4) *holds and equation* (1.1.1) *is nonoscillatory, then* $\mathbb{M}^+_\infty = \mathbb{M}^+_B = \mathbb{M}^-_B = \emptyset$, $\mathbb{M}^-_0 \neq \emptyset$.

If (C5) *holds, then equation* (1.1.1) *is nonoscillatory and* $\mathbb{M}^+_\infty = \mathbb{M}^+_B = \emptyset$, $\mathbb{M}^-_B \neq \emptyset$, $\mathbb{M}^-_0 \neq \emptyset$.

If (C6) *holds, then equation* (1.1.1) *is nonoscillatory and* $\mathbb{M}^+_\infty = \emptyset$, $\mathbb{M}^+_B \neq \emptyset$, $\mathbb{M}^-_B \neq \emptyset$, $\mathbb{M}^-_0 \neq \emptyset$.

Proof. The proof follows from the previous statements of this section.

(C2) From Lemma 4.1.3-(ii) and Lemma 4.1.7-(i) we have $\mathbb{M}^+_B = \mathbb{M}^-_B = \mathbb{M}^-_0 = \emptyset$. Since (1.1.1) is nonoscillatory, we obtain $\mathbb{M}^+ = \mathbb{M}^+_\infty \neq \emptyset$.

(C3) The assertion follows from Lemma 4.1.3-(ii), Lemma 4.1.7-(ii), Lemma 4.1.8.

(C4) From Lemma 4.1.3-(i) and Theorem 4.1.9-(ii) we have $\mathbb{M}^+_\infty = \mathbb{M}^+_B = \mathbb{M}^-_B = \emptyset$. Since (1.1.1) is nonoscillatory, we obtain $\mathbb{M}^- = \mathbb{M}^-_0 \neq \emptyset$.

(C5) The assertion follows from Lemma 4.1.3-(i), Theorem 4.1.9-(i), Lemma 4.1.9.

(C6) From Lemma 4.1.4-(iii), Lemma 4.1.8, Lemma 4.1.9, we obtain $\mathbb{M}^+_\infty = \emptyset$, $\mathbb{M}^+_B \neq \emptyset$, $\mathbb{M}^-_0 \neq \emptyset$. Finally the existence in $\mathbb{M}^-_B$ can be proved using an argument similar to that given in the proof of Theorem 4.1.1 (see also [364, Theorem 3.3] and its proof). □

Taking into account that the possible cases concerning the convergence or divergence of J_r, J_c, J_1, J_2 are the cases (C1) – (C6), from Theorem 4.1.10 we easily obtain the following result, which gives a necessary and sufficient condition for the existence of nonoscillatory solutions of (1.1.1) in the classes $\mathbb{M}^+_\infty, \mathbb{M}^+_B, \mathbb{M}^-_B, \mathbb{M}^-_0$.

Theorem 4.1.11. *Let $c(t) > 0$ for large t.*

(i) *Assume* (1.1.1) *nonoscillatory. The class* $\mathbb{M}_\infty^+$ *is nonempty if and only if* $J_r = \infty$.

(ii) *The class* $\mathbb{M}_B^+$ *is nonempty if and only if* $J_2 < \infty$.

(iii) *Assume* (1.1.1) *nonoscillatory. The class* $\mathbb{M}_0^-$ *is nonempty if and only if* $J_r < \infty$.

(iv) *The class* $\mathbb{M}_B^-$ *is nonempty if and only if* $J_1 < \infty$.

4.1.5 Generalized Fubini's theorem and its applications

In this subsection, we will show that half-linear equation (1.1.1) exhibits a "surprising" difference in asymptotic behavior of nonoscillatory solutions comparing with linear equation (1.1.2). In particular, differences between the three cases $p = 2$, $p < 2$, and $p > 2$ will be shown, especially as regards a possible coexistence of nonoscillatory solutions with various asymptotics.

We consider equation (1.1.1) under the assumptions $c(t) > 0$ and

$$\int^\infty r^{1-q}(t)\,dt = \infty, \qquad \int^\infty c(t)\,dt < \infty. \tag{4.1.17}$$

When (1.1.1) is nonoscillatory and (4.1.17) holds, then by Theorem 4.1.10 every nontrivial solution x of (1.1.1) satisfies $x(t)x'(t) > 0$ for large t, i.e., it belongs to the class $\mathbb{M}^+$. The more detailed description of asymptotic properties of solutions is given by the limit behavior of the quasiderivative $x^{[1]} = r\Phi(x')$. Hence, in addition to the previous classification of nonoscillatory solutions, we introduce two subclasses in $\mathbb{M}_\infty^+$

$$\begin{aligned}
\mathbb{M}_{\infty,0}^+ &:= \{x \in \mathbb{M}_\infty^+ : \lim_{t\to\infty} x^{[1]}(t) = 0\},\\
\mathbb{M}_{\infty,B}^+ &:= \{x \in \mathbb{M}_\infty^+ : \lim_{t\to\infty} x^{[1]}(t) = \ell_x \in (0,\infty)\}.
\end{aligned}$$

Then every solution of (1.1.1) belongs to one of the subclasses $\mathbb{M}_B^+$, $\mathbb{M}_{\infty,B}^+$ and $\mathbb{M}_{\infty,0}^+$.

Following the terminology introduced in [180], solutions in the class $\mathbb{M}_B^+$ are called *subdominant solutions*, in $\mathbb{M}_{\infty,0}^+$ *intermediate solutions* and in $\mathbb{M}_{\infty,B}^+$ are called *dominant solutions.* It is not known (see [180] where this problem has been formulated), whether intermediate solutions can coexist with dominant and/or subdominant solutions, and it has been shown in this paper that the existence of such solutions is determinated by the divergence of the integrals

$$\begin{aligned}
I &= \int_0^\infty c(t)\left(\int_0^t r^{1-q}(s)\,ds\right)^{p-1} dt,\\
J &= \int_0^\infty r^{1-q}(t)\left(\int_t^\infty c(s)\,ds\right)^{q-1} dt.
\end{aligned}$$

Hence the principal role is played by an extension of the classical Fubini theorem on the change of integration in a double integral. In particular, the question whether

$$J = \infty \Longleftrightarrow I = \infty \tag{4.1.18}$$

plays an important role. If $p = 2$, then $J = I$ (Fubini theorem) and so (4.1.18) holds. The answer whether (4.1.18) remains to hold when $p \neq 2$ (under assumption (4.1.17)) has been given in [96] and the result reads as follows.

Theorem 4.1.12. *Let A, b be continuous nonnegative functions such that the integral $\int_0^\infty b(t)\,dt < \infty$.*

(a) If $\alpha \geq 1$, then

$$\int_0^\infty b(t)\Big(\int_0^t A(s)\,ds\Big)^\alpha dt \leq 2\left(\int_0^\infty A(t)\Big(\int_t^\infty b(s)\,ds\Big)^{1/\alpha} dt\right)^\alpha. \tag{4.1.19}$$

(b) If $0 < \alpha \leq 1$, then

$$\int_0^\infty A(t)\Big(\int_t^\infty b(s)\,ds\Big)^{1/\alpha} dt \leq \left(\int_0^\infty b(t)\Big(\int_0^t A(s)\,ds\Big)^\alpha dt\right)^{1/\alpha}. \tag{4.1.20}$$

In particular,

(a) If $\alpha \geq 1$, then the condition

$$\int_0^\infty b(t)\left(\int_0^t A(s)\,ds\right)^\alpha dt = \infty \tag{4.1.21}$$

implies

$$\int_0^\infty A(t)\left(\int_t^\infty b(s)\,ds\right)^{1/\alpha} = \infty. \tag{4.1.22}$$

(b) If $0 < \alpha \leq 1$, then condition (4.1.22) *implies* (4.1.21).

The following examples show that the opposite statements to the second part of the previous theorem does not hold for $p \neq 2$, i.e., in general, the following cases are possible, see [96, 180]:

$$\begin{array}{llll} (K_1) & J = \infty, & I = \infty, & \\ (K_2) & J = \infty, & I < \infty, & \text{when } p > 2, \\ (K_3) & J < \infty, & I = \infty, & \text{when } p < 2, \\ (K_4) & J < \infty, & I < \infty. & \end{array}$$

Example 4.1.1. (a) Let

$$\alpha > 1, \quad A(t) = e^t, \quad b(t) = \frac{e^{-\alpha t}}{1 + t^\alpha}.$$

In this case $I < \infty$ and $J = \infty$.

(b) Let

$$A(t) = 1, \quad b(t) = (t+1)^{-\alpha-1}\big(\log(t+1)\big)^{-\mu}, \quad 0 < \alpha < \mu < 1.$$

In this case $I = \infty$ and $J < \infty$.

Now we can answer the question whether intermediate solutions may coexist with dominant or subdominant solutions and whether

$$\mathbb{M}^+_{\infty,B} \neq \emptyset \quad \Longleftrightarrow \quad \mathbb{M}^+_B \neq \emptyset. \tag{4.1.23}$$

We start with the following result.

Lemma 4.1.10. *The following statements hold:*

(i) If $J < \infty$, then (1.1.1) *is nonoscillatory and $\mathbb{M}^+_B \neq \emptyset$.*

(ii) If $\mathbb{M}^+_B \neq \emptyset$, then $J < \infty$.

(iii) If $I < \infty$, then (1.1.1) *is nonoscillatory and $\mathbb{M}^+_{\infty,B} \neq \emptyset$.*

(iv) If $\mathbb{M}^+_{\infty,B} \neq \emptyset$, then $I < \infty$.

(v) If $J = \infty$ and $I < \infty$, then (1.1.1) *is nonoscillatory and $\mathbb{M}^+_{\infty,0} \neq \emptyset$.*

(vi) If (1.1.1) *is nonoscillatory, then $\mathbb{M}^+_{\infty,0} \cup \mathbb{M}^+_{\infty,B} \neq \emptyset$.*

Proof. Claims (i), (ii), (iii), (iv) can be found in [180, Theorems 4.1,4.2], see also some results of [159] obtained for a more general equation.

Claim (v): By (ii) and (iii) we have $\mathbb{M}^+_B = \emptyset$ and $\mathbb{M}^+_{\infty,B} \neq \emptyset$. Assume $\mathbb{M}^+_{\infty,0} = \emptyset$, i.e., all nontrivial solutions belong to $\mathbb{M}^+_{\infty,B}$. Then for any pair of linearly independent solutions $x, \tilde{x}$ we have $\lim_{t\to\infty} \tilde{x}^{[1]}(t)/x^{[1]}(t) = L \neq 0$ and by L'Hospital's rule

$$\lim_{t\to\infty} \frac{\tilde{x}(t)}{x(t)} = \lim_{t\to\infty} \frac{\tilde{x}'(t)}{x'(t)} = L^{1/p-1} \neq 0$$

which yields a contradiction with the later given Theorem 4.2.7. Hence $\mathbb{M}^+_{\infty,0} \neq \emptyset$.

Claim (vi) also follows from Theorem 4.2.7. □

Concerning the linear equation (1.1.2), i.e., the case $p = 2$, from the statements (i)–(iv) of Lemma 4.1.10 and the equality $I = J$, we have that solutions in $\mathbb{M}^+_{\infty,0}$ *cannot coexist* with other types of solutions and (4.1.23) holds, see also [63, Theorems 1,2].

Using Theorem 4.1.12 and statements (i)–(iv) of Lemma 4.1.10 this result can be extended to the half-linear equation (1.1.1) as follows.

Theorem 4.1.13. *Assume* (4.1.17). *We have the following statements.*

(i) Let $p \geq 2$. If $\mathbb{M}^+_B \neq \emptyset$, then $\mathbb{M}^+_{\infty,B} \neq \emptyset$.

(ii) Let $p \in (1,2]$. If $\mathbb{M}^+_{\infty,B} \neq \emptyset$, then $\mathbb{M}^+_B \neq \emptyset$.

The following result describes the coexistence of solutions in cases (K_3) and (K_4) and shows the discrepancy between the linear and half-linear equations.

Theorem 4.1.14. *Assume* (4.1.17).

(a) *A necessary and sufficient condition for* (1.1.1) *to have solutions satisfying*

$$\mathbb{M}_B^+ = \emptyset, \quad \mathbb{M}_{\infty,0}^+ \neq \emptyset, \quad \mathbb{M}_{\infty,B}^+ \neq \emptyset$$

is $p > 2$, $J = \infty$ *and* $I < \infty$.

(b) *A necessary and sufficient condition for* (1.1.1) *to have solutions with*

$$\mathbb{M}_B^+ \neq \emptyset, \quad \mathbb{M}_{\infty,0}^+ \neq \emptyset, \quad \mathbb{M}_{\infty,B}^+ = \emptyset$$

is $p \in (1, 2)$, $J < \infty$ *and* $I = \infty$.

Proof. The proof of the statement (a) and (b) follows from Lemma 4.1.10, items (i)-(v) and items (i)-(iv), (vi), respectively. □

The following continuation of Example 4.1.1 illustrates the previous statements.

Example 4.1.2. Consider the half-linear differential equation

$$\left(e^{-(p-1)t}\Phi(x')\right)' + \frac{e^{-(p-1)t}}{1+t^{p-1}}\Phi(x) = 0.$$

The coefficients of this equation satisfies (4.1.17) (observe that $A(t) = r^{1-q}(t) = e^{-(p-1)(1-q)t} = e^t$) and, as it was shown in Example 4.1.1, $J = \infty$ and $I < \infty$. By Theorem 4.1.14-(a), equation (1.1.1) possesses nonoscillatory solutions satisfying (4.1.23).

Remark 4.1.2. (i) Note also when

$$\int^\infty r^{1-q}(t)\,dt < \infty, \quad \int^\infty c(t)\,dt = \infty,$$

then all solutions of (1.1.1) are in the class $\mathbb{M}^-$, as we have shown in Subsection 4.1.4. This case can be treated by applying the previous results to the reciprocal equation (4.1.1).

(ii) Finally observe that when both integrals $\int^\infty r^{1-q}(t)\,dt$, $\int^\infty c(t)\,dt$ are convergent, integrals I, J are convergent as well, and so $\mathbb{M}_B^+ \neq \emptyset$ and $\mathbb{M}_{\infty,B}^+ \neq \emptyset$. It is an open problem whether $\mathbb{M}_{\infty,0}^+ \neq \emptyset$.

4.2 Principal solution

The concept of the principal solution of the linear second order differential equation (1.1.2) was introduced in 1936 by Leighton and Morse [236], and plays an important role in the oscillation and asymptotic theory of (1.1.2). In this section we show that this concept can be introduced also for (nonoscillatory) half-linear equation (1.1.1).

4.2.1 Principal solution of linear equations

First we recall basic properties of the principal solution of linear equation (1.1.2). Suppose that this equation is nonoscillatory, i.e., any solution of this equation is eventually positive or negative. Then, using the below described method, one can distinguish among all solutions of this equation a solution $\tilde{x}$, called the *principal solution* (determined uniquely up to a multiplicative factor), which is near ∞ less than any other solution of this equation in the sense that

$$\lim_{t\to\infty} \frac{\tilde{x}(t)}{x(t)} = 0$$

for any solution x which is linearly independent of $\tilde{x}$.

Let x, y be eventually positive linearly independent solutions of (1.1.2), then

$$r(t)[x'(t)y(t) - x(t)y'(t)] =: \omega,$$

where $\omega \neq 0$ is a real constant. This means that the function x/y is monotonic (since $(x/y)' = \omega/(ry^2)$) and hence there exists (finite or infinite) limit $\lim_{t\to\infty} x(t)/y(t) = L$. If $L = 0$, x is the principal solution of (1.1.2), if $L = \infty$, the solution y is principal. If $0 < L < \infty$, we set $\tilde{x} = x - Ly$. Then obviously $\lim_{t\to\infty} \tilde{x}(t)/y(t) = 0$ and $\tilde{x}$ is the principal solution. Observe that this construction of the principal solution is based on the linearity of the solution space of (1.1.2).

Using the Wronskian identity, the principal solution $\tilde{x}$ of (1.1.2) can be equivalently characterized as a solution satisfying

$$\int^{\infty} \frac{dt}{r(t)\tilde{x}^2(t)} = \infty. \tag{4.2.1}$$

Indeed, let y be a solution linearly independent of $\tilde{x}$. Then by the previous argument $y/\tilde{x}$ tends monotonically to ∞ as $t \to \infty$, hence

$$\lim_{t\to\infty} \int^{t} \frac{ds}{r(s)\tilde{x}^2(s)} = \lim_{t\to\infty} \frac{y(t)}{\tilde{x}(t)} = \infty. \tag{4.2.2}$$

Another characterization of the principal solution of (1.1.2) is via the eventually minimal solution of the associated Riccati equation

$$w' + c(t) + \frac{w^2}{r(t)} = 0. \tag{4.2.3}$$

Let $\tilde{x}$, x be linearly independent solutions of (1.1.2), the solution $\tilde{x}$ being principal, and let $\tilde{w} = r\tilde{x}'/\tilde{x}$, $w = rx'/x$ be the solutions of the associated Riccati equation. Without loss of generality we may suppose that $\tilde{x}$ and x are eventually positive. We have

$$w(t) - \tilde{w}(t) = \frac{r(t)x'(t)}{x(t)} - \frac{r(t)\tilde{x}'(t)}{\tilde{x}(t)} = \frac{r(t)[x'(t)\tilde{x}(t) - \tilde{x}'(t)x(t)]}{\tilde{x}(t)x(t)}.$$

The numerator of the last fraction is a constant and this constant is positive since we have

$$\frac{r(t)[x'(t)\tilde{x}(t) - \tilde{x}'(t)x(t)]}{r(t)\tilde{x}^2(t)} = \left(\frac{x(t)}{\tilde{x}(t)}\right)' > 0.$$

This follows from the fact that $\tilde{x}$ is the principal solution, i.e., $x/\tilde{x}$ tends monotonically to ∞. Hence, the solution $\tilde{w}$ of the Riccati equation (4.2.3) given by the principal solution of (1.1.2) is less than any other solution of (4.2.3) near ∞, i.e., a solution $\tilde{x}$ of (1.1.2) is principal if and only if

$$\frac{\tilde{x}'(t)}{\tilde{x}(t)} < \frac{x'(t)}{x(t)} \quad \text{for large } t \tag{4.2.4}$$

for any solution x linearly independent of $\tilde{x}$. Conversely, let $\tilde{w} = r\tilde{x}'/\tilde{x}$ be the minimal solution of (4.2.3), i.e., $\tilde{w}(t) < w(t)$ for large t, for any solution w of (4.2.3), and suppose that the solution $\tilde{x}$ of (1.1.2) is not principal, i.e., the integral in (4.2.1) is convergent. Let $T \in \mathbb{R}$ be such that $\int_T^\infty r^{-1}(t)\tilde{x}^{-2}(t)\,dt < 1$ and consider the solution w of (4.2.3) given by the initial condition $w(T) = \tilde{w}(T) - 1/(2\tilde{x}^2(T))$. Put $v = \tilde{x}^2(\tilde{w} - w)$. Then $v(T) = 1/2$ and by a direct computation we have

$$v' = \frac{v^2}{r(t)\tilde{x}^2(t)}.$$

Hence

$$v(t) = \frac{1}{2 - \int_T^t r^{-1}(s)\tilde{x}^{-2}(s)\,ds} \le \frac{1}{2 - \int_T^\infty r^{-1}(s)\tilde{x}^{-2}(s)\,ds} \le 1.$$

This means that v is extensible up to ∞ and hence w has the same property and $w(t) < \tilde{w}(t)$. This contradiction shows that the eventual minimality of $\tilde{w}$ implies that (4.2.1) holds, i.e., the associated solution $\tilde{x}$ of (1.1.1) is principal.

The last construction of the principal solution of (1.1.2) which we present here requires (in addition to nonoscillation of (1.1.2)) the assumption that for any t_0, the solution of (1.1.2) given by the initial condition $x(t_0) = 0$, $x'(t_0) \neq 0$ has a zero point to the right of t_0. We denote this zero point by $\eta(t_0)$. The function η is nondecreasing according to the Sturmian theory, hence there exists $\lim_{t\to\infty} \eta(t) =: T$ and $T < \infty$ since we suppose that (1.1.2) is nonoscillatory. Now, the solution $\tilde{x}$ given by the initial condition $\tilde{x}(T) = 0$, $\tilde{x}'(T) \neq 0$ is the principal solution of (1.1.2). This construction is used in the original paper of Leighton and Morse [236]. Concerning other papers dealing with the principal solution of (1.1.2) and its properties we refer to [174] and the references given therein.

4.2.2 Mirzov's construction of principal solution

This construction defines the principal solution of half-linear equation (1.1.1) via the minimal solution of the associated Riccati equation (1.1.21). Nonoscillation of (1.1.1) implies that there exist $T \in \mathbb{R}$ and a solution $\hat{w}$ of (1.1.21) which is defined in the whole interval $[T, \infty)$, i.e., such that (1.1.1) is disconjugate on $[T, \infty)$. Let $d \in (T, \infty)$ and let w_d be the solution of (1.1.21) determined by the solution

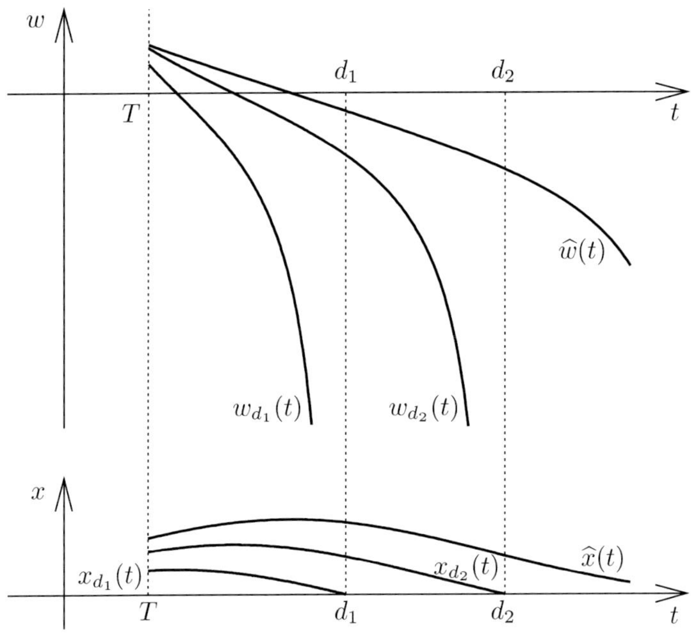

Figure 4.2.1: Mirzov's construction of principal solution

x_d of (1.1.1) satisfying the initial condition $x(d) = 0$, $r(d)\Phi(x'(d)) = -1$. Then $w_d(d-) = -\infty$ and $w_d(t) < \hat{w}(t)$ for $t \in (T, d)$. See also Figure 4.2.1. Moreover, if $T < d_1 < d_2$, then

$$w_{d_1}(t) < w_{d_2}(t) < \hat{w}(t) \quad \text{for } t \in (T, d_1).$$

This implies that for $t \in (T, \infty)$ there exists the limit $w_\infty(t) := \lim_{d\to\infty} w_d(t)$ and monotonicity of this convergence (with respect to the "subscript" variable) implies that this convergence is uniform on every compact subinterval of $[T, \infty)$. Consequently, the limit function w_∞ solves (1.1.21) as well and any solution w of this equation which is extensible up to ∞ satisfies the inequality $w(t) > w_\infty(t)$ near ∞. Indeed, if a solution $\bar{w}$ satisfies the inequality $\bar{w}(t) < w_\infty(t)$ on some interval (T_1, ∞), then for $\bar{t} \in (T_1, \infty)$ and d sufficiently large we would have $\bar{w}(\bar{t}) < w_d(\bar{t}) < w_\infty(\bar{t})$. But this contradicts the fact that $w_d(d-) = -\infty$ and that graphs of solutions of (1.1.21) cannot intersect (because of unique solvability of this equation). Summarizing, we have found a solution w_∞ of (1.1.21) with the property that

$$w_\infty(t) < w(t), \quad \text{for large } t, \tag{4.2.5}$$

for any other solution w of (1.1.21).

Now, having defined the *minimal solution* w_∞ of (1.1.21), i.e., a solution w_∞ such that (4.2.5) holds for any other solution w of (1.1.21), we define a *principal solution* of (1.1.1) at ∞ as a (nontrivial) solution of the first order equation

$$x' = \Phi^{-1}\left(\frac{w_\infty(t)}{r(t)}\right) x, \tag{4.2.6}$$

i.e., the principal solution of (1.1.1) at ∞ is determined uniquely up to a multiplicative factor by the formula

$$x(t) = x(T)\exp\left\{\int_T^t r^{1-q}(s)\Phi^{-1}(w_\infty(s))\,ds\right\}.$$

Consequently, a solution $\tilde{x}$ of (1.1.1) is principal if and only if

$$\frac{r(t)\Phi(\tilde{x}'(t))}{\Phi(\tilde{x}(t))} < \frac{r(t)\Phi(x'(t))}{\Phi(x'(t))} \quad \text{for large } t, \tag{4.2.7}$$

which is equivalent to (4.2.4).

Remark 4.2.1. (i) Mirzov actually used a slightly different approach in his paper [291] which can be briefly explained as follows. Suppose that (1.1.1) is nonoscillatory and let $\hat{w}$ be a solution of the associated Riccati equation which exists on some interval $[T, \infty)$ and let $W := \hat{w}(T)$. Denote by

$$\mathcal{W} = \{v \in (-\infty, W) : \text{ the solution } w \text{ of (1.1.21) given by the initial condition } w(T) = v \text{ is not extensible up to } \infty\},$$

i.e., $\mathcal{W}$ are initial values of solutions of (1.1.21) at $t = T$ which blow down to $-\infty$ at some finite time $t > T$. Note that the set $\mathcal{W}$ is nonempty what can be seen as follows. Let $T_1 > T$ be arbitrary and consider a solution x of (1.1.1) given by $x(T_1) = 0$, $x'(T_1) \neq 0$. Disconjugacy of (1.1.1) on $[T, \infty)$ implies that $x(t) \neq 0$ on $[T, T_1)$ and the value of the associated solution of the Riccati equation $w = r\Phi(x')/\Phi(x)$ at $t = T$ clearly belongs to $\mathcal{W}$. Now, let $\tilde{v} := \sup \mathcal{W}$ and let w_∞ be the solution of (1.1.21) given by the initial condition $w(T) = \tilde{v}$. Then this solution is extensible up to ∞ (supposing that this is not the case, we get a contradiction with the definition of the number $\tilde{v}$) and the principal solution of (1.1.1) is defined again by (4.2.6).

(ii) Let b be a *regular* point of equation (1.1.1) in the sense that for any $A, B \in \mathbb{R}$ the initial condition $x(b) = A$, $r(b)\Phi(x'(b)) = B$ determines uniquely a solution of (1.1.1). Let x_b be a solution given by $x_b(b) = 0$, $x_b'(b) \neq 0$. Replacing in the above construction the point $t = \infty$ by $t = b$, i.e., $w_b(t) := \lim_{d\to b-} w_d(t)$, it is not difficult to see that $w_b = r\Phi(x_b')/\Phi(x_b)$. Consequently, what we call the *principal solution* x_b *of* (1.1.1) *at a regular point* $b \in \mathbb{R}$ is a nontrivial solution satisfying the condition $x_b(b) = 0$.

4.2.3 Construction of Elbert and Kusano

This construction was introduced (independently of Mirzov's approach) in the paper [145] and it is based on the half-linear Prüfer transformation.

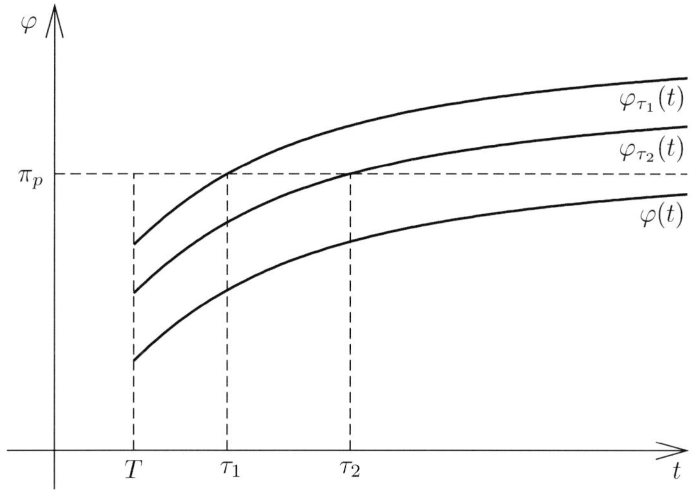

Figure 4.2.2: Construction of principal solution by Elbert and Kusano

Let (1.1.1) be nonoscillatory and let T be such that this equation is disconjugate on $[T,\infty)$. Take a solution x which is positive on $[T,\infty)$. By the generalized Prüfer transformation (see Subsection 1.1.3) this solution can be expressed in the form

$$x(t)=\rho(t)\sin_p\varphi(t),\quad r^{q-1}(t)x'(t)=\rho(t)\cos_p\varphi(t), \tag{4.2.8}$$

where ρ is a positive function, the half-linear sine and cosine functions $\sin_p$, $\cos_p$ were defined in Subsection 1.1.3 and the function φ is a solution of the first order equation

$$\varphi'=r^{1-q}(t)|\cos_p\varphi(t)|^p+\frac{c(t)}{p-1}\,|\sin_p\varphi(t)|^p. \tag{4.2.9}$$

The fact that $x(t)>0$ for $t\in[T,\infty)$ implies that $\varphi(t)\in(k\pi_p,(k+1)\pi_p)$ for some even $k\in\mathbb{Z}$ and without loss of generality we can suppose that $k=0$. Now, let $\tau\in(T,\infty)$ and let φ_τ be the solution of (4.2.9) given by the initial condition $\varphi_\tau(\tau)=\pi_p$. Since any solution of (4.2.9) satisfies $\varphi'(t)>0$ whenever $\varphi(t)=0 \pmod{\pi_p}$, the unique solvability of (4.2.9) (compare again with Subsection 1.1.3) implies that $\varphi(t)<\varphi_{\tau_2}(t)<\varphi_{\tau_1}(t)$ for $t\ge T$ whenever $T<\tau_1<\tau_2$ (see Figure 4.2.2). Consequently, the monotonicity of φ_τ with respect to τ implies that there exists a finite limit

$$\lim_{\tau\to\infty}\varphi_\tau(T)=\varphi^*.$$

Now, the *principal solution* is the solution of (1.1.1) given by the initial condition

$$\tilde{x}(T)=\sin_p\varphi^*,\quad \tilde{x}'(T)=r^{1-q}(T)\cos_p\varphi^*.$$

This means that we take $\rho(T)=1$ in the definition of $\tilde{x}$, this can be done according to the homogeneity of the solution space of (1.1.1).

Theorem 4.2.1. *A solution $\tilde{x}$ of a nonoscillatory equation* (1.1.1) *is principal in the sense of Mirzov's construction if and only if it is principal in the sense of Elbert and Kusano.*

Proof. Let x_τ be a nontrivial solution of (1.1.1) satisfying $x_\tau(\tau) = 0$. This solution can be expressed in the form

$$x_\tau(t) = \rho(t)\sin_p \varphi_\tau(t), \quad r^{1-q}(t)x'_\tau(t) = \rho(t)\cos_p \varphi_\tau(t),$$

where φ_τ is the solution of (4.2.9) satisfying $\varphi_\tau(\tau) = 0$. The corresponding solution of the associated Riccati equation (1.1.21)

$$w_\tau(t) = \frac{r(t)\Phi(x'_\tau(t))}{\Phi(x_\tau(t))} = \Phi(\cot_p \varphi_\tau(t))$$

satisfies $w_\tau(\tau-) = -\infty$. The minimal solution of (1.1.21) (which defines the principal solution of (1.1.1) in Mirzov's definition) is given by $\tilde{w}(t) = \lim_{\tau\to\infty} w_\tau(t)$, i.e., it is just the solution satisfying $\tilde{w}(T) = \Phi(\cot_p \varphi^*)$ and this is the solution of Riccati equation (1.1.21) given by the principal solution obtained by Elbert-Kusano's construction. □

We finish this subsection with some examples of equations whose principal solution can be computed explicitly.

Example 4.2.1. (i) Consider the one-term half-linear equation

$$(r(t)\Phi(x'))' = 0. \tag{4.2.10}$$

As we have mentioned in Section 1.4, the solution space of this equation is a two-dimensional linear space with the basis $x_1(t) \equiv 1$, $x_2(t) = \int^t r^{1-q}(s)\,ds$. The Riccati equation associated with (4.2.10) is $w' + (p-1)r^{1-q}|w|^q = 0$ and the general solution of this equation is

$$w(t) = \frac{1}{\Phi\left(C + \int_T^t r^{1-q}(s)\,ds\right)}, \qquad w(t) \equiv 0. \tag{4.2.11}$$

If $\int^\infty r^{1-q}(t)\,dt = \infty$, then by an easy computation one can verify that $\tilde{w}(t) \equiv 0$ is the eventually minimal solution of this equation and hence $\tilde{x}(t) = 1$ is the principal solution of (1.1.1). If $\int^\infty r^{1-q}(t)\,dt < \infty$, then $\tilde{w}(t) = -\left(\int_t^\infty r^{1-q}(s)\,ds\right)^{1-p}$ is the eventually minimal solution of the Riccati equation (we take $C = -\int_T^\infty r^{1-q}(s)\,ds$ in formula (4.2.11)) and $\tilde{x}(t) = \int_t^\infty r^{1-q}(s)\,ds$ is the principal solution of (4.2.10).

(ii) The nonoscillatory equation $(\Phi(x'))' - (p-1)\Phi(x) = 0$ investigated in Subsection 1.4.1 has solutions $x(t) = \mathrm{e}^{\pm t}$ and all other solutions are asymptotically equivalent to e^t. Consequently, the solution $\tilde{x}(t) = \mathrm{e}^{-t}$ is the principal solution at ∞.

(iii) The Euler type equation

$$(\Phi(x'))' + \frac{\gamma}{t^p}\Phi(x) = 0 \tag{4.2.12}$$

is nonoscillatory if and only if $\gamma \le \gamma_p = [(p-1)/p]^p$, see Subsection 1.4.2. If $\gamma = \gamma_p$, then (4.2.12) has a solution $x(t) = t^{\frac{p-1}{p}}$ and all linearly independent solutions are asymptotically equivalent to $t^{\frac{p-1}{p}} \log^{\frac{2}{p}} t$. Consequently, $\tilde{x}(t) = t^{\frac{p-1}{p}}$ is the principal solution of (4.2.12). If $\gamma < \gamma_p$, then $x_1(t) = e^{\lambda_1 t}$, $x_2(t) = e^{\lambda_2 t}$, where $\lambda_1 < \lambda_2$ are the roots of the algebraic equation $(p-1)[|\lambda|^p - \Phi(\lambda)] + \gamma = 0$, are solutions of (4.2.12), and all other linearly independent solutions are asymptotically equivalent to $x_2(t)$. Consequently, $\tilde{x}(t) = x_1(t) = e^{\lambda_1 t}$ is the principal solution of (4.2.12) in this case.

4.2.4 Comparison theorem for eventually minimal solutions of Riccati equations

Similarly as in the linear case we have the following inequalities for solutions of a pair of Riccati equations corresponding to nonoscillatory half-linear equations.

Theorem 4.2.2. *Consider a pair of half-linear equations* (1.1.1),

$$(R(t)\Phi(y'))' + C(t)\Phi(y) = 0 \tag{4.2.13}$$

and suppose that (4.2.13) *is a Sturmian majorant of* (1.1.1) *for large* t, *i.e., there exists* $T \in \mathbb{R}$ *such that* $0 < R(t) \le r(t)$, $c(t) \le C(t)$ *for* $t \in [T, \infty)$. *Suppose that the majorant equation* (4.2.13) *is nonoscillatory and denote by* $\tilde{w}$, $\tilde{v}$ *eventually minimal solutions of* (1.1.21) *and of*

$$v' + C(t) + (p-1)R^{1-q}(t)|v|^q = 0, \tag{4.2.14}$$

respectively. Then $\tilde{w}(t) \le \tilde{v}(t)$ *for large* t.

Proof. Nonoscillation of (4.2.13) implies the existence of $T \in \mathbb{R}$ such that $\tilde{w}$ and $\tilde{v}$ exist on $[T, \infty)$. Suppose that there exists $t_1 \in [T, \infty)$ such that $\tilde{w}(t_1) > \tilde{v}(t_1)$. Let w be the solution of (1.1.21) given by the initial condition $w(t_1) = \tilde{v}(t_1)$. Then according to the standard theorem on differential inequalities (see e.g. [232]) we have $w(t) \ge \tilde{v}(t)$ for $t \ge t_1$, i.e., w is extensible up to ∞. At the same time $w(t) < \tilde{w}(t)$ for $t \ge t_1$ since graphs of solutions of (1.1.21) cannot intersect (because of the unique solvability). But this contradicts the eventual minimality of $\tilde{w}$. ☐

In some oscillation criteria, we will need the following immediate consequence of the previous theorem.

Corollary 4.2.1. *Let* $\int^\infty r^{1-q}(t)\,dt = \infty$, $c(t) \ge 0$ *for large* t *and suppose that* (1.1.1) *is nonoscillatory. Then the eventually minimal solution of the associated Riccati equation* (1.1.21) *satisfies* $\tilde{w}(t) \ge 0$ *for large* t.

Proof. Under the assumptions of corollary, (1.1.1) is the majorant of the one-term equation $(r(t)\Phi(y'))' = 0$. Since $\int^\infty r^{1-q}(t)\,dt = \infty$, $\tilde{y} \equiv 1$ is the principal solution of this equation (compare Example 4.2.1). Hence $\tilde{v}(t) = 0$ is the eventually minimal solution of the associated Riccati equation which implies the required statement. ☐

4.2.5 Sturmian property of the principal solution

In this short subsection we briefly show that the principal solution of (1.1.1) has a Sturmian type property and that the largest zero point of this solution (if any) behaves like the left conjugate point of ∞, in a certain sense.

Theorem 4.2.3. *Suppose that equation* (1.1.1) *is nonoscillatory and its principal solution* $\tilde{x}$ *has a zero point and let* T *be the largest of them. Further suppose that equation* (4.2.13) *is a Sturmian majorant of* (1.1.1) *on* $[T,\infty)$, *i.e.,* $0 < R(t) \le r(t)$ *and* $C(t) \ge c(t)$ *for* $t \in [T,\infty)$. *Then any solution* y *of* (4.2.13) *has a zero point in* (T,∞) *or it is a constant multiple of* $\tilde{x}$. *The latter possibility is excluded if one of the inequalities between* r, R *and* c, C, *respectively, is strict on a nondegenerate subinterval of* $[T,\infty)$.

Proof. If (4.2.13) is oscillatory, the statement of theorem trivially holds, so suppose that (4.2.13) is nonoscillatory and let $\tilde{y}$ be its principal solution. Denote by $\tilde{w}$ and $\tilde{v}$ the minimal solutions of corresponding Riccati equations (1.1.21) and (4.2.14), respectively. According to the comparison theorem for minimal solutions of Riccati equations presented in the previous subsection, we have $\tilde{w}(t) \ge \tilde{v}(t)$ on the interval of existence of $\tilde{w}$. Since we suppose that $\tilde{x}(T) = 0$, this implies that $\tilde{w}(T+) = \infty$, so the interval of existence of $\tilde{v}$ must be a subinterval of $[T,\infty)$, say $[T_1,\infty)$, i.e., T_1 is the largest zero of the principal solution $\tilde{y}$ of (4.2.13). If one of the inequalities between r, R and c, C is strict, it can be shown that the possibility $T = T_1$ is excluded. Now let y be any nontrivial solution of (4.2.13). If $y(t) \ne 0$ for $t \in [T_1,\infty)$, then the solution $v = R\Phi(y')/\Phi(y)$ of the associated Riccati equation (4.2.14) exists on $[T_1,\infty)$ and satisfies there the inequality $v(t) < \tilde{v}(t)$ on $[T_1,\infty)$ (since $\tilde{v}(T_1+) = \infty$ and $v(T-) < \infty$) and this is a contradiction with minimality of $\tilde{v}$. □

Remark 4.2.2. (i) Let T be the largest zero of the principal solution $\tilde{x}$ of (1.1.1), i.e., the same as in the previous theorem, and suppose that $R(t) = r(t)$ and $C(t) = c(t)$ for $t \in [T,\infty)$. Then Theorem 4.2.3 shows that T plays the role of the left conjugate point of ∞ in the sense that any nonprincipal solution of (1.1.1), i.e., a solution linearly independent of $\tilde{x}$, has exactly one zero in (T,∞). Note that the fact that this zero is exactly one (and not more) follows from the classical Sturm separation theorem (Theorem 1.2.3).

(ii) In Remark 1.2.2 we have pointed out that the disconjugacy of (1.1.1) on a bounded interval $I = [a,b]$ (which, by definition, means that the solution x given by $x(a) = 0$, $x'(a) \ne 0$ has no zero in $(a,b]$) is actually equivalent to the existence of a solution *without any zero* in $[a,b]$. Theorem 4.2.3 shows that we have the same situation with unbounded intervals or an interval whose endpoints are singular points of (1.1.1). For example, if $I = \mathbb{R} = (-\infty,\infty)$, then disconjugacy of (1.1.1) on this interval (i.e., no nontrivial solution has two or more zeros on $\mathbb{R}$) is equivalent to the existence of a solution *without any zero on* $\mathbb{R}$, the solution having this property is e.g. the principal solution (at ∞).

4.2.6 Principal solution of reciprocal equation

In this subsection we suppose that $c(t) > 0$ for large t. We will show that under certain additional conditions a solution x of (1.1.1) is principal if and only if its quasiderivative $x^{[1]} = r\Phi(x')$ is the principal solution of the reciprocal equation

$$\left(\frac{\Phi^{-1}(u')}{c^{q-1}(t)}\right)' + \frac{\Phi^{-1}(u)}{r^{q-1}(t)} = 0. \tag{4.2.15}$$

Recall that $\Phi^{-1}(u) = |u|^{q-2}u$, $q = p/(p-1)$, is the inverse function of Φ. We will need the following auxiliary statement which is already partially hidden in Lemma 4.1.3. Here we reformulate the statement into the form directly applicable in the proof of the main result of this subsection (and also in the proof of the main result of Subsection 4.2.8 given below).

Theorem 4.2.4. *Let* (1.1.1) *be nonoscillatory and* $J_r + J_c = \infty$. *Then a solution* $\tilde{x}$ *of* (1.1.1) *is principal solution if and only if its quasiderivative* $\tilde{x}^{[1]}$ *is principal solution of* (4.2.15).

Proof. Assume that $\tilde{x}$ is the principal solution of (1.1.1). Then $\tilde{u} = \tilde{x}^{[1]}(t)$ is a solution of (4.2.15); let u be another solution of (4.2.15) such that $\tilde{u} \neq \lambda u$ for any $\lambda \in \mathbb{R}$. Since reciprocity is a mutual property, the function $x = \Phi^{-1}(u'/c)$ is a solution of (1.1.1) and clearly $x \neq \mu\tilde{x}$ for any $\mu \in \mathbb{R}$. Because $\tilde{x}$ is principal solution, taking into account that Φ is increasing, from Riccati equation (1.1.21) we obtain for large t

$$\frac{\tilde{x}^{[1]}(t)}{c(t)\Phi(\tilde{x}(t))} < \frac{x^{[1]}(t)}{c(t)\Phi(\tilde{x}(t))}. \tag{4.2.16}$$

From the relationship between $\tilde{x}, \tilde{x}^{[1]}$ and between $x, x^{[1]}$, we have

$$\frac{\tilde{u}(t)}{\tilde{u}'(t)} > \frac{u(t)}{u'(t)}. \tag{4.2.17}$$

From Lemma 4.1.3, as claimed, all the solutions of (4.2.15) are in the same class ($\mathbb{M}^+$ or $\mathbb{M}^-$). Then either $u(t)/u'(t) > 0$ or $\tilde{u}(t)/\tilde{u}'(t) < 0$ and from (4.2.17) we obtain $\tilde{u}'(t)/\tilde{u}(t) < u'(t)/u(t)$ which means that $\tilde{u}$ is the principal solution of (4.2.15). The converse can be proved by using a similar argument. □

The following example shows that Theorem 4.2.4 does not hold if the assumption $J_r + J_c = \infty$ is violated. This example is "linear", i.e., it concerns linear equation (1.1.2), but it can be modified to half-linear equations (1.1.1) and (4.2.15).

Example 4.2.2. Consider the linear equation

$$\left((t+1)\log^2(t+2)x'(t)\right)' + \frac{\log(t+2)}{(t+2)^2}x(t) = 0. \tag{4.2.18}$$

Clearly, in view of property (4.2.1), $\tilde{x}(t) = (\log(t+2))^{-1}$ is a principal solution of (4.2.18). It is easy to verify, again from (4.2.1), that the quasiderivative of $\tilde{x}$

$$\tilde{u}(t) = \tilde{x}^{[1]}(t) = -\frac{t+1}{t+2}$$

is a nonprincipal solution of the reciprocal equation

$$\left(\frac{(t+2)^2}{\log(t+2)}u'(t)\right)' + \frac{1}{(t+1)\log^2(t+2)}u(t) = 0.$$

Observe that Theorem 4.2.4 cannot be applied since $J_r < \infty$ and $J_c < \infty$.

4.2.7 Integrals associated with eventually minimal solution of Riccati equation

In this subsection we present some necessary and/or sufficient integral conditions for a solution $\tilde{w}$ to be the minimal solution of (1.1.21). Similarly to some previous parts of the book, we reformulate the Mirzov results originally stated for first order system (1.1.8) (see [292, Sec. 15]) to (1.1.1). We will use the following notation:

$$F(x,a) := \begin{cases} \frac{|a|^q - |x|^q}{a-x} + \Phi^{-1}(a-x) & \text{for } x \neq a, \\ q\Phi^{-1}(a) & \text{for } x = a, \end{cases}$$

$$\begin{aligned} \alpha(v,z) &:= \min\left\{F(x,z) : v \le x \le z\right\}, \\ \beta(v,z) &:= \max\left\{F(x,z) : v \le x \le z\right\}, \end{aligned}$$

and

$$m_* = \min\{F(x,1) : 0 \le x \le 1\}, \quad m^* = \max\{F(x,1) : 0 \le x \le 1\}.$$

An important role is played by the following statement.

Theorem 4.2.5. *Let $\tilde{w}$ be the minimal solution of Riccati equation* (1.1.21) *defined on an interval $[t_0,\infty)$, v be a continuous function on $[t_0,\infty)$ such that $v(t) < \tilde{w}(t)$ for $t \ge t_0$, and suppose that the inequality*

$$\text{(4.2.19)} \quad \liminf_{t\to\infty} \big(\tilde{w}(t) - v(t)\big)^{-(q-1)} \exp\left(-\int_{t_0}^t r^{1-q}(\tau)\alpha\big(v(\tau),\tilde{w}(\tau)\big)\,d\tau\right) < \infty$$

holds. Then

$$\text{(4.2.20)} \quad \int_{t_0}^\infty r^{1-q}(t)\exp\left(-\int_{t_0}^t r^{1-q}(\tau)\alpha\big(v(\tau),\tilde{w}(\tau)\big)\,d\tau\right)dt = \infty.$$

Proof. We assume the contrary. In view of (4.2.19), there exists a sequence $\{t_k\}$ such that $t_k < t_{k+1}$, $\lim_{k\to\infty} t_k = \infty$ and

$$\text{(4.2.21)} \quad \lim_{k\to\infty} \big(\tilde{w}(t_k) - v(t_k)\big)^{-(q-1)} \exp\left(-\int_{t_0}^{t_k} r^{1-q}(\tau)\alpha\big(v(\tau),\tilde{w}(\tau)\big)\,d\tau\right) < \infty.$$

Consider a sequence of solutions w_n of equation (1.1.21), which are defined by the initial conditions $w_n(t_0) = \tilde{w}(t_0) + (v(t_0) - \tilde{w}(t_0))/(n+1)$. Since the solution

$\tilde{w}$ is minimal, every solution w_n of (1.1.21) blows down to $-\infty$ at some $t = T_n$, i.e., $\lim_{t\to T_n} w_n(t) = -\infty$ and $v(t) < w_n(t) < \tilde{w}(t)$ for $t_0 \le t < T_n$. Obviously, $T_n < T_{n+1}$ and $\lim_{n\to\infty} T_n = \infty$. Choose a subsequence of the intervals $[T_{n_i}, T_{n_i+1})$ such that each interval contains at least one point of the sequence $\{t_k\}$. Set $t_{k_i} = \max\{t_1, t_2, \dots, t_k, \dots\} \cap [T_{n_i}, T_{n_i+1})$. By virtue of the continuous dependence of a solution on the initial data, there exists a solution $\bar{w}_{k_i}$ of (1.1.21) such that

$$\begin{gathered}\bar{w}_{k_i}(t_{k_i}) = v(t_{k_i}), \quad w_{n_i}(t_0) \le \bar{w}_{k_i}(t_0) < \tilde{w}_{n_i+1}(t_0),\\ v(t) < \bar{w}_{k_i}(t) < \tilde{w}(t) \text{ for } t_0 \le t < t_{k_i}.\end{gathered} \tag{4.2.22}$$

From (1.1.21) it can be easily obtained that for $t_0 \le t \le t_{k_i}$,

$$\big(\tilde{w}(t) - \bar{w}_{k_i}(t)\big)^{q-1} = d_{k_i} \exp\left(-\int_{t_0}^t r^{1-q}(\tau)\frac{|\tilde{w}(\tau)|^q - |\bar{w}_{k_i}(\tau)|^q}{\tilde{w}(\tau) - \bar{w}_{k_i}(\tau)}\, d\tau\right), \tag{4.2.23}$$

where $d_{k_i} = (\tilde{w}(t_0) - \bar{w}_{k_i}(t_0))^{q-1}$. Let us multiply (4.2.23) by

$$-r^{1-q}(t)\exp\left(-\int_{t_0}^t r^{1-q}(s)\big(\tilde{w}(s) - \bar{w}_{k_i}(s)\big)^{q-1}\, ds\right)$$

and integrate the obtained equality from t_0 to t. Then we get

$$\begin{aligned}&\exp\left(-\int_{t_0}^t r^{1-q}(\tau)\big(\tilde{w}(\tau) - \bar{w}_{k_i}(\tau)\big)^{\lambda_1}\, d\tau\right)\\ &\quad = 1 - d_{k_i}\int_{t_0}^t r^{1-q}(\tau)\exp\left(-\int_{t_0}^\tau r^{1-q}(s)F\big(\bar{w}_{k_i}(s), \tilde{w}(s)\big)\, ds\right) d\tau,\end{aligned} \tag{4.2.24}$$

Equalities (4.2.23) and (4.2.24) imply

$$\big(\tilde{w}(t) - \bar{w}_{k_i}(t)\big)^{q-1} = \frac{d_{k_i}\exp\big(-\int_{t_0}^t r^{1-q}(\tau)F(\bar{w}_{k_i}(\tau), \tilde{w}(\tau))\, d\tau\big)}{1 - d_{k_i}\int_{t_0}^t r^{1-q}(\tau)\exp\big(-\int_{t_0}^\tau r^{1-q}(s)F(\bar{w}_{k_i}(s), \tilde{w}(s))\, ds\big)\, d\tau}. \tag{4.2.25}$$

For $t = t_{k_i}$, from (4.2.25) in view of (4.2.22) and the definition of the function α we obtain

$$\big(\tilde{w}(t_{k_i}) - v(t_{k_i})\big)^{q-1} \le \frac{d_{k_i}\exp\big(-\int_{t_0}^{t_{k_i}} r^{1-q}(\tau)\alpha(v(\tau), \tilde{w}(\tau))\, d\tau\big)}{1 - d_{k_i}\int_{t_0}^{t_{k_i}} r^{1-q}(\tau)\exp\big(-\int_{t_0}^\tau r^{1-q}(s)\alpha(v(s), \tilde{w}(s))\, ds\big)\, d\tau}.$$

If we pass to the limit in this inequality, in view of (4.2.21) and the equality $\lim_{i\to\infty} d_{k_i} = 0$, we obtain the contradiction. □

Using the previous statement, we can now prove the following necessary condition for a solution w of Riccati equation (1.1.21) to be the minimal one.

Corollary 4.2.2. *Let*

$$\int_{t_0}^\infty r^{1-q}(t)\, dt = \infty \tag{4.2.26}$$

and let $\tilde{w}(t) > 0$ for $t \geq t_0$ be the minimal solution of Riccati equation (1.1.21). *Then*

$$\int_{t_0}^{\infty} r^{1-q}(t) \exp\left(- m_* \int_{t_0}^{t} r^{1-q}(\tau)\tilde{w}^{q-1}(\tau)\, d\tau \right) dt = \infty. \tag{4.2.27}$$

In particular, if $\tilde{x}$ is the principal solution of (1.1.1), *then*

$$\int^{\infty} \frac{dt}{r^{q-1}(t)|\tilde{x}(t)|^{m_*}} = \infty. \tag{4.2.28}$$

Proof. We assume the contrary. Then in view of (4.2.26) we have

$$\int_{t_0}^{\infty} r^{1-q}(t)\tilde{w}^{q-1}(t)\, dt = \infty$$

and hence

$$\liminf_{t\to\infty} \tilde{w}^{-(q-1)}(t) \exp\left(-m_* \int_{t_0}^{t} r^{1-q}(s)\tilde{w}^{q-1}(s)\, ds\right) = 0.$$

Indeed, if this is not the case, then there exists a number $d > 0$ such that

$$d\tilde{w}^{q-1}(t) < \exp\left(-m_* \int_{t_0}^{t} r^{1-q}(\tau)\tilde{w}^{q-1}(\tau)\, d\tau\right)$$

for $t \geq t_0$. Consequently,

$$d \int_{t_0}^{t} r^{1-q}(\tau)\tilde{w}^{q-1}(\tau)\, d\tau \leq \int_{t_0}^{t} r^{1-q}(\tau) \exp\left(- m_* \int_{t_0}^{\tau} r^{1-q}(s)\tilde{w}^{q-1}(s)\, ds\right) d\tau.$$

Since we have assumed that (4.2.27) is violated, the right-hand side of the last inequality tends to a finite number as $t \to \infty$, while the left-hand side tends to ∞, i.e., we obtain a contradiction. Now, if we set $v(t) \equiv 0$ in Theorem 4.2.5 and note that $\alpha(0, z) = m_* z^{q-1}$, then we conclude that (4.2.27) is valid by virtue of Theorem 4.2.5. Divergence of the integral in (4.2.28) follows from the relationship between solutions of (1.1.1) and (1.1.21). □

Now we turn our attention to a sufficient integral condition for a solution of (1.1.21) to be the minimal one.

Theorem 4.2.6. *Let a continuous function v, defined on $[t_0, \infty)$, be such that for any solution w of Riccati equation* (1.1.21), *the inequality $v(t) \leq w(t)$ is valid for $t \geq t_0$, and a solution $\tilde{w}$ of* (1.1.21), *defined on $[t_0, \infty)$, satisfies the condition*

$$\int_{t_0}^{\infty} r^{1-q}(t) \exp\left(- \int_{t_0}^{t} r^{1-q}(\tau)\beta\big(v(\tau), \tilde{w}(\tau)\big)\, d\tau \right) dt = \infty. \tag{4.2.29}$$

Then $\tilde{w}$ is the minimal solution of Riccati equation (1.1.21).

Proof. First we note that if $\tilde{w}(t_1) = v(t_1)$ for some $t_1 \in [t_0, \infty)$, then $\tilde{w}$ is the minimal solution. Indeed, if not, then there exists a solution w of (1.1.21), defined on $[t_0, \infty)$ and satisfying the inequality $w(t) < \tilde{w}(t)$ for $t \geq t_0$. But then $w(t_1) < \tilde{w}(t_1) = v(t_1)$, which contradicts the fact how the function v is chosen.

Now we consider the case $v(t) < \tilde{w}(t)$ for $t \geq t_0$ and suppose that the conclusion of the theorem is not true. Then there exists a solution w of (1.1.21) satisfying the inequalities

$$v(t) \leq w(t) < \tilde{w}(t) \quad \text{for } t \geq t_0. \tag{4.2.30}$$

Since for $t \geq t_0$,

$$\big(\tilde{w}(t) - w(t)\big)^{q-1} = \frac{d_0 \exp\big(- \int_{t_0}^t r^{1-q}(\tau) F(w(\tau), \tilde{w}(\tau))\, d\tau \big)}{1 - d_0 \int\limits_{t_0}^{t} r^{1-q}(\tau) \exp\big(- \int_{t_0}^{\tau} r^{1-q}(s) F(w(s), \tilde{w}(s))\, ds\big)\, d\tau},$$

where $d_0 = (\tilde{w}(t_0) - w(t_0))^{q-1}$, by virtue of (4.2.30) we obtain the inequality

$$\big(\tilde{w}(t) - w(t)\big)^{q-1} \geq \frac{d_0 \exp\big(- \int_{t_0}^t r^{1-q}(\tau) \beta(v(\tau), \tilde{w}(\tau))\, d\tau \big)}{1 - d_0 \int_{t_0}^{t} r^{1-q}(\tau) \exp\big(- \int_{t_0}^{\tau} r^{1-q}(s) \beta(v(s), \tilde{w}(s))\, ds\big)\, d\tau}.$$

The last inequality yields the contradiction in view of (4.2.29). □

As a consequence of the previous theorem we have the following statement.

Corollary 4.2.3. *Let* (4.2.26) *hold, the integral* $\int^\infty c(t)\, dt$ *converge with*

$$C(t) := \int_t^\infty c(s)\, ds \geq 0 \quad \textit{for large } t, \tag{4.2.31}$$

and let w *be an eventually nonnegative solution of Riccati equation* (1.1.21) *satisfying*

$$\int_{t_0}^\infty r^{1-q}(t) \exp\left(- m^* \int_{t_0}^t r^{1-q}(\tau) w^{q-1}(\tau)\, d\tau \right) dt = \infty. \tag{4.2.32}$$

Then w *is the minimal solution of* (1.1.21). *In particular, if*

$$\int^\infty \frac{dt}{r^{q-1}(t) x^{m^*}(t)} = \infty, \tag{4.2.33}$$

then x *is the principal solution of* (1.1.1).

Proof. According to our assumptions, the function w solves the Riccati integral equation

$$w(t) = (p-1) \int_t^\infty r^{1-q}(\tau) |w(\tau)|^q\, d\tau + C(t), \tag{4.2.34}$$

which implies that $w(t) \geq 0$ for large t. Now, the statement follows from the previous theorem with $v(t) \equiv 0$ taking into account that $\beta(0, w) = m^* w^{q-1}$. □

4.2.8 Limit characterization of the principal solution

The "most characteristic" property of the principal solution in the linear case is the limit characterization (4.2.2). In this subsection we show that the limit characterization holds provided $c(t) \neq 0$ for large t also for half-linear equation (1.1.1), i.e., we prove that $\tilde{x}$ is the principal solution of (1.1.1) if and only if

$$\lim_{t\to\infty} \frac{\tilde{x}(t)}{x(t)} = 0 \tag{4.2.35}$$

for any solution x of (1.1.1) linearly independent of $\tilde{x}$.

First we reformulate and complement some statements of the previous section to be directly applicable in the proof of the limit characterization of the principal solution.

When $c(t) < 0$ for large t, we denote conditions involving integrals J_1, J_2 as follows:

$$(\mathrm{C}_1)\ J_2 = \infty; \qquad (\mathrm{C}_2)\ J_1 = \infty,\ J_2 < \infty; \qquad (\mathrm{C}_3)\ J_1 < \infty,\ J_2 < \infty,$$

where the integrals J_1, J_2 are defined by (4.1.3)

Lemma 4.2.1. *Suppose that $c(t) < 0$ for large t.*

(i) If (C_1) holds, then $\mathbb{M}^- = \mathbb{M}_0^- \neq \emptyset$.

(ii) If (C_2) holds, then $\mathbb{M}^- = \mathbb{M}_B^- \neq \emptyset$.

(iii) If (C_3) holds, then $\mathbb{M}_0^- \neq \emptyset$ and $\mathbb{M}_B^- \neq \emptyset$.

In addition, when (C_1) or (C_3) holds, there exists a unique solution x of (1.1.1) in the class $\mathbb{M}_0^-$ such that $x(t_0) = \mu$ for any $(t_0, \mu) \in [0, \infty) \times \mathbb{R} \setminus \{0\}$. When (C_2) holds, there exists a unique solution x of (1.1.1) in the class $\mathbb{M}_B^-$ such that $x(t_0) = \mu$ for any $(t_0, \mu) \in (0, \infty) \times \mathbb{R} \setminus \{0\}$.

Proof. Claims (i)–(iii) follow from [59, Theorem 9], see also Theorem 4.1.4. The existence of solutions in $\mathbb{M}^-$ has been shown in [71]. The uniqueness can be obtained by using a similar argument as given in the final part of the proof of [57, Theorem 4]. □

Lemma 4.2.2. *Let $c(t) < 0$ for large t and assume the case (C_3) of the previous lemma. If x is a solution of (1.1.1) in the class $\mathbb{M}_0^-$, then*

$$\lim_{t\to\infty} r(t)\Phi(x'(t)) \neq 0.$$

Proof. Assume $\lim_{t\to\infty} r(t)\Phi(x'(t)) = 0$ and, without loss of generality, suppose $0 < x(t) \leq 1$, $x'(t) < 0$ for $t \geq T \geq 0$. Integrating (1.1.1) over (t, ∞), $t \geq T$, we obtain

$$-r(t)\Phi(x'(t)) = \int_t^\infty c(s)\Phi(x(s))\, ds \leq \Phi(x(t)) \int_t^\infty c(s)\, ds.$$

Since $x(t) \le 1$, we have

$$-\frac{\Phi(x'(t))}{\Phi(x(t))} \le \frac{1}{r(t)} \int_t^\infty c(s)\, ds$$

or

$$-\frac{x'(t)}{x(t)} < \Phi^{-1}\left(\frac{1}{r(t)} \int_t^\infty c(s)\, ds\right).$$

Integrating on $(T, t), t \ge T$ we obtain

$$-\log \frac{x(t)}{x(T)} < \int_T^t \Phi^{-1}\left(\frac{1}{r(\tau)} \int_\tau^\infty c(s)\, ds\right) d\tau,$$

that gives a contradiction as $t \to \infty$. □

Lemma 4.2.3. *Let $c(t) > 0$ for large t.*

(i) Assume $J_r = \infty$. Then any bounded nonoscillatory solution x of (1.1.1) satisfies $\lim_{t\to\infty} x^{[1]}(t) = 0$.

(ii) Assume that (1.1.1) is nonoscillatory and $J_r < \infty$. Then (1.1.1) has a solution x such that $\lim_{t\to\infty} x(t) = 0$.

(iii) Assume $J_r < \infty, J_c < \infty$. Then any solution x of (1.1.1) in $\mathbb{M}^-$ satisfies

$$\lim_{t\to\infty} x^{[1]}(t) = \ell_x,$$

where $0 < |\ell_x| < \infty$.

Proof. Claim (i): Let x be a bounded nonoscillatory solution of (1.1.1). In view of Lemma 4.2.3, $x \in \mathbb{M}^+$ and, without loss of generality, suppose $x(t) > 0, x'(t) > 0$ for $t \ge t_x \ge 0$. Because $x^{[1]}$ is (positive) decreasing on (t_x, ∞), if $x^{[1]}(\infty) > 0$, we have

$$x(t) > x(t_x) + \Phi^{-1}(x^{[1]}(\infty)) \int_{t_x}^t r^{1-q}(s)\, ds$$

that gives a contradiction as $t \to \infty$.

Claim (ii): If $J_c < \infty$, the assertion follows, as a particular case, from [286, Theorem 2.2]. Assume now $J_c = \infty$ and consider the reciprocal equation (4.2.15) (its solutions are of the form $y = r\Phi(x') = x^{[1]}$). Taking into account the reciprocity principle and the fact that $y^{[1]} = \Phi^{-1}(y'/c)$, it is sufficient to show that there exists a solution y of (4.2.15) such that $y^{[1]}(\infty) = 0$. Assume that all (nonoscillatory) solutions y of (4.2.15) satisfy $y^{[1]}(\infty) \ne 0$. Let v be a principal solution of (4.2.15) and, without loss of generality, suppose $v(t) \ne 0, v^{[1]}(t) \ne 0$ for $t \ge t_v \ge 0$. Because $|v^{[1]}|$ is decreasing, we have

$$A_v := \lim_{t\to\infty} \int_{t_v}^t \frac{v'(s)}{v^2(s) v^{[1]}(s)} ds \le \frac{1}{|v^{[1]}(\infty)|} \lim_{t\to\infty} \left(\frac{1}{v(t_v)} - \frac{1}{v(t)}\right).$$

Because $v \in \mathbb{M}^+$, we have $A_v < \infty$, i.e., a contradiction with Theorem 4.2.8 given in the next subsection.

Claim (iii): Without loss of generality assume $x(t) > 0$, $x'(t) < 0$ for $t \geq t_x \geq 0$. Integrating (1.1.1) on (t_x, t) we obtain

$$x^{[1]}(t) - x^{[1]}(t_x) = -\int_{t_x}^{t} c(s)\Phi(x(s))ds \geq -\Phi(x(t_x))\int_{t_x}^{t} c(s)ds.$$

Taking into account that $x^{[1]}$ is negative decreasing, as $t \to \infty$ the assertion follows. □

Now, we are ready to formulate the main result of this subsection.

Theorem 4.2.7. *Suppose that* (1.1.1) *is nonoscillatory and* $c(t) \neq 0$ *for large* t. *Then a solution* $\tilde{x}$ *is principal if and only if the limit characterization* (4.2.35) *holds for every solution* x *linearly independent of* $\tilde{x}$.

Proof. The proof follows different ideas in cases $c(t) < 0$ and $c(t) > 0$, so we divide it into two parts.

(I) *The case* $c(t) < 0$ *for large* t.

"⇒": Let $\tilde{x}$ be a principal solution of (1.1.1). Since in case $c(t) < 0$ both classes $\mathbb{M}^+$, $\mathbb{M}^-$ are nonempty, we have $\tilde{x} \in \mathbb{M}^-$ and there are two possibilities: (i) $\tilde{x}(\infty) = 0$, (ii) $\tilde{x}(\infty) \neq 0$. Suppose case (i); if $x \in \mathbb{M}^+$, (4.2.35) is clearly satisfied; if $x \in \mathbb{M}^-$, in view of the homogeneity property, the function

$$w(t) = \frac{\tilde{x}(0)}{x(0)}x(t)$$

is a solution of (1.1.1), such that $w(0) = \tilde{x}(0) \neq 0$. By uniqueness in $\mathbb{M}_0^-$, it holds $x(\infty) \neq 0$, that yields the assertion. Suppose case (ii) and, without loss of generality, assume $\tilde{x}(t) > 0$, $\tilde{x}'(t) < 0$. Integrating (4.2.15) on $(T, t), T$ large, we obtain

$$0 < \tilde{x}(\infty) < \tilde{x}(T)\frac{x(t)}{x(T)}.$$

Then no solution of (1.1.1) converges to zero as $t \to \infty$ and, from Lemma 4.2.1, the case (C_2) holds. Using again homogeneity property, it is easy to show that $x \in \mathbb{M}^+$. In view of [59, Theorem 4] and also Theorem 4.1.4, x is unbounded and (4.2.35) follows.

"⇐": Without loss of generality, assume there exists an eventually positive solution $\tilde{x}$ such that (4.2.35) holds for any solution x of (1.1.1) such that $x \neq \lambda\tilde{x}$, $\lambda \in \mathbb{R}$. Clearly $\tilde{x} \in \mathbb{M}^-$, otherwise, if $\tilde{x} \in \mathbb{M}^+$, by choosing $x \in \mathbb{M}^-$, condition (4.2.35) fails. When $x \in \mathbb{M}^+$, condition (4.2.5) is satisfied. Now assume $x \in \mathbb{M}^-$, and, without loss of generality, suppose $x(t) > 0$, $x'(t) < 0$. Taking into account the homogeneity property and the uniqueness in $\mathbb{M}_0^-$, we obtain $x(\infty) > 0$ and so, from Lemma 4.2.1, the case (C_3) holds. Applying Lemma 4.2.2 and taking into account that $r(t)\Phi(x'(t))$ is negative increasing, there exists a positive constant m such that

$$\Phi\left(\frac{\tilde{x}'(t)}{x'(t)}\right) = \frac{r(t)\Phi(\tilde{x}'(t))}{r(t)\Phi(x'(t))} > m > 0, \tag{4.2.36}$$

i.e.,

$$\frac{\tilde{x}'(t)}{x'(t)} > \Phi^{-1}(m) > 0.$$

Taking into account (4.2.35), we obtain for large t

$$\frac{\tilde{x}(t)}{x(t)} < \frac{\tilde{x}'(t)}{x'(t)},$$

i.e., (4.2.5) holds with $\tilde{w}$ instead of w_∞.

(II) *The case $c(t) > 0$ for large t.*

"$\Leftarrow$": Assume that (4.2.35) holds for any solution x of (1.1.1) such that $x \neq \lambda\tilde{x}$, $\lambda \in \mathbb{R}$. Suppose that $\tilde{x}$ is a nonprincipal solution of (1.1.1) and let z be a principal solution of (1.1.1). Without loss of generality, assume $\tilde{x}(t) > 0$, $z(t) > 0$ for $t \geq t_1 \geq 0$. Then for $t \geq t_1$ we obtain

$$z'(t)/z(t) < \tilde{x}'(t)/\tilde{x}(t) \tag{4.2.37}$$

and, because $\tilde{x} \neq \mu z$ for any $\mu \in \mathbb{R}$, from (4.2.35) we have

$$\lim_{t\to\infty} \tilde{x}(t)/z(t) = 0. \tag{4.2.38}$$

In view of (4.2.37), the ratio $\tilde{x}(t)/z(t)$ is positive increasing, that gives a contradiction with (4.2.38).

"$\Rightarrow$": Assume now that $\tilde{x}$ is the principal solution of (1.1.1) and let us show that (4.2.35) holds for any solution x of (1.1.1) such that $x \neq \lambda\tilde{x}$, $\lambda \in \mathbb{R}$. Without loss of generality, assume $\tilde{x}(t) > 0$, $x(t) > 0$ for $t \geq t_1 \geq 0$. In view of (4.2.37), the ratio $\tilde{x}(t)/x(t)$ is (positive) decreasing. Three cases are possible: (A) $J_r = \infty, J_c < \infty$; (B) $J_r < \infty, J_c = \infty$; (C) $J_r < \infty, J_c < \infty$.

Case (A): There are two possibilities: either (A_1) all solutions of (1.1.1) are unbounded (as $t \to \infty$), or (A_2) (1.1.1) has a bounded solution.

Assume (A_1): In view of Lemma 4.2.3, $\tilde{x}$, $x \in \mathbb{M}^+$. From Theorem 4.2.4, $\tilde{x}^{[1]}$ is a principal solution of (4.2.15) and so the ratio $\tilde{x}^{[1]}(t)/x^{[1]}(t)$ is (positive) decreasing. Then there exists the limit

$$\lim_{t\to\infty} \frac{\tilde{x}^{[1]}(t)}{x^{[1]}(t)} = \lim_{t\to\infty} \Phi\left(\frac{\tilde{x}'(t)}{x'(t)}\right) = L,\ L \geq 0.$$

By L'Hospital's rule we obtain

$$\lim_{t\to\infty} \frac{\tilde{x}(t)}{x(t)} = \lim_{t\to\infty} \frac{\tilde{x}'(t)}{x'(t)} = \Phi^{-1}(L).$$

If $L = 0$, the assertion follows. If $L > 0$, we have

$$\lim_{t\to\infty} \left[\frac{\tilde{x}'(t)}{\tilde{x}^2(t)\tilde{x}^{[1]}(t)}\right]\left[\frac{x'(t)}{x^2(t)x^{[1]}(t)}\right]^{-1} = \lim_{t\to\infty} \frac{\tilde{x}'(t)}{x'(t)} \cdot \frac{x^2(t)}{\tilde{x}^2(t)} \cdot \frac{x^{[1]}(t)}{\tilde{x}^{[1]}(t)} = \frac{1}{L\Phi^{-1}(L)} > 0$$

Hence both integrals

$$\int_{t_1}^{t} \frac{\tilde{x}'(s)}{\tilde{x}^2(s)\tilde{x}^{[1]}(s)}\, ds, \quad \int_{t_1}^{t} \frac{x'(s)}{x^2(s)x^{[1]}(s)}\, ds$$

have the same behavior as $t \to \infty$, that contradicts Theorem 4.2.8 given in the next subsection, because $\tilde{x}$ is the principal solution of (1.1.1) and x is a nonprincipal solution.

Assume (A_2): From (4.2.15) we obtain, for $t > t_1$, $\tilde{x}(t) < [\tilde{x}(t_1)/x(t_1)]x(t)$, that yields the boundedness of $\tilde{x}$. If x is unbounded, the assertion immediately follows. Now assume x bounded and let $\lim_{t\to\infty}[\tilde{x}(t)/x(t)] = d \geq 0$. In view of Lemma 4.2.3-(i), $\lim_{t\to\infty} \tilde{x}^{[1]}(t) = \lim_{t\to\infty} x^{[1]}(t) = 0$ and, by L'Hospital's rule, we obtain

$$\lim_{t\to\infty} \frac{\tilde{x}^{[1]}(t)}{x^{[1]}(t)} = \lim_{t\to\infty} \Phi\left(\frac{\tilde{x}(t)}{x(t)}\right) = \Phi(d).$$

Using again a similar argument to that given in the final part of the proof of claim (A_1), we obtain a contradiction and in case (A) the proof is complete.

Case (B): Taking into account Lemma 4.1.3 and Lemma 4.2.3-(ii), clearly $\tilde{x}(\infty) = 0$. If $x(\infty) > 0$, the assertion follows. Now assume $x(\infty) = 0$. In view of Theorem 4.2.4, applying the same argument as in case (A) to the reciprocal equation, we obtain

$$\lim_{t\to\infty} \frac{\tilde{x}^{[1]}(t)}{x^{[1]}(t)} = 0,$$

that implies $\lim_{t\to\infty}[\tilde{x}'(t)/x'(t)] = 0$. Hence the assertion follows by using the L'Hospital's rule.

Case (C): In view of Lemma 4.2.3 (ii)-(iii), we have $\tilde{x}(\infty) = 0$, $\tilde{x}^{[1]}(\infty) \neq 0$. If $x \in \mathbb{M}^+$, or $x \in \mathbb{M}^-$ and $x(\infty) \neq 0$, then the assertion follows. Now suppose $x \in \mathbb{M}^-$, $x(\infty) = 0$. In view of Lemma 4.2.3-(iii), we have $x^{[1]}(\infty) \neq 0$. Taking into account the homogeneity property, we can suppose $\tilde{x}^{[1]}(\infty) = x^{[1]}(\infty)$. Now define $v = \tilde{x}^{[1]}(t)$, $y = x^{[1]}(t)$. Then v and y are solutions of the reciprocal equation (4.1.1) and $v, y \in \mathbb{M}^+$. Because $v(\infty) = y(\infty) \neq 0$ and $v^{[1]}(\infty) = y^{[1]}(\infty) = 0$, we obtain a contradiction with Theorem 4.1.7 and the proof is complete. □

4.2.9 Integral characterization of the principal solution

Among all (equivalent) characterizations of the principal solution of linear equation (1.1.2), the most suitable seems be the integral one (4.2.1), since it needs to know just one solution and according to the divergence/convergence of the characterizing integral it is possible to decide whether or not it is the principal solution. The remaining characterizations require to know other solutions since they are of comparison type. In the linear case, this is not serious disadvantage because of the reduction of order formula which enables to compute all solutions (at least locally) of the linear second order equation when one solution is already known. However, in the half-linear case we have no reduction of order formula as pointed in Section 1.3, so some kind of the integral characterization would be very useful. An attempt to find an integral characterization of the principal solution of (1.1.1) has already been presented in Subsection 4.2.7, see (4.2.28), (4.2.33). However, since $m_* < m^*$ if $p \neq 2$, these formulas do not provide the *equivalent* integral characterization of the principal solution. In the main result of this subsection we present one candidate for the equivalent integral characterization of the principal solution of (1.1.1).

First we prove an auxiliary statement concerning the function P introduced in Subsection 1.2.1.

Lemma 4.2.4. *The function* $P(u,v)$ *defined in* (1.2.2) *satisfies the following inequalities*

$$P(u,v) \gtrless \frac{1}{2}|u|^{2-p}\left(\Phi(u)-v\right)^2, \quad p \lessgtr 2, \quad \Phi(u)\neq v,\ u\neq 0, \tag{4.2.39}$$

and

$$P(u,v) \lessgtr \frac{1}{2(p-1)}|u|^{2-p}(\Phi(u)-v)^2, \quad p\lessgtr 2, \quad |\Phi(u)|>|v|,\ uv>0. \tag{4.2.40}$$

More generally, for every $T>0$ *there exists a constant* $K=K(T)$ *such that*

$$P(u,v) \gtrless K(T)|u|^{2-p}\left(\Phi(u)-v\right)^2, \quad p\gtrless 2, \quad \Phi(u)\neq v, \ \left|\frac{v}{\Phi(u)}\right|<T. \tag{4.2.41}$$

Proof. We present an outline of the proof only, for details we refer to [108, 117]. We have

$$P(u,v)=|u|^p\left\{\frac{1}{q}\left|\frac{v}{\Phi(u)}\right|^q-\frac{v}{\Phi(u)}+\frac{1}{p}\right\}$$

and

$$|u|^{2-p}\left(v-\Phi(u)\right)^2=|u|^p\left(\frac{v}{\Phi(u)}-1\right)^2.$$

Denote $F(t)=|t|^q/q-t+1/p$, $G(t)=(t-1)^2/2$. The function $H=F-G$ satisfies $H(-1)=0=H(1)$, $H(0)=1/p-1/2\gtrless 0$ for $p\lessgtr 2$ and a closer investigation of the graph of this function shows that (4.2.39) and (4.2.41) really hold. □

We will also need the following partial result concerning asymptotics of the nonprincipal solution of (1.1.1).

Lemma 4.2.5. *Assume that* (1.1.1) *is nonoscillatory and* $J_r<\infty$, $J_c=\infty$. *Let* x *be a nonprincipal solution of* (1.1.1). *Then* $\lim_{t\to\infty}|x^{[1]}(t)|=\infty$.

Proof. Let $\tilde{x}$ be a principal solution of (1.1.1). From Lemma 4.1.3 and Lemma 4.2.3-(ii) we have $\tilde{x}, x\in \mathbb{M}^-$ and $\tilde{x}(\infty)=0$. Assume $|x^{[1]}(\infty)|<\infty$. From Theorem 4.2.7, $\lim_{t\to\infty}[\tilde{x}(t)/x(t)]=0$. In view of Theorem 4.2.4, $\tilde{x}^{[1]}$ is principal solution of (4.2.15)) and so Theorem 4.2.7 yields $\lim_{t\to\infty}[\tilde{x}^{[1]}(t)/x^{[1]}(t)]=0$, that implies $\tilde{x}^{[1]}(\infty)=0$. This is a contradiction because $|\tilde{x}^{[1]}|$ is eventually positive increasing. □

Now we are ready to formulate and prove the main result of this subsection.

Theorem 4.2.8. *Suppose that equation* (1.1.1) *is nonoscillatory and* $\tilde{x}$ *is its solution such that* $\tilde{x}'(t)\neq 0$ *for large* t.

(i) Let $p \in (1,2)$. If

$$I(\tilde{x}) := \int^{\infty} \frac{dt}{r(t)\tilde{x}^2(t)|\tilde{x}'(t)|^{p-2}} = \infty, \tag{4.2.42}$$

then $\tilde{x}$ is the principal solution.

(ii) Let $p > 2$. If $\tilde{x}$ is the principal solution, then (4.2.42) holds.

(iii) Suppose that $\int^{\infty} r^{1-q}(t)\,dt = \infty$, the function $\gamma(t) := \int_t^{\infty} c(s)\,ds$ exists and $\gamma(t) \geq 0$, but $\gamma(t) \not\equiv 0$ eventually. Then $\tilde{x}(t)$ is the principal solution if and only if (4.2.42) holds.

(iv) Let $c(t) > 0$ for large t, $\int^{\infty} r^{1-q}(t)\,dt < \infty$ and $\int^{\infty} c(t)\,dt = \infty$. Then $\tilde{x}(t)$ is the principal solution if and only if (4.2.42) holds.

Proof. (i) Suppose, by contradiction, that a (positive) solution x of (1.1.1) satisfying (4.2.42) is not principal. Then the corresponding solution $w_x = r\Phi(x'/x)$ of the associated Riccati equation (1.1.21) is not eventually minimal. Hence, there exists another nonoscillatory solution y of (1.1.1) such that

$$w_y = r\Phi(y'/y) < w_x \quad \text{eventually.} \tag{4.2.43}$$

Due to the Picone identity given in Subsection 1.2.1 we have

$$r(t)|x'|^p - c(t)x^p = [x^p w_y]' + pr^{1-q}(t)x^p P\left(\Phi^{-1}(w_x), w_y\right)$$

and at the same time

$$r(t)|x'|^p - c(t)x^p = (x^p w_x)' - x\left[(r(t)\Phi(x'))' + c(t)\Phi(x)\right] = (x^p w_x)'.$$

Subtracting the last two equalities, we get

$$[x^p(w_x - w_y)]' = pr^{1-q}(t)x^p P(\Phi^{-1}(w_x), w_y).$$

Let $f(t) = x^p(w_x - w_y)$. By (4.2.43) there exists T sufficiently large such that $f(t) > 0$ for $t \geq T$. Then by Lemma 4.2.4 we have

$$\begin{aligned} \frac{f'}{f^2} &= \frac{p}{f^2} r^{1-q}(t)x^p P(\Phi_q(w_x), w_y) \\ &> \frac{p}{2}\frac{x^p r^{1-q}(t)}{[x^p(w_x - w_y)]^2} \left|r^{q-1}(t)x'/x\right|^{2-p} (w_x - w_y)^2 \\ &= \frac{p}{2r(t)x^2|x'|^{p-2}}. \end{aligned}$$

Integrating the last inequality from T to T_1 ($T_1 > T$), we have

$$\frac{1}{f(T)} > \frac{1}{f(T)} - \frac{1}{f(T_1)} \geq \frac{p}{2}\int_T^{T_1} \frac{dt}{r(t)x^2(t)|x'(t)|^{p-2}}$$

and letting $T_1 \to \infty$ we are led to contradiction. Hence a solution satisfying (4.2.42) is principal.

(ii) We proceed again by contradiction. Suppose that $\tilde{x}$ is the principal solution and $I(\tilde{x}) < \infty$. Let T be chosen so large that $\tilde{x}(t) > 0$ for $t \geq T$ and

$$\int_T^\infty \frac{dt}{r(t)\tilde{x}^2(t)|\tilde{x}'(t)|^{p-2}} < \frac{1}{p}.$$

Consider the solution $\overline{w}(t)$ of Riccati equation (1.1.21) given by the initial condition

$$\overline{w}(T) = \tilde{w}(T) - \frac{1}{2\tilde{x}^p(T)},$$

where $\tilde{w} = r\Phi(\tilde{x}'/\tilde{x})$, i.e., $\overline{w}(T) < \tilde{w}(T)$. We want to show that $\overline{w}(t)$ is extensible up to ∞. To this end, denote $f(t) = \tilde{x}^p(t)(\tilde{w}(t) - \overline{w}(t))$. Then $f(T) = 1/2$ and using the Picone identity, we have

$$\frac{f'}{f^2} = \frac{pr^{1-q}(t)}{f^2} P(r^{q-1}(t)\tilde{x}', \overline{w}\Phi(\tilde{x})),$$

hence, integrating this identity from T to t

$$\frac{1}{f(T)} - \frac{1}{f(t)} = p\int_T^t \frac{r^{1-q}(s)\tilde{x}^p(s)}{f^2(s)} P\left(r^{q-1}(s)\frac{\tilde{x}'(s)}{\tilde{x}(s)}, \overline{w}(s)\right) ds. \tag{4.2.44}$$

By (4.2.39) of Lemma 4.2.4 we have

$$P\left(\frac{r^{q-1}\tilde{x}'}{\tilde{x}}, \overline{w}\right) \leq \frac{1}{2}\left|\frac{r^{q-1}\tilde{x}'}{\tilde{x}}\right|^{2-p} (\tilde{w} - \overline{w})^2,$$

which means, using (4.2.44) and taking into account that $f(T) = 1/2$,

$$\begin{aligned} f(t) &\leq \left(2 - p\int_T^t \frac{r^{1-q}(s)\tilde{x}^p(s)}{f^2(s)} P(r^{q-1}(s)\tilde{x}'(s)/\tilde{x}(s), \overline{w}(s))\, ds\right)^{-1} \\ &\leq \left(2 - p\int_T^\infty \frac{dt}{2r(t)\tilde{x}^2(t)|\tilde{x}'(t)|^{p-2}}\right)^{-1} \leq 1. \end{aligned}$$

Consequently, $1/2 \leq f(t) \leq 1$ and $f(t)$ can be continued to ∞, hence $\overline{w}(t)$ is a continuable up to infinity solution of (1.1.21) and $\overline{w}(t) < \tilde{w}(t)$ for $t \geq T$, i.e., $\tilde{w}(t)$ is not minimal. Thus, the solution $\tilde{x}(t)$ is not principal, which was to be proved.

(iii) The principal solution $\tilde{x}(t)$ of (1.1.1) is associated with the minimal solution $\tilde{w}(t)$ of (1.1.21) and hence it is also the minimal solution of the Riccati integral equation (the convergence of $\int^\infty r^{1-q}(t)|w(t)|^q\, dt$ follows from Theorem 2.2.3)

$$\tilde{w}(t) = \gamma(t) + (p-1)\int_t^\infty r^{1-q}(s)|\tilde{w}(s)|^q ds, \qquad t \geq T_1,$$

and by the assumptions on $\gamma(t)$, there exists $T \in \mathbb{R}$ such that $\tilde{w}(t) > 0$ for $t \geq T$. Since $\tilde{w}$ is the minimal solution, for any other solution w of (1.1.21) we have

$w(t) > \tilde{w}(t) > 0$ for $t \geq T_1 \geq T$, and hence the associated solutions $x(t)$ and $\tilde{x}(t)$ satisfy the inequalities $x'(t) > 0$, $\tilde{x}'(t) > 0$ for $t \geq T_1$.

Now the proof goes in different ways according to $1 < p < 2$ or $p \geq 2$.

Case A ($1 < p < 2$): By the part (i) it is sufficient to show that the integral in (4.2.42) is really divergent. Suppose the contrary, i.e.,

$$\int^{\infty} \frac{dt}{r(t)\tilde{x}^2(t)|\tilde{x}'(t)|^{p-2}} < \infty.$$

Let $T_2 \geq T_1$ be chosen so large that

$$\int_{T_2}^{\infty} \frac{dt}{r(t)\tilde{x}^2(t)|\tilde{x}'(t)|^{p-2}} \leq \frac{2(p-1)}{p}.$$

Consider the solution $\bar{w}(t)$ of (1.1.21) with the initial condition

$$\bar{w}(T_2) = \tilde{w}(T_2) - \frac{1}{2\tilde{x}^2(T_2)},$$

and accordingly, let the function $f(t)$ be defined by

$$f(t) = \tilde{x}^p(t)[\tilde{w}(t) - \bar{w}(t)].$$

Clearly, $f(T_2) = 1/2$. Following the computation in the proof of the claim (i), we find

$$\frac{f'(t)}{f^2(t)} = p\frac{r^{1-q}(t)\tilde{x}^p(t)}{f^2(t)} P(\Phi^{-1}(\tilde{w}(t)), \bar{w}(t)),$$

hence by (4.2.40)

$$\frac{f'(t)}{f^2(t)} < \frac{p}{2(p-1)} \frac{1}{r(t)\tilde{x}^2(t)|\tilde{x}'(t)|^{p-2}}$$

and integrating this inequality over $[T_2, t]$ we find

$$\begin{aligned} \frac{1}{f(T_2)} - \frac{1}{f(t)} &< \frac{1}{2}\frac{p}{p-1} \int_{T_2}^{t} \frac{ds}{r(s)\tilde{x}^2(s)|\tilde{x}'(s)|^{p-2}} \\ &< \frac{p}{2(p-1)} \int_{T_2}^{\infty} \frac{ds}{r(s)\tilde{x}^2(s)|\tilde{x}'(s)|^{p-2}} \leq 1, \end{aligned}$$

consequently, $1/2 \leq f(t) \leq 1$ for $t \geq T_2$. Thus, the function $\bar{w}(t)$ exists on $[T_2, \infty)$ and $\bar{w}(t) < \tilde{w}(t)$, i.e., $\tilde{w}(t)$ is not minimal solution of (1.1.21), hence $\tilde{x}(t)$ is not the principal solution, and this contradiction proves the first case.

Case B ($p \geq 2$): By the claim (ii), it is sufficient to show that if the solution x is not principal then the corresponding integral in (4.2.42) is convergent. Let $w(t) = r(t)\Phi(x'(t)/x(t))$ be the associated solution of (1.1.21). Then $w(t)$ is not minimal solution of (1.1.21) and let $\tilde{w}(t)$ be the minimal solution of this equation. Then we have $w(t) > \tilde{w}(t)$ for $t \geq T_2$ with T_2 sufficiently large. Consider the function $f(t)$ given again by

$$f(t) = x^p(t)[w(t) - \tilde{w}(t)] > 0 \quad \text{for } t \geq T_2.$$

By inequality (4.2.40) given in Lemma 4.2.4 we have again

$$\frac{f'}{f^2} = \frac{p}{f^2} r^{1-q} x^p P(\Phi^{-1}(w), \tilde{w}) > \frac{p}{2(p-1)} \frac{1}{r x^2 |x'|^{p-2}}, \qquad t \geq T_2,$$

hence

$$\frac{1}{f(T_2)} > \frac{1}{f(T_2)} - \frac{1}{f(t)} \geq \frac{p}{2(p-1)} \int_{T_2}^{t} \frac{ds}{r x^2 |x'|^{p-2}},$$

then letting $t \to \infty$ we obtain the desired result

$$\int^{\infty} \frac{dt}{r(t) x^2(t) |x'(t)|^{p-2}} < \infty.$$

(iv) In view of Lemma 4.1.3, let $\tilde{x}$ be a principal solution of (1.1.1) such that $\tilde{x}(t) > 0, \tilde{x}^{[1]}(t) < 0$ for $t \geq T$. Using Theorem 4.2.4 and applying the statement of the part (iii) to the reciprocal equation (4.2.15) we obtain

$$\int_{T}^{\infty} \frac{c(t)\Phi(\tilde{x}(t))}{\tilde{x}(t)(\tilde{x}^{[1]}(t))^2}\, dt = \infty.$$

We have

$$\begin{aligned} I(\tilde{x}) &= \int_{T}^{t} \frac{\tilde{x}'(s)}{\tilde{x}^2(s)\tilde{x}^{[1]}(s)}\, ds = -\frac{1}{\tilde{x}(t)\tilde{x}^{[1]}(t)} + L + \int_{T}^{t} \frac{c(s)\Phi(\tilde{x}(s))}{\tilde{x}(s)(\tilde{x}^{[1]}(s))^2}\, ds \\ &> L + \int_{T}^{t} \frac{c(s)\Phi(\tilde{x}(s))}{\tilde{x}(s)(\tilde{x}^{[1]}(s))^2}\, ds, \end{aligned}$$

where $L = [\tilde{x}(T) u^{[1]}(T)]^{-1}$, i.e., (4.2.42) holds.

Vice versa, let x be a nonprincipal solution of (1.1.1) and let us show that

$$I(x) = \int^{\infty} \frac{x'(s)}{x^2(s) x^{[1]}(s)}\, ds < \infty. \tag{4.2.45}$$

Without loss of generality suppose $x(t) > 0$, $x^{[1]}(t) < 0$ for $t \geq t_x \geq 0$. By using Theorem 4.2.4 and applying again the part (iii) to (4.2.15) we obtain

$$\int_{t_x}^{\infty} \frac{c(t)\Phi(x(t))}{x(t)(x^{[1]}(t))^2}\, dt < \infty. \tag{4.2.46}$$

Hence the function

$$F_x(t) = \int_{t_x}^{t} \frac{c(s)\Phi(x(s))}{x(s)(x^{[1]}(s))^2}\, ds - \int_{t_x}^{t} \frac{x'(s)}{x^2(s) x^{[1]}(s)}\, ds$$

admits limit as $t \to \infty$ and $\lim_{t\to\infty} F_x(t) = L_x$, $-\infty \leq L_x < \infty$. Put $d_x = [x(t_x) x^{[1]}(t_x)]^{-1}$. From the identity

$$x(t)x^{[1]}(t) = \left[d_x + \int_{t_x}^{t} \frac{c(s)\Phi(x(s))}{x(s)(x^{[1]}(s))^2}\, ds - \int_{t_x}^{t} \frac{x'(s)}{x^2(s)x^{[1]}(s)}\, ds \right]^{-1} \tag{4.2.47}$$

also the function $x(t)x^{[1]}(t)$ has a limit as $t \to \infty$ and $\lim_{t\to\infty} x(t)x^{[1]}(t) = \ell_x$, $-\infty \le \ell_x \le 0$. We claim that

$$\lim_{t\to\infty} x(t)x^{[1]}(t) = -\infty. \tag{4.2.48}$$

Otherwise there exists a positive constant K such that $x(t)x^{[1]}(t) > -K$ and

$$\int_{t_x}^{t} \frac{c(s)\Phi(x(s))}{x(s)(x^{[1]}(s))^2}\, ds > \frac{1}{K}\int_{t_x}^{t} \frac{c(s)\Phi(x(s))}{|x^{[1]}(s)|} ds = \frac{1}{K}\log\frac{|x^{[1]}(t)|}{|x^{[1]}(t_x)|},$$

that gives a contradiction with (4.2.46) as $t \to \infty$, because, from Lemma 4.2.5 we have $\lim_{t\to\infty} |x^{[1]}(t)| = \infty$. Hence (4.2.48) holds and, from (4.2.47) we obtain (4.2.45). □

Remark 4.2.3. The equivalent integral characterization of the principal solution of (1.1.1) is stated in the parts (iii) and (iv) of the previous theorem is proved under some restriction on the functions r, c in (1.1.1). In order to better understand these restrictions, the concept of the *regular* half-linear equation has been introduced in [117] as follows. A nonoscillatory equation (1.1.1) is said to be regular if there exists a constant $K \ge 0$ such that

$$\limsup_{t\to\infty} \left|\frac{w_1(t)}{w_2(t)}\right| \le K$$

for any pair of solutions w_1, w_2 of the associated Riccati equation (1.1.21) such that $w_2(t) > w_1(t)$ eventually. It was shown that for regular half-linear equation, (4.2.42) holds if and only if the solution $\tilde{x}$ is principal and that under assumptions of (iii) and (iv) of the previous theorem equation (1.1.1) is regular.

4.2.10 Another integral characterization

The integral characterization (4.2.42) of the principal solution of (1.1.1) reduces to the usual integral characterization of the principal solution of linear equation (4.2.1) if $p = 2$. However, this characterization applies in case $p > 2$ only to solutions x for which $x'(t) \ne 0$ eventually. Moreover, in [60, 61] examples of half-linear equations are given which show that if the assumptions of the parts (iii) and (iv) of Theorem 4.2.8 are violated, (4.2.42) is *no longer equivalent characterization* of the principal solution of (1.1.1). For this reason, another integral characterization of the principal solution was suggested in [60].

Theorem 4.2.9. *Suppose that either*

(i) $c(t) < 0$ *for large* t, *or*

(ii) $c(t) > 0$ *for large* t *and both integrals* $\int^\infty r^{1-q}(t)\, dt$, $\int^\infty c(t)\, dt$ *are convergent.*

Then a solution $\tilde{x}$ *of* (1.1.1) *is principal if and only if*

$$\int^\infty \frac{dt}{r^{q-1}(t)\tilde{x}^2(t)} = \infty. \tag{4.2.49}$$

Proof. (i) Suppose that $c(t) < 0$ for large t.

"$\Rightarrow$": Let $\tilde{x}$ be a principal solution of (1.1.1). As already claimed, $\tilde{x} \in \mathbb{M}^-$ and there are two possibilities: (a) $\tilde{x}(\infty) = 0$, (b) $\tilde{x}(\infty) \neq 0$. Suppose case (a); if $\int_T^\infty r^{1-q}(t)\,dt = \infty$, the assertion clearly follows. Assume $\int_T^\infty r^{1-q}(t)\,dt < \infty$ and, without loss of generality, suppose $\tilde{x}(t) > 0, \tilde{x}'(t) < 0$. Since, for $t \geq T$, it holds

$$r(t)\Phi(\tilde{x}'(t)) > r(T)\Phi(\tilde{x}'(T)),$$

integrating on (t, ∞) with $t > T$, we obtain

$$\tilde{x}(t) < K \int_t^\infty r^{1-q}(s)\,ds,$$

where $K = \Phi^{-1}(r(T))|\tilde{x}'(T)|$. Then

$$\begin{aligned}\mathcal{I}_{\tilde{x}} = \int_T^\infty \frac{r^{1-q}(t)}{\tilde{x}^2(t)}\,dt &> \frac{1}{K^2}\int_T^\infty \frac{r^{1-q}(t)}{[\int_t^\infty r^{1-q}(s)\,ds]^2}\,dt \\ &= \frac{1}{\int_t^\infty r^{1-q}(s)\,ds}\Big|_T^\infty = \infty,\end{aligned}$$

i.e., condition (4.2.49) is verified. Now assume the case (b) when $\tilde{x}(\infty) \neq 0$. Integrating (1.1.21) on (T, t), $t \geq T \geq 0$, and using the same argument as given in Theorem 4.2.7 (the only if part), the case (C_2) defined at the beginning of Subsection 4.2.8 holds. From [59, Lemma 2], we have $\int_T^\infty r^{1-q}(t)\,dt = \infty$ and the condition (4.2.49) follows.

"$\Leftarrow$": Let x be a nonprincipal solution of (1.1.1) and let us show that for T large it holds

$$\int_T^\infty \frac{dt}{r^{q-1}(t)x^2(t)} < \infty. \tag{4.2.50}$$

Suppose $x \in \mathbb{M}^+$ and, without loss of generality, assume $x(t) > 0, x'(t) > 0$ for $t \geq T \geq 0$. Then there exists $K > 0$ such that for $t \geq T$ it holds $r(t)\Phi(x'(t)) > K > 0$, i.e., $x'(t) > \Phi^{-1}(K)r^{1-q}(t)$. Integrating over (T, t) with $t > T$ we obtain

$$x(t) > x(T) + \Phi^{-1}(K)\int_T^t r^{1-q}(s)\,ds \geq \Phi^{-1}(K)\int_T^t r^{1-q}(s)\,ds.$$

Then for $t_2 > t_1 > T$ it holds

$$\begin{aligned}\int_{t_1}^{t_2} \frac{r^{1-q}(t)}{x^2(t)}\,dt &< -\frac{1}{K^{2(q-1)}}\int_{t_1}^{t_2} \frac{r^{1-q}(t)}{[\int_T^t r^{1-q}(s)\,ds]^2}\,dt \\ &= -\frac{1}{K^{2(q-1)}\int_T^t r^{1-q}(s)\,ds}\Big|_{t_1}^{t_2} < K^{2(1-q)}\frac{1}{\int_T^{t_1} r^{1-q}(t)\,dt} < \infty.\end{aligned}$$

As $t_2 \to \infty$ we obtain the assertion.

It remains to consider the case $x \in \mathbb{M}^-$. Let $\tilde{x}$ be a principal solution of (1.1.1). Because x is a nonprincipal solution, we obtain $\lim_{t\to\infty}[\tilde{x}(t)/x(t)] = 0$

in view of Theorem 4.2.7. Then $\tilde{x} \in \mathbb{M}_0^-$ and, using the homogeneity property and the uniqueness in $\mathbb{M}_0^-$, we obtain $x \in \mathbb{M}_B^-$. In view of Lemma 4.2.1, the case (C_3) holds, that yields $\int^\infty r^{1-q}(t)\,dt < \infty$. Hence (4.2.50) is verified because x^2 is asymptotic to a positive constant.

(ii) Suppose that $c(t) > 0$ for large t and $J_r + J_c < \infty$.

"$\Rightarrow$": Let $\tilde{x}$ be a principal solution of (1.1.1). In view of Lemma 4.2.3 (ii)-(iii), we have $\tilde{x}(\infty) = 0$, $\tilde{x}^{[1]}(\infty) = \ell \neq 0$. Without loss of generality, suppose $\tilde{x}(t) > 0$, $\tilde{x}'(t) < 0$ for $t \geq T$. Because

$$\lim_{t\to\infty} \frac{\tilde{x}(t)}{\int_t^\infty r^{q-1}(s)\,ds} = \Phi^{-1}(\ell),$$

there exists a positive constant k such that $\tilde{x}(t) < k\int_t^\infty \Phi^{-1}\Big(r^{-1}(s)\Big)\,ds$ for $t \geq T$. Then

$$\begin{aligned}\int_T^\infty \frac{1}{r^{q-1}(s)\tilde{x}^2(s)}\,ds &> \frac{1}{k^2}\int_T^\infty r^{1-q}(s)\left[\int_s^\infty r^{1-q}(\tau)\,d\tau\right]^{-2} ds \\ &= \left[\int_t^\infty r^{-1}(s)\,ds\right]^{-1}\Bigg|_T^\infty = \infty,\end{aligned}$$

i.e., condition (4.2.49) is verified.

"$\Leftarrow$": Let x be a nonprincipal solution of (1.1.1) and let us show that (4.2.50) holds. If $x \in \mathbb{M}^+$, then clearly the condition (4.2.49) follows. It remains to consider the case $x \in \mathbb{M}^-$. In this case, by using the same argument as given in the proof of Theorem 4.2.7 – Case C, we obtain $x(\infty) \neq 0$ and so the assertion again follows. □

4.2.11 Oscillation criteria and (non)principal solution

In this subsection we use the integral characterizations of principal and nonprinipal solutions of (1.1.1) given at the end of Subsection 4.2.7 in order to establish sufficient conditions for oscillation of (1.1.1). We also present one oscillation criterion based on the integral characterization given in Subsection 4.2.9.

Theorem 4.2.10. *Let* (4.2.26) *holds and suppose that*

$$C(t) := \int_t^\infty c(s)\,ds \quad \textit{is convergent} \tag{4.2.51}$$

and $C(t) \geq 0$ *for large* t. *Then* (1.1.1) *is oscillatory provided one of the following conditions is fulfilled:*

(i) $p > 2$, $\int^\infty r^{1-q}(t)[C(t)]^{\frac{q}{2}}\,dt < \infty$ *and*

$$\int_{t_0}^\infty r^{1-q}(t)\exp\left(-q\int_{t_0}^t r^{1-q}(s)\big(a(s)+C(s)\big)^{q-1}\,ds\right)dt < \infty,$$

where

$$a(t) = \left[2^{q-1}(p-2)\int_t^\infty r^{1-q}(s)[C(s)]^{\frac{q}{2}}\,ds\right]^{\frac{2}{2-q}};$$

(ii) $p \in (1,2)$, $\int^\infty r^{1-q}(t)[C(t)]^{\frac{q}{2}}\,dt = \infty$ *and*

$$\int_{t_0}^\infty r^{1-q}(t)\exp\left(-q\int_{t_0}^t r^{1-q}(s)\big(\gamma(s)+C(s)\big)^{q-1}\,ds\right)dt = \infty,$$

where

$$\gamma(t) = \left[2^{q-1}(2-p)\int_{t_0}^t r^{1-q}(s)[C(s)]^{\frac{q}{2}}\,ds\right]^{\frac{2}{2-q}}.$$

Proof. Assume, by contradiction, that (1.1.1) has a nonoscillatory solution x and let w be the associated solution of Riccati equation (1.1.21). Then w solves also the Riccati integral equation (4.2.34). Set $\rho(t) = (p-1)\int_t^\infty r^{1-q}(s)|w(s)|^q\,ds$. Then

(4.2.52) $$w(t) = \rho(t) + C(t).$$

Using the inequality $\rho + C \geq 2\sqrt{\rho C}$, we find

(4.2.53) $$\rho'(t) = -(p-1)r^{1-q}(t)\big|\rho(t)+C(t)\big|^q \leq -2^q(p-1)r^{1-q}(t)[C(t)]^{\frac{q}{2}}\rho^{\frac{q}{2}}(t)$$

for large t.

We will prove the statement (i) only, the proof of (ii) is similar (and it is also based on Corollary 4.2.3). From (4.2.53) we have $\rho(t) \geq a(t)$ for large t, say $t \geq t_0$. Therefore, due to (4.2.52), $w(t) \geq a(t) + C(t)$. Hence, we obtain

$$|x(t)| \geq |x(t_0)|\exp\left(\int_{t_0}^t r^{1-q}(s)\big(a(s)+C(s)\big)^{q-1}\,ds\right) \quad \text{for } t \geq t_0.$$

Thus we have the inequality

$$\int_{t_0}^\infty r^{1-q}(t)|x(t)|^{-q}\,dt$$
$$\leq |x(t_0)|^{-q}\int_{t_0}^\infty r^{1-q}(t)\exp\left(-q\int_{t_0}^t r^{1-q}(s)\big(a(s)+C(s)\big)^{q-1}\,ds\right)dt < \infty,$$

which, by virtue of Corollary 4.2.2 and the condition $m_* = q$ for $1 < q < 2$, contradicts the behavior of principal solutions.

The sufficiency of the condition (ii) for the oscillation of (1.1.1) can be proved by a similar way, using the fact that $m^* = q$ for $q > 2$. □

Now we give some integral conditions for the oscillation, in which no additional restrictions are imposed on the value of p.

Theorem 4.2.11. *Let* (4.2.26), (4.2.51) *hold,* $C(t) \geq 0$ *for* $t \geq t_0$ *and*

$$\int_{t_0}^\infty r^{1-q}(t)[C(t)]^q\,dt < \infty.$$

Then (1.1.1) *is oscillatory provided one of the following conditions holds.*

(i) Either

$$\int_{t_0}^{\infty} r^{1-q}(t) \exp\Big(- m_* \int_{t_0}^{t} r^{1-q}(s)\big(\delta(s) + C(s)\big)^{q-1}\, ds\Big)\, dt < \infty,$$

where

$$\delta(t) = (p-1) \int_{t}^{\infty} r^{1-q}(s)[C(s)]^q\, ds;$$

(ii) or

$$\int_{t_1}^{\infty} r^{1-q}(t) \exp\Big(- m^* \int_{t_1}^{t} r^{1-q}(s)\big(\omega(s) + C(s)\big)^{q-1}\, ds\Big)\, dt = \infty,$$

where $t_1 > t_0$*, and*

$$\omega(t) = \Big(\int_{t_0}^{t} r^{1-q}(s)\, ds\Big)^{1-p}.$$

Proof. We assume the contrary. Then we have (4.2.52), which yields the inequality $w(t) \geq C(t)$ for $t \geq t_0$. This means that

$$\rho(t) \geq (p-1) \int_{t}^{\infty} r^{1-q}(s)[C(s)]^q\, ds \tag{4.2.54}$$

for $t \geq t_0$. On the other hand, from (4.2.52) it follows

$$\rho'(t) = -(p-1) r^{1-q}(t)\big(\rho(t) + C(t)\big)^q \leq (p-1) r^{1-q}(t) \rho^q(t)$$

for $t \geq t_0$. Consequently,

$$\rho(t) \leq \Big(\int_{t_0}^{t} r^{1-q}(s)\, ds\Big)^{1-p} \tag{4.2.55}$$

for $t \geq t_0$. Relations (4.2.52), (4.2.54), and (4.2.55) imply

$$|u_1(t_1)| \exp\Big(\int_{t_1}^{t} r^{1-q}(s)\big(\delta(s) + C(s)\big)^{(q-1)}\, ds\Big) \leq |u_1(t)|$$

and

$$|u_1(t_1)| \leq |u_1(t_1)| \exp\Big(\int_{t_1}^{t} r^{1-q}(s)\big(\omega(s) + C(s)\big)^{(q-1)}\, ds\Big)$$

for $t \geq t_1$, where $t_1 > t_0$. These inequalities contradict the behavior of principal and nonprincipal solutions of (1.1.1). □

We conclude this subsection by the oscillation criterion for (1.1.1) whose proof is based on the integral characterization of the principal solution of (1.1.1) given in Theorem 4.2.8. This criterion is a half-linear extension of the oscillation criterion of Wintner (see [341, Theorem 2.17]) which claims that (1.1.2) with $r(t) \equiv 1$ is oscillatory provided

$$\int^{\infty} \exp\Big\{-2\int^{t} \Big[\int_{s}^{\infty} c(\tau)\, d\tau\Big]\, ds\Big\}\, dt < \infty.$$

Theorem 4.2.12. *Suppose that $p \geq 2$, $\int^\infty r^{1-q}(t)\,dt = \infty$, the integral $\int^\infty c(t)\,dt$ is convergent and $\int_t^\infty c(s)\,ds > 0$ for large t. If*
(4.2.56)

$$\int^\infty r^{1-q}(t)\left(\int_t^\infty c(s)\,ds\right)^{q-2} \exp\left\{-p\int^t r^{1-q}(s)\left(\int_s^\infty c(\tau)\,d\tau\right)^{q-1} ds\right\} dt < \infty,$$

then (1.1.1) *is oscillatory.*

Proof. Suppose, by contradiction, that (1.1.1) is nonoscillatory and let $\tilde{x}$ be its principal solution at ∞. Put $\tilde{w} = r\Phi(\tilde{x}')/\Phi(\tilde{x})$. Then $\tilde{w}$ is a solution of Riccati integral equation (2.2.17). Then, since $\tilde{x}' = \Phi^{-1}\left(\tilde{w}(t)/r(t)\right)x$, we have

$$\tilde{x}(t) = \exp\left\{\int^t \frac{\Phi^{-1}(\tilde{w}(s)}{\Phi^{-1}(r(s))}\,ds\right\} > \exp\left\{\int^t \frac{\Phi^{-1}(\int_s^\infty c(\tau)\,d\tau)}{\Phi^{-1}(r(s))}\,ds\right\}$$

and

$$\begin{aligned} |\tilde{x}'(t)|^{p-2} &= \left|\frac{\tilde{w}(t)}{r(t)}\right|^{(q-1)(p-2)} |\tilde{x}(t)|^{p-2} \\ &= \left|\frac{\tilde{w}(t)}{r(t)}\right|^{2-q} |\tilde{x}(t)|^{p-2} \\ &> r^{q-2}(t)\left|\int_t^\infty c(s)\,ds\right|^{2-q} \\ &\quad \times \exp\left\{(p-2)\int^t r^{1-q}(s)\Phi^{-1}\left(\int_s^\infty c(\tau)\,d\tau\right)\right\} ds. \end{aligned}$$

Consequently,

$$\begin{aligned} \frac{1}{r(t)\tilde{x}^2(t)|\tilde{x}'(t)|^{p-2}} &< \frac{1}{r^{q-1}(t)\left|\int_t^\infty c(s)\,ds\right|^{2-q}} \\ &\quad \times \exp\left\{-p\int^t r^{1-q}(s)\left(\int_s^\infty c(\tau)\,d\tau\right)^{q-1} ds\right\} \\ &= r^{1-q}(t)\left(\int_t^\infty c(s)\,ds\right)^{q-2} \\ &\quad \times \exp\left\{-p\int^t r^{1-q}(s)\left(\int_s^\infty c(\tau)\,d\tau\right)^{q-1} ds\right\}. \end{aligned}$$

This implies that

$$\int^\infty \frac{dt}{r(t)\tilde{x}^2(t)|\tilde{x}'(t)|^{p-2}} < \infty,$$

and this contradiction with the part (ii) of Theorem 4.2.8 completes the proof. □

4.3 Half-linear differential equations and Karamata functions

In this part we establish the criteria, which are natural generalizations of the results for linear equations (see e.g. [181] and other works, in particular those of Marić). Since the purpose is to develop nonoscillation theory of the equation

$$(\Phi(x'))' + c(t)\Phi(x) = 0, \tag{4.3.1}$$

where $c : [0, \infty) \to \mathbb{R}$ is a continuous function for which

$$\int_0^\infty c(t)\, dt = \lim_{t\to\infty} \int_0^t c(s)\, ds \text{ is convergent},$$

in the framework of regularly varying functions in the sense of Karamata, let us recall briefly those basic definitions and properties of such functions which will be indispensable later. Note that the detailed presentation of the theory of regularly varying functions can be found in the books [44, 338]

Definition 4.3.1. A positive measurable function $L(t)$ defined on $(0, \infty)$ is said to be *slowly varying function* if it satisfies

$$\lim_{t\to\infty} \frac{L(\lambda t)}{L(t)} = 1 \text{ for any } \lambda > 0.$$

The next result is a type of representation theorem.

Theorem 4.3.1. *A positive measurable function $L(t)$ defined on $(0, \infty)$ is slowly varying if and only if it can be written in the form*

$$L(t) = \varphi(t) \exp\left\{\int_{t_0}^t \frac{\psi(s)}{s}\, ds\right\}, \quad t \geq t_0, \tag{4.3.2}$$

for some $t_0 > 0$, where $\varphi(t)$ and $\psi(t)$ are measurable functions such that

$$\lim_{t\to\infty} \varphi(t) = a \in (0, \infty) \quad \textit{and} \quad \lim_{t\to\infty} \psi(t) = 0.$$

Definition 4.3.2. If, in particular, $\varphi(t)$ is identically a positive constant in (4.3.2), then $L(t)$ is called a *normalized slowly varying function.*

The totality of slowly varying functions (resp. normalized slowly varying functions) is denoted by $\mathcal{SV}$ (resp. $\mathcal{NSV}$).

Definition 4.3.3. Let ϱ be a fixed real number. A positive measurable function $f(t)$ is said to be a *regularly varying function with index* ϱ if it satisfies

$$\lim_{t\to\infty} \frac{f(\lambda t)}{f(t)} = \lambda^\varrho \text{ for any } \lambda > 0.$$

Just defined function can be represented as follows.

Theorem 4.3.2. *A positive measurable function f defined on $(0,\infty)$ is regularly varying with index ϱ if and only if it can be written in the form (4.3.2) where φ and ψ are measurable functions such that*

$$\lim_{t\to\infty}\varphi(t) = a \in (0,\infty) \quad \textit{and} \quad \lim_{t\to\infty}\psi(t) = \varrho. \tag{4.3.3}$$

Definition 4.3.4. A positive measurable function f defined on $(0,\infty)$ satisfying (4.3.2) and (4.3.3) with $\varphi(t) \equiv$ const is called a *normalized regularly varying function with index ϱ.*

Theorem 4.3.3. *A positive measurable function f defined on $(0,\infty)$ is a regularly varying function (resp. a normalized regularly varying function) with index ϱ if and only if $f(t)$ is expressed as $f(t) = t^{\varrho}L(t)$, where $L(t) \in \mathcal{SV}$ (resp. $L(t) \in \mathcal{NSV}$).*

The totality of regularly varying functions (resp. normalized regularly varying functions) with index ϱ is denoted by $\mathcal{RV}(\varrho)$ (resp. $\mathcal{NRV}(\varrho)$).

Definition 4.3.5. A positive measurable function f is said to be an *O-regularly varying function* if it satisfies

$$0 < \liminf_{t\to\infty}\frac{f(\lambda t)}{f(t)} \le \limsup_{t\to\infty}\frac{f(\lambda t)}{f(t)} < \infty \quad \text{for any} \quad \lambda \ge 1.$$

Theorem 4.3.4. *A positive measurable function f is O-regularly varying if and only if*

$$f(t) = \exp\left\{\xi(t) + \int_{t_0}^{t}\frac{\eta(s)}{s}\,ds\right\}, \quad t \ge t_0, \tag{4.3.4}$$

for some $t_0 > 0$, where ξ and η are bounded measurable functions on $[t_0,\infty)$.

Definition 4.3.6. If, in particular, $\xi(t) \equiv$ const in (4.3.4), then $f(t)$ is referred to as a *normalized O-regularly varying function.*

The totality of O-regularly varying functions (resp. normalized O-regularly varying functions) is denoted by $\mathcal{OR}$ (resp. $\mathcal{NOR}$).

We finish this part, which deals with the theory of Karamata functions, by the following simple but important property.

Theorem 4.3.5. *Let L be any slowly varying function. Then, for any $\gamma > 0$,*

$$\lim_{t\to\infty} t^{\gamma}L(t) = \infty \quad \textit{and} \quad \lim_{t\to\infty} t^{-\gamma}L(t) = 0.$$

4.3.1 Existence of regularly varying solutions

Now we prove nonoscillation theorems for (4.3.1) asserting the existence of solutions in the classes of regularly varying functions. We start with an auxiliary statement.

Lemma 4.3.1. *Put* $\sigma(t) = \int_t^\infty c(s)\,ds$ *and suppose that there exists a continuous function* $P : [t_0, \infty) \to (0, \infty)$, $t_0 \geq 0$, *such that* $\lim_{t\to\infty} P(t) = 0$, $|\sigma(t)| \leq P(t)$, $t \geq t_0$, *and*

$$\int_t^\infty P^q(s)\,ds \leq \frac{1}{p-1} a^{q-1} P(t), \quad t \geq t_0, \tag{4.3.5}$$

for some positive constant

$$a < \frac{1}{p}\left(\frac{p-1}{p}\right)^{p-1}. \tag{4.3.6}$$

Then equation (4.3.1) *is nonoscillatory and has a solution of the form*

$$y(t) = \exp\left\{\int_{t_0}^t \Phi^{-1}[v(s) + \sigma(s)]\,ds\right\}, \quad t \geq t_0, \tag{4.3.7}$$

where v *is a solution of the integral equation*

$$v(t) = (p-1)\int_t^\infty |v(s) + \sigma(s)|^q ds, \quad t \geq t_0, \tag{4.3.8}$$

satisfying

$$v(t) = \mathcal{O}(P(t)) \quad as \quad t \to \infty. \tag{4.3.9}$$

Proof. Consider the function y defined by (4.3.7). Recall that if w is a solution of generalized Riccati equation (3.1.2), then the function $\exp\left\{\int_{t_0}^t \Phi^{-1}(w(s))\,ds\right\}$ is a (nonoscillatory) solution of (4.3.1). Hence, y is a solution of (4.3.1) if v is chosen in such a way that $w = v + \sigma$ satisfies (3.1.2) on $[t_0, \infty)$. The differential equation for v then reads $v' + (p-1)|v + \sigma(t)|^q = 0$, which upon integration under the additional requirement that $\lim_{t\to\infty} v(t) = 0$, yields (4.3.8). We denote by $C_P[t_0, \infty)$ the set of all continuous functions v on $[t_0, \infty)$ such that

$$\|v\|_P = \sup_{t\geq t_0} \frac{|v(t)|}{P(t)} < \infty.$$

Clearly, $C_P[t_0, \infty)$ is a Banach space equipped with the norm $\|v\|_P$. Let Ω be a subset of $C_P[t_0, \infty)$ defined by $\Omega = \{v \in C_P[t_0, \infty) : |v(t)| \leq (p-1)P(t), t \geq t_0\}$ and define the mapping $\mathcal{T} : \Omega \to C_P[t_0, \infty)$ by

$$\mathcal{T}v(t) = (p-1)\int_t^\infty |v(s) + \sigma(s)|^q ds, \quad t \geq t_0. \tag{4.3.10}$$

If $v \in \Omega$, then

$$|\mathcal{T}v(t)| \leq (p-1)p^q \int_t^\infty P^q(s)\,ds \leq p^q a^{q-1} P(t),$$

$t \geq t_0$, which implies that

$$\|\mathcal{T}v\|_P \leq p^q a^{q-1} < p^q \left(\frac{(p-1)^{p-1}}{p^p} \right)^{q-1} = p-1. \tag{4.3.11}$$

Thus $\mathcal{T}$ maps Ω into itself. If $v_1, v_2 \in \Omega$, then, using the Mean Value Theorem, we see that

$$\begin{aligned}
|\mathcal{T}v_1(t) - \mathcal{T}v_2(t)| &\leq (p-1) \int_t^\infty \big| |v_1(s) + \sigma(s)|^q - |v_2(s) + \sigma(s)|^q \big| \, ds \\
&\leq (p-1)q \int_t^\infty (pP(s))^{q-1} |v_1(s) - v_2(s)| \, ds \\
&= p^q \int_t^\infty P^q(s) \frac{|v_1(s) - v_2(s)|}{P(s)} \, ds \\
&\leq p^q (q-1) a^{q-1} P(t) \|v_1 - v_2\|_P,
\end{aligned}$$

$t \geq t_0$, from which it follows that

$$\|\mathcal{T}v_1 - \mathcal{T}v_2\|_P \leq (q-1) p^q a^{q-1} \|v_1 - v_2\|_P.$$

In view of (4.3.6) (or (4.3.11)), this implies that $\mathcal{T}$ is a contraction mapping on Ω. Therefore, by the contraction mapping principle, there exists an element $v \in \Omega$ such that $v = \mathcal{T}v$, that is, a solution of integral equation (4.3.8). Thus the function $y(t)$ defined by (4.3.7) with this $v(t)$ gives a solution of (4.3.1) on $[t_0, \infty)$. The fact that v satisfies (4.3.9) is a consequence of $v \in \Omega$. This completes the proof. □

Theorem 4.3.6. *If*

$$\begin{aligned}
-\frac{1}{p} \left(\frac{p-1}{p} \right)^{p-1} &< \liminf_{t \to \infty} t^{p-1} \int_t^\infty c(s) \, ds \\
&\leq \limsup_{t \to \infty} t^{p-1} \int_t^\infty c(s) \, ds < \frac{1}{p} \left(\frac{p-1}{p} \right)^{p-1}
\end{aligned}$$

holds, then equation (4.3.1) *is nonoscillatory and has a normalized O-regularly varying solution.*

Proof. If the assumption holds, then there exist positive constants t_0 and a satisfying (4.3.6) such that $|t^{p-1}\sigma(t)| \leq a$, $t \geq t_0$. Put $P(t) = a t^{1-p}$. Then it can be easily verified that P satisfies $|\sigma(t)| \leq P(t)$ and (4.3.5). So, by Lemma 4.3.1, (4.3.1) has a nonoscillatory solution of the form

$$y(t) = \exp\left\{ \int_{t_0}^t \Phi^{-1}(v(s) + \sigma(s)) \, ds \right\} = \exp\left\{ \int_{t_0}^t \frac{\Phi^{-1}[s^{p-1}(v(s) + \sigma(s))]}{s} \, ds \right\}, \tag{4.3.12}$$

$t \geq t_0$, with v satisfying (4.3.9). Since

$$\left| \Phi^{-1}[t^{p-1}(v(t) + \sigma(t))] \right| \leq p^{q-1} (t^{p-1} P(t))^{q-1} = (ap)^{q-1},$$

$t \geq t_0$, the solution y is a normalized O-regularly varying function by Theorem 4.3.4. □

Now let us turn to the case where the coefficient c satisfies the condition

$$(4.3.13)\qquad -\infty < \lim_{t\to\infty} t^{p-1}\int_t^\infty c(s)\,ds =: a < \frac{1}{p}\left(\frac{p-1}{p}\right)^{p-1}$$

or

$$(4.3.14)\qquad \lim_{t\to\infty} t^{p-1}\int_t^\infty c(s)\,ds = \frac{1}{p}\left(\frac{p-1}{p}\right)^{p-1},$$

and investigate how these conditions affect the regularity indices of nonoscillatory solutions of (4.3.1) considered as Karamata functions. Suppose first that (4.3.13) holds. Let λ_1 and λ_2, $\lambda_1 < \lambda_2$, denote the two real roots of the equation

$$(4.3.15)\qquad |\lambda|^q - \lambda + a = 0.$$

It is easy to see that (4.3.15) has two distinct real roots if and only if $a < \frac{1}{p}\left(\frac{p-1}{p}\right)^{p-1}$. Clearly, $\lambda_1 < 0 < \lambda_2$ if $a < 0$, and $0 < \lambda_1 < \lambda_2$ if $0 < a < \frac{1}{p}\left(\frac{p-1}{p}\right)^{p-1}$. It should be noticed that $p\Phi^{-1}(\lambda_1) < p-1 < p\Phi^{-1}(\lambda_2)$.

Theorem 4.3.7. *Equation* (4.3.1) *is nonoscillatory and has two solutions* y_1 *and* y_2 *such that* $y_1 \in \mathcal{NRV}(\Phi^{-1}(\lambda_1))$ *and* $y_2 \in \mathcal{NRV}(\Phi^{-1}(\lambda_2))$ *if and only if* (4.3.13) *holds.*

Proof. The "only if" part: Let y_i be solutions belonging to $\mathcal{NRV}(\Phi^{-1}(\lambda_i))$, $i = 1, 2$. From Theorem 4.3.2 it follows that

$$(4.3.16)\qquad \lim_{t\to\infty} t\frac{y_i'(t)}{y_i(t)} = \Phi^{-1}(\lambda_i), \quad \text{so that} \quad \lim_{t\to\infty}\frac{y_i'(t)}{y_i(t)} = 0, \quad i = 1, 2.$$

Put $w_i = \Phi(y'/y)$, $i = 1, 2$. Then w_i satisfies generalized Riccati equation (3.1.2), from which, integrating on $[t,\infty)$ and noting that $\lim_{t\to\infty} w_i(t) = 0$, we have

$$(4.3.17)\quad t^{p-1}w_i(t) = (p-1)t^{p-1}\int_t^\infty \frac{|s^{p-1}w_i(s)|^q}{s^p}\,ds + t^{p-1}\int_t^\infty c(s)\,ds, \quad i = 1, 2,$$

for all sufficiently large t. Let $t\to\infty$ in (4.3.17). Using (4.3.16), we then conclude that

$$\lim_{t\to\infty} t^{p-1}\int_t^\infty c(s)\,ds = \lambda_i - |\lambda_i|^q = a, \quad i = 1, 2.$$

The "if" part: Assume that (4.3.13) holds. Put $\omega(t) = t^{p-1}\int_t^\infty c(s)\,ds - a$ and consider the functions

$$(4.3.18)\qquad y_i(t) = \exp\left\{\int_{t_i}^t \Phi^{-1}\left(\frac{\lambda_i + \omega(s) + v_i(s)}{s^{p-1}}\right)ds\right\}, \quad i = 1, 2.$$

Then the function y_i is a solution of (4.3.1) on $[t_i, \infty)$ if v_i is chosen in such a way that $w_i = (\lambda_i + \omega + v_i)/t^{p-1}$ satisfies (3.1.2) on $[t_i,\infty)$, $i = 1, 2$. The differential equation for v_i then reads

$$(4.3.19)\qquad v_i' - \frac{p-1}{t}v_i + \frac{p-1}{t}\left(|\lambda_i + \omega(t) + v_i|^q - |\lambda_i|^q\right) = 0, \quad i = 1, 2.$$

We rewrite (4.3.19) as

$$v_i' + \frac{p\Phi^{-1}(\lambda_i + \omega(t)) - (p-1)}{t} v_i + \frac{p-1}{t}\Big[|\lambda_i + \omega(t) + v_i|^q - q\Phi^{-1}(\lambda_i + \omega(t))v_i - |\lambda_i|^q\Big] = 0$$

and transform it into

$$(r_i(t)v_i)' + \frac{p-1}{t} r_i(t) F_i(t, v_i) = 0, \tag{4.3.20}$$

where

$$r_i(t) = \exp\left\{\int_1^t \frac{p\Phi^{-1}(\lambda_i + \omega(s)) - (p-1)}{s}\, ds\right\}$$

and

$$F_i(t,v) = |\lambda_i + \omega(t) + v|^q - q\Phi^{-1}(\lambda_i + \omega(t))v - |\lambda_i|^q, \quad i = 1, 2.$$

It is convenient to express $F_i(t,v)$ as $F_i(t,v) = G_i(t,v) + h_i(t)$, with $G_i(t,v)$ and $h_i(t)$ defined by

$$G_i(t,v) = |\lambda_i + \omega(t) + v|^q - q\Phi^{-1}(\lambda_i + \omega(t))v - |\lambda_i + \omega(t)|^q$$

and $h_i(t) = |\lambda_i + \omega(t)|^q - |\lambda_i|^q$, $i = 1, 2$. Now we suppose that $a \neq 0$ in (4.3.13), which implies $\lambda_i \neq 0$ for $i = 1, 2$. Let $t_0 > 0$ be such that $|\omega(t)| \le |\lambda_i|/4$ for $t \ge t_0$, $i = 1, 2$. This is possible because $\omega(t) \to 0$ as $t \to \infty$ by hypothesis. It follows that $\frac{3}{4}|\lambda_i| \le |\lambda_i + \omega(t)| \le \frac{5}{4}|\lambda_i|$ for $t \ge t_0$, $i = 1, 2$. We observe that there exist positive constants $K_i(p), L_i(p)$ and $M_i(p)$ such that $|G_i(t,v)| \le K_i(p)v^2$,

$$\left|\frac{\partial G_i}{\partial v}(t,v)\right| \le L_i(p)|v| \tag{4.3.21}$$

and $|h_i(t)| \le M_i(p)|\omega(t)|$ for $t \ge t_0$ and $|v| \le |\lambda_i|/4$, $i = 1, 2$. In fact, the last two estimations follow from the Mean Value Theorem, while the estimation for G_i is a consequence of the L'Hospital rule applied to G_i:

$$\lim_{v\to 0} \frac{G_i(t,v)}{v^2} = \frac{1}{2}\lim_{v\to 0} \frac{\partial^2 G_i(t,v)}{\partial v^2} = \frac{q}{2(p-1)}|\lambda_i + \omega(t)|^{q-2}.$$

Let us examine equation (4.3.20) with $i = 1$. The following properties of r_1 are needed: $r_1 \in \mathcal{NRV}(\Phi^{-1}(p) - (p-1))$, $\lim_{t\to\infty} r_1(t) = 0$,

$$\lim_{t\to\infty} \frac{p-1}{r_1(t)} \int_t^\infty \frac{r_1(s)}{s}\, ds = (p-1)\lim_{t\to\infty} \frac{-r_1(t)}{tr_1'(t)} = \frac{p-1}{p-1-p\Phi^{-1}(\lambda_1)}, \tag{4.3.22}$$

$$\lim_{t\to\infty} \frac{p-1}{r_1(t)} \int_t^\infty \frac{r_1(s)}{s} h(s)\, ds = 0 \quad \text{if} \quad h \in C[0,\infty), \quad \text{and} \quad \lim_{t\to\infty} h(t) = 0. \tag{4.3.23}$$

Let ε_1 be a positive constant such that $\varepsilon_1 < \min\{1, |\lambda_1|/4\}$ and

$$\frac{2(p-1)}{p-1-p\Phi^{-1}(\lambda_1)}[K_1(p)+L_1(p)+M_1(p)]\varepsilon_1 \le 1, \tag{4.3.24}$$

and choose $t_1 \ge t_0$ so that

$$|\omega(t)| \le \varepsilon_1^2, \quad t \ge t_1, \tag{4.3.25}$$

and

$$\frac{p-1}{r_1(t)}\int_t^\infty \frac{r_1(s)}{s}\,ds \le \frac{2(p-1)}{p-1-p\Phi^{-1}(\lambda_1)}, \quad t \ge t_1. \tag{4.3.26}$$

Note that (4.3.26) is an immediate consequence of (4.3.22).

Let $C_0[t_1,\infty)$ denote the set of all continuous functions on $[t_1,\infty)$ that tend to zero as $t \to \infty$. Then $C_0[t_1,\infty)$ is a Banach space with the sup-norm $\|v\| = \sup\{|v(t)| : t \ge t_1\}$. Consider the set $\Omega_1 \subset C_0[t_1,\infty)$ defined by $\Omega_1 = \{v \in C_0[t_1,\infty) : |v(t)| \le \varepsilon_1, t \ge t_1\}$ and define the integral operator $\mathcal{T}_1$ by

$$(\mathcal{T}_1 v)(t) = \frac{p-1}{r_1(t)}\int_t^\infty \frac{r_1(s)}{s}F_1(s,v(s))\,ds,$$

$t \ge t_1$. It can be shown that $\mathcal{T}_1$ is a contraction mapping on Ω_1. In fact, if $v \in \Omega_1$, then using the above inequalities we see that

$$\begin{aligned}
|(\mathcal{T}_1 v)(t)| &\le \frac{p-1}{r_1(t)}\int_t^\infty \frac{r_1(s)}{s}(|G_1(s,v(s))| + |h_1(s)|)\,ds \\
&\le \frac{p-1}{r_1(t)}\int_t^\infty \frac{r_1(s)}{s}(K_1(p)v^2(s) + M_1(p)|\omega(s)|)\,ds \\
&\le \frac{p-1}{r_1(t)}\int_t^\infty \frac{r_1(s)}{s}(K_1(p)+M_1(p))\varepsilon_1^2\,ds \\
&\le \frac{2(p-1)}{p-1-p\Phi^{-1}(\lambda_1)}[K_1(p)+M_1(p)]\varepsilon_1^2,
\end{aligned}$$

$t \ge t_1$. Since $F_1(t,v(t)) \to 0$ as $t \to \infty$, we have $\lim_{t\to\infty}(\mathcal{T}_1 v)(t) = 0$ by (4.3.23). It follows that $\mathcal{T}_1 v \in \Omega_1$, and so $\mathcal{T}_1$ maps Ω_1 into itself. Furthermore, if $u, v \in \Omega_1$, then using (4.3.21) and (4.3.24), we obtain

$$\begin{aligned}
|(\mathcal{T}_1 v)(t) - (\mathcal{T}_1 u)(t)| &\le \frac{p-1}{r_1(t)}\int_t^\infty \frac{r_1(s)}{s}|F_1(s,v(s)) - F_1(s,u(s))|\,ds \\
&= \frac{p-1}{r_1(t)}\int_t^\infty \frac{r_1(s)}{s}|G_1(s,v(s)) - G_1(s,u(s))|\,ds \\
&\le \frac{2(p-1)}{p-1-p\Phi^{-1}(\lambda_1)}L_1(p)\varepsilon_1\|v-u\|,
\end{aligned}$$

$t \ge t_1$, which implies that

$$\|\mathcal{T}_1 v - \mathcal{T}_1 u\| \le \frac{2(p-1)L_1(p)}{p-1-p\Phi^{-1}(\lambda_1)}\varepsilon_1\|v-u\|.$$

In view of (4.3.24), this shows that $\mathcal{T}_1$ is a contraction mapping on Ω_1. The contraction mapping principle then ensures the existence of a unique fixed element $v_1 \in \Omega_1$ such that $v_1 = \mathcal{T}_1 v_1$, which is equivalent to the integral equation

$$v_1(t) = \frac{p-1}{r_1(t)} \int_t^\infty \frac{r_1(s)}{s} F_1(s, v_1(s))\, ds, \tag{4.3.27}$$

$t \geq t_1$. Differentiation of (4.3.27) shows that v_1 satisfies differential equation (4.3.20) with $i = 1$ on $[t_1, \infty)$ and so substitution of this v_1 into (4.3.18) gives rise to a solution y_1 of half-linear equation (4.3.1) defined on $[t_1, \infty)$. Since $\lim_{t\to\infty} v_1(t) = 0$, from the representation theorem we have $y_1 \in \mathcal{NRV}(\Phi^{-1}(\lambda_1))$.

Our next task is to solve equation (4.3.20) for $i = 2$ in order to construct a larger solution y_2 of (4.3.1) via formula (4.3.18). It is easy to see that $r_2 \in \mathcal{NRV}(p\Phi^{-1}(\lambda_2) - p + 1)$, $\lim_{t\to\infty} r_2(t) = \infty$ and that for any fixed $t_2 > 0$,

$$\lim_{t\to\infty} \frac{p-1}{r_2(t)} \int_{t_2}^t \frac{r_2(s)}{s}\, ds = (p-1) \lim_{t\to\infty} \frac{r_2(t)}{t r_2'(t)} = \frac{p-1}{p\Phi^{-1}(\lambda_2) - p + 1},$$

$$\lim_{t\to\infty} \frac{p-1}{r_2(t)} \int_{t_2}^t \frac{r_2(s)h(s)}{s}\, ds = 0 \quad \text{if} \quad h \in C[t_2, \infty) \quad \text{and} \quad \lim_{t\to\infty} h(t) = 0.$$

Let $\varepsilon_2 > 0$ be small enough so that

$$\frac{2(p-1)}{p\Phi^{-1}(\lambda_2) - p + 1} [K_2(p) + L_2(p) + M_2(p)]\varepsilon_2 \leq 1,$$

and choose $t_2 > 0$ so large that $\omega(t) \leq \varepsilon_2^2$, $t \geq t_2$, and

$$\frac{p-1}{r_2(t)} \int_{t_2}^t \frac{r_2(s)}{2}\, ds \leq \frac{2(p-1)}{p\Phi^{-1}(\lambda_2) - p + 1},$$

$t \geq t_2$. Define the set $\Omega_2 \subset C_0[t_2, \infty)$ and the integral operator $\mathcal{T}_2$ by $\Omega_2 = \{v \in C_0[t_2, \infty) : |v(t)| \leq \varepsilon_2, t \geq t_2\}$, and

$$(\mathcal{T}_2 v)(t) = -\frac{p-1}{r_2(t)} \int_{t_2}^t \frac{r_2(s)}{s} F_2(s, v(s))\, ds,$$

$t \geq t_2$. It is a matter of easy calculation to verify that $\mathcal{T}_2$ is a contraction mapping on Ω_2. Therefore there exists a unique fixed element $v_2 \in \Omega_2$ of $\mathcal{T}_2$, which satisfies the integral equation

$$v_2(t) = -\frac{p-1}{r_2(t)} \int_{t_2}^t \frac{r_2(s)}{s} F(s, v_2(s))\, ds,$$

$t \geq t_2$, and hence differential equation (4.3.20) with $i = 2$. Then the function y_2 defined by (4.3.18) with this v_2 is a nonoscillatory solution of (4.3.1) on $[t_2, \infty)$. The fact that $y_2 \in \mathcal{NRV}(\Phi^{-1}(\lambda_2))$ follows from the representation theorem. This finishes the proof of the "if" part of the theorem for the case $a \neq 0$.

It remains to consider the case $a = 0$ in (4.3.13). Then equation (4.3.15) has the two real roots $\lambda_1 = 0$, $\lambda_2 = 1$. The solution $y_1 \in \mathcal{NRV}(0) = \mathcal{NSV}$ of (4.3.1) corresponding to λ_1 has already been constructed in Theorem 4.3.9. The existence of the solution $y_2 \in \mathcal{NRV}(1)$ corresponding to λ_2 can be proved in exactly the same manner as developed for the case $a \neq 0$. □

Let us consider equation (4.3.1) for which the condition

$$\lim_{t\to\infty} t^{p-1} \int_t^\infty c(s)\, ds = \frac{1}{p}\left(\frac{p-1}{p}\right)^{p-1} \tag{4.3.28}$$

is satisfied. Such an equation can be regarded as a perturbation of generalized Euler equation (1.4.20) with $\gamma = \bar{\gamma} = \left(\frac{p-1}{p}\right)^p$. Although (1.4.20) is nonoscillatory, because it has a solution $y(t) = t^{(p-1)/p}$, its perturbation may be oscillatory or nonoscillatory depending on the asymptotic behavior of the perturbed term as $t \to \infty$, see Section 5.2. Our purpose here is to show the existence of a class of perturbations which preserve the nonoscillation character of (1.4.20).

Theorem 4.3.8. *Suppose that (4.3.28) holds. Put*

$$\Upsilon(t) = t^{p-1} \int_t^\infty c(s)\, ds - \frac{1}{p}\left(\frac{p-1}{p}\right)^{p-1}$$

and suppose that

$$\int^\infty \frac{|\Upsilon(t)|}{t}\, dt < \infty \tag{4.3.29}$$

and

$$\int^\infty \frac{\Psi(t)}{t}\, dt < \infty, \quad \text{where } \Psi(t) = \int_t^\infty \frac{|\Upsilon(s)|}{s}\, ds. \tag{4.3.30}$$

Then equation (4.3.1) is nonoscillatory and has a normalized regularly varying solution with index $(p-1)/p$ of the form $y(t) = t^{(p-1)/p} L(t)$ with $L(t) \in \mathcal{NSV}$ and $\lim_{t\to\infty} L(t) = l \in (0, \infty)$.

Proof. The solution is sought in the form

$$y(t) = \exp \int_T^t \Phi^{-1}\left(\frac{\bar{\gamma} + \Upsilon(s) + v(s)}{s^{p-1}}\right) ds, \quad \bar{\gamma} = \left(\frac{p-1}{p}\right)^p, \tag{4.3.31}$$

for some $T > 0$ and $v : [T, \infty) \to \mathbb{R}$. The same argument as in the proof of the "if" part of the previous theorem leads to the differential equation for v

$$(r(t)v)' + \frac{p-1}{t} r(t) F(t, v) = 0, \tag{4.3.32}$$

where

$$r(t) = \exp\left(\int_1^t \frac{p\Phi^{-1}(\bar{\gamma} + \Upsilon(s)) - p + 1}{s}\, ds\right)$$

and

$$F(t,v) = |\bar{\gamma} + \Upsilon(t) + v|^q - q\Phi^{-1}(\bar{\gamma} + \Upsilon(t))v - \bar{\gamma}^q. \tag{4.3.33}$$

Choose $t_0 > 0$ so that

$$|\Upsilon(t)| \le \frac{\bar{\gamma}}{4}, \tag{4.3.34}$$

$t \ge t_0$. Since

$$|\Phi^{-1}(\bar{\gamma} + \Upsilon(t)) - p + 1| = p|(\bar{\gamma} + \Upsilon(t))^{q-1} - \bar{\gamma}^{q-1}| \le pm(p)|\Upsilon(t)|,$$

$t \ge t_0$, for some constant $m(p) > 0$, we see in view of (4.3.29) that r is a slowly varying function and tends to a finite positive limit as $t \to \infty$. It follows that there exists $t_1 \ge t_0$ such that

$$r(s)/r(t) \le 2 \tag{4.3.35}$$

for $s \ge t \ge t_1$. We rewrite the function $F(t,v)$ defined by (4.3.33) as $F(t,v) = G(t,v) + h(t)$, where

$$G(t,v) = |\bar{\gamma} + \Upsilon(t) + v|^q - q\Phi^{-1}(\bar{\gamma} + \Upsilon(t))v - |\bar{\gamma} + \Upsilon(t)|^q$$

and $h(t) = |\bar{\gamma} + \Upsilon(t)|^q - \bar{\gamma}^q$. As it is easily seen, there exist positive constants $K(p), L(p)$ and $M(p)$ such that

$$|G(t,v)| \le K(p)v^2, \tag{4.3.36}$$

$$\left| \frac{\partial G(t,v)}{\partial v} \right| \le L(p)|v| \tag{4.3.37}$$

and $|h(t)| \le M(p)|\Upsilon(t)|$ for $t \ge t_1$ and $|v| \le \bar{\gamma}/4$. Let $T > t_1$ be large enough so that

$$4(p-1)M(p)\Psi(t) \le \frac{\bar{\gamma}}{4}, \tag{4.3.38}$$

$t \ge T$,

$$16(p-1)^2 K(p)M(p) \int_T^\infty \frac{\Psi(s)}{s}\, ds \le 1,$$

and

$$16(p-1)^2 L(p)M(p) \int_T^\infty \frac{\Psi(s)}{s}\, ds \le 1. \tag{4.3.39}$$

We want to solve the integral equation

$$v(t) = \frac{p-1}{r(t)} \int_t^\infty \frac{r(s)}{s} F(s, v(s))\, ds, \tag{4.3.40}$$

$t \geq T$, which follows from (4.3.32), subject to the condition $\lim_{t\to\infty} v(t) = 0$. Let $C_\Psi[T,\infty)$ denote the set of all continuous functions v on $[T,\infty)$ such that

$$\|v\|_\Psi = \sup_{t\geq T} \frac{|v(t)|}{\Psi(t)} < \infty.$$

Clearly, $C_\Psi[T,\infty)$ is a Banach space equipped with the norm $\|v\|_\Psi$. Consider the set $\Omega \subset C_\Psi[T,\infty)$ and the mapping $\mathcal{T} : \Omega \to C_\Psi[T,\infty)$ defined by

$$\Omega = \{v \in C_\Psi[T,\infty) : |v(t)| \leq 4(p-1)M(p)\Psi(t),\ t \geq T\} \tag{4.3.41}$$

and

$$\mathcal{T}v(t) = \frac{p-1}{r(t)} \int_t^\infty \frac{r(s)}{s} F(s,v(s))\, ds = \frac{p-1}{r(t)} \int_t^\infty \frac{r(s)}{s} [G(s,v(s)) + h(s)]\, ds,$$

$t \geq T$. Using (4.3.35),(4.3.36) and (4.3.37), we see that

$$\frac{p-1}{r(t)} \int_t^\infty \frac{r(s)}{s} |h(s)|\, ds \leq 2(p-1) \int_t^\infty \frac{M(p)|\Upsilon(s)|}{s}\, ds = 2(p-1)M(p)\Psi(t),$$

$t \geq T$, and that

$$\begin{aligned}
&\frac{p-1}{r(t)} \int_t^\infty \frac{r(s)}{s} |G(s,v(s))|\, ds \leq 2(p-1) \int_t^\infty \frac{K(p)[4(p-1)M(p)\Psi(s)]^2}{s}\, ds \\
&= 32(p-1)^3K(p)M^2(p) \int_t^\infty \frac{\Psi^2(s)}{s}\, ds \leq 32(p-1)^3K(p)M^2(p)\Psi(t) \int_t^\infty \frac{\Psi(s)}{s}\, ds \\
&\qquad \leq 32(p-1)^3K(p)M^2(p)\Psi(t) \int_T^\infty \frac{\Psi(s)}{s}\, ds \leq 2(p-1)M(p)\Psi(t),
\end{aligned}$$

$t \geq T$. This shows that $v \in \Omega$ implies $\mathcal{T}v \in \Omega$, and hence $\mathcal{T}$ maps Ω into itself. If $u, v \in \Omega$, then using (4.3.37) we have

$$\begin{aligned}
|\mathcal{T}v(t) - \mathcal{T}u(t)| &\leq \frac{p-1}{r(t)} \int_t^\infty \frac{r(s)}{s} |G(s,v(s)) - G(s,u(s))|\, ds \\
&\leq 2(p-1) \int_t^\infty \frac{4(p-1)L(p)M(p)\Psi(s)|v(s)-u(s)|}{s}\, ds \\
&= 8(p-1)^2L(p)M(p) \int_t^\infty \frac{\Psi^2(s)|v(s)-u(s)|}{s\Psi(s)}\, ds \\
&\leq 8(p-1)^2L(p)M(p)\Psi(t)\|v-u\|_\Psi \int_t^\infty \frac{\Psi(s)}{s}\, ds,
\end{aligned}$$

$t \geq T$, from which, in view of (4.3.39), we conclude that $\mathcal{T}$ is a contraction mapping: $\|\mathcal{T}v - \mathcal{T}u\|_\Psi \leq \|v-u\|_\Psi/2$. Let $v \in \Omega$ be a unique fixed element of of $\mathcal{T}$. Then v satisfies (4.3.40), and hence (4.3.32), on $[T,\infty)$, and the function y defined by (4.3.31) provides a nonoscillatory solution of (4.3.1) on $[T,\infty)$. Since

$\Phi^{-1}(\bar{\gamma}+\Upsilon(t)+v(t)) \to (p-1)/p$ as $t \to \infty$, $y \in \mathcal{NRV}((p-1)/p)$, and y is expressed as $y(t) = t^{(p-1)/p}L(t)$, where

$$L(t) = \exp\left\{\int_1^t \frac{\Phi^{-1}(\bar{\gamma}+\Upsilon(s)+v(s)) - \bar{\gamma}^{q-1}}{s}\,ds\right\},$$

$t \geq T$. Noting that $|\Upsilon(t)+v(t)| \leq \bar{\gamma}/2$, $t \geq T$, by (4.3.34) and (4.3.38), and applying the Mean Value Theorem, we see with the use of (4.3.41) that $|\Phi^{-1}(\bar{\gamma}+\Upsilon(t)+v(t)) - \bar{\gamma}^{q-1}| \leq N(p)(|\Upsilon(t)|+|\Psi(t)|)$, $t \geq T$, for some constant $N(p) > 0$. This, combined with the hypotheses (4.3.29) and (4.3.30), guarantees that $L(t)$ tends to a finite positive limit as $t \to \infty$. □

Now we give some examples illustrating the results developed in this section. Consider the function $c_1(t) = kt^{\beta}\sin(t^{\gamma})$, where $k \neq 0$, $\beta \geq 0$ and $\gamma > 0$ are constants. Note that $\liminf_{t\to\infty} c_1(t) = -\infty$, $\limsup_{t\to\infty} c_1(t) = \infty$ if $\beta > 0$, and $\liminf_{t\to\infty} c_1(t) > -\infty$, $\limsup_{t\to\infty} c_1(t) < \infty$ if $\beta = 0$. If $\gamma \geq p+\beta$, we have by integration by parts

$$t^{p-1}\int_t^\infty s^{\beta}\sin(s^{\gamma})\,ds = \frac{1}{\gamma}t^{p+\beta-\gamma}\cos(t^{\gamma})$$
$$-\frac{1+\beta-\gamma}{\gamma^2}t^{p+\beta-2\gamma}\sin(t^{\gamma}) - \frac{(1+\beta-\gamma)(1+\beta-2\gamma)}{\gamma^2}t^{p-1}\int_t^\infty s^{\beta-2\gamma}\sin(s^{\gamma})\,ds,$$

$t > 0$, from which it follows that

$$t^{p-1}\int_t^\infty c_1(s)\,ds = \frac{k}{\gamma}t^{p+\beta-\gamma}\cos(t^{\gamma}) + o(t^{-\gamma}) \quad \text{as} \quad t \to \infty, \tag{4.3.42}$$

if $\gamma > p+\beta$, and

$$t^{p-1}\int_t^\infty c_1(s)\,ds = \frac{k\cos(t^{p+\beta})}{p+\beta} + \mathcal{O}(t^{-\gamma}) \quad \text{as} \quad t \to \infty,$$

if $\gamma = p+\beta$. The first example is a consequence of Theorem 4.3.6.

Example 4.3.1. If

$$\frac{|k|}{p+\beta} < \frac{1}{p}\left(\frac{p-1}{p}\right)^{p-1},$$

then the equation $(\Phi(y'))' + kt^{\beta}\sin(t^{p+\beta})\Phi(y) = 0$ is nonoscillatory and has an O-regularly varying solution.

An example illustrating Theorem 4.3.7 is derived from the observation that

$$\lim_{t\to\infty} t^{p-1}\int_t^\infty \left(c_1(s) + \frac{(p-1)a}{s^p}\right) = a,$$

for any constant a, if $\gamma > p+\beta$.

Example 4.3.2. Let $a \in \left(-\infty, \frac{1}{p}\left(\frac{p-1}{p}\right)^{p-1}\right)$, $a \neq 0$, and suppose that $\gamma > p + \beta$. Then the equation

$$(\Phi(y'))' + \left(\frac{(p-1)a}{t^p} + kt^\beta \sin(t^\gamma)\right)\Phi(y) = 0$$

is nonoscillatory for any k and has two normalized regularly varying solutions with nonzero indices $\Phi^{-1}(\lambda_1)$ and $\Phi^{-1}(\lambda_2)$, where λ_1, λ_2 denote the two real roots of the equation $|\lambda|^q - \lambda + a = 0$.

To give an example to which Theorem 4.3.8 is applicable, consider the function

$$c(t) = c_1(t) + \left(\frac{p-1}{p}\right)^p \frac{1}{t^p}, \quad \gamma > p + \beta.$$

Put

$$\Upsilon_1(t) = t^{p-1}\int_t^\infty c(s)\,ds - \frac{1}{p}\left(\frac{p-1}{p}\right)^{p-1} = t^{p-1}\int_t^\infty c_1(s)\,ds.$$

Using (4.3.42), we have

$$\frac{\Upsilon_1}{t} = \frac{k}{\gamma} t^{p-1+\beta-\gamma}\cos(t^\gamma) + o(t^{-\gamma-1}) \quad \text{as} \quad t \to \infty,$$

which implies that

$$\int^\infty \frac{|\Upsilon_1(t)|}{t}\,dt < \infty$$

and

$$\Psi_1(t) := \int_t^\infty \frac{|\Upsilon_1(s)|}{s}\,ds = \mathcal{O}(t^{p+\beta-\gamma}) \quad \text{as} \quad t \to \infty.$$

Since $p + \beta - \gamma < 0$, we see that

$$\int^\infty \frac{\Psi_1(t)}{t}\,dt < \infty.$$

Example 4.3.3. If $\gamma > p + \beta$, then the equation

$$(\Phi(y'))' + \left(\left(\frac{p-1}{p}\right)^p \frac{1}{t^p} + kt^\beta \sin(t^\gamma)\right)\Phi(y) = 0$$

is nonoscillatory for any k and has a normalized regularly varying solution y with the index $(p-1)/p$ of the form $y(t) = t^{(p-1)/p}L(t)$, $L \in \mathcal{NSV}$, where L satisfies $\lim_{t\to\infty} L(t) = l \in (0, \infty)$.

Let $M : [0, \infty) \to [-1, 1]$ be a continuous function which takes both positive and negative values in any neighborhood of infinity. The functions $\sin t$, $\sin(\log t)$, $\sin(e^t)$ are simple examples of such a function M. Consider the equation

$$(\Phi(y'))' + \left(\frac{(p-1)aM(t)}{t^p} + kt^\beta \sin(t^\gamma)\right)\Phi(y) = 0. \tag{4.3.43}$$

Noting that $aM(t) \leq |a|$, $t \geq 0$, for any $a \in \mathbb{R}$, and applying the Sturm comparison theorem for half-linear equations, we conclude from Examples 4.3.2 and 4.3.3 that (4.3.43) is nonoscillatory if $|a| \leq \frac{1}{p}\left(\frac{p-1}{p}\right)^{p-1}$ and $\gamma > p + \beta$.

4.3.2 Existence of slowly varying solutions

In this subsection we present conditions guaranteeing the existence of solutions belonging to the class of slowly varying functions.

Theorem 4.3.9. *If*

$$\lim_{t\to\infty} t^{p-1}\int_t^\infty c(s)\,ds = 0 \tag{4.3.44}$$

holds, then equation (4.3.1) *is nonoscillatory and has a normalized slowly varying solution.*

Proof. Suppose that the condition from the theorem holds. Put

$$\varphi(t) = \sup_{s\geq t}\left| s^{p-1}\int_s^\infty c(\tau)\,d\tau\right|. \tag{4.3.45}$$

Then φ is nonincreasing and tends to zero as $t\to\infty$. Let $t_0 > 0$ be such that

$$\varphi(t) < \frac{1}{p}\left(\frac{p-1}{p}\right)^{p-1} \quad \text{and} \quad |\sigma(t)| \leq \frac{\varphi(t)}{t^{p-1}}$$

for $t\geq t_0$, where $\sigma(t)=\int_t^\infty c(s)\,ds$. Put $P(t)=\varphi(t)t^{1-\alpha}$. Then $|\sigma(t)|\leq P(t)$ holds and

$$\int_t^\infty P^q(s)\,ds = \int_t^\infty\left(\frac{\varphi(s)}{s^{p-1}}\right)^q ds \leq \frac{\varphi^q(t)}{(p-1)t^{p-1}} = \frac{1}{p-1}\varphi^{q-1}(t)P(t)$$

$t\geq t_0$. Consequently, by Lemma 4.3.1, (4.3.1) has a nonoscillatory solution of the form (4.3.12) on $[t_0,\infty)$ with v satisfying (4.3.9). Since

$$t^{p-1}v(t) = \mathcal{O}(t^{p-1}P(t)) = o(1) \quad \text{and} \quad t^{p-1}\sigma(t) = \mathcal{O}(t^{p-1}P(t)) = o(1)$$

as $t\to\infty$, we conclude that y is a normalized slowly varying function. □

Remark 4.3.1. It can be shown that (4.3.44) is a necessary condition for (4.3.1) to possess a normalized slowly varying solution

$$y(t) = \exp\left\{\int_{t_0}^t \frac{\psi(s)}{s}\,ds\right\}, \tag{4.3.46}$$

with ψ satisfying $\lim_{t\to\infty}\psi(t)=0$. In fact, (4.3.46) implies that $ty'(t)/y(t) = \psi(t)\to 0$ as $t\to\infty$. From generalized Riccati equation (3.1.2), satisfied by $w = \Phi(y'/y)$, we have

$$t^{p-1}w(t) = (p-1)t^{p-1}\int_t^\infty \frac{|s^{p-1}w(s)|^q}{s^p}\,ds + t^{p-1}\int_t^\infty c(s)\,ds,$$

$t\geq t_0$. Letting $t\to\infty$ and noting that $t^{p-1}w(t)\to 0$ as $t\to\infty$, we see that (4.3.44) holds. Consequently, this condition is sufficient and necessary for the existence of a normalized slowly varying solution.

For the next result we will need the statement of Lemma 4.3.1, with a slight modification, namely that condition (4.3.5) is replaced by the more general one:

$$\int_t^\infty P^q(s)\,ds \le \frac{1}{p-1} a^{q-1}(t)P(t), \quad t \ge t_0,$$

where $a(t)$ is a continuous nonincreasing function satisfying

$$0 < a(t) \le a < \frac{1}{p}\left(\frac{p-1}{p}\right)^{p-1}$$

for some constant a. The proof of such modified lemma is almost the same as that of the original one, and so it is omitted.

Theorem 4.3.10. *Suppose that the hypotheses of Lemma 4.3.1 with the above modification are satisfied. Let there exist a positive integer n such that*

$$\text{(4.3.47)} \qquad \int^\infty a^{n(q-1)}(t)P^{q-1}(t)\,dt < \infty \quad \text{if} \quad 1 < p \le 2,$$

$$\text{(4.3.48)} \qquad \int^\infty a^{n(q-1)^2}(t)P^{q-1}(t)\,dt < \infty \quad \text{if} \quad p > 2.$$

Then, for the solution (4.3.7) of (4.3.1), the following asymptotic formula holds for $t \to \infty$

$$\text{(4.3.49)} \qquad y(t) \sim A\exp\left\{\int_{t_0}^t \Phi^{-1}[v_{n-1}(s) + \sigma(s)]\,ds\right\},$$

where A is a positive constant. Here the sequence $\{v_n(t)\}$ of successive approximations is defined by

$$\text{(4.3.50)} \qquad v_0(t) = 0, \quad v_n(t) = (p-1)\int_t^\infty |v_{n-1}(s) + \sigma(s)|^q\,ds, \quad n = 1, 2, \ldots.$$

Proof. Let y be the solution (4.3.7). Recall that the function v used in (4.3.7) has been constructed as the fixed element in $C_P[t_0, \infty)$ of the contractive mapping $\mathcal{T}$ defined by (4.3.10). The standard proof of the contraction mapping principle shows that the sequence $\{v_n(t)\}$ defined by (4.3.50) converges to $v(t)$ uniformly on $[t_0, \infty)$. To see how fast $v_n(t)$ approaches $v(t)$ we proceed as follows. First, note that $|v_n(t)| \le (p-1)P(t)$, $t \ge t_0$, $n = 1, 2, \ldots$. By definition, we have

$$|v_1(t)| = (p-1)\int_t^\infty |\sigma(s)|^q\,ds \le (p-1)\int_t^\infty P^q(s)\,ds \le a^{q-1}(t)P(t),$$

and

$$
\begin{aligned}
|v_2(t) - v_1(t)| &\leq (p-1)\int_t^\infty \big||v_1(s)+\sigma(s)|^q - |\sigma(s)|^q\big|\,ds \\
&\leq (p-1)q\int_t^\infty [pP(s)]^q|v_1(s)|\,ds \\
&\leq p^q\int_t^\infty a^{q-1}(s)P^q(s)\,ds \leq p^q a^{q-1}(t)\int_t^\infty P^q(s)\,ds \\
&\leq (q-1)p^q a^{2(q-1)}(t)P(t) \leq \gamma_p^{q-1}\left(\frac{a(t)}{\gamma_p}\right)^{2(q-1)} P(t)
\end{aligned}
$$

for $t \geq t_0$, where $\gamma_p = \frac{1}{p}\left(\frac{p-1}{p}\right)^{p-1}$. Assuming that

$$
|v_n(t) - v_{n-1}(t)| \leq \gamma_p^q\left(\frac{a(t)}{\gamma_p}\right)^{n(q-1)} P(t), \tag{4.3.51}
$$

$t \geq t_0$, for some $n \in \mathbb{N}$, we compute

$$
\begin{aligned}
|v_{n+1}(t) - v_n(t)| &\leq (p-1)\int_t^\infty \big||\sigma(s)+v_n(s)|^q - |\sigma(s)+v_{n-1}(s)|^q\big|\,ds \\
&\leq (p-1)q\int_t^\infty [pP(s)]^{q-1}|v_n(s)-v_{n-1}(s)|\,ds \\
&= p^q\int_t^\infty \gamma_p^{q-1}\left(\frac{a(s)}{\gamma_p}\right)^{n(q-1)} P^q(s)\,ds \\
&\leq p^q\gamma_p^{q-1}\left(\frac{a(t)}{\gamma_p}\right)^{n(q-1)}\int_t^\infty P^q(s)\,ds \\
&\leq p^q\gamma_p^{q-1}\left(\frac{a(t)}{\gamma_p}\right)^{n(q-1)} (q-1)a^{q-1}(t)P(t) \\
&= \gamma_p^{q-1}\left(\frac{a(t)}{\gamma_p}\right)^{n(q-1)} P(t),
\end{aligned}
$$

$t \geq t_0$, which establishes the truth of (4.3.51) for all integers $n \in \mathbb{N}$. Now we have

$$
v(t) = v_{n-1}(t) + \sum_{k=n}^\infty [v_k(t) - v_{k-1}(t)],
$$

from which, due to (4.3.51), it follows that

$$
\begin{aligned}
|v(t) - v_{n-1}(t)| &\leq \gamma_p^{q-1}\left(\frac{a(t)}{\gamma_p}\right)^{k(q-1)} P(t) \\
&\leq \gamma_p^{q-1}\left(\frac{a(t)}{\gamma_p}\right)^{n(q-1)}\sum_{k=0}^\infty\left(\frac{a}{\gamma_p}\right)^k P(t) \qquad (4.3.52) \\
&= \gamma_p\left(\frac{a(t)}{\gamma_p}\right)^{n(q-1)}\frac{\gamma_p}{\gamma_p - c}P(t) = Ka^{n(q-1)}(t)P(t),
\end{aligned}
$$

where K is a constant depending only on p and n. Using (4.3.7) and (4.3.52), we obtain

$$(4.3.53)\quad y(t)\left(\exp\left\{\int_{t_0}^{t}\Phi^{-1}[\sigma(s)+v_{n-1}(s)]\,ds\right\}\right)^{-1} = \exp\left\{\int_{t_0}^{t}\left(\Phi^{-1}[\sigma(s)+v(s)]-\Phi^{-1}[\sigma(s)+v_{n-1}(s)]\right)ds\right\}.$$

Let $1<p\le 2$. Then, by the Mean Value Theorem and (4.3.52),

$$\begin{aligned}\left|\Phi^{-1}[\sigma(t)+v(t)]-\Phi^{-1}[\sigma(t)+v_{n-1}(t)]\right| &\le (q-1)[pP(t)]^{q-2}|v(t)-v_{n-1}(t)|\\ (4.3.54)\qquad &\le La^{n(q-1)}(t)P^{q-1}(t),\end{aligned}$$

$t\ge t_0$, where L is a constant depending on p and n. Let $p>2$. Then, using (4.3.52) and the inequality $|a^\lambda-b^\lambda|\le 2|a-b|^\lambda$ holding for $\lambda\in(0,1)$ and $a,b\in\mathbb{R}$, we see that

$$\begin{aligned}\left|\Phi^{-1}[\sigma(t)+v(t)]-\Phi^{-1}[\sigma(t)+v_{n-1}(t)]\right| &\le 2|v(t)-v_{n-1}(t)|^{q-1}\\ (4.3.55)\qquad &\le Ma^{n(q-1)^2}(t)P^{q-1}(t),\end{aligned}$$

$t\ge t_0$, where M is a constant depending on p and n. Combining (4.3.53) with (4.3.54) or (4.3.55) according as $1<p\le 2$ or $p>2$, and using (4.3.47) or (4.3.48), we conclude that the right-hand side of (4.3.53) tends to a constant $A>0$ as $t\to\infty$, which implies that $y(t)$ has the desired asymptotic behavior (4.3.49). □

Theorem 4.3.11. *Suppose that* (4.3.44) *holds and that the function* $a(t)=\varphi(t)$, *where* $\varphi(t)$ *defined by* (4.3.45), *satisfies*

$$(4.3.56)\qquad \int^{\infty}\frac{a^{(n+1)(q-1)}(t)}{t}\,dt<\infty \quad \textit{if}\quad 1<p\le 2,$$

$$(4.3.57)\qquad \int^{\infty}\frac{a^{(n+p-1)(q-1)^2}}{t}\,dt<\infty \quad \textit{if}\quad p>2.$$

Then the formula (4.3.49) *holds for the slowly varying solution* $y(t)$ *of* (4.3.1).

Proof. The conclusion follows from the previous theorem combined with the observation that in this case

$$a^{n(q-1)}(t)P^{q-1}(t)=\frac{a^{(n+1)(q-1)}(t)}{t}\quad\text{and}\quad a^{n(q-1)^2}(t)P^{q-1}(t)=\frac{a^{(n+p-1)(q-1)^2}(t)}{t}$$

according to whether $1<p\le 2$ or $p>2$. □

Example 4.3.4. Consider the equation

$$(4.3.58)\qquad (\Phi(y'))'+kt^\beta\sin(t^\gamma)\Phi(y)=0,$$

$t\ge 1$, where k,β and γ are positive constants satisfying

$$(4.3.59)\qquad \gamma>p+\beta.$$

Since

$$\int_t^\infty s^\beta \sin(s^\gamma)\,ds = \frac{1}{\gamma} t^{1+\beta-\gamma}\cos(t^\gamma) + \frac{1+\beta-\gamma}{\gamma}\int_t^\infty s^{\beta-\gamma}\cos(s^\gamma)\,ds,$$

there exists a positive constant K such that

$$\left|\int_t^\infty k s^\beta \sin(s^\gamma)\,ds\right| \le K t^{1+\beta-\gamma}, \tag{4.3.60}$$

which, in view of (4.3.59), implies that

$$\lim_{t\to\infty} t^{p-1}\int_t^\infty k s^\beta \sin(s^{p-1})\,ds = 0.$$

Therefore, equation (4.3.58) has a slowly varying solution by Theorem 4.3.9. In this case, the function $\varphi(t)$ defined by (4.3.45) can be taken to be $\varphi(t) = Kt^{p+\beta-\gamma}$. Since $\varphi(t)$ satisfies both (4.3.56) and (4.3.57) for any $n \in \mathbb{N}$, because of (4.3.59), from Theorem 4.3.11 for $n = 1$ we conclude that the slowly varying solution $y(t)$ of (4.3.58) has the asymptotic behavior

$$y(t) \sim A\exp\left\{\int_{t_0}^t \Phi^{-1}\left(\int_s^\infty k\tau^\beta \sin(\tau^\gamma)\,d\tau\right) ds\right\} \quad \text{as } t\to\infty, \tag{4.3.61}$$

which is equivalent to $y(t) \sim A_0$ (A_0 being a constant), since the integral in the braces in (4.3.61) converges as $t \to \infty$, because of (4.3.60).

4.4 Notes and references

The asymptotic behavior of nonoscillatory solutions of (1.1.1) (or of more general equations) has been deeply studied by the Georgian and Russian mathematical school, wee refer at last to the papers and book of Chanturia, Kiguradze, Kvinikadze and Rabtsevich [71, 72, 200, 201, 202, 203, 229, 319]. A comprehensive treatment of this topic can be found in Mirzov's book [292], see also updated english translation of this book [293].

The classification of nonoscillatory solutions of linear equation (1.1.2) into the classes $\mathbb{M}^+_\infty, \mathbb{M}^\infty_B, \mathbb{M}^-_B, \mathbb{M}^-_0$ treated in Section 4.1 was suggested in the paper of Marini and Zezza [285], and later it was extended to various directions in the papers of Cecchi, Marini and Villari [63, 64, 65, 66], Chen, Huang and Erbe [74], Nagabuchi and Yamamoto [300]. The results of Subsection 4.1.2 are taken from the papers of Cecchi, Došlá and Marini [57, 59]. Asymptotic behavior of nonoscillatory solutions of (1.1.1) (or of more general equations) when the function c is eventually positive has been considered by several authors. Here we refer in particular to the papers of Fan, Li and Zhong [159], J. Wang [359], Li and Cheng [251], Matucci [286], Wong and Agarwal [364] and to the papers of the Japan mathematicians Hoshino, Imabayashi, Kitano, Kusano, M. Naito, Ogata, Tanigawa and Usami [180, 212, 224], see also references given therein. The main part of the results of Subsections 4.1.3 and 4.1.4 can be found in the papers of Cecchi, Došlá and Marini

[57, 61]. Subsection 4.1.5 is a part of the paper of Došlá and Vrkoč [96], this paper contains also the application of Theorem 4.1.12 to oscillation of Emden-Fowler equation (1.1.4).

Mirzov's construction of the principal solution of (1.1.1) is established in his paper [291]. The construction described in Subsection 4.2.3 is taken from the paper of Elbert and Kusano [145], this paper contains also the main statement of Subsection 4.2.4. The main result of Subsection 4.2.5, in the form presented here, is a new result, but implicitly it is hidden in both the papers of Mirzov [290] and of Elbert and Kusano [145]. Theorem 4.2.4 is taken from the paper of Cecchi, Došlá and Marini [61]. Subsection 4.2.7 is the substantial part of [293, Section 15], the remaining part of this section of Mirzov's book is also presented in Subsection 4.2.11. The limit characterization of the principal solution of (1.1.1) is established in the papers of Cecchi, Došlá and Marini [60] (the case $c(t) < 0$ eventually) and in [61] (the case $c(t) > 0$ eventually). These two papers also contain the integral characterization of the principal solution suggested in Subsection 4.2.10, while the characterization suggested in Theorem 4.2.8 of Subsection 4.2.9 is taken from the paper of Došlý and Elbert [108] (parts (i) – (iii)) and from the paper of Cecchi, Došlá and Marini [61] – part (iv). A related paper containing the relationship between the integral characterization of Subsection 4.2.9 and the concept of regular half-linear equations is the paper of Došlý and Řezníčková [117]. Finally, as mentioned above, the most of the results of Subsection 4.2.11 is taken from Mirzov's book [293], the last statement of this subsection (Theorem 4.2.12) is taken from Došlý [106].

A detailed tratment of regularly varying functions and their relationship to differential equations can be found in the book of Marić [269]. The results of Section 4.3 are taken form the papers of Jaroš, Kusano and Tanigawa [190] and of Kusano, Marić and Tanigawa [215]. A related paper on this subject is Kusano, Marić [214].

CHAPTER 5

VARIOUS OSCILLATION PROBLEMS

In this rather long chapter we present various results concerning oscillation theory of equation (1.1.1). Among others, we will see that the methods developed in the first chapter play very important role in proving subsequent results. We start with extensions of Lyapunov and Vallée-Poussin inequality. Then we present focal point and conjugacy criteria. In the second section we describe the so-called perturbation principle in which the main idea lies in comparison of (1.1.1) with two-term equation. Disconjugacy/nonoscillation domains and equations with (almost) periodic coefficients are studied in the third section. Section 5.4 deals with strongly and conditionally oscillatory equations. In Section 5.5 we show that nonoscillation of (1.1.1) can be characterized by means of certain function sequences. Many applications will be given as well. Asymptotic formula for distance of zeros and conditions guaranteeing the existence of quick/slow oscillatory solutions are presented in the sixth section of this chapter. Various aspects of half-linear Sturm-Liouville problem are treated in Section 5.7. Energy functionals considered on classes of functions satisfying general boundary conditions are studied in Section 5.8. The last section is devoted to generalized Hartman-Wintner Theorem, Milloux Theorem, Armellini-Tonelli-Sansone Theorem and interval oscillation criteria. Many of the statements will be formulated, for simplicity, for (1.1.1) with $r(t) \equiv 1$, i.e.,

$$(\Phi(y'))' + c(t)\Phi(y) = 0, \tag{5.1.1}$$

c being a continuous function on an interval under consideration. An extension to a general r satisfying $\int^{\infty} r^{1-q}(t)\,dt = \infty$ is straightforward; see also Subsection 1.2.7.

5.1 Conjugacy and disconjugacy

The first section of this chapter presents (dis)conjugacy and focal point criteria which are based on various half-linear extensions of the Lyapunov and Vallée-Poussin type inequalities.

5.1.1 Lyapunov inequality

The classical Lyapunov inequality (see e.g. [174, Chap. XI]) for the linear differential equation (1.1.2) states that if a, b, $a < b$, are consecutive zeros of a nontrivial solution of this equation, then

$$\int_a^b c_+(t)\, dt > \frac{4}{\int_a^b r^{-1}(t)\, dt}, \qquad c_+(t) = \max\{0, c(t)\}.$$

This inequality has been extended in many directions and its half-linear extension reads as follows.

Theorem 5.1.1. *Let a, b, $a < b$, be consecutive zeros of a nontrivial solution of* (1.1.1). *Then*

$$\int_a^b c_+(t)\, dt > \frac{2^p}{\left(\int_a^b r^{1-q}(t)\, dt\right)^{p-1}}. \tag{5.1.2}$$

Proof. According to homogeneity of the solution space of (1.1.1), we can suppose that $x(t) > 0$ on (a, b). Let $c \in (a, b)$ be the least point of the local maxima of x in (a, b), i.e., $x'(c) = 0$ and $x'(t) > 0$ on $[a, c)$. By the Hölder inequality, we have

$$\begin{aligned} x^p(c) &= \left(\int_a^c x'(t)\, dt\right)^p = \left(\int_a^c r^{-\frac{1}{p}}(t) r^{\frac{1}{p}}(t) x'(t)\, dt\right)^p \\ &\le \left(\int_a^c r^{-\frac{q}{p}}(t)\, dt\right)^{\frac{p}{q}} \left(\int_a^b r(t)(x'(t))^p\, dt\right). \end{aligned}$$

Multiplying (1.1.1) by $x(t)$ and integrating from a to c by parts,we get

$$\begin{aligned} \int_a^c r(t)(x'(t))^p\, dt &= \int_a^c c(t) x^p(t)\, dt \le \int_a^c c_+(t) x^p(t)\, dt \\ &\le x^p(c) \int_a^c c_+(t)\, dt, \end{aligned}$$

hence

$$x^p(c) \le \left(\int_a^c r^{1-q}(t)\, dt\right)^{p-1} \left(\int_a^c c_+(t)\, dt\right) x^p(c),$$

which yields

$$\left(\int_a^c r^{1-q}(t)\, dt\right)^{1-p} \le \int_a^c c_+(t)\, dt.$$

Similarly, if d is the greatest point of local maxima of x in (a,b), i.e., $x'(d)=0$ and $x'(t)<0$ on (d,b), we have

$$\left(\int_d^b r^{1-q}(t)\,dt\right)^{1-p} \le \int_d^b c_+(t)\,dt.$$

Consequently,

$$\int_a^b c_+(t)\,dt \ge \left(\int_a^c r^{1-q}(t)\,dt\right)^{1-p} + \left(\int_d^b r^{1-q}(t)\,dt\right)^{1-p}.$$

Finally, since the function $f(u)=u^{1-p}$ is convex for $u>0$, the Jensen inequality $f\left((u+v)/2\right) \le [f(u)+f(v)]/2$ with $u=\int_a^c r^{1-q}(t)\,dt$, $v=\int_d^b r^{1-q}(t)\,dt$ implies

$$\begin{aligned}\left(\int_a^c r^{1-q}(t)\,dt\right)^{1-p} &+ \left(\int_d^b r^{1-q}(t)\,dt\right)^{1-p}\\ &\ge 2\left[\frac{1}{2}\left(\int_a^c r^{1-q}(t)\,dt + \int_d^b r^{1-q}(t)\,dt\right)\right]^{1-p} \ge \frac{2^p}{\left(\int_a^b r^{1-q}(t)\,dt\right)^{p-1}},\end{aligned}$$

what completes the proof. □

A closer examination of the proof shows that Theorem 5.1.1 can be modified in the following way.

Theorem 5.1.2. *If there exists* $0 \not\equiv \xi \in W_0^{1,p}(a,b)$ *such that* $\mathcal{F}(\xi;a,b) \le 0$ *(*$\mathcal{F}$ *is defined in Subsection 1.2.2), then* (5.1.2) *holds.*

Remark 5.1.1. Combining the last theorem with the variational principle (see Section 2.1), we get the following disconjugacy criterion: If

$$\left(\int_a^b r^{1-q}(t)\,dt\right)^{p-1} \int_a^b c_+(t)\,dt < 2^p,$$

then (1.1.1) is disconjugate on $[a,b]$.

5.1.2 Vallée-Poussin type inequality

Another important inequality concerning disconjugacy of the linear differential equation

$$x'' + a(t)x' + b(t)x = 0 \tag{5.1.3}$$

was introduced by Vallée-Poussin [354] in 1929 and reads as follows. Suppose that $t_1 < t_2$ are consecutive zeros of a nontrivial solution x of (5.1.3), then

$$2\int_0^\infty \frac{dt}{t^2+At+B} \le t_2 - t_1, \quad A := \max_{t\in[t_1,t_2]} |a(t)|, \; B := \max_{t\in[t_1,t_2]} |b(t)|.$$

The half-linear version was originally proved in a somewhat different form, but here we prefer to formulate this criterion in a simplified form, to underline similarity with the original criterion of Vallée-Poussin. For the same reason we consider the equation

$$(\Phi(x'))' + a(t)\Phi(x') + b(t)\Phi(x) = 0 \tag{5.1.4}$$

instead of (1.1.1) (if the function r in (1.1.1) is differentiable, then this equation can be easily reduced to (5.1.4)).

Theorem 5.1.3. *Suppose that $t_1 < t_2$ are consecutive zeros of a nontrivial solution x of* (5.1.4). *Then*

(5.1.5)
$$2\int_0^\infty \frac{dt}{(p-1)t^q + At + B} \le t_2 - t_1, \quad A = \max_{t\in[t_1,t_2]} |a(t)|, \; B = \max_{t\in[t_1,t_2]} |b(t)|.$$

Proof. Suppose that $x(t) > 0$ in (t_1, t_2). In case $x(t) < 0$ in (t_1, t_2), the proof is analogical. Let $c, d \in (t_1, t_2)$, $c \le d$, be the least and the greatest points of the local maximum of x in (a, b), respectively, i.e., $x'(t) > 0$ for $t \in (t_1, c)$, $x'(t) < 0$ for $t \in (d, t_2)$ and $x'(c) = 0 = x'(d)$. The Riccati variable $v = \Phi(x'/x)$ satisfies $v(t_1+) = \infty$, $v(c) = 0$, $v(t) > 0$, $t \in (t_1, c)$ and

$$v' = -b(t) - a(t)v - (p-1)v^q \ge -B - Av - (p-1)v^q. \tag{5.1.6}$$

Hence,

$$\int_0^\infty \frac{dv}{(p-1)v^q + Av + B} \le c - t_1 \tag{5.1.7}$$

Concerning the interval (d, t_2), we set $v = -\frac{\Phi(x')}{\Phi(x)} > 0$ for $t \in (d, t_2)$, and similarly as for $t \in (t_1, c)$ we have

$$\int_0^\infty \frac{dv}{(p-1)v^q + Av + B} \le t_2 - d. \tag{5.1.8}$$

The summation of (5.1.7) and (5.1.8) gives

$$2\int_0^\infty \frac{dv}{(p-1)v^q + Av + B} \le c - t_1 + t_2 - d \le t_2 - t_1,$$

what we needed to prove. □

Remark 5.1.2. (i) Since (5.1.5) is a necessary condition for conjugacy of (5.1.4) in $[t_1, t_2]$, the opposite inequality is a disconjugacy criterion: If

$$2\int_0^\infty \frac{dv}{(p-1)v^q + Av + B} > t_2 - t_1,$$

then (5.1.4) is disconjugate in $[t_1, t_2]$.

(ii) A more general Riccati substitution $v = \alpha(t)\Phi(x'/x)$, $t \in (t_1, c]$, $v = -\beta(t)\Phi(x'/x)$, $t \in [d, t_2)$, where α, β are suitable positive functions, enables to formulate the Vallée-Poussin type criterion in a more general form than presented in Theorem 5.1.3, we refer to [111] for details. Concerning the various extensions of the linear Vallée-Poussin criterion we refer to the survey paper [22] and the references given therein.

5.1.3 Focal point criteria

Recall that a point b is said to be the *first right focal point* of $c < b$ with respect to (1.1.1) if there exists a nontrivial solution x of this equation such that $x'(c) = 0 = x(b)$ and $x(t) \neq 0$ for $t \in [c, b)$. See also below given Remark 5.8.1. The *first left focal point* a of c is defined similarly by $x(a) = 0 = x'(c)$, $x(t) \neq 0$ on $(a, c]$. Equation (1.1.1) is said to be *right disfocal* on $[c, b)$ if there exists no right focal point of c relative to (1.1.1) in (c, b), the *left disfocality* on $(a, c]$ is defined in a similar way. Consequently, (1.1.1) is conjugate on an interval (a, b) if there exists $c \in (a, b)$ such that this equation is neither right disfocal on $[c, b)$ nor left disfocal on $(a, c]$. This idea is illustrated in the next statement for (5.1.1) considered on $(a, b) = (-\infty, \infty)$. The extension of this statement to general half-linear equation (1.1.1) is immediate.

Theorem 5.1.4. *Suppose that the function* $c(t) \not\equiv 0$ *for* $t \in (0, \infty)$ *and there exist constants* $\alpha \in (-1/p, p-2]$ *and* $T \geq 0$ *such that*

$$\int_0^t s^\alpha \left(\int_0^s c(\tau)\, d\tau \right) ds \geq 0 \quad \text{for } t \geq T. \tag{5.1.9}$$

Then the solution x *of* (5.1.1) *satisfying the initial conditions* $x(0) = 1$, $x'(0) \leq 0$ *has a zero in* $(0, \infty)$.

Proof. Suppose, by contradiction, that the solution x has no zero on $(0, \infty)$, i.e., $x(t) > 0$. Let $w = -\Phi(x'/x)$ be the solution of the Riccati equation

$$w' = c(t) + (p-1)|w|^q.$$

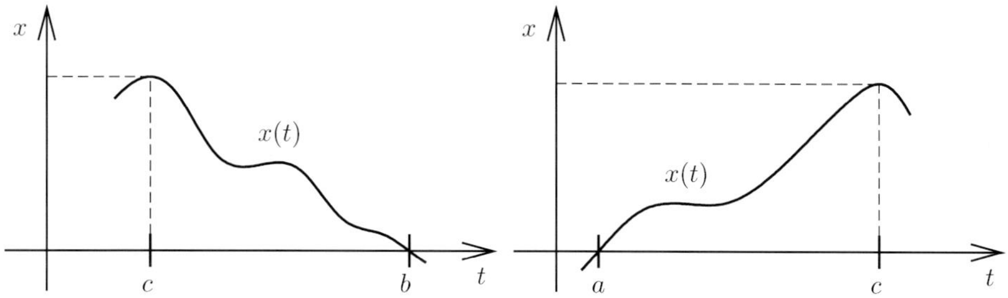

Figure 5.1.1: First right focal point and first left focal point

Since $w(0) \geq 0$, we have

$$
\begin{aligned}
(5.1.10) \qquad w(t) &= w(0) + \int_0^t c(s)\, ds + (p-1)\int_0^t |w(s)|^q\, ds \\
&\geq \int_0^t c(s)\, ds + (p-1)\int_0^t |w(s)|^q\, ds
\end{aligned}
$$

and

$$\int_0^t s^\alpha w(s)\, ds \geq \int_0^t \left(s^\alpha \int_0^s c(\tau)\, d\tau \right) ds + G(t),$$

where

$$G(t) = (p-1)\int_0^t s^\alpha \left(\int_0^s |w(\tau)|^q\, d\tau \right) ds.$$

Then

$$(5.1.11) \qquad G'(t) = (p-1)t^\alpha \int_0^t |w(\tau)|^q\, d\tau \geq 0 \quad \text{for } t \geq 0,$$

and according to (5.1.9),

$$(5.1.12) \qquad G(t) \leq \int_0^t s^\alpha w(s)\, ds \qquad \text{for } t \geq T.$$

By the Hölder inequality we have

$$\int_0^t s^\alpha w(s)\, ds \leq \left[\int_0^t s^{p\alpha}\, ds \right]^{\frac{1}{p}} \left[\int_0^t |w(s)|^q\, ds \right]^{\frac{1}{q}} = \left[\frac{t^{1+p\alpha}}{1+p\alpha} \right]^{\frac{1}{p}} \left[\int_0^t |w(s)|^q\, ds \right]^{\frac{1}{q}},$$

hence by (5.1.12)

$$(5.1.13) \qquad \frac{t^{(1+p\alpha)\frac{q}{p}}}{(1+p\alpha)^{\frac{q}{p}}} \int_0^t |w(s)|^q\, ds \geq G^q(t).$$

Here we need the relation $G(t) > 0$ for sufficiently large t. By (5.1.11) $G(t)$ is nondecreasing function of t and $G(0) = 0$. The equality $G(t) = 0$ for all $t \geq 0$ would imply that $w(t) \equiv 0$, consequently by (5.1.11) $x'(t) \equiv 0$ for $t \geq 0$. But this may happen only if $c(t) \equiv 0$, which case has been excluded. Hence we may suppose that T is already chosen so large that the inequality $G(t) > 0$ holds for $t \geq T$.

Denote $\beta = \alpha - (1+p\alpha)\frac{q}{p}$ and $K = (p-1)(1+p\alpha)^{\frac{q}{p}} > 0$. Then by (5.1.11), inequality (5.1.13) yields $G'G^{-q} \geq Kt^\beta$. Integrating this inequality from T to t, we get

$$\frac{1}{q-1} G^{1-q}(T) > \frac{1}{q-1}\left[G^{1-q}(T) - G^{1-q}(t) \right] \geq K \int_T^t s^\beta\, ds,$$

where the integral on the right-hand side tends to ∞ as $t \to \infty$ because an easy computation shows that $\alpha \leq p-2$ implies $\beta \geq -1$. This contradiction proves that x must have a positive zero. □

Remark 5.1.3. (i) Clearly, in Theorem 5.1.4 the starting point $t_0 = 0$ can be shifted to any other value $t_0 \in \mathbb{R}$ if the condition (5.1.9) is modified to

$$\int_{t_0}^{t} (s - t_0)^{\alpha} \left(\int_{t_0}^{s} c(\tau)\, d\tau \right) ds \geq 0 \quad \text{for } t \geq T \geq t_0.$$

A similar statement can be formulated on the interval $(-\infty, t_0)$, too.

(ii) In the previous theorem we have used the weight function s^{α}, $\alpha \in (-1/p, p-2]$. The results of Subsection 3.2.1 suggest to use a more general weight functions. This research is a subject of the present investigation.

Using the just established focal point criterion we can prove the following conjugacy criterion for (5.1.1).

Theorem 5.1.5. *Suppose that $c(t) \not\equiv 0$ both in $(-\infty, 0)$ and $(0, \infty)$ and there exist constants $\alpha_1, \alpha_2 \in (-1/p, p-2]$ and $T_1, T_2 \in \mathbb{R}$, $T_1 < 0 < T_2$, such that*

(5.1.14)
$$\int_{t}^{0} |s|^{\alpha_1} \left(\int_{s}^{0} c(\tau)\, d\tau \right) ds \geq 0,\ t \leq T_1, \quad \int_{0}^{t} s^{\alpha_2} \left(\int_{0}^{s} c(\tau)\, d\tau \right) ds \geq 0,\ t \geq T_2.$$

Then equation (5.1.1) is conjugate in $\mathbb{R}$, more precisely, there exists a solution of (5.1.1) having at least one positive and one negative zero.

Proof. The statement follows immediately from Theorem 5.1.4 since by this theorem the solution x given by $x(0) = 1$, $x'(0) = 0$ has a positive zero. Using the same argument as in Theorem 5.1.4 and the second condition in (5.1.14) we can show the existence of a negative zero. □

Remark 5.1.4. (i) Assumptions of the previous theorem are satisfied if

$$\liminf_{s_1 \downarrow -\infty, s_2 \uparrow \infty} \int_{s_1}^{s_2} c(t)\, dt > 0. \tag{5.1.15}$$

This conjugacy criterion for the linear Sturm-Liouville equation (1.1.2) with $r(t) \equiv 1$ is proved in [350] and the extension to (5.1.1) can be found in [310].

(ii) Several conjugacy criteria for linear equation (1.1.2) (in terms of its coefficients r, c) are proved using the fact that this equation is conjugate on (a, b) if and only if

$$\int_{a}^{b} \frac{dt}{r(t)[x_1^2(t) + x_2^2(t)]} > \pi \tag{5.1.16}$$

for any pair of solutions of (1.1.2) for which $r(x_1' x_2 - x_1 x_2') \equiv \pm 1$. This statement is based on the trigonometric transformation of (1.1.2), see [97, 98] and also Section 1.3. However, since we have in disposal no half-linear analogue of the trigonometric transformation, conjugacy criteria of this kind for (1.1.1) are (till now) missing.

5.1.4 Lyapunov-type focal points and conjugacy criteria

The next results concern again equation (5.1.1).

Theorem 5.1.6. *Let x be a nontrivial solution of (5.1.1) satisfying $x'(d) = 0 = x(b)$ and $x(t) \neq 0$ for $t \in [d, b)$. Then*

$$(b-d)^{p-1} \sup_{d \le t \le b} \left| \int_d^t c(s)\, ds \right| > 1. \tag{5.1.17}$$

Moreover, if there is no extreme value of x in (d, b), then

$$(b-d)^{p-1} \sup_{d \le t \le b} \int_d^t c(s)\, ds > 1. \tag{5.1.18}$$

Proof. Suppose that $x(t) > 0$ on $[d, b)$, if $x(t) < 0$ we proceed in the same way. Let $v = -\Phi(x'/x)$ and $V(t) = (p-1)\int_d^t |v(s)|^q\, ds$. Then we have

$$v(t) = \int_d^t c(s)\, ds + V(t). \tag{5.1.19}$$

Thus, $v(d) = 0 = V(d)$ and $\lim_{t \to b-} v(t) = \lim_{t \to b-} V(t) = \infty$. Set

$$C^* := \sup_{d \le t \le b} \left| \int_d^t c(s)\, ds \right|$$

and observe that $|v(t)| \le C^* + V(t)$, so that

$$V'(t) = (p-1)|v(t)|^q \le (p-1)(C^* + V(t))^q,$$

and

$$\frac{V'(t)}{(p-1)(C^* + V(t))^q} \le 1.$$

Integrating this inequality from d to b and using $\lim_{t \to b-} V(t) = \infty$, we obtain

$$-\frac{1}{(C^* + V(t))^{q-1}} \bigg|_d^b \le b - d,$$

which implies that $(b-d)^{p-1} C^* \le 1$. We remark that the equality cannot hold, for otherwise $|\int_d^t c(s)\, ds| = C^*$ on $[d, b)$ which implies that $c(t) \equiv 0$, a contradiction, thus (5.1.17) holds.

If d is the largest extreme point of x in (a, b), then $x'(t) \le 0$ and hence $v(t) \ge 0$ on $[d, b)$. Set $C_* = \sup_{d \le t \le b} \int_d^t c(s)\, ds$. Then we also have $C_* > 0$ since the assumption $C_* \le 0$ contradicts to $V(d) = 0$, $\lim_{t \to b-} V(t) = \infty$. Hence, by (5.1.19), $0 \le v(t) \le C_* + V(t)$. The remaining part of the proof is similar to the first one. □

The following theorem can be proved similarly as the previous one, and hence we omit its proof.

Theorem 5.1.7. *Let x be a nontrivial solution of (5.1.1) satisfying $x(a) = 0 = x'(c)$ and $x(t) \neq 0$ for $t \in (a, c]$. Then*

$$(c-a)^{p-1} \sup_{a \le t \le c} \left| \int_t^c c(s)\, ds \right| > 1.$$

Moreover, if there is no extreme value of x in (a, c), then

$$(c-a)^{p-1} \sup_{a \le t \le c} \int_t^c c(s)\, ds > 1.$$

Corollary 5.1.1. *If*

$$(b-d)^{p-1} \sup_{d \le t \le b} \left| \int_d^t c(s)\, ds \right| \le 1,$$

then (5.1.1) is right disfocal on $[d, b)$. If

$$(c-a)^{p-1} \sup_{a \le t \le c} \left| \int_t^c c(s)\, ds \right| \le 1,$$

then (5.1.1) is left disfocal on $(a, c]$.

Theorem 5.1.8. *Let $a < b$ be consecutive zeros of a nontrivial solution x of (5.1.1). Then there exist two disjoint subintervals $I_1, I_2 \subset [a, b]$ such that*

$$(b-a)^{p-1} \int_{I_1 \cup I_2} c(s)\, ds > \min\{4, 4^{p-1}\} \tag{5.1.20}$$

and

$$\int_{[a,b] \setminus (I_1 \cup I_2)} c(s)\, ds \le 0. \tag{5.1.21}$$

Proof. Let c and d denote the smallest and largest extreme points of x on $[a, b]$, respectively. Without loss of generality, we may assume that $c < d$ (if there is only one zero of x' in (a, b), then c and d coincide). Thus, $x'(d) = 0$, $x(b) = 0$ and $x'(t) \neq 0$ for $t \in (d, b]$. By Theorem 5.1.6, inequality (5.1.18) holds. Then there exists $b_1 \in (d, b]$ such that

$$(b-d)^{p-1} \int_d^{b_1} c(s)\, ds > 1, \quad \text{and} \quad \int_d^{b_1} c(s)\, ds \ge \int_d^b c(s)\, ds. \tag{5.1.22}$$

Similarly, it follows from Theorem 5.1.7 that there exists $a_1 \in [a, c)$ such that

$$(c-a)^{p-1} \int_{a_1}^c c(s)\, ds > 1, \quad \text{and} \quad \int_{a_1}^c c(s)\, ds \ge \int_a^c c(s)\, ds. \tag{5.1.23}$$

Let $I_1 = [d, b_1]$ and $I_2 = [a_1, c]$. Now, we divide the proof into the following two cases.

Case 1. If $p \geq 2$, then

$$\begin{aligned}
(b-a)^{p-1} \int_{I_1 \cup I_2} c(s)\, ds &= [(b-d)+(c-a)]^{p-1} \left(\int_d^{b_1} c(s)\, ds + \int_{a_1}^{c} c(s)\, ds \right) \\
&> [(b-d)^{p-1} + (c-a)^{p-1}] \left(\frac{1}{(b-d)^{p-1}} + \frac{1}{(c-a)^{p-1}} \right) \\
&\geq \left(\frac{\sqrt{(b-d)^{p-1}}}{\sqrt{(b-d)^{p-1}}} + \frac{\sqrt{(c-a)^{p-1}}}{\sqrt{(c-a)^{p-1}}} \right)^2 = 4.
\end{aligned}$$

Case 2. If $1 < p < 2$, then

$$\begin{aligned}
(b-a) \left(\int_{I_1 \cup I_2} c(s)\, ds \right)^{q-1} &\geq (b-a) \left(\int_d^{b_1} c(s)\, ds + \int_{a_1}^{c} c(s)\, ds \right)^{q-1} \\
&> (b-a) \left[\left(\int_d^{b_1} c(s)\, ds \right)^{q-1} + \left(\int_{a_1}^{c} c(s)\, ds \right)^{q-1} \right] \\
&\geq [(b-d)+(c-a)] \left(\frac{1}{b-d} + \frac{1}{c-a} \right) \geq 4.
\end{aligned}$$

It follows from Cases 1 and 2 that (5.1.20) holds.

By (5.1.22) and (5.1.23), $\int_{b_1}^{b} c(s)\, ds \leq 0$ and $\int_a^{a_1} c(s)\, ds \leq 0$. In order to verify (5.1.21), it is sufficient to show that $\int_c^d c(s)\, ds \leq 0$. Since $y'(c) = y'(d) = 0$, we have $v(c) = v(d) = 0$, where $v = -\Phi(y'/y)$. Hence

$$0 = v(d) - v(c) = \int_c^d c(s)\, ds + (p-1) \int_c^d |v(s)|^q\, ds.$$

This means that $\int_c^d c(s)\, ds \leq 0$, which implies that (5.1.21) holds. □

Corollary 5.1.2. *Suppose that for every two disjoint subintervals $I_1, I_2 \subset [a,b]$,*

$$(b-a)^{p-1} \int_{I_1 \cup I_2} c(s)\, ds \leq \min\{4, 4^{p-1}\}$$

Then (5.1.1) is disconjugate on $[a,b]$.

Proof. Suppose by a contradiction that there exists a nontrivial solution x of (5.1.1) with $x(c) = x(d) = 0$ for $a \leq c < d \leq b$. Without loss of generality, we may assume that $x(t) \neq 0$ for $t \in (c,d)$. By Theorem 5.1.8, there exist two disjoint subintervals $I_1, I_2 \subset [c,d] \subseteq [a,b]$ such that

$$(d-c)^{p-1} \int_{I_1 \cup I_2} c(s)\, ds > \min\{4, 4^{p-1}\}.$$

Thus $(b-a)^{p-1} \int_{I_1 \cup I_2} c(s)\, ds > \min\{4, 4^{p-1}\}$, a contradiction. □

Corollary 5.1.3. *Suppose that a nontrivial solution x of (5.1.1) has N zeros on $[a,b]$, where $N \geq 2$. Then there exist $2(N-1)$ disjoint subintervals $I_{ij} \subset [a,b]$, $i = 1, \ldots, N-1$, $j = 1, 2$, such that*

$$N < \left(\frac{(b-a)^{p-1}}{\min\{4, 4^{p-1}\}} \int_I c(s)\, ds \right)^{1/p} + 1, \tag{5.1.24}$$

and

$$\int_{[a,b]\setminus I} c(s)\, ds \leq 0, \tag{5.1.25}$$

where $I = \bigcup_{i=1}^{N-1} (I_{i1} \cup I_{i2})$.

Proof. Let t_i, $i = 1, \ldots, N$, be the zeros of x on $[a,b]$. By Theorem 5.1.8, for each $i = 1, \ldots, N-1$, there are two disjoint subintervals I_{i1} and I_{i2} of $[t_i, t_{i+1}]$ such that

$$\int_{I_{i1}\cup I_{i2}} c(s)\, ds > \frac{\min\{4, 4^{p-1}\}}{(t_{i+1} - t_i)^{p-1}} \tag{5.1.26}$$

and

$$\int_{[t_i, t_{i+1}]\setminus(I_{i1}\cup I_{i2})} c(s)\, ds \leq 0. \tag{5.1.27}$$

Summing (5.1.26) for i from 1 to $N-1$,

$$\begin{aligned}
\int_I c(s)\, ds &> \min\{4, 4^{p-1}\} \sum_{i=1}^{N-1} \frac{1}{(t_{i+1} - t_i)^{p-1}} \\
&\geq \min\{4, 4^{p-1}\}(N-1) \sqrt[N-1]{\frac{1}{(t_2 - t_1)^{p-1}} \cdots \frac{1}{(t_N - t_{N-1})^{p-1}}} \\
&= \min\{4, 4^{p-1}\}(N-1) \frac{1}{\left(\sqrt[N-1]{(t_2 - t_1) \cdots (t_N - t_{N-1})} \right)^{p-1}} \\
&\geq \min\{4, 4^{p-1}\}(N-1) \frac{(N-1)^{p-1}}{[(t_2 - t_1) + \cdots + (t_N - t_{n-1})]^{p-1}} \\
&= \min\{4, 4^{p-1}\}(N-1)^p \frac{1}{(t_N - t_1)^{p-1}} \\
&\geq \min\{4, 4^{p-1}\}(N-1)^p \frac{1}{(b-a)^{p-1}},
\end{aligned}$$

which implies

$$(N-1)^p < \frac{(b-a)^{p-1}}{\min\{4, 4^{p-1}\}} \int_I c(s)\, ds.$$

This implies that (5.1.24) holds. From (5.1.27), it is easy to deduce (5.1.25). □

5.1.5 Further related results

We start with the following Opial type inequality (proved in [32]), which will be useful in proving the subsequent statement. Note that another Opial type inequality, which is directly related to equation (1.1.1), is presented in Section 9.5.

Lemma 5.1.1. *Let $\alpha\beta > 0$, $\alpha + \beta > 1$, be constants and $f(t)$ be a nonnegative, measurable function on (a, b), and*

$$K(\alpha,\beta) = K_1(d,\alpha,\beta) = K_2(d,\alpha,\beta) < \infty, \tag{5.1.28}$$

where

$$K_1(d,\alpha,\beta) = \left(\frac{\beta}{\alpha+\beta}\right)^{\beta/(\alpha+\beta)} \left\{\int_a^d f^{(\alpha+\beta)/\alpha}(t)(t-a)^{\alpha+\beta-1}\,dt\right\}^{\alpha/(\alpha+\beta)},$$

$$K_2(d,\alpha,\beta) = \left(\frac{\beta}{\alpha+\beta}\right)^{\beta/(\alpha+\beta)} \left\{\int_d^b f^{(\alpha+\beta)/\alpha}(t)(b-t)^{\alpha+\beta-1}\,dt\right\}^{\alpha/(\alpha+\beta)},$$

and d is the (unique) solution of (5.1.28). If x is absolutely continuous function on $[a, b]$, with $x(a) = x(b) = 0$, then

$$\int_a^b f(t)|x(t)|^\alpha |x'(t)|^\beta\,dt \le K(\alpha,\beta)\int_a^b |x'(t)|^{\alpha+\beta}\,dt. \tag{5.1.29}$$

Theorem 5.1.9. *Let (5.1.1) have a solution x satisfying $x(a) = x(b) = 0$, $x(t) \neq 0$ for $t \in (a, b)$. Let $Q(t) = \int_a^t c(s)\,ds$ or $Q(t) = \int_t^b c(s)\,ds$. Then there exists a (unique) $d \in (a, b)$ such that*

$$\int_a^d |Q(t)|^q (t-a)^{p-1}\,dt = \int_d^b |Q(t)|^q (b-t)^{p-1}\,dt \ge \frac{1}{p}, \tag{5.1.30}$$

where $1/p + 1/q = 1$.

Proof. Multiplying (5.1.1) by $x(t)$ and integrating by parts over $[a, b]$ gives

$$\begin{aligned} \int_a^b |x'(t)|^p\,dt &= \int_a^b c(t)|x(t)|^p\,dt = -p\int_a^b Q(t)p\Phi(x(t))x'(t)\,ds \\ &\le p\int_a^b |Q(t)||x(t)|^{p-1}|x'(t)|\,dt \\ &\le pK(p-1,1)\int_a^b |x'(t)|^p\,dt, \end{aligned} \tag{5.1.31}$$

where we have used (5.1.29) of Lemma 5.1.1, since $x(t)$ is a nontrivial solution of (5.1.1), $\int_a^b |x'(t)|\,dt > 0$, (5.1.30) follows from (5.1.31). □

5.2 Perturbation principle

In this section we investigate oscillatory properties of (1.1.1) viewed as a perturbation of another half-linear equation of the same form. We present modified versions of several previously formulated criteria like Leighton conjugacy criterion, Leighton-Wintner and Nehari type (non)oscillation criteria. The last two subsections contain criteria where oscillatory properties of (1.1.1) and (5.1.1) are studied via certain associated *linear* second order differential equations of the form (1.1.2).

5.2.1 General idea

In the previous sections devoted to oscillation and nonoscillation criteria for (1.1.1), this equation was essentially viewed as a perturbation of the *one-term* equation

$$(r(t)\Phi(x'))' = 0. \tag{5.2.1}$$

As we have already mentioned, for oscillation (nonoscillation) of (1.1.1), the function c must be "sufficiently positive" ("not too positive") comparing with the function r. In this section we use a more general approach: Equation (1.1.1) is investigated as a perturbation of another (nonoscillatory) *two-term* half-linear equation

$$(r(t)\Phi(x'))' + \tilde{c}(t)\Phi(x) = 0 \tag{5.2.2}$$

with a continuous function $\tilde{c}$, i.e., (1.1.1) is written in the form

$$(r(t)\Phi(x'))' + \tilde{c}(t)\Phi(x) + (c(t) - \tilde{c}(t))\,\Phi(x) = 0. \tag{5.2.3}$$

The main idea is essentially the same as before. If the difference $(c-\tilde{c})$ is sufficiently positive (not too positive), then (5.2.3) becomes oscillatory (remains nonoscillatory).

Note that in the linear case $p = 2$, the idea to investigate the linear Sturm-Liouville equation (1.1.2) as a perturbation of the nonoscillatory two-term equation

$$(r(t)x')' + \tilde{c}(t)x = 0 \tag{5.2.4}$$

(and not only as a perturbation of the one-term equation $(r(t)x')' = 0$) brings essentially no new idea. Indeed, let us write (1.1.2) in the "perturbed" form

$$(r(t)x')' + \tilde{c}(t)x + (c(t) - \tilde{c}(t))\,x = 0. \tag{5.2.5}$$

Further, let h be a solution of (5.2.4) and consider the transformation $x = h(t)u$. This transformation transforms (5.2.5) into the equation

$$(r(t)h^2(t)u')' + [c(t) - \tilde{c}(t)]h^2(t)u = 0 \tag{5.2.6}$$

(compare (1.3.14)) and this equation, whose oscillatory properties are the same as those of (1.1.2), can be again investigated as a perturbation of the one-term equation $(r(t)h^2(t)u')' = 0$. In the half-linear case we have in disposal no transformation which reduces nonoscillatory two-terms equation into an one-term equation, so we have to use different methods. This "perturbation principle" then becomes a very useful tool which enables to prove new results in the half-linear case.

5.2.2 Singular Leighton's theorem

In this subsection we show that if the points a, b are singular points of considered equations, in particular, $a = -\infty$, $b = \infty$ (or finite singularities, i.e., the points where the unique solvability is violated), Leighton type comparison theorem (Theorem 2.3.5) still holds if we replace the solution satisfying $y(a) = 0 = y(b)$ by the principal solution at a and b.

Theorem 5.2.1. *Suppose that $\tilde{c}$ is a continuous function such that the equation*

$$(r(t)\Phi(y'))' + \tilde{c}(t)\Phi(y) = 0 \tag{5.2.7}$$

has the property that the principal solutions at a and b coincide and denote by h this simultaneous principal solution at these points. If

$$\liminf_{s_1 \downarrow a, s_2 \uparrow b} \int_{s_1}^{s_2} (c(t) - \tilde{c}(t))\, |h(t)|^p \, dt \geq 0, \quad c(t) \not\equiv \tilde{c}(t) \text{ in } (a, b), \tag{5.2.8}$$

then (1.1.1) is conjugate in $I = (a, b)$, i.e., there exists a nontrivial solution of this equation having at least two zeros in I.

Proof. Our proof is based on the relationship between nonpositivity of the energy functional $\mathcal{F}$ and conjugacy of (1.1.1) given in Theorem 1.2.2. We construct a nontrivial function, piecewise of the class C^1, with a compact support in I, such that $\mathcal{F}(y; a, b) < 0$.

Continuity of the functions $c, \tilde{c}$ and (5.2.8) imply the existence of $\bar{t} \in I$ and $d, \varrho > 0$ such that $(c(t) - \tilde{c}(t))|h(t)|^p > d$ for $(\bar{t} - \varrho, \bar{t} + \varrho)$. Let Λ be any positive differentiable function with the compact support in $(\bar{t} - \varrho, \bar{t} + \varrho)$. Further, let $a < t_0 < t_1 < \bar{t} - \varrho < \bar{t} + \varrho < t_2 < t_3 < b$ and let f, g be the solutions of (5.2.7) satisfying the boundary conditions

$$f(t_0) = 0, \; f(t_1) = h(t_1), \quad g(t_2) = h(t_2), \; g(t_3) = 0.$$

Note that such solutions exist if t_0, t_1 and t_2, t_3 are sufficiently close to a and b, respectively, due to nonoscillation of (5.2.7) near a and b (this is implied by the existence of principal solutions at these points) and the fact that the solution space of this equation is homogeneous. Define the function y as follows

$$y(t) = \begin{cases} 0 & t \in (a, t_0], \\ f(t) & t \in [t_0, t_1], \\ h(t) & t \in [t_1, t_2] \setminus [\bar{t} - \varrho, \bar{t} + \varrho], \\ h(t)(1 + \delta\Lambda(t)) & t \in [\bar{t} - \varrho, \bar{t} + \varrho], \\ g(t) & t \in [t_2, t_3], \\ 0 & t \in [t_3, b), \end{cases}$$

where δ is a real parameter. Then we have

$$\begin{aligned}
\mathcal{F}(y;t_0,t_3) &= \int_{t_0}^{t_3} [r(t)|y'|^p - c(t)|y|^p]\,dt \\
&= \int_{t_0}^{t_3} [r(t)|y'|^p - \tilde{c}(t)|y|^p]\,dt - \int_{t_0}^{t_3} [c(t)-\tilde{c}(t)]\,|y|^p\,dt \\
&= \int_{t_0}^{t_1} [r(t)|f'|^p - \tilde{c}(t)|f|^p]\,dt - \int_{t_0}^{t_1} [c(t)-\tilde{c}(t)]\,|f|^p\,dt \\
&\quad + \int_{t_1}^{t_2} [r(t)|y'|^p - \tilde{c}(t)|y|^p]\,dt - \int_{t_1}^{t_2} [c(t)-\tilde{c}(t)]\,|y|^p\,dt \\
&\quad + \int_{t_2}^{t_3} [r(t)|g'|^p - \tilde{c}(t)|g|^p]\,dt - \int_{t_2}^{t_3} [c(t)-\tilde{c}(t)]\,|g|^p\,dt.
\end{aligned}$$

Denote by w_f, w_g, w_h the solutions of the Riccati equation associated with (5.2.7)

$$w' + \tilde{c}(t) + (p-1)r^{1-q}(t)|w|^q = 0 \tag{5.2.9}$$

generated by f, g and h, respectively, i.e.,

$$w_f = \frac{r\Phi(f')}{\Phi(f)}, \quad w_g = \frac{r\Phi(g')}{\Phi(g)}, \quad w_h = \frac{r\Phi(h')}{\Phi(h)}.$$

Then using Picone's identity (1.2.3),

$$\begin{aligned}
\int_{t_0}^{t_1} [r(t)|f'|^p - \tilde{c}(t)|f|^p]\,dt &= w_f|f|^p\,\Big|_{t_0}^{t_1} + p\int_{t_0}^{t_1} r^{1-q}(t)P(r^{q-1}f', \Phi(f)w_f)\,dt \\
&= w_f|f|^p\,\Big|_{t_0}^{t_1},
\end{aligned}$$

where $P(u,v) = |u|^p/p - uv + |v|^q/q$ (see Subsection 1.2.1). Similarly,

$$\int_{t_2}^{t_3} [r(t)|g'|^p - \tilde{c}(t)|g|^p]\,dt = w_g|g|^p\,\Big|_{t_2}^{t_3}.$$

Concerning the interval $[t_1,t_2]$, we have (again by identity (1.2.3))

$$\begin{aligned}
\tilde{\mathcal{F}}(y;t_1,t_2) &:= \int_{t_1}^{t_2} [r(t)|y'|^p - \tilde{c}(t)|y|^p]\,dt \\
&= w_h|h|^p\,\Big|_{t_1}^{t_2} + p\int_{t_1}^{t_2} r^{1-q}(t)P(r^{q-1}y', \Phi(y)w_h)\,dt = w_h|h|^p\,\Big|_{t_1}^{t_2} \\
&\quad + \int_{\bar{t}-\varrho}^{\bar{t}+\varrho} \Big\{ r(t)|h' + \delta(\Lambda h)'|^p - pr(t)\frac{\Phi(h')}{\Phi(h)}y'h^{p-1}(1+\delta\Lambda)^{p-1} \\
&\quad + (p-1)r^{1-q}(t)\left|\frac{r(t)\Phi(h')}{\Phi(h)}\right|^q h^p(1+\delta\Lambda)^p \Big\}\,dt \\
&= w_h|h|^p\,\Big|_{t_1}^{t_2} + \int_{\bar{t}-\varrho}^{\bar{t}+\varrho} r(t)\,\{|h'|^p + p\delta(\Lambda h)'\Phi(h') + o(\delta)
\end{aligned}$$

$$
\begin{aligned}
& \quad - p(h' + \delta(\Lambda h)')\Phi(h')(1 + (p-1)\delta\Lambda + o(\delta)) \\
& \quad + (p-1)|h'|^p(1 + p\delta\Lambda + o(\delta))\} \, dt \\
&= w_h|h|^p \,\Big|_{t_1}^{t_2} + \int_{\bar t-\varrho}^{\bar t+\varrho} r(t)\big\{|h'|^p + p\delta(\Lambda h)'\Phi(h') \\
& \quad + p|h'|^p - p\delta\Phi(h')(\Lambda h)' - p(p-1)\delta\Lambda|h'|^p \\
& \quad + (p-1)|h'|^p + (p-1)p\delta\Lambda|h'|^p + o(\delta)\big\} \, dt \\
&= w_h|h|^p \,\Big|_{t_1}^{t_2} + o(\delta)
\end{aligned}
$$

as $\delta \to 0+$ Consequently,

$$
\begin{aligned}
\tilde{\mathcal{F}}(y; t_0, t_3) &= w_f|f|^p \,\Big|_{t_0}^{t_1} + w_h|h|^p \,\Big|_{t_1}^{t_2} + w_g|g|^p \,\Big|_{t_2}^{t_3} + o(\delta) \\
&= |h(t_1)|^p(w_f(t_1) - w_h(t_1)) + |h(t_2)|^p(w_h(t_2) - w_g(t_2)) + o(\delta)
\end{aligned}
$$

as $\delta \to 0+$. Further, observe that the function f/h is monotonically increasing in (t_0, t_1) since $f/h(t_0) = 0$, $f/h(t_1) = 1$ and $(f/h)' = (f'h - fh')/h^2 \neq 0$ in (t_0, t_1). Indeed, if $f'h - fh' = 0$ at some point $\tilde t \in (t_0, t_1)$, i.e., $(f'/f)(\tilde t) = (h'h)(\tilde t)$, then $w_f(\tilde t) = w_h(\tilde t)$ which contradicts the unique solvability of the generalized Riccati equation. By the second mean value theorem of integral calculus there exists $\xi_1 \in (t_0, t_1)$ such that

$$
\begin{aligned}
\int_{t_0}^{t_1} (c(t) - \tilde c(t))\, |f|^p \, dt &= \int_{t_0}^{t_1} (c(t) - \tilde c(t))\, |h|^p \frac{|f|^p}{|h|^p} \, dt \\
&= \int_{\xi_1}^{t_1} (c(t) - \tilde c(t))\, |h|^p \, dt.
\end{aligned}
$$

By the same argument the function g/h is monotonically decreasing in (t_2, t_3) and

$$
\int_{t_2}^{t_3} (c(t) - \tilde c(t))\, |g|^p \, dt = \int_{t_2}^{\xi_2} (c(t) - \tilde c(t))\, |h|^p \, dt
$$

for some $\xi_2 \in (t_2, t_3)$.

Concerning the interval (t_1, t_2) we have

$$
\begin{aligned}
& \int_{t_1}^{t_2} (c(t) - \tilde c(t))\, |y|^p \, dt \\
&= \int_{t_1}^{\bar t-\varrho} (c(t) - \tilde c(t))\, |h|^p dt \\
& \quad + \int_{\bar t-\varrho}^{\bar t+\varrho} (c(t) - \tilde c(t))\, |h|^p (1 + \delta\Lambda)^p \, dt + \int_{\bar t+\varrho}^{t_2} (c(t) - \tilde c(t))\, |h|^p \, dt \\
&= \int_{t_1}^{t_2} (c(t) - \tilde c(t))\, |h|^p \, dt + \delta \int_{\bar t-\varrho}^{\bar t+\varrho} (c(t) - \tilde c(t))\, |h|^p \Lambda(t) \, dt + o(\delta) \\
&\geq \int_{t_1}^{t_2} (c(t) - \tilde c(t))\, |h|^p \, dt + \delta K + o(\delta)
\end{aligned}
$$

as $\delta \to 0+$, where $K = d\int_{\bar{t}-\varrho}^{\bar{t}+\varrho} \Lambda(t)\,dt > 0$. Therefore,

$$\int_{t_0}^{t_3} (c(t) - \tilde{c}(t))\,|y|^p\,dt \ge \int_{\xi_1}^{\xi_2} (c(t) - \tilde{c}(t))\,|h|^p\,dt + K\delta + o(\delta).$$

Summarizing our computations, we have

$$\begin{aligned}\mathcal{F}(y; t_0, t_3) \;\le\;& |h(t_1)|^p\,(w_f(t_1) - w_h(t_1)) + |h(t_2)|^p\,(w_h(t_2) - w_g(t_2)) \\ & - \int_{\xi_1}^{\xi_2} (c(t) - \tilde{c}(t))\,|h|^p\,dt - (K\delta + o(\delta))\end{aligned}$$

with a positive constant K.

Now, let $\delta > 0$ (sufficiently small) be such that $K\delta + o(\delta) =: \varepsilon > 0$. According to (5.2.8) the points t_1, t_2 can be chosen in such a way that

$$\int_{s_1}^{s_2} (c(t) - \tilde{c}(t))\,|h|^p\,dt > -\frac{\varepsilon}{4}$$

whenever $s_1 \in (a, t_1)$, $s_2 \in (t_2, b)$. Further, since w_h is generated by the solution h of (5.2.7) which is principal both at $t = a$ and $t = b$, according to the Mirzov construction of the principal solution, we have (for t_1, t_2 fixed for a moment)

$$\lim_{t_0 \to a+} [w_f(t_1) - w_h(t_1)] = 0, \qquad \lim_{t_3 \to b-} [w_g(t_2) - w_h(t_2)] = 0.$$

Hence

$$|h(t_1)|^p\,[w_f(t_1) - w_h(t_1)] < \frac{\varepsilon}{4}, \quad |h(t_2)|^p\,[w_h(t_2) - w_g(t_2)] < \frac{\varepsilon}{4}$$

if $t_0 < t_1$, $t_3 > t_2$ are sufficiently close to a and b, respectively. Consequently, for the above specified choice of $t_0 < t_1 < t_2 < t_3$ we have

$$\begin{aligned}\mathcal{F}(y; t_0, t_3) \;=\;& \int_{t_0}^{t_3} [r(t)|y'|^p - \tilde{c}(t)|y|^p]\,dt - \int_{t_0}^{t_3} (c(t) - \tilde{c}(t))\,|y|^p\,dt \\ \le\;& |h(t_1)|^p\,[w_f(t_1) - w_h(t_1)] + |h(t_2)|^p\,[w_h(t_2) - w_g(t_2)] \\ & - \int_{\xi_1}^{\xi_2} (c(t) - \tilde{c}(t))\,|h|^p\,dt - (K\delta + o(\delta)) < \frac{\varepsilon}{4} + \frac{\varepsilon}{4} + \frac{\varepsilon}{4} - \varepsilon < 0.\end{aligned}$$

The proof is now complete. □

5.2.3 Leighton-Wintner type oscillation criterion

Recall that if $\int^\infty r^{1-q}(t)\,dt = \infty$ and $\int^\infty c(t)\,dt = \infty$, then (1.1.1) is oscillatory. This direct extension of the classical linear Leighton-Wintner criterion has been proved in Section 1.2. This criterion characterizes exactly what means that for oscillation of (1.1.1) the function c must be sufficiently positive comparing with the function r in one-term equation (5.2.1). Here we extend this result to the situation when (1.1.1) is investigated as a perturbation of (5.2.2).

Theorem 5.2.2. *Suppose that h is the principal solution of (nonoscillatory) equation (5.2.2) and*

$$\int^{\infty} (c(t) - \tilde{c}(t))\, h^p(t)\, dt := \lim_{b\to\infty} \int^{b} (c(t) - \tilde{c}(t))\, h^p(t)\, dt = \infty. \tag{5.2.10}$$

Then equation (1.1.1) is oscillatory.

Proof. According to the variational principle (see Section 2.1), it suffices to find (for any $T \in \mathbb{R}$) a function $y \in W^{1,p}(T,\infty)$, with a compact support in (T,∞), such that $\mathcal{F}(y;T,\infty) < 0$. Hence, let $T \in \mathbb{R}$ be arbitrary and $T < t_0 < t_1 < t_2 < t_3$ (these points will be specified later). Define the test function y as follows:

$$y(t) = \begin{cases} 0 & T \le t \le t_0, \\ f(t) & t_0 \le t \le t_1, \\ h(t) & t_1 \le t \le t_2, \\ g(t) & t_2 \le t \le t_3, \\ 0 & t_3 \le t < \infty, \end{cases}$$

where f, g are solutions of (5.2.2) given by the boundary conditions $f(t_0) = 0$, $f(t_1) = h(t_1)$, $g(t_2) = h(t_2)$, $g(t_3) = 0$. Denote

$$w_f = \frac{r\Phi(f')}{\Phi(f)}, \qquad w_h = \frac{r\Phi(h')}{\Phi(h)}, \qquad w_f = \frac{r\Phi(g')}{\Phi(g)},$$

i.e., w_f, w_g, w_h are solutions of the Riccati equation associated with (5.2.2) generated by f, g, h respectively. Using exactly the same computations as in the proof of Theorem 5.2.1, one can show that

$$\mathcal{F}(y;T,\infty) = K - \int_{t_1}^{\xi} (c(t) - \tilde{c}(t))\, h^p(t)\, dt + h^p(t_2)\left[w_h(t_2) - w_g(t_2)\right], \tag{5.2.11}$$

where

$$K := h^p(t_1)\left[w_f(t_1) - w_h(t_1)\right] - \int_{t_0}^{t_1} (c(t) - \tilde{c}(t)) f^p(t)\, dt$$

and $\xi \in (t_2, t_3)$. Now, if $\varepsilon > 0$ is arbitrary and $T < t_0 < t_1$ are fixed, then, according to (5.2.10), t_2 can be chosen in such a way that $\int_{t_1}^{t} (c(s) - \tilde{c}(s)) h^p(s)\, ds > K + \varepsilon$ whenever $t > t_2$. Finally, again using the same argument as in [100] we have (observe that w_g actually depends also on t_3)

$$\lim_{t_3\to\infty} h^p(t_2)\left[w_h(t_2) - w_g(t_2)\right] = 0,$$

hence the last summand in (5.2.11) is less than ε if t_3 is sufficiently large. Consequently, $\mathcal{F}(y; t_0, t_3) < 0$ if t_0, t_1, t_2, t_3 are chosen in the above specified way. □

If $r(t) \equiv 1$ in (1.1.1) and $\tilde{c}(t) = \frac{\tilde{\gamma}}{t^p}$, $\tilde{\gamma} = \left(\frac{p-1}{p}\right)^p$, i.e., (5.2.2) is the generalized Euler equation with the critical coefficient (1.4.20), then the previous theorem reduces to the oscillation criterion given by Elbert [141].

5.2.4 Hille-Nehari type oscillation criterion

The results of this subsection can be viewed as an extension of some criteria given in Section 3.1.

Theorem 5.2.3. *Let $\int^\infty r^{1-q}(t)\,dt = \infty$ and $c(t) \geq 0$ for large t. Further suppose that equation (5.2.2) is nonoscillatory and possesses a positive solution h satisfying the following conditions:*

(i) $h'(t) > 0$ for large t;

(ii) it holds

$$\int^\infty r(t)(h'(t))^p\,dt = \infty; \tag{5.2.12}$$

(iii) there exists a finite limit

$$\lim_{t\to\infty} r(t)h(t)\Phi(h'(t)) =: L > 0. \tag{5.2.13}$$

Denote by

$$G(t) = \int^t \frac{ds}{r(s)h^2(s)(h'(s))^{p-2}} \tag{5.2.14}$$

and suppose that the integral

$$\int^\infty (c(t) - \tilde{c}(t))\, h^p(t)\,dt = \lim_{b\to\infty} \int^b (c(t) - \tilde{c}(t))\, h^p(t)\,dt \tag{5.2.15}$$

is convergent. If

$$\liminf_{t\to\infty} G(t) \int_t^\infty (c(s) - \tilde{c}(s))\, h^p(s)\,ds > \frac{1}{2q} \tag{5.2.16}$$

then equation (1.1.1) is oscillatory.

Proof. Suppose, by contradiction, that (1.1.1) is nonoscillatory, i.e., there exists an eventually positive principal solution x of this equation. Denote by $\rho := r(t)\Phi(x'/x)$. Then ρ satisfies Riccati equation (1.1.21) and using the Picone identity for half-linear equations (1.2.1), we have

$$\int_T^t (r(s)|y'|^p - c(s)|y|^p)\,ds = \rho(s)|y|^p\,\Big|_T^t + p\int_T^t r^{1-q}(s)P\left(r^{q-1}y', \rho\Phi(y)\right)\,ds$$

for any differentiable function y, where P is given by (1.2.2), and integration by parts yields

$$\begin{aligned}\int_T^t [r(s)|y'|^p - c(s)|y|^p]\,ds &= \int_T^t [r(s)|y'|^p - \tilde{c}(s)|y|^p]\,ds - \int_T^t (c(s) - \tilde{c}(s))|y|^p\,ds \\ &= r(s)y\Phi(y')\,\Big|_T^t - \int_T^t y\left[(r(s)\Phi(y'))' - \tilde{c}(s)\Phi(y)\right]ds \\ &\quad - \int_T^t (c(s) - \tilde{c}(s))|y|^p\,ds.\end{aligned}$$

Substituting $y = h$ into the last two equalities (h being a solution of (5.2.2) satisfying the assumptions (i) – (iii) of the theorem), we get

$$\text{(5.2.17)}\quad h^p(\tilde\rho-\rho)\,|_T^t = \int_T^t (c(s)-\tilde c(s))h^p\,ds + p\int_T^t r^{1-q}(s)P(r^{q-1}h',\rho\Phi(h))\,ds,$$

where $\tilde\rho = r\Phi(h'/h)$. Since $\int^\infty r^{1-q}(t)\,dt = \infty$, $w \equiv 0$ is the distinguished solution of the Riccati equation corresponding to the equation $(r(t)\Phi(x'))' = 0$ and since $c(t) \ge 0$, by Theorem 4.2.2 $\rho(t) \ge 0$ eventually. Hence, with L given by (5.2.13), we have

$$L + h^p(T)(\rho(T)-\tilde\rho(T)) \ge \int_T^t (c(s)-\tilde c(s))h^p\,ds + p\int_T^t r^{1-q}(s)P(r^{q-1}h',\rho\Phi(h))\,ds,$$

and since $P(u,v) \ge 0$, this means that

$$\text{(5.2.18)}\qquad \int^\infty r^{1-q}(t)P(r^{q-1}(t)h'(t),\rho(t)\Phi(h(t)))\,dt < \infty.$$

Now, since (5.2.13), (5.2.15), (5.2.18) hold, from (5.2.17) it follows that there exists a finite limit

$$\lim_{t\to\infty} h^p(t)\,(\rho(t)-\tilde\rho(t)) =: \beta$$

and also the limit

$$\text{(5.2.19)}\qquad \lim_{t\to\infty}\frac{\rho(t)}{\tilde\rho(t)} = \lim_{t\to\infty}\frac{h^p(t)\rho(t)}{h^p(t)\tilde\rho(t)} = \frac{L+\beta}{L}.$$

Therefore,

$$h^p(t)(\rho(t)-\tilde\rho(t)) - \beta = C(t) + p\int_t^\infty r^{1-q}(s)P(r^{q-1}h',\rho\Phi(h))\,ds,$$

where $C(t) = \int_t^\infty (c(s)-\tilde c(s))h^p(s)\,ds$.

Concerning the function $P(u,v)$, we have for $u, v > 0$

$$\text{(5.2.20)}\quad P(u,v) = \frac{u^p}{p} - uv + \frac{v^q}{q} = u^p\left(\frac{v^q}{qu^p} - vu^{1-p} + \frac{1}{p}\right) = u^pQ\left(vu^{1-p}\right),$$

where $Q(\lambda) = \lambda^q/q - \lambda + 1/p \ge 0$ for $\lambda \ge 0$ with equality if and only if $\lambda = 1$ and

$$\text{(5.2.21)}\qquad \lim_{\lambda\to 1}\frac{Q(\lambda)}{(\lambda-1)^2} = \frac{q-1}{2}.$$

Hence for every $\varepsilon > 0$ there exists $\delta > 0$ such that

$$\text{(5.2.22)}\qquad P(u,v) \ge \left(\frac{q-1}{2} - \varepsilon\right)u^p\left(\frac{v}{u^{p-1}} - 1\right)^2,$$

whenever $|vu^{1-p} - 1| < \delta$. This implies that $\beta = 0$ in (5.2.19) since the case $\beta \ne 0$ contradicts the divergence of $\int^\infty r(t)(h'(t))^{p-1}\,dt$. If we denote

$$f(t) := h^p(t)(\rho(t)-\tilde\rho(t)), \quad H(t) := \frac{1}{r(t)h^2(t)(h'(t))^{p-2}},$$

then using

$$f(t) \geq C(t) + \left(\frac{p(q-1)}{2} - \tilde{\varepsilon}\right) \int_t^\infty r(s)\,(h'(s))^p \left(\frac{\rho(s)}{\tilde{\rho}(s)} - 1\right)^2 ds$$

$$= C(t) + \left(\frac{q}{2} - \tilde{\varepsilon}\right) \int_t^\infty H(s) f^2(s)\, ds, \tag{5.2.23}$$

where $\tilde{\varepsilon} = p\varepsilon$. Multiplying (5.2.23) by $G(t)$ we get

$$G(t)f(t) \geq G(t)C(t) + \left(\frac{q}{2} - \tilde{\varepsilon}\right) G(t) \int_t^\infty H(s) f^2(s)\, ds. \tag{5.2.24}$$

Inequality (5.2.24) together with (5.2.16) imply that there exists a $\tilde{\delta} > 0$ such that

$$G(t)f(t) \geq \frac{1}{2q} + \tilde{\delta} + \left(\frac{q}{2} - \tilde{\varepsilon}\right) G(t) \int_t^\infty \frac{H(s)}{G^2(s)} [G(s)f(s)]^2\, ds \tag{5.2.25}$$

for large t.

Suppose first that $\liminf_{t\to\infty} G(t)f(t) =: c < \infty$. Then for every $\bar{\varepsilon} > 0$, we have $[G(t)f(t)]^2 > (1-\bar{\varepsilon})c^2$ for large t and (5.2.25) implies

$$c \geq \frac{1}{2q} + \tilde{\delta} + \left(\frac{q}{2} - \tilde{\varepsilon}\right)(1-\bar{\varepsilon})c^2.$$

Now, letting $\tilde{\varepsilon}, \bar{\varepsilon} \to 0$ we have

$$c \geq \frac{1}{2q} + \tilde{\delta} + \frac{q}{2}c^2,$$

but this is impossible since the discriminant $1 - 2q(1/(2q) + \tilde{\delta}) < 0$.

Finally, if

$$\liminf_{t\to\infty} G(t)f(t) = \infty, \tag{5.2.26}$$

denote by $m(t) = \inf_{t\leq s}\{G(s)f(s)\}$. Then m is nondecreasing and (5.2.25) implies that

$$G(t)f(t) \geq K + \left(\frac{q}{2} - \tilde{\varepsilon}\right) m^2(t),$$

where $K = 1/(2q) + \tilde{\delta}$. Since m is nondecreasing, we have for $s > t$

$$G(s)f(s) \geq K + \left(\frac{q}{2} - \tilde{\varepsilon}\right) m^2(s) \geq K + \left(\frac{q}{2} - \tilde{\varepsilon}\right) m^2(t),$$

and hence

$$m(t) \geq K + \left(\frac{q}{2} - \tilde{\varepsilon}\right) m^2(t)$$

which contradicts (5.2.26). The proof is complete. □

Let $r(t) \equiv 1$, $c(t) \geq 0$ and

$$\tilde{c}(t) = \frac{\gamma_p}{t^p}, \quad \gamma_p = \left(\frac{p-1}{p}\right)^p.$$

Then (5.2.2) reduces to the generalized Euler equation with the critical coefficient

$$(\Phi(y'))' + \frac{\gamma_p}{t^p}\Phi(y) = 0, \tag{5.2.27}$$

and the solution $h(t) = t^{\frac{p-1}{p}}$ of this equation satisfies all assumptions of Theorem 5.2.3 with

$$G(t) = \left(\frac{p}{p-1}\right)^{p-2} \log t.$$

Thus we get the following corollary.

Corollary 5.2.1. *Equation* (5.1.1) *is oscillatory provided*

$$\liminf_{t\to\infty} \log t \int_t^\infty s^{p-1}\left[c(s) - \frac{\gamma_p}{s^p}\right] ds > \frac{1}{2}\left(\frac{p-1}{p}\right)^{p-1}.$$

5.2.5 Hille-Nehari type nonoscillation criterion

Now we turn our attention to a nonoscillation criterion which is proved under no sign restriction on the function c and also under no assumption concerning divergence of the integral $\int^\infty r^{1-q}(t)\,dt$ (compare with Theorem 5.2.3). The result of this subsection can be viewed as an extension of Theorems 2.2.9 and 2.3.2 to the situation when (1.1.1) (or (5.1.1)) is viewed as a perturbation of a two-term equation.

Theorem 5.2.4. *Suppose that equation* (5.2.2) *is nonoscillatory and possesses a solution h satisfying (i), (iii) of Theorem 5.2.3. Moreover, suppose that*

$$\int^\infty \frac{dt}{r(t)h^2(t)(h'(t))^{p-2}} = \infty. \tag{5.2.28}$$

If $G(t)$ is the same as in Theorem 5.2.3, now satisfying

$$\limsup_{t\to\infty} G(t)\int_t^\infty (c(s) - \tilde{c}(s))\, h^p(s)\, ds < \frac{1}{2q} \tag{5.2.29}$$

and

$$\liminf_{t\to\infty} G(t)\int_t^\infty (c(s) - \tilde{c}(s))\, h^p(s)\, ds > -\frac{3}{2q}, \tag{5.2.30}$$

then (1.1.1) *is nonoscillatory.*

Proof. Denote again

$$C(t) = \int_t^\infty (c(s) - \tilde{c}(s))\, h^p(s)\, ds.$$

To prove that (1.1.1) is nonoscillatory, according to Theorem 2.2.1 it suffices to find a differentiable function ρ which verifies differential inequality (2.2.7) for large t. This inequality can be written in the form (with $w = h^{-p}(\rho + C)$)

$$\begin{aligned}
\rho' &\le -p\left[\frac{1}{q}\left|\frac{\rho+C}{h}\right|^q r^{1-q} - h'\left(\frac{\rho+C}{h}\right) + \frac{r(h')^p}{p}\right] + r(h')^p - \tilde{c}(t)h^p \\
&= -pr^{1-q}\left[\frac{1}{q}\left|\frac{\rho+C}{h}\right|^q - r^{q-1}h'\left(\frac{\rho+C}{h}\right) + \frac{1}{p}r^q(h')^p\right] + r(h')^p - \tilde{c}h^p \\
&= -pr^{1-q}P\left(r^{q-1}h', \frac{\rho+C}{h}\right) + r(h')^p - \tilde{c}h^p.
\end{aligned}$$

We will show that the function

$$\rho(t) = r(t)h(t)\Phi(h'(t)) + \frac{1}{2qG(t)} \tag{5.2.31}$$

satisfies the last inequality for large t. To this end, let $v = (\rho + C)/h$, $u = r^{q-1}h'$. The fact that the solution h of (5.2.2) is increasing together with (5.2.28), (5.2.29), (5.2.30) and the assumption (iii) of Theorem 5.2.3 imply that

$$\frac{v}{\Phi(u)} = \frac{\rho(t) + C(t)}{h(t)r(t)\Phi(h'(t))} = 1 + \frac{1 + 2qC(t)G(t)}{2qG(t)r(t)h(t)\Phi(h'(t)))} \to 1$$

as $t \to \infty$. Hence, using (5.2.20) and the same argument as in the proof of the Theorem 5.2.3, for any $\varepsilon > 0$, we have (with Q satisfying (5.2.21))

$$\begin{aligned}
pr^{1-q}&\left[\frac{1}{q}\left|\frac{\rho+C}{h}\right|^q - h'r^{q-1}\left(\frac{\rho+C}{h}\right) + \frac{r^q(h')^p}{p}\right] \\
&= pr^{1-q}r^q(h')^pQ\left(\frac{\rho+C}{hr\Phi(h')}\right) \\
&\le p\left(\frac{q-1}{2} + \varepsilon\right) r(h')^p \frac{(1+2qGC)^2}{4q^2r^2h^2(h')^{2p-2}G^2} \\
&= \left(\frac{q}{2} + p\varepsilon\right)\frac{1}{rh^2(h')^{p-2}}\frac{(1+2qGC)^2}{4G^2q^2}
\end{aligned}$$

for large t.

Now, since (5.2.29), (5.2.30) hold, there exists $\delta > 0$ such that

$$\frac{-3+\delta}{2q} < G(t)C(t) < \frac{1-\delta}{2q} \iff |1 + 2qG(t)C(t)| < 2 - \delta$$

for large t, hence $\varepsilon > 0$ can be chosen in such a way that

$$\left(\frac{q}{2} + \varepsilon\right)\frac{(1 + 2qG(t)C(t))^2}{4q^2} < \frac{1}{2q}$$

for large t. Consequently (using the fact that h solves (5.2.2)), we have

$$\begin{aligned} - pr^{1-q}&\left[\frac{1}{q}\left|\frac{\rho+C}{h}\right|^q - r^{q-1}h'\left(\frac{\rho+C}{h}\right)+\frac{r(h')^p}{p}\right] + r(h')^p - \tilde{c}(t)h^p \\ &\geq -\left(\frac{q}{2}+\varepsilon\right)\frac{1}{G^2rh^2(h')^{p-2}}\frac{(1+2qGC)^2}{4q^2} + r(h')^p - \tilde{c}(t)h^p \\ &> -\frac{1}{2qG^2rh^2(h')^{p-2}} + [rh\Phi(h')]' = \left[rh\Phi(h')+\frac{1}{2qG}\right]' = \rho'. \end{aligned}$$

The proof is complete. □

If (5.2.2) reduces to Euler type equation (5.2.27), then we obtain the following corollary from the previous statement.

Corollary 5.2.2. *Equation* (5.1.1) *is nonoscillatory provided*

$$\limsup_{t\to\infty} \log t \int_t^\infty \left(c(s) - \frac{\gamma_p}{s^p}\right) s^{p-1}\, ds < \frac{1}{2}\left(\frac{p-1}{p}\right)^{p-1}$$

and

$$\liminf_{t\to\infty} \log t \int_t^\infty \left(c(s) - \frac{\gamma_p}{s^p}\right) s^{p-1}\, ds > -\frac{3}{2}\left(\frac{p-1}{p}\right)^{p-1}.$$

5.2.6 Perturbed Euler equation

If we distinguish the cases $p \in (1,2]$ and $p \geq 2$, the following refinement of oscillation and nonoscillation criteria from the previous subsection can be proved.

Theorem 5.2.5. *Consider the half-linear equation*

$$(5.2.32) \qquad (\Phi(x'))' + \frac{\gamma_p}{t^p}\Phi(x) + 2\left(\frac{p-1}{p}\right)^{p-1}\delta(t)\Phi(x) = 0, \quad \gamma_p = \left(\frac{p-1}{p}\right)^p,$$

where δ is a continuous function, and the linear second order equation

$$(5.2.33) \qquad (ty')' + \frac{\delta(t)}{t}y = 0.$$

Suppose that the integral

$$(5.2.34) \qquad \sigma(t) := \int_t^\infty \frac{\delta(s)}{s}\, ds \geq 0$$

for large t, (in particular, we suppose that $\int^\infty \delta(t)/t\, dt$ is convergent).

(i) *Let $p \geq 2$ and linear equation* (5.2.33) *be nonoscillatory. Then* (5.2.32) *is also nonoscillatory.*

(ii) *Let $p \in (1,2]$ and half-linear equation* (5.2.32) *be nonoscillatory. Then linear equation* (5.2.33) *is also nonoscillatory.*

Proof. We only present the proof of the part (i), the proof of the claim (ii) is analogical. Since (5.2.33) is nonoscillatory, the same is true for the differential equation

$$(\Lambda z')' + \Lambda\delta(e^s)z = 0, \quad \Lambda = 2\left(\frac{p-1}{p}\right)^{p-1}, \tag{5.2.35}$$

which results from (5.2.33) upon the transformation $x(t) = z(s)$, $s = \log t$. Since the integral $\int^\infty \delta(t)/t\,dt = \int^\infty \delta(e^s)\,ds < \infty$, the function $u(s) = \Lambda z'(s)/z(s)$ is a solution of the Riccati integral equation

$$u(s) = \Lambda \int_s^\infty \delta(e^\tau)\,d\tau + \frac{1}{\Lambda}\int_s^\infty u^2(\tau)\,d\tau, \tag{5.2.36}$$

and hence $u(s) > 0$ for large s according to the assumption on $\sigma(t)$. Now we use similar ideas as in Subsection 1.4.2. We denote

$$F(\varrho) := |\varrho|^q - \varrho + \tilde{\gamma}, \quad \tilde{\gamma} := \frac{1}{p}\left(\frac{p-1}{p}\right)^{p-1}. \tag{5.2.37}$$

Recall that this function is nonnegative, attains its minimum at $\varrho_0 = \left(\frac{p-1}{p}\right)^{p-1}$ and this minimum is $F(\varrho_0) = 0$. By a direct computation we have

$$F''(\varrho_0) = \frac{2}{(p-1)\Lambda}, \quad F'''(\varrho) = q(q-1)(q-2)|\varrho|^{q-3}\operatorname{sgn}\varrho. \tag{5.2.38}$$

Hence, by the Taylor formula

$$F(\varrho) < \frac{1}{(p-1)\Lambda}(\varrho - \varrho_0)^2 \quad \text{for} \quad p > 2, \tag{5.2.39}$$

provided $\varrho \neq \varrho_0$ and $\varrho \geq 0$, (for $p \in (1,2)$ we have the opposite inequality). Hence $F(\varrho_0 + u) < u^2/[(p-1)\Lambda]$ for $u > 0$. Applying the last inequality in (5.2.36), we obtain

$$u(s) > \int_s^\infty \Lambda\delta(e^\tau)\,d\tau + (p-1)\int_s^\infty F(u(\tau))\,d\tau. \tag{5.2.40}$$

After some computation, one can verify that (5.2.40) is the Riccati integral inequality associated with the half-differential equation

$$(\Phi(z'))' - \Phi(z') + \gamma_p\Phi(z) + 2\varrho_0\delta(t)\Phi(z) = 0 \tag{5.2.41}$$

which is, in turn, nonoscillatory and this means that (5.2.32) is nonoscillatory as well since the transformation $x(t) = z(s)$, $s = \log t$, transforms (5.2.41) into (5.2.32). □

The proofs of the next two theorems are based on the detailed investigation of the function F from the previous proof. Since computations are rather complicated, we skip the proof, we refer to the original paper [149].

Theorem 5.2.6. *Suppose that (5.2.34) holds and (5.2.33) is nonoscillatory. Assume that there exists a constant $\theta \in (0,1)$ such that for a solution z of (5.2.33) the functions $\zeta = z'/z$ and μ defined by*

$$\mu(s) = \int_s^\infty \zeta^3(\tau)\, d\tau,$$

satisfy the relations

$$\int^\infty \mu(\tau)\, d\tau < \infty, \quad \int_s^\infty \zeta(\tau)\mu(\tau)\, d\tau \le \frac{\theta}{2}\mu(s) \text{ for large } s. \tag{5.2.42}$$

Then (5.2.32) is nonoscillatory and it has a solution with the asymptotics

$$\begin{aligned} x(t) &= t^{\frac{p-1}{p}} z^{\frac{2}{p}}(\log t)[C + o(1)], \quad C \neq 0, \\ t\frac{x'(t)}{x(t)} &= \frac{p-1}{p} + \frac{2}{p}\zeta(\log t) + o(\zeta(\log t)) \end{aligned} \tag{5.2.43}$$

as $t \to \infty$. Moreover, if the assumptions of the previous part of theorem are satisfied for the principal solution $\tilde{z}$ of the equation

$$z'' + \delta(s)z = 0 \tag{5.2.44}$$

and the quantities $\tilde{\zeta}$, $\tilde{\mu}$ are defined accordingly via $\tilde{z}$, then the solution $\tilde{x}$ given by (5.2.43) with $\tilde{\zeta}$ instead of ζ is the principal solution of (5.2.44). Let x be any nonprincipal solution of (5.2.32), then

$$\int^\infty \frac{dt}{x^2(t)|x'(t)|^{p-2}} < \infty, \quad \lim_{t\to\infty} \frac{\tilde{x}(t)}{x(t)} = 0,$$

while for the principal solution $\tilde{x}$

$$\int^\infty \frac{dt}{\tilde{x}^2(t)|\tilde{x}'(t)|^{p-2}} = \infty.$$

Finally, if x_1, x_2 are two nonprincipal solutions of (5.2.32), then there exists the limit

$$\lim_{t\to\infty} \frac{x_1(t)}{x_2(t)} = C \in \mathbb{R} \setminus \{0\}.$$

The previous statement, applied to the half-linear differential equation with the iterated logarithms

$$(\Phi(x'))' + \left(\frac{\gamma_p}{t^p} + \frac{1}{2}\left(\frac{p-1}{p}\right)^{p-1} \sum_{k=1}^n \frac{1}{\log^2 t \cdot \log_2^2 t \cdot \log_k^2 t} \right) \Phi(x) = 0, \tag{5.2.45}$$

where $\log_2 t = \log(\log t)$, $\log_k t = \log(\log_{k-1} t)$, gives the following result.

Corollary 5.2.3. *Each nontrivial solution x of (5.2.45) has the asymptotic form either*

$$x(t) = t^{\frac{p-1}{p}} (\log t \log_2 t \cdots \log_n t)^{\frac{1}{p}} \left(C + O(\log^{-1} t)\right),$$

or

$$x(t) = t^{\frac{p-1}{p}} (\log t \log_2 t \cdots \log_n t)^{\frac{1}{p}} \left(\log_{n+1} t\right)^{\frac{2}{p}} \left(C + O(\log^{-1})t\right),$$

as $t \to \infty$, where $C \neq 0$ is a real constant and $n \in \mathbb{N}$.

We finish this subsection with a statement where we apply linear Hille-Nehari (non)oscillation criterion to linear equation (5.2.33).

Theorem 5.2.7. *Suppose that (5.2.34) holds. Then (5.2.32) is oscillatory if*

$$\liminf_{s\to\infty} s \int_s^\infty \delta(e^\tau)\, d\tau > \frac{1}{4},$$

and it is nonoscillatory if

$$s \int_s^\infty \delta(e^\tau)\, d\tau \leq \frac{1}{4}, \quad \textit{for large } s.$$

In particular, the differential equation

$$(\Phi(x'))' + \left(\frac{\gamma_p}{t^p} + \frac{\lambda}{t^p \log^2 t}\right) \Phi(x) = 0$$

is oscillatory for $\lambda > \frac{1}{2}\left(\frac{p-1}{p}\right)^{p-1}$ and nonoscillatory in the opposite case.

5.2.7 Linearization method in half-linear oscillation theory

In this subsection we extend some results of the previous subsection to a general situation. We show that oscillatory properties of (1.1.1) can be studied via properties of a certain associated linear equation of the form (1.1.2). In particular, we show that this linear equation plays a role of the Sturmian majorant for $p \geq 2$ and the role of the Sturmian minorant for $p \in (1, 2]$.

Along with (1.1.1) we consider the equation (5.2.2), we consider both (1.1.1) and (5.2.2) on an interval $I \subseteq \mathbb{R}$, and we suppose that

(5.2.46) there exists a solution h of (5.2.2) such that $h(t) > 0$ and $r(t)\Phi(h'(t)) \neq 0$ on I.

We denote

(5.2.47) $$R(t) := r(t)h^2(t)|h'(t)|^{p-2}, \quad C(t) := (c(t) - \tilde{c}(t))h^p(t).$$

and we compare oscillatory behavior of half-linear equation (1.1.1) with the behavior of the linear Sturm-Liouville equation

(5.2.48) $$(R(t)y')' + \frac{p}{2}C(t)y = 0.$$

Theorem 5.2.8. *Suppose that* (5.2.46) *holds.*

(i) *Suppose that* $p \geq 2$ *and that linear equation* (5.2.48) *is disconjugate in an interval* I. *Then half-linear equation* (1.1.1) *is also disconjugate in this interval.*

(ii) *Suppose that* $1 < p \leq 2$ *and that half-linear equation* (1.1.1) *is disconjugate in an interval* I. *Then linear equation* (5.2.48) *is also disconjugate in* I.

Proof. We will outline the proof of the part (i), the proof of the part (ii) is similar. Disconjugacy of (5.2.48) implies the existence of a solution u of the associated Riccati equation

$$u' + C(t) + \frac{pu^2}{2R(t)} = 0 \tag{5.2.49}$$

on I, where $u = 2Ry'/(yp)$, y being a positive solution to (5.2.48) on I. Fix a $t_0 \in I$ and consider the solution of generalized Riccati equation (1.1.21) satisfying the initial condition

$$w(t_0) = h^{-p}(t_0)u(t_0) + w_h(t_0),$$

where $w_h = r\Phi(h'/h)$. We will show that this solution w exists on the whole interval I. This then implies the required result – disconjugacy of (1.1.1) on I, since $x(t) = \exp\left\{\int_{t_0}^t r^{1-q}(s)\Phi^{-1}(w(s))\,ds\right\}$ is a solution of (1.1.1) which is positive on I. Let $v = h^p(w - w_h)$. Then $v(t_0) = u(t_0)$ and by a direct computation, similar to that in the proof of Theorem 5.2.3, one can verify that

$$v' + C(t) + pr^{1-q}(t)h^p(t)P(w_h, w) = 0,$$

where P is defined by (1.2.2). Now we use the inequality from Lemma 4.2.4 to obtain

$$P(\alpha, \beta) \leq \frac{1}{2}|\alpha|^{2-p}(\alpha - \beta)^2, \quad p \geq 2, \quad \alpha, \beta \in \mathbb{R}, \quad \alpha \neq 0. \tag{5.2.50}$$

For $p \in (1, 2]$ we have the opposite inequality, which is then used in the proof of the part (ii). By (5.2.50), after a short computation, we get

$$v' + C(t) + \frac{p}{2}\frac{v^2}{R(t)} \geq 0.$$

Now, the standard statement for differential inequalities (see, e.g. [174]) claims that $v(t) \geq u(t)$ for $t \geq t_0$ and $v(t) \leq u(t)$ for $t \leq t_0$, and this implies that v exists on the whole interval I which means that w exists on I as well. □

Remark 5.2.1. (i) If $p = 2$, i.e., (1.1.1) reduces to the linear equation (1.1.2), then (5.2.48) reduces to the equation

$$\left[r(t)h^2(t)y'\right]' + h^2(t)\left[c(t) - \tilde{c}(t)\right]y = 0$$

and this is just the equation which results from (1.1.2) upon the transformation $x = h(t)y$, where h is a solution (5.2.2), compare (1.3.14). From this point of view, the statement of Theorem 5.2.8 can be regarded as an extension of the linear transformation method to half-linear equations.

(ii) If we substitute $r \equiv 1$, $\tilde{c}(t) = \gamma_p t^{-p}$, $h(t) = t^{\frac{p-1}{p}}$ in Theorem 5.2.8, oscillatory properties of (5.1.1) are related to oscillatory properties of the linear equation

$$(t\,y')' + \frac{p}{2}\left(\frac{p}{p-1}\right)^{p-2} t^{p-1}\left(c(t) - \gamma_p t^{-p}\right) y = 0. \tag{5.2.51}$$

The constant $\frac{p}{2}\left(\frac{p}{p-1}\right)^{p-2}$ in (5.2.51) is worse than the corresponding constant in Theorem 5.2.5 (it is $(p-1)$-times bigger). The explanation of the fact that the constant in Theorem 5.2.5 is better is the following. In the proof of Theorem 5.2.5, a modified version of Lemma 4.2.4 has been used. In this modified version, the variables u, v are restricted to the region $0 \le v \le \Phi(u)$ and under this restriction the constant $1/2$ in (4.2.39), (4.2.40) can be replaced by a better constant $(q-1)/2$. Note that this restriction on u, v is enabled by additional assumption of the convergence and nonnegativity of the integral given by (5.2.34), see the previous section and also [149] for details. However, if no additional restriction on u, v is available, the constant $1/2$ in Lemma 4.2.4 is exact since (4.2.39) reduces to the equality if $v = -\Phi(u)$.

5.3 Nonoscillation domains and (almost) periodicity

The aim of this section is to study the properties (like closedness, convexity, but also many others) of the nonoscillation and disconjugacy domains of the half-linear equations with two parameters $(\Phi(y'))' + (-\alpha A(t) + \beta B(t))\Phi(y) = 0$. Various eigenvalue problems may be cast in this form. For example, setting $A(t) \equiv 1$, fixing α and allowing β to be the parameter we obtain weighted Sturm-Liouville equation with a possibly sign indefinite B. Since the results on equation (5.1.1) with an (almost) periodic coefficient are closely related to these problems we present them within this section as well. This is contained in the second and the third subsection.

5.3.1 Disconjugacy domain and nonoscillation domain

In the paper [270], it was studied the disconjugacy domain for the linear Hill type equation

$$y'' + (-\alpha + \beta B(t))y = 0,$$

where α, β are real parameters and $B(t)$ is a real almost periodic function (in a Bohr sense); see below for the definition of disconjugacy domain and Bohr almost periodic functions. This complemented the paper [294], where the same equation was studied under the condition of the periodicity of B with a period one and a mean value equal to zero. The theory was further extended to the equation

$$y'' + (-\alpha A(t) + \beta B(t))y = 0, \tag{5.3.1}$$

where, in particular, A, B do not need to be (almost) periodic, see the monograph [289]. What we offer here is a generalization of some aspects of the above mentioned theory (in particular that in [289]) to the half-linear case. Consider the equation

$$(\Phi(y'))' + (-\alpha A(t) + \beta B(t))\Phi(y) = 0 \tag{5.3.2}$$

on an interval I. The interval I is mostly taken to be $[0, \infty)$, although many results herein are valid also on $(-\infty, \infty)$ (or on finite open or half-open intervals). We assume that α, β are real parameters and A, B are continuous functions, but it is not difficult to see that the results are still valid if A, B are merely Lebesgue integrable on every compact subset of I (this may be important in the connection with generalizations of the concept of almost periodicity – in contrast to the almost periodic function in the sense of Bohr, almost periodic functions in a more general sense, e.g. of Weyl, may not be continuous); sometimes, for comparison purposes, we will present the results in this more general setting. Also, instead of (5.3.2) we may consider the equation $(r(t)\Phi(y'))' + (-\alpha A(t) + \beta B(t))\Phi(y) = 0$, r being a positive continuous function, and, with some little adjustments, the theory works as well. The so-called *parameter space*, i.e., the $\alpha\beta$-plane (which is equal to $\mathbb{R}^2$) plays an important role. Before we give the definition of its two significant subsets, recall that equation (5.3.2) is said to be disconjugate on I if every its nontrivial solution has at most one zero in I, while nonoscillation of equation (5.3.2) means that every its nontrivial solution has at most a finite number of zeros in I. Also recall that (5.3.2) is disconjugate on I if it has a solution without zeros on I. For a compact or an open interval I this condition is also necessary, see Subsection 1.2.6. Now we give two important definitions.

Definition 5.3.1. The collection of all $(\alpha, \beta) \in \mathbb{R}^2$ for which (5.3.2) is disconjugate (resp. nonoscillatory) on I is called the *disconjugacy domain* (resp. *nonoscillation domain*) of (5.3.2), and denoted by $\mathcal{D}$ (resp. $\mathcal{N}$). The set of points (α, β) for which (5.3.2) is oscillatory is the *oscillation domain*, denoted by $\mathcal{O}$.

Definition 5.3.2. We say that a function f is *Bohr almost periodic* (or *uniformly almost periodic*) if for any $\varepsilon > 0$ there exists $L = L(\varepsilon)$ such that in each interval of length L there exists at least one number τ such that $|f(t + \tau) - f(t)| < \varepsilon$, $-\infty < t < \infty$. In other words, if f is continuous and for any $\varepsilon > 0$ there exists a relatively dense set of ε-almost periods of this function. Recall that a number $\tau = \tau_f(\varepsilon)$ is called an ε-almost period of f if for all t, $|f(t + \tau) - f(t)| < \varepsilon$.

We emphasize that any Bohr almost periodic function is continuous. We are interested in closedness, convexity and boundendess of $\mathcal{D}$ (and/or $\mathcal{N}$). We start with the simple statement which in fact says that nonoscillation and disconjugacy coincides when c in (5.1.1) is a Bohr almost periodic function. In particular, $\mathcal{D} = \mathcal{N}$ if $c(t) = -\alpha A(t) + \beta B(t)$, where A, B are Bohr almost periodic. Later this statement will be extended to the Stepanoff-almost periodic case. For comparison purposes, note that for more general equation (1.1.1), the statement works as well provided r is assumed to be Bohr almost periodic.

Theorem 5.3.1. *Let c be a Bohr almost periodic function. If (5.1.1) is not disconjugate on $(-\infty, \infty)$, then it is oscillatory at both $\pm\infty$.*

Proof. Suppose that a nontrivial solution y of (5.1.1) vanishes at two distinct points $t = t_1$ and $t = t_2$. For each $\varepsilon > 0$ there are arbitrarily large ε-almost periods, say τ_n, of $c(t)$. Consider the translated equations $(\Phi(y'))' + c(t+\tau_n)\Phi(y) = 0$ with solutions y_n which attain the same initial values at $t = t_1$ as does y. Also there is a solution z_n of (5.1.1) for which $z_n(t + \tau_n) = y_n(t)$. For each $\xi > 0$ there exists an $\varepsilon > 0$ such that y_n vanishes on $t_2 - \xi < t < t_2 + \xi$. Then z_n vanishes at $t = t_1 + \tau_n$ and also near $t_2 + \tau_n$. Hence every solution of (5.1.1) must vanish on $t_1 + \tau_n \le t \le t_2 + \tau_n + \xi$. Since the translation numbers τ_n are arbitrarily large (or small), (5.1.1) is oscillatory. □

The lack of periodicity-type assumptions on A, B usually has the effect of splitting $\mathcal{D}$ and $\mathcal{N}$. However, later we will see that $\mathcal{D} = \mathcal{N}$ may occur even in the "non-periodic" case.

Next we present the results concerning the basic properties of the disconjugacy domain. The central role in this subsection is played by the following lemma.

Lemma 5.3.1. *Let $c(t)$ be a continuous function on I. The equation*

$$(\Phi(y'))' + \lambda c(t)\Phi(y) = 0 \tag{5.3.3}$$

is disconjugate on I for each $\lambda \in \mathbb{R}$ if and only if $c(t) \equiv 0$ on I.

Proof. The sufficiency is trivial since $(\Phi(y'))' = 0$ is certainly disconjugate on I. In order to prove the necessity, let (5.3.3) be disconjugate on I for every $\lambda \in \mathbb{R}$. The variational principle now says that whenever $\eta \not\equiv 0$ is in $W_0^{1,p}(a,b)$, where $[a,b] \subset I$,

$$\int_a^b [|\eta'|^p - \lambda c|\eta|^p](t)\, dt > 0 \tag{5.3.4}$$

for every $\lambda \in \mathbb{R}$. Now let $[a,b]$ be a given subinterval of I and fix $\eta \not\equiv 0$ in $W_0^{1,p}(a,b)$. Since $\eta' \in L^p(a,b)$ it follows from (5.3.4) that

$$\lambda \int_a^b (c|\eta|^p)(t)\, dt < \|\eta'\|_p^p, \tag{5.3.5}$$

where $\|\cdot\|_p$ is the usual norm in $L^p(a,b)$. We emphasize that (5.3.5) is valid for every $\lambda \in \mathbb{R}$. Since $|\lambda|$ can be chosen arbitrarily large it follows from (5.3.5) that

$$\int_a^b (c|\eta|^p)(t)\, dt = 0. \tag{5.3.6}$$

We find that (5.3.6) holds for every $\eta \in W_0^{1,p}(a,b)$ and for every $[a,b] \subset I$. Define the test function ϕ_ε by

$$\phi_\varepsilon(t) = \begin{cases} t-a & \text{for } a \le t \le a+\varepsilon, \\ \varepsilon & \text{for } a+\varepsilon \le t \le b-\varepsilon, \\ b-t & \text{for } b-\varepsilon \le t \le b. \end{cases}$$

Clearly, ϕ_ε is in $W_0^{1,p}(a,b)$ for each $\varepsilon > 0$. Inserting this into (5.3.6) and passing to the limit as $\varepsilon \to 0+$, it is easily seen that

$$\int_a^b c(t)\,dt = 0. \tag{5.3.7}$$

Thus (5.3.7) holds for every compact subinterval $[a,b] \subset I$. Hence $c(t) = 0$ on I. □

An interesting formulation of Lemma 5.3.1 is its contrapositive.

Corollary 5.3.1. *Let $c(t)$ be a continuous function on I with $c(t) \not\equiv 0$ in I. Then there exists at least one value of $\lambda \in \mathbb{R}$ such that (5.3.3) is not disconjugate on I.*

Corollary 5.3.2. *Let c_i be continous functions on I for $i = 1, \ldots, n$. If the equation*

$$(\Phi(y'))' + (\lambda_1 c_1(t) + \cdots + \lambda_n c_n(t))\Phi(y) = 0 \tag{5.3.8}$$

is disconjugate on I for every point $(\lambda_1, \ldots, \lambda_n)$ in $\mathbb{R}^n$, then $c_i(t) = 0$ on I for $i = 1, \ldots, n$.

Note, once again, that the contrapositive of the last corollary is of interest. If none of the functions c_i vanishes identically, then there exists at least one point $(\lambda_1, \ldots, \lambda_n)$ in $\mathbb{R}^n$ for which (5.3.8) is not disconjugate on I.

From the last corollary we get the following theorem.

Theorem 5.3.2. *The disconjugacy domain $\mathcal{D}$ of equation (5.3.2) is the whole space $\mathbb{R}^2$ if and only if $A(t) = B(t) = 0$ on I.*

Corollary 5.3.3. *If at least one of the functions A, B is not identically equal to zero, then $\mathcal{D}$ is a proper subset of $\mathbb{R}^2$.*

The question of the boundedness or non-boundedness is a more difficult one. The next result gives a necessary and sufficient condition for $\mathcal{D}$ to contain a full ray through the origin of $\mathbb{R}^2$, and thus a sufficient condition for non-boundedness of $\mathcal{D}$.

Theorem 5.3.3. *The set $\mathcal{D}$ contains a proper subspace of the vector space $\mathbb{R}^2$ (other than the subspaces formed by the coordinate axes) if and only if the function A is a constant multiple of the function B over I.*

Proof. Let A be a constant multiple of B. Then there exists $k \neq 0$ in $\mathbb{R}$ such $A(t) = kB(t)$ on I. Equation (5.3.2) then becomes $(\Phi(y'))' + (-\alpha k + \beta)B(t)\Phi(y) = 0$. Hence $\mathcal{D}$ contains the subspace $\{(\alpha, \beta) : \beta = k\alpha\}$.

Conversely, let $\mathcal{D}$ contain a proper subspace of $\mathbb{R}^2$ other than the coordinate axes. Then $S = \{(\alpha, \beta) : \beta = k\alpha\}$ for some $k \neq 0$. Hence, on this subspace, we must have $(\Phi(y'))' + (-A(t) + kB(t))\alpha\Phi(y) = 0$ disconjugate on I for every $\alpha \in \mathbb{R}$. Applying Lemma 5.3.1 with $c(t) = -A(t) + kB(t)$, we find that $A(t) = kB(t)$ on I, so that A is a constant multiple of B. □

Note that the assumption of linear independence excludes the possibility that either one of A, B vanishes on I. Hence, form the previous theorem, we have the following statement.

Corollary 5.3.4. *Whenever A, B are linearly independent functions on I, $\mathcal{D}$ cannot contain any full ray through the origin of parameter space.*

Remark 5.3.1. Note that if $\mathcal{D}$ contains two proper subspaces of $\mathbb{R}^2$, then $\mathcal{D} = \mathbb{R}^2$. Thus whenever $\mathcal{D} \subset \mathbb{R}^2$ (which is generally the case), $\mathcal{D}$ contains at most one full ray through $(0, 0)$. It is important to point out that $\mathcal{D}$ need not always contain a full ray through $(0, 0)$. In fact, in some cases, $\mathcal{D}$ may be even a bounded set (see Example 1 on p. 5 of [289] concerning equation (5.3.1)). However note that, for example, $\mathcal{D}$ is always unbounded whenever A, B are linearly dependent, or if $A(t) = 1$ on I and $B(t)$ is arbitrary, as, in this case, $\mathcal{D} \supset \{(\alpha, 0) : \alpha \geq 0\}$.

We have seen that $\mathcal{D}$ may contain various full rays through the origin of $\mathbb{R}^2$. It should be interesting to determine whether or not $\mathcal{D}$ may contain curves of higher (lower) order than one.

For example, a glance at the geometry of the curve $\beta = \alpha^3$ shows that, since $\mathcal{D}$ is convex (this will be shown later), the line segment joining the points (α_1, α_1^3) and (α_2, α_2^3) on this curve must belong to $\mathcal{D}$. However as α varies over $\mathbb{R}$, the convex hull of the set $\{(\alpha, \beta) : \beta = \alpha^3\}$ is in fact all of $\mathbb{R}^2$. Thus $\mathcal{D} = \mathbb{R}^2$ and so the coefficients must vanish on I (Theorem 5.3.2). From this simple argument it follows that the equation $(\Phi(y'))' + (-\alpha A(t) + \alpha^3 B(t))\Phi(y) = 0$, $t \in [0, \infty)$, must be nondisconjugate on $[0, \infty)$ for at least one value of $\alpha \in \mathbb{R}$ (if A and/or B are not zero on I). A similar argument applies for the general cubic $\beta = a_1\alpha^3 + a_2\alpha^2 + a_3\alpha + a_4$ where $a_1 \neq 0$, and for the general odd degree polynomial equation (with nonzero leading coefficient).

The case when $\beta = \alpha^2$ (or any even degree polynomial equation) is very different. This is because the convex hull of the set $\{(\alpha, \beta) : \beta = \alpha^2\}$ as α varies over $\mathbb{R}$ is not all of $\mathbb{R}^2$, in contrast with the cubic case mentioned above. The idea of the proof of Theorem 5.3.2 may be used to derive necessary conditions for $\mathcal{D}$ to contain the full parabola $\beta = \alpha^2$. For example, we have the following statement.

Theorem 5.3.4. *Let A, B be such that $A^2(t) + B^2(t) > 0$ on a set of positive measure on I. Then a necessary condition for the disconjugacy domain $\mathcal{D}$ to contain the full parabola $\beta = \alpha^2$ is that $B(t) \leq 0$ on I.*

Proof. We proceed as in the proof of Theorem 5.3.2. Since $\mathcal{D}$ contains the parabola $\{(\alpha, \alpha^2) : \alpha \in \mathbb{R}\}$ it follows that for each interval $[a, b] \subset I$ and each $\eta \neq 0$ in

$W_0^{1,p}(a,b)$ $\int_a^b [|\eta'|^p - (-\alpha A + \alpha^2 B)|\eta|^p](t)\,dt > 0$ for each $\alpha \in \mathbb{R}$. From this it is readily derived that, for a fixed interval $[a,b] \subset I$ and a given $\eta \not\equiv 0$ in $W_0^{1,p}(a,b)$,

$$\alpha^2 \int_a^b (B|\eta|^p)(t)\,dt - \alpha \int_a^b (A|\eta|^p)(t)\,dt < \|\eta'\|_p^p \tag{5.3.9}$$

for each $\alpha \in \mathbb{R}$. Since $\eta' \in L^p(a,b)$ it follows that

$$\int_a^b (B|\eta|^p)(t)\,dt \le 0 \tag{5.3.10}$$

for our η, otherwise the left-hand side of (5.3.9) will exceed the right-hand side for all sufficiently large $\alpha > 0$. Since $\eta \not\equiv 0$ and $[a,b]$ are arbitrary it follows that (5.3.10) holds for each $\eta \not\equiv 0$ in $W_0^{1,p}$ (and clearly for $\eta \equiv 0$) and each $[a,b] \subset I$. Let $\eta \equiv \phi_\varepsilon$ be the test function appearing in the proof of Theorem 5.3.2 (which is in fact based on Lemma 5.3.1). Then, arguing as in that proof, we find $\int_a^b B(t)\,dt \le 0$ for each $[a,b] \subset I$. The conclusion now follows. □

Remark 5.3.2. Note that the above type of proof may be used to show that $B(t) \le 0$ is a necessary condition for $\mathcal{D}$ to contain the graph of any even degree polynomial equation with nonzero leading coefficient.

Other results in the same vein are the following two statements.

Theorem 5.3.5. *Let A, B satisfy the hypotheses in Theorem 5.3.4. A necessary condition for $\mathcal{D}$ to contain the parabolic segment $\{(\alpha,\beta) : \beta = k\sqrt{\alpha}, k > 0\}$ for all sufficiently large α (depending on k) is that $A(t) \ge 0$ on I.*

Proof. This is analogous to the proof of Theorem 5.3.4, and so will be omitted. □

Now we give a special type of converse. As we have already said above, the coefficients do not need to be continuous in our theory but we make the assumption of their continuity because of simpler formulation. In fact, there is almost no difference when we relax the assumption of continuity appropriately. For comparison purposes, let A, B be locally Lebesgue integrable in I in the following theorem.

Theorem 5.3.6. *Let $A(t) \ge 0$ almost everywhere on I and* $\operatorname{ess\,inf} A(t) > 0$. *In addition, let $B \in L^\infty(I)$. If $\alpha \ge \alpha_0$ where*

$$\alpha_0 = k^2 \|B\|_\infty^2 / (\operatorname{ess\,inf} A(t))^2, \tag{5.3.11}$$

then $\mathcal{D}$ contains the parabolic segment $\{(\alpha,\beta) : \beta = k\sqrt{\alpha}\}$ for such α. The lower bound α_0 in (5.3.11) *is sharp when A, B are constant functions almost everywhere on I.*

Proof. Note that for $\alpha \ge \alpha_0$ we have $\sqrt{\alpha}(\operatorname{ess\,inf} A(t)) - k\|B\|_\infty \ge 0$, and so $\sqrt{\alpha}A(t) - kB(t) \ge 0$ almost everywhere on I for $\alpha \ge \alpha_0$. Thus

$$\alpha A(t) - k\sqrt{\alpha}B(t) \ge 0 \tag{5.3.12}$$

almost everywhere on I for $\alpha \geq \alpha_0$. Now for a given fixed $[a, b] \subset I$ and $\eta \not\equiv 0$ in $W_0^{1,p}(a, b)$ we have

$$\int_a^b [|\eta'|^p - (-\alpha A + k\sqrt{\alpha}B)|\eta|^p](t)\, dt \geq 0 \tag{5.3.13}$$

on account of (5.3.12), with equality holding in (5.3.13) if and only if $\eta \equiv 0$. Hence this p-degree functional is positive definite on $W_0^{1,p}(a, b)$ for each $[a, b] \subset I$ and consequently $(\Phi(y'))' + (\alpha A(t) + k\sqrt{\alpha}B(t))\Phi(y) = 0$ must be disconjugate on $[0, \infty)$ for $\alpha \geq \alpha_0$ by the variational principle. Therefore $\{(\alpha, k\sqrt{\alpha}) : \alpha \geq \alpha_0\} \subset \mathcal{D}$ which is what we needed to show.

Next, let $A(t) \equiv a$, $B(t) \equiv b$ be constant functions almost everywhere on I. Then it is trivial that $\mathcal{D} = \{(\alpha, \beta) : -\alpha a + \beta b \leq 0\}$. Now the boundary of $\mathcal{D}$ is the ray $\beta b - \alpha a = 0$, or $\alpha = (b/a)\beta$ (since $a > 0$). The point of intersection of the parabolic segment $\beta = k\sqrt{\alpha}$ for $\alpha \geq \alpha_0 = k^2 b^2/a^2$ with the boundary of $\mathcal{D}$ is given by equating $(b/a)\beta = \beta^2/k^2$ which yields $\beta = bk^2/a$, i.e., $\alpha = \alpha_0$. Hence the said parabolic segment originates at the boundary of $\mathcal{D}$ for each $k > 0$. □

Remark 5.3.3. Similarly as in Theorem 5.3.5 it is possible to show that $\mathcal{D}$ contains the parabolic segment $\{(\alpha, k\alpha^{1/\lambda}), \alpha \geq \alpha_0\}$ where $\lambda > 1$, $k > 0$ if $\alpha_0 = k^\lambda \|B\|_\infty^\lambda / (\operatorname{ess\,inf} A(t))^\lambda$.

Next two lemmas will serve to prove very important general properties of $\mathcal{D}$.

Lemma 5.3.2. *Let c_i be continuous functions on I for $i = 1, 2$. Assume that each one of the equations $(\Phi(y'))' + c_i(t)\Phi(y) = 0$, $i = 1, 2$, is disconjugate on I. Then the equation $(\Phi(y'))' + ((1-\lambda)c_1(t) + \lambda c_2(t))\Phi(y)$ is disconjugate on I for each $\lambda \in [0, 1]$.*

Proof. Let $[a, b] \subset I$ be a compact subinterval. Then the functional $\int_a^b [|\eta'|^p - c_i|\eta|^p](t)\, dt$, $i = 1, 2$, is positive on $W_0^{1,p}(a, b)$ by the variational principle. On account of the same principle, it suffices to show that $\int_a^b [|\eta'|^p - ((1-\lambda)c_1 + \lambda c_2)|\eta|^p](t)\, dt$ is positive on $W_0^{1,p}(a, b)$. Now if $\lambda \in [0, 1]$ and $\eta \not\equiv 0$ is in $W_0^{1,p}(a, b)$, then

$$(1-\lambda)\int_a^b [|\eta'|^p - c_1|\eta|^p](t)\, dt + \lambda \int_a^b [|\eta'|^p - c_2|\eta|^p](t)\, dt > 0 \tag{5.3.14}$$

However the left side of (5.3.14) is equal to $\int_a^b [|\eta'|^p - ((1-\lambda)c_1 + \lambda c_2)|\eta|^p](t)\, dt$ as a simple calculation shows. Hence the latter is positive definite on $W_0^{1,p}(a, b)$. This is valid for every $[a, b]$, therefore the result follows. □

Before presenting the following lemma, let us recall that throughout this subsection, the interval I is assumed to be equal to $[0, \infty)$ or $(-\infty, \infty)$.

Lemma 5.3.3. *Let c_i, $i = 1, 2, \ldots$, be a sequence of continuous functions on I such that*

$$(\Phi(y'))' + c_n(t)\Phi(y) = 0 \tag{5.3.15}$$

is disconjugate on I for each $i = 1, 2, \ldots$. If $c_n(t)$ converge uniformly on each compact subinterval of I to $c(t)$, then the limit equation (5.1.1) *is also disconjugate on I.*

Proof. Assume, on the contrary, that (5.1.1) has two distinct zeros t_1, t_2. Consider the solutions y_n of (5.3.15) with initial data $y_n(t_1) = 0$, $y_n'(t_1) = y'(t_1)$. Let J be a compact interval containing t_1 and t_2 in its interior. For sufficiently large n, $|c_n(t) - c(t)|$ and $|y_n(t) - y(t)|$ are smaller than any prescribed $\varepsilon > 0$ for $t \in J$. Therefore y_n vanishes near t_2. But this contradicts the hypothesis that (5.3.15) is disconjugate. Thus (5.1.1) is necessarily disconjugate. □

Theorem 5.3.7. *In the usual topology of $\mathbb{R}^2$, the disconjugacy domain $\mathcal{D}$ of* (5.3.2) *is a closed set.*

Proof. Let (α_0, β_0) be a limit point of the sequence $(\alpha_n, \beta_n) \in \mathcal{D}$, $n = 1, 2, \ldots$. Then for each $\varepsilon > 0$ there exists an n such that $|\alpha_n - \alpha_0| < \varepsilon$, $|\beta_n - \beta_0| < \varepsilon$ and

$$(\Phi(y'))' + (-\alpha_n A(t) + \beta_n B(t))\Phi(y) = 0 \tag{5.3.16}$$

is disconjugate. Now let y be any nontrivial solution of (5.3.2) for $(\alpha, \beta) = (\alpha_0, \beta_0)$. Either y never vanishes in which case $(\alpha_0, \beta_0) \in \mathcal{D}$, or $y(t_0) = 0$ for some t_0. In the latter case let y_n be the solution of (5.3.16) which satisfies $y_n(t_0) = 0$, $y_n'(t_0) = y'(t_0)$. Then, by assumption, $y_n(t) \neq 0$ for $t \neq t_0$. However $\{y_n\}$ uniformly approximates y on each interval $[t_0, t_0 + \xi]$ (by Lemma 5.3.3) for $\xi > 0$ if ε is sufficiently small. Hence y can only change sign at $t = t_0$, and so $y(t) \neq 0$ for $t \neq t_0$ in I. Thus every solution y has at most one zero in I. □

Theorem 5.3.8. *When viewed as a subset of parameter space $\mathbb{R}^2$, the disconjugacy domain of* (5.3.2) *is a convex set.*

Proof. We must show that if $(\alpha_i, \beta_i) \in \mathcal{D}$, $i = 1, 2$, then the line segment joining these two points is also in $\mathcal{D}$, i.e., that $(1 - \lambda)(\alpha_1, \beta_1) + \lambda(\alpha_2, \beta_2) \in \mathcal{D}$ for each $\lambda \in [0, 1]$. This is equivalent to showing that $(\Phi(y'))' + [(-(1 - \lambda)\alpha_1 - \lambda\alpha_2)A(t) + ((1 - \lambda)\beta_1 + \lambda\beta_2)B(t)]\Phi(y) = 0$ is disconjugate on I for each $\lambda \in [0, 1]$. Simplifying and rearranging terms in the potential of the last equation, we may rewrite it in the equivalent form $(\Phi(y'))' + [(1-\lambda)(-\alpha_1 A(t) + \beta_1 B(t)) + \lambda(-\alpha_2 A(t) + \beta_2 B(t))]\Phi(y) = 0$ for $\lambda \in [0, 1]$. Since $(\alpha_i, \beta_i) \in \mathcal{D}$ for $i = 1, 2$, Lemma 5.3.2 yields the conclusion. □

Example 5.3.1. Let $A(t) = B(t) = t^{-p}$ on $I = [1, \infty)$. Since A, B are linearly dependent, $\mathcal{D}$ must contain full ray through $(0, 0)$. Moreover we know that $\mathcal{D}$ is closed and convex. Note that (5.3.2), with the above identification, becomes an Euler type equation of the form

$$(\Phi(y'))' + \frac{-\alpha + \beta}{t^p}\Phi(y) = 0$$

for $t \in I$. Thus if $-\alpha + \beta \leq [(p - 1)/p]^p$, the equation is disconjugate, see Subsection 1.4.2, and so $\{(\alpha, \beta) : \beta \leq [(p - 1)/p]^p + \alpha\} \subseteq \mathcal{D}$. On the other hand, if $-\alpha+\beta > [(p-1)/p]^p$, the equation is oscillatory. Hence $\mathcal{D} = \{(\alpha,\beta) : \beta \leq [(p-1)/p]^p + \alpha\}$. Note that $\mathcal{D}$ contains precisely one subspace S of $\mathbb{R}^2$, namely, $S = \{(\alpha,\beta) : \alpha = \beta\}$ (cf. Theorem 5.3.3).

Remark 5.3.4. If $c(t) = -\alpha A(t) + \beta B(t)$ satisfies the assumption of Theorem 5.3.1, then the oscillation domain $\mathcal{O}$ is open, since $\mathcal{D}$ and $\mathcal{O}$ are complementary in the $\alpha\beta$-plane.

Now we turn our attention to the nonoscillation domain $\mathcal{N}$ (of (5.3.2)). First note that $\mathcal{N} \neq \emptyset$ since $\mathcal{D} \subseteq \mathcal{N}$ and $\mathcal{D} \neq \emptyset$. The following examples show the interplay between $\mathcal{D}$ and $\mathcal{N}$.

Example 5.3.2. In contrast with Theorem 5.3.2 one can have $\mathcal{N} = \mathbb{R}^2$ without either of A, B being equal to zero on I. A simple way of seeing this is by choosing $A, B \in \{f \in C(I) : f(t)$ vanishes identically outside a closed and bounded subinterval of $I\}$ For such a choice of A, B it is readily seen that (5.3.2) reduces to $(\Phi(y'))' = 0$ for all sufficiently large t, independently of α, β. Hence every solution must have finitely many zeros. So $\mathcal{N} = \mathbb{R}^2$ for such potentials.

Example 5.3.3. We now give an example where $\{(0,0)\} \subset \mathcal{D} \subseteq \mathcal{N} \subset \mathbb{R}^2$. To this end, let $A(t) = 1$ and $B(t) = t^{-p}$ on $I = [1,\infty)$. Then (5.3.2) takes the form $(\Phi(y'))' + (-\alpha + \beta t^{-p})\Phi(y) = 0$ for $t \in I$. Now it is clear that $\{(\alpha,\beta) : \alpha \geq 0, \beta \leq [(p-1)/p]^p\} \subset \mathcal{D}$ by a simple application of Sturm's comparison theorem with a generalized Euler equation. Moreover the region $\{(\alpha,\beta) : \alpha < 0\}$ is certainly part of oscillation domain (i.e., the complement in $\mathbb{R}^2$ of the nonoscillation domain) by Leighton-Wintner type criterion (Theorem 1.2.9). Thus $\mathcal{D} \neq \{(0,0)\}$ and $\mathcal{N} \neq \mathbb{R}^2$.

Example 5.3.4. The phenomenon $\mathcal{N} = \mathcal{D} = \{(0,0)\}$ may also occur. Indeed, we already know (see Theorem 5.3.1) that for the class of potentials which are uniformly (i.e., Bohr) almost periodic the notions of disconjugacy and nonoscillation coincide. Moreover, below mentioned Corollary 5.3.6 says that (5.1.1) is oscillatory provided $M\{c\} = 0$, where the mean value $M\{c\}$ is defined by

$$M\{c\} = \lim_{T\to\infty} \frac{1}{T} \int_0^T c(s+a)\, ds,$$

$a \in \mathbb{R}$. Now let A, B be Bohr almost periodic functions with $M\{A\} = M\{B\} = 0$. Then $-\alpha A + \beta B$ is also Bohr almost periodic with $M\{-\alpha A + \beta B\} = 0$ for each $(\alpha,\beta) \in \mathbb{R} \setminus \{(0,0)\}$. Hence (5.3.2) is oscillatory for each $(\alpha,\beta) \neq (0,0)$. Thus $\mathcal{D} = \mathcal{N} = \{(0,0)\}$.

Example 5.3.5. In contrast with Theorem 5.3.7 we show that, in general, $\mathcal{N}$ is not a closed set. It suffices to find a sequence $\{c_n(t)\}$ such that $c_n(t) \to c(t)$ uniformly on compact subsets of I as $n \to \infty$, and $(\Phi(y'))' + c_n(t)\Phi(y) = 0$ is nonoscillatory (for each n) with $(\Phi(y'))' + c(t)\Phi(y) = 0$ being oscillatory at ∞. Let $A(t) = t^p$, $B(t) = t^{-p}$ on $I = [1,\infty)$, and consider the equation

$$(\Phi(y'))' + (-\alpha t^p + \beta t^{-p})\Phi(y) = 0 \tag{5.3.17}$$

on I. Let $(\alpha_n, \beta_n) = (1/n, 1+1/n)$, $n = 1, 2, \dots$. Then (5.3.17), with α_n, β_n as parameters, is nonoscillatory at ∞ for each $n = 1, 2, \dots$, since $-\alpha_n t^p + \beta_n t^{-p} \leq 0$

for each $t \geq T_n := \sqrt[2p]{n+1}$. Hence for each n, (5.3.17) with $(\alpha, \beta) = (\alpha_n, \beta_n)$ is disconjugate on $[T_n, \infty)$ and hence must be nonoscillatory on $[1, \infty)$. However $-\alpha_n t^p + \beta_n t^{-p} \to t^{-p}$ (uniformly on compact subintervals of $[0, \infty)$), and the limit equation $(\Phi(y'))' + t^{-p}\Phi(y) = 0$ is an oscillatory Euler type equation. Thus $(\alpha_n, \beta_n) \in \mathcal{N}$ but $\lim(\alpha_n, \beta_n) \notin \mathcal{N}$. Hence $\mathcal{N}$ cannot be closed in this case. Note that since $\mathcal{D}$ is closed, it follows that $\mathcal{D} \neq \mathcal{N}$ for equation (5.3.17).

Remark 5.3.5. Let us summarize what we have seen so far. Recall that equation (5.3.2) is considered on the interval I, where I equals $(-\infty, \infty)$ or $[0, \infty)$. In contrast with $\mathcal{D}$, the set $\mathcal{N}$ is not always closed. Furthermore, there may occur $\mathcal{N} = \mathbb{R}^2$ without either one of A, B vanishing on I. In general $\mathcal{D} \subset \mathcal{N}$ for equation (5.3.2), however there exists various classes of potentials A, B, for which $\mathcal{D} = \mathcal{N}$. Further, the oscillation domain $\mathcal{O}$ may be an open set (e.g., when $\mathcal{D} = \mathcal{N}$) and it may be empty (e.g., when $\mathcal{N} = \mathcal{R}^2$). For comparison purposes, it is not difficult to see that whenever $A(t) = 1$, $B(t)$ is uniformly almost periodic, then $\mathcal{O}$ is open, connected and nonempty. Indeed, if $(\alpha_0, \beta_0) \in \mathcal{O}$, then $(\alpha_0 - \xi, \beta_0) \in \mathcal{O}$ for each $\xi > 0$ by the Sturm type comparison theorem. Moreover, $\mathcal{O}$ contains the sector $\alpha < -|\beta| \sup_{-\infty < t < \infty} c(t)$. The sets $\mathcal{D}, \mathcal{N}$ may be unbounded (e.g. when A, B are linearly dependent) and $\mathcal{D}$ may even be a bounded set (so that $\mathcal{D}$ is nonempty, convex and compact), $\mathcal{D} \neq \{(0,0)\}$. However if $A(t) = 1$, then $\mathcal{D}$ is unbounded. It is possible that $\mathcal{N}$ may be bounded (as $\mathcal{N} = \{(0,0)\}$ can occur). However it is an open question whether or not $\mathcal{N}$ may be bounded if $\mathcal{N} \neq \{(0,0)\}$ (even in the linear case). We have also seen that $\mathcal{O} = \mathbb{R}^2 \setminus \{(0,0)\}$ is a possibility so that $\mathcal{O}$ is as large as possible in such a case. Moreover it is immediate that the boundary of $\mathcal{D}$ and $\mathcal{N}$ is a continuous curve (since each of these sets is convex).

Already at the beginning of this section, the problem on the equivalence of $\mathcal{D}$ and $\mathcal{N}$ was mentioned. There is an interesting question: For which class of potentials A, B does equation (5.3.2) have the property that $\mathcal{D} = \mathcal{N}$? We already know that this question is related to oscillatory potentials. Indeed, the desired property is guaranteed by Bohr almost periodicity of the coefficients of the equation, see Theorem 5.3.1. In Subsection 5.3.3 we extend this statement to the Stepanoff case. However there are classes of nonoscillating potentials for which $\mathcal{N} = \mathcal{D}$ also. Recall Example 5.3.1, where $\mathcal{D} = \mathcal{N}$ provided $A(t) = B(t) = t^{-p}$ on $[1, \infty)$. Here is another example.

Example 5.3.6. Let $A(t) \geq 0$ on $I = [0, \infty)$ and $A(t) \equiv B(t)$. Assume further that

$$\lim_{t \to \infty} \int_0^t A(s)\, ds = \infty. \tag{5.3.18}$$

Then the equation $(\Phi(y'))' + (-\alpha + \beta)A(t)\Phi(y) = 0$ is oscillatory whenever $\beta > \alpha$, on account of (5.3.18) and the Leighton-Wintner type criterion, see Theorem 1.2.9. Moreover, if $\beta \leq \alpha$, we have $(\beta - \alpha)A(t) \leq$ on I and so (5.3.2) is disconjugate. Hence $\mathcal{O} = \{(\alpha, \beta) : \beta > \alpha\}$ and so $\mathcal{D} = \mathcal{N}$ $(= \{(\alpha, \beta) : \beta \leq \alpha\})$.

Now any one of a multitude of oscillation criteria for half-linear equations with positive coefficients may be used instead of (5.3.18) to obtain still wider classes of potentials for which $\mathcal{D} = \mathcal{N}$.

We finish this subsection with applications to the (extended) weighted Sturm-Liouville type equation

$$(\Phi(y'))' + (\lambda B(t) - A(t))\Phi(y) = 0, \tag{5.3.19}$$

where $t \in I$ $(= [0, \infty))$ and $\lambda \in \mathbb{R}$ is a parameter. We turn our attention to the collection of those $\lambda \in \mathbb{R}$ for which (5.3.19) is nonoscillatory/oscillatory. The collection of such λ is a vertical line $\mathcal{L}$ through $\alpha = 1$ in parameter space $\mathbb{R}^2$. Hence we will investigate the number of possible ways in which $\mathcal{L}$ may intersect $\mathcal{N}$ (or $\mathcal{D}$).

Using the technique developed in the proof of Lemma 5.3.1, we can show: If equation (5.3.19) is disconjugate on $[0, \infty)$ for every real λ, then $B(t) = 0$ on I (and $(\Phi(y'))' = A(t)\Phi(y)$ is disconjugate on I). Therefore it follows that if $B(t) \not\equiv 0$, then there exists at least one value of λ for which (5.3.19) is not disconjugate on I.

Furthermore we have: Let A, B be (nontrivial) Bohr almost periodic functions with $M\{A\} = M\{B\} = 0$. Then (5.3.19) is oscillatory at infinity for every value of λ (provided the case when $\lambda B(t) - A(t) \equiv 0$ eventually is excluded).

Theorem 5.3.9. *Precisely one of the following five cases occurs for each equation of the form* (5.3.19)*:*

(i) It is oscillatory for every real λ.

(ii) It is oscillatory for every real value of λ except at some unique point $\lambda = \lambda_0$.

(iii) There exists a finite interval (λ_1, λ_2) in $\mathbb{R}$ such that (5.3.19) *is oscillatory for $\lambda \in (-\infty, \lambda_1) \cup (\lambda_2, \infty)$ and nonoscillatory for $\lambda \in (\lambda_1, \lambda_2)$.*

(iv) There exists a point $\lambda_3 \in \mathbb{R}$ such that (5.3.19) *is oscillatory (resp. nonoscillatory) for $\lambda \in (-\infty, \lambda_3)$ and nonoscillatory (resp. oscillatory) for $\lambda \in (\lambda_3, \infty)$.*

(v) It is nonoscillatory for every real λ.

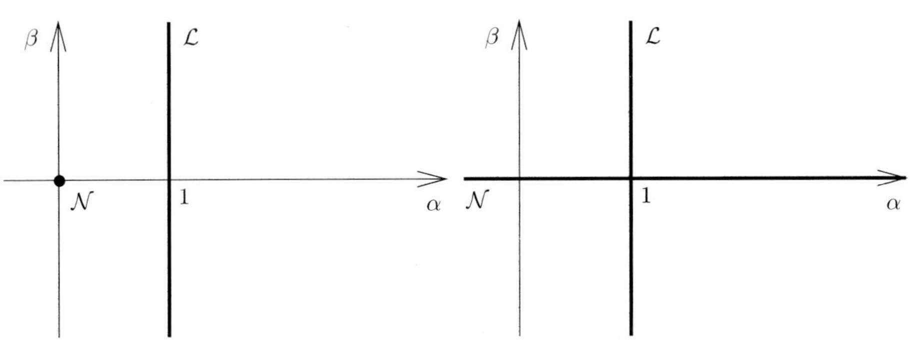

Figure 5.3.1: Intersection of $\mathcal{L}$ with $\mathcal{N}$: Cases (i) and (ii)

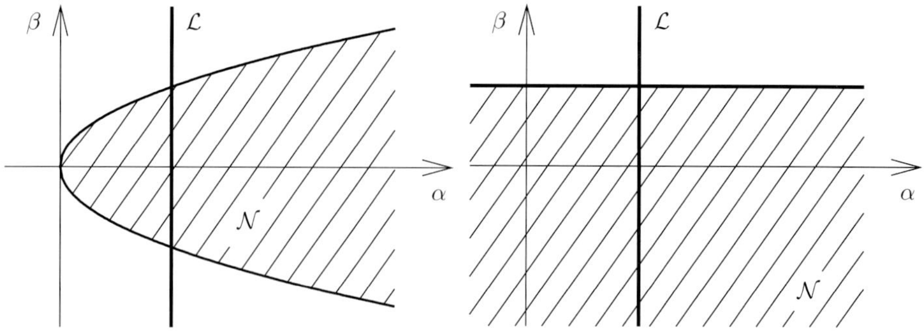

Figure 5.3.2: Intersection of $\mathcal{L}$ with $\mathcal{N}$: Cases (iii) and (iv)

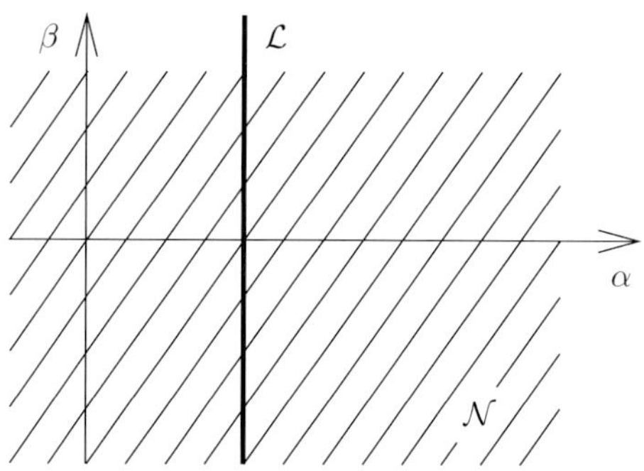

Figure 5.3.3: Intersection of $\mathcal{L}$ with $\mathcal{N}$: Case (v)

Proof. Let $\mathcal{L} = \{(1, \lambda) : \lambda \in \mathbb{R}\}$ be the ray mentioned above. Since $\mathcal{L}$ is a ray and $\mathcal{N}$ is convex, the claims (i)-(v) are consequences of the geometrical nature of the intersection of $\mathcal{L}$ with $\mathcal{N}$. Recall that $\mathcal{N}$ is a convex set in general position in $\mathbb{R}^2$ (but, as always, containing $(0,0)$). Enumerating the possibilities is now an easy matter. There are only five distinct ways in which $\mathcal{L}$ may intersect $\mathcal{N}$: (i) $\mathcal{L} \cap \mathcal{N} = \emptyset$, (ii) $\mathcal{L} \cap \mathcal{N} = \{\text{a single point}\}$, (iii) $\mathcal{L} \cap \mathcal{N} = \{\text{a finite line segment}\}$, (iv) $\mathcal{L} \cap \mathcal{N} = \{\text{a "half-ray"}\}$, (v) $\mathcal{L} \cap \mathcal{N} = \{\text{a full ray}\} = \mathcal{L}$. See also Figures 5.3.1-5.3.3. Each of the possibilities (i)-(v) listed here corresponds to the stated claims, respectively, in the theorem, as it is not difficult to see. □

Example 5.3.7. We show here that, in fact, each of the possibilities (i)–(v) stated in Theorem 5.3.9 may occur.

On (i): See the note before Theorem 5.3.9.

On (ii): See the note before Theorem 5.3.9 but now set $A \equiv 0$ there. Then $(\Phi(y'))' + \lambda B(t)\Phi(y) = 0$ is oscillatory for each $\lambda \neq 0$, and so $\mathcal{L} \cap \mathcal{N} = \{(1,0)\}$ in this case.

On (iii): It is known that at least in the special case $p = 2$, such an example exists, see [289, p. 21], where the results on the structure of $\mathcal{N}$, given in [294], are utilized.

On (iv): Let $B(t) = (t+1)^{-p}$ on $[0, \infty)$ and set $A(t) \equiv 0$ on $[0, \infty)$. Then for $\lambda \in (-\infty, [(p-1)/p]^p)$ (5.3.19) is nonoscillatory (in fact, disconjugate) while for $\lambda \in ([(p-1)/p]^p, \infty)$, (5.3.19) is oscillatory at ∞ as the equation is of generalized Euler type. Replacing B by $-B$ and $[(p-1)/p]^p$ by $-[(p-1)/p]^p$ interchanges the words "nonoscillatory" and "oscillatory".

On (v): See Example 5.3.2.

5.3.2 Equations with periodic coefficients

In this subsection we give an oscillation criterion for equation (5.1.1), where the coefficient c is periodic.

Theorem 5.3.10. *Suppose that the function $c(t)$ in (5.1.1) is a periodic function with the period ω, $c(t) \not\equiv 0$, and*

$$\int_0^\omega c(t)dt \geq 0.$$

Then (5.1.1) is oscillatory both at $t = \infty$ and at $t = -\infty$.

Proof. To prove oscillation of (5.1.1), it is sufficient to find a solution of this equation with at least two zeros. Indeed, the periodicity of the function c implies that if x is a solution of (5.1.1) then $x(t \pm \omega)$ is a solution as well and hence any solution with two zeros has actually infinitely many of them, tending both to ∞ and $-\infty$.

The statement of theorem is clearly true if c is a positive constant function since then $x(t) = \sin_p \mu t$ is a solution of this equation, where μ is a constant depending on c and p. So we need to consider the cases when $c(t)$ is not a constant only. Also, it is sufficient to deal with the case when $\int_0^\omega c(t)dt = 0$ because otherwise we can define $c_0 = \frac{1}{\omega}\int_0^\omega c(t)dt > 0$ and $\tilde{c}(t) = c(t) - c_0$. Clearly, we have $c(t) > \tilde{c}(t)$. If we prove (5.1.1) with $\tilde{c}$ instead of c to be oscillatory then by the Sturmian comparison theorem equation (5.1.1) is also oscillatory.

Now let

$$C(t) = \int_0^t c(s)ds.$$

This is a continuous periodic function with the period ω. Let γ and δ be defined by

$$C(\delta) = \max_{0 \leq t \leq \omega} C(t), \qquad C(\gamma) = \min_{\delta \leq t \leq \delta + \omega} C(t).$$

Then $0 \leq \delta < \gamma < \delta + \omega$ and

$$\int_\gamma^t c(s)ds \geq 0, \qquad \int_t^\delta c(s)ds \geq 0 \quad \text{for } t \in \mathbb{R}.$$

Now, by Theorem 5.1.4 and the remark given below this theorem, the solution of (5.1.1) given by the initial condition $x(\delta) = 1$, $x'(\delta) = 0$ has a zero in $(-\infty, \delta)$. Indeed, $C(t) \not\equiv 0$ and

$$\int_t^\delta |s - \delta|^\alpha \left(\int_t^\delta c(\tau)\, d\tau \right) ds \geq 0 \quad \text{for } t \geq \delta,$$

with any $\alpha \in (-1/p, p-2]$. Now we need to show that this solution has a zero on (δ, ∞) as well. We proceed by contradiction, suppose that $x(t) > 0$ for $t \geq \delta$. Consider the function $w = -\Phi(x'/x)$ on $[\delta, \infty)$. This function satisfies the Riccati differential equation

$$w' = c(t) + (p-1)|w|^q \tag{5.3.20}$$

and by integration we have

$$w(t+\omega) - w(t) = (p-1)\int_t^{t+\omega} |w(s)|^q ds, \tag{5.3.21}$$

hence $w(t+\omega) > w(t)$. Consider now the sequence

$$w(\gamma),\ w(\gamma+\omega),\ w(\gamma+2\omega),\ \ldots.$$

By Theorem 5.1.4 and by our indirect assumption on the solution $x(t)$, this sequence consists of negative terms:

$$w(\gamma) < w(\gamma+\omega) < w(\gamma+2\omega) < \cdots < 0.$$

Indeed, if $w(\gamma + k\omega) \geq 0$ for some $k \in \mathbb{N}$, then by Theorem 5.1.4 the solution $x(t)$ would have a zero in $(\gamma + k\omega, \infty)$. Hence $\lim_{k\to\infty} w(\gamma + k\omega) \leq 0$, consequently by (5.3.21)

$$w(\gamma) + (p-1)\int_\gamma^\infty |w(s)|^q ds \leq 0,$$

i.e., the integral $\int_\gamma^\infty |w(s)|^q ds$ is convergent. This implies by (5.3.20) that

$$w(t) = w(\gamma) + \int_\gamma^t c(s)\, ds + (p-1)\int_\gamma^t |w(s)|^q\, ds$$

and the function $w(t)$ is bounded. Again by (5.3.20) we find that w' is also bounded, say, $|w'(t)| < L$. Then

$$\left|\frac{|w(t_2)|^{q+1} - |w(t_1)|^{q+1}}{q+1}\right| = \left|\int_{t_1}^{t_2} w'(s)|w(s)|^q \operatorname{sgn} w(s)\, ds\right| \leq L\int_{t_1}^{t_2} |w(s)|^q\, ds,$$

$\gamma < t_1 < t_2$, hence $\lim_{t\to\infty} |w(t)|^{q+1}$ exists. Clearly, we have $\lim_{t\to\infty} w(t) = 0$.

On the other hand, $w(\delta) = 0$, and by (5.3.21) we have $\lim_{k\to\infty} w(\delta + k\omega) > 0$ and this contradicts the fact that $\lim_{t\to\infty} w(t) = 0$. □

5.3.3 Equations with almost periodic coefficients

We start with recalling the concept of almost periodic function. Bohr almost periodic function has already been defined at the beginning of this section. We emphasize that any Bohr almost periodic function is continuous, but the following generalized almost periodic functions are not necessarily so. In the sequel we assume that f is locally Lebesgue integrable. One of the first generalizations of Bohr almost periodicity was due to Stepanoff. We say that f is *Stepanoff almost periodic* if there exists a real number $L > 0$ such that for each $\varepsilon > 0$ there exists a relatively dense set of Stepanoff translation numbers $\tau_f(\varepsilon)$, i.e., numbers such that

$$\sup_{t\in\mathbb{R}} \left\{ \frac{1}{L} \int_t^{t+L} |f(s+\tau_f(\varepsilon)) - f(s)|\, ds \right\} < \varepsilon.$$

This definition is classical. It is to be noted however that the class of Stepanoff almost periodic functions may be found by taking the closure of the class of all finite trigonometric polynomials relative to the metric $\mathcal{D}_S$ defined by

$$\mathcal{D}_S[f,g] = \sup_{t\in\mathbb{R}} \left\{ \frac{1}{L} \int_t^{t+L} |f(s) - g(s)|\, ds \right\},$$

L being a positive real number. The notion of a *Weyl almost periodic function* is defined analogously, the only difference being in the definition of the metric. Thus, the Weyl metric is given by

$$\mathcal{D}_W[f,g] = \lim_{L\to\infty} \sup_{t\in\mathbb{R}} \left\{ \frac{1}{L} \int_t^{t+L} |f(s) - g(s)|\, ds \right\},$$

where the limit may be shown to exist. The completion of the class of all finite trigonometric polynomials relative to this metric gives the space of the Weyl almost periodic functions. The generalization of almost periodic functions undertaken by Besicovitch is as follows. The Besicovitch metric is defined by

$$\mathcal{D}_B[f,g] = \limsup_{T\to\infty} \frac{1}{2T} \int_{-T}^{T} |f(s) - g(s)|\, ds$$

and the space of all *Besicovitch almost periodic functions* is obtained by completing the space of all finite trigonometric polynomials relative to the Besicovitch metric. However, we will rather work with the Besicovitch seminorm $\|f\|_B = \limsup_{T\to\infty} \frac{1}{2T} \int_{-T}^{T} |f(s)|\, ds$. For each one of these notions of almost periodic functions, the *mean value* $M\{f\}$ always exists, is finite, and is uniform with respect to a, $a \in \mathbb{R}$, where

$$M\{f\} = \lim_{T\to\infty} \frac{1}{T} \int_{t_0}^{T} f(s+a)\, ds.$$

It is immediate from the translation properties of such functions that $M\{|f|\}$ always exists, since $|f|$ enjoys the same almost periodic properties as does f, and is finite. Finally note that if we consider almost periodic functions in the sense of Bohr, Stepanoff, Weyl and Besicovitch, then each such class is included in the next one. The monographs [35, 51] are very good sources for finding further information on almost periodic functions.

Now we give an extension of Theorem 5.3.1 from the Bohr almost periodic case to the Stepanoff almost-periodic case, as we promised in the previous subsection. Recall that the function c in (5.1.1) may be considered to be merely locally Lebesgue integrable. This nicely matches the fact that generalized almost periodic functions do not need to be continuous.

Theorem 5.3.11. *Let c be a Stepanoff almost periodic function. If equation (5.1.1) is not disconjugate on $(-\infty, \infty)$, then it is oscillatory at both $\pm\infty$.*

Proof. Let y be a nontrivial solution of (5.1.1) such that for some $a < b$, $y(a) = y(b) = 0$. Let $\varepsilon = |n|^{-1}$, $n = \pm 1, \pm 2, \ldots$. Then there exists (by definition) arbitrarily large positive and arbitrarily large negative $\{\tau_n\}_{n=-\infty}^{\infty}$ (arranged in increasing order) so that, for each n,

$$\sup_{t\in\mathbb{R}} \left\{ \frac{1}{L} \int_t^{t+L} |c(s+\tau_n) - c(s)|\, ds \right\} < \frac{1}{|n|}.$$

Hence

$$\frac{1}{L} \int_{a+iL}^{a+(i+1)L} |c(s+\tau_n) - c(s)|\, ds < \frac{1}{|n|} \tag{5.3.22}$$

for each i, $i = 0, 1, 2, \ldots$. Since $L > 0$ is fixed, let $m \in \mathbb{N}$ be chosen so that $mL > b - a$. Fix such an m. Then (5.3.22) holds for $i = 0, 1, \ldots, m-1$, and so

$$\frac{1}{L} \int_a^{a+mL} |c(s+\tau_n) - c(s)|\, ds < \frac{m}{|n|}$$

holds for each n, $n = \pm 1, \pm 2, \ldots$. Now there exists some $\eta > 0$, which we now fix, such that

$$\frac{1}{L} \int_a^{b+\eta} |c(s+\tau_n) - c(s)|\, ds < \frac{m}{|n|} \tag{5.3.23}$$

for each n. It follows from (5.3.23) that given $\nu > 0$ there exists $N > 0$ such that

$$\int_a^{b+\eta} |c(s+\tau_n) - c(s)|\, ds < \nu \tag{5.3.24}$$

provided $|n| \geq N$. Now consider the equations $(\Phi(z_n'))' + c(t+\tau_n)\Phi(z_n) = 0$, $t \in \mathbb{R}$, and let z_n be the solution corresponding to $z_n(a) = y(a) = 0$, $z_n'(a) = y'(a) (\neq 0)$. Then z_n is defined and continuous on $[a, b+\eta]$ and since $c(t+\tau_n)$ approximates $c(t)$ in the L^1-sense over $[a, b+\eta]$ (by (5.3.24)), it follows by the continuous dependence of solutions (see [171]) that there holds $\sup_{t\in[a,b+\eta]} |z_n(t) - y(t)| < \varepsilon_n$, where $\varepsilon_n \to 0$ as $|n| \to \infty$. Since $y(b) = 0$, there exists n such that z_n vanishes in some neighborhood $(b-\delta, b+\delta)$ of b where δ may be chosen less than or equal to $\min\{b-a, \eta\}$. Hence z_n vanishes at two points for all $|n| \geq N$. But $z_n(t) \equiv y_n(t+\tau_n)$ where $y_n \not\equiv 0$ satisfies (5.1.1) and $y_n(a+\tau_n) = 0$, $y_n'(a+\tau_n) = y'(a)$. Hence y_n vanishes at $t = a + \tau_n$ and near $t = b + \tau_n$ for each n. The Sturm type separation theorem now implies that every solution of (5.1.1) must vanish between $a+\tau_n$ and $b + \tau_n + \delta$. Thus (5.1.1) is oscillatory at $\pm\infty$. □

Thus we see that if c is Bohr almost periodic or, more generally, Stepanoff almost periodic, then (5.1.1) is either oscillatory at ∞ and $-\infty$ or else is disconjugate on $\mathbb{R}$. We show now that this is, in general, false for Weyl almost periodic functions.

Example 5.3.8. Let c be defined on $\mathbb{R}$ as follows: On $[0,\infty)$, $c(t)=\gamma_1(t+1)^{-p}$, while on $(-\infty,0]$, $c(t)=\gamma_2(t-1)^{-p}$, where $\gamma_1>[(p-1)/p]^p$ and $0<\gamma_2<[(p-1)/p]^p$ are some constants. Recall that the function c in (5.1.1) may be considered to be merely locally Lebesgue integrable. Since each of these c gives rise to a generalized Euler equation on the respective half-axes, we see that the resulting equation is oscillatory on $[0,\infty)$ and nonoscillatory on $(-\infty,0]$. Finally note that c is Weyl almost periodic as it is in $L(\mathbb{R})$.

Corollary 5.3.5. *Let A, B be Stepanoff almost periodic functions. Then $\mathcal{D}=\mathcal{N}$ for equation* (5.3.2).

Proof. If A, B are Stepanoff almost periodic functions, then $\beta B-\alpha A$ is a Stepanoff almost periodic function, see [35], since α,β are real constants. Hence either $(\alpha,\beta)\in\mathcal{D}$ or $(\alpha,\beta)\in\mathcal{O}$, by the previous theorem. □

If, in addition to almost periodicity of the potential, suitable conditions on the mean value are required, then the oscillatory properties are affected in an interesting way. Before we give the main statement of this part, recall the history of the problem. In [172], it was shown that under the assumption that c is Bohr almost periodic, c is not identically zero, the linear equation $y''+\lambda c(t)y=0$ is oscillatory at both ∞ and $-\infty$ for every real nonzero λ if and only if $M\{c\}=0$. Note that the proof of sufficiency actually follows also from [270, Theorems 2 and 6] or from [80]. This was extended slightly in [289] to include the case of generalized almost periodic functions in the sense of Stepanoff, under the tacit assumption that either c or its indefinite integral is uniformly bounded on $\mathbb{R}$. In each of the cited works use is made of the Bohr uniqueness theorem which states that a nonnegative almost periodic function c with $M\{c\}=0$ must vanish identically, if it is Bohr almost periodic, or vanish almost everywhere with respect to Lebesgue measure, if it is Stepanoff almost periodic. It is a classical result that this uniqueness theorem fails in the Weyl or Besicovitch case. It turns out that a natural condition, namely, that $M\{|c|\}>0$, is needed in addition to the original one, that is, $M\{c\}=0$, in order to extend the result of [172] to the Stepanoff, Weyl and Besicovitch cases, and that the former condition is necessary. This is seen by the example: Let $c(t)=\exp(-x^2)$. Then c is Weyl almost periodic [35, p. 77] and $M\{c\}=0=M\{|c|\}$. But $y''+\lambda c(t)y=0$ is nonoscillatory on $\mathbb{R}$ if $\lambda>0$. In [136], it was proved the following result: Let c be Besicovitch almost periodic and assume that $M\{|c|\}>0$. Then the equation $y''+\lambda c(t)y=0$ is oscillatory at both ∞ and $-\infty$ for every real nonzero λ if and only if $M\{c\}=0$. What we offer here is a half-linear extension of the "sufficient part". Note that in the linear case the proof of necessity is valid for any class of almost periodic functions with no change. However, it is based on the linear transformation which has no half-linear analogue. Therefore, it is an open problem to prove an extension of the "necessary part". We start with an auxiliary statement which can be viewed as a certain variant of Hartman-Wintner type theorem.

Lemma 5.3.4. *Suppose that $c:[t_0,\infty)\to\mathbb{R}$ is a locally Lebesgue integrable function with $M\{c\}=0$ and* (5.1.1) *is nonoscillatory. If $y(t)\neq 0$ is a solution of*

(5.1.1) *on* $[t_0, \infty)$, *then*

$$\lim_{t\to\infty} \frac{1}{t} \int_{t_0}^{t} |w(s)|^q \, ds = 0,$$

where w *is given by the Riccati substitution* $w = -\Phi(y'/y)$.

Proof. Let w be defined as in the lemma. Then it satisfies the generalized Riccati equation

$$w' - (p-1)|w|^q - c(t) = 0 \tag{5.3.25}$$

on $[t_0, \infty)$. Since $|w|^q \geq 0$, it suffices to show that

$$\limsup_{t\to\infty} \frac{1}{t} \int_{t_0}^{t} |w(s)|^q ds = 0.$$

Assume to the contrary that

$$\limsup_{t\to\infty} \frac{1}{t} \int_{t_0}^{t} |w(s)|^q ds > 0. \tag{5.3.26}$$

Integrating (5.3.25) from t_0 to t and dividing it by t, we have

$$\frac{w(t)}{t} = \frac{w(t_0)}{t} + \frac{1}{t} \int_{t_0}^{t} c(s)\, ds + \frac{p-1}{t} \int_{t_0}^{t} |w(s)|^q ds \tag{5.3.27}$$

for all $t > t_0$. It follows from (5.3.26), (5.3.27) and $M\{c\} = 0$ that there exist a positive constant m and an increasing sequence $\{t_n\}_{n=1}^{\infty}$ of (t_0, ∞) with $\lim_{n\to\infty} t_n = \infty$ such that

$$\frac{w(t_n)}{t_n} > (p-1)m^p \quad \text{for all } n \text{ large enough.} \tag{5.3.28}$$

It follows from $M\{c\} = 0$ that there exists t^* large enough such that

$$\left| \int_{t_0}^{t} c(s)\, ds \right| < (p-1)(m/2)^p t \tag{5.3.29}$$

for $t \geq t^*$. Using (5.3.29) we have

$$\int_{t_n}^{t} c(s)\, ds = \int_{t_0}^{t} c(s)\, ds - \int_{t_0}^{t_n} c(s)\, ds < (p-1)(m/2)^p (t + t_n) \tag{5.3.30}$$

for all $t \geq t_n \geq t^*$. It follows from (5.3.28) and (5.3.30) that

$$\begin{aligned} w(t_n) - \int_{t_n}^{t} c(s)\, ds &> (p-1)m^p t_n - (p-1)(m/2)^p (t + t_n) \\ &\geq (p-1)m^p t_n - (p-1)(m/2)^p [(2^p - 1)t_n + t_n] = 0 \end{aligned} \tag{5.3.31}$$

for all $t \in [t_n, (2^p - 1)t_n] \subset [t^*, \infty)$. From the existence of solutions to the initial value problem, the differential equation

$$(\Phi(y_n'))' - (p-1)(m/2)^p \Phi(y_n) = 0 \tag{5.3.32}$$

has a solution y_n on $[t_n, (2^p - 1)t_n]$ satisfying $y_n(t_n) = y(t_n)$ and

$$-\frac{\Phi(y_n'(t_n))}{\Phi(y_n(t_n))} = w(t_n) - 2(p-1)(m/2)^p t_n.$$

It follows from (5.3.30) and (5.3.31) that

$$\begin{aligned} -\frac{\Phi(y'(t_n))}{\Phi(y(t_n))} - \int_{t_n}^t c(s)\, ds &= w(t_n) - \int_{t_n}^t c(s)\, ds \\ &> w(t_n) - (p-1)(m/2)^p (t + t_n) \\ &= w(t_n) - 2(p-1)(m/2)^p t_n - (p-1)(m/2)^p (t - t_n) \\ &= -\frac{\Phi(y_n'(t_n))}{\Phi(y_n(t_n))} - \int_{t_n}^t (p-1)(m/2)^p ds \geq 0 \end{aligned}$$

on $[t_n, (2^p - 1)t_n] \subset [t^*, \infty)$. Using the Leighton-Levin type comparison theorem (see Subsection 5.8.3 below), we have

$$-\frac{\Phi(y'(t_n))}{\Phi(y(t_n))} > \left| -\frac{\Phi(y_n'(t))}{\Phi(y_n(t))} \right| \tag{5.3.33}$$

on $[t_n, (2^p - 1)t_n] \subset [t^*, \infty)$. Now define $w_n = -\Phi(y_n'/y_n)$. It is clear that w_n is a solution of the equation

$$w_n' - (p-1)|w_n|^q + (p-1)(m/2)^p = 0 \tag{5.3.34}$$

on $[t_n, (2^p - 1)t_n] \subset [t^*, \infty)$ with $w_n(t_n) = w(t_n) - 2(p-1)(m/2)^p t_n$. Let $s_n = [w_n(t_n) - (m/2)^{p/q}]^{1-q}$ and $v_n(t) = (m/2)^{p/q} + (t_n - t + s_n)^{1-p}$ on $[t_n, t_n + s_n) \subset [t^*, \infty)$, where n is large enough such that $w_n(t_n) > (m/2)^{p/q}$. Then $v_n(t_n) = w_n(t_n)$. From the inequality $(a+b)^\gamma > a^\gamma + b^\gamma$, which holds for every $\gamma > 1$, $a > 0$, $b > 0$, we get

$$\begin{aligned} v_n'(t) &= (p-1)(t_n - t + s_n)^{-p} \\ &= (p-1)[(t_n - t + s_n)^{-p} + (m/2)^p] - (p-1)(m/2)^p \\ &< (p-1)[(m/2)^{p/q} + (t_n - t + s_n)^{1-p}]^q - (p-1)(m/2)^p \\ &= (p-1)|v_n(t)|^q - (p-1)(m/2)^p \end{aligned}$$

on $[t_n, t_n + s_n) \subset [t^*, \infty)$. Thus

$$v_n'(t) - (p-1)|v_n(t)|^q + (p-1)(m/2)^p < 0 = w_n'(t) - (p-1)|w_n(t)|^q + (p-1)(m/2)^p$$

for all $t \in [t_n, (2^p - 1)t_n] \cap [t_n, t_n + s_n) \subset [t^*, \infty)$. A simple comparison argument shows that $v_n(t) \leq w_n(t)$ on this interval. It follows from

$$w_n(t_n) = w(t_n) - 2(p-1)(m/2)^p t_n > (p-1)(1 - 2^{1-p}) m^p t_n$$

that $t_n + s_n \in [t_n, (2^p - 1)t_n]$ for n large enough. By the definition of v_n we see that $\lim_{t \to (t_n+s_n)-} v_n(t) = \infty$ for n large enough. Hence

$$\lim_{t \to (t_n+s_n)-} w_n(t) = \infty \quad \text{for } n \text{ large enough.} \tag{5.3.35}$$

Now we take k large enough such that $t_k + s_k \in [t_k, (2^p - 1)t_k]$. Clearly, there exists a positive constant M such that

$$-\frac{\Phi(y'(t_n))}{\Phi(y(t_n))} \le M < \infty$$

on $[t_k, (2^p - 1)t_k] \subset [t^*, \infty)$. It follows from (5.3.33) and (5.3.35) that

$$\infty = \lim_{t \to (t_k+s_k)-} w_n(t) \le \lim_{t \to (t_k+s_k)-} \left\{ -\frac{\Phi(y'(t_n))}{\Phi(y(t_n))} \right\} \le M < \infty,$$

which is a contradiction. □

Theorem 5.3.12. *Suppose that c is a Besicovitch almost periodic function with the mean value $M\{c\} = 0$ and $M\{|c|\} > 0$. Then (5.3.3) is oscillatory at ∞ and $-\infty$ for every $\lambda \neq 0$.*

Proof. Without loss of generality we only show that (5.3.3) is oscillatory ∞. Suppose, by a contradiction, that (5.3.3) is nonoscillatory (at ∞) for some λ with a solution $y(t) > 0$ for large t, say $t \ge t_0$. Then $w = -\Phi(y'/y)$ is the corresponding solution of the associated Riccati equation

$$w' = \lambda c(t) + (p-1)|w(t)|^q. \tag{5.3.36}$$

Integrating (5.3.36) (with any fixed $\delta > 0$ and $t \ge t_0$) we get

$$\frac{\lambda}{\delta}\int_t^{t+\delta} c(s)\,ds = \frac{w(t+\delta)}{\delta} - \frac{w(t)}{\delta} - \frac{p-1}{\delta}\int_t^{t+\delta} |w(s)|^q. \tag{5.3.37}$$

Applying the Besicovitch seminorm $\|\cdot\|_{B'}$ defined by

$$\|f\|_{B'} = \limsup_{t \to \infty} \frac{1}{t}\int_{t_0}^t |f(s)|\,ds$$

(this is essentially restriction of the seminorm $\|\cdot\|_B$ to the interval $[t_0, \infty)$) to (5.3.37), we find

$$0 \le \left\| \frac{\lambda}{\delta}\int_t^{t+\delta} c(s)\,ds \right\|_{B'} \le \left\| \frac{p-1}{\delta}\int_t^{t+\delta} |w(s)|^q\,ds \right\|_{B'} + \left\| \frac{w(t+\delta)}{\delta} \right\|_{B'} + \left\| \frac{w(t)}{\delta} \right\|_{B'}$$

for all $\delta > 0$. From Lemma 5.3.4 it follows that $M\{|w|^q\} = 0$, thus $\|w\|_{B'} =$

$\|w(t+\delta)\|_{B'} = 0$ for all $\delta > 0$. Using the Fubini theorem we have for some $t_0 > 0$

$$\begin{aligned}
\frac{1}{\delta t}\int_{t_0}^{t}\int_{s}^{s+\delta} |w(\tau)|^q \, d\tau \, ds &= \frac{1}{\delta t}\int_{t_0}^{t}\int_{0}^{\delta} |w(\tau+s)|^q \, d\tau \, ds \\
&= \frac{1}{\delta t}\int_{0}^{\delta}\int_{t_0}^{t} |w(\tau+s)|^q \, ds \, d\tau \\
&\le \frac{1}{\delta t}\int_{0}^{\delta}\int_{t_0}^{t+\delta} |w(s)|^q \, ds \, d\tau \\
&= \frac{1}{t}\int_{t_0}^{t+\delta} |w(s)|^q \, ds
\end{aligned}$$

for any fixed $\delta > 0$. Using the last computation and Lemma 5.3.4 we have

$$\left\| \frac{p-1}{\delta}\int_{t}^{t+\delta} |w(s)|^q \, ds \right\|_{B'} = 0.$$

Applying the last equality coupled with the fact that $\|w(t)\|_{B'} = 0$ to the previous computation, we see that

$$\left\| \frac{\lambda}{\delta}\int_{t}^{t+\delta} c(s) \, ds \right\|_{B'} = 0 \tag{5.3.38}$$

for every $\delta > 0$. Since c is almost periodic, it follows from [35, p. 97] that

$$\lim_{\delta \to 0+} \left\| c(t) - \frac{1}{\delta}\int_{t}^{t+\delta} c(s) \, ds \right\|_{B'} = 0.$$

This and (5.3.38) imply $M\{|c|\} = \|c\|_{B'} = 0$ which is a contradiction. □

The following statement is an immediate consequence of the Stepanoff uniqueness theorem, see e.g. [289], and the last theorem.

Corollary 5.3.6. *Let c be a Stepanoff almost periodic function, which is not almost everywhere zero, with $M\{c\} = 0$. Then (5.3.3) is oscillatory at ∞ and $-\infty$ for every $\lambda \neq 0$.*

Since every Bohr almost periodic function is Stepanoff almost periodic, this corollary includes an extension of the above mentioned result from [172].

5.4 Strongly and conditionally oscillatory equation

Let us now discuss the problem of strong (non)oscillation of the equation

$$(r(t)\Phi(y'))' + \lambda c(t)\Phi(y) = 0, \tag{5.4.1}$$

where r, c are subject to the usual conditions with c *positive* and λ is a positive parameter. This concept was introduced by Nehari in [304] for linear differential equation and its extension is obvious. Also, it is natural to expect that the situation for strong (non)oscillation can be completely characterized. Differential equation (5.4.1) is said to be *conditionally oscillatory* if there exists a constant $\lambda_0 > 0$ such that (5.4.1) is oscillatory for $\lambda > \lambda_0$ and nonoscillatory if $\lambda < \lambda_0$. The value λ_0 is called the *oscillation constant* of (5.4.1). Since this constant depends on the coefficients of the equation, sometimes we speak about *oscillation constant of the function c (with respect to r)*. If equation is oscillatory (resp. nonoscillatory) for every $\lambda > 0$, then equation is said to be *strongly oscillatory* (resp. *strongly nonoscillatory*).

5.4.1 Strong (non)oscillation criteria

We start with the assumption $\int^\infty r^{1-q}(t)\,dt = \infty$. A complementary case will be discussed later. Then we need only to consider the case where the function $c(t)$ is integrable on $[t_0, \infty)$. Indeed, if $\int^\infty c(t)\,dt = \infty$, then (1.1.1) is oscillatory by Leighton-Wintner type criterion, and hence (5.4.1) is strongly oscillatory.

Theorem 5.4.1. *Suppose that $\int^\infty c(t)\,dt < \infty$ and $\int^\infty r^{1-q}(t)\,dt = \infty$. Equation* (5.4.1) *is strongly oscillatory if and only if*

$$\limsup_{t\to\infty} \left(\int^t r^{1-q}(s)\,ds\right)^{p-1} \int_t^\infty c(s)\,ds = \infty, \tag{5.4.2}$$

and it is strongly nonoscillatory if and only if

$$\lim_{t\to\infty} \left(\int^t r^{1-q}(s)\,ds\right)^{p-1} \int_t^\infty c(s)\,ds = 0. \tag{5.4.3}$$

Proof. Suppose that (5.4.2) holds. Then

$$\limsup_{t\to\infty} \left(\int^\infty r^{1-q}(s)\,ds\right)^{p-1} \int_t^\infty \lambda c(s)\,ds > 1$$

for every $\lambda > 0$, and so (5.4.1) oscillatory for every $\lambda > 0$ by Theorem 3.1.2. This implies strong oscillation of (5.4.1). Conversely, suppose that (1.1.1) is strongly oscillatory and (5.4.2) fails to hold. Then, by Theorem 3.1.3 we get, for every $\lambda > 0$,

$$\limsup_{t\to\infty} \left(\int^t r^{1-q}(s)\,ds\right)^{p-1} \int_t^\infty \lambda c(s)\,ds \ge \frac{1}{p}\left(\frac{p-1}{p}\right)^{p-1}. \tag{5.4.4}$$

This implies (5.4.2), otherwise (5.4.4) would be violated for sufficiently small λ.

Now suppose that (5.4.3) hold. Then

$$\lim_{t\to\infty} \left(\int^t r^{1-q}(s)\,ds\right)^{p-1} \int_t^\infty \lambda c(s)\,ds = 0$$

for every $\lambda > 0$, and from Theorem 3.1.3, (5.4.1) is nonoscillatory for every $\lambda > 0$, which implies strong nonoscillation of (5.4.1). Conversely, let (5.4.1) be strongly nonoscillatory. From Theorem 3.1.2, we have

$$\limsup_{t\to\infty} \left(\int^t r^{1-q}(s)\,ds\right)^{p-1} \int_t^\infty \lambda c(s)\,ds \le 1$$

for every $\lambda > 0$. The arbitrariness of λ then implies

$$\limsup_{t\to\infty} \left(\int^t r^{1-q}(s)\,ds\right)^{p-1} \int_t^\infty c(s)\,ds = 0,$$

which is equivalent to (5.4.3). □

Example 5.4.1. The examples illustrating these concepts we have already seen in the previous sections. For instance, if $\int^\infty r^{1-q}(t)\,dt = \infty$, the equation

$$(5.4.5)\qquad (r(t)\Phi(x'))' + \frac{\lambda r^{1-q}(t)}{\left(\int^t r^{1-q}(s)\,ds\right)^\mu}\Phi(x) = 0$$

is conditionally oscillatory if $\mu = p$, strongly oscillatory if $\mu < p$ and strongly nonoscillatory if $\mu > p$. This follows from the fact that the transformation of independent variable $t \mapsto \int^t r^{1-q}(s)\,ds$ transforms (5.4.5) into the equation

$$(5.4.6)\qquad (\Phi(x'))' + \frac{\lambda}{t^\mu}\Phi(x) = 0,$$

and (5.4.6) is compared with the Euler equation (1.4.20). It is then easy to see that if $\mu = p$, then $\lambda = [(p-1)/p]^p$ is the oscillation constant of t^{-p}.

Now we will discuss the complementary case to the previous one, i.e., the case when $\int^\infty r^{1-q}(t)\,dt < \infty$. We will present the theorem without proof, since the idea is the same to that of the proof of Theorem 5.4.1, with the only difference that instead of Theorem 3.1.2 and Theorem 3.1.3 we use Theorem 3.1.6. Also, similarly to the previous case, we need only to consider the case where c satisfies

$$(5.4.7)\qquad \int^\infty \left(\int_t^\infty r^{1-q}(s)\,ds\right)^p c(t)\,dt < \infty,$$

since otherwise (1.1.1) is oscillatory by Theorem 2.2.11, so that (5.4.1) is strongly oscillatory.

Theorem 5.4.2. *Suppose that* (5.4.7) *holds and* $\int^\infty r^{1-q}(t)\,dt < \infty$. *Equation* (5.4.1) *is strongly oscillatory if and only if*

$$\limsup_{t\to\infty} \left(\int_t^\infty r^{1-q}(s)\,ds\right)^{-1} \int_t^\infty \left(\int_s^\infty r^{1-q}(\tau)\,d\tau\right)^p c(s)\,ds = \infty,$$

and it is strongly nonoscillatory if and only if

$$\lim_{t\to\infty} \left(\int_t^\infty r^{1-q}(s)\,ds\right)^{-1} \int_t^\infty \left(\int_s^\infty r^{1-q}(\tau)\,d\tau\right)^p c(s)\,ds = 0.$$

Remark 5.4.1. In view of the criteria in Section 3.1, it is not difficult to see that using similar ideas as in the above two theorems, we can obtain corresponding statements (under appropriate assumptions) involving the functions $\bar{\mathcal{A}}(t), \bar{\mathcal{B}}(t), \mathcal{C}(t)$ and $\bar{\mathcal{C}}(t)$, defined at the beginning of Section 3.1. The same holds for the results of the next subsection.

5.4.2 Oscillation constant

The following theorems provide information about the oscillation constants of conditionally oscillatory equations of the form (5.4.1). From Theorem 5.4.1 we conclude that under the assumption $\int^\infty r^{1-q}(t)\, dt = \infty$ equation (5.4.1) is conditionally oscillatory if and only if either

$$0 < \lim_{t\to\infty} \left(\int^t r^{1-q}(s)\, ds\right)^{p-1} \int_t^\infty c(s)\, ds < \infty$$

or

$$0 \le \liminf_{t\to\infty} \left(\int^t r^{1-q}(s)\, ds\right)^{p-1} \int_t^\infty c(s)\, ds < \limsup_{t\to\infty} \left(\int^t r^{1-q}(s)\, ds\right)^{p-1} \int_t^\infty c(s)\, ds < \infty.$$

We introduce the notation

$$M_* = \liminf_{t\to\infty} \left(\int^t r^{1-q}(s)\, ds\right)^{p-1} \int_t^\infty c(s)\, ds,$$

$$M^* = \limsup_{t\to\infty} \left(\int^t r^{1-q}(s)\, ds\right)^{p-1} \int_t^\infty c(s)\, ds.$$

Theorem 5.4.3. *Suppose that $\int^\infty c(t)\, dt < \infty$ and $\int^\infty r^{1-q}(t)\, dt = \infty$. Let $0 < M_* \le M^* < \infty$. Then the oscillation constant λ_0 of (5.4.1) satisfies*

$$\frac{1}{pM^*}\left(\frac{p-1}{p}\right)^{p-1} \le \lambda_0 \le \min\left\{\frac{1}{M^*}, \frac{1}{pM_*}\left(\frac{p-1}{p}\right)^{p-1}\right\}.$$

In particular, if $M_* = M^*$, then

$$\lambda_0 = \frac{1}{pM^*}\left(\frac{p-1}{p}\right)^{p-1} = \frac{1}{pM_*}\left(\frac{p-1}{p}\right)^{p-1}.$$

Proof. Let $\lambda \in (0, \lambda_0)$. Then (5.4.1) is nonoscillatory, and so by Theorem 3.1.1 and Theorem 3.1.2, we have

$$\liminf_{t\to\infty} \left(\int^t r^{1-q}(s)\, ds\right)^{p-1} \int_t^\infty \lambda c(s)\, ds = \lambda M_* \le \frac{1}{p}\left(\frac{p-1}{p}\right)^{p-1}$$

and

$$\limsup_{t\to\infty}\left(\int^t r^{1-q}(s)\,ds\right)^{p-1}\int_t^\infty \lambda c(s)\,ds = \lambda M^* \le 1$$

whence, letting $\lambda \to \lambda_0-$, we obtain

$$\lambda_0 \le \frac{1}{pM_*}\left(\frac{p-1}{p}\right)^{p-1} \quad \text{and} \quad \lambda_0 \le \frac{1}{M^*}.$$

Let $\lambda \in (\lambda_0, \infty)$. Since (5.4.1) is oscillatory, from Theorem 3.1.3 we see that

$$\limsup_{t\to\infty}\left(\int^t r^{1-q}(s)\,ds\right)^{p-1}\int_t^\infty \lambda c(s)\,ds = \lambda M^* \ge \frac{1}{p}\left(\frac{p-1}{p}\right)^{p-1},$$

which, as $\lambda \to \lambda_0+$, yields

$$\lambda_0 \ge \frac{1}{pM^*}\left(\frac{p-1}{p}\right)^{p-1}.$$

□

Now we will discuss the complementary case to the previous one, i.e., the case when $\int^\infty r^{1-q}(t)\,dt < \infty$ (and (5.4.7) holds). Introduce the notation

$$N_* = \liminf_{t\to\infty}\left(\int_t^\infty r^{1-q}(s)\,ds\right)^{-1}\int_t^\infty\left(\int_s^\infty r^{1-q}(\tau)\,d\tau\right)^p c(s)\,ds,$$

$$N^* = \limsup_{t\to\infty}\left(\int_t^\infty r^{1-q}(s)\,ds\right)^{-1}\int_t^\infty\left(\int_s^\infty r^{1-q}(\tau)\,d\tau\right)^p c(s)\,ds.$$

From Theorem 5.4.2, it is clear that (5.4.1) is conditionally oscillatory (provided (5.4.7) holds) if and only if either

$$0 < \lim_{t\to\infty}\left(\int_t^\infty r^{1-q}(s)\,ds\right)^{-1}\int_t^\infty\left(\int_s^\infty r^{1-q}(\tau)\,d\tau\right)^p c(s)\,ds < \infty$$

or $0 \le N_* < N^* < \infty$.

Theorem 5.4.4. *Suppose that (5.4.7) and $\int^\infty r^{1-q}(t)\,dt < \infty$. Let $0 < N_* \le N^* < \infty$. Then the oscillation constant λ_0 of (5.4.1) satisfies*

$$\frac{1}{N^*}\left(\frac{p-1}{p}\right)^p \le \lambda_0 \le \min\left\{\frac{1}{N^*}, \frac{1}{N_*}\left(\frac{p-1}{p}\right)^p\right\}.$$

If, in particular, $N_* = N^*$, then

$$\lambda_0 = \frac{1}{N^*}\left(\frac{p-1}{p}\right)^p = \frac{1}{N_*}\left(\frac{p-1}{p}\right)^p.$$

We conclude this section with comparison type results involving oscillation constants.

Theorem 5.4.5. *Let $c(t)$ and $d(t)$ be two nonnegative (and eventually nontrivial) integrable functions, and let λ_d, $0 < \lambda_d < \infty$, be the oscillation constant of d (with respect to r). Assume that $\int^\infty r^{1-q}(s)\,ds = \infty$ and the limit*

$$L_d = \lim_{t\to\infty} \left(\int^t r^{1-q}(s)\,ds \right)^{p-1} \int_t^\infty d(s)\,ds$$

exists. If

$$\Psi := \liminf_{t\to\infty} \frac{\int_t^\infty c(s)\,ds}{\int_t^\infty d(s)\,ds} > \lambda_d,$$

then (1.1.1) *is oscillatory.*

Proof. The equation $(r(t)\Phi(y'))' + \lambda d(t)\Phi(y) = 0$ is oscillatory if $\lambda > \lambda_d$. Thus, by Theorem 3.1.3, we have $L_d \geq \gamma_p/\lambda$, where $\gamma_p = \frac{1}{p}\left(\frac{p-1}{p}\right)^{p-1}$, so that for any $0 < \varepsilon < \gamma_p/\lambda_d$ there exists T such that

$$\left(\int^t r^{1-q}(s)\,ds \right)^{p-1} \int_t^\infty d(s)\,ds > \frac{1}{\lambda}\gamma_p - \varepsilon$$

for $t \geq T$. As $\lambda \to \lambda_d$, we have

$$\left(\int^t r^{1-q}(s)\,ds \right)^{p-1} \int_t^\infty d(s)\,ds > \frac{1}{\lambda_d}\gamma_p - \varepsilon$$

for $t \geq T$, so that

$$\begin{aligned} \Psi &\leq \liminf_{t\to\infty} \frac{\left(\int^t r^{1-q}(s)\,ds\right)^{p-1} \int_t^\infty c(s)\,ds}{\gamma_p/\lambda_d - \varepsilon} \\ &= \frac{1}{\gamma_p/\lambda_d - \varepsilon} \liminf_{t\to\infty} \left(\int^t r^{1-q}(s)\,ds \right)^{p-1} \int_t^\infty c(s)\,ds. \end{aligned}$$

Since this implies

$$\liminf_{t\to\infty} \left(\int^t r^{1-q}(s)\,ds \right)^{p-1} \int_t^\infty c(s)\,ds > \gamma_p,$$

equation (1.1.1) is oscillatory by Theorem 3.1.1. □

Similarly we can prove the following theorem.

Theorem 5.4.6. *Let $c(t)$ and $d(t)$ be two nonnegative (and eventually nontrivial) functions, and let λ_d, $0 < \lambda_d < \infty$, be the oscillation constant of d (with respect to r). Assume that $\int^\infty r^{1-q}(s)\,ds < \infty$. Further, suppose that* (5.4.7) *and the same condition, with d instead of c, hold. Let the limit*

$$\lim_{t\to\infty} \left(\int_t^\infty r^{1-q}(s)\,ds \right)^{-1} \int_t^\infty \left(\int_s^\infty r^{1-q}(\tau)\,d\tau \right)^p d(s)\,ds$$

exist. If

$$\liminf_{t\to\infty} \frac{\int_t^\infty \left(\int_s^\infty r^{1-q}(\tau)\,d\tau\right)^p c(s)\,ds}{\int_t^\infty \left(\int_s^\infty r^{1-q}(\tau)\,d\tau\right)^p d(s)\,ds} > \lambda_d,$$

then (1.1.1) *is oscillatory.*

5.5 Function sequence technique

In this section, we will see how nonoscillation of (1.1.1) can be expressed in terms of suitable function sequences, which will be proved to be strictly related to solvability of the Riccati integral equation (inequality). Several new criteria will be given and some of the already presented ones will be proved alternatively. The Hille-Wintner comparison theorem and its extension will be mentioned in this framework as well.

Let us consider equation (1.1.1), where r, c are continuous on $[a, \infty)$ with $r(t) > 0$. Throughout this section we assume that

$$\int_a^\infty r^{1-q}(s)\,ds = \infty \tag{5.5.1}$$

and

$$\int_t^\infty c(s)\,ds \quad \text{exists and is nonnegative} \tag{5.5.2}$$

for large t, say $t \in [a, \infty)$, without loss of generality.

5.5.1 Function sequences and Riccati integral equation

First we give a characterization of nonoscillation in terms of the sequence $\{\varphi_k(t)\}$ defined by

$$\varphi_0(t) = \int_t^\infty c(s)\,ds, \quad \varphi_k(t) = \int_t^\infty S(\varphi_{k-1}, r)(s)\,ds + \varphi_0(t), \quad k = 1, 2, \dots, \tag{5.5.3}$$

where the function S is defined by (2.2.2). By induction, it is not difficult to see that

$$\varphi_{k+1}(t) \geq \varphi_k(t), \quad k = 0, 1, 2, \dots, \tag{5.5.4}$$

i.e., this function sequence is nondecreasing on $[a, \infty)$.

Theorem 5.5.1. *Equation* (1.1.1) *is nonoscillatory if and only if there exists* $t_0 \in [a, \infty)$ *such that*

$$\lim_{k\to\infty} \varphi_k(t) = \varphi(t) \tag{5.5.5}$$

for $t \geq t_0$, *i.e., the sequence* $\{\varphi_k(t)\}$ *is well defined and pointwise convergent.*

Proof. Let (1.1.1) be nonoscillatory. Then there is w satisfying (2.2.17) for large t, say $t \geq t_0 \geq a$, by Theorem 2.2.4. Hence $w(t) \geq \varphi_0(t)$, and so

$$w(t) = \varphi_0(t) + \int_t^\infty S(w, r)(s)\, ds \geq \varphi_0(t) + \int_t^\infty S(\varphi_0, r)(s)\, ds = \varphi_1(t).$$

By induction, $w(t) \geq \varphi_k(t) \geq 0$, $k = 0, 1, 2, \ldots$, $t \geq t_0$. Recall that $\{\varphi_k(t)\}$ is nondecreasing. Now we can easily see that this sequence is bounded above on $[t_0, \infty)$, and so (5.5.5) holds.

Conversely, if (5.5.5) holds, then it follows from (5.5.4) and (5.5.5) that $\varphi_k(t) \leq \varphi(t)$, $k = 0, 1, 2, \ldots$, $t \in [t_0, \infty)$. Applying the Lebesgue Monotone Convergence Theorem to the second equation in (5.5.3) we have

$$\varphi(t) = \varphi_0(t) + \int_t^\infty S(\varphi, r)(s)\, ds.$$

Now, (1.1.1) is nonoscillatory by Theorem 2.2.4. □

Corollary 5.5.1. *Equation* (1.1.1) *is oscillatory if and only if either*

(i) there is a positive integer m such that $\varphi_k(t)$ is defined for $k = 1, 2, \ldots, m-1$, but $\varphi_m(t)$ does not exists, i.e., $\int_t^\infty S(\varphi_{m-1}, r)(s)\, ds = \infty$,

or

(ii) $\varphi_k(t)$ is defined for $k = 1, 2, \ldots$, but for arbitrarily large $t_0 \geq a$, there is $t^* \geq t_0$ such that $\lim_{k\to\infty} \varphi_k(t^*) = \infty$.

The (approximating) sequence $\{\varphi_k(t)\}$ defined by (5.5.3) is not the only one that is available. Let us define the sequence $\{\psi_k(t)\}$ by

$$\psi_0(t) = \int_t^\infty c(s)\, ds, \quad \psi_1(t) = \int_t^\infty S(\psi_0, r)(s)\, ds, \quad \psi_{k+1}(t) = \int_t^\infty S(\psi_k + \psi_0, r)(s)\, ds,$$

$k = 1, 2, \ldots$. Now we proceed similarly as above. Indeed, nonoscillation of (1.1.1) implies

$$\psi_0(t) \leq \psi_0(t) + \int_t^\infty S(w, r)(s)\, ds = w(t).$$

Hence

$$\psi_0(t) + \psi_1(t) = \psi_0(t) + \int_t^\infty S(\psi_0, r)(s)\, ds \leq w(t).$$

Similarly,

$$\psi_0(t) + \psi_k(t) \leq w(t), \quad k = 1, 2, \ldots, \tag{5.5.6}$$

which implies $0 < \psi_0(t) \leq \psi_1(t) \leq \cdots \leq w(t)$. The converse is obvious. Hence we have the following statement. Note that later we show another slightly modified approaches.

Theorem 5.5.2. *Equation* (1.1.1) *is nonoscillatory if and only if there exists* $t_0 \in [a, \infty)$ *such that*

$$\lim_{k\to\infty} \psi_k(t) = \psi(t) \tag{5.5.7}$$

for $t \geq t_0$, *i.e., the sequence* $\{\psi_k(t)\}$ *is well defined and pointwise convergent.*

Also in this case we have a corollary, which is literatim the same as Corollary 5.5.1, except of φ is replaced by ψ.

Remark 5.5.1. Observe that in the proof of Theorem 5.5.1 we use the fact that nonoscillation of (1.1.1) implies solvability of the *Riccati integral equation*

$$w(t) = \int_t^\infty c(s)\,ds + \int_t^\infty S(w,r)(s)\,ds,$$

and then we make estimations which are based on this equation. A closer examination shows that if this equation is replaced by the *Riccati integral inequality*

$$w(t) \geq \int_t^\infty c(s)\,ds + \int_t^\infty S(w,r)(s)\,ds,$$

then everything works as well. Recall that we have already shown that solvability of this inequality is sufficient for nonoscillation of (1.1.1), see Theorem 2.2.5. What we show here now is that this fact can be proved alternatively, by means of the sequence approach (without using Theorem 2.2.1): simply combine the proof of Theorem 5.5.1 with the observation at the beginning of this remark. Altogether, we see that under conditions (5.5.1) and (5.5.2), in view of our observations, the following statements are equivalent:

(i) Equation (1.1.1) is nonoscillatory.

(ii) The equation $w(t) = \int_t^\infty c(s)\,ds + \int_t^\infty S(w,r)(s)\,ds$ is solvable in a neighborhood of ∞.

(iii) The inequality $w(t) \geq \int_t^\infty c(s)\,ds + \int_t^\infty S(w,r)(s)\,ds$ is solvable in a neighborhood of ∞.

(iv) The sequence $\{\varphi_k(t)\}$ satisfies (5.5.5).

(v) The sequence $\{\psi_k(t)\}$ satisfies (5.5.7).

Recall that throughout this section we assume (5.5.1) and (5.5.2).

Theorem 5.5.3. *Let* $c(t) \geq 0$ *for large* t. *If*

$$\limsup_{t\to\infty} \left(\int_a^t r^{1-q}(s)\,ds\right)^{p-1} \varphi_k(t) > 1 \tag{5.5.8}$$

for some $k \in \mathbb{N} \cup \{0\}$, *then* (1.1.1) *is oscillatory.*

Proof. If the conclusion is not true, then as in the proof of Theorem 5.5.1 we have $\varphi_k(t) \le w(t)$, $k = 0, 1, 2, \dots$, $t \ge t_0$ and hence, by Lemma 2.2.6, $\varphi_k(t) \le \left(\int_{t_0}^t r^{1-q}(s)\,ds\right)^{1-p}$, $k = 0, 1, 2, \dots$, $t \ge t_0$. But then

$$\limsup_{t\to\infty} \left(\int_{t_0}^t r^{1-q}(s)\,ds\right)^{p-1} \varphi_k(t) \le 1,$$

which contradicts (5.5.8). □

Remark 5.5.2. In particular, the condition

$$\limsup_{t\to\infty} \left(\int_a^t r^{1-q}(s)\,ds\right)^{p-1} \int_t^\infty c(s)\,ds > 1$$

guarantees oscillation of (1.1.1), which is in fact the statement of (Hille-Nehari type) Theorem 3.1.2.

The "ψ-variant" of the last theorem is the following criterion.

Theorem 5.5.4. *Let* $c(t) \ge 0$ *for large* t. *If*

$$\limsup_{t\to\infty} \left(\int_a^t r^{1-q}(s)\,ds\right)^{p-1} \left[\int_t^\infty c(s)\,ds + \psi_k(t)\right] > 1 \tag{5.5.9}$$

for some $k \in \mathbb{N}$, *then* (1.1.1) *is oscillatory.*

Proof. If not, then we have inequality (5.5.6). From Lemma 2.2.6 we get

$$\int_t^\infty c(s)\,ds + \psi_k(t) \le \left(\int_{t_0}^t r^{1-q}(s)\,ds\right)^{1-p}, \quad k = 1, 2, \dots,$$

which contradicts (5.5.9). □

Remark 5.5.3. In particular, the condition

$$\limsup_{t\to\infty} \left(\int_a^t r^{1-q}(s)\,ds\right)^{p-1} \left[\int_t^\infty c(s)\,ds + (p-1)\int_t^\infty r^{1-q}(s)\left(\int_s^\infty c(\tau)\,d\tau\right)^q ds\right] > 1$$

guarantees oscillation of (1.1.1). Its consequence is again Theorem 3.1.2.

Next we give an alternative proof of the already presented Hille-Nehari type criterion (Theorem 3.1.1).

Theorem 5.5.5. *If*

$$\left(\int_a^t r^{1-q}(s)\,ds\right)^{p-1} \int_t^\infty c(s)\,ds \ge \gamma_0 \tag{5.5.10}$$

for large t, where

$$\gamma_0 > \frac{1}{p}\left(\frac{p-1}{p}\right)^{p-1}, \tag{5.5.11}$$

then (1.1.1) *is oscillatory.*

Proof. Condition (5.5.10) can be rewritten as $\varphi_0(t) \ge \gamma_0 \left(\int_a^t r^{1-q}(s)\,ds\right)^{1-p}$. Then

$$\begin{aligned}
\varphi_1(t) &= \int_t^\infty S(\varphi_0, r)(s)\,ds + \varphi_0(t) \\
&\ge (p-1)\gamma_0^q \int_t^\infty r^{1-q}(s)\left(\int_a^s r^{1-q}(\tau)\,d\tau\right)^{-p} ds + \gamma_0\left(\int_a^t r^{1-q}(s)\,ds\right)^{1-p} \\
&= \gamma_0^q\left(\int_a^t r^{1-q}(s)\,ds\right)^{1-p} + \gamma_0\left(\int_a^t r^{1-q}(s)\,ds\right)^{1-p} \\
&= \gamma_1\left(\int_a^t r^{1-q}(s)\,ds\right)^{1-p},
\end{aligned}$$

where $\gamma_1 = \gamma_0^q + \gamma_0$. By induction,

$$\varphi_k(t) \ge \gamma_k\left(\int_a^t r^{1-q}(s)\,ds\right)^{1-p},$$

$k = 1, 2, \ldots,$ where

$$\gamma_k = \gamma_{k-1}^q + \gamma_0. \tag{5.5.12}$$

Clearly, $\gamma_k < \gamma_{k+1}$, $k = 0, 1, 2, \ldots$. Now we claim that $\lim_{k\to\infty}\gamma_k = \infty$. If not, let $\lim_{k\to\infty}\gamma_k = L < \infty$. Then from (5.5.12) we have $L = L^q + \gamma_0$. It is not difficult to see that, in view of (5.5.11), the latter algebraic equation has no real solution. Hence we must have $\gamma_k \to \infty$ as $k \to \infty$, which implies $\varphi_k(t) \to \infty$ as $k \to \infty$, and so (1.1.1) is oscillatory by Corollary 5.5.1. □

Remark 5.5.4. Condition (5.5.10) can be rewritten as

$$\liminf_{t\to\infty}\left(\int_a^t r^{1-q}(s)\,ds\right)^{p-1}\int_t^\infty c(s)\,ds > \frac{1}{p}\left(\frac{p-1}{p}\right)^{p-1},$$

which is exactly (3.1.5).

The following Willet type criterion is based on the idea similar to the last one, but with the use of the sequence $\{\psi_k(t)\}$. See Remark 3.1.5 for an alternative approach. Another approach can be found in Theorem 5.5.11.

Theorem 5.5.6. *If*

$$\liminf_{t\to\infty}\left(\int_t^\infty c(s)\,ds\right)^{-1}\int_t^\infty r^{1-q}(s)\left(\int_s^\infty c(\tau)\,d\tau\right)^q ds > p^{-q}, \tag{5.5.13}$$

then (1.1.1) *is oscillatory.*

Proof. Condition (5.5.13) can be rewritten as

$$\int_t^\infty r^{1-q}(s)\left(\int_s^\infty c(\tau)\,d\tau\right)^q ds \ge \frac{\delta_0}{p-1}\int_t^\infty c(s)\,ds$$

for large t, where $\delta_0 > (p-1)p^{-q}$. This is equivalent to $\psi_1(t) = \int_t^\infty S(\psi_0, r)(s)\,ds \ge \delta_0\psi_0(t)$. Hence, in view of the definition of $\{\psi_k(t)\}$ we have

$$\begin{aligned}\psi_2(t) &= \int_t^\infty S(\psi_1+\psi_0, r)(s)\,ds \ge \int_t^\infty S(\psi_0(1+\delta_0), r)(s)\,ds \\ &= (1+\delta_0)^q\int_t^\infty S(\psi_0, r)(s)\,ds \ge \delta_1\psi_0,\end{aligned}$$

where $\delta_1 = (1+\delta_0)^q\delta_0$. By induction, $\psi_{k+1}(t) \ge \delta_k\psi_0(t)$, where $\delta_k = (1+\delta_{k-1})^q\delta_0$, $k = 1, 2, \dots$. Clearly, the sequence $\{\delta_k\}$ is increasing. We claim that it is unbounded. Otherwise, $\delta_k \to M < \infty$ as $k \to \infty$ would imply $M = (1+M)^q\delta_0$. However, as it is not difficult to show, this algebraic equation cannot have a real solution if $\delta_0 > (p-1)p^{-q}$. Hence $\delta_k \to \infty$ as $k \to \infty$, implying that $\{\psi_k(t)\}$ is not convergent, and so the assumptions of Corollary 5.5.1 are satisfied. □

The sequence approach can be also used to show nonoscillatory counterparts of the above oscillation criteria (see also Section 3.1 where a different technique is used). First we give a counterpart to Theorem 5.5.6 (i.e., Willet type criterion, see also Theorem 5.5.11 and the text right before it). Observe that this approach enables to prove the nonoscillatory criteria where the inequality in the conditions like (5.5.14) or (5.5.16) does not need to be strict, in the sense that the lim sup of the expressions depending on t (being of the left-hand side) can be equal to the critical constant (compare with the results of Section 3.1). Recall that we assume (5.5.1) and (5.5.2).

Theorem 5.5.7. *If*

$$\int_t^\infty r^{1-q}(s)\left(\int_s^\infty c(\tau)\,d\tau\right)^q ds \le \frac{\delta_0}{p-1}\int_t^\infty c(s)\,ds \tag{5.5.14}$$

for large t, where $\delta_0 \le (p-1)p^{-q}$, then (1.1.1) *is nonoscillatory.*

Proof. Consider the sequence $\{\psi_k(t)\}$. Recall that $\psi_0(t) = \int_t^\infty c(s)\,ds$. We have

$$\psi_1(t) = (p-1)\int_t^\infty r^{1-q}(s)\psi_0^q(s)\,ds \le \delta_0\psi_0(t)$$

by (5.5.14). Further,

$$\begin{aligned}\psi_2(t) &= (p-1)\int_t^\infty r^{1-q}(s)(\psi_1(s)+\psi_0(s))^q ds \\ &\le (p-1)(1+\delta_0)^q\int_t^\infty r^{1-q}(s)\psi_0^q(s)\,ds \le \delta_1\psi_0(t),\end{aligned}$$

where $\delta_1 = (1+\delta_0)^q\delta_0$. Inductively, we see that

$$\psi_{k+1}(t) \le \delta_k\psi_0(t), \ \text{where } \delta_k = (1+\delta_{k-1})^q\delta_0, \ k = 1, 2, \dots. \tag{5.5.15}$$

Clearly, $\{\delta_k\}$ is nondecreasing. We claim that it converges. Indeed, consider the fixed point problem $x = g(x)$, where $g(x) = \delta_0(1+x)^q$. In fact, it is more convenient to consider $g(x) = \delta_0|1+x|^q$, which is not a restriction in view of the form of the function S. Moreover, we are particularly interested in the first quadrant. We find fixed points by means of the iteration scheme $x_k = \delta_0|1+x_{k-1}|^q$, $k = 1, 2, \dots$. Note that when $\delta_0 = (p-1)p^{-q}$, the graph of g is a parabola like curve which has a unique minimum at $x = -1$ and touches the line $y = x$ at $(x, y) = (p-1, p-1)$. Therefore, if we choose $x_0 = \delta_0$, then we see that the approximating sequence $\{x_k\}$ is strictly increasing and converges to $x = p-1$. If $\delta_0 < (p-1)p^{-q}$, then clearly $\delta_k < x_k < p-1$ for all k. This shows that $\{\delta_k\}$ is bounded and hence converges. Thus $\{\psi_k\}$ converges by (5.5.15), and so equation (1.1.1) is nonoscillatory in view of Theorem 5.5.2. □

The counterpart to Theorem 5.5.5 is already stated (Hille-Nehari type) Theorem 2.3.2, but here (5.5.2) holds, and so we have:

Theorem 5.5.8. *If*

$$\left(\int_a^t r^{1-q}(s)\,ds\right)^{p-1} \int_t^\infty c(s)\,ds \le \frac{1}{p}\left(\frac{p-1}{p}\right)^{p-1} \tag{5.5.16}$$

for large t, then (1.1.1) *is nonoscillatory.*

Concerning the sequence approach (utilizing $\{\varphi_k(t)\}$), the idea of the proof is similar to that of the previous theorem and so it is omitted. Note only that the assumptions of the criterion imply the inequality

$$\varphi_k(t) \le \gamma_k\left(\int_a^t r^{1-q}(s)\,ds\right)^{1-p},$$

where $\gamma_k = \gamma_{k-1}^q + \gamma_0$, $k = 1, 2, \dots$, $\{\gamma_k\}$ being a nondecreasing and bounded sequence, provided γ_0 is assumed to be less than or equal to $\frac{1}{p}\left(\frac{p-1}{p}\right)^{p-1}$. Nevertheless, in the next subsection we show a modified approach.

We conclude this subsection by an alternative proof of Hille-Wintner type theorem (see Subsection 2.3.1). Along with (1.1.1) consider the equation

$$(\tilde r(t)\Phi(x'))' + \tilde c(t)\Phi(x) = 0, \tag{5.5.17}$$

where $\tilde r, \tilde c$ are subject to the same conditions as imposed on r, c, respectively. Let the sequence $\{\tilde\varphi_k(t)\}$ be defined by means of $\tilde r, \tilde c$ like $\{\varphi_k(t)\}$ is defined by means of r, c.

Theorem 5.5.9. *Assume that*

$$\tilde r(t) \le r(t), \quad \int_t^\infty c(s)\,ds \le \int_t^\infty \tilde c(s)\,ds \quad \text{for large } t. \tag{5.5.18}$$

If (5.5.17) *is nonoscillatory, then* (1.1.1) *is nonoscillatory.*

Proof. Suppose that (5.5.17) is nonoscillatory. It follows from Theorem 5.5.1 that there is $t_0 \geq a$ such that $\lim_{t\to\infty} \tilde{\varphi}_k(t) = \tilde{\varphi}(t) < \infty$, $t \geq t_0$. By (5.5.18), $0 \leq \varphi_0(t) \leq \tilde{\varphi}_0(t)$. This implies

$$\varphi_1(t) = \int_t^\infty S(\varphi_0, r)(s)\, ds + \varphi_0(t) \leq \int_t^\infty S(\tilde{\varphi}_0, \tilde{r})(s)\, ds + \tilde{\varphi}_0(t) = \tilde{\varphi}_1(t).$$

By induction, $0 \leq \varphi_k(t) \leq \tilde{\varphi}_k(t)$, $k = 0, 1, 2, \dots$, $t \geq t_0$. Therefore, $\varphi_k(t) \leq \tilde{\varphi}(t)$, $k = 0, 1, 2, \dots$, $t \geq t_0$. This and (5.5.4) imply $\lim_{k\to\infty} \varphi_k(t) = \varphi(t)$, $t \geq t_0$. By Theorem 5.5.1, (1.1.1) is nonoscillatory. □

Remark 5.5.5. (i) It is easy to see that the approach based on Theorem 5.5.2 works in the last proof as well. Also, the sequence $\{\vartheta_k(t)\}$, defined in the next subsection, can be utilized to prove the Hille-Wintner theorem, even under the conditions of Theorem 2.3.1, see [180].

(ii) It is not difficult to see that higher order iterated comparison theorems can be established by using the nonoscillatory characterizations via function sequences. For instance, instead of the second condition in (5.5.18), we may assume the following one: There exists $m \in \mathbb{N} \cup \{0\}$ such that $\varphi_k(t) - \tilde{\varphi}_k(t)$ changes sign on $[T, \infty)$ for arbitrarily large T and $k = 0, 1, \dots, m-1$, but there is t_0 such that $\varphi_m(t) \leq \tilde{\varphi}_m(t)$, $t \geq t_0$.

5.5.2 Modified approaches

In this subsection we introduce another function sequences, which can be used in our theory as well. We start with the alternative proof of Theorem 5.5.8.

Proof. (Proof of Theorem 5.5.8) Suppose that (5.5.16) holds for $t \geq t_0$. Introduce the integral operator

$$(\mathcal{H}v)(t) = (p-1)\left(\int_a^t r^{1-q}(s)\, ds\right)^{p-1} \int_t^\infty r^{1-q}(s) \left(\int_a^t r^{1-q}(\tau)\, d\tau\right)^{-p} |v(s)|^q ds$$

and define the sequence of functions $\{\vartheta_k(t)\}$, $t \geq t_0$, by

$$\begin{aligned}\vartheta_0(t) &= \left(\int_a^t r^{1-q}(s)\, ds\right)^{p-1} \int_t^\infty c(s)\, ds,\\ \vartheta_k(t) &= \left(\int_a^t r^{1-q}(s)\, ds\right)^{p-1} \int_t^\infty c(s)\, ds + (\mathcal{H}\vartheta_{k-1})(t), \quad k = 1, 2, \dots.\end{aligned} \tag{5.5.19}$$

From (5.5.16) we have

$$0 < \vartheta_0(t) \leq \frac{1}{p}\left(\frac{p-1}{p}\right)^{p-1} < \left(\frac{p-1}{p}\right)^{p-1}$$

and

$$0 < \vartheta_0(t) < \vartheta_1(t) < \frac{1}{p}\left(\frac{p-1}{p}\right)^{p-1} + \left\{\mathcal{H}\left(\left(\frac{p-1}{p}\right)^{p-1}\right)\right\}(t) = \left(\frac{p-1}{p}\right)^{p-1},$$

$t \geq t_0$. By induction,

$$0 < \vartheta_{k-1}(t) < \vartheta_k(t) < \left(\frac{p-1}{p}\right)^{p-1},$$

$n \in \mathbb{N}$, $t \geq t_0$. Let $\vartheta(t)$ denote the limit $\lim_{k\to\infty} \vartheta(t)$, $t \geq t_0$. Applying the Lebesgue Monotone Convergence Theorem to (5.5.19), we see that $\vartheta(t)$ satisfies the integral equation

$$\begin{aligned} u(t) = & \left(\int_a^t r^{1-q}(s)\, ds\right)^{p-1} \int_t^\infty c(s)\, ds \\ & + (p-1)\left(\int_a^t r^{1-q}(s)\, ds\right)^{p-1} \int_t^\infty r^{1-q}(s) \left(\int_a^s r^{1-q}(\tau)\, d\tau\right)^{-p} |u(s)|^q ds \end{aligned}$$

for $t \geq t_0$. It then follows that the function $w(t) = \vartheta(t)\left(\int_a^t r^{1-q}(s)\, ds\right)^{1-p}$ is a solution of (2.2.17), so that Theorem 2.2.4 implies that (1.1.1) is nonoscillatory. □

Next we describe another approach, where a slightly modified sequence appears, which enables to state new criteria. Let (1.1.1) be nonoscillatory. Then there is w such that

$$w(t) = \varphi_0(t) + \int_t^\infty S(w, r)(s)\, ds$$

for large t, say $t \geq t_0$, by Theorem 2.2.4, where $\varphi_0(t) = \int_t^\infty c(s)\, ds$. Let us define the sequence $\{\omega_k(t)\}$ as follows:

$$\omega_0(t) = \varphi_0(t), \quad \omega_{k+1}(t) = \varphi_0(t) + (p-1)\int_t^\infty r^{1-q}(s)\varphi_0^{q-1}(s)\omega_k(s)\, ds,$$

$k = 0, 1, 2, \ldots$, $t \geq t_0$. Clearly, $\varphi_0(t) = \omega_0(t) \leq w(t)$, and hence

$$\omega_0(t) \leq \omega_1(t) \leq \varphi_0(t) + (p-1)\int_t^\infty r^{1-q}(s)\varphi_0^{q-1}(s) w(s)\, ds \leq w(t).$$

Similarly, by induction, it is easy to see that $\{\omega_k(t)\}$ is a nondecreasing sequence such that

$$\omega_k(t) \leq w(t), \quad k = 0, 1, 2, \ldots, \tag{5.5.20}$$

$t \geq t_0$. From

$$\int_t^\infty r^{1-q}(s)\varphi_0^{q-1}(s)\omega_k(s)\, ds \leq \int_t^\infty r^{1-q}(s)\varphi_0^{q-1}(s) w(s)\, ds < \infty$$

we see that the family $\{\omega_0, \omega_1, \omega_2, \ldots\}$ is equicontinuous, so (5.5.20) says that there is a subsequence $\{\omega_{k_n}(t)\}$ with a locally uniform limit on $[t_0, \infty)$ (i.e., the convergence is uniform in each compact subinterval). In view of the monotonicity,

$\{\omega_k(t)\}$ has a locally uniform limit, say $\omega^*(t)$, on $[t_0, \infty)$. Obviously, $\omega^*(t) \le w(t)$, so that

$$\int_t^\infty r^{1-q}(s)\varphi_0^{q-1}(s)\omega^*(s)\,ds < \infty.$$

Now the Lebesgue Dominated Convergence Theorem yields

$$\int_t^\infty r^{1-q}(s)\varphi_0^{q-1}(s)\omega^*(s)\,ds = \lim_{k\to\infty} \int_t^\infty r^{1-q}(s)\varphi_0^{q-1}(s)\omega_k(s)\,ds$$

for $t \ge t_0$. It therefore makes clear now that

$$\omega^*(t) = \varphi_0(t) + (p-1)\int_t^\infty r^{1-q}(s)\varphi_0^{q-1}(s)\omega^*(s)\,ds, \tag{5.5.21}$$

$t \ge t_0$.

The above observation serves to prove the following necessary condition for nonoscillation of (1.1.1).

Theorem 5.5.10. *If* (1.1.1) *is nonoscillatory, then*

$$\int_a^\infty c(s)\exp\left\{(p-1)\int_a^s r^{1-q}(\tau)\left(\int_\tau^\infty c(\xi)\,d\xi\right)^{q-1} d\tau\right\} ds < \infty \tag{5.5.22}$$

and

$$\begin{aligned}\int_a^\infty r^{1-q}(s)\left(\int_s^\infty c(\xi)\,d\xi\right)^{q} \exp\Bigg\{(p-1)\int_a^s r^{1-q}(\tau) \\ \times\left(\int_\tau^\infty c(\xi)\,d\xi\right)^{q-1} d\tau\Bigg\}\,ds < \infty.\end{aligned} \tag{5.5.23}$$

Proof. In view of the above discussion, there is a number $t_0 \ge a$ and a function ω^* satisfying (5.5.21). This implies that

$$(\omega^*)'(t) = -c(t) - (p-1)r^{1-q}(t)\varphi_0^{q-1}(t)\omega^*(t). \tag{5.5.24}$$

Equation (5.5.24) is the first order linear equation, and from the formula for its solution we get

$$\begin{aligned}\omega^*(t_0) - \int_{t_0}^t c(s)\exp\left\{(p-1)\int_{t_0}^s r^{1-q}(\tau)\varphi_0^{q-1}(\tau)\,d\tau\right\} ds \\ = \omega^*(t)\exp\left\{(p-1)\int_{t_0}^s r^{1-q}(\tau)\varphi_0^{q-1}(\tau)\,d\tau\right\} > 0.\end{aligned}$$

Hence

$$\omega^*(t_0) > \int_{t_0}^t c(s)\exp\left\{(p-1)\int_{t_0}^s r^{1-q}(\tau)\varphi_0^{q-1}(\tau)\,d\tau\right\} ds.$$

This implies (5.5.22). Let z be a function given on $[t_0, \infty)$ by

$$z(t) = \int_t^\infty r^{1-q}(s)\varphi_0^{q-1}(s)\omega^*(s)\, ds.$$

Then, in view of (5.5.21),

$$z'(t) = -r^{1-q}(t)\varphi_0^{q-1}(t)\omega^*(t) = -r^{1-q}(t)\varphi_0^q(t) - (p-1)r^{1-q}(t)\varphi_0^{q-1}(t)z(t),$$

in view of (5.5.21), which implies (5.5.23), using a similar computation as for ω^*. □

Corollary 5.5.2. *If the integral in* (5.5.22) *or in* (5.5.23) *is divergent, then* (1.1.1) *is oscillatory.*

We conclude this section by showing another possibility, how to define a suitable sequence, which then finds an application. The difference, when comparing it with the above approach, is that we do not work directly with the functions, which are defined by means of (known) coefficients, but with suitable general functions. Nevertheless, the conclusion then leads to (un)solvability of Riccati type integral equation (inequality). This approach can be directly used to show the Willet type criteria (already proved in the previous subsection, see Theorem 5.5.6 and Theorem 5.5.7). The reason for this terminology is the original linear version, see the paper of Willet [362], where the function sequence technique was used as well.

Theorem 5.5.11. *Suppose that $B(s)$ and $Q(t,s)$ are nonnegative continuous functions on $[T, \infty)$ and $[T, \infty) \times [T, \infty)$, respectively.*

(i) If

$$\int_t^\infty Q(t,s)B^q(s)\, ds \le p^{-q}B(t), \quad t \ge T, \tag{5.5.25}$$

then the equation

$$v(t) = B(t) + (p-1)\int_t^\infty Q(t,s)|v(s)|^q\, ds \tag{5.5.26}$$

has a solution on $[T, \infty)$.

(ii) If there exists $\varepsilon > 0$ such that

$$\int_t^\infty Q(s,t)B^q(s)\, ds \ge p^{-q}(1+\varepsilon)B(t) \not\equiv 0, \quad t \ge T, \tag{5.5.27}$$

then the inequality

$$v(t) \ge B(t) + (p-1)\int_t^\infty Q(t,s)|v(s)|^q\, ds \tag{5.5.28}$$

possesses no solution on $[T, \infty)$.

Proof. (i) Let $v_1(t) = B(t)$ and define

$$v_{k+1}(t) = B(t) + (p-1)\int_t^\infty Q(t,s)|v_k(s)|^q\, ds, \quad k \in \mathbb{N}.$$

Then by (5.5.25)

$$v_2(t) = B(t) + (p-1)\int_t^\infty Q(t,s)B^q(s)\, ds \le B(t) + (p-1)p^{-q}B(t) \le pB(t),$$

and $v_1(t) \le v_2(t)$. Suppose, by induction, that $v_1(t) \le v_2(t) \le \cdots \le v_n(t) \le pB(t)$. Then

$$\begin{aligned} v_{n+1}(t) &\le B(t) + (p-1)\int_t^\infty Q(t,s)|v_n(s)|^q ds \\ &\le B(t) + (p-1)p^q\int_t^\infty Q(t,s)B^q(s)\, ds \\ &\le B(t) + (p-1)p^q \cdot p^{-q}B(t) = pB(t). \end{aligned}$$

Thus, the sequence $\{v_n\}$ is nondecreasing and bounded above. Hence, it converges uniformly to a continuous function v, which is a solution of (5.5.26).

(ii) Suppose, to the contrary, that v is a continuous function satisfying (5.5.28). Then $v(t) \ge B(t) \ge 0$, which implies $v^q(t) \ge B^q(t) \ge 0$. Thus

$$v(t) \ge B(t) + (p-1)\int_t^\infty Q(t,s)B^q(s)\, ds \ge [1 + (p-1)(1+\varepsilon)p^{-q}]B(t).$$

Continuing in this way, we obtain $v(t) \ge a_n B(t)$, where $a_1 = 1$, $a_n < a_{n+1}$ and

(5.5.29) $$a_{n+1} = 1 + (p-1)a_n^q p^{-q}(1+\varepsilon).$$

We claim that $\lim_{n\to\infty} a_n = \infty$. Assume, to the contrary, that $\lim_{n\to\infty} a_n = a < \infty$. Then $a \ge 1$ and from (5.5.29)

$$a = 1 + (p-1)(1+\varepsilon)a^q p^{-q},$$

but this is the contradiction since the equation $\lambda = 1 + (p-1)(1+\varepsilon)\lambda^q p^{-q}$ has no solution for which $\lambda \ge 0$. This contradiction proves that $\lim_{n\to\infty} a_n = \infty$ and hence $B(t) \equiv 0$. This contradiction with (5.5.27) proves the lemma. □

Finally note that a certain variant of the function sequence technique has been used also in the part devoted to the existence of slowly varying solutions, see Subsection 4.3.2.

5.6 Distance between zeros of oscillatory solutions

In the first part of this section we present an asymptotic formula for number of zeros. The second, respectively third subsection deal with conditions guaranteeing the existence of quickly, respectively slowly oscillating solutions.

5.6.1 Asymptotic formula for distribution of zeros

Here we present an asymptotic formula for number of zeros of oscillatory solutions of the equation

$$(\Phi(x'))' + (p-1)c(t)\Phi(x) = 0. \tag{5.6.1}$$

It is supposed that $c(t) > 0$ for large t and the results are based on the generalized Prüfer transformation from Section 1.2. In this transformation, a nontrivial solution and its derivative are expressed via the generalized half-linear sine and cosine functions. Recall that the half-linear sine function, denoted by $\sin_p t$, is the solution of the equation

$$(\Phi(x'))' + (p-1)\Phi(x) = 0$$

satisfying the initial condition $x(0) = 0$, $x'(0) = 1$ and the half-linear cosine function is defined by $\cos_p t = \sin_p' t$. Generalized π, denoted by π_p, is introduced in Subsection 1.1.2.

Theorem 5.6.1. *Suppose that c is a differentiable function such that $c(t) > 0$ on an interval $[T, \infty)$, and*

$$\lim_{t\to\infty} c'(t)[c(t)]^{-\frac{p+1}{p}} = 0 \tag{5.6.2}$$

holds. Then (5.6.1) is oscillatory. Moreover, if $N[x;T]$ denotes the number of zeros of a solution x of (5.6.1) in the interval $[a, T]$, then

$$N[x;T] = P[x;T] + R[x;T], \tag{5.6.3}$$

where $P[x;T]$ is the principal term given by

$$P[x;T] = \frac{1}{\pi_p}\int_a^T [c(s)]^{\frac{1}{p}}\, ds$$

and $R[x;T]$ is the remainder which is of smaller order than $P[x;T]$ as $T \to \infty$ and satisfies

$$|R[x;T]| \le \frac{1}{p\pi_p}\int_a^T \frac{|c'(s)|}{c(s)}\, ds + O(1).$$

Proof. Set $C(t) := c'(t)[c(t)]^{-\frac{p+1}{p}}$ and define

$$C^*(t) = \sup\{|C(s)| : s \ge t\}, \ t \ge a. \tag{5.6.4}$$

Then $C^*(t)$ is nonincreasing and satisfies $\lim_{t\to\infty} C^*(t) = 0$ by (5.6.2). We have

$$\left|[c(t+h)]^{-\frac{1}{p}} - [c(t)]^{-\frac{1}{p}}\right| = \frac{1}{p}\left|\int_t^{t+h} C(s)\, ds\right| \le \frac{|h|}{p} C^*(t),$$

which implies that

$$\limsup_{h\to\infty} \frac{[c(t+h)]^{-\frac{1}{p}}}{t+h} \le \frac{C^*(t)}{p}.$$

It follows that $\lim_{t\to\infty} t^{-1}[c(t)]^{-\frac{1}{p}} = 0$, or equivalently, $\lim_{t\to\infty} t^p c(t) = \infty$. This implies, by Theorem 1.4.5, that (5.6.1) is oscillatory.

Now we turn our attention to the proof of the asymptotic formulas for numbers of zeros. By the Sturmian comparison theorem (Theorem 1.2.4) we have that $N[x_1;T]$ and $N[x_2;T]$ differ at most by one for any solutions x_1 and x_2 of (5.6.1), so we may restrict our attention to the solution x_0 of (5.6.1) determined by the initial conditions $x_0(a) = 0$, $x_0'(a) = 1$. This solution is oscillatory by the first part of the our theorem.

We introduce the polar coordinates $\rho(t)$, $\varphi(t)$ for $x_0(t)$ by setting

$$[c(t)]^{\frac{1}{p}} x_0(t) = \rho(t) \sin_p \varphi(t), \quad x_0'(t) = \rho(t)\cos_p\varphi(t). \tag{5.6.5}$$

It can be shown without difficulty that $\rho(t)$ and $\varphi(t)$ are continuously differentiable on $[a,\infty)$ and satisfy the differential equations

$$\begin{aligned} \frac{\rho'}{\rho} &= \frac{c'(t)}{pc(t)} |\sin_p \varphi|^p, \\ \varphi' &= [c(t)]^{\frac{1}{p}} + \frac{c'(t)}{pc(t)} \sin_p \varphi \Phi(\cos_p \varphi). \end{aligned} \tag{5.6.6}$$

We use the notation

$$g(\varphi) = \sin_p \varphi \Phi(\cos_p \varphi),$$

in terms of which (5.6.6) is written as

$$\varphi' = [c(t)]^{\frac{1}{p}} + \frac{c'(t)}{pc(t)} g(\varphi). \tag{5.6.7}$$

From the first equation in (5.6.5) we see that $x_0(t) = 0$ if and only if $\varphi(t) = j\pi_p$, $j \in \mathbb{Z}$. We may suppose that $\varphi(a) = 0$. In view of (5.6.2) there is no loss of generality in assuming that

$$C^*(t) < p \quad \text{for} \quad t \geq a,$$

where $C^*(t)$ is defined by (5.6.4). Since

$$|g(\varphi)| \leq 1 \quad \text{for all } \varphi, \tag{5.6.8}$$

we have

$$[c(t)]^{\frac{1}{p}} + \frac{c'(t)}{pc(t)} g(\varphi(t)) \geq [c(t)]^{\frac{1}{p}} \left(1 - \frac{1}{p} C^*(t)\right) > 0,$$

which implies that $\varphi'(t) > 0$, so that $\varphi(t)$ is increasing for $t \geq a$.

We now integrate (5.6.7) over $[a, T]$, obtaining

$$\varphi(T) = \int_a^T [c(s)]^{\frac{1}{p}} ds + \frac{1}{p} \int_a^T \frac{c'(s)}{c(s)} g(\varphi(s)) ds = F(T) + G(T), \tag{5.6.9}$$

where

$$F(T) := \int_a^T [c(s)]^{\frac{1}{p}} ds, \quad G(T) := \frac{1}{p} \int_a^T \frac{c'(s)}{c(s)} g(\varphi(s)) ds$$

From (5.6.8) it is clear that

$$(5.6.10) \qquad |G(T)| \le \frac{1}{p}\int_a^T \frac{|c'(s)|}{c(s)}ds.$$

Noting that the number of zeros of $x_0(t)$ in $[a,T]$ is given by

$$N[x_0;T] = \left[\frac{\varphi(T)}{\pi_p}\right] + 1,$$

where $[u]$ denotes the greatest integer not exceeding u, we see from (5.6.9) and (5.6.10) that the conclusion of the theorem holds with the choice

$$P[x_0;T] = \frac{1}{\pi_p}F(T) = \frac{1}{\pi_p}\int_a^T [c(s)]^{\frac{1}{p}}ds.$$

That the term $R[x_0;T] = N[x_0;T] - P[x_0;T]$ is of smaller order than $P[x_0;T]$ follows from the observation that

$$\begin{aligned}\int_a^T \frac{|c'(s)|}{c(s)}ds &= \int_a^T |C(s)|[c(s)]^{\frac{1}{p}}ds \\ &\le \int_a^T C^*(s)[c(s)]^{\frac{1}{p}}ds = o\left(\int_a^T [c(s)]^{\frac{1}{p}}ds\right) \quad \text{as } T \to \infty.\end{aligned}$$

This completes the proof. □

Example 5.6.1. Consider the equation

$$(5.6.11) \qquad (\Phi(x'))' + (p-1)t^\beta\Phi(x) = 0, \ t \ge 1,$$

where β is a constant with $p + \beta > 0$. The function $c(t) = t^\beta$ satisfies

$$\int_1^T [c(s)]^{\frac{1}{p}}ds = \frac{p}{p+\beta}\left(T^{\frac{p+\beta}{p}} - 1\right),$$

$$\int_1^T \frac{|c'(s)|}{c(s)}ds = |\beta|\log T,$$

and so we conclude from Theorem 5.6.1 that the quantity $P[x;T]$ can be taken to be

$$P[x;T] = \frac{p}{(p+\beta)\pi_p}T^{\frac{p+\beta}{p}}$$

and (5.6.3) holds with this $P[x;T]$ and $R[x;T]$ satisfying

$$R[x;T] = \frac{|\beta|}{p\pi_p}\log T + O(1).$$

Remark 5.6.1. (i) The results of this subsection cannot be applied to the generalized Euler equation 1.4.20, since the function $c(t) = \lambda(p-1)t^{-p}$ does not satisfy (5.6.2). A calculation of $P[x;T]$ and $R[x;T]$ for the generalized Euler equation

$$(\Phi(x'))' + \lambda(p-1)t^{-p}\Phi(x) = 0$$

shows that both of them are of the same logarithmic order as $T \to \infty$.

(ii) In [317], M. Piros has investigated a similar problem under a more stringent restriction on $c(t)$, namely he supposed that $c^\nu(t)$ is a concave function of t for some $\nu > 0$. Then he proved that the error term $R[x;T]$ in (5.6.3) is $O(1)$. Exactly, the differential equation (5.6.11) with $\beta = 1/\nu$ plays the exceptional role in determining the precise value of $R[x;T]$.

(iii) It is not difficult to see how the results can be extended to the case of a general r in (1.1.1), which satisfies $\int^\infty r^{1-q}(s)\,ds = \infty$, by using the transformation of independent variable, see Subsection 1.2.7.

5.6.2 Quickly oscillating solution

We start with the definition of quick oscillation.

Definition 5.6.1. A function $h : \mathbb{R} \to \mathbb{R}$ is said to be *quickly oscillating* if it is defined in a neighborhood of ∞ and if there exists a sequence $\{t_n\}_{n=1}^\infty$ with $\lim_{n\to\infty} t_n = \infty$ such that $h(t_n) = 0$, $n = 1, 2, 3, \ldots$, $t_{n+1} > t_n$ and $\lim_{n\to\infty}(t_{n+1} - t_n) = 0$.

The following theorem is based on the Lyapunov inequality (Theorem 5.1.1).

Theorem 5.6.2. *If equation* (5.1.1) *has a quickly oscillating solution, then*

$$\int^\infty c_+(t)\,ds = \infty, \tag{5.6.12}$$

where $c_+(t) = \max\{0, c(t)\}$, *and*

$$\limsup_{t\to\infty} c(t) = \infty.$$

Proof. Let y be a quickly oscillating solution of (5.1.1) with zeros t_n such that $t_n \to \infty$ and $t_{n+1} - t_n \to 0$ as $n \to \infty$. Consider the consecutive zeros $t_{n+1} > t_n > 0$ and the interval $[t_n, t_{n+1}]$. It follows from Theorem 5.1.1 that

$$\int_0^\infty c_+(t)\,dt > \int_{t_n}^{t_{n+1}} c_+(t)\,dt > 2^p(t_{n+1} - t_n)^{1-p} \to \infty \quad \text{as } t \to \infty.$$

Hence (5.6.12) holds. Applying the Mean Value Theorem for integrals, we have

$$c_+(\xi_n)(t_{n+1} - t_n) = \int_{t_n}^{t_{n+1}} c_+(t)\,dt > 2^p(t_{n+1} - t_n)^{1-p},$$

where $t_n < \xi_n < t_{n+1}$. This implies $c_+(\xi_n) > 2^p(t_{n+1} - t_n)^{-p}$, and it follows that $\limsup_{t\to\infty} c(t) = \infty$. □

Remark 5.6.2. (i) It is not difficult to see how the statement can be extended to equation (1.1.1).

(ii) In Subsection 9.2.2 we show how quick oscillation of all solutions of (1.1.1) and some additional conditions guarantee oscillation of a more general forced non-linear equation.

5.6.3 Slowly oscillating solution

Also in the results of this subsection, a crucial role is played by the Lyapunov inequality. We start with the definition.

Definition 5.6.2. A function $h : \mathbb{R} \to \mathbb{R}$ is said to be *slowly oscillating* if it is defined in a neighborhood of ∞ and if there exists a sequence $\{t_n\}_{n=1}^{\infty}$ with $\lim_{n\to\infty} t_n = \infty$ such that $h(t_n) = 0$, $n = 1, 2, 3, \ldots$, $t_{n+1} > t_n$ and $\lim_{n\to\infty}(t_{n+1} - t_n) = \infty$.

Theorem 5.6.3. *Suppose that*

$$\limsup_{t\to\infty} \delta^{p-1} \int_t^{t+\delta} c_+(s)\, ds < 2^p \tag{5.6.13}$$

for all $\delta > 0$, *where* $c_+(t) = \max\{0, c(t)\}$. *If* (5.1.1) *is oscillatory, then any its nontrivial solution* y *is slowly oscillating.*

Proof. Suppose, by contradiction, that (5.1.1) has a solution y with its sequence of zeros $\{t_n\}_{n=1}^{\infty}$, which has a subsequence $\{t_{n_k}\}_{k=1}^{\infty}$ such that $0 < t_{n_k+1} - t_{n_k} \le \delta < \infty$ for some δ and for all k. Then Theorem 5.1.1 implies that

$$\int_{t_{n_k}}^{t_{n_k+1}} c_+(s)\, ds > \frac{2^p}{(t_{n_k+1} - t_{n_k})^{p-1}} \ge \frac{2^p}{\delta^{p-1}} > 0$$

for any k. Thus we have that

$$\delta^{p-1} \int_{t_{n_k}}^{t_{n_k}+\delta} c_+(s)\, ds \ge \delta^{p-1} \int_{t_{n_k}}^{t_{n_k+1}} c_+(s)\, ds > 2^p$$

for all k. Therefore,

$$\limsup_{t\to\infty} \delta^{p-1} \int_t^{t+\delta} c_+(s)\, ds \ge 2^p,$$

which contradicts (5.6.13). □

We will need the following a lemma, which is proved in [369].

Lemma 5.6.1. *Suppose that a nonnegative function* $f(t)$ *is locally integrable on* $[a, \infty)$. *If there is a* $\delta_0 > 0$ *such that*

$$\lim_{t\to\infty} \int_t^{t+\delta_0} f(s)\, ds = 0,$$

then for any $\delta > 0$,

$$\lim_{t\to\infty} \int_t^{t+\delta} f(s)\, ds = 0.$$

Theorem 5.6.4. *Suppose that* $0 < \lambda < \infty$, *and for a constant* $\delta_0 > 0$,

$$\lim_{t\to\infty} \int_t^{t+\delta_0} c_+^\lambda(s)\, ds = 0. \tag{5.6.14}$$

If (5.1.1) *is oscillatory, then any its nontrivial solution* y *is slowly oscillating.*

Proof. Suppose to the contrary that (5.1.1) has a solution y with its sequence of zeros $\{t_n\}_{n=1}^\infty$, which has a subsequence $\{t_{n_k}\}_{k=1}^\infty$ such that $0 < t_{n_k+1} - t_{n_k} \le \delta < \infty$ for some δ and for all k. Now, we distinguish two cases:
Case 1. $[\lambda > 1]$ By Theorem 5.1.1 and the Hölder inequality,

$$\begin{aligned}
2^p &< (t_{n_k+1} - t_{n_k})^{p-1} \int_{t_{n_k}}^{t_{n_k+1}} c_+(s)\, ds \\
&< (t_{n_k+1} - t_{n_k})^{p-1+(\lambda-1)/\lambda} \left(\int_{t_{n_k}}^{t_{n_k+1}} c_+^\lambda(s)\, ds \right)^{1/\lambda} \\
&\le \delta^{p-1+(\lambda-1)/\lambda} \left(\int_{t_{n_k}}^{t_{n_k+1}} c_+^\lambda(s)\, ds \right)^{1/\lambda},
\end{aligned}$$

which contradicts (5.6.14).
Case 2. $[0 < \lambda \le 1]$ Let

$$\begin{aligned}
E_1(t) &= \{s : c_+(s) \le 1, s \in (t, t+\delta_0)\}, \\
E_2(t) &= \{s : c_+(s) > 1, s \in (t, t+\delta_0)\}.
\end{aligned}$$

Then

$$\int_t^{t+\delta_0} c_+(s)\, ds = \int_{E_1(t)} c_+(s)\, ds + \int_{E_2(t)} c_+(s)\, ds.$$

Then (5.6.14) implies that $\lim_{t\to\infty} \operatorname{mes} E_2(t) = 0$. Using this fact, we see that for any $\varepsilon > 0$, there exists T_1 such that for all $t > T_1$,

$$\int_{E_2(t)} c_+(s)\, ds < \frac{\varepsilon}{2}.$$

On the other hand, it follows from (5.6.14) that there exists T_2 such that

$$\int_{E_1(t)} c_+(s)\, ds \le \int_t^{t+\delta_0} c_+^\lambda(s)\, ds < \frac{\varepsilon}{2}$$

for all $t > T_2$. Thus we obtain that

$$0 \le \int_t^{t+\delta_0} c_+(s)\, ds \le \varepsilon$$

for every $t \ge \max\{T_1, T_2\}$. Then, by Lemma 5.6.1, we get

$$\lim_{t\to\infty} \delta^{p-1} \int_t^{t+\delta} c_+(s)\, ds = 0$$

for any $\delta > 0$. Using Theorem 5.6.3, we obtain the statement. □

Remark 5.6.3. If $\int^{\infty} r^{1-q}(s)\,ds = \infty$, then the above two theorems can be easily extended to (1.1.1) using the transformation of the independent independent variable (Section 1.2.7). Condition (5.6.13) then remains the same, while (5.6.14) is replaced by

$$\lim_{t\to\infty} \int_t^{t+\delta_0} \frac{[r^{q-1}(s)c_+(s)]^{\lambda}}{r^{q-1}(s)}\,ds = 0.$$

5.7 Half-linear Sturm-Liouville problem

In this section we show that the solutions of the Sturm-Liouville problem for half-linear equation (1.1.1) have similar properties as in the linear case. Of course, we cannot consider the problem of orthogonality of eigenfunctions since this concept has no meaning in L^p, $p \neq 2$.

5.7.1 Basic Sturm-Liouville problem

We start with the problem

$$(\Phi(x'))' + \lambda c(t)\Phi(x) = 0, \quad x(a) = 0 = x(b), \tag{5.7.1}$$

under the assumption that $c(t) \geq 0$ and $c(t) \not\equiv 0$. The value λ is called the *eigenvalue* if there exists a nontrivial solution x of (5.7.1). The solution x is said to the *eigenfunction* corresponding to the eigenvalue λ. Clearly, according to the assumption $c(t) \geq 0$ and the Sturm comparison theorem, only values $\lambda > 0$ can be eigenvalues.

Theorem 5.7.1. *The eigenvalue problem* (5.7.1) *has infinitely many eigenvalues* $0 < \lambda_1 < \lambda_2 < \cdots < \lambda_n < \dots$, $\lambda_n \to \infty$ *as* $n \to \infty$. *The* n*-th eigenfunction has exactly* $n-1$ *zeros in* (a,b). *Moreover, if the function* c *is supposed to be positive in the whole interval* (a,b), *the eigenvalues satisfy the asymptotic relation*

$$\lim_{n\to\infty} \frac{\sqrt[p]{\lambda_n}}{n} = \frac{\pi_p}{\int_a^b \sqrt[p]{c(t)}\,dt}. \tag{5.7.2}$$

Proof. The proof of the first part of the theorem is a special case of the problem treated in the next subsection, so we present only its main idea. Let $x(t;\lambda)$ be the solution of (5.7.1) given by the initial condition $x(a;\lambda) = 0$, $x'(a;\lambda) = 1$ and let $\varphi(t;\lambda)$ be the continuous function given at all points where $x'(t;\lambda) \neq 0$ by the formula

$$\varphi(t;\lambda) = \operatorname{arctan}_p \frac{x(t;\lambda)}{x'(t;\lambda)},$$

i.e., $\varphi(t;\lambda)$ is the angular variable in the half-linear Prüfer transformation for $x(t;\lambda)$. This means that $\varphi(t;\lambda)$ satisfies the differential equation

$$\varphi' = |\cos_p \varphi|^p + \frac{\lambda c(t)}{p-1} |\sin_p \varphi|^p, \quad \varphi(a;\lambda) = 0. \tag{5.7.3}$$

The proof is based on the fact that $\varphi(b;\lambda)$ is a continuous function of λ and $\varphi(b;\lambda)\to\infty$ as $\lambda\to\infty$. The continuity property follows from the general theory of continuous dependence of solutions of first order differential equations on the right-hand side. The limit property of $\varphi(b;\lambda)$ is proved via the comparison of (5.7.1) with the "minorant" problem with a constant coefficient

$$(\Phi(y'))'+\lambda\bar{c}\Phi(y)=0,\quad y(a')=0=y(b'), \tag{5.7.4}$$

where $[a',b']\subset[a,b]$ is such that $c(t)>0$ on $[a',b']$ and $\bar{c}=\min_{t\in[a',b']}c(t)>0$. The eigenvalues and eigenfunctions of (5.7.4) can be computed explicitly and if $\theta(t;\lambda)$ is defined for this problem in the same way as $\varphi(t;\lambda)$ for (5.7.1), we have

$$\varphi(b;\lambda)\geq\theta(b;\lambda)\to\infty,\quad \text{as } \lambda\to\infty.$$

Now, the eigenvalues are those $\lambda=\lambda_n$ for which $\varphi(b;\lambda_n)=n\pi_p$ and taking into account that $\varphi(t;\lambda)$ is increasing in t (this follows from (5.7.3)), zero points of the associated eigenfunction $x_n(t)=x(t;\lambda_n)$ are those t_k, $k=1,\dots,n-1$, for which $\varphi(t_k,\lambda_n)=k\pi_p$, $k=1,\dots,n-1$.

Concerning the proof of the asymptotic formula (5.7.2), first suppose that the function $c(t)\equiv c_1>0$ is a constant function and consider the problem

$$(\Phi(z'))'+\lambda c_1\Phi(z)=0,\quad z(a)=0=z(b). \tag{5.7.5}$$

A nontrivial solution of the half-linear equation in this problem satisfying $z(a)=0$ is $z=\sin_p(\sqrt[p]{\lambda c_1}(t-a))$ and hence the k-th eigenvalue is given by $\sqrt[p]{\lambda_k c_1}(b-a)=k\pi_p$, thus

$$\frac{\sqrt[p]{\lambda_k}}{k}=\frac{\pi_p}{\int_a^b\sqrt[p]{c_1}\,dt},$$

i.e., the asymptotic formula (5.7.2) is automatically satisfied in this case.

Now let us consider the original Sturm-Liouville problem (5.7.1) and its k-th eigenfunction $x(t;\lambda_k)$. This function has zeros at $t_0=a<t_1<t_2<\cdots<t_{k-1}<t_k=b$. Put $\lambda=\lambda_k$ in (5.7.1) and in (5.7.5) and define

$$c_{1,i}:=\min_{t_{i-1}\leq t\leq t_i}c(t),\quad c_{2,i}:=\max_{t_{i-1}\leq t\leq t_i}c(t),\quad i=1,\dots,k.$$

Then the differential equation in (5.7.1) is a Sturmian majorant of the differential equation in (5.7.5) with $c_1=c_{1,i}$ on the interval $[t_{i-1},t_i]$. Hence the solution $\sin_p(\sqrt[p]{\lambda_k c_{1,i}}(t-t_{i-1}))$ of (5.7.5) has no zero on (t_{i-1},t_i) so that

$$\sqrt[p]{\lambda_k c_{1,i}}(t_i-t_{i-1})\leq\pi_p. \tag{5.7.6}$$

By a similar argument we have $\pi_p\leq\sqrt[p]{\lambda_k c_{2,i}}(t_i-t_{i-1})$. On the other hand,

$$\int_{t_{i-1}}^{t_i}\sqrt[p]{\lambda_k c_{1,i}}\,dt\leq\int_{t_{i-1}}^{t_i}\sqrt[p]{\lambda_k c(t)}\,dt\leq\int_{t_{i-1}}^{t_i}\sqrt[p]{\lambda_k c_{2,i}}\,dt,$$

and consequently

$$\left|\pi_p-\int_{t_{i-1}}^{t_i}\sqrt[p]{\lambda_k c(t)}\right|\leq\int_{t_{i-1}}^{t_i}\sqrt[p]{\lambda_k}\left(\sqrt[p]{c_{2,i}}-\sqrt[p]{c_{1,i}}\right)\,dt. \tag{5.7.7}$$

Let $\omega(f,\delta)$ be defined for any continuous function f on $[a,b]$ by

$$\omega(f,\delta) = \max\{|f(\tau_1) - f(\tau_2)| : |\tau_1 - \tau_2| \le \delta,\ \tau_1, \tau_2 \in [a,b]\}.$$

Making use of this definition we deduce from (5.7.7) that

$$\left| \frac{\pi_p}{\sqrt[p]{\lambda_k}} - \int_{t_{i-1}}^{t_i} \sqrt[p]{c(t)}\, dt \right| \le \omega(\sqrt[p]{c}, t_i - t_{i-1}) \cdot |t_i - t_{i-1}|.$$

Let $c_1 = \min_{i=1,\dots,k} c_{1,i}$. Then by (5.7.6)

$$\left| \frac{k\pi_p}{\sqrt[p]{\lambda_k}} - \int_a^b \sqrt[p]{c(t)}\, dt \right| \le \omega\left(\sqrt[p]{c}, \frac{\pi_p}{\sqrt[p]{\lambda_k c_1}} \right)(b-a).$$

By the first part of the proof $\lambda_k \to \infty$ as $k \to \infty$. Therefore the continuity of the function c yields the formula which has to be proved. □

5.7.2 Regular problem with indefinite weight

We consider the Sturm-Liouville problem

$$\begin{cases} (r(t)\Phi(x'))' + \lambda c(t)\Phi(x) = 0, \\ Ax(a) - A'x'(a) = 0, \quad Bx(b) + B'x'(b) = 0. \end{cases} \tag{5.7.8}$$

It is supposed that r, c are continuous in $[a,b]$ and $r(t) > 0$ in this interval. No sign restriction on the function c is supposed, A, A', B, B' are real numbers such that $A^2 + A'^2 > 0$, $B^2 + B'^2 > 0$, λ is a real-valued eigenvalue parameter.

Theorem 5.7.2. *Suppose that $AA' \ge 0$, $BB' \ge 0$ and $A^2 + B^2 > 0$. Further suppose that the function c takes both positive and negative values in $[a,b]$. Then the totality of eigenvalues of (5.7.8) consists of two sequences $\{\lambda_n^+\}_{n=0}^\infty$ and $\{\lambda_n^-\}_{n=0}^\infty$ such that*

$$\dots < \lambda_n^- < \dots < \lambda_1^- < \lambda_0^- < 0 < \lambda_0^+ < \lambda_1^+ < \dots < \lambda_n^+ < \dots$$

and

$$\lim_{n\to\infty} \lambda_n^+ = \infty, \qquad \lim_{n\to\infty} \lambda_n^- = -\infty.$$

The eigenfunctions $x = x(t;\lambda_n^+)$ and $x = x(t;\lambda_n^-)$ associated with $\lambda = \lambda_n^+$ and λ_n^- have exactly n zeros in (a,b).

Proof. The proof is again based on the half-linear Prüfer transformation. Let $\lambda \in \mathbb{R}$ and let $x(t;\lambda)$ be the solution of

$$(r(t)\Phi(x'))' + \lambda c(t)\Phi(x) = 0 \tag{5.7.9}$$

satisfying the initial conditions $x(a) = A', x'(a) = A$. Note that this solution satisfies the boundary condition $Ax(a) - A'x'(a) = 0$. According to the continuous dependence of solutions on a perturbation of the functions r, c in (1.1.1), the function $x(t;\lambda)$ depends continuously on λ. In particular, if $\lambda_i \to \lambda$ as $i \to \infty$, then $x(t;\lambda_i) \to x(t;\lambda)$ uniformly on $[a,b]$ as $i \to \infty$. If $x(t;\lambda)$ satisfies the second part of the boundary conditions, i.e., $Bx(b) + B'x'(b) = 0$ for some $\lambda \in \mathbb{R}$, then λ is an eigenvalue and $x(t;\lambda)$ is the corresponding eigenfunction.

For $\lambda = 0$ we can compute $x(t;\lambda)$ explicitly as follows

$$x(t;0) = A' + r^{q-1}(a)A \int_a^t r^{1-q}(s)\, ds$$

and it is easy to see that this solution does not satisfy the condition at $t = b$, so $\lambda = 0$ is not an eigenvalue.

In what follows we suppose that $\lambda > 0$. For the solution $x(t;\lambda)$ we perform slightly modified Prüfer transformation, we express $x(t;\lambda)$ and its quasiderivative in the form

(5.7.10)

$$x(t;\lambda) = \rho(t;\lambda)\sin_p(\varphi(t;\lambda)), \quad r^{q-1}(t)x'(t;\lambda) = \lambda^{q-1}\rho(t;\lambda)\cos_p(\varphi(t;\lambda)).$$

The function $\rho(t;\lambda)$ is given by

$$\rho(t;\lambda) = \left[|x(t;\lambda)|^p + \left(\frac{r(t)}{\lambda}\right)^q |x'(t;\lambda)|^p\right]^{\frac{1}{p}}.$$

The functions ρ and φ satisfy the first order system

$$\begin{aligned} \varphi' &= \left(\frac{\lambda}{r(t)}\right)^{q-1} |\cos_p \varphi|^p + \frac{c(t)}{p-1} |\sin_p \varphi|^p, \\ \rho' &= \rho \left[\left(\frac{\lambda}{r(t)}\right)^{q-1} - \frac{c(t)}{p-1}\right] \Phi(\sin_p \varphi)\cos_p \varphi \end{aligned} \tag{5.7.11}$$

with the initial conditions

$$\begin{aligned} \rho(a;\lambda) &= \left[|A'|^p + \left(\frac{r(a)}{\lambda}\right)^q |A|^p\right]^{\frac{1}{p}}, \\ \varphi(a;\lambda) &= \operatorname{arctan}_p\left(\left(\frac{\lambda}{r(a)}\right)^{\frac{1}{p}} \frac{A'}{A}\right). \end{aligned} \tag{5.7.12}$$

Since $AA' \geq 0$, we may assume without loss of generality that

$$0 \leq \varphi(a;\lambda) < \frac{\pi_p}{2} \quad \text{if } A \neq 0, \tag{5.7.13}$$

$$\varphi(a,\lambda) = \frac{\pi_p}{2} \quad \text{if } A = 0. \tag{5.7.14}$$

Observe that as soon as $\varphi(t;\lambda)$ is known, $\rho = \rho(t;\lambda)$ can be computed explicitly and

$$\rho(t;\lambda) = \rho(a,\lambda)\exp\left\{\int_a^t \left[\left(\frac{\lambda}{r(s)}\right)^{q-1} - \frac{c(s)}{p-1}\right] \Phi(\sin_p(\varphi(s;\lambda)))\cos_p(\varphi(s;\lambda))\, ds\right\}.$$

Thus, it is important to discuss the initial value problem (5.7.11), (5.7.12). We denote by $f(t,\varphi,\lambda)$ the right-hand side of (5.7.11). It is clear that, for each $\lambda > 0$, the function $f(t,\varphi,\lambda)$ is bounded for $t \in [a,b]$ and $\varphi \in \mathbb{R}$. In view of the Pythagorean identity (1.1.13) the function $f(t,\varphi,\lambda)$ can be written in the form

$$f(t,\varphi,\lambda) = \left(\frac{\lambda}{r(t)}\right)^{q-1} + \left\{-\left(\frac{\lambda}{r(t)}\right)^{q-1} + \frac{c(t)}{p-1}\right\} |\sin_p \varphi|^p.$$

Similarly as in the standard half-linear Prüfer transformation, $f(t,\varphi,\lambda)$ is Lipschitzian in φ, hence unique solvability is guaranteed and the solution $\varphi = \varphi(t;\lambda)$ depends continuously on $(t,\lambda) \in [a,b] \times (0,\infty)$.

It is easy to see that $\lambda > 0$ is an eigenvalue of (5.7.8) if and only if λ satisfies

$$\varphi(b;\lambda) = \arctan_p \left(-\left(\frac{\lambda}{r(b)}\right)^{q-1} \frac{B'}{B}\right) + (n+1)\pi_p \tag{5.7.15}$$

for some $n \in \mathbb{Z}$. Here, by virtue of $BB' \geq 0$, we assume without loss of generality that the value of the function $\arctan_p$ in (5.7.15) is in $(-(\pi_p/2), 0]$ if $B \neq 0$ and equals $-(\pi_p/2)$ if $B = 0$.

Observe that the function $\varphi(b;\lambda)$ is strictly increasing for $\lambda \in (0,\infty)$. Indeed, denote as before $f(t,\varphi,\lambda)$ the right-hand side of (5.7.11). Clearly, $f(t,\varphi,\lambda)$ is nondecreasing function of $\lambda \in (0,\infty)$, and, since $AA' \geq 0$, the initial value $\varphi(a;\lambda)$ given by (5.7.12) is also nondecreasing for $\lambda \in (0,\infty)$. Then a standard comparison theorem for the first order scalar differential equations implies that $\varphi(t;\lambda)$ is a nondecreasing function of $\lambda \in (0,\infty)$ for each fixed $t \in [a,b]$. Now, let $0 < \lambda < \mu$ be fixed. Since the function $\varphi(t;\lambda)$ is nondecreasing with respect to λ, we have $\varphi(t;\lambda) \leq \varphi(t;\mu)$. Assume that $\varphi(t;\lambda) \equiv \varphi(t;\mu)$ for all $t \in (a,b)$. Then $\varphi'(t;\lambda) \equiv \varphi'(t;\mu)$, and so we have $f(t,\varphi(t;\lambda),\lambda) \equiv f(t,\varphi(t;\mu),\mu)$ from which it follows that $\cos_p(\varphi(t;\lambda)) \equiv \cos_p(\varphi(t;\mu)) \equiv 0$. This implies that $\varphi(t;\lambda) \equiv (m+1/2)\pi_p$ for some integer $m \in \mathbb{Z}$, and hence, by equation (5.7.11), $c(t) \equiv 0$ for $t \in (a,b)$. This is a contradiction to the assumption that $c(t) > 0$ for some $t \in [a,b]$. Therefore we have $\varphi(t_0;\lambda) < \varphi(t_0;\mu)$ for some $t_0 \in (a,b)$. Then applying a standard comparison theorem again, we conclude that $\varphi(b;\lambda) < \varphi(b;\mu)$.

Now we claim that $x(t;\lambda)$ has no zeros in the interval $(a,b]$ for all sufficiently small $\lambda > 0$. As stated before, $x(t;\lambda) \to x(t;0)$ as $\lambda \to 0+$ uniformly on $[a,b]$. We note that $x(t;\lambda)$ satisfies

$$x(t;\lambda) = A' + \int_a^t \left|\frac{r(a)}{r(s)}\Phi(A) - \frac{\lambda}{r(s)}I(s;\lambda)\right|^{q-2} \left\{\frac{r(a)}{r(s)}\Phi(A) - \frac{\lambda}{r(s)}I(s;\lambda)\right\} ds,$$

for all $a \leq t \leq b$, where

$$I(s;\lambda) = \int_a^s c(\tau)\Phi(x(\tau;\lambda))\, d\tau, \quad a \leq s \leq b.$$

Then it is easy to find that if $A = 0$ or $AA' > 0$, then $x(t;\lambda)$ has no zero in the closed interval $[a,b]$ for all sufficiently small $\lambda > 0$, and that if $A \neq 0$ and $A' = 0$, then $x(t;\lambda)$ has no zero in the interval $(a,b]$ for all sufficiently small $\lambda > 0$. Further, since

$$r(t)\Phi(x'(t;\lambda)) = r(a)\Phi(A) - \lambda \int_a^t c(s)\Phi(x(s;\lambda))\,ds$$

for $a \le t \le b$, we see that if $A \ne 0$, then $x'(t;\lambda)$ has no zeros in $[a,b]$ for all sufficiently small $\lambda > 0$.

Next we claim that the number of zeros of $x(t;\lambda)$ in $[a,b]$ can be made as large as possible if $\lambda > 0$ is chosen sufficiently large. To this end, we consider the equation

$$(\Phi(x'))' + (p-1)\mu^p\Phi(x) = 0,$$

where $\mu > 0$ is a constant. Clearly, $\sin_p(\mu t)$ is a solution of this equation, and has zeros $t = j\pi_p/\mu$, $j \in \mathbb{Z}$. Since c is supposed to be positive at some $t \in [a,b]$, there exists $[a',b'] \subset [a,b]$ such that $c(t) > 0$ on $[a',b']$. Let $k \in \mathbb{N}$ be any given positive integer and take $\mu > 0$ so that $\sin_p(\mu t)$ has at least $k+1$ zeros in $[a',b']$. Let $r^* > 0$ and $\lambda_* > 0$ be numbers such that

$$r^* = \max_{t\in[a',b']} r(t), \qquad \lambda_* \min_{t\in[a',b']} c(t) = (p-1)r^*\mu^p.$$

Then, comparing the half-linear equation in (5.7.8) with $\lambda > \lambda_*$ and the equation

$$(r^*\Phi(x'))' + (p-1)r^*\mu^p\Phi(x) = 0, \quad a' \le t \le b',$$

we conclude by the Sturm comparison theorem that all solution of the equation in (5.7.8) with $\lambda > \lambda_*$ have at least k zeros in $[a,b]$. Since k was arbitrary, this shows that the number of zeros of $x(t;\lambda)$ in $[a,b]$ can be made as large as possible if $\lambda > 0$ is chosen sufficiently large.

Since the radial variable $\rho(t;\lambda) > 0$, it follows from (5.7.10) that $x(t;\lambda)$ has a zero at $t = c$ if and only if there exists $j \in \mathbb{Z}$ such that $\varphi(c;\lambda) = j\pi_p$. Moreover, if $\varphi(c;\lambda) = j\pi_p$, then by (5.7.11) we have $\varphi'(c;\lambda) = (\lambda/r(c))^{q-1} > 0$. Therefore we easily see that if $\varphi(c;\lambda) = j\pi_p$, then $\varphi(t;\lambda) > j\pi_p$ for $c < t \le b$. Consequently, we have: (i) For all $\lambda > 0$ sufficiently small

$$0 < \varphi(b;\lambda) < \frac{\pi_p}{2} \quad \text{if } A \ne 0,$$
$$0 < \varphi(b;\lambda) < \pi_p \quad \text{if } A = 0;$$

(ii) $\lim_{\lambda\to\infty} \varphi(b;\lambda) = \infty$.

Now we seek $\lambda > 0$ satisfying (5.7.15) for some $n \in \mathbb{Z}$. The left-hand side $\varphi(b;\lambda)$ of (5.7.15) is a continuous function of $\lambda \in (0,\infty)$, and it is strictly increasing for $\lambda \in (0,\infty)$, moreover, it has the following properties

$$0 \le \lim_{\lambda\to 0+} \varphi(b;\lambda) < \frac{\pi_p}{2}, \quad \text{if } A \ne 0,$$
$$0 \le \lim_{\lambda\to 0+} \varphi(b;\lambda) < \pi_p, \quad \text{if } A = 0,$$

and $\lim_{\lambda\to\infty} \varphi(b;\lambda) = \infty$. On the other hand, by virtue of $BB' \ge 0$, the right-hand side of (5.7.15) is a nonincreasing function of $\lambda \in (0,\infty)$ for each $n \in \mathbb{Z}$. More precisely, in case $BB' > 0$, it is strictly decreasing and varies from $(n+1)\pi_p$ to $(n+1/2)\,\pi_p$ as λ varies from 0 to ∞. In the case $B' = 0$, it is the constant function $(n+1/2)\,\pi_p$.

From what was observed above we find that, for each $n = 0, 1, 2, \dots$, there exists a unique $\lambda_n^+ > 0$ such that

$$\varphi(b; \lambda_n^+) = \operatorname{arctan}_p \left(-\left(\frac{\lambda_n^+}{r(b)}\right)^{q-1} \frac{B'}{B} \right) + (n+1)\pi_p.$$

Then, each λ_n^+ is an eigenvalue of (5.7.8), and the associated eigenfunction $x(t; \lambda_n^+)$ has exactly n zeros in the open interval (a, b), where $n = 0, 1, 2, \dots$. It is clear that

$$\lambda_0^+ < \lambda_1^+ < \cdots < \lambda_n^+ < \dots, \qquad \lim_{n\to\infty} \lambda_n^+ = \infty.$$

The proof concerning the sequence of negative eigenvalues λ_n^- and the number of zeros of associated eigenfunctions can be proved in the same way. □

5.7.3 Singular Sturm-Liouville problem

We consider the equation

(5.7.16) $$(\Phi(x'))' + \lambda c(t)\Phi(x) = 0, \quad t \in [a, \infty),$$

where $\lambda > 0$ is a real-valued parameter and c is a nonnegative piecewise continuous eventually nonvanishing function. A solution $x_0 = x_0(t; \lambda)$ of (5.7.16) is said to be *subdominant* if

(5.7.17) $$\lim_{t\to\infty} x_0(t; \lambda) = k_0,$$

for some constant $k_0 \neq 0$, and a solution $x_1 = x_1(t; \lambda)$ of (5.7.16) is said to be *dominant* if

(5.7.18) $$\lim_{t\to\infty} [x_1(t; \lambda) - k_1(t-a)] = 0$$

for some constant $k_1 \neq 0$. We will show that the subdominant and dominant solutions are essentially unique in the sense that if $\tilde{x}_0(t; \lambda)$ and $\tilde{x}_1(t; \lambda)$ denote the solutions of (5.7.16) satisfying

(5.7.19) $$\lim_{t\to\infty} \tilde{x}_0(t; \lambda) = 1$$

and

(5.7.20) $$\lim_{t\to\infty} [\tilde{x}_1(t; \lambda) - (t-a)] = 0,$$

then $x_0(t; \lambda) = k_0 \tilde{x}_0(t; \lambda)$ and $x_1(t; \lambda) = k_1 \tilde{x}_1(t; \lambda)$. According to the results presented in Section 4.1, any eventually positive solution (5.7.16) has one of the following properties:

(i) $\lim_{t\to\infty} \Phi(x'(t;\lambda)) = \text{const} > 0$;

(ii) $\lim_{t\to\infty} \Phi(x'(t;\lambda)) = 0$, $\lim_{t\to\infty} x(t;\lambda) = 0$;

(iii) $\lim_{t\to\infty} \Phi(x'(t;\lambda)) = 0$, $\lim_{t\to\infty} x(t;\lambda) = \text{const} > 0$.

In view of this result, the dominant and subdominant solutions investigated in this subsection correspond to cases (i) and (iii), respectively.

The proofs of three statements presented in this subsection are rather complex, so we skip them and we refer to the paper [146]. We note only that these proofs are again based on the half-linear Prüfer transformation, this time combined with a detailed asymptotic analysis of solutions of (5.7.16).

Theorem 5.7.3. *Suppose that*

$$\int^{\infty} \left(\int_t^{\infty} c(s)\, ds \right)^{q-1} dt < \infty.$$

Then for every λ equation (5.7.16) *has a unique solution $\tilde{x}_0(t;\lambda)$ satisfying* (5.7.19) *and there exists a sequence $\{\lambda_n^{(0)}\}_{n=0}^{\infty}$ of positive parameters with the properties that*

(i) $0 = \lambda_0^{(0)} < \lambda_1^{(0)} < \cdots < \lambda_n^{(0)} < \dots$, $\lim_{n\to\infty} \lambda_n^{(0)} = \infty$;

(ii) for $\lambda \in (\lambda_{n-1}^{(0)}, \lambda_n^{(0)})$, $n = 1, 2, \dots$, $\tilde{x}_0(t;\lambda)$ has exactly $n-1$ zeros in (a,∞) and $\tilde{x}_0(a;\lambda) \neq 0$;

(iii) for $\lambda = \lambda_n^{(0)}$, $n = 1, 2, \dots$, $\tilde{x}_0(t;\lambda)$ has exactly $n-1$ zeros in (a,∞) and $\tilde{x}_0(a;\lambda) = 0$.

Theorem 5.7.4. *Let the sequence $\{\lambda_n^{(0)}\}_{n=0}^{\infty}$ be defined as in the previous theorem. Then the number of zeros of any nontrivial solution $x(t;\lambda)$ on $[a,\infty)$ can be*

(i) exactly n if $\lambda = \lambda_n^{(0)}$, $n = 1, 2, \dots$;

(ii) either $n-1$ or n if $\lambda_{n-1}^{(0)} < \lambda < \lambda_n^{(0)}$, and both cases occur.

Theorem 5.7.5. *Suppose that*

$$\int^{\infty} t^p c(t)\, dt < \infty.$$

Then for every $\lambda > 0$ equation (5.7.16) *has a unique solution $\tilde{x}_1(t;\lambda)$ satisfying* (5.7.20) *and there exists a sequence $\{\lambda_n^{(1)}\}_{n=0}^{\infty}$ of positive parameters with the properties that*

(i) $0 = \lambda_0^{(1)} < \lambda_1^{(1)} < \cdots < \lambda_n^{(1)} < \dots$, $\lim_{n\to\infty} \lambda_n^{(1)} = \infty$;

(ii) for $\lambda \in (\lambda_{n-1}^{(1)}, \lambda_n^{(1)})$, $n = 1, 2, \dots$, the solution $\tilde{x}_1(t;\lambda)$ has exactly n zeros in (a,∞) and $\tilde{x}_1(t;\lambda) \neq 0$;

(iii) *for* $\lambda = \lambda_n^{(1)}$, $n = 1, 2, \dots$, *the solution* $\tilde{x}_1(t;\lambda)$ *has exactly* n *zeros and* $\tilde{x}_1(a;\lambda) = 0$;

(iv) *the parameters* $\{\lambda_n^{(0)}\}$ *and* $\{\lambda_n^{(1)}\}$ *have the interlacing property* $0 = \lambda_0^{(1)} = \lambda_0^{(0)} < \lambda_1^{(1)} < \lambda_1^{(0)} < \dots \lambda_n^{(1)} < \lambda_n^{(0)} < \dots$.

5.7.4 Singular eigenvalue problem associated with radial p-Laplacian

As we have mentioned in Section 1.1, radially symmetric p-Laplacian can be expressed in the form

$$\Delta_p u(r) = r^{1-N}(r^{N-1}\Phi(u'))' = 0, \quad ' = \frac{d}{dr}.$$

Motivated by this fact, in this short subsection we investigate the (singular) BVP

$$L^\alpha u + [c(t) + \lambda w(t)]\Phi(u), \quad u'(0) = 0, \ u(b) = 0, \tag{5.7.21}$$

where $L^\alpha u := t^{-\alpha}(t^\alpha \Phi(u'))'$, $\alpha \geq 0$, $b > 0$ and w is a continuous positive function.

Similarly as in the previous subsections, the proof of the main statement of this subsection is based on the (modified) Prüfer transformation in the form

$$\Phi(u) = \rho(t)\sin_p \varphi(t), \quad t^\alpha \Phi(u') = \rho(t)\cos_p \varphi(t).$$

In this modification of the Prüfer transformation, ρ, φ are solutions of the system

$$\begin{aligned} (5.7.22)\ \varphi' &= [c(t) + \lambda w(t)]t^\alpha \sin_p^2 \varphi + (p-1)t^{\alpha(1-q)}|\cos_p \varphi|^q |\sin_p \varphi|^{2-q}, \\ (5.7.23)\ \rho' &= \rho\big\{ - [c(t) + \lambda w(t)]t^\alpha \sin_p \varphi \cos_p \varphi \\ &\quad + (p-1)t^{\alpha(1-q)}\Phi^{-1}(\cos_p \varphi)|\sin_p \varphi|^{3-q}\operatorname{sgn}(\sin_p \varphi)\big\}. \end{aligned}$$

For the sake of later applications, we denote the right-hand side of (5.7.22) by $f(t, \varphi, \lambda)$.

Concerning the existence and unique solvability, it can be shown that this system, has the same properties as system (1.1.20). We present the majority of the statements of this subsection without proofs, we refer to [356] for details. The difference with respect to the case treated in Subsection 5.7.2 is that $t = 0$ is generally a *singular point* of (5.7.21). However, the Sturmian theory (which is the basic "ingredience" of the proof of the properties of the below defined argument function $\varphi(t, \lambda)$) extends also to the singular case treated in this subsection (the argument is similar to that of Subsection 4.2.5), see also [331, 332].

First we present a statement which concerns the existence and unique solvability of the singular initial value problem

$$L^\alpha u + c(t)\Phi(u) = 0, \quad u(0) = u_0, \ u'(0) = 0. \tag{5.7.24}$$

Theorem 5.7.6. *Suppose that* c *is continuous in* $(0, \infty)$. *Then the initial value problem* (5.7.24) *has a unique solution in* $[0, \infty)$ *which depends continuously on* u_0 *in the sense of* C^1 *convergence on compact subintervals of* $[0, \infty)$.

Let us denote by $u(t,\lambda)$ the solution of

$$L^\alpha u + [c(t) + \lambda w(t)]\Phi(u) = 0, \quad u(0) = 1, \; u'(0) = 0, \tag{5.7.25}$$

and let $\varphi(t,\lambda)$ be a function satisfying differential equation (5.7.22). The next lemma will be used in the proof of the last statement of this subsection.

Lemma 5.7.1. *The function $\varphi(t,\lambda)$ is continuous in $[0,b] \times \mathbb{R}$, it is strictly increasing in λ for $t > 0$, and it has the properties*

$$\varphi(b,\lambda) \to 0 \text{ as } \lambda \to -\infty, \quad \varphi(b,\lambda) \to \infty \text{ as } \lambda \to \infty.$$

Theorem 5.7.7. *Suppose that the functions c, w are continuous in $J = [0,b]$ and $w(t) > 0$ for $t \in J$. Then the eigenvalue problem (5.7.21) has a countable number of simple eigenvalues $\lambda_1 < \lambda_2 < \cdots < \lambda_n \to \infty$, as $n \to \infty$. The eigenfuction u_n corresponding to λ_n has exactly $n-1$ zeros in the open interval $(0,b)$. Between $t = 0$ and the first zero of u_n, between two consecutive zeros of u_n and between the last zero of u_n and $t = b$ there is exactly one zero of u_{n+1}.*

Proof. The solution $u(t,\lambda)$ of (5.7.25) vanishes at b if and only if $\varphi(b,\lambda) = k\pi_p$, where φ is a solution of (5.7.22) corresponding to $u(t,\lambda)$. Hence, all eigenvalues λ_n and eigenfunctions u_n are obtained from

$$\varphi(b,\lambda_n) = n\pi_p, \quad u_n(t) = u(t,\lambda_n), \quad n \in \mathbb{N},$$

where $\lambda_n < \lambda_{n+1}$ and $\lim \lambda_n = \infty$ by the similar argument as in Lemma 5.7.1. The argument function $\varphi(t,\lambda_n)$ crosses each line $\varphi = k\pi_p$ with $k = 1,\dots,n-1$, only once, but never other values of k. Hence u_n has $n-1$ interior zeros $0 < t_1 < \cdots < t_{n-1} < b$. The Sturm comparison theorem shows that u_{n+1} has a zero in each of the open intervals (t_i, t_{i+1}), $i = 1,\dots,n-2$. From the strict monotonicity of $\varphi(t,\lambda)$ in λ the inequality $\varphi(t_1,\lambda_{n+1}) > \pi_p$ follows; hence u_{n+1} has a zero in $(0,t_1)$. Finally, if u_{n+1} has no zero in (t_{n-1},b), then $\varphi(t_{n-1},\lambda_{n+1}) \geq n\pi_p$. The function $\bar\varphi(t) = \varphi(t,\lambda_{n+1}) - \pi_p$ is also a solution of $\bar\varphi' = f(t,\varphi,\lambda_{n+1})$, and it has an initial value $\bar\varphi(t_{n-1}) \geq \varphi(t_{n-1},\lambda_n)$, which implies $\bar\varphi(b) > \varphi(b,\lambda_n) = n\pi_p$, which leads to $\varphi(b,\lambda_{n+1}) > (n+1)\pi_p$, a contradiction. □

5.7.5 Rotation index and periodic potential

In this subsection we extend the so-called *rotation index* technique of the investigation of linear differential equations to half-linear equations. We consider the equation

$$(\Phi(x'))' + (\lambda + c(t))\Phi(x) = 0 \tag{5.7.26}$$

together with the periodic boundary condition

$$x(0) - x(2\pi_p) = x'(0) - x'(2\pi_p) = 0, \tag{5.7.27}$$

or with the anti-periodic boundary condition

$$x(0) + x(2\pi_p) = x'(0) + x'(2\pi_p) = 0 \tag{5.7.28}$$

where c is a periodic function such that $c \in L^1_{\mathrm{loc}}(\mathbb{R})$. We always assume that the period of c is $T = 2\pi_p$.

Denote by $\mathcal{P}$ and $\mathcal{A}$ the sets of eigenvalues of the problem (5.7.26), (5.7.27) and the problem (5.7.26), (5.7.28), respectively. Let $\mathcal{P}_\bullet = \mathcal{P} \cup \mathcal{A}$. Before explaining the difficulty in understanding the structure of $\mathcal{P}_\bullet$, let us recall some classical results for the linear counterpart of (5.7.26), that is, the periodic and anti-periodic eigenvalues of the following linear Sturm-Liouville operator

$$(Lx)(t) := -x''(t) - c(t)x(t) = \lambda x(t), \tag{5.7.29}$$

where $c(t)$ is 2π-periodic and $c \in L^1(0, 2\pi)$. Then one has the following classical result, which is a part of [264, Theorem 2.1].

Theorem 5.7.8. *There exist two sequences* $\{\underline{\lambda}_n(c) : n \in \mathbb{N}\}$ *and* $\{\overline{\lambda}_n(c) : n \in \mathbb{Z}^+\}$ *of the reals with the following properties:*

(i) They have the following order:

$$\begin{aligned}\overline{\lambda}_0(c) < \underline{\lambda}_1(c) \le \overline{\lambda}_1(c) < \underline{\lambda}_2(c)& \\ \le \overline{\lambda}_2(c) < \cdots < \underline{\lambda}_n(c) \le \overline{\lambda}_n(c) < \ldots .&\end{aligned} \tag{5.7.30}$$

(ii) λ *is an eigenvalue of* (5.7.29), (5.7.27) *if only if* $\lambda = \underline{\lambda}_n(c)$ *or* $\overline{\lambda}_n(c)$ *for some even integer* n*; and* λ *is an eigenvalue of* (5.7.29), (5.7.28) *if and only if* $\lambda = \underline{\lambda}_n(c)$ *or* $\overline{\lambda}_n(c)$ *for some odd integer* n*.*

The proof of Theorem 5.7.8 is essentially based on the Floquet theory for *linear* equations with periodic coefficients and on the classification of symplectic 2×2 matrices. Concerning the periodic and anti-periodic problem for (5.7.26), we know that there is no hope of giving the complete structure of $\mathcal{P}_\bullet$, because a Floquet-like theory for periodic equations (5.7.26) is not available. On the other hand, unlike the Dirichlet problem, there would be a coexistence problem for eigenvalues $\underline{\lambda}_n(c)$ and $\overline{\lambda}_n(c)$:

$$\underline{\lambda}_n(c) = \overline{\lambda}_n(c).$$

Such a coexistence problem is extraordinarily difficult for general potentials $c(t)$. This also adds to the difficulty in understanding the structure of $\mathcal{P}_\bullet$.

In this subsection, we try to give a partial generalization of Theorem 5.7.8 to eigenvalues $\mathcal{P}_\bullet$ of (5.7.26) for using a more geometric approach, that is, the rotation index approach, which has been well developed for linear periodic systems; see [195] and [295]. This approach applies also to (5.7.29) with quasi-periodic or almost periodic potentials and is very useful in understanding the spectrum and the dynamics aspect of (5.7.29).

For linear case (5.7.29) with periodic potential c, the rotation index function in [195] is an analytic function on the upper half-plane. When real parameters are considered, the rotation index function $\rho(\lambda)$ is defined as follows. Let $x = r\cos\theta$ and $x' = -r\sin\theta$ in (5.7.29), i.e., we consider the classical Prüfer transformation. Then θ satisfies

$$\theta' = (\lambda + c(t))\cos^2\theta + \sin^2\theta =: \Phi(t, \theta; \lambda). \tag{5.7.31}$$

As $\Phi(t, \theta; \lambda)$ is 2π-periodic in both t and θ, it is known that the rotation index of (5.7.31)

$$\rho(\lambda) = \rho(\lambda; c) = \lim_{t\to\infty} \frac{\theta(t; \theta_0, \lambda) - \theta_0}{t}$$

exists and is independent of θ_0, where $\theta(t; \theta_0, \lambda)$ is the solution of (5.7.31) satisfying the initial condition $\theta(0; \theta_0, \lambda) = \theta_0$. Using the function $\rho(\lambda)$ all eigenvalues $\underline{\lambda}_n(c)$ and $\overline{\lambda}_n(c)$ can be characterized in the following way [295, Theorem 4.3].

Theorem 5.7.9. *Let c and $\rho(\lambda; c)$ be as above. Then*

$$\underline{\lambda}_n(c) = \min\{\lambda : \rho(\lambda; q) = n/2\}, \quad n \in \mathbb{N}, \tag{5.7.32}$$

$$\overline{\lambda}_n(c) = \max\{\lambda : \rho(\lambda; q) = n/2\}, \quad n \in \mathbb{N}. \tag{5.7.33}$$

In fact, the graph of $\rho(\lambda)$ looks like a staircase which is continuous and is strictly increasing except on the possible platforms $\rho^{-1}(n/2), n \in \mathbb{N}$, while the endpoints of the platforms $\rho^{-1}(n/2)$ are exactly the periodic and anti-periodic eigenvalues of (5.7.29).

In this subsection, when c is a periodic potential, we follow the idea in [295] of introducing a rotation index function $\rho(\lambda)$ for (5.7.26). Then, we use (5.7.32) and (5.7.33) to define two sequences $\{\underline{\lambda}_n(c) : n \in \mathbb{N}\}$, $\{\overline{\lambda}_n(c) : n \in \mathbb{N}\}$. It will be proved that $\{\underline{\lambda}_n(c), \overline{\lambda}_n(c) : n \text{ is even}\}$ are the periodic eigenvalues of (5.7.26), (5.7.27) and $\{\underline{\lambda}_n(c), \overline{\lambda}_n(c) : n \text{ is odd}\}$ are the anti-periodic eigenvalues of (5.7.26), (5.7.28). Namely, the "if" part of Theorem 5.7.8 (ii) holds for such a half-linear problem (below presented Theorem 5.7.13).

Differently from the variational structure of (5.7.26) in defining the variational eigenvalues of higher dimensional p-Laplacian, we extensively make use of the Hamiltonian structure of (5.7.26) for the half-linear problems (5.7.26), (5.7.27) and (5.7.26), (5.7.28). However, as in the higher dimensional Dirchlet problem for p-Laplacian (see Chapter 7), it remain open as to whether the "only if" part of Theorem 5.7.8 (ii) also holds for the one-dimensional p-Laplacian, namely, whether those eigenvalues $\underline{\lambda}_n(c), \overline{\lambda}_n(c)$ represent a *complete* list of eigenvalues of (5.7.26), (5.7.27) and (5.7.26), (5.7.28). For this reason, we will call these periodic and anti-periodic eigenvalues $\underline{\lambda}_n(c), \overline{\lambda}_n(c)$, constructed using the rotation index function, the *rotational periodic eigenvalues* of (5.7.26).

In accordance with the original paper [382] (compare also Subsection 1.1.1) we introduce the planar system

$$x' = \Phi^{-1}(y), \; y' = \Phi(x), \tag{5.7.34}$$

which is an integrable Hamiltonian system with the Hamiltonian

$$H(x, y) = \frac{|x|^p}{p} + \frac{|y|^q}{q}.$$

Let $(C_p(t), S_p(t))$ be the unique solution of equation (5.7.34) with the initial value $(x(0), y(0)) = (1, 0)$. Some properties of $C_p(t)$ and $S_p(t)$ are summarized in the following lemma, its proof follows essentially ideas introduced in Section 1.1.

Lemma 5.7.2. *The functions $C_p(t)$ and $S_p(t)$ have the following properties:*

(i) *Both $C_p(t)$ and $S_p(t)$ are $2\pi_p$-periodic.*

(ii) *$C_p(t)$ is even in t and $S_p(t)$ is odd in t.*

(iii) *$C_p(t+\pi_p) = -C_p(t)$, $S_p(t+\pi_p) = -S_p(t)$.*

(iv) *$C_p(t) = 0$ if and only if $t = \pi_p/2 + m\pi_p$, $m \in \mathbb{Z}$; $S_p(t) = 0$ if and only if $t = m\pi_p$, $m \in \mathbb{Z}$.*

(v) *$C_p'(t) = -\Phi^{-1}(S_p(t))$ and $S_p'(t) = \Phi(C_p(t))$.*

(vi) *$|C_p(t)|^p/p + |S_p(t)|^q/q \equiv 1/p$.*

Let us define the polar coordinates in $\mathbb{R}^2$ by

$$x = r^{2/p} C_p(\theta), \quad y = r^{2/q} S_p(\theta). \tag{5.7.35}$$

Equation (5.7.26) is equivalent to the following Hamiltonian system

$$x' = -\Phi^{-1}(y) = -\frac{\partial H(t,x,y)}{\partial y}, \quad y' = (\lambda + q(t))\Phi(x) = \frac{\partial H(t,x,y)}{\partial x}, \tag{5.7.36}$$

where the Hamiltonian is

$$H(t,x,y;\lambda) = (\lambda + c(t))\frac{|x|^p}{p} + \frac{|y|^q}{q}.$$

In the polar coordinates (5.7.35), it is not difficult to check that r and θ satisfy the following equations, see also Section 1.1,

$$r' = \frac{p}{2}(\lambda + c(t) - 1)\Phi(C_p(\theta))\Phi^{-1}(S_p(\theta))r =: \Psi(t,\theta,r;\lambda), \tag{5.7.37}$$

$$\theta' = p((\lambda + c(t))|C_p(\theta)|^p/p + |S_p(\theta)|^q/q) =: \Xi(t,\theta;\lambda). \tag{5.7.38}$$

For any $(x_0, y_0) \in \mathbb{R}^2$, let $(x(t;x_0,y_0,\lambda)), y(t;x_0,y_0,\lambda))$ be the unique solution of (5.7.36) satisfying the initial condition $(x(0),y(0)) = (x_0,y_0)$. Similarly, for $\theta_0 \in \mathbb{R}$, let $(r(t;\theta_0,\lambda), \theta(t;\theta_0,\lambda))$ be the unique solution of (5.7.37),(5.7.38) satisfying the initial conditions $r(0) = 1$, $\theta(0) = \theta_0$. As $\Xi(t,\theta;\lambda)$ – the right-hand side of (5.7.38) – is bounded in θ, we know that $r(t;\theta_0,\lambda)$ and $\theta(t;\theta_0,\lambda)$ are defined on the whole line $\mathbb{R}$. Due to the oddness of (5.7.36) in (x,y) and the homogeneity of (5.7.37) in r, we have the following relation between the solutions $(x(t;x_0,y_0,\lambda), y(t;x_0,y_0,\lambda))$ and $(r(t;\theta_0,\lambda), \theta(t;\theta_0,\lambda))$:

$$\begin{aligned}&[x(t;\varphi_p(r_0)C_p(\theta_0), \varphi_q(r_0)S_p(\theta_0),\lambda), y(t;\varphi_p(r_0)C_p(\theta_0), \varphi_q(r_0)S_p(\theta_0),\lambda)]\\ &\equiv [\varphi_p(r_0)r^{2/p}(t;\theta_0,\lambda)C_p(\theta(t;\theta_0,\lambda)), \varphi_q(r_0)r^{2/q}(t;\theta_0,\lambda)S_p(\theta(t;\theta_0,\lambda))]\end{aligned} \tag{5.7.39}$$

for all $r_0, \theta_0 \in \mathbb{R}$, where

$$\varphi_p(r_0) = |r_0|^{2/p}\operatorname{sgn}(r_0), \quad r_0 \in \mathbb{R}.$$

In particular, the Poincaré map P_λ of (5.7.36) is given by

(5.7.40)

$$\begin{aligned}&P_\lambda(\varphi_p(r_0)C_p(\theta_0),\varphi_q(r_0)S_p(\theta_0))\\&\quad=(x(2\pi_p;\varphi_p(r_0)C_p(\theta_0),\varphi_q(r_0)S_p(\theta_0),\lambda),y(2\pi_p;\varphi_p(r_0)C_p(\theta_0),\varphi_q(r_0)S_p(\theta_0),\lambda))\\&\quad=(\varphi_p(r_0)R_\lambda^{2/p}(\theta_0)C_p(\Theta_\lambda(\theta_0)),\varphi_q(r_0)R_\lambda^{2/q}(\theta_0)S_p(\Theta_\lambda(\theta_0)))\end{aligned}$$

for $r_0, \theta_0 \in \mathbb{R}$, where $R_\lambda(\theta_0) := r(2\pi_p; \theta_0, \lambda)$ and

$$\Theta_\lambda(\theta_0) := \theta(2\pi_p; \theta_0, \lambda). \tag{5.7.41}$$

Thus $\Theta_\lambda(\theta_0)$ is the corresponding Poincaré map of (5.7.38).

Equation (5.7.38) is a family of $2\pi_p$-periodic equations on the circle $\mathbb{S} := \mathbb{R}/\pi_p\mathbb{Z}$. By the periodicity of $\Xi(t, \theta; \lambda)$, we have the following equalities on the solutions $\theta(t; \theta_0, \lambda)$ of (5.7.38)

$$\theta(t + 2m\pi_p; \theta_0, \lambda) \equiv \theta(t; \theta(2m\pi_p; \theta_0, \lambda), \lambda), \tag{5.7.42}$$

$$\theta(t; \theta_0 + m\pi_p, \lambda) \equiv \theta(t; \theta_0, \lambda) + m\pi_p \tag{5.7.43}$$

for all $\theta_0 \in \mathbb{R}$ and all $m \in \mathbb{Z}$. In particular, the *rotation index* of (5.7.38)

$$\begin{aligned}\rho(\lambda) = \rho(\lambda; c) :&= \lim_{|t|\to\infty} \frac{\theta(t; \theta_0, \lambda) - \theta_0}{t}\\&= \lim_{|m|\to\infty} \frac{\Theta_\lambda^m(\theta_0) - \theta_0}{2m\pi_p}\end{aligned} \tag{5.7.44}$$

exists and is independent of θ_0; see [171, Theorem 2.1, Chapter 2].

Next, we give without proof an auxiliary statement concerning the behavior of the function θ. In this statement, $w \in L^1(0, 2\pi_p)$ is a 2π-periodic function, and θ is a solution of

$$\theta' = p\left[w(t)\frac{|C_p(\theta)|^p}{p} + \frac{|S_p(\theta)|^q}{q}\right] =: \tilde{\Xi}(t, \theta). \tag{5.7.45}$$

The solution of this equation satisfying $\theta(0) = \theta_0$ we denote by $\theta(t; t_0)$.

Lemma 5.7.3. *The following statements hold.*

(i) *If $\theta_0 \geq \pi_p/2 + m\pi_p$ for some $m \in \mathbb{Z}$, then $\theta(t; \theta_0) > \pi_p/2 + m\pi_p$ for all $t > 0$.*

(ii) *Assume that $w(t) < 0$. If $\theta_0 \leq m\pi_p$ for some $m \in \mathbb{Z}$, then $\theta(t; \theta_0) < m\pi_p$ for all $t > 0$.*

Now we give some properties of the rotation index function $\rho(\lambda)$.

Theorem 5.7.10. *The rotation index has the following properties:*

(i) *$\rho(\lambda)$ is continuous in $\lambda \in \mathbb{R}$.*

(ii) *$\rho(\lambda)$ is non-decreasing in $\lambda \in \mathbb{R}$.*

(iii) $\rho(\lambda) = 0$ *if* $\lambda \ll -1$ *and* $\lim_{\lambda\to+\infty} \rho(\lambda) = +\infty$.

Proof. (i) Note that the vector field $\Xi(t, \theta; \lambda)$ depends on λ continously. Then so does the Poincaré map $\Theta_\lambda(\theta_0)$. Now the continuity of $\rho(\lambda)$ follows from [171, Corollary 2.1, Chapter 2].

(ii) The monotonicity of $\rho(\lambda)$ follows simply from the monotonicity of the right-hand side of (5.7.38).

(iii) Let us first prove that $\lim_{\lambda\to+\infty} \rho(\lambda) = +\infty$. To this end, assume that $\lambda > 0$. Let us consider differential equation (5.7.26) again. Let $\hat{x} = x, \hat{y} = -\lambda^{-1/q}\Phi(x')$. Then (5.7.26) is equivalent to the following system:

$$\hat{x}' = -\lambda^{1/p}\Phi^{-1}(\hat{y}), \quad \hat{y}' = (\lambda^{1/p} + \lambda^{-1/q}c(t))\Phi(\hat{x}). \tag{5.7.46}$$

Let $\hat{x} = \hat{r}^{2/p}C_p(\hat{\theta})$, $\hat{y} = \hat{r}^{2/q}S_p(\hat{\theta})$ in (5.7.46). Then $\hat{r}, \hat{\theta}$ satisfy the equations

$$\hat{r}' = \frac{p}{2}\lambda^{-1/q}c(t)\Phi(C_p(\hat{\theta}))\Phi^{-1}(S_p(\hat{\theta}))\hat{r}, \tag{5.7.47}$$

$$\hat{\theta}' = \lambda^{1/p} + \lambda^{-1/q}c(t)\frac{|C_p(\hat{\theta})|^p}{p}. \tag{5.7.48}$$

As before, let $\hat{\theta}(t; \hat{\theta}_0, \lambda)$ denote the solution of (5.7.48) with the initial condition $\hat{\theta}(0) = \hat{\theta}_0$. By (5.7.48), one has

$$\lambda^{1/p}t - \lambda^{-1/q}\int_0^t c_-(s)ds \le \hat{\theta}(t; \hat{\theta}_0, \lambda) - \hat{\theta}_0 \le \lambda^{1/p}t + \lambda^{-1/q}\int_0^t c_+(s)ds \tag{5.7.49}$$

for all $t \ge 0$ and all $\hat{\theta}_0 \in \mathbb{R}$, where

$$c_+(t) = \max\{c(t), 0\}, \quad c_-(t) = \max\{-c(t), 0\}.$$

Comparing (5.7.46) with (5.7.36), we have the relation

$$(x, y) = (\hat{x}, \lambda^{1/q}\hat{y}).$$

In the corresponding polar coordinates, we have

$$r^{2/p}C_p(\theta) = \hat{r}^{2/p}C_p(\hat{\theta}), \quad r^{2/q}S_p(\theta) = \lambda^{1/q}\hat{r}^{2/q}S_p(\hat{\theta}).$$

As a result,

$$C_p(\theta) = \frac{C_p(\hat{\theta})}{(|C_p(\hat{\theta})|^p + (p-1)\lambda|S_p(\hat{\theta})|^q)^{1/p}}, \tag{5.7.50}$$

$$S_p(\theta) = \frac{\lambda^{1/q}C_p(\hat{\theta})}{(|C_p(\hat{\theta})|^p + (p-1)\lambda|S_p(\hat{\theta})|^q)^{1/p}}, \tag{5.7.51}$$

Let us define a homeomorphism $\mathcal{H}_\lambda : \mathbb{R} \to \mathbb{R}$, where $\theta = \mathcal{H}_\lambda(\hat{\theta})$ is determined by (5.7.50) and (5.7.51). Note that $\mathcal{H}_\lambda$ fixes the points $\{m\pi_p, \pi_p/2 + m\pi_p : m \in \mathbb{Z}\} \subset \mathbb{R}$ and

$$\lim_{|\hat{\theta}|\to\infty} \frac{\mathcal{H}_\lambda(\hat{\theta})}{\hat{\theta}} = 1. \tag{5.7.52}$$

Comparing (5.7.48) with (5.7.38), $\mathcal{H}_\lambda$ preserves solutions

$$\theta(t; \mathcal{H}_\lambda(\hat{\theta}_0), \lambda) \equiv \mathcal{H}_\lambda(\hat{\theta}(t; \hat{\theta}_0, \lambda)). \tag{5.7.53}$$

By (5.7.52) and (5.7.53) we have

$$\begin{aligned}\rho(\lambda) &= \lim_{t\to\infty} \frac{\theta(t; \mathcal{H}_\lambda(\hat{\theta}_0), \lambda)}{t} = \lim_{t\to\infty} \frac{\mathcal{H}_\lambda(\hat{\theta}(t; \hat{\theta}_0, \lambda))}{t} \\ &= \lim_{t\to\infty} \frac{\hat{\theta}(t; \hat{\theta}_0, \lambda)\mathcal{H}_\lambda(\hat{\theta}(t; \hat{\theta}_0, \lambda))}{t\hat{\theta}(t; \hat{\theta}_0, \lambda)} = \lim_{t\to\infty} \frac{\hat{\theta}(t; \hat{\theta}_0, \lambda)}{t},\end{aligned} \tag{5.7.54}$$

because, when $\lambda \gg 1$,

$$\hat{\theta}(t; \hat{\theta}_0, \lambda) \geq \hat{\theta}_0 + \lambda^{1/p} t - \lambda^{-1/q} \int_0^t c_-(s) ds \to \infty \quad \text{as } t \to \infty.$$

Next, by (5.7.49), we get from (5.7.54)

$$\rho(\lambda) = \lim_{t\to\infty} \frac{\hat{\theta}(t; \hat{\theta}_0, \lambda)}{t} \geq \lambda^{1/p} - \bar{c}_- \lambda^{-1/q},$$

where

$$\bar{c}_- = \frac{1}{2\pi_p} \int_0^{2\pi_p} c_-(s) ds$$

is the mean value. Now it is obvious that $\rho(\lambda) \to \infty$ when $\lambda \to \infty$.

Next let us prove that $\rho(\lambda) \geq 0$ for all λ. Applying Lemma 5.7.3 (i) to $w(t) = \lambda + c(t)$, we have $\theta(t; \pi_p/2, \lambda) > \pi_p/2$ for all $t > 0$. Thus

$$\rho(\lambda) = \lim_{t\to\infty} \frac{\theta(t; \pi_p/2, \lambda)}{t} \geq 0.$$

Finally we prove that $\rho(\lambda) = 0$ if $\lambda \ll -1$. For simplicity, let us assume that the potential $c(t)$ is bounded on $\mathbb{R}$, that is, there is some $M_0 > 0$ such that $|c(t)| \leq M_0, t \in \mathbb{R}$. Suppose that $\lambda < -M_0$. Then $w(t) = \lambda + q(t) < 0$ for all t. It follows from Lemma 5.7.3-(ii) that $\theta(t; 0, \lambda) < 0$ for all $t > 0$. Thus

$$\rho(\lambda) = \lim_{t\to\infty} \frac{\theta(t; 0, \lambda)}{t} \leq 0.$$

This, together with the conclusion $\rho(\lambda) \geq 0$, shows that $\rho(\lambda) = 0$ for all $\lambda \ll -1$. For general periodic potentials $c(t) \in L^1_{\text{loc}}(\mathbb{R})$, the result can be proved similarly by using a similar transformation as in the proof of the conclusion $\lim_{\lambda\to\infty} \rho(\lambda) = \infty$. □

Remark 5.7.1. It follows from (5.7.49) that the rotation index function has the following estimates:

$$\lambda^{1/p} - \bar{c}_- \lambda^{-1/q} \leq \rho(\lambda) \leq \lambda^{1/p} + \bar{c}_+ \lambda^{-1/q}, \ \lambda > 0, \tag{5.7.55}$$

where $\bar{c}_\pm$ are the mean values of $2\pi_p$-periodic functions $c_\pm(t)$.

Now we use the rotation index function $\rho(\lambda)$ to study the main problem in this subsection, that is, the periodic eigenvalues $\mathcal{P}$ of (5.7.26), (5.7.27) and the anti-periodic eigenvalues $\mathcal{A}$ of (5.7.26), (5.7.28).

Note that $(x(t; x_0, y_0, \lambda), y(t; x_0, y_0, \lambda))$ is a solution of (5.7.26), (5.7.27) or (5.7.26), (5.7.28) if and only if (x_0, y_0) is a fixed point of the Poincaré map P_λ, i.e., it satisfies

$$P_\lambda(x_0, y_0) = (x_0, y_0),$$

or an anti-fixed point of P_λ:

$$P_\lambda(x_0, y_0) = -(x_0, y_0).$$

Using expression (5.7.40) for P_λ the following result is obvious.

Theorem 5.7.11. *We have $\lambda \in \mathcal{P}_\bullet$ if and only if there exist some $\theta_0 \in \mathbb{R}$ and some $n \in \mathbb{Z}^+$ such that*

$$\theta(2\pi_p; \theta_0, \lambda) = \theta_0 + n\pi_p \quad \text{and} \quad r(2\pi_p; \theta_0, \lambda) = 1. \tag{5.7.56}$$

Moreover, the case n is even (n is odd, respectively) in (5.7.56) *corresponds to the periodic eigenvalues (the anti-periodic eigenvalues, respectively).*

Note that the vector field $\Xi(t, \theta; \lambda)$ is differentiable in θ. Thus the Poincaré map $\Theta_\lambda(\theta_0)$ of (5.7.38) (see (5.7.41)) is also differentiable in θ_0. The following statement, which is a result of the area-preserving property of P_λ, is fundamental in obtaining some eigenvalues in $\mathcal{P}_\bullet$. Using (5.7.37) and (5.7.38) we obtain the following result.

Lemma 5.7.4. *It holds*

$$\frac{d\Theta_\lambda(\theta_0)}{d\theta_0} = \frac{1}{R_\lambda^2(\theta_0)}, \qquad \theta_0 \in \mathbb{R}, \tag{5.7.57}$$

where R_λ is defined right after (5.7.40).

Proof. Consider the space $S = (0, \infty) \times (\mathbb{R}/2\pi_p\mathbb{Z})$. Let θ_0 be any fixed real number. For any θ_1 ($> \theta_0$) which is near θ_0, consider the following domain in S:

$$\mathcal{D} = \{(r, \theta) : 0 \le r \le 1, \theta_0 \le \theta \le \theta_1\}.$$

The domain $\mathcal{D}$ has the area

$$|\mathcal{D}| = \int_0^1 \int_{\theta_0}^{\theta_1} r\, dr\, d\theta = \frac{1}{2}(\theta_1 - \theta_0).$$

Note that the Poincaré map P_λ, in the (r, θ)-coordinates, is

$$P_\lambda(r, \theta) = (rR_\lambda(\theta), \Theta_\lambda(\theta)).$$

The image $\widetilde{\mathcal{D}}$ of $\mathcal{D}$ is the following domain in S:

$$\widetilde{\mathcal{D}} = \{(\tilde{r}, \tilde{\theta}) : 0 \le \tilde{r} \le R_\lambda(\Theta_\lambda^{-1}(\tilde{\theta})), \Theta_\lambda(\theta_0) \le \tilde{\theta} \le \Theta_\lambda(\theta_1)\},$$

see (5.7.40). Here $\Theta_\lambda^{-1}(\cdot)$ is the inverse of $\Theta_\lambda(\cdot)$. Thus $\tilde{\mathcal{D}}$ has the area

$$|\tilde{\mathcal{D}}| = \int_{\Theta_\lambda(\theta_0)}^{\Theta_2(\theta_1)} d\tilde{\theta} \int_0^{R_\lambda(\Theta_\lambda^{-1}(\tilde{\theta}))} \tilde{r} d\tilde{r} = \frac{1}{2} \int_{\Theta_\lambda(\theta_0)}^{\Theta_2(\theta_1)} R_\lambda^2(\Theta_\lambda^{-1}(\tilde{\theta})) d\tilde{\theta}$$
$$= \frac{1}{2} \int_{\theta_0}^{\theta_1} R_\lambda^2(\theta) \frac{d\Theta_\lambda(\theta)}{d\theta} d\theta.$$

As P_λ preserves the expression $r\,dr\,d\theta$, we have

$$\frac{1}{2}(\theta_1 - \theta_0) \equiv \frac{1}{2} \int_{\theta_0}^{\theta_1} R_\lambda^2(\theta) \frac{d\Theta_\lambda(\theta)}{d\theta} d\theta.$$

Differentiating this equality with respect to θ_1 at $\theta_1 = \theta_0$, we get the desired equality (5.7.57). □

Remark 5.7.2. Equality (5.7.57) can be also proved using equations (5.7.37) and (5.7.38).

By Lemma 5.7.4, we can state an equivalent form of Theorem 5.7.11.

Theorem 5.7.12. *It holds that $\lambda \in \mathcal{P}_\bullet$ if and only if there exist some $\theta_0 \in \mathbb{R}$ and some $n \in \mathbb{N}$ such that*

$$\Theta_\lambda(\theta_0) = \theta_0 + n\pi_p \quad \textit{and} \quad \left.\frac{d\Theta_\lambda(\theta)}{d\theta}\right|_{\theta=\theta_0} = 1. \tag{5.7.58}$$

Moreover, the case n is even (n is odd, respectively) in (5.7.58) *corresponds to the periodic eigenvalues (the anti-periodic eigenvalues, respectively).*

We will establish some relation between condition (5.7.58) with the rotation index function $\rho(\lambda)$. To this end, we need the following result.

Lemma 5.7.5. *Let h be a homeomorphism of $\mathbb{R}$ satisfying*

$$h(\theta + m\pi_p) = h(\theta) + m\pi_p, \quad \theta \in \mathbb{R},\ m \in \mathbb{Z}. \tag{5.7.59}$$

Define the rotation index of h by

$$\rho(h) = \lim_{|m|\to\infty} \frac{h^m(\theta_0) - \theta_0}{2m\pi_p}.$$

Suppose that n is an integer. Then

(i) $\rho(h) \geq n/2$ if and only if $\max_{\theta_0 \in \mathbb{R}}(h(\theta_0) - (\theta_0 + n\pi_p)) \geq 0$;

(ii) $\rho(h) \leq n/2$ if and only if $\max_{\theta_0 \in \mathbb{R}}(h(\theta_0) - (\theta_0 + n\pi_p)) \leq 0$.

Now we introduce for (5.7.26) two sequences $\{\underline{\lambda}_n(c) : n \in \mathbb{N}\}$ and $\{\overline{\lambda}_n(c) : n \in \mathbb{Z}^+\}$ using the rotation index function $\rho(\lambda)$.

Let $c(t)$ be a $2\pi_p$-periodic potential with $c(t) \in L^1(0, 2\pi_p)$. Define

$$\underline{\lambda}_n(c) = \min\{\lambda \in \mathbb{R} : \rho(\lambda; c) = n/2\} \text{ for } n \in \mathbb{N}, \tag{5.7.60}$$

$$\overline{\lambda}_n(c) = \max\{\lambda \in \mathbb{R} : \rho(\lambda; c) = n/2\} \text{ for } n \in \mathbb{N}. \tag{5.7.61}$$

Note that these sequences are well-defined by Theorem 5.7.10. Now we prove the main result in this subsection.

Theorem 5.7.13. *If $n \in \mathbb{Z}^+$ is even, then $\underline{\lambda}_n(c)$ and $\overline{\lambda}_n(c)$are eigenvalues of* (5.7.26), (5.7.27). *If $n \in \mathbb{N}$ is odd, then $\underline{\lambda}_n(c)$ and $\overline{\lambda}_n(c)$ are eigenvalues of* (5.7.26), (5.7.28).

Proof. Let us consider the family of Poincaré maps Θ_λ of (5.7.38). By (5.7.43) Θ_λ satisfies (5.7.59) for each λ. By Lemma 2.2, for any fixed θ_0, the function $\Theta_\lambda(\theta_0) - \theta_0$ is strictly increasing in λ. Thus the functions

$$\max_{\theta_0 \in \mathbb{R}}(\Theta_\lambda(\theta_0) - \theta_0), \quad \text{and} \quad \min_{\theta_0 \in \mathbb{R}}(\Theta_\lambda(\theta_0) - \theta_0),$$

are strictly increasing in λ. Now it follows from (5.7.60) and (5.7.61) and from Lemma 5.7.5 that $\lambda = \underline{\lambda}_n(c)$ if and only if λ satisfies

$$\max_{\theta_0 \in \mathbb{R}}(\Theta_\lambda(\theta_0) - \theta_0) = n\pi_p, \tag{5.7.62}$$

and that $\lambda = \overline{\lambda}_n(c)$ if and only if λ satisfies

$$\min_{\theta_0 \in \mathbb{R}}(\Theta_\lambda(\theta_0) - \theta_0) = n\pi_p. \tag{5.7.63}$$

As a result, if $\lambda = \underline{\lambda}_n(c)$ or $\overline{\lambda}_n(c)$, we know from (5.7.62) or (5.7.63) that there must be some $\theta_0 \in \mathbb{R}$ such that (5.7.58) is satisfied. Now the result follows from Theorem 5.7.12. □

Remark 5.7.3. (i) We do not know if the converse of Theorem 5.7.13 also holds, that is, whether all eigenvalues of (5.7.26), (5.7.27) and (5.7.26), (5.7.28) are necessarily given by t$\underline{\lambda}_n(c)$ and $\overline{\lambda}_n(c)$.

(ii) Since $\rho(\lambda)$ is nondecreasing, one can see that the ordering (5.7.30) holds also for the p-Laplacian. As $\rho(\underline{\lambda}_n(c)) = \rho(\overline{\lambda}_n(c)) = n/2$, it follows from the estimates (5.7.55) that one has the following asymptotic formula for the rotational periodic eigenvalues $\underline{\lambda}_n(c)$ and $\overline{\lambda}_n(c)$:

$$\underline{\lambda}_n(c), \overline{\lambda}_n(c) \sim \left(\frac{n}{2}\right)^p \quad \text{as } n \to \infty. \tag{5.7.64}$$

We will finish this subsection with another characterization of rotational periodic eigenvalues using the eigenvalues of (5.7.26) with respect to certain two-point boundary conditions.

The general two-point boundary conditions are of the form

$$\xi x(0) + \eta x'(0) = 0, \quad \sigma x(2\pi_p) + \tau x'(2\pi_p) = 0, \tag{5.7.65}$$

where ξ, η, σ, τ are constants such that $\xi^2+\eta^2>0$ and $\sigma^2+\tau^2>0$. The conditions (5.7.65) can be rewritten as the following concise form:

$$\begin{aligned}\Phi^{-1}(S_p(\alpha))x(0)+C_p(\alpha)x'(0)=0,\\ \Phi^{-1}(S_p(\beta))x(2\pi_p)+C_p(\beta)x'(2\pi_p)=0,\end{aligned} \tag{5.7.66}$$

where $\alpha, \beta \in [0,\pi_p)$. In the following, we are only interested in the case $\beta=\alpha\in[0,\pi_p)$ in (5.7.66), i.e.,

$$\begin{aligned}\Phi^{-1}(S_p(\alpha))x(0)+C_p(\alpha)x'(0)=0,\\ \Phi^{-1}(S_p(\alpha))x(2\pi_p)+C_p(\alpha)x'(2\pi_p)=0.\end{aligned} \tag{5.7.67}$$

Hence λ is an eigenvalue of (5.7.26), (5.7.67) if and only if

$$\theta(2\pi_p;\alpha,\lambda)=\alpha+n\pi_p, \quad n\in\mathbb{N}. \tag{5.7.68}$$

Remark 5.7.4. Let us denote by λ_n^α eigenvalues of (5.7.26), (5.7.67). It follows from (5.7.68) that $\rho(\lambda_n^\alpha(c))=n/2$. Thus one has the following relation between $\lambda_n^\alpha(c)$ and the rotational eigenvalues, which is well known for the linear case $p=2$:

$$\underline{\lambda}_n(c)\le\lambda_n^\alpha(c)\le\overline{\lambda}_n(c), \quad \alpha\in[0,\pi_p),\ n\in\mathbb{N}. \tag{5.7.69}$$

By the asymptotic formula (5.7.64) for $\underline{\lambda}_n(c)$ and $\overline{\lambda}_n(c)$, one has the following asymptotic formula $\lambda_n^\alpha(c)$:

$$\lambda_n^\alpha(c)\sim(n/2)^p \text{ as } n\to\infty.$$

Now we give a characterization of rotational periodic eigenvalues using the eigenvalues of Sturm-Liouville problems.

Theorem 5.7.14. *We have for any $n\in\mathbb{N}$*

$$\underline{\lambda}_n(c)=\min_{s\in\mathbb{R}}\lambda_n^\alpha(c_s), \tag{5.7.70}$$

$$\overline{\lambda}_n(c)=\max_{s\in\mathbb{R}}\lambda_n^\alpha(c_s). \tag{5.7.71}$$

The eigenvalue $\overline{\lambda}_0(c)$ can be characterized using the Neumann eigenvalues

$$\overline{\lambda}_0(c)=\max_{s\in\mathbb{R}}\lambda_0^0(c_s). \tag{5.7.72}$$

Here $c_s(t)$ are translations of $c(t)$, that is, $c_s(t)=c(t+s)$.

Proof. For any fixed $s\in\mathbb{R}$, let $\theta(t;\theta_0,\lambda,c_s)$ be the solution of the following equation

$$\theta'=p\Big((\lambda+c_s(t))\frac{|C_p(\theta)|^p}{p}+\frac{|S_p(\theta)|^q}{q}\Big)$$

satisfying the initial condition $\theta(0)=\theta_0$. Then one has the equality

$$\theta(t;\theta(-s;\theta_0,\lambda,c_s),\lambda,c)\equiv\theta(t-s;\theta_0,\lambda,c_s)$$

for all t, s, θ_0 and λ. Hence $\rho(\lambda; c) = \rho(\lambda; c_s)$ for all λ and all s. Consequently, $\underline{\lambda}_n(c_s) = \underline{\lambda}_n(c)$ and $\overline{\lambda}_n(c_s) = \overline{\lambda}_n(c)$ for all s. By (5.7.69), one has

$$\underline{\lambda}_n(c) \le \lambda_n^\alpha(c_s) \le \overline{\lambda}_n(c), \quad \text{for all } s \in \mathbb{R}.$$

Moreover, the eigenvalues $\lambda_n^\alpha(c_s)$, as functions of s, are continuous in s, see (5.7.68).

Now we are going to prove (5.7.71). Assume first that $n \in \mathbb{N}$ is even. Then $\lambda = \overline{\lambda}_n(c)$ is an eigenvalue of the periodic problem (5.7.26), (5.7.27), that is, equation (5.7.26) has a nonzero $2\pi_p$-periodic function $x(t)$. We claim that there exists $s_0 \in \mathbb{R}$ such that

$$\Phi^{-1}(S_p(\alpha))x(s_0) + C_p(\alpha)x'(s_0) = 0. \tag{5.7.73}$$

Case 1: $\alpha = 0$. As $x(t)$ is $2\pi_p$-periodic, there exists s_0 such that $x'(s_0) = 0$. Thus (5.7.73) is satisfied.

Case 2: $\alpha = \pi_p/2$. If (5.7.73) does not hold, then either $x(t) > 0$ for all t or $x(t) < 0$ for all t. Let r_0 ($\neq 0$) and θ_0 be such that

$$x(0) = \varphi_p(r_0)C_p(\theta_0), \quad y(0) = -\Phi(x'(0)) = \varphi_q(r_0)S_p(\theta_0),$$

where φ_p is defined right after (5.7.39). Then $\theta(t; \theta_0, \lambda, c)$ is bounded because $x(t) \neq 0$ for all t. As a result, we have $\rho(\lambda) = 0$. This is a contradiction because $\rho(\lambda) = \rho(\overline{\lambda}_n(c)) = n/2 > 0$.

Case 3: $\alpha \in (0, \pi_p/2) \cup (\pi_p/2, \pi_p)$. If (5.7.73) does not hold, then the function

$$\xi(t) = x'(t) + \gamma x(t) \tag{5.7.74}$$

is a $2\pi_p$-periodic function and satisfies $\xi(t) \neq 0$ for all t, where

$$\gamma = \frac{\Phi(S_p(\alpha))}{C_p(\alpha)} \neq 0.$$

Without loss of generality, we assume that $\xi(t) > 0$ for all t. It is easy to prove that for any given $2\pi_p$-periodic function $\xi(t)$, linear equation (5.7.74) has a unique $2\pi_p$-periodic solution $x(t)$. In fact, such a $2\pi_p$-periodic solution is given by

$$x(t) = \int_{\pm\infty}^{0} \xi(s+t)\exp(\gamma s)ds,$$

where the sign $-$ is for $\alpha < \pi_p/2$ while the sign $+$ is for $\alpha > \pi_p/2$. As $\xi(t) > 0$ for all t, we have $x(t) > 0$ for all t if $\alpha < \pi_p/2$, or $x(t) < 0$ for all t if $\alpha > \pi_p/2$. Now we can use a similar argument as in Case 2 to prove that (5.7.73) holds for some $s_0 \in \mathbb{R}$.

Let now $z(t) = x(t + s_0)$, where s_0 satisfies (5.7.73). Then $z(t)$ satisfies the differential equation

$$(\Phi(z'(t)))' + (\lambda + c_{s_0}(t))\Phi(z(t)) = 0. \tag{5.7.75}$$

Moreover, by (5.7.73) and $2\pi_p$-periodicity of $x(t)$, $z(t)$ satisfies the boundary condition (5.7.67). This means that $\lambda = \overline{\lambda}_n(c)$ is an eigenvalue of (5.7.71), (5.7.67).

As we have shown, $n/2 = \rho(\overline{\lambda}_n(c); c) = \rho(\overline{\lambda}_n(c); c_{s_0})$, therefore $\overline{\lambda}_n(c) = \lambda_n^\alpha(c_{s_0})$. This proves that the maximum in (5.7.71) can be attained when $n > 0$ is even. The characterization (5.7.72) also holds because we have (5.7.73) in this case.

Now we consider (5.7.71) when $n \in \mathbb{N}$ is odd. In this case, for $\lambda = \overline{\lambda}_n(c)$, (5.7.26) has a nonzero solution $x(t)$ satisfying the anti-periodic boundary condition (5.7.28). As (5.7.26) is odd in x, it is easy to check that $x(t)$ satisfies $x(t + 2\pi_p) \equiv -x(t)$ and $x(t)$ is a $4\pi_p$-periodic solution of (5.7.26). Now a similar argument shows that (5.7.71) holds also in this case.

Equality (5.7.70) can be proved similarly. □

5.8 Energy functional and various boundary conditions

The aim of this section is to establish necessary and sufficient conditions for non-negativity and positivity of the energy functional associated to (1.1.1) over the class of functions satisfying various boundary conditions. These conditions are formulated in terms of the so-called coupled points and also in terms of solutions of the generalized Riccati equation. As applications, the comparison theorems of Leighton-Levin type will be given.

Along with (1.1.1) consider the p-degree functional

$$\mathcal{J}(\eta; a, b) = \alpha|\eta(a)|^p + \beta|\eta(b)|^p + \int_a^b \left[r(t)|\eta'(t)|^p - c(t)|\eta(t)|^p \right] dt \tag{5.8.1}$$

over the class of functions $\eta \in W^{1,p}(a, b)$ satisfying the boundary condition

$$D \begin{pmatrix} \eta(a) \\ \eta(b) \end{pmatrix} = \begin{pmatrix} 0 \\ 0 \end{pmatrix}, \tag{5.8.2}$$

where D is a diagonal 2×2 matrix. This condition covers the cases, in which the values at the endpoints of the interval $[a, b]$ are independent. Since the matrix D is diagonal, it can be considered in one of the following forms

$$D_1 = \begin{pmatrix} 1 & 0 \\ 0 & 1 \end{pmatrix}, \quad D_2 = \begin{pmatrix} 0 & 0 \\ 0 & 1 \end{pmatrix}, \quad D_3 = \begin{pmatrix} 1 & 0 \\ 0 & 0 \end{pmatrix}, \quad D_4 = \begin{pmatrix} 0 & 0 \\ 0 & 0 \end{pmatrix}.$$

The class of functions η will be said to be the class of *admissible functions.*

In [379, 380] the concept of *coupled points* and *regularity condition* has been introduced for the study of *quadratic* functionals with general boundary conditions. This concept will be used throughout this chapter. The following definitions are motivated by the definition of the coupled point in the linear case.

Definition 5.8.1. A point $d \in (a, b]$ is said to be the *semicoupled point* with the point a relative to the functional $\mathcal{J}(\cdot; a, b)$ if there exists a nontrivial solution $y(t)$

of (1.1.1) and $\sigma \in \mathbb{R}^2$ such that

$$D \begin{pmatrix} y(a) \\ y(d) \end{pmatrix} = \begin{pmatrix} 0 \\ 0 \end{pmatrix}, \tag{5.8.3}$$

$$\begin{pmatrix} \alpha\Phi(y(a)) \\ \beta\Phi(y(d)) \end{pmatrix} + \begin{pmatrix} -r(a)\Phi(y'(a)) \\ r(d)\Phi(y'(d)) \end{pmatrix} - \begin{pmatrix} 0 \\ \Phi(y(d)) \int_d^b c(t)\,dt \end{pmatrix} = D\sigma. \tag{5.8.4}$$

A point $d \in (a,b)$ is said to be the *coupled point* with the point a relative to the functional $\mathcal{J}(\cdot; a, b)$ if it is semicoupled with a and the solution $y(\cdot)$ from the definition of semicoupled point satisfies

$$y(\cdot) \not\equiv y(d) \text{ on } [d,b]. \tag{5.8.5}$$

A point a is said to be *its own coupled point* if $y(\cdot) \equiv \text{const} \neq 0$ is admissible, is not a solution of (1.1.1) on $[a,b]$ and

$$\alpha + \beta - \int_a^b c(t)\,dt = 0.$$

If d is a coupled point with a and the interval $[a,d)$ does not contain any other point coupled with a, then d is said to be the *first coupled point* with a.

Definition 5.8.2. The functional $\mathcal{J}(\eta; a, b)$ is said to satisfy the *regularity condition* if it is nonnegative for any admissible constant function.

Recall that a point d is said to be the *conjugate point* to the point a if there exists a nontrivial solution y of (1.1.1) satisfying $y(a) = 0 = y(d)$. Let $I_0 \subset I$. Recall that equation (1.1.1) is said to be *disconjugate* on I_0 if there exists no pair of conjugate points in the interval I_0.

It is easy to see that if $D = D_1$, then the definition of the point coupled with the point a coincides with the definition of point conjugate to the point a. Observe that a can be its own coupled point only in the case of free end points, i.e., boundary condition $D = D_4$.

The regularity condition is trivially satisfied in the cases of boundary conditions D_1, D_2, D_3, since the only constant admissible function is the zero function. In case of free endpoints the functional $\mathcal{J}(\cdot; a, b)$ satisfies the regularity condition if and only if

$$\mathcal{K} := \alpha + \beta - \int_a^b c(t)\,dt \geq 0. \tag{5.8.6}$$

For comparison purposes, let us recall Roundabout theorems (Theorem 1.2.7 and Theorem 1.2.2, respectively), now slightly reformulated into the forms which are more consistent with what we are going to study.

Theorem 5.8.1. *The following statements are equivalent:*

(i) The functional $\mathcal{J}(\cdot; a, b)$ *is nonnegative on* $W_0^{1,p}(a, b)$.

(ii) Equation (1.1.1) *is disconjugate on* (a, b).

(iii) The solution w *of Riccati equation* (1.1.21) *which satisfies* $w(a+) = \infty$ *is defined on the entire interval* (a, b).

(iv) The solution w *of Riccati equation* (1.1.21) *which satisfies* $w(b-) = -\infty$ *is defined on the entire interval* (a, b).

Theorem 5.8.2. *The following statements are equivalent:*

(i) The functional $\mathcal{J}(\cdot; a, b)$ *is positive on* $W_0^{1,p}(a, b)$.

(ii) Equation (1.1.1) *is disconjugate on* $[a, b]$.

(iii) The solution w *of Riccati equation* (1.1.21) *which satisfies* $w(a+) = \infty$ *is defined on* $(a, b]$.

(iv) The solution w *of Riccati equation* (1.1.21) *which satisfies* $w(b-) = -\infty$ *is defined on* $[a, b)$.

We close this introductory part with the technical lemma, which actually was already used in the proof of Theorem 1.2.4. We present it here for an easy reference.

Lemma 5.8.1. *Let* $\eta \in W^{1,p}(a, b)$ *and let* w *be a solution of* (1.1.21) *defined on* (a, b). *If* $\eta(a) = 0$, $\eta(b) = 0$, *then*

$$\lim_{t \to a^+} w(t)|\eta(t)|^p = 0, \quad \lim_{t \to b^-} w(t)|\eta(t)|^p = 0,$$

respectively.

5.8.1 Disfocality

Let us study the functional $\mathcal{J}(\cdot; a, b)$ under the boundary condition $\eta(b) = 0$, i.e.,

$$\mathcal{J}(\eta; a, b) = \alpha|\eta(a)|^p + \int_a^b \Big[r(t)|\eta'(t)|^p - c(t)|\eta(t)|^p \Big] \, dt$$

over the set of admissible functions

$$U_{*0} = \{\eta \in W^{1,p}(a, b) : \eta(b) = 0\}.$$

In this case, $D = D_2$ in (5.8.2) and the definition of coupled point corresponds to the definition of a *focal point*, which is defined as follows.

Definition 5.8.3. A point $d \in (a, b]$ is said to be *focal point* (or *right focal point*) *to* a *relative to the functional* $\mathcal{J}(\cdot; a, b)$ if the nontrivial solution $y(t)$ of (1.1.1) satisfying

$$\alpha \Phi(y(a)) - r(a) \Phi(y'(a)) = 0 \tag{5.8.7}$$

has a zero at $t = d$.

Remark 5.8.1. In many works, especially when an equation under consideration is not considered in the variational context (cf. Subsection 5.1.3), the focal point is defined with $\alpha = 0$ in Definition 5.8.3, i.e., a point $t_1 \in (t_0, b)$ is said to be a focal point to the point t_0 if there exists a nontrivial solution of the equation such that $y'(t_0) = 0 = y(t_1)$.

Note that the solution of (1.1.1) is by the condition (5.8.7) given uniquely up to a constant multiple.

First we give a condition for nonnegativity of $\mathcal{J}(\cdot; a, b)$ in terms of solutions of Riccati equation (1.1.21). Denote by $y_b(t)$ the solution of (1.1.1) which satisfies the initial conditions $y_b(b) = 0$, $y_b'(b) = -1$, and let $w_b(t)$ be the corresponding solution of Riccati equation. The function w_b satisfies $w_b(b-) = -\infty$.

Lemma 5.8.2. *Functional $\mathcal{J}(\eta; a, b)$ is nonnegative on the class of admissible functions $\eta \in U_{*0}$ if and only if $w_b(t)$ is defined on $[a, b)$ and satisfies $\alpha - w_b(a) \ge 0$.*

Proof. "$\Rightarrow$": Let $\mathcal{J}(\cdot; a, b)$ be nonnegative on the class of admissible functions. Suppose that $w_b(t)$ is not defined at a point $d \in [a, b)$ and it is defined on (d, b), this means that $y_b(d) = 0$ and $y_b(t) \ne 0$ for every $t \in (d, b)$. Let $\lambda \in (d, b)$ be a real number and define the admissible function

$$\eta_\lambda(t) = \begin{cases} y_b(\lambda) & t \in [a, \lambda], \\ y_b(t) & t \in [\lambda, b], \end{cases}$$

see Figure 5.8.1. From the definition of the function $\eta_\lambda(t)$ it follows

$$\begin{aligned} \mathcal{J}(\eta_\lambda; a, b) &= \alpha |\eta_\lambda(a)|^p + \int_a^\lambda \Big[r(t) |\eta_\lambda'(t)|^p - c(t) |\eta_\lambda(t)|^p \Big] \, dt \\ &\quad + \int_\lambda^b \Big[r(t) |y_b'(t)|^p - c(t) |y_b(t)|^p \Big] \, dt \\ &= |y_b(\lambda)|^p \Big(\alpha - \int_a^\lambda c(t) \, dt \Big) + r(t) \Phi(y_b'(t)) y_b(t) \Big|_\lambda^b \\ &\quad - \int_\lambda^b y_b(t) \Big[\Big(r(t) \Phi(y_b'(t)) \Big)' + c(t) \Phi(y_b(t)) \Big] \, dt \end{aligned}$$

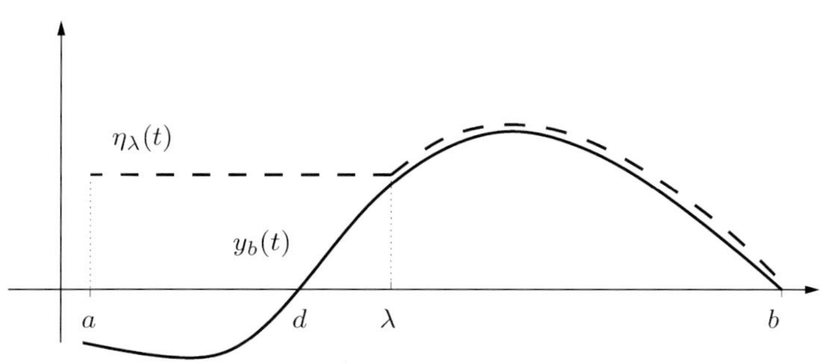

Figure 5.8.1: Construction of the admissible function $\eta_\lambda(t)$

$$= \quad |y_b(\lambda)|^p \Big(\alpha - r(\lambda)\Phi(y_b'(\lambda)/y_b(\lambda)) - \int_a^\lambda c(t)\,dt\Big),$$

where the second integral was computed using integration by parts. If $\lambda \to d+$, then the expression in parenthesis tends to $-\infty$ and clearly there exists $\lambda_0 \in (d,b)$ such that $\mathcal{J}(\eta_{\lambda_0}; a, b) < 0$, a contradiction. This means that $w_b(t)$ is defined on $[a,b)$. The function y_b is admissible and a similar computation as above gives

$$0 \le \mathcal{J}(y_b; a, b) = |y_b(a)|^p \Big(\alpha - w_b(a)\Big).$$

It holds $y_b(a) \neq 0$, hence $\alpha - w_b(a) \ge 0$.

"$\Leftarrow$": Let $w_b(t)$ be defined on $[a,b)$, $\alpha - w_b(a) \ge 0$ and η be an admissible function. Integrating Picone's identity (1.2.3) from a to $b - \varepsilon$, letting $\varepsilon \to 0+$ and using Lemma 5.8.1 we get

$$\mathcal{J}(\eta; a, b) = |\eta(a)|^p \Big(\alpha - w_b(a)\Big) + \int_a^b pP(r^{1/p}y', r^{-1/p}w\Phi(\eta))\,dt \ge 0.$$

□

The next theorem gives the conditions that are equivalent to right disfocality of (1.1.1) (i.e., the condition (ii)).

Theorem 5.8.3. *The following statements are equivalent:*

(i) The functional $\mathcal{J}(\cdot; a, b)$ is nonnegative on U_{*0}.

(ii) There exists no focal point to a in (a,b).

(iii) The solution w of Riccati equation (1.1.21) satisfying $w(a) = \alpha$ is defined on $[a,b)$.

(iv) The solution $w_b(t)$ of Riccati equation (1.1.21) satisfying $w_b(b-) = -\infty$ is defined on $[a,b)$ and $\alpha - w_b(a) \ge 0$ holds.

Proof. (i)$\Rightarrow$(ii): Suppose, by contradiction, that the functional is nonnegative on the class of admissible functions and $d \in (a,b)$ is the first focal point to a. Then there exists a nontrivial solution $u(t)$ of (1.1.1) such that $\alpha\Phi(u(a)) - r(a)\Phi(u'(a)) = 0$, $u(d) = 0$, and $u(t) \neq 0$ for $t \in (a,d)$. Let $\lambda \in (a,d)$ be a real number and $\eta_\lambda(t)$ be an admissible function defined by

$$\eta_\lambda(t) := \begin{cases} u(t) & t \in [a,\lambda], \\ u(\lambda)(b-t)/(b-\lambda) & t \in [\lambda, b], \end{cases} \tag{5.8.8}$$

see Figure 5.8.2. Then

$$\begin{aligned} \mathcal{J}(\eta_\lambda; a, b) &= \alpha|\eta_\lambda(a)|^p + \int_a^\lambda \big(r(t)|u'(t)|^p - c(t)|u(t)|^p\big)\,dt \\ &\quad + \Big|\frac{u(\lambda)}{b-\lambda}\Big|^p \int_\lambda^b \big(r(t) - c(t)|b-t|^p\big)\,dt \\ &= |u(\lambda)|^p \Big[r(\lambda)\Phi\big(u'(\lambda)/u(\lambda)\big) + \frac{1}{|b-\lambda|^p}\int_\lambda^b \big(r(t) - c(t)|b-t|^p\big)\,dt\Big]. \end{aligned}$$

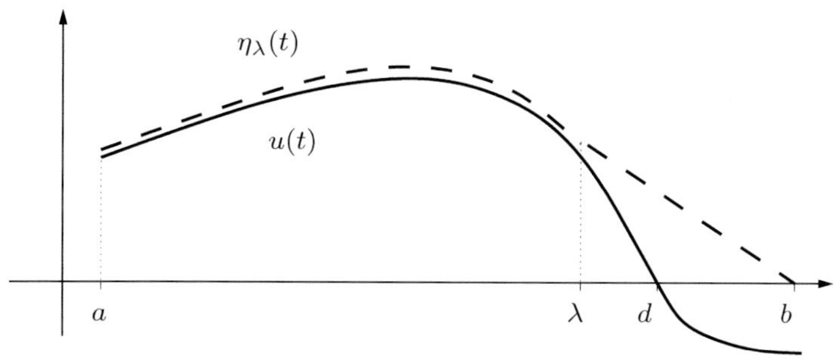

Figure 5.8.2: Construction of the admissible function $\eta_\lambda(t)$

The first term in the parenthesis tends to $-\infty$ if $\lambda \to d-$ and the second one is bounded. Hence there exists λ_0 such that $\mathcal{J}(\eta_{\lambda_0}; a, b) < 0$, a contradiction.

(ii)$\Rightarrow$(iii): This follows immediately from the definition of focal point.

(iii)$\Rightarrow$(i): If a solution w of (1.1.21) satisfying $w(a) = \alpha$ is defined on $[a, b)$, then integrating Picone's identity from a to $b - \varepsilon$, letting $\varepsilon \to 0$ and using Lemma 5.8.1 we have

$$\mathcal{J}(\eta; a, b) = \int_a^b P(r^{1/p}\eta', r^{-1/p} w\Phi(\eta))\, dt \geq 0,$$

and so the functional is nonnegative on U_{*0}.
(i)$\Leftrightarrow$(iv): See Lemma 5.8.2. □

An analogous method yields the following modification of Theorem 5.8.3.

Theorem 5.8.4. *The following statements are equivalent:*

(i) The functional $\mathcal{J}(\cdot; a, b)$ *is positive definite on* U_{*0}.

(ii) There exists no focal point to a *on* $(a, b]$.

(iii) The solution w *of Riccati equation* (1.1.21) *satisfying* $w(a) = \alpha$ *is defined on* $[a, b]$.

(iv) The solution $w_b(t)$ *of Riccati equation* (1.1.21) *satisfying* $w_b(b-) = -\infty$ *is defined on* $[a, b)$ *and* $\alpha - w_b(a) > 0$ *holds.*

Proof. (i)$\Rightarrow$(ii): Let $d \in (a, b]$ be a focal point to a, and $y(t)$ be the solution of (1.1.1) which realizes this focal point. Let η be an admissible function, which is equal to $y(t)$ on $[a, d]$ and to zero on $[d, b]$. Then integrating by parts we have $\mathcal{J}(\eta; a, b) = 0$, but $\eta \not\equiv 0$, a contradiction.

(ii)$\Rightarrow$(iii): This follows immediately from the definition of focal point.

(iii)$\Rightarrow$(i): This follows from the Picone identity and from the properties of the function $P(\cdot, \cdot)$.

(i)$\Leftrightarrow$(iv): Analogously to Lemma 5.8.2. □

Definition 5.8.4. A point $d \in [a,b)$ is said to be *left focal point* to the point b relative to the functional $\mathcal{J}(\cdot;a,b)$ if the nontrivial solution y of (1.1.1) satisfying

$$r(b)\Phi(y'(b)) + \beta\Phi(y(b)) = 0$$

has the zero at $t = d$.

In this case $D = D_3$, the functional $\mathcal{J}(\cdot;a,b)$ takes the form

$$\mathcal{J}(\eta;a,b) = \beta|\eta(b)|^p + \int_a^b \Big[r(t)|\eta'(t)|^p - c(t)|\eta(t)|^p\Big]\,dt$$

over the set of admissible functions

$$U_{0*} = \{\eta \in W^{1,p}(a,b) : \eta(a) = 0\}$$

and Definition 5.8.1 yields: A point d is a coupled point to the point a relative to the functional $\mathcal{J}(\cdot;a,b)$ if there exists a nontrivial solution $y_a(t)$ of (1.1.1) such that

$$y_a(a) = 0, \tag{5.8.9}$$

$$r(d)\Phi(y_a'(d)) + \Phi(y_a(d))\Big(\beta - \int_d^b c(t)\,dt\Big) = 0, \tag{5.8.10}$$

$$y_a(\cdot) \not\equiv y_a(d) \text{ on } [d,b]. \tag{5.8.11}$$

Before presenting the next theorem note that the condition (iv) of that statement is in fact left disfocality of (1.1.1).

Theorem 5.8.5. *Let us consider the boundary condition D_3. The following statements are equivalent:*

(i) The functional $\mathcal{J}(\cdot;a,b)$ is nonnegative on U_{0*}.

(ii) There exists no coupled point with a in the interval $[a,b)$.

(iii) The solution $w_a(t)$ of Riccati equation (1.1.21) satisfying $w_a(a+) = +\infty$ is defined on $(a,b]$ and $\beta + w_a(b) \ge 0$ holds.

(iv) The solution w of Riccati equation (1.1.21) satisfying $w(b) + \beta = 0$ is defined on $(a,b]$, i.e., there exists no left focal point to the point b in (a,b).

Proof. (i)⇒(ii): Suppose that there exists a point $d \in (a,b)$ coupled with a and y_a is a nontrivial solution of (1.1.1) satisfying (5.8.9). Then $y_a(d) \ne 0$. Indeed, if $y_a(d) = 0$, then from (5.8.10) it follows $y_a'(d) = 0$ and $y_a \equiv 0$, a contradiction. Then in view of (5.8.11) there exists a point $e \in (d,b]$ such that $y_a(t) \ne 0$ for all $t \in [d,e]$ and $y_a(\cdot) \not\equiv \text{const}$ on $[d,e]$, hence $w_a(\cdot) \not\equiv 0$ on $[d,e]$. Define the function $\eta(t)$ by

$$\eta(t) = \begin{cases} y_a(t) & t < e, \\ y_a(e) & t \ge e. \end{cases} \tag{5.8.12}$$

Then

$$\begin{aligned}\mathcal{J}(\eta; a, b) &= r(t)\Phi(y_a'(t))y_a(t)|_a^e + \beta|y_a(e)|^p - \int_e^b c(t)|y_a(e)|^p\,dt \\ &= |y_a(e)|^p\Big(r(e)\Phi\Big(\frac{y_a'(e)}{y_a(e)}\Big) + \beta - \int_e^b c(t)\,dt\Big) \\ &= |y_a(e)|^p\Big(w_a(e) + \beta - \int_e^b c(t)\,dt\Big).\end{aligned}$$

The function w_a is defined on $[d, e]$. From (1.1.21) and (5.8.10) it follows

$$w_a(e) - w_a(d) + \int_d^e c(t)\,dt = -(p-1)\int_d^e r^{1-q}(t)|w_a(t)|^q\,dt$$

$$\beta = -w_a(d) + \int_d^b c(t)\,dt.$$

Combining these computations, we get

$$\mathcal{J}(\eta; a, b) = -|y_a(e)|^p(p-1)\int_d^e r^{1-q}(t)|w_a(t)|^q\,dt.$$

In view of the fact that $w_a \not\equiv 0$ on $[d, e]$, this integral is negative, a contradiction.

(ii)⇒(iii): Assume that the solution $w_a(t)$ is not defined on $(a, b]$. Then there exists a point $e \in (a, b]$ such that $w_a(t)$ is defined on (a, e) and $w_a(e-) = -\infty$. Define the function

$$f(t) = w_a(t) + \beta - \int_t^b c(t)\,dt$$

for $t \in (a, e)$. The function $f(t)$ is continuous and satisfies $f(a+) = \infty$ and $f(e-) = -\infty$. Hence there exists a point $d \in [a, e)$ such that $f(d) = 0$, i.e., d satisfies (5.8.10). Moreover,

$$y_a(e) = 0 \neq y_a(d),$$

(5.8.11) is satisfied and d is a coupled point with a. Hence $w_a(t)$ is defined on $(a, b]$. Suppose $w_a(b) + \beta < 0$. Then the function $f(t)$ satisfies $f(a+) = \infty$ and $f(b) < 0$ and again there exists a point d such that $f(d) = 0$, i.e., (5.8.10) holds. We claim that (5.8.11) holds, too. Indeed, if $y_a(\cdot) \equiv$ const on $[d, b]$, then $c(t) \equiv 0 \equiv w_a(t)$ on $[d, b]$ and

$$w_a(d) + \beta - \int_d^b c(t)\,dt = w_a(b) + \beta = f(b) < 0,$$

which contradicts (5.8.10).

(iii)⇒(iv): Let $w(t)$, $w_a(t)$ be the solutions of (1.1.21) given by the initial conditions $w(b) = -\beta$, $w_a(a+) = \infty$, respectively. In view of (iii) it holds $w_a(b) \geq -\beta = w(b)$. Due to the fact that the graphs of two distinct solutions of (1.1.21) cannot intersect it holds $w(t) \leq w_a(t)$ for every $t \in (a, b]$. Clearly there cannot exist a point $e \in (a, b)$ such that $w(e+) = \infty$, hence $w(t)$ is defined on $(a, b]$.

(iv)⇒(i): Let $w(t)$ be the solution of Riccati equation (1.1.21) given by the initial condition $w(b) = -\beta$. From the Picone identity we have for every admissible function η

$$\begin{aligned}\mathcal{J}(\eta; a, b) &= \lim_{\varepsilon\to 0+}\Big[\big(\alpha - w(a+\varepsilon)\big)|\eta(a+\varepsilon)|^p \\ &\qquad + \big(\beta + w(b)\big)|\eta(b)|^p + \int_{a+\varepsilon}^{b} P(r^{1/p}\eta', r^{-1/p}w\Phi(\eta))\,dt\Big].\end{aligned}$$

The first term tends to zero by Lemma 5.8.1, the second one is vanishing and the third one is nonnegative, hence the functional is nonnegative. □

Theorem 5.8.6. *Let us consider the boundary condition D_3. The following statements are equivalent:*

(i) The functional $\mathcal{J}(\cdot; a, b)$ is positive definite on U_{0*}.

(ii) There exists no semicoupled point with a in the interval $[a, b]$.

(iii) The solution $w_a(t)$ of Riccati equation (1.1.21) *satisfying $w_a(a+) = +\infty$ is defined on $(a, b]$ and satisfies $\beta + w_a(b) > 0$.*

(iv) The solution w of Riccati equation (1.1.21) *satisfying $w(b)+\beta = 0$ is defined on $[a, b]$, i.e., there exists no left focal point to the point b in $[a, b)$.*

Proof. (i)⇒(ii): If $d \in [a, b]$ is semicoupled with a, then the function η defined by $\eta \equiv y_a$ on $[a, d]$, $\eta \equiv y_a(d)$ on $[d, b]$ satisfies $\mathcal{J}(\eta; a, b) = 0$ and $y_a \not\equiv 0$, a contradiction.

(ii)⇒(iii): The existence of w_a follows from Theorem 5.8.3. If $w_a(b) + \beta \le 0$, then there exists a point $d \in [a, b]$ such that $f(d) = 0$, where the function f is defined in the proof of Theorem 5.8.5. The point d is semicoupled with a.

(iii)⇒(iv): This follows from the fact that two different solutions of generalized Riccati equation cannot intersect.

(iv)⇒(i): From Theorem 5.8.5, it follows nonnegativity of $\mathcal{J}(\cdot; a, b)$ over the class of admissible functions. The equality holds only if η is a constant multiple of y, where y is the solution of (1.1.1) corresponding to the solution of generalized Riccati equation given by the initial condition $w(b) = -\beta$. Since $\eta(a) = 0 \neq y(a)$, the only possible case is $\eta \equiv 0$ and the functional is positive definite. □

5.8.2 Nonexistence of coupled points

Let us discuss the case with both free endpoints. In this case, the coupled point is defined in the following way: a point $d \in [a, b)$ is said to be coupled point with a if there exists a nontrivial solution $y(t)$ of (1.1.1) satisfying

$$\alpha\Phi(y(a)) - r(a)\Phi(y'(a)) = 0, \tag{5.8.13}$$

$$r(d)\Phi(y'(d)) + \Phi(y(d))(\beta - \int_d^b c(t)\,dt) = 0, \tag{5.8.14}$$

$$y(\cdot) \not\equiv y(d) \text{ on } [d, b]. \tag{5.8.15}$$

To simplify the proof of our theorem we first give an alternative representation of the coupled points in Lemmas 5.8.3 and 5.8.4. In the following part we suppose that the function y is a solution of (1.1.1) satisfying (5.8.13). This solution is defined up to a constant multiple and the corresponding solution $w(t)$ of generalized Riccati equation satisfies the initial condition $w(a) = \alpha$. Denote

$$\theta(t) = \int_a^t (p-1)r^{1-q}(s)|w(s)|^q\, ds,$$

where $w(s)$ is defined on $[a,t)$. From Riccati equation (1.1.21) it follows that $\theta(t) = \alpha - w(t) - \int_a^t c(s)\, ds$.

Lemma 5.8.3. *Let $w(t)$ be defined on $[a,e] \subset [a,b)$, $d \in [a,e)$ be such that $\theta(d) = \mathcal{K}$ and $\theta(e) > \mathcal{K}$, where $\mathcal{K}$ is defined in (5.8.6). Then d is a coupled point with a.*

Proof. It holds $y(d) \neq 0$. Because of

$$\begin{aligned} 0 &= \mathcal{K} - \theta(d) = \alpha + \beta - \int_a^b c(t)\, dt - \alpha + w(d) + \int_a^d c(t)\, dt \\ &= \beta + r(d)\Phi\Big(\frac{y'(d)}{y(d)}\Big) - \int_d^b c(t)\, dt, \end{aligned}$$

relation (5.8.14) holds. Since

$$0 < \theta(e) - \mathcal{K} = \theta(e) - \theta(d) = \int_d^e (p-1)r^{1-q}(s)|w(s)|^q\, ds,$$

one has $w(\cdot) \not\equiv 0$ on $[d,e]$, hence $y(\cdot) \not\equiv \text{const}$ on $[d,e]$ and d is a coupled point with a. $\square$

Lemma 5.8.4. *Let $\mathcal{J}(\eta; a, b)$ satisfy the regularity condition. If $d \in [a,b)$ is the first coupled point with the point a, then there exists a point $e \in (d,b)$ such that the solution $w(t)$ of Riccati equation (1.1.21) given by the initial condition $w(a) = \alpha$ is defined on $[a,e]$, $\theta(d) = \mathcal{K}$ and $\theta(e) > \mathcal{K}$.*

Proof. Suppose, that $w(t)$ is not defined on $[a,d]$, i.e., there exists a point $\tau \in [a,d]$ such that $w(t)$ is defined for $t \in [a,\tau)$ and $\lim_{t\to\tau-} w(t) = -\infty$. Then

$$\theta(\tau-) = \int_a^{\tau-} (p-1)r^{1-q}(t)|w(t)|^q\, dt = \infty$$

and clearly there exist points $d_0 \in (a,\tau)$ and $e_0 \in (d_0,\tau)$ such that $\theta(d_0) = \mathcal{K}$ and $\theta(e_0) > \mathcal{K}$. By the previous lemma $d_0 < d$ is a focal point, a contradiction.

In the same way as in the proof of the previous lemma we can show that if d is a coupled point, then (5.8.14) implies $\theta(d) = \mathcal{K}$. The point d is a coupled point, hence $y(d) \neq 0$. Indeed, if $y(d) = 0$, then it follows from (5.8.14) that $y'(d) = 0$ and in view of the uniqueness of solution of (1.1.1) it holds $y(\cdot) \equiv 0$, a contradiction. Because of $y(\cdot) \neq \text{const}$ on $[d,b]$ and $y(d) \neq 0$, there exists a point $e \in (d,b)$ such that $y(t) \neq 0$ for all $t \in [d,e]$ and $y(\cdot) \not\equiv$ const on $[d,e]$. Hence $w(t)$ is defined on $[a,e]$ and $w(\cdot) \not\equiv 0$ on $[d,e]$. From here

$$\theta(e) = \theta(d) + \int_d^e (p-1) r^{1-k}(t) |w(t)|^k \, dt > \theta(d) = \mathcal{K},$$

which completes the proof. □

Theorem 5.8.7. *Consider the boundary conditions $D = D_4$, i.e., both endpoints are free. The following statements are equivalent:*

(i) The functional $\mathcal{J}(\cdot; a, b)$ is nonnegative over $W^{1,p}(a,b)$.

(ii) There exists no coupled point with a in the interval $[a,b)$ and the regularity condition is satisfied.

(iii) The solution $w(t)$ of Riccati equation (1.1.21) given by the initial condition $w(a) = \alpha$ is defined on $[a,b]$ and satisfies $\beta + w(b) \geq 0$.

(iv) The solution $w(t)$ of Riccati equation (1.1.21) given by the initial condition $w(b) + \beta = 0$ is defined on $[a,b]$ and satisfies $\alpha - w(a) \geq 0$.

Proof. (i)⇒(ii): The validity of the regularity condition follows immediately from the nonnegativity of the functional $\mathcal{J}(\eta; a, b)$. Now let $y(t)$ be a solution of (1.1.1) satisfying (5.8.13) and $d \in [a,b)$ be the first coupled point with a. From Lemma 5.8.4 it follows that there exists $e \in (d,b)$ such that the function $y(t)$ has no zero on $[a,e]$ and $\theta(e) > \mathcal{K}$. Define an admissible function

$$\eta(t) = \begin{cases} y(t) & t \in [a,e), \\ y(e) & t \in [e,b]. \end{cases}$$

Then it holds

$$\begin{aligned} \mathcal{J}(\eta; a, b) &= \alpha |y(a)|^p + \beta |y(e)|^p + r(t) \Phi\Big(\frac{y'(t)}{y(t)}\Big)\Big|_a^e - |y(e)|^p \int_e^b c(t)\, dt \\ &= |y(e)|^p \left[\beta - \int_e^b c(t)\, dt + r(e) \Phi\Big(\frac{y'(e)}{y(e)}\Big) \right] = |y(e)|^p \Big(\mathcal{K} - \theta(e)\Big) < 0, \end{aligned}$$

a contradiction.

(ii)⇒(iii): Assume that (iii) does not hold, there exists no coupled point on $[a,b)$ and the regularity condition is satisfied. If $w(t)$ is not defined on $[a,b]$, then following the same way as in the first part of the proof of Lemma 5.8.4 we can show that there exists a coupled point with a on $[a,b)$, a contradiction. Hence $w(t)$ really exists on $[a,b]$. Assume that $w(b) + \beta < 0$. Then in view of the relations $w(b) + \beta = \mathcal{K} - \theta(b) < 0$, $\theta(a) = 0$, $\mathcal{K} \geq 0$ and the fact that θ is a continuous function, there exist points d, e which satisfies the conditions from Lemma 5.8.3. Then the point d is a coupled point, a contradiction.

(iii)⇒(iv): Denote by w and $\tilde{w}$ the solutions of (1.1.21) given by the initial conditions $w(a) = \alpha$ and $\tilde{w}(b) = -\beta$, respectively. Then $w(b) \geq -\beta = \tilde{w}(b)$ and

$w(t)$ is defined on $[a,b]$. Due to the fact that two distinct solutions of (1.1.21) cannot intersect it holds $\tilde{w}(t) \le w(t)$ on $[a,b]$ and there cannot exist a point $e \in [a,b]$ such that $\tilde{w}(e+) = \infty$. Hence $\tilde{w}(t)$ is defined on $[a,b]$ and $\tilde{w}(a) \le w(a) = \alpha$.

(iv)⇒(i): Since $w(t)$ is defined on $[a,b]$ then from the Picone identity it follows

$$\mathcal{J}(\eta; a, b) = |\eta(a)|^p(\alpha - w(a)) + |\eta(b)|^p(\beta + w(b)) + \int_a^b P(r^{1/p}\eta', r^{-1/p}w\Phi(\eta))\,dt.$$

The second term equals zero, and the first and the third ones are nonnegative, hence the functional is nonnegative. □

We close this section with summarizing all cases.

Theorem 5.8.8. *Let the matrix D in (5.8.2) be diagonal. The functional $\mathcal{J}(\eta; a, b)$ is nonnegative for every admissible function η if and only if there exists no coupled point with a in the interval $[a,b)$ and the regularity condition is satisfied.*

Remark 5.8.2. When $p = 2$, Theorem 5.8.8 coincides in the scalar case with Theorem 3 from [95], where general boundary conditions and n-dimensional problem are considered.

Since $W_0^{1,p}(a,b) \subset U_{*0} \subset W^{1,p}(a,b)$, the nonnegativity of functional $\mathcal{J}(\cdot; a, b)$ as a functional with free endpoints (with one free endpoint) implies nonnegativity of this functional as a functional with one free endpoint (with zero boundary conditions). Hence we have the result for ordering of coupled points according to different boundary conditions.

Theorem 5.8.9. *Let d_f, d_l, d_0 be the first coupled point with a relative to the functional $\mathcal{J}(\cdot; a, b)$ with free endpoints, free left endpoint and zero boundary conditions, respectively. Then*

$$d_f \le d_l \le d_0.$$

The same statement holds if "left endpoint" and "d_l" are replaced by "right endpoint" and "d_r", respectively.

5.8.3 Comparison theorems of Leighton-Levin type

The aim of this subsection is to establish comparison theorems for focal points of half-linear ordinary differential equations. Our main tool is the relationship between nonnegativity of an appropriate functional and the nonexistence of focal points, established in the previous subsections. Thus, at the same time, we give complementary results to those in Subsection 2.3.2 and an application of the above theory. For historical reasons, we speak about comparison theorems of Leighton-Levin type, see [235, 341, 378] for the linear case.

Let us consider two differential equations

$$\mathcal{L}_{r,c}[x] \equiv (r(t)\Phi(x'))' + c(t)\Phi(x) = 0, \tag{5.8.16}$$

$$\mathcal{L}_{R,C}[y] \equiv (R(t)\Phi(y'))' + C(t)\Phi(y) = 0, \tag{5.8.17}$$

where $r(t)$, $R(t)$, $c(t)$ and $C(t)$ are continuous on $[a,b]$, $R(t)$ and $r(t)$ are positive. In the classical Sturmian comparison theorem we assume

$$R(t) \le r(t) \text{ for } t \in [a,b] \tag{5.8.18}$$

$$c(t) \le C(t) \text{ for } t \in [a,b]. \tag{5.8.19}$$

Recall that equation (5.8.17) is then called the *Sturmian majorant* of (5.8.16). Here we relax these assumptions.

Recall that in Theorem 2.3.5 we have assumed the existence of a solution to (5.8.16) satisfying $x(a) = 0 = x(b)$. For the case $x(a) \neq 0 = x(b)$ using Theorems 5.8.3 and 5.8.4 we obtain the comparison theorem for focal points.

Theorem 5.8.10. *Let x be a solution of* (5.8.16) *such that $x(b) = 0 \neq x(a)$, y be a solution of* (5.8.17) *such that $y(a) \neq 0$. Denote*

$$\alpha = r(a)\Phi\Big(\frac{x'(a)}{x(a)}\Big) \quad \text{and} \quad A = R(a)\Phi\Big(\frac{y'(a)}{y(a)}\Big).$$

If

$$\mathcal{V}[x] \equiv (\alpha - A)|x(a)|^p + \int_a^b \Big[\Big(r(t) - R(t)\Big)|x'(t)|^p + \Big(C(t) - c(t)\Big)|x(t)|^p\Big]\, dt \ge 0,$$

then y has a zero in $(a,b]$. If, in addition, the inequality is strict, then y has a zero in (a,b).

Proof. Denote

$$\mathcal{J}_\alpha(\eta; a,b) = \alpha|\eta(a)|^p + \int_a^b (r|\eta'|^p - c|\eta|^p)\, dt,$$

$$\mathcal{J}_A(\eta; a,b) = A|\eta(a)|^p + \int_a^b (R|\eta'|^p - C|\eta|^p)\, dt.$$

To prove the theorem, it is sufficient to find a function $\eta \in C^1[a,b]$ such that $\eta(b) = 0$ and $\mathcal{J}_A(\eta; a,b)$ is nonpositive (negative) and then Theorems 5.8.3, 5.8.4 yield the conclusion. Integration by parts shows $\mathcal{J}_\alpha(x; a,b) = 0$. Hence

$$\mathcal{J}_A(x; a,b) = \mathcal{J}_A(x; a,b) - \mathcal{J}_\alpha(x; a,b) = -\mathcal{V}[x] \le 0.$$

Therefore the functional $\mathcal{J}_A(\eta; a,b)$ is not positive definite and by Theorem 5.8.4 there exists a focal point to a relative to the functional $\mathcal{J}_A(\cdot; a,b)$ on $(a,b]$, or equivalently, $y(t)$ has a zero in $(a,b]$. If the strict inequality holds, the functional $\mathcal{J}_A(\cdot; a,b)$ is not positive semidefinite and by Theorem 5.8.3 there exists a focal point to a in (a,b), i.e., y has a zero in (a,b). Theorem is proved. □

Corollary 5.8.1. *Suppose* (5.8.18) *and* (5.8.19)*. Let x and y be solutions of* (5.8.16) *and* (5.8.17)*, respectively. Let $x(a) \neq 0 = x(b)$. If*

$$R(a)\Phi\Big(\frac{y'(a)}{y(a)}\Big) \le r(a)\Phi\Big(\frac{x'(a)}{x(a)}\Big), \tag{5.8.20}$$

then the function y has a zero in $(a,b]$. Moreover, y has a zero in (a,b) in the case when any of the following conditions is satisfied:

(i) The sharp inequality in (5.8.20) *is satisfied.*

(ii) $c(\cdot) \not\equiv C(\cdot)$ *on* (a,b).

(iii) $R(t_0) < r(t_0)$ *and* $c(t_0) \neq 0$ *at some* $t_0 \in (a,b)$.

Proof. This follows immediately from Theorem 5.8.10. □

Under additional condition on the nonnegativity of the function c the inequalities (5.8.19) and (5.8.20) can be replaced by a more general integral inequality, as the following theorem shows.

Corollary 5.8.2. *Suppose* (5.8.18). *Let $x(t)$ and $y(t)$ be solutions of* (5.8.16) *and* (5.8.17), *respectively. Let $c(t) \geq 0$, $x'(a) \leq 0 = x(b)$, $x(t) > 0$ for $t \in [a,b)$, $y(a) \neq 0$. If*

$$r(a)\Phi\Big(\frac{x'(a)}{x(a)}\Big) - R(a)\Phi\Big(\frac{y'(a)}{y(a)}\Big) + \int_a^t \Big[C(s) - c(s)\Big]\, ds \geq 0 \quad \text{for } t \in [a,b], \tag{5.8.21}$$

then the function y has a zero in $(a,b]$. Moreover, y has a zero in (a,b) if the sharp inequality in (5.8.21) *is satisfied, or if the condition* (iii) *of Corollary 5.8.1 holds.*

Proof. Let α, A be the numbers from Theorem 5.8.10. It holds

$$\int_a^b \Big[C(t) - c(t)\Big] |x(t)|^p\, dt = \int_a^b \int_0^{|x(t)|^p} \Big[C(t) - c(t)\Big]\, ds\, dt.$$

From (5.8.16) it follows

$$r(t)\Phi(x'(t)) = \alpha\Phi(x(a)) - \int_a^t c(s)\Phi(x(s))\, ds$$

and x is decreasing or it is constant in some interval $[a,c] \subset [a,b)$ and decreasing on $[c,b]$. The same is true also for $|x|^p$. Then we can interchange the order of the integration and we get

$$\begin{aligned}\int_a^b \Big[C(t) - c(t)\Big] |x(t)|^p\, dt &= \int_0^{|x(a)|^p} \int_0^{\varphi(s)} \Big[C(t) - c(t)\Big]\, dt\, ds \\ &\geq \int_0^{|x(a)|^p} [A - \alpha]\, ds = [A - \alpha]|x(a)|^p\end{aligned}$$

by (5.8.21). Note that $\varphi(s)$ is well-defined and continuous for $0 \leq s \leq |x(a)|^p$. By this computation we get

$$\mathcal{V}[x] \geq \int_a^b \Big[r(t) - R(t)\Big] |x'(t)|^p\, dt \geq 0,$$

and by Theorem 5.8.10 y has a zero in $(a,b]$. If the strict inequality is satisfied, then $\mathcal{V}[x] > 0$ and y has a zero in (a,b). □

The next comparison theorem is an alternative to Theorem 5.8.10.

Theorem 5.8.11. *Let r and R be positive functions which are continuously differentiable on (a,b). Let x be a solution of (5.8.16) such that $x(b)=0\neq x(a)$ and y be a solution of (5.8.17) such that $y(a)\neq 0$. Denote*

$$\alpha = r(a)\Phi\Big(\frac{x'(a)}{x(a)}\Big) \quad \textit{and} \quad A = R(a)\Phi\Big(\frac{y'(a)}{y(a)}\Big).$$

If

$$\mathcal{W}[x] := \Big(\frac{R(a)}{r(a)}\alpha - A\Big)|x(a)|^p + \int_a^b \Big[\Big(C(t) - \frac{R(t)}{r(t)}c(t)\Big)|x(t)|^p + r(t)\Big(\frac{R(t)}{r(t)}\Big)' x(t)\Phi\big(x'(t)\big)\Big]\, dt \geq 0,$$

then y has a zero in $(a,b]$. If, in addition, the sharp inequality holds, then y has a zero in (a,b).

Proof. Let x be a solution of (5.8.16) such that $x(b)=0$. Then

$$(5.8.22) \qquad \begin{aligned} \mathcal{L}_{R,C}[x] &= \Big(\frac{R}{r} r\Phi(x')\Big)' + C\Phi(x) \\ &= \Big(\frac{R}{r}\Big)' r\Phi(x') + \frac{R}{r}\big(r\Phi(x')\big)' + C\Phi(x) \\ &= \Big[C - \frac{R}{r}c\Big]\Phi(x) + \Big(\frac{R}{r}\Big)' r\Phi(x'), \end{aligned}$$

and since integration by parts shows that

$$\begin{aligned} \int_a^b \Big(\frac{R}{r}\Big)' r\Phi(x')x\, dt &= \big[R\Phi(x')x\big]_a^b - \int_a^b \Big[\frac{R}{r}x\big(r\Phi(x')\big)' + \frac{R}{r}r\Phi(x')x'\Big]\, dt \\ &= -R(a)\Phi\big(x'(a)\big)x(a) - \int_a^b \Big[R|x'|^p - \frac{R}{r}c|x|^p\Big]\, dt, \end{aligned}$$

it holds

$$\int_a^b x\mathcal{L}_{R,C}[x]\, dt = -\frac{R(a)}{r(a)}\alpha|x(a)|^p - \int_a^b \Big[R|x'|^p - C|x|^p\Big]\, dt.$$

Hence

$$\begin{aligned} \mathcal{J}_A(x;a,b) &= \Big(A - \frac{R(a)}{r(a)}\alpha\Big)|x(a)|^p + \int_a^b \Big[R|x'|^p - C|x|^p\Big]\, dt + \frac{R(a)}{r(a)}\alpha|x(a)|^p \\ &= \Big(A - \frac{R(a)}{r(a)}\alpha\Big)|x(a)|^p - \int_a^b x\mathcal{L}_{R,C}[x]\, dt \end{aligned}$$

and by (5.8.22),

$$\begin{aligned}\mathcal{J}_A(x;a,b) &= \Big(A-\frac{R(a)}{r(a)}\alpha\Big)|x(a)|^p-\int_a^b\Big[\Big(C-\frac{R}{r}c\Big)\Phi(x)+\Big(\frac{R}{r}\Big)' r\Phi(x')\Big]x\,dt\\ &= -\mathcal{W}[x].\end{aligned}$$

Now the statement follows from Theorems 5.8.3 and 5.8.4. □

An immediate consequence of this theorem is the following corollary. It extends the result of Reid, proved for the linear equation, see [341, p. 29].

Corollary 5.8.3. *Let $r, R \in C^1[a,b]$ be positive and $c \in C[a,b]$ be nonnegative. Let x and y be solutions of* (5.8.16) *and* (5.8.17)*, respectively, such that $x'(a) \le 0 = x(b)$, x is positive on $[a,b)$, $y(a) \ne 0$. If*

$$\frac{x'(a)}{x(a)} \ge \frac{y'(a)}{y(a)}, \quad \Big(\frac{R(t)}{r(t)}\Big)' \le 0 \quad \text{and} \quad \frac{C(t)}{R(t)} \ge \frac{c(t)}{r(t)},$$

then y has a zero in $(a,b]$.

If we apply the method from the proof of Theorem 5.8.2, we get the following corollary.

Corollary 5.8.4. *Let $r, R \in C^1[a,b]$ be positive and $c \in C[a,b]$ be nonnegative. Further let x, y be solutions of* (5.8.16), (5.8.17) *such that $x'(a) \le 0 = x(b)$, x is positive on $[a,b)$, $y(a) \ne 0$, respectively. If $(R/r)' \le 0$ and*

$$\frac{R(a)}{r(a)}\alpha - A + \int_a^t \Big[C(t) - \frac{R(t)}{r(t)}c(t)\Big]\,dt \ge 0 \quad \text{for } t \in [a,b],$$

then y has a zero in $(a,b]$.

Remark 5.8.3. As we have already pointed out, the focal point is sometimes defined with $\alpha = 0$ in Definition 5.8.3, i.e., a point $t_1 \in (t_0, b)$ is said to be the focal point to the point t_0 if there exists a nontrivial solution of (5.8.16) such that $y'(t_0) = 0 = y(t_1)$. If there exists in every neighborhood of ∞ such a point t_0 and its focal point t_1, then equation (5.8.16) is called *focal oscillatory*. This is motivated by the property following from the Sturmian type theorem of separation of zeros which states, that (5.8.16) is oscillatory if and only if there exists a pair of conjugate points in every neighborhood of ∞. Due to the separation of zeros of two solutions of (5.8.16), oscillation implies focal oscillation. The reverse statement does not hold, which can be showed on the example of Euler equation, which is nonoscillatory on $[1,\infty)$, but it is focal oscillatory on this interval. For more details see [303], where the linear systems are studied. Corollary 5.8.1 states, that if equation (5.8.16) is focal oscillatory, then its Sturmian majorant is focal oscillatory as well.

The same method as in Theorem 5.8.10 can be applied to the case of the functional with other boundary conditions. Then we obtain the following results.

Theorem 5.8.12. *Let x be solution of* (5.8.16) *such that $x(a) = 0 \neq x(b)$, y be a solution of* (5.8.17) *such that $y(b) \neq 0$. Denote*

$$\beta = -r(b)\Phi\Big(\frac{x'(b)}{x(b)}\Big) \quad \textit{and} \quad B = -R(b)\Phi\Big(\frac{y'(b)}{y(b)}\Big).$$

If

$$\mathcal{V}_b[x] \equiv (\beta - B)|x(b)|^p + \int_a^b \Big[\Big(r(t) - R(t)\Big)|x'(t)|^p + \Big(C(t) - c(t)\Big)|x(t)|^p\Big]\, dt \geq 0,$$

then y has a zero in $[a, b)$. If, in addition, the sharp inequality is satisfied, then y has a zero in (a, b).

Proof. The statement follows from the fact that

$$\mathcal{J}_B(x; a, b) \equiv B|x(b)|^p + \int_a^b \Big[R(t)|x'(t)|^p - C(t)|x(t)|^p\Big]\, dt = -\mathcal{V}_b[x] \leq 0$$

and from Theorems 5.8.5 and 5.8.6. □

Corollary 5.8.5. *Suppose* (5.8.18) *and $c(t) \geq 0$. Let x and y be solutions of* (5.8.16) *and* (5.8.17), *respectively. Let $x'(b) \geq 0 = x(a)$, $x(t) > 0$ for $t \in [a, b)$, $y(b) \neq 0$. Let β, B be the numbers from the preceding theorem. If*

$$\beta - B + \int_t^b \Big[C(s) - c(s)\Big]\, ds \geq 0 \quad \textit{for } t \in [a, b], \tag{5.8.23}$$

then the function $y(t)$ has a zero in $[a, b)$. Moreover, $y(t)$ has a zero in (a, b) if the sharp inequality in (5.8.23) *is satisfied, or if the condition* (iii) *of Theorem 5.8.1 holds.*

Proof. The proof is an obvious modification of the proof of Corollary 5.8.2. □

5.9 Miscellaneous

The topics treated in this section have no immediate unifying point. The results presented here do not fit into any of the previous sections of this chapter, but we consider these results sufficiently important to present them in this book.

5.9.1 Extended Hartman-Wintner criterion

In Section 2.1 we have shown that (1.1.1) with $\int^\infty r^{1-q}(t)\, dt = \infty$ is oscillatory provided

$$-\infty < \liminf_{t\to\infty} \frac{\int_T^t r^{1-q}(s) \int_T^s c(\tau) d\tau\, ds}{\int_T^t r^{1-q}(s)\, ds} < \limsup_{t\to\infty} \frac{\int_T^t r^{1-q}(s) \int_T^s c(\tau) d\tau\, ds}{\int_T^t r^{1-q}(s)\, ds}$$

or

$$\lim_{t\to\infty} \frac{\int_T^t r^{1-q}(s) \int_T^s c(\tau)d\tau\, ds}{\int_T^t r^{1-q}(s)\, ds} = \infty.$$

In this subsection we present one extension of this criterion. We formulate this criterion for (5.1.1), its extension to the equation with a general r satisfying $\int^\infty r^{1-q}(t)\, dt = \infty$ is immediate. For the sake of convenience, we introduce the linear operator $A : C[0,\infty) \to C[0,\infty)$ defined by

$$(5.9.1) \qquad (Af)(t) := \frac{1}{t}\int_0^t f(s)\, ds, \quad t > 0, \quad (Af)(0) := 0.$$

By A^n we denote the n-th iteration of A.

Theorem 5.9.1. *Let $C(t) := \int_0^t c(s)\, ds$. If there exists $n \in \mathbb{N}$ such that*

$$(5.9.2) \qquad -\infty < \liminf_{t\to\infty} (A^n C)(t) < \limsup_{t\to\infty} (A^n C)(t),$$

or

$$(5.9.3) \qquad \lim_{t\to\infty} (A^n C)(t) = \infty,$$

then (5.1.1) is oscillatory.

Proof. The proof is similar to that of Theorem 2.2.3 and of Theorem 2.2.10. Suppose, by contradiction, that (5.1.1) is nonoscillatory and let w be a solution of the associated Riccati equation (3.1.2). For convenience, we suppose that this solution is defined on $[0,\infty)$, this is no loss of generality since the lower integration limit 0 in the next computation can be replaced by any T sufficiently large. Integrating equation (3.1.2) from 0 to t we get

$$(5.9.4) \qquad w(t) - w(0) + C(t) + (p-1)\int_0^t |w(s)|^q\, ds = 0.$$

The application of the operator A^n to the previous equation yields

$$(5.9.5) \qquad (A^n w)(t) + (A^n C)(t) + (p-1)A^n\left(\int_0^t |w(s)|^q\, ds\right) - w(0) = 0.$$

Each of the conditions (i), (ii) implies the existence of $K \geq 0$ such that $(A^n C)(t) \geq -K$ for large t. This implies that $\int^\infty |w(t)|^q\, dt < \infty$. The proof of this claim goes by contradiction in the same way as in the proof of Theorem 2.2.3. Having proved the convergence of $\int^\infty |w(t)|^q\, dt$, again in the same way as in the proof of Theorem 2.2.3, we prove that $\lim_{t\to\infty}(A^n C)(t)$ exists finite, a contradiction with (i) or (ii). □

The following example presents a construction of the function c for which the classical Hartman-Wintner criterion (i.e., the case $n = 1$ in the previous theorem) does not apply, while the previous theorem with $n = 2$ does.

Example 5.9.1. Let $\{a_n\}_{n=1}^\infty$, $\{b_n\}_{n=1}^\infty$ be two sequences of real numbers defined by $a_n = n - 2^{-n}$, $b_n = n + 2^{-n}$, $n \in \mathbb{N}$. Let $\{g_n\}_{n=1}^\infty$ denote a sequence of functions $g_n : [0, \infty) \to \mathbb{R}$ of the class C^2 such that $g_n(t) > 0$ if $t \in (a_n, b_n)$ and $g_n(t) = 0$ otherwise. We also ask that

$$\int_0^\infty g_n(t)\,dt = n. \tag{5.9.6}$$

Next define $g : [0, \infty) \to \mathbb{R}$ by

$$g(t) = \sum_{n=1}^\infty (-1)^n g_n(t). \tag{5.9.7}$$

The function g is also of the class C^2 and using this function we define

$$c(t) = (tg(t))'', \quad C(t) := \int_0^t c(s)\,ds. \tag{5.9.8}$$

The reason for defining c in this form is two-fold. From (5.9.8) we find that

$$(AC)(t) = \frac{1}{t}\int_0^t \int_0^s c(\tau)\,d\tau = g(t), \tag{5.9.9}$$

and from (5.9.6) and the Mean Value Theorem

$$\max_{t\in[0,\infty)} g_n(t) \geq n2^{n-1}. \tag{5.9.10}$$

Thus, from (5.9.9) and (5.9.10) we obtain $\liminf_{t\to\infty}(AC)(t) = -\infty$, so the Hartman-Wintner theorem does not apply. Let us consider $p = 2$ for a moment (so we actually consider the linear equation) in (5.1.1) and we note that the Lebesgue measure of the set $\left\{t : \int_0^t c(s)\,ds \neq 0\right\}$ is finite. This implies that

$$\lim_{t\to\infty} \text{approx} \int_0^t c(s)\,ds = 0,$$

so the criterion of Olech, Opial and Wazewski [307] does not apply either. Recall that

$$\lim_{t\to\infty} \text{approx}\, f(t) = l$$

if and only if, by definition,

$$l = \sup\{\ell : \text{mes}\{t : f(t) > \ell\} = \infty\} = \inf\{\ell : \text{mes}\{t : f(t) < \ell\} = \infty\}.$$

Here $\text{mes}\{\cdot\}$ denotes the Lebesgue measure of the set indicated. Recall also that the oscillation criterion of [307] states that the equation $x'' + c(t)x = 0$ is oscillatory provided

$$\lim_{t\to\infty} \text{approx} \int_0^t c(s)\,ds = \infty.$$

Nevertheless, by Theorem 5.9.1, equation (5.1.1) with any $p > 1$ is oscillatory. Indeed, from (5.9.10) we have $\limsup_{t\to\infty}(AC)(t) = \infty$. Also, for $t \in (b_n, a_{n+1}) = (n + 2^{-n}, n + 1 - 2^{-(n+1)})$, we have

$$(A^2C)(t) = \begin{cases} n/2t & n \text{ even}, \\ -(n+1)/2t & n \text{ odd}. \end{cases}$$

From the last equality it can be easily shown that $(A^2C)(t)$ is bounded for $t \in (0, \infty)$. Hence from (5.9.2) equation (5.1.1) is oscillatory.

Now we give another example, where the classical Hartman-Wintner criterion fails, but its improvement applies.

Example 5.9.2. Consider the equation

$$(\Phi(y'))' + t^\lambda g(t)\Phi(y) = 0, \tag{5.9.11}$$

where $\lambda > 1$ is a constant and $g : [0, \infty) \to \mathbb{R}$ is a T-periodic function with the mean value zero. We claim that then (5.9.11) is oscillatory. This will be shown as a consequence of the next lemma, which is stated without proof.

Lemma 5.9.1. *Let* $[\lambda]$ *denote the greatest positive integer less than or equal to* λ. *Then* $C(t) = \int_0^t s^\lambda g(s)\, ds$ *can be written as*

$$C(t) = \sum_{i=0}^{[\lambda]} t^{\lambda - i} p_i(t) + \Theta(t), \tag{5.9.12}$$

where each p_i *is a* T*-periodic continuous function defined on* $[0, \infty)$ *with the mean value zero and* $\Theta : [0, \infty) \to \mathbb{R}$ *is a bounded continuous function.*

From (5.9.12), we can find that for $j \in \mathbb{N}$

$$(A^jC)(t) = \begin{cases} \sum_{i=j}^{[\lambda]} t^{\lambda-i} p_i^j(t) + \Theta^j(t) & \text{if } j \le [\lambda], \\ \Theta^j(t) & \text{if } j > [\lambda], \end{cases} \tag{5.9.13}$$

where the functions p_i^j, Θ^j satisfy the same properties as p_i, Θ, respectively. Thus from (5.9.13), $\limsup_{t\to\infty}(AC)(t) = \infty$ and for $j > [\lambda]$,

$$-\infty < \liminf_{t\to\infty}(A^j)(t) < M$$

for a certain constant M. Hence, from Theorem 5.9.1 the claim follows. Note that in this example we have $\liminf_{t\to-\infty}(A^jC)(t) = -\infty$ if $j \le [\lambda]$, thus the oscillatory nature of (5.9.11) does not follow from the classical Hartman-Wintner theorem.

5.9.2 Half-linear Milloux and Armellini-Tonelli-Sansone theorems

Recall that the classical Armellini-Tonelli-Sansone theorem concerns the convergence to zero of all solutions of the second order linear differential equation

$$x'' + c(t)x = 0. \tag{5.9.14}$$

In particular, by the theorem of Milloux [288], if the function c is continuously differentiable, nondecreasing, and

$$\lim_{t\to\infty} c(t) = \infty \tag{5.9.15}$$

then (5.9.14) has at least one solution satisfying

$$\lim_{t\to\infty} x(t) = 0. \tag{5.9.16}$$

Note that condition (5.9.15) guarantees oscillation of (5.9.14) by the Leighton-Wintner criterion. The theorem of Armellini-Tonelli-Sansone deals with the situation when *all* solutions of (5.9.14) satisfy (5.9.16). This happens when c goes to infinity "regularly" (the exact definition is given below). Regular growth means, roughly speaking, that a function does not increase fast on intervals of short length.

Here we show that both theorems extend verbatim to (5.1.1). First we present some definitions. Let $S := \{(\alpha_k, \beta_k)\}$ be a sequence of intervals such that

$$0 \le \alpha_1 < \beta_1 < \alpha_2 < \beta_2 < \cdots < \beta_k \to \infty \quad \text{as } k \to \infty. \tag{5.9.17}$$

Then

$$\limsup_{k\to\infty} \frac{\sum_{i=1}^{k}(\beta_i - \alpha_i)}{\beta_k} =: \delta(S) = \delta \tag{5.9.18}$$

is called the *density* of the sequence of intervals S. A nondecreasing positive function f tends to infinity *intermittently* (an alternative terminology is *quasi-jumping*) as $t \to \infty$, provided to every $\varepsilon > 0$ there exists a sequence of intervals S satisfying (5.9.17) such that $\delta(S) \le \varepsilon$ and the increase of f on $\mathbb{R}_+ \setminus S$ is finite, i.e.,

$$\mathcal{S}(f; S) := \sum_{k=1}^{\infty} [f(\alpha_k) - f(\beta_{k-1})] < \infty. \tag{5.9.19}$$

In the opposite case we say that $f(t) \to \infty$ *regularly* as $t \to \infty$.

First we give an extension of Milloux theorem.

Theorem 5.9.2. *Suppose that c is a nondecreasing continuously differentiable function satisfying* (5.9.15). *Then* (5.6.1) *possesses at least one nontrivial solution satisfying* (5.9.16).

Proof. From the variety of proofs we present that one based on the modified Prüfer transformation (compare with Subsection 1.1.3). An alternative approach to the problem is presented in [176, 177, 198].

For any nontrivial solution x of (5.6.1) there exist a positive function ϱ given by the formula

$$\varrho = \left[|x|^p + \frac{1}{c}|x'|^p\right]^{\frac{1}{p}}$$

and a continuous function ϑ such that x can be expressed in the form

$$x(t) = \varrho(t)\sin_p(\vartheta(t)), \quad x'(t) = c^{\frac{1}{p}}(t)\varrho(t)\cos_p(\vartheta(t)).$$

The functions ϑ and ϱ satisfy the differential system

$$\vartheta' = c^{\frac{1}{p}}(t) + \frac{c'(t)}{c(t)} f(\vartheta), \quad \frac{\varrho'}{\varrho} = -\frac{c'(t)}{c(t)} g(\vartheta(t)), \tag{5.9.20}$$

where

$$f(\vartheta) = \frac{1}{p}\Phi(\cos_p \vartheta)\sin_p \vartheta, \quad g(\vartheta) = \frac{1}{p}|\cos_p \vartheta|^p.$$

The right-hand side of (5.9.20) is Lipschitzian in ϑ, hence a solution of (5.9.20) is uniquely determined by the initial conditions. We denote by $\vartheta(t,\varphi)$, $\varrho(t,\varphi)$ the solution given by the initial conditions $\vartheta(0) = \varphi$, $\varrho(0) = 1$. Then

$$\varrho(t,\varphi) = \exp\left\{-\int_0^t \frac{c'(s)}{c(s)} g(\vartheta(s,\varphi))\,ds\right\},$$

and since $g(\vartheta) \geq 0$, the function $\varrho(t,\vartheta)$ is nonincreasing and tends to a nonnegative limit $\varrho(\infty,\varphi)$ as $t \to \infty$. Obviously, $\varrho(\infty,\varphi) = 0$ implies that $x(t) \to 0$ as $t \to \infty$. The converse is also true because x is oscillatory.

There are the following two possibilities: (i) We have $\varrho(\infty,\varphi) = 0$, the corresponding solution x satisfies $x(t) \to 0$ as $t \to \infty$, and

$$\int_0^\infty \frac{c'(t)}{c(t)} g(\vartheta(t,\varphi))\,dt = \infty.$$

(ii) $\varrho(\infty,\varphi) > 0$, the solution x oscillates, its amplitude tends to a positive limit, and

$$\int_0^\infty \frac{c'(t)}{c(t)} g(\vartheta(t,\varphi))\,dt < \infty. \tag{5.9.21}$$

Now, the proof is based on the behavior as $t \to \infty$ of the function $\psi(t,\varphi_1,\varphi_2) = \vartheta(t,\varphi_2) - \vartheta(t,\varphi_1)$ which is described in the next two auxiliary statements. Here $\mathcal{X}$ denotes the set of φ's such that (5.9.21) holds, this means that the corresponding solution does not tend to zero as $t \to \infty$. The proof can be found in [26].

Lemma 5.9.2. *Let $\varphi_1, \varphi_2 \in \mathcal{X}$ and $\varphi_1 < \varphi_2 < \varphi_1 + \pi_p$. Then*

$$\psi(\infty,\varphi_1,\varphi_2) := \lim_{t\to\infty}[\vartheta(t,\varphi_2) - \vartheta(t,\varphi_1)]$$

exists and equals 0 *or* π_p.

Lemma 5.9.3. *Let $\varphi_0 \in \mathcal{X}$. Then for any $\varepsilon > 0$ there exists $\eta \in (0,\pi_p)$ such that if $|\varphi - \varphi_0| < \eta$, then*

$$|\vartheta(t,\varphi) - \vartheta(t,\varphi_0)| < \varepsilon \quad \text{for } t \geq 0. \tag{5.9.22}$$

Now, returning to the proof of our theorem, suppose that $\mathcal{X} = \mathbb{R}$. Then the function ψ given by $\psi(\infty, 0, \varphi) = 0$ is nondecreasing as φ increases in $[0, \pi_p]$. It must go from 0 to π_p, taking on only these two values, by Lemma 5.9.2. But this is impossible since by Lemma 5.9.3 this function is continuous, so the assumption $\mathcal{X} = \mathbb{R}$ was false and the theorem is proved. $\square$

Now we turn our attention to the extension of the Armellini-Tonelli-Sansone theorem.

Theorem 5.9.3. *Let c be a continuously differentiable function for large t. If the function $\log c(t) \to \infty$ regularly, then every solution of* (5.6.1) *satisfies* (5.9.16).

Proof. Consider the function

$$H(t) = |x(t)|^p + \frac{|x'(t)|^p}{c(t)},$$

where x is a nontrivial solution of (5.6.1). The function H is nonincreasing since

$$H'(t) = -\frac{c'(t)}{c^2(t)}|x'(t)|^p \leq 0. \tag{5.9.23}$$

Consequently, there exists a (finite or infinite) limit $H = \lim_{t\to\infty} H(t)$ and $H \geq 0$.

Suppose, by contradiction, that there exists a solution x of (5.6.1) which does not tend to zero. For this solution, obviously, $H > 0$. By (5.9.23)

$$\begin{aligned} H(t) &= H(0) - \int_0^t \frac{c'(t)}{c^2(s)}|x'(s)|^p\, ds \\ &= H(0) - \int_0^t \frac{c'(s)}{c(s)}(A(s) - |x(s)|^p)\, ds \\ &= H(0) - \int_0^t (A(s) - |x(s)|^p)\frac{dc(s)}{c(s)}. \end{aligned}$$

Let $\varepsilon > 0$ be a number such that for every sequence S of intervals with $\delta(S) \leq \varepsilon$ one has

$$\mathcal{S} = \sum_{i=1}^{k} [\log c(\alpha_{i+1}) - \log c(\beta_i)] = \sum_{i=1}^{k} \log \frac{c(\alpha_{i+1})}{c(\beta_i)} \to \infty \tag{5.9.24}$$

as $k \to \infty$.

In the remaining part of the proof we use the following statement.

Lemma 5.9.4. *For every $\varepsilon_0 > 0$ there exists $\eta > 0$ such that the density of the sequence S of all intervals, where*

$$H(t) - |x(t)|^p \leq \eta, \tag{5.9.25}$$

is less than ε_0.

Since the proof of this lemma is rather technical and follows essentially the original linear idea, we skip it and return to the proof of theorem.

Denote by (α_i, β_i) intervals, where (5.9.25) holds. On the intervals (β_i, α_{i+1}), we have

$$H(t) - |x(t)|^p > \eta,$$

therefore

$$H(\alpha_k) \le H(0) - \sum_{i=1}^{k-1} \int_{\beta_i}^{\alpha_{i+1}} (H(t) - |x(t)|^p) \frac{dc(t)}{c(t)} < H(0) - \eta \sum_{i=1}^{k} \log \frac{c(\alpha_{i+1})}{c(\beta_i)},$$

which implies by (5.9.24) that $H(\alpha_k)$ becomes negative for large k. This is a contradiction with $H = \lim_{t\to\infty} H(t) > 0$. □

5.9.3 Interval oscillation criteria

The main idea used in this subsection is based on the fact that oscillation of (1.1.1) is defined as conjugacy on any interval of the form $[T, \infty)$. More precisely, equation (1.1.1) is oscillatory if and only if there exists a sequence of intervals $[a_n, b_n]$, $a_n \to \infty$, such that (1.1.1) is conjugate on $[a_n, b_n]$, no matter how "bad" functions r, c are on complements of these intervals. The (dis)conjugacy and (dis)focality criteria of this subsection are closely related, in a certain sense, to the results of Subsection 3.2.3 which are based on the H-function averaging technique.

We introduce the following notation. We define $D := \{(t, s) : -\infty < s \le t < \infty\}$ and the class of functions $\mathcal{H} \subset C(D, [0, \infty))$ satisfying $H(t, t) = 0$, $H(t, s) > 0$ for $t > s$ and having partial derivatives $\partial H/\partial t$, $\partial H/\partial s$ on D such that

$$\frac{\partial H}{\partial t} = h_1(t, s)[H(t, s)]^{1/q} \quad \text{and} \quad \frac{\partial H}{\partial s} = -h_2(t, s)[H(t, s)]^{1/q}, \tag{5.9.26}$$

where h_1, h_2 are functions locally integrable on D.

Lemma 5.9.5. *Assume that x is a solution of (1.1.1) such that $x(t) > 0$ on an interval $[d, b)$ and let $w = -r\Phi(x'/x)$. Then for any $H \in \mathcal{H}$,*

$$\int_d^b H(b, s)c(s)\, ds \le -H(b, d)w(d) + \frac{1}{p^p} \int_d^b r(s)h_2^p(b, s)\, ds. \tag{5.9.27}$$

Proof. The function w is a solution of the Riccati equation

$$w' = c(t) + (p - 1)r^{1-q}(t)|w|^q. \tag{5.9.28}$$

Multiplying (5.9.28) by $H(t, s)$, integrating it with respect to s from d to t and using properties of the function H, we obtain

$$\int_d^t H(t, s)c(s)\, ds = \int_d^t H(t, s)w'(s)\, ds - (p - 1) \int_d^t r^{1-q}(s)H(t, s)|w(s)|^q\, ds$$

$$= -H(t, d)w(d) + \int_d^t \left((h_2(t, s)H^{1/q}(t, s)w(s) - (p-1)r^{1-q}(s)H(t, s)|w(s)|^q\right) ds.$$

Now, the inequality

$$Bw - A|w|^q \le \frac{1}{p}\left(\frac{p-1}{p}\right)^{p-1} |B|^p A^{-(p-1)}$$

(that can be written in the form

$$\frac{(p-1)B}{pA} \le \frac{1}{p}\left(\frac{p-1}{p}\right)^{p} \left|\frac{B}{A}\right|^p + \frac{1}{q}|w|^q$$

and follows from (1.2.2) with $u = B(p-1)/pA$, $v = w$), which holds for fixed $A > 0$, $B \in \mathbb{R}$ and any $w \in \mathbb{R}$, applied to the function

$$f(w) = h_2(t,s)H^{1/q}(t,s)w - (p-1)r^{1-q}(s)H(t,s)|w|^q$$

yields

$$\int_d^t H(t,s)c(s) \le -H(t,d)w(d) + \frac{1}{p^p}\int_d^t r(s)h_2^p(t,s)\,ds.$$

Letting $t \to b-$ we have the required inequality. □

Corollary 5.9.1. *Assume that* (1.1.1) *is right disfocal on* $[d,b)$, *i.e., there is no solution of this equation satisfying* $x'(d) = 0 = x(b_1)$ *for any* $b_1 \in (d,b)$. *Then for any* $H \in \mathcal{H}$,

$$\int_d^b H(b,s)c(s)\,ds \le \frac{1}{p^p}\int_d^b r(s)h_2^p(b,s)\,ds.$$

Remark 5.9.1. (i) Similarly as in Lemma 5.9.5, if x is a solution of (1.1.1) such that $x(t) > 0$ on $[a,d)$ and w is the same as in Lemma 5.9.5, then

$$\int_a^d H(s,a)c(s)\,ds \le -H(d,a)w(d) + \frac{1}{p^p}\int_a^d r(s)h_1^p(s,a)\,ds, \tag{5.9.29}$$

Corollary 5.9.1 can be reformulated in a similar way, in particular, if (1.1.1) is left-difocal on $(a,d]$, then for any $H \in \mathcal{H}$

$$\int_a^d H(s,a)c(s)\,ds \le \frac{1}{p^p}\int_a^d r(s)h_1^p(s,a)\,ds. \tag{5.9.30}$$

(ii) Combining (5.9.29), (5.9.30) we have the following statement. If equation (1.1.1) is disconjugate on an interval $[a,b]$, then for any $H \in \mathcal{H}$ there exists $d \in [a,b]$ such that (5.9.29) and (5.9.30) hold.

Another corollaries of Lemma 5.9.5 can be formulated as follows.

Corollary 5.9.2. *Assume that for some* $d \in (a,b)$ *and some* $H \in \mathcal{H}$,

$$\begin{aligned} &\frac{1}{H(d,a)}\int_a^d H(s,a)c(s)\,ds + \frac{1}{H(b,d)}\int_d^b H(b,s)c(s)\,ds \\ &> \frac{1}{p^p}\left(\frac{1}{H(d,a)}\int_a^d r(s)h_1^p(s,a)\,ds + \frac{1}{H(b,d)}\int_d^b r(s)h_2^p(b,s)\,ds\right). \end{aligned} \tag{5.9.31}$$

Then every solution of (1.1.1) *has at last one zero in* (a,b).

Corollary 5.9.3. *Assume that there exists $H \in \mathcal{H}$ such that for any $d \in [a,b]$ at least one of the following inequalities holds:*

$$\int_a^d H(s,a)c(s)\,ds > \frac{1}{p^p}\int_a^d r(s)h_1^p(s,a)\,ds \tag{5.9.32}$$

or

$$\int_d^b H(b,d)c(s)\,ds > \frac{1}{p^p}\int_d^b r(s)h_2^p(b,s)\,ds. \tag{5.9.33}$$

Then every solution of (1.1.1) *has at least one zero in* (a,b).

As an immediate consequence of the previous remark we have the following oscillation criterion.

Theorem 5.9.4. *Equation* (1.1.1) *is oscillatory provided for each $T \in \mathbb{R}$ there exists $H \in \mathcal{H}$ and either*

(i) *there exists $a,b \in [T,\infty)$, $a < b$, and $d \in [a,b]$ such that* (5.9.31) *holds, or*

(ii) *there exist $a,b \in [T,\infty)$, $a < b$, and for any $d \in [a,b]$ at least one of conditions* (5.9.32) *or* (5.9.33) *holds.*

Corollary 5.9.4. *Assume that for some $H \in \mathcal{H}$ and for each T sufficiently large,*

$$\limsup_{t\to\infty}\int_T^t \left[H(s,T)c(s) - \frac{1}{p^p}r(s)h_1^p(s,T)\,ds\right] ds > 0 \tag{5.9.34}$$

and

$$\limsup_{t\to\infty}\int_T^t \left[H(t,s)c(s) - \frac{1}{p^p}r(s)h_2^p(t,s)\,ds\right] ds > 0. \tag{5.9.35}$$

Then (1.1.1) *is oscillatory.*

Proof. Let $T \in \mathbb{R}$ be arbitrary (sufficiently large). By (5.9.34) there exists $d > T$ such that

$$\int_T^d \left[H(s,T)c(s) - \frac{1}{p^p}r(s)h_1^p(s,T)\,ds\right] ds > 0, \tag{5.9.36}$$

similarly, substituting $T = d$ in (5.9.35), by this inequality there exists $b > d$ such that

$$\int_d^b \left[H(b,s)c(s) - \frac{1}{p^p}r(s)h_2^p(b,s)\,ds\right] ds > 0. \tag{5.9.37}$$

Combining (5.9.36) and (5.9.37) we obtain (5.9.31). □

Now, choose $H(t,s) = (t-s)^\lambda$ for $\lambda > p-1$, then we can take $h_1(t,s) = h(t-s) = \lambda(t-s)^{(\lambda/p)-1}$. Then we have the following Kamenev type oscillation criterion for (5.1.1) (compare with Subsection 3.2.2). We skip the proof since it is similar to those of previous statements of this subsection.

Corollary 5.9.5. *Suppose that for each $T \in \mathbb{R}$ sufficiently large the following inequalities hold:*

$$\limsup_{t\to\infty} \frac{1}{t^{\lambda-p+1}} \int_T^t (s-T)^\lambda c(s)\,ds > \frac{\lambda^p}{p^p(\lambda-p+1)}$$

and

$$\limsup_{t\to\infty} \frac{1}{t^{\lambda-p+1}} \int_T^t (t-s)^\lambda c(s)\,ds > \frac{\lambda^p}{p^p(\lambda-p+1)},$$

then (5.1.1) *is oscillatory.*

Remark 5.9.2. The method of this subsection has been extended in several directions: To forced equations, damped equations, half-linear functional differential equations and to various nonlinear differential equations (not only of the second order), we refer to [249, 250] and the references given therein. As an example we present one result of [250]. There is given an interval oscillation criterion for the half-linear functional differential equation

$$(r(t)\Phi(x'(t)))' + c(t)\Phi(x(\tau(t))) = 0, \tag{5.9.38}$$

where $\tau'(t) > 0$ and $\lim_{t\to\infty} \tau(t) = \infty$. We present this statement without proof, we refer to [250] for details. This proof is similar to that of Corollary 5.9.4. See also Section 9.3 for other results concerning half-linear functional differential equations.

In the next theorem, H, h_1, h_2 are the same as in the previous part of this subsection, only (5.9.26) is replaced by

$$\frac{\partial H}{\partial t} = h_1(t,s)H^{1/2}H(t,s), \qquad \frac{\partial H}{\partial s} = -h_2(t,s)H^{1/2}H(t,s).$$

Theorem 5.9.5. *Suppose that for every $T \in \mathbb{R}$ (sufficiently large) there exists a nondecreasing function $\rho \in C^1[T,\infty)$ such that*

$$\limsup_{t\to\infty} \int_T^t \left[H(s,T)c(s) - \frac{r(\tau(s))\rho(s)\left(h_1(s,T) + \frac{\rho'(s)}{\rho(s)}\sqrt{H(s,T)}\right)^p}{p^p(\tau'(s))^{p-1}[H(s,T)]^{(p-2)/2}} \right] ds > 0$$

and

$$\limsup_{t\to\infty} \int_T^t \left[H(t,s)c(s) - \frac{r(\tau(s))\rho(s)\left(h_2(t,s) + \frac{\rho'(s)}{\rho(s)}\sqrt{H(s,T)}\right)^p}{p^p(\tau'(s))^{p-1}[H(t,s)]^{(p-2)/2}} \right] ds > 0.$$

Then (5.9.38) *possesses no nonoscillatory solution which is extensible up to ∞.*

Taking $H(t,s) = [R(t) - R(s)]^\lambda$, $\lambda > p-1$, where $R(t) = \int^t r^{1-q}(s)\,ds$, and $\rho(t) \equiv 1$, then applying the previous theorem to the Euler type equation with deviated argument

$$(\Phi(x'(t)))' + \frac{\gamma}{t^p}\Phi(x(t\pm\tau)) = 0 \tag{5.9.39}$$

gives the following result showing that the constant argument deviation does not affect oscillatory nature of Euler type equations.

Corollary 5.9.6. *If* $\gamma > \left(\frac{p-1}{p}\right)^p$, *then* (5.9.39) *possesses no nonoscillatory solution which is extensible up to* ∞.

Remark 5.9.3. Finally note that one of the advantages of the interval oscillation criteria is that they can apply in cases where some of the classical criteria fail. This is, for instance, the case when $\int^\infty c(t)\,dt = -\infty$. Consider equation (1.1.1) with a continuous function $0 < r(t) \le 1$, $t \in [0,\infty)$, and

$$c(t) = \begin{cases} (t-3k)\eta & 3k \le t \le 3k+1, \\ (-t+3k+2)\eta & 3k+1 < t \le 3k+2, \\ -k \text{ or } -k|\sin \pi t| & 3k+2 < t < 3k+3, \end{cases}$$

where

$$\eta > \frac{(\lambda+2)\lambda^p}{(\lambda-p+1)p^p},$$

is a constant for fixed $\lambda > \max\{1, p-1\}$, $k \in \mathbb{N}_0$ It is not difficult to verify (see [361]) that such an equation is oscillatory by Theorem 5.9.4, and $\int^\infty c(t)\,dt = -\infty$.

5.10 Notes and references

Lyapunov type inequality was proved for the first time by Elbert [139] for (1.1.1) with $r(t) \equiv 1$, but the extension to general (1.1.1) in Theorem 5.1.1 is straightforward. Half-linear Lyapunov inequality has been rediscovered in several later papers, e.g., in [245] by Li and Yeh and in [373] by X. Yang. Vallée-Poussin type inequality is taken from Došlý and Lomtatidze [111], while focal point criteria were proved by Elbert and Došlý [107]. Hong, Lian and Yeh [179] showed Lyapunov type focal points and conjugacy criteria presented in Subsection 5.1.4. Similar results can be found also in Peña's paper [310]. Conjugacy criterion based on Opial's inequality (Theorem 5.1.9) is taken from X. Yang [373].

Subsection 5.2.3 is based on Došlý and Lomtatidze [112]. More precisely, Theorem 5.2.3 from Subsection 5.2.4 is [112, Theorem 3.2] and Theorem 5.2.4 of Subsection 5.2.5 is [112, Theorem 3.3]. Singular Leighton's Theorem is formulated in a simplified form, as can be found in Došlý [101], a more general formulation is presented in Došlý and Jaroš [110]. The results on perturbed Euler equation in Subsection 5.2.6 are due to Elbert and Schneider [149]. A similar idea, applied to the nonlinear equation of the form $(\Phi(x'))' + t^{-p}g(x) = 0$, with a nonlinearity g satisfying certain additional conditions, can be found in the recent interesting paper of Sugie and Yamaoka [340]. A related result can be found in the paper of Řezníčková [333]. Finally, the linearization method presented in Subsection 5.2.7 is the main result of the paper of Došlý and Peña [114].

The theory of disconjugacy and nonoscillation domain in Section 5.3.1 is an original extension of linear results, as described at the beginning of that section. The same holds for the results of Subsection 5.3.3 except of Lemma 5.3.4 and Theorem 5.3.12 that were presented in H. J. Li and Yeh [240]. Half-linear equations with periodic coefficients (see Subsection 5.3.2) were studied in Došlý and Elbert [107].

The results concerning strong and conditional oscillation (Section 5.4) are taken from Kusano, Naito, Ogata [218, 219], and supplemented by some new ones (which seem to be original even in the linear case).

The discussion on the function sequence technique in Section 5.5 is based on the papers [180, 182, 244, 246, 247, 253] by Fan, Hoshino, Hsu, Imabayashi, Kusano, H. J. Li, W. T. Li, Tanigawa and Yeh, and it is extended by numerous new observations. Note that the function sequence was also used by Mirzov [293] for the investigation of half-linear systems.

The results of Subsection 5.6.1 are taken from Elbert, Kusano, Tanigawa [148], see also [147]. Theorem 5.6.2 dealing with quick oscillation was proved by H. J. Li and Yeh [245], while Subsection 5.6.3 concerning slow oscillation is based on Lian, Yeh and H. J. Li [255].

The main statement of Subsection 5.7.1 is taken from the classical paper of Elbert [139]. Regular problem with indefinite weight (see Subsection 5.7.2) was studied in Kusano and Naito [216] and the results concerning singular Sturm-Liouville problem (see Subsection 5.7.3) are due to Elbert, Kusano and Naito [146]. For related results we refer to Kusano, Naito and Tanigawa [217, 220]. Subsection 5.7.4 devoted to singular eigenvalue problem associated radial p-Laplacian is based on [356]. The results of Subsection 5.7.5 are taken from the paper of Zhang [382]. Related results concerning half-linear Sturm-Liouville BVP's are presented in the papers of Binding and Drábek [42], Eberhard, Elbert, [137] and Huang [183].

Energy functionals considered on classes of functions satisfying various boundary conditions (see Section 5.8) were studied by Mařík [271, 274, 276], see also the paper of Yeh [377] concerning Levin type comparison theorem. Note that singular energy p-degree functionals were investigated in [272] also by Mařík.

Generalized Hartman-Wintner's criterion, as well as related examples (see Subsection 5.9.1) are due to Del Pino, Elgueta and Manasevich [90]. Results of Subsection 5.9.2 are taken from Atkinson, Elbert [26] and Bihari [40]. Interval oscillation criteria (see Subsection 5.9.3) come from Kong [207], see also Wang, Yang [361] for similar results and Jiang [194] for interval oscillation criteria of a different nature.

CHAPTER 6

BOUNDARY VALUE PROBLEMS FOR HALF-LINEAR DIFFERENTIAL EQUATIONS

In this chapter we deal with boundary value problems associated with half-linear differential equations. This problem has already been partially discussed in Section 5.7, but here we focus our attention to different aspects of this problem.

The main part of this chapter deals with the BVP of the form

$$(\Phi(x'))' + \lambda\Phi(x) = f(t), \quad x(0) = 0 = x(\pi_p), \tag{6.1.1}$$

where λ is a spectral parameter and the function f satisfies various smoothness assumptions depending on the particular investigated problem. In the first section we study the so-called *nonresonant case*, i.e., the situation when λ is not an eigenvalue of the unforced problem ($f \equiv 0$ in (6.1.1)). The second section is devoted mainly to the half-linear version of the classical Fredholm alternative for linear BVP's. This is perhaps the most interesting part of the qualitative theory of half-linear differential equations since one meets there phenomena which are completely different in comparison with the linear case. The last section contains results concerning solvability of the resonant BVP's associated with the equation

$$(\Phi(x'))' + \lambda_n\Phi(x) + g(x) = f(t),$$

λ_n being an eigenvalue of the unforced BVP (6.1.1). In particular, the classical (linear) Landesman-Lazer and Ambrosetti-Prodi type results are extended to half-linear equations.

6.1 Eigenvalues, existence, and nonuniqueness problems

This section is mainly devoted to nonresonant BVP's, i.e., to (6.1.1) where λ is not an eigenvalue of the below given BVP (6.1.2). First we present a variational characterization of eigenvalues and then we deal with the existence and multiplicity results for nonresonant problems.

6.1.1 Basic boundary value problem

Under the "basic" boundary value problem we understand the problem

$$(\Phi(x'))' + \lambda\Phi(x) = 0, \quad x(0) = 0 = x(\pi_p), \tag{6.1.2}$$

where λ is the eigenvalue parameter. Here π_p is the same as in the Section 1.1 and its value is defined by the formula

$$\pi_p = 2\int_0^1 \frac{ds}{\sqrt[p]{1-s^p}}. \tag{6.1.3}$$

Eigenvalue problem (6.1.2) is a special case of the general Sturm-Liouville problem for half-linear equations treated in Section 5.7, and its simple structure enables to determine completely the eigenvalues and eigenfunctions. The situation is essentially the same as in the linear case where the eigenvalues are $\lambda_n = n^2$ with the associated eigenfunctions $x_n(t) = \sin nt$.

Theorem 6.1.1. *The eigenvalues of* (6.1.2) *are* $\lambda_n = (p-1)n^p$, $n \in \mathbb{N}$, *and the corresponding eigenfunctions are (up to a nonzero multiplicative factor)* $x_n(t) = \sin_p(nt)$, *where the half-linear sine function* $\sin_p$ *is defined in Section 1.1.*

Proof. The proof of this statement follows immediately from the homogeneity of the solution space of half-linear equations and from the unique solvability the initial value problem for these equations. The function $x_1(t) = \sin_p t$ is a solution of (6.1.2) with $\lambda = (p-1)$ and satisfies $x(0) = 0 = x(\pi_p)$ (by the definition of this function in Section 1.1), and $x_n(t) = \sin_p(nt)$ is a solution of (6.1.2) with $\lambda_n = (p-1)n^p$. □

6.1.2 Variational characterization of eigenvalues

In the linear case, the Courant-Fischer minimax principle provides a variational characterization of eigenvalues of the classical Sturm-Liouville eigenvalue problem. This characterization is based on the orthogonality of the eigenfunctions corresponding to different eigenvalues. In the half-linear case, the meaning of orthogonality is lost, but eigenvalues can be described using the Lusternik-Schnirelmann procedure, for general facts concerning this approach we refer to [168].

Let us introduce the functionals over $W_0^{1,p}(0,\pi_p)$, endowed with the norm $\|x\| = \left(\int_0^{\pi_p} |x'|^p\,dt\right)^{\frac{1}{p}}$, as follows:

(6.1.4)
$$A(x) = \frac{1}{p}\int_0^{\pi_p} |x'|^p\,dt, \quad B(x) = \frac{1}{p}\int_0^{\pi_p} |x|^p\,dt, \quad C(x) = \int_0^{\pi_p} (F(x) - f(t)x)\,dt,$$

where $F(t) = \int_0^t f(s)\,ds$. Eigenfunctions and eigenvalues of (6.1.2) are equivalent to critical points and critical values of the functional

$$E(x) = \frac{A(x)}{B(x)}.$$

The proof of the next statement (which concerns the differentials A', B', C' of the operators A, B, C) can be found in [118, Lemma 10.3], here $\|\cdot\|_*$ denotes the norm in the dual space $(W_0^{1,p}(0,\pi_p))^*$.

Lemma 6.1.1. *The operators $A', B', C' : W_0^{1,p}(0,\pi_p) \to (W_0^{1,p}(0,\pi_p))^*$ have the following properties:*

(a) A' is $(p-1)$-homogeneous, odd, continuously invertible, and $\|A'(u)\|_* = \|u\|^{p-1}$ for any $u \in W_0^{1,p}(0,\pi_p)$.

(b) B' is $(p-1)$-homogeneous, odd and compact.

(c) C' is bounded and compact.

We also introduce the notation (with $c \in \mathbb{R}$)

$$\begin{aligned} \mathcal{S} &= \{x \in W_0^{1,p}(0,\pi_p) : B(x) = 1\}, \\ \mathcal{K}_c &= \{u \in W_0^{1,p}(0,\pi_p) \setminus \{0\},\ E(u) = c,\ E'(u) = 0\} \end{aligned}$$

(hence $E(x) = A(x)$, $E'(x) = A'(x) - A(x)B'(x)$ for $x \in \mathcal{S}$).

In our variation characterization of eigenvalues (or, more generally, in all "min-max" procedures), an important role is played by the so-called *Palais-Smale* condition.

Lemma 6.1.2. *The functional $E|_{\mathcal{S}}$ satisfies the Palais-Smale condition, i.e., if $\{u_k\} \subset \mathcal{S}$ is a sequence such that $E(u_k)$ is bounded and $E'(u_k) \to 0$ in $(W_0^{1,p}(0,\pi_p))^*$, then $\{u_k\}$ contains a convergent subsequence.*

Proof. Since $E(u_k) = A(u_k) = \frac{1}{p}\|u_k\|^p$ for $u_k \in \mathcal{S}$, it is clear that $\{u_k\}$ is bounded in $W_0^{1,p}(0,\pi_p)$, so we may assume, without loss of generality, that $u_k \rightharpoonup u_0$ in $W_0^{1,p}(0,\pi_p)$, where $\rightharpoonup$ denotes the weak convergence, and that $A(u_k) \to A_0 \in \mathbb{R}$. By compactness, $B'(u_k) \to B'(u_0)$ in $(W_0^{1,p}(0,\pi_p))^*$. Since $E'(u_k) \to 0$, we have $A'(u_k) - A(u_k)B'(u_k) \to 0$ in $(W_0^{1,p}(0,\pi_p))^*$, and thus $A'(u_k) \to A_0 B'(u_0)$. Applying Lemma 6.1.1, $u_k \to u_0 = (A')^{-1}(A_0 B'(u_0))$ in $W_0^{1,p}(0,\pi_p)$. $\square$

Let us recall also the definition of the Krasnoselskii genus of a symmetric set $\mathcal{A} \subset W_0^{1,p}(0, \pi_p)$. Let

$$\mathcal{F} := \{\mathcal{A} \subset W_0^{1,p}(0, \pi_p) : \mathcal{A} \text{ is closed and } \mathcal{A} = -\mathcal{A}\}$$

and let

$$\mathcal{M} = \{m \in \mathbb{N} : \exists h \in C(\mathcal{A}; \mathbb{R}^m \setminus \{0\}) \text{ such that } h(-x) = -h(x)\}.$$

Then the Krasnoselskii genus of $\mathcal{A}$ is defined by

$$\gamma(\mathcal{A}) := \begin{cases} \inf \mathcal{M} & \text{if } \mathcal{M} \neq \emptyset, \\ \infty & \text{if } \mathcal{M} = \emptyset. \end{cases}$$

Intuitively, γ provides a measure of the dimension of a symmetric set. For example, if Ω is a bounded symmetric neighborhood of the origin in $\mathbb{R}^m$, then $\gamma(\partial\Omega) = m$.

Using the above given concepts we can now present the formulas for the variational characterization of *all* eigenvalues of (6.1.2).

Theorem 6.1.2. *Let*

$$\mathcal{F}_k := \{\mathcal{A} \in \mathcal{F} : 0 \notin \mathcal{A},\ \gamma(\mathcal{A}) \geq k\}, \quad \tilde{\mathcal{F}}_k := \{\mathcal{A} \in \mathcal{F}_k : \mathcal{A} \subset \mathcal{S},\ \mathcal{A} \textit{ is compact}\},$$

and let

$$\beta_k = \min_{\mathcal{A} \in \tilde{\mathcal{F}}_k} \max_{x \in \mathcal{A}} E(x). \tag{6.1.5}$$

Then $\beta_n = \lambda_n = (p-1)n^p$ *for* $n \in \mathbb{N}$.

Proof. Let $k \in \mathbb{N}$ be fixed. Clearly, $\beta_k = \lambda_n$ for some $n \in \mathbb{N}$. Thus $\mathcal{K}_{\beta_k} \cap \mathcal{S} = \{\pm p^{1/p}\varphi_n\}$, where φ_n is the normalized eigenfunction corresponding to λ_n (i.e., $\|\varphi_n\|_p = \left(\int_0^{\pi_p} |\varphi_n(t)|^p\, dt\right)^{1/p} = 1$), and so $\gamma(\mathcal{K}_{\beta_k} \cap \mathcal{S}) = 1$. Moreover, it is known that if $\beta_j = \beta_{j+1} = \cdots = \beta_{j+m}$, then $\gamma(\mathcal{K}_{\beta_j} \cap \mathcal{S}) \geq m+1$, see [339, Lemma 5.6]. Thus $m = 0$ and $\{\beta_n\}$ must be an increasing sequence. It follows that $\beta_k \geq \lambda_k$.

Now, consider the functions $\varphi_{k,i} = \chi_{[(i-1)\pi_p/k, i\pi_p/k]}\varphi_k$, for $i = 1, \dots, k$, where $\chi_{[\cdot]}$ is a characteristic function of the interval indicated, and let

$$\Lambda_k = \{\alpha_1\varphi_{k,1} + \cdots + \alpha_k\varphi_{k,k},\ |\alpha_1|^p B(\varphi_{k,1}) + \cdots + |\alpha|^p B(\varphi_{k,k}) = 1\}, \tag{6.1.6}$$

where α_i are real constants. Each $\varphi_{k,i}$ is in $W_0^{1,p}(0, \pi_p)$ and it is a principal eigenfunction, with the eigenvalue λ_k, for the differential equation restricted to the appropriate subinterval. Observe that Λ_k is symmetric and homeomorphic to the unit sphere in $\mathbb{R}^n$. Thus Λ_k is the compact with $\gamma(\Lambda_k) = k$. Moreover, observe that for $u \in \Lambda_k$

$$\begin{aligned} B(u) &= B(\alpha_1\varphi_{k,1} + \cdots + \alpha_k\varphi_{k,k}) \\ &= B(\alpha_1\varphi_{k,1}) + \cdots + B(\alpha_k\varphi_{k,k}) \\ &= |\alpha_1|^p B(\varphi_{k,1}) + \cdots + |\alpha_k|^p B(\varphi_{k,k}) \\ &= 1. \end{aligned}$$

Thus, $\Lambda_k \subset \mathcal{S}$ and so $\Lambda_k \in \tilde{\mathcal{F}}_k$. A similar computation shows that $E(u) = A(u) = \lambda_k$ for all $u \in \Lambda_k$. This implies that

$$\beta_k = \inf_{\mathcal{A} \in \mathcal{F}_k} \sup_{u \in \mathcal{A}} E(u) \leq \lambda_k,$$

so $\lambda_k = \beta_k$. Moreover, the inf sup has been achieved on a set in $\tilde{\mathcal{F}}_k$. Using the compactness of the sets in $\tilde{\mathcal{F}}_k$ we replace sup by max, and since the inf is achieved we replace inf by min. This completes the proof. □

6.1.3 Existence and (non)uniqueness below the first eigenvalue

In this subsection we consider the problem (6.1.1), i.e., the nonhomogeneous problem

$$(\Phi(x'))' + \lambda\Phi(x) = f(t), \quad x(0) = 0 = x(\pi_p). \tag{6.1.7}$$

Consider the energy functional

$$\mathcal{J}_f^\lambda(x) := \frac{1}{p}\int_0^{\pi_p} |x'(t)|^p\,dt - \frac{\lambda}{p}\int_0^{\pi_p} |x(t)|^p\,dt + \int_0^{\pi_p} f(t)x(t)\,dt \tag{6.1.8}$$

and suppose that λ is not an eigenvalue, i.e., $\lambda \neq \lambda_k$. For simplicity we deal with $f \in C[0, \pi_p]$ and solution of (6.1.7) is understood in the classical sense, i.e., it is a function x such that $\Phi(x') \in C^1[0, \pi_p]$ and the equation with the boundary conditions in (6.1.7) are satisfied. The critical points of $\mathcal{J}_f^\lambda$ are in one to one correspondence with solutions of (6.1.7).

Due to the variational characterization of the least eigenvalue

$$\lambda_1 = \min \frac{\int_0^{\pi_p} |x'(t)|^p\,dt}{\int_0^{\pi_p} |x(t)|^p\,dt}, \tag{6.1.9}$$

where the minimum is taken over all nonzero elements of $W_0^{1,p}(0, \pi_p)$, and due to the monotonicity of the operators $A', B' : W_0^{1,p}(0, \pi_p) \to \left(W_0^{1,p}(0, \pi_p)\right)^*$ defined by

$$\langle A'u, v\rangle = \int_0^{\pi_p} \Phi(u'(t))v'(t)\,dt, \quad \langle B'u, v\rangle = \int_0^{\pi_p} \Phi(u(t))v(t)\,dt$$

(here $\langle \cdot, \cdot \rangle$ is the duality pairing between $\left(W_0^{1,p}(0, \pi_p)\right)^*$ and $W_0^{1,p}(0, \pi_p)$) it is easy to prove that for $\lambda \leq 0$ the energy functional $\mathcal{J}_f^\lambda$ has a unique minimizer in $W_0^{1,p}(0, \pi_p)$ for arbitrary $f \in \left(W_0^{1,p}(0, \pi_p)\right)^*$. In particular, it follows from here that given arbitrary $f \in C[0, \pi_p]$, the problem (6.1.7) has a unique solution. So, from this point of view, the situation is the same for $p = 2$ (linear case) and $p \neq 2$.

The case $\lambda > 0$ is different. It is well known that for $p = 2$ and $\lambda \neq \lambda_k$, $k = 1, 2, \ldots$, for any $f \in C[0, \pi_p]$ the problem (6.1.7) has a unique solution, which follows e.g. from the Fredholm alternative. Let us consider now $p \neq 2$ and

$0 < \lambda < \lambda_1$. Due to the variational characterization of λ_1 given by (6.1.9), the energy functional is still coercive but the monotone operators A', B' "compete" because of the positivity of λ. While in the linear case $p = 2$ this fact does not affect the uniqueness, for $p \neq 2$ the following interesting phenomenon is observed.

Theorem 6.1.3. *Let $0 < \lambda < \lambda_1 = p - 1$ and $p \neq 2$. There exists a function $f \in C[0, \pi_p]$ such that $\mathcal{J}_f^\lambda$ has at least two critical points. One of them corresponds to the global minimizer of $\mathcal{J}_f^\lambda$ on $W_0^{1,p}(0, \pi_p)$ (which does exist due to $\lambda < \lambda_1$) and the other is a critical point of saddle type.*

Proof. As we have mentioned before, for $0 < \lambda < \lambda_1$ the functional $\mathcal{J}_f^\lambda$ is coercive and hence (together with its weak lower semicontinuity) possesses a global minimum. Hence it suffices to construct a critical point of this functional which is not the global minimizer.

First consider the case $p > 2$, and let $\tilde{x}$ be a $C^2[0, \pi_p]$ function which is equal to a nonzero constant for $t \in [\varepsilon, \pi_p - \varepsilon]$ and $\tilde{x}(0) = 0 = \tilde{x}(\pi_p)$, where $\varepsilon > 0$ is sufficiently small. Define the function f by

$$f(t) = (\Phi(\tilde{x}'(t)))' + \lambda\Phi(\tilde{x}(t)).$$

Then $\tilde{x}$ is a solution of (6.1.7). We will show that this solution is not the above mentioned global minimizer of $\mathcal{J}_f^\lambda$, i.e., (6.1.7) has at least two solutions. To show this, note that for $p > 2$ the functional $\mathcal{J}$ (we skip the sub/superscripts f, λ if these values are not important at the moment) is twice Fréchet differentiable and its second derivative at $\tilde{x}$ is given by

$$\langle \mathcal{J}''(\tilde{x})v, v\rangle = (p-1)\left(\int_0^{\pi_p} |\tilde{x}|^{p-2} v'^2\, dt - \lambda \int_0^{\pi_p} |\tilde{x}|^{p-2} v^2\, dt\right) \tag{6.1.10}$$

for $v \in W_0^{1,p}$. Next, let $z \in C_0^\infty$ be such that $\operatorname{supp} z \subset (\varepsilon, \pi_p - \varepsilon)$. We find from (6.1.10) and the definition of $\tilde{x}$ that

$$\langle J''(\tilde{x})z, z\rangle = -(p-1)\lambda \int_0^{\pi_p} |\tilde{x}|^{p-2} z^2\, dt < 0$$

which together with the fact that $J'(\tilde{x}) = 0$ shows that $\tilde{x}$ is not a local minimum.

Now we deal with the case $p \in (1, 2)$. We will follow the presentation of [164], where the interval $[-1, 1]$ instead of $[0, \pi_p]$ is considered, i.e., BVP (6.1.7) is considered with the boundary condition $u(-1) = 0 = u(1)$. In this modification, the construction of the forcing term f is more transparent.

Fix the numbers $0 < \varepsilon < \varepsilon_1 < \varepsilon_2 < 1/2$ and let $m > q = p/(p-1)$ be a real constant which will be specified later. Define a function $u_0 \in C^2[-1, 1]$ such that

$$\begin{cases} u_0(t) = |t|^m & \text{for } |t| \le \varepsilon_1, \\ tu_0'(t) > 0 & \text{for } \varepsilon_1 \le |t| \le \varepsilon_2, \\ u_0(t) = \frac{1}{2^m} - \left|\frac{1}{2} - |t|\right|^m & \text{for } \varepsilon_2 \le |t| \le 1. \end{cases}$$

Notice that the condition $tu_0'(t) > 0$ for $\varepsilon_1 \le |t| \le \varepsilon_2$ can be satisfied because $u_0(\pm\varepsilon_1) < u_0(\pm\varepsilon_2)$ follows from $\varepsilon_1^m + |(1/2) - \varepsilon_2|^m < 2^{-m}$. Clearly, we have

$u_0(\pm 1) = 0$ and $(\Phi(u_0'))' \in C[-1,1]$. Similarly as in the case $p > 2$, we define the right-hand side f of (6.1.7) by

$$f(t) = (\Phi(u_0'(t)))' + \lambda\Phi(u_0(t))$$

and we will show that the functional $\mathcal{J}$ does not attain its minimum at u_0 (over $W_0^{1,p}(-1,1)$).

To prove this claim, we note that $\mathcal{J}$ is Fréchet differentiable and its derivative is given by

$$\langle \mathcal{J}'(u), v\rangle = \int_{-1}^{1} \Phi(u')v'\,dt - \lambda\int_{-1}^{1} \Phi(u)v\,dt + \int_{-1}^{1} fv\,dt \tag{6.1.11}$$

for all $v \in W_0^{1,p}(-1,1)$. Next, let $z \in C^1[-1,1]$ be any function such that

$$\begin{cases} z(t) = 1 & \text{for } |t| \le \varepsilon, \\ 0 \le z(t) \le 1 & \text{for } \varepsilon \le |t| \le \varepsilon_1, \\ z(t) = 0 & \text{for } \varepsilon_1 \le |t| \le 1. \end{cases}$$

We find from (6.1.11) and the definition of z that

$$\langle \mathcal{J}'(u), z\rangle = \int_{-\varepsilon_1}^{\varepsilon_1} \Phi(u')z'\,dt - \lambda\int_{-\varepsilon_1}^{\varepsilon_1} \Phi(u)z\,dt + \int_{-\varepsilon_1}^{\varepsilon_1} fz\,dt.$$

Now we investigate the function

$$\zeta(\alpha) := \frac{1}{\alpha}(\langle \mathcal{J}'(u_0+\alpha z), z\rangle - \langle \mathcal{J}'(u_0), z\rangle) \quad \text{for } \alpha \in [-1,1]. \tag{6.1.12}$$

Observe that $\zeta(\alpha) = J(\alpha) + J_1(\alpha)$, where

$$J(\alpha) = -\frac{\lambda}{\alpha}\int_{-\varepsilon}^{\varepsilon} (\Phi(u_0+\alpha) - \Phi(u_0))\,dt$$

and

$$J_1(\alpha) = \frac{1}{\alpha}\int_{\varepsilon\le|t|\le\varepsilon_1} [\Phi(u_0'+\alpha z') - \Phi(u_0')]z'\,dt - \frac{\lambda}{\alpha}\int_{\varepsilon\le|t|\le\varepsilon_1} [\Phi(u_0+\alpha z) - \Phi(u_0)]\,dt.$$

Since $u_0(t) = |t|^m$ for $|t| \le \varepsilon_1$, there exists (possibly infinite) nonpositive limit

$$J(0) = \lim_{\alpha\to 0} J(\alpha) = -(p-1)\lambda\int_{-\varepsilon}^{\varepsilon} |u_0|^{p-2}\,dt = -(p-1)\lambda\int_{-\varepsilon}^{\varepsilon} |t|^{m(p-2)}\,dt, \tag{6.1.13}$$

and the finite limit

$$J_1(0) = \lim_{\alpha\to 0} J_1(\alpha) = (p-1)\int_{\varepsilon\le|t|\le\varepsilon_1} \left(|u_0|^{p-2}|z'|^2 - \lambda|u_0|^{p-2}z^2\right)\,dt. \tag{6.1.14}$$

Combining (6.1.12) with (6.1.13) and (6.1.14), we arrive at

$$\lim_{\alpha\to 0}\zeta(\alpha) = -(p-1)\lambda\int_{-\varepsilon}^{\varepsilon} |t|^{m(p-2)}\,dt + J_1(0).$$

and we find that $\zeta(0) = -\infty$ provided $m(p-2) \le -1$, which shows that at u_0 the functional $\mathcal{J}$ does not attain its minimum. □

The last result of this subsection presents a nonuniqueness result regardless of the value of the eigenparameter λ. We skip the proof, but its idea is similar as above, only the construction of the forcing term f is technically more complicated when $\lambda > \lambda_1$.

Theorem 6.1.4. *Let $p \neq 2$ and $\lambda > 0$. Then there exists $f \in C[0,T]$ such that (6.1.7) has at least two distinct solutions.*

6.1.4 Homotopic deformation along p and Leray-Schauder degree

The Leray-Schauder degree of a mapping associated with the investigated BVP is one of the most frequently used methods when dealing with this problem. First consider the associated problem

$$(\Phi_p(x'))' = f(t), \quad x(0) = 0 = x(\pi_p), \quad \Phi_p(s) = |s|^{p-1} \operatorname{sgn} s \tag{6.1.15}$$

(note that $\Phi_p = \Phi$), where $f \in L^\beta(0,\pi_p)$ for some $\beta > 1$, and the energy functional corresponding to this problem

$$\Psi_p(x) = \frac{1}{p}\int_0^{\pi_p} |x'(t)|^p \, dt + \int_0^{\pi_p} f(t)x(t) \, dt. \tag{6.1.16}$$

Here we use the index p by Φ and Ψ to emphasize their dependence on the power p. The functional Ψ_p is coercive, continuous and convex over $W_0^{1,p}(0,\pi_p)$, and hence it possesses the unique global minimum which is the critical point and hence a solution of (6.1.15). This means that we have correctly defined mapping $G_p : L^\beta(0,\pi_p) \to C^1[0,\pi_p]$ which assigns to the right-hand side f of (6.1.15) the solution x of this problem. This mapping is completely continuous. Moreover, if p_n is a real sequence such that $p_n \to p$ and $f_n \in L^\beta(0,\pi_p)$ is such that $f_n \rightharpoonup f \in L^\beta(0,\pi_p)$ ($\rightharpoonup$ denotes again the weak convergence), then $\lim_{n\to\infty} G_{p_n}(f_n) = G_p(f)$ as it is shown in [91].

Now, for fixed $p > 1$, we define $T_p : C[0,\pi_p] \to C[0,\pi_p]$ by $T_p(x) = x - G_p(\lambda\Phi_p(x))$ with $\lambda \in \mathbb{R}$. Obviously, the equation $T_p(x) = 0$ has a nontrivial solution if and only if $\lambda = \lambda_n(p) = (p-1)n^p$ and this solution is $x_n(t) = \alpha \sin_p(nt)$, $\alpha \in \mathbb{R}$, $\alpha \neq 0$.

The following statement concerns a homotopic deformation along the power p of the Leray-Schauder degree of the mapping T_p with $\lambda \neq \lambda_n$. Note that the classical result of the linear theory is that the Leray-Schauder degree of T_2 with respect to the ball

$$B(0,r) := \left\{ u \in C[0,\pi_p] : \|u\|_C = \max_{t\in[0,\pi_p]} |u(t)| \le r \right\}$$

is

$$d(T_2, B(0,r), 0) = (-1)^n, \tag{6.1.17}$$

where n is the number of the eigenvalues of problem (6.1.2) with $p = 2$ which are less than λ.

Theorem 6.1.5. *Let $p > 1$ be arbitrary, $\lambda \neq \lambda_n(p) = (p-1)n^p$, $n \in \mathbb{N}$. Then for every $r > 0$, the Leray-Schauder degree $d(T_p, B(r,0), 0)$ is well defined and satisfies*

$$d(T_p, B(r,0), 0) = (-1)^n, \tag{6.1.18}$$

where n is the number of eigenvalues of (6.1.2) *which are less than λ.*

Proof. Suppose that $p \geq 2$ and $\lambda > \lambda_1 = (p-1)$, i.e., $\lambda = (p-1)(n+s)^p$ for some $s \in (0,1)$ and $n \in \mathbb{N}$. In the remaining cases the idea of the proof is the same. We will show that $d(T_p, B(r,0), 0) = (-1)^n$ for every $r > 0$.

Let $\Lambda : [p, \infty) \to \mathbb{R}$ be defined by $\Lambda(\alpha) = [(n+s)\pi_\alpha/\pi_p]^\alpha$, where π_α is given by (6.1.3) with α instead of p. Obviously, π_α depends continuously on α and hence Λ is continuous. Next, define the mapping

$$T(\alpha, x) = x - G_\alpha(\Lambda(\alpha)\Phi_\alpha(x)).$$

The mapping $\tilde{G}(\alpha, x) := G_\alpha(\Lambda(\alpha)\Phi_\alpha(x))$ is completely continuous and $T(\alpha, x) \neq 0$ for all $\alpha \in [p, \infty)$ (for details see [91, Theorem 4.1]). Hence, from the invariance of the degree under homotopies and from (6.1.17) we obtain the required statement. ☐

Now we apply the previous statement to derive solvability conditions for the BVP

$$(\Phi(x'))' + g(t,x) = 0, \quad x(0) = 0 = x(\pi_p), \tag{6.1.19}$$

where $g : [0, \pi_p] \times \mathbb{R} \to \mathbb{R}$ is a continuous function.

Theorem 6.1.6. *Suppose that there exists $n \in \mathbb{N}$ such that the nonlinearity g in* (6.1.19) *satisfies*

$$\lambda_n \leq a(t) := \liminf_{|s|\to\infty} \frac{g(t,s)}{\Phi(s)} \leq \limsup_{|s|\to\infty} \frac{g(t,s)}{\Phi(s)} =: b(t) \leq \lambda_{n+1} \tag{6.1.20}$$

uniformly on $[0, \pi_p]$, the first and the last inequalities being strict on a subset of positive measure in $[0, \pi_p]$. Then the BVP (6.1.19) *has a solution.*

Proof. Let $\nu \in (\lambda_n, \lambda_{n+1})$. According to Theorem 6.1.5, it suffices to construct a homotopic bridge connecting (6.1.19) with the problem

$$(\Phi(x'))' + \nu\Phi(x) = 0, \quad x(0) = 0 = x(\pi_p).$$

The degree of the mapping associated with this problem has been computed in Theorem 6.1.5. This homotopy is defined as follows

$$H(\tau, x) = G_p(\tau\nu\Phi(x) + (1-\tau)g(t, x(t))).$$

Using the standard method it can be proved that there exists $r > 0$ such that $x - H(\tau, x) \neq 0$ for $x \in \partial B(r, 0)$ for every $\tau \in [0,1]$ if $r > 0$ is sufficiently large. This proof goes by contradiction. Supposing that there exists $x_n \in C[0, \pi_p]$ and with $\|x_n\|_C \to \infty$ and $\tau_n \in [0,1]$ such that $x_n = H(\tau_n, x_n)$, functions v and c are constructed (using essentially the same construction as in the linear case) in such a way that the half-linear equation

$$(\Phi(v'))' + c(t)\Phi(v) = 0, \quad v(0) = 0 = v(\pi_p),$$

with $\lambda_n \le a(t) \le c(t) \le b(t) \le \lambda_{n+1}$ has a nontrivial solution. Since the first and the last of the previous inequalities is strict on the set of positive measure (since inequalities in (6.1.20) are of the same character), we have a contradiction with the Sturmian comparison theorem. □

6.1.5 Multiplicity nonresonance results

In order to illustrate one of the basic methods for the investigation of the generally *nonresonant* case, we consider the following simple BVP in this subsection. We investigate the number of solutions of the BVP

$$(\Phi(u'))' + \lambda\Phi(u) = 1, \quad u(0) = 0 = u(\pi_p). \tag{6.1.21}$$

We denote the number of solutions of (6.1.21) by $N_p(\lambda)$. Note that in the linear case $p = 2$ we have

$$N_2(\lambda) = \begin{cases} 1 & \text{if } \lambda \neq k^2 \text{ for all } k \in \mathbb{N}, \\ 0 & \text{if } \lambda = (2k-1)^2 \text{ for all } k \in \mathbb{N}, \\ \infty & \text{if } \lambda = (2k)^2 \text{ for all } k \in \mathbb{N}. \end{cases}$$

In this section we show, among others, that

$$\lim_{\lambda\to\infty} N_p(\lambda) = \infty. \tag{6.1.22}$$

An important role in the proof of the main result of this subsection is played by the quantity $t_\lambda(\alpha)$, which is the first positive zero of the derivative u' of the initial problem

$$(\Phi(u'))' + \lambda\Phi(u) = 1, \quad u(0) = 0, \ u'(0) = \alpha. \tag{6.1.23}$$

Multiplying both sides of (6.1.23) by u' and integrating the obtained equality from 0 to t, we get the identity

$$\frac{|u'(t)|^p}{q} + \lambda\frac{|u(t)|^p}{p} = \frac{|\alpha|^p}{q} + u(t), \quad q = \frac{p}{p-1}.$$

Thus, for $t \in (0, t_\lambda(\alpha))$

$$t = \int_0^{u(t)} \frac{ds}{(\alpha^p + qs - \lambda(q-1)s^p)^{1/p}} \tag{6.1.24}$$

if $\alpha \ge 0$, and

$$t = \int_0^{-u(t)} \frac{ds}{(|\alpha|^p - qs - \lambda(q-1)s^p)^{1/p}} \tag{6.1.25}$$

if $\alpha < 0$. Hence, we consider the function

$$F(t) = \int_0^t \frac{ds}{(|\alpha|^p + q\,\mathrm{sgn}(\alpha)s - \lambda(q-1)s^p)^{1/p}}, \tag{6.1.26}$$

here we adjust the usual sgn function as follows

$$\mathrm{sgn}(\alpha) = \begin{cases} 1 & \text{if } \alpha \geq 0, \\ -1 & \text{if } \alpha < 0. \end{cases}$$

It follows that

$$t_\lambda(\alpha) = F(h(\alpha)), \tag{6.1.27}$$

where $h(\alpha)$ is the unique positive root of the equation

$$\lambda(q-1)x^p - q\,\mathrm{sgn}(\alpha)x = |\alpha|^p. \tag{6.1.28}$$

Also, from (6.1.24)-(6.1.26), for $t \in (t, t_\lambda(\alpha)]$, we have

$$u(t) = \begin{cases} F^{-1}(t) & \text{if } \alpha \geq 0, \\ -F^{-1}(t) & \text{if } \alpha < 0. \end{cases} \tag{6.1.29}$$

Conversely, if we have a function u of the form (6.1.29), it can be directly verified that u is a solution of the differential equation in (6.1.23) and hence the (unique) solution of this initial value problem on the interval $[0, t_\lambda(\alpha)]$.

Next, we extend u to obtain a global solution $u_\lambda(\alpha, t)$ of (6.1.23). Define $u_\lambda(\alpha, t)$ as the $2[t_\lambda(\alpha) + t_\lambda(-\alpha)]$ periodic extension of the function

$$\tilde{u}_\lambda(\alpha, t) = \begin{cases} u(t) & t \in [0, t_\lambda(\alpha)], \\ u(2t_\lambda(\alpha) - t) & t \in [t_\lambda(\alpha), 2t_\lambda(\alpha)], \\ \tilde{u}_\lambda(-\alpha, 2(t_\lambda(\alpha) + t_\lambda(-\alpha) - t)) & t \in [2t_\lambda(\alpha), 2(t_\lambda(\alpha) + t_\lambda(-\alpha))]. \end{cases}$$

Note that the zeros of $u_\lambda(\alpha, t)$ are the numbers $2kt_\lambda(\alpha) + 2(k-\nu)t_\lambda(-\alpha)$, $k \in \mathbb{N}$, $\nu \in \{0, 1\}$. The basic properties of the function $t_\lambda(\alpha)$, which we will need in proving the main result of this subsection, are summarized in the next lemma.

Lemma 6.1.3. *The function $t_\lambda(\alpha)$ has the following properties*

(i) t_λ is strictly decreasing and continuous on $(-\infty, 0)$ and on $[0, \infty)$.

(ii) t_λ has a jump discontinuity at $\alpha = 0$; more precisely

$$0 = \lim_{\alpha \to 0-} t_\lambda(\alpha) < t_\lambda(0) = \frac{q\pi_p}{2}\left[\frac{p-1}{\lambda}\right]^{1/p}.$$

(iii) It holds

$$\lim_{\alpha \to \pm\infty} t_\lambda(\alpha) = \frac{\pi_p}{2}\left[\frac{p-1}{\lambda}\right]^{1/p}.$$

Proof. From (6.1.26), (6.1.27) it follows

$$t_\lambda(\alpha) = \int_0^{h(\alpha)} \frac{ds}{(|\alpha|^p + q\,\mathrm{sgn}(\alpha)s - \lambda(q-1)s^p)^{1/p}}. \tag{6.1.30}$$

Substituting $s \to sq(\alpha)$ in (6.1.30) and calling on (6.1.28) with $x = q(\alpha)$, we obtain

$$t_\lambda(\alpha) = \int_0^1 \frac{ds}{\left[\lambda(q-1)(1-s^p)^{1/p} - h^{1-p}(\alpha)q\,\mathrm{sgn}(\alpha)(1-s)\right]^{1/p}}. \tag{6.1.31}$$

To examine the behavior of $h(\alpha)$ with respect to α we note that the Implicit Function Theorem together with (6.1.28) imply the $h(\alpha)$ is of the class C^1 on $\mathbb{R} \setminus \{0\}$ and

$$\frac{dh}{d\alpha}(\alpha) = \frac{p\Phi(\alpha)}{q[\lambda h^{p-1}(\alpha) - \mathrm{sgn}(\alpha)]}. \tag{6.1.32}$$

From the definition of $h(\alpha)$, we easily see that the denominator on the right-hand side of (6.1.32) is positive, hence $h(\alpha)$ is strictly increasing and continuous for $\alpha \in [0, \infty)$ and strictly decreasing and continuous on $(-\infty, 0)$. Thus, the statement (i) immediately follows from (6.1.31).

To show (ii), we first observe that $h(0-) = 0$ and hence from (6.1.30) we conclude that $\lim_{\alpha\to 0-} t_\lambda(\alpha) = 0$. Next, setting $\alpha = 0$ in (6.1.31) and using the fact that $h(0) = (p/\lambda)^{1/(p-1)} = (p/\lambda)^{q-1}$, from (6.1.31), we obtain

$$t_\lambda(0) = \left(\frac{p-1}{\lambda}\right)^{1/p} \int_0^1 \frac{ds}{(s - s^p)^{1/p}}. \tag{6.1.33}$$

The substitution $s = \tau^q$ in (6.1.33) yields

$$t_\lambda(0) = \left(\frac{p-1}{\lambda}\right)^{1/p} \int_0^1 \frac{q\,d\tau}{(1-\tau^p)^{1/p}} = \frac{q\pi_p}{2}\left[\frac{p-1}{\lambda}\right]^{1/p}.$$

This shows the statement (ii). Finally, to show (iii), we note that $h(\alpha) \to \infty$ as $|\alpha| \to \infty$. Letting $\alpha \to \pm\infty$ in (6.1.31), it follows from the Lebesgue Dominated Convergence Theorem that

$$\lim_{\alpha\to\pm\infty} t_\lambda(\alpha) = \left(\frac{p-1}{\lambda}\right)^{1/p} \int_0^1 \frac{ds}{(1-s^p)^{1/p}} = \frac{\pi_p}{2}\left[\frac{p-1}{\lambda}\right]^{1/p}.$$

This concludes the proof of lemma. □

Now we are ready to prove the main result of this subsection and some of its consequences. Together with the sequence of $\lambda_k = (p-1)k^p$ of (6.1.2) we will consider also the sequence $\mu_k = (p-1)(qk)^p$. We observe that $\mu_k < (=)(>)\lambda_{2k}$ if $p > (=)(<)2$. The numbers μ_k will play an important role in our results, the reason is their nonuniform distribution with respect to the λ_k's for $p \neq 0$ which gives rise to the existence of a large number of solutions of (6.1.21) for large λ.

We say that a function $u \in C^1[0, \pi_p]$ belongs to the classes E_k^+ (E_k^0) (E_k^-) if u possesses exactly $k-1$ zeros on $(0, \pi_p)$ and $u'(0) >(=0)(<)0$.

Theorem 6.1.7. *The following statements hold:*

(a) If $\lambda \in (0, \lambda_1)$, then (6.1.21) has exactly one solution u, and $u \in E_1^-$.

(b) If $\lambda = \lambda_1$, then (6.1.21) has no solution.

(c) If λ is strictly between λ_{2k-1} and μ_k, then (6.1.21) possesses at least one solution in $u \in E_{2k-1}^+$.

(d) If λ is strictly between μ_k and λ_{2k+1}, then (6.1.21) possesses at least one solution in E_{2k+1}^+.

(e) If $\lambda = \mu_k$, then (6.1.21) possesses a solution $u \in E_k^0$.

(f) If λ is strictly between μ_k and λ_{2k}, then (6.1.21) possesses a solution in E_{2k}^+ and a solution in E_{2k}^-.

Proof. (a) Assume that $0 < \lambda < \lambda_1 = p - 1$. From Lemma 6.1.3 we find that $2t_\lambda(\alpha) > \pi_p$ for $\alpha \geq 0$. Hence, there is no solution of (6.1.21) with nonnegative derivative at $t = 0$. Again, from Lemma 6.1.3, but for $\alpha < 0$, we see that there exists a unique $\alpha^* < 0$ such that $2t_\lambda(\alpha^*) = \pi_p$. It follows that $u_\lambda(\alpha^*, t)$ is the unique solution of (6.1.21) and belongs to E_1^-.

(b) Suppose that $\lambda = \lambda_1 = p - 1$. The absence of solutions to (6.1.21) follows in this case directly from the fact that $2t_\lambda(\alpha) > \pi_p$ for $\alpha \geq 0$ and $2t_\lambda(\alpha) < \pi_p$ for $\alpha < 0$.

(c) We assume λ strictly between λ_{2k-1} and μ_k. Consider the function

$$f(\alpha) = 2(k-1)[t_\lambda(\alpha) + t_\lambda(-\alpha)] + 2t_\lambda(\alpha). \tag{6.1.34}$$

From Lemma 6.1.3 we see that f is continuous on $(0, \infty)$, $f(0+) = qk(\pi_p/\lambda)^{1/p}$ and we have $\lim_{\alpha\to\infty} f(\alpha) = (2k - 1)$. This implies that there exists $\bar{\alpha} > 0$ such that $f(\bar{\alpha}) = \pi_p$ and hence $u_\lambda(\bar{\alpha}, t)$ is a solution of (6.1.21) with exactly $2k - 2$ zeros in $(0, \pi_p)$. Clearly $u \in E_{2k-1}^+$.

(d) This proof is analogous to the previous one, only instead of f defined by (6.1.34) we consider the function

$$f(\alpha) = 2k[t_\lambda(\alpha) + t_\lambda(-\alpha)] + 2t_\lambda(\alpha) \tag{6.1.35}$$

on the interval $(-\infty, 0)$.

(e) If $\lambda = \mu_k$, we have $2kt_\lambda(0) = \pi_p$ and, therefore, $u(t) = u_\lambda(0, t)$ is a solution of (6.1.21) with exactly $k - 1$ zeros in $(0, \pi_p)$. Clearly, $u \in E_k^0$. Note that all zeros of u are double in this case.

(f) Finally, suppose that λ is between λ_{2k} and μ_k. In this case, we define the function

$$f(\alpha) = 2k[t_\lambda(\alpha) + t_\lambda(-\alpha)].$$

The function f is continuous on $(-\infty, 0)$ and $(0, \infty)$, $f(0+) = qk\pi_p/\left((p-1)/\lambda\right)^{1/p}$ and again we have $\lim_{\alpha\to\pm\infty} f(\alpha) = 2k\pi_p\left((p-1)/\lambda\right)^{1/p}$. Reasoning as in the previous parts of the proof, we obtain the existence of solutions u, v of (6.1.21) with $u \in E_{2k}^+$, $v \in E_{2k}^-$. □

Corollary 6.1.1. *Let $p \neq 2$. Then (6.1.21) is solvable for all $\lambda > 0$ except $\lambda = \lambda_1$ and, eventually, those numbers λ of the form $\lambda = \lambda_{2k-1}$ for $k < 1/|q-2|$.*

From the previous corollary we obtain, in particular, that (6.1.21) is solvable for all large positive λ. Furthermore, as $\lambda \to \infty$, the number of solutions $N_p(\lambda)$ tends ∞, as the following statement shows.

Theorem 6.1.8. *Let $p \neq 2$. Then the number of solutions $N_p(\lambda)$ of (6.1.21) satisfies*

$$N_p(\lambda) \geq 3\left(\frac{\lambda}{p-1}\right)^{1/p}\left|\frac{1}{q}-\frac{1}{2}\right| - 3 \tag{6.1.36}$$

for all $\lambda > 0$. In particular, $\lim_{\lambda\to\infty} N_p(\lambda) = \infty$.

Proof. We will assume that $p > 2$, the case $p \in (1,2)$ can be treated in a similar way. Let us fix $\lambda > 0$ and denote by M_f the number of positive integers such that (f) of Theorem 6.1.7 holds, i.e.,

$$M_f = \operatorname{card}\left\{k \in \mathbb{N} : (\mu_k)^{1/p} < \lambda^{1/p} < (\lambda_{2k})^{1/p}\right\}.$$

Clearly,

$$M_f \geq \max\left\{k \in \mathbb{N} : k < \left(\frac{\lambda}{\mu_k}\right)^{1/p}\right\} - \min\left\{k \in \mathbb{N} : \left(\frac{\lambda}{\lambda_{2k}}\right)^{1/p} < k\right\} + 1,$$

and hence

(6.1.37)

$$M_f \geq \left[\left(\frac{\lambda}{\mu_k}\right)^{1/p} - 1\right] - \left[\left(\frac{\lambda}{\lambda_{2k}}\right)^{1/p} + 1\right] + 1 = \left(\frac{\lambda}{p-1}\right)^{1/p}\left(\frac{1}{q}-\frac{1}{2}\right) - 1.$$

Next, let us denote by M_c and M_d the number of positive integers such that (c) and (d) of Theorem 6.1.7 holds, respectively. Estimates similar to those for M_f yield

$$M_c \geq \left(\frac{\lambda}{(p-1)}\right)^{1/p}\left(\frac{1}{q}-\frac{1}{2}\right) - \frac{3}{2}, \tag{6.1.38}$$

$$M_d \geq \left(\frac{\lambda}{(p-1)}\right)^{1/p}\left(\frac{1}{q}-\frac{1}{2}\right) - \frac{1}{2}. \tag{6.1.39}$$

From (6.1.37)-(6.1.39) we obtain

$$N_p(\lambda) \geq M_c + M_d + M_f \geq 3\left(\frac{\lambda}{(p-1)}\right)^{1/p}\left(\frac{1}{q}-\frac{1}{2}\right) - 3,$$

what we needed to prove. □

We finish this subsection with a statement which extends and complements the previous results of this subsection. We prefer to present the result without proof because of its technical complexity.

Theorem 6.1.9. *Let $p \neq 2$ and set*

$$k(p) = \begin{cases} (p-1)/(2-p) & \text{if } p \in (1,2), \\ 1/(p-2) & \text{if } p > 2. \end{cases}$$

Let either $\lambda = \lambda_{2k}$ with $k \geq 1$, or $\lambda = \lambda_{2k+1}$ with $k > k(p)$, $k \in \mathbb{N}$. Then there exists a number $0 < \delta < 1$ such that (6.1.7) *has at least one solution for any $f = 1 + h$ with $\|h\|_{L^\infty(0,\pi_p)} < \delta$.*

6.2 Fredholm alternative for one-dimensional p-Laplacian

As we have mentioned at the beginning of this chapter, the investigation of the BVP (6.1.7) when $\lambda = \lambda_k$ is an eigenvalue is perhaps the most interesting part of the qualitative theory of half-linear differential equations, since in comparison with the classical Fredholm alternative for the linear boundary value problem

$$u'' + m^2 u = h(t), \quad u(0) = 0 = u(\pi),$$

which has a solution if and only if

$$\int_0^\pi h(t) \sin mt \, dt = 0, \tag{6.2.1}$$

in case of the half-linear BVP (6.1.7), the situation is very different.

In this section we discuss a possible extension of the Fredholm alternative to (6.1.7). We suppose that $\lambda = \lambda_k$ for some $k \in \mathbb{N}$, so the problem (6.1.2) possesses a nontrivial solution $x(t) = \sin_p(kt)$. The half-linear version of (6.2.1) when $\lambda = \lambda_1$ and $m = 1$ is is the orthogonality condition

$$\int_0^{\pi_p} h(t) \sin_p t \, dt = 0. \tag{6.2.2}$$

In the first subsection we describe the situation when $\lambda = \lambda_1 = p - 1$, the second subsection deals briefly with resonance at higher eigenvalues.

6.2.1 Resonance at the first eigenvalue

First we present (without proofs) some technical statements which concern certain initial value problem, and present essentially asymptotic formulas for zero points of the solution of this IVP (compare with Lemma 6.1.3 of the previous subsection). These auxiliary statements are used in the proof of the three main results of this subsection which concern the BVP

$$(\Phi(u'))' + (p-1)\Phi(u) = \frac{1}{q} h(t), \quad u(0) = 0 = u(\pi_p). \tag{6.2.3}$$

The factor $1/q = (p-1)/p$ is inserted here for convenience, some computations are then formally slightly easier.

Lemma 6.2.1. *Let u_α be a solution of*

$$(\Phi(u'))' + (p-1)\Phi(u) = \frac{1}{q}h(t), \quad u(0) = 0, \ u'(0) = \alpha, \quad q = \frac{p}{p-1}, \tag{6.2.4}$$

Then $u_\alpha(t)/\alpha \to \sin_p t$ as $|\alpha| \to \infty$ in $C^1[0,K]$-sense, for every $K > 0$. In particular, for large $|\alpha|$, u_α has a first positive zero t_1^α and so does u'_α at a point $t(\alpha) > 0$. Moreover, for $\alpha > 0$ ($\alpha < 0$), u_α is strictly incerasing (decreasing) for $t \in (0, t(\alpha))$ and strictly decreasing (increasing) in $(t(\alpha), t_1^\alpha)$, and $t(\alpha) \to \pi_p/2$, $t_1^\alpha \to \pi_p$ as $|\alpha| \to \infty$. For a fixed $M > 0$, all these convergences are uniform in h with $\|h\|_{L^\infty[0,K]} \le M$.

Lemma 6.2.2. *Assume that $h \in L^\infty_{loc}[0, 2\pi_p)$ and denote*

$$I_h := \int_0^{\pi_p} h(t) \sin_p t \, dt.$$

Then

$$t_1^\alpha = \pi_p + \frac{I_h}{p\Phi(\alpha)} + o(|\alpha|^{1-p}) \quad \text{as } |\alpha| \to \infty, \tag{6.2.5}$$

where $o(|\alpha|^{1-p})$ is uniform with respect to all h such that $\|h\|_{L^\infty(0,2\pi_p)} < H$ for a fixed constant $H > 0$.

Lemma 6.2.3. *Let $h \in C^1[0, 2\pi_p]$, and denote*

$$J_h := \frac{1}{2p^3} \int_0^{\pi_p/2} (\cos_p t)^{-p} \left[\left(\int_t^{\pi_p/2} h(s) \cos_p s \, ds \right)^2 + \left(\int_t^{\pi_p/2} h(\pi_p - s) \cos_p s \, ds \right)^2 \right] dt.$$

If $I_h = 0$, then

$$t_1^\alpha = \pi_p + (p-2) J_h |\alpha|^{2(1-p)} + o(|\alpha|^{2(1-p)}), \quad \text{as } |\alpha| \to \infty, \tag{6.2.6}$$

where $o(|\alpha|^{2(1-p)})$ is uniform with respect to all h with $\|h\|_{C^1[0,2\pi_p]} < H$ for some fixed positive constant H. Moreover, if $|\alpha|$ is sufficiently large, $t_1^\alpha < \pi_p$ for $p \in (1,2)$ and $t_1^\alpha > \pi_p$ for $p > 2$.

The next statement concerns the topological degree of the mapping associated with BVP (6.2.3). For $h \in L^\infty(0, \pi_p)$ and $\lambda \in [0, \infty)$ define an operator $T_{\lambda,h}$ by $T_{\lambda,h}(v) = u$ if and only if

$$(\Phi(u'))' = \lambda \left[\frac{1}{q} h(\lambda^{\frac{1}{p}} t) - (p-1)\Phi(v) \right], \quad u(0) = u(\pi_p) = 0. \tag{6.2.7}$$

Now we define $T_h := T_{1,h}$.

Lemma 6.2.4. *Assume that for $h \in L^\infty(0,\pi_p)$ one has $I_h \neq 0$. Then solutions to the boundary value problem* (6.2.3) *are a-priori bounded and there exists $R > 0$ such that*

$$\deg[I - T_h; B_R(0), 0] = 0. \tag{6.2.8}$$

The first theorem of this subsection shows that (6.2.2) is sufficient but generally not necessary for solvability of the BVP (6.2.4).

Theorem 6.2.1. *Let us assume that $h \in C^1[0,\pi_p]$, $h \not\equiv 0$, and* (6.2.2) *holds. Then* (6.2.3) *has at least one solution. Moreover, if $p \neq 2$, then the set of possible solutions is bounded in $C^1[0,\pi_p]$.*

Proof. Set $X := C_0^1[0,\pi_p] = \{u \in C^1[0,\pi_p]; u(0) = u(\pi_p) = 0\}$, and let the operator $T_{\lambda,h}$ be defined by (6.2.7).

Standard arguments based on the Ascoli-Arzelá Theorem imply that $T_{\lambda,h}$ is a well defined operator which is compact from X into X^*. Moreover, $T_{\lambda,h}$ depends continuously (in the operator norm) on the perturbations of $h \in L^\infty(0,\pi_p)$ and $\lambda \in \mathbb{R}$. A formula for the change of the index for $T_{\lambda,0}$, when the spectral parameter $\lambda \in \mathbb{R}$ crosses the first eigenvalue $\lambda_1 = 1$ can be found in the proof of Theorem 6.1.5 or [118, Theorem 14.9]. Adapting that result to our case, we have that for small $\varepsilon > 0$, and any $R > 0$,

$$\deg[I - T_{1-\varepsilon,0}; B_R(0), 0] = 1, \quad \deg[I - T_{1+\varepsilon,0}; B_R(0), 0] = -1, \tag{6.2.9}$$

where $B_R(0) := \{u \in X; \|u\|_X < \mathbb{R}\}$. Using the homogeneity in equation (6.2.7) and the boundary conditions in this BVP we have that for fixed $h \in L^\infty(0,\pi_p)$ we can take $R > 0$ so large that (6.2.9) extend to

$$\deg[I - T_{1-\varepsilon,h}; B_R(0), 0] = 1, \quad \deg[I - T_{1+\varepsilon,h}; B_R(0), 0] = -1. \tag{6.2.10}$$

We distinguish between the two cases $1 < p < 2$ and $p > 2$.

Case $1 < p < 2$. Let $h \in C^1[0,\pi_p]$ be such that orthogonality condition (6.2.2) holds. For $t \geq \pi_p$, let us extend h to $[0,\infty)$ as a C^1-function (e.g., as a linear function $h(t) = h'(\pi_p)t + h(\pi_p)$). We claim that there exists a constant $R > 0$ such that for any $\lambda \in [1, 2^{p-1}]$ the boundary value problem

$$(\Phi(u'))' + \lambda(p-1)\Phi(u) = \frac{\lambda}{q} h(\lambda^{\frac{1}{p}} t), \quad u(0) = u(\pi_p) = 0, \tag{6.2.11}$$

has no solution with $\|u\|_{C^1[0,\pi_p]} \geq R$.

To prove this claim we argue by contradiction. Thus we suppose there exist sequences $\{u_n\}_{n=1}^\infty \subset C^1[0,\pi_p]$, $\{\lambda_n\}_{n=1}^\infty \subset [1, 2^{p-1}]$, such that $\lambda_n \to \bar\lambda \in [1, 2^{p-1}]$, and $\|u_n\|_{C^1[0,\pi_p]} \to \infty$, and u_n, λ_n satisfy (6.2.11). By Lemma 6.2.1 it is not difficult to see that $|\alpha_n| \to \infty$, where as before $\alpha_n = u_n'(0)$. So, assume that $\alpha_n \to \infty$ (the case $\alpha_n \to -\infty$ is similar). Then u_n is the solution of the initial value problem

$$(\Phi(u_n'))' + \lambda_n(p-1)\Phi(u_n) = \frac{\lambda_n}{q} h(\lambda_n^{1/p} t), \quad u_n(0) = 0, u_n'(0) = \alpha_n,$$

on $[0,\infty)$, and hence $v_n(t) := u_n(t\lambda_n^{-(1/p)})$ solves the initial value problem

$$(\Phi(v_n'))' + (p-1)\Phi(v_n) = \frac{1}{q}h(t), \quad v_n(0) = 0,\ v_n'(0) = \tilde{\alpha}_n,$$

where $\tilde{\alpha}_n = \alpha_n\lambda_n^{1/p} \to \infty$. By Lemma 6.2.1 the first positive zero point $t_1^{\tilde{\alpha}_n}$ of v_n satisfies $t_1^{\tilde{\alpha}_n} \to \pi_p$ as $n \to \infty$ and similarly the second positive zero point approaches $2\pi_p$. Then condition (6.2.2) and Lemma 6.2.3 imply that $t_1^{\tilde{\alpha}_n} < \pi_p$ for n large enough. But this contradicts the fact that $0 = u_n(\pi_p) = v_n(\pi_p\lambda_n^{1/p})$ because $1 \le \lambda_n \le 2^{p-1}$ for any $n \in \mathbb{N}$. Thus the claim is proved.

From this claim we have that for $\varepsilon > 0$ small the homotopy $\mathcal{H} : [1, 1+\varepsilon] \times X \to X$ defined by $\mathcal{H}(u,\lambda) = u - T_{\lambda,h_\lambda}(u)$, where $h_\lambda(t) = h(\lambda^{\frac{1}{p}}t)$, satisfies $\mathcal{H}(u,\lambda) \ne 0$ for all $\lambda \in [1, 1+\varepsilon]$ and $\|u\|_{C^1[0,\pi_p]} \ge R$. Thus, from the homotopy invariance property of the Leray-Schauder degree, we obtain that

$$\deg[I - T_{1,h}; B_R(0), 0] = \deg[I - T_{1+\varepsilon,h_{1+\varepsilon}}; B_R(0), 0] = -1,$$

by (6.2.9). This proves that for given $h \in C^1[0,\pi_p]$ satisfying $I_h = 0$ the boundary value problem (6.2.3) has at least one solution. Moreover, it follows from our considerations that all possible solutions of (6.2.3) are a-priori bounded in the $C^1[0,\pi_p]$ norm.

Case $p > 2$. Let h be as in the previous ase. We claim now that there exists a constant $R > 0$ such that for any $\lambda \in [1/2, 1]$ the boundary value problem (6.2.11) has no solution with $\|u\|_{C_1[0,\pi_p]} \ge R$.

The proof of this claim follows the same steps as in the previous case, we obtain now

$$\deg[I - T_{1,h}; B_R(0), 0] = \deg[I - T_{1-\varepsilon,h_{1-\varepsilon}}; B_R(0), 0] = 1,$$

by (6.2.10). Thus the proof is completed. □

The eigenvalue problem (6.1.7) with $\lambda = \lambda_1$ and $f \equiv 0$ is closely related to the L^p-Poincaré inequality

$$\int_0^{\pi_p} |x'(t)|^p\,dt \ge C\int_0^{\pi_p} |x(t)|^p\,dt, \quad \text{for all } x \in W_0^{1,p}(0,\pi_p). \tag{6.2.12}$$

The constant $C = \lambda_1$ is precisely the largest $C > 0$ for which (6.2.12) holds. Then $\int_0^{\pi} |x'|^p - \lambda_1\int_0^{\pi_p} |x|^p \ge 0$ for all $x \in W_0^{1,p}(0,\pi_p)$ while it minimizes and equals 0 exactly on the ray generated by the first eigenfunction $\sin_p t$. Now we consider the following question: What is the sensitivity of this optimal Poincaré's inequality under a linear perturbation? We consider then the energy functional

$$E(u) = \frac{1}{p}\int_0^{\pi_p} |u'|^p - \frac{p-1}{p}\int_0^{\pi_p} |u|^p + \frac{1}{q}\int_0^{\pi_p} hu, \tag{6.2.13}$$

and ask whether E is *bounded from below*. It is easy to see that a necessary condition for this is that h satisfies the orthogonality condition (6.2.2), for otherwise E is unbounded below along the ray generated by the first eigenfunction. If $p = 2$, an

L^2-orthogonal expansion into the Fourier series yields that this condition is also sufficient for the boundedness from below. However, this approach seems to be of no use when $p \neq 2$. Under the additional assumption $h \in C^1[0, \pi_p]$, the result answering the sufficiency is provided by the following statement. Note that some of its conclusions are already implied by the previous theorem.

Theorem 6.2.2. *Assume that $h \in C^1[0, \pi_p]$, $h \not\equiv 0$, and (6.2.2) holds.*

(i) For $1 < p < 2$ the functional E is unbounded from below. The set of its critical points is nonempty and bounded.

(ii) For $p > 2$ the functional E is bounded from below and has a global minimizer. The set of its critical points is bounded, however E does not satisfy the Palais-Smale condition at the level 0.

Proof. Let us consider the energy functional $E : W_0^{1,p}(0, \pi_p) \to \mathbb{R}$ given by (6.2.13) whose critical points are solutions of boundary value problem (6.2.3). We will distinguish between the case $1 < p < 2$ and the case $p > 2$.

(i) *Case* $1 < p < 2$. For $\alpha \gg 1$, say $\alpha_n \to \infty$, $n \in \mathbb{N}$, consider the solutions to the initial value problem (6.2.4) given by $u_n(t) = u_{\alpha_n}(t)$ for $t \in [0, t_1^{\alpha_n})$, $u_n(t) = 0$ for $t \in [t_1^{\alpha_n}, \pi_p]$ (recall that $t_1^{\alpha_n} < \pi_p$ by Lemma 6.2.3). Clearly $u_n \in W_0^{1,p}(0, \pi_p), n \in \mathbb{N}$.

By Lemma 6.2.3, it follows that

$$\delta_n := \pi_p - t_1^{\alpha_n} = \frac{(2-p)J_h}{\alpha_n^{2(p-1)}} + o(\alpha_n^{2(1-p)}) \quad \text{as} \quad n \to \infty. \tag{6.2.14}$$

We shall prove that the energy functional E defined in (6.2.13) satisfies

$$\lim_{n \to \infty} E(u_n) = -\infty. \tag{6.2.15}$$

By definition,

$$E(u_n) = \frac{1}{p} \int_0^{\pi_p - \delta_n} |u_n'|^p - \frac{p-1}{p} \int_0^{\pi_p - \delta_n} |u_n|^p + \frac{1}{q} \int_0^{\pi_p - \delta_n} h u_n. \tag{6.2.16}$$

Multiplying $(\Phi(u_{\alpha_n}'))' + (p-1)\Phi(u_{\alpha_n}) = \frac{1}{q} h$ by u_{α_n} and integrating over $[0, \pi_p - \delta_n]$, we find

$$-\int_0^{\pi_p - \delta_n} |u_n'|^p + (p-1) \int_0^{\pi_p - \delta_n} |u_n|^p = \frac{1}{q} \int_0^{\pi_p - \delta_n} h u_n. \tag{6.2.17}$$

Then, from (6.2.16) and (6.2.17),

$$E(u_n) = -\frac{1}{q} \left[\int_0^{\pi_p - \delta_n} |u_n'|^p - (p-1) \int_0^{\pi_p - \delta_n} |u_n|^p \right]. \tag{6.2.18}$$

On the other hand from the Poincaré inequality, we have that

$$\int_0^{\pi_p - \delta_n} |u_n'|^p \geq (p-1) \left(\frac{\pi_p}{\pi_p - \delta_n} \right)^p \int_0^{\pi_p - \delta_n} |u_n|^p,$$

and then, from (6.2.18)

$$E(u_n) \le -p\left[\left(1-\frac{\delta_n}{\pi_p}\right)^{-p}-1\right]\int_0^{\pi_p-\delta_n}|u_n'|^p. \tag{6.2.19}$$

Now since by (6.2.14),

$$\left[\left(1-\frac{\delta_n}{\pi_p}\right)^{-p}-1\right] = \frac{p\delta_n}{\pi_p}+o(\delta_n) = \frac{p(2-p)J_h}{\pi_p}\,\alpha_n^{2(1-p)}+o\left(\alpha_n^{2(1-p)}\right),$$

as $n\to\infty$, we have $u_n(t)=\alpha_n \sin_p t + o(\alpha_n)$ for $t\in[0,\pi_p-\delta_n]$, and it follows from (6.2.19) that

$$\begin{aligned} E(u_n) \;&\le\; -\frac{p^2(2-p)J_h}{\pi_p}\,\alpha_n^{2(1-p)}\left[\alpha_n^p\int_0^{\pi_p-\delta_n}\sin_p^p t\,dt + o(\alpha_n^p)\right] \\ &\quad + o(\alpha^{2(1-p)})\left[\alpha_n^p\int_0^{\pi_p-\delta_n}\sin_p^p t\,dt + o(\alpha_n^p)\right] \\ &= -\frac{p^2(2-p)J_h}{\pi_p}\,\alpha_n^{2-p}\int_0^{\pi_p}\sin_p^p t\,dt + o(\alpha_n^{2-p}) \end{aligned} \tag{6.2.20}$$

for $n\to\infty$. Thus (6.2.15) follows from (6.2.20).

(ii) *Case* $p>2$. For a large positive number α, let us consider the solutions u_α and $u_{-\alpha}$ of the initial value problem (6.2.4). Then from Lemma 6.2.3, $u_{-\alpha}$ and u_α are respectively, lower and upper solutions (see later Subsection 6.3.1 for precise definitions) of the boundary value problem (6.2.3). Now, from Lemma 6.3.2 given in Subsection 6.3.1 we obtain that E attains its minimum on the set of functions between $u_{-\alpha}$ and u_α, at a (global) critical point of E. The set of such global critical points is compact, from Theorem 6.2.1. Let $-K$ be the minimum value of E on this set. Since given any $\psi\in C_0^\infty(0,\pi_p)$ and sufficiently large α,ψ lies between $u_{-\alpha}$ and u_α, so that we get $E(\psi)\ge -K$. Finally, by density of $C_0^\infty(0,\pi_p)$ in $W_0^{1,p}(0,\pi_p)$ we get $E(u)\ge -K$ for any $u\in W_0^{1,p}(0,\pi_p)$. Moreover, E minimizes precisely on the (nonempty) set of its critical points.

Finally we will exhibit an example which shows that the *Palais-Smale condition fails* for $p>2$ at the level zero, recall that this condition is defined in Lemma 6.1.2. Let $h\in C^1[0,\pi_p]$ be such that orthogonality condition (6.2.2) holds for $p>2$. Consider the solutions $u_n=u_n(t)$, $n\in\mathbb{N}$, of the initial value problem (6.2.4) with $u_n'(0)=\alpha_n\to\infty$ as $n\to\infty$. Then, from the fact that (6.2.4) is equivalent to its first integral

$$|u_\alpha'(t)|^p+|u_\alpha(t)|^p=\alpha^p+W(u_\alpha),\quad W(s):=\int_0^s h(\tau_\alpha(\xi))\,d\xi,$$

where $\tau_\alpha=u_\alpha^{-1}$ is the inverse function of u_α. Substituting $t=t(\alpha)$ in the last identity we obtain

$$\left(\frac{\alpha}{\beta}\right)^p = 1-\frac{W(\beta)}{\beta^p},\quad \beta:=u_\alpha(t(\alpha)),$$

and after a short computation we have $\alpha = \beta + O(\beta^{2-p})$ and thus for $t \in [0, 2\pi_p]$, we get

$$u_n(t) = \alpha_n \sin_p t + O(\alpha_n^{2-p}) \quad \text{as} \quad n \to \infty. \tag{6.2.21}$$

Since u_n solves the initial value problem (6.2.4), the function $v_n(t) := u_n(t_1^{\alpha_n} t/\pi_p)$ solves the boundary value problem

$$(\Phi(v_n'))' + (p-1)\left(\frac{\pi_p}{t_1^{\alpha_n}}\right)^p \Phi(v_n) = \frac{1}{q}\left(\frac{\pi_p}{t_1^{\alpha_n}}\right)^p \tilde{h}, \quad v_n(0) = v_n(\pi_p) = 0, \tag{6.2.22}$$

where $\tilde{h}(t) = h(t_1^{\alpha_n}/\pi_p\, t)$. From (6.2.21) and $\|h\|_{C^1[0,2\pi_p]} < H$, it follows that

$$v_n(t) = \alpha_n \sin_p t + O(\alpha_n^{2-p}) \quad \text{as} \quad n \to \infty, \tag{6.2.23}$$

and

$$\tilde{h}(t) = h(t) + o(\alpha_n^{2(1-p)}) \quad \text{as} \quad n \to \infty. \tag{6.2.24}$$

Also, from Hölder inequality and (6.2.22), we have that

$$\begin{aligned} &\sup_{\|\psi\|_{W_0^{1,p}} \le 1} |\langle E'(v_n), \psi\rangle| \\ &= \sup_{\|\psi\|_{W_0^{1,p}} \le 1} \left| \int_0^{\pi_p} \Phi(v_n')\psi' - (p-1)\int_0^{\pi_p} \Phi(v_n)\psi \right. \\ &\quad \left. + \frac{1}{q}\int_0^{\pi_p} h\psi \right| \le (p-1)\left|\left(\frac{\pi_p}{t_1^{\alpha_n}}\right)^p - 1\right| \sup_{\|\psi\|_{W_0^{1,p}} \le 1} \int_0^{\pi_p} \Phi(v_n)\psi \\ &\quad + \frac{1}{q}\int_0^{\pi_p} h\psi - \frac{1}{q}\left(\frac{\pi_p}{t_1^{\alpha_n}}\right)^p \int_0^{\pi_p} \tilde{h}\psi \\ &\le (p-1)\left|\left(\frac{\pi_p}{t_1^{\alpha_n}}\right)^p - 1\right| \left(\int_0^{\pi_p} |v_n|^p\right)^{\frac{1}{q}} + \frac{1}{q}\left(\int_0^{\pi_p} |h - \tilde{h}|^q\right)^{\frac{1}{q}} \\ &\quad + \frac{1}{q}\left|1 - \left(\frac{\pi_p}{t_1^{\alpha_n}}\right)^p\right| \left(\int_0^{\pi_p} |\tilde{h}|^q\right)^{\frac{1}{q}}. \end{aligned} \tag{6.2.25}$$

By (6.2.6),

$$\left|\left(\frac{\pi_p}{t_1^{\alpha_n}}\right)^p - 1\right| = \frac{p(p-2)}{\pi_p} J_h \alpha_n^{2(1-p)} + o\left(\alpha_n^{2(1-p)}\right), \tag{6.2.26}$$

and by (6.2.24),

$$|h(t) - \tilde{h}(t)| = o(\alpha_n^{2(1-p)}), \tag{6.2.27}$$

as $n \to \infty$, then from (6.2.26) and (6.2.27), we find that

$$\sup_{\|\psi\|_{W_0^{1,p}} \le 1} |\langle E'(v_n), \psi\rangle| \le \frac{p(p-1)(p-2)}{\pi_p} J_h \alpha_n^{1-p} + o(\alpha_n^{1-p})$$

as $n \to \infty$, i.e., $\lim_{n\to\infty} E'(v_n) = 0$.

From (6.2.2), (6.2.22), (6.2.23), (6.2.25), (6.2.26) and (6.2.27) we obtain that

$$\begin{aligned}
|E(v_n)| &= \left| \frac{1}{p} \int_0^{\pi_p} |v_n'|^p - \frac{p-1}{p} \int_0^{\pi_p} |v_n|^p + \frac{1}{q} \int_0^{\pi_p} h v_n \right| \\
&\le \frac{p-1}{p} \left| \left(\frac{\pi_p}{t_1^{\alpha_n}} \right)^p - 1 \right| \int_0^{\pi_0} |v_n|^p \\
&\quad + \frac{1}{q} \left| \int_0^{\pi_p} h v_n \right| + \frac{1}{pq} \left(\frac{\pi_p}{t_1^{\alpha_n}} \right)^p \left| \int_0^{\pi_p} \tilde{h} v_n \right| \\
&\le \frac{(p-1)(p-2)}{\pi_p} J_h \alpha_N^{2-p} + o(\alpha_n^{2-p}) \quad \text{as } n \to \infty,
\end{aligned}$$

i.e., $\lim_{n\to\infty} E(v_n) = 0$. Hence $\{v_n\}_{n=1}^{\infty} \subset W_0^{1,p}(0, \pi_p)$ is an unbounded Palais-Smale sequence. □

Our next result shows, in particular, another interesting difference with the linear case $p = 2$. If $p \ne 2$, then the set of functions f for which (6.1.7) with $\lambda = \lambda_1$ is solvable has *nonempty interior* in $L^\infty(0, \pi_p)$.

Theorem 6.2.3. *Let $p \neq 2$. Then there exists an open cone $\mathcal{C} \subset L^\infty(0, \pi_p)$ such that for any $h \in \mathcal{C}$ problem (6.2.3) with $\lambda = \lambda_1$ has at least two solutions. Moreover*

$$\int_0^{\pi_p} h(t) \sin_p t \, dt \neq 0 \tag{6.2.28}$$

for all $h \in \mathcal{C}$.

Proof. With the notation of the proof of Theorem 6.2.1 let us define $T_h : X \to X$ by $T_h = T_{1,h}$. Thus for each $v \in X$ and $h \in L^\infty(0, \pi_p)$, $u = T_h(v)$ satisfies

$$(\Phi(u'))' = \frac{1}{q} h - (p-1)\Phi(v), \;\; u(0) = u(\pi_p) = 0.$$

As mentioned before, T_h is a well defined compact operator and depends continuously (in the operator norm) on the perturbations of the (parameter) function h (with respect to the $L^\infty(0, \pi_p)$-norm).

Let us first construct an auxilliary function $h_0 \in C^2[0, \pi_p]$ for which the boundary value problem (6.2.3) (with $h = h_0$) has a solution and, moreover,

$$\int_0^{\pi_p} h_0(t) \sin_p t \, dt \neq 0.$$

For $0 < \varepsilon \ll 1$, set

$$u_\varepsilon(t) = \begin{cases} \frac{p-1}{p+1} - \frac{p-1}{p+1} \frac{(\varepsilon - t)^{\frac{p+1}{p-1}}}{\varepsilon^{\frac{p+1}{p-1}}} & \text{for } t \in [0, \varepsilon), \\ \frac{p-1}{p+1} & \text{for } t \in [\varepsilon, \pi_p/2], \\ u_\varepsilon(\pi_p - t) & \text{for } t \in (\pi_p/2, \pi_p]. \end{cases}$$

Let us define $h_\varepsilon := q[(\Phi(u'_\varepsilon))' + (p-1)\Phi(u_\varepsilon)]$. Straightforward calculation yields $h_\varepsilon \in C^2[0,\pi_p]$ and, by definition, $u_\varepsilon \in X$ is a positive solution of the boundary value problem (6.2.3) with $h = h_\varepsilon$. On the other hand, the following asymptotic estimates for $\varepsilon \to 0+$ hold

$$\begin{aligned}
\frac{1}{q} I_\varepsilon &:= \frac{1}{q}\int_0^{\pi_p} h_\varepsilon(t)\sin_p t\,dt \\
&= 2\int_0^{\frac{\pi_p}{2}} (\Phi(u'_\varepsilon))'\sin_p t\,dt + 2(p-1)\int_0^{\frac{\pi_p}{2}} \Phi(u_\varepsilon)\sin_p t\,dt \\
&= -2\int_0^{\frac{\pi_p}{2}} |u'_\varepsilon|^{p-2}u'_\varepsilon\cos_p t\,dt + 2(p-1)\int_0^{\frac{\pi_p}{2}} |u_\varepsilon|^{p-2}u_\varepsilon\sin_p t\,dt \\
&= -2\int_0^{\varepsilon} \frac{(\varepsilon-t)^2}{\varepsilon^{p+1}}\cos_p t\,dt + 2(p-1)\int_0^{\varepsilon} |u_\varepsilon|^{p-2}u_\varepsilon\sin_p t\,dt \\
&\quad +2(p-1)\int_\varepsilon^{\frac{\pi_p}{2}} |u_\varepsilon|^{p-2}u_\varepsilon\sin_p t\,dt.
\end{aligned}$$

Using the facts that $\sin_p\varepsilon = \varepsilon + o(\varepsilon)$ and $\cos_p\varepsilon = 1 + o(1)$, we obtain

$$\begin{aligned}
\frac{1}{q} I_\varepsilon &= -2\int_0^{\varepsilon} \frac{(\varepsilon-t)^2}{\varepsilon^{p+1}}(1+o(1))\,dt + 2O(\varepsilon^2) \\
&\quad + \left(\frac{p-1}{p+1}\right)^{p-1}(p-1)\int_0^{\pi_p}\sin_p t\,dt + o(1) \\
&= -\frac{2}{3}\varepsilon^{2-p} + \left(\frac{p-1}{p+1}\right)^{p-1}(p-1)\int_0^{\pi_p}\sin_p t\,dt + o(1).
\end{aligned}$$

Hence, for $1<p<2$, we have $I_\varepsilon > 0$ while for $p>2$, we have $I_\varepsilon < 0$, if $0<\varepsilon\le\varepsilon_0$ with ε_0 small enough. So we can take $h_0 := h_{\varepsilon_0}$. We have to distinguish between the cases $1<p<2$ and $p>2$.

Case $1<p<2$. In this case we have

$$\int_0^{\pi_p} h_0(t)\sin_p t\,dt > 0, \tag{6.2.29}$$

and the boundary value problem (6.2.3) with $h = h_0$ has a positive solution $u_0 := u_{\varepsilon_0} \in X$. By (6.2.29) there exists $\delta > 0$ so small that for $u_{-\delta}(t) := u_0(t) - \delta$ and for

$$h_{-\delta} := q[(\Phi(u'_{-\delta}))' + (p-1)\Phi(u_{-\delta})]$$

we also have

$$\int_0^{\pi_p} h_{-\delta}(t)\sin_p t\,dt > 0. \tag{6.2.30}$$

Fix such a δ. Clearly we can choose a small number $\rho > 0$ such that for any h with $\|h - h_{-\delta}\|_\infty < \rho$ one has $h < h_{-\delta/2}$ on $(0,\pi_p)$, and also $\int_0^{\pi_p} h(t)\sin_p t\,dt > 0$. Let us fix such an h and extend it e.g. by zero for $t > \pi_p$. Then, from Lemma 6.2.1, it follows that for α sufficiently large and positive, the solution u_α of the initial value problem

$$(\Phi(u'_\alpha))' + (p-1)\Phi(u_\alpha) = \frac{1}{q}h,$$
$$u_\alpha(0) = 0, \; u'_\alpha(0) = \alpha$$

satisfies $u_\alpha \geq u_{-\delta/2}, u_\alpha(\pi_p) > 0$. Since $h < h_{-\delta/2}$, it then follows that $u_{-\delta/2}$ and u_α are respectively lower and upper solutions of the boundary value problem (6.2.3) with this h.

Hence setting $\underline{u} = u_{-\delta/2}$ and $\bar{u} = u_\alpha$ in Lemma 6.3.2 of later Subsection 6.3.1 we obtain the existence of at least one solution u, which lies between $\underline{u}$ and $\bar{u}$. We claim that there exists at least a second solution. Assume the opposite, namely that only one solution exists. Then, from Lemma 6.3.2, it follows that for a certain bounded open set Ω in $C^1[0, \pi_p]$ which contains u, we have

$$\deg[I - T_h; \Omega, 0] = 1. \tag{6.2.31}$$

On the other hand, Lemma 6.2.4 guarantees that for $R > 0$ so large that $\Omega \subset B_R(0)$, we have

$$\deg[I - T_h; B_R(0), 0] = 0. \tag{6.2.32}$$

Now, (6.2.31), (6.2.32) and the additivity of the Leray–Schauder degree yield that the boundary value problem (6.2.3) has a second solution in $B_R(0)\backslash\Omega$. Hence the boundary value problem (6.2.3) has at least two distinct solutions for any $h \in B_\rho(h_{-\delta})$. The existence of an open cone $\mathcal{C}$ with the desired property is now a consequence of the homogeneity of the boundary value problem (6.2.3).

Case $p > 2$. Now we have

$$\int_0^{\pi_p} h_0(t) \sin_p t \, dt < 0, \tag{6.2.33}$$

and the boundary value problem (6.2.3) with $h = h_0$ has a positive solution u_0. By (6.2.33) there exists $\delta > 0$ so small that for $u_\delta(t) := u_0(t) + \delta$ and $h_\delta := q[(\Phi(u'_\delta))' + (p-1)\Phi(u_\delta)]$ we also have

$$\int_0^{\pi_p} h_\delta(t) \sin_p t \, dt < 0. \tag{6.2.34}$$

It follows from (6.2.34) and Lemma 6.2.1 that for large and positive α, the solution of the initial value problem

$$(\Phi(u'_{-\alpha}))' + (p-1)\Phi(u_{-\alpha}) = \frac{1}{q}h_\delta, \quad u_{-\alpha}(0) = 0, \; u'_{-\alpha}(0) = -\alpha$$

satisfies $u_{-\alpha} \leq u_\delta, u_{-\alpha}(\pi_p) < 0$. Then we are in a position to proceed symmetrically to the previous case. In this form we have completed the proof of our statement. $\square$

A by-product of the proof of this theorem is the following general fact. For any $h \in L^\infty(0, \pi_p)$ such that (6.2.2) holds, one has that the set of all possible solutions of (6.1.7) is bounded and the degree of the associated fixed point operator equals 0. Combining this and Theorem 6.2.1 yields, in particular, that for any $h \not\equiv 0$ of the class C^1 and $p \neq 2$, there are *a-priori estimates* for the solution set.

6.2.2 Resonance at higher eigenvalues

In this subsection we briefly deal with the solvability of (6.1.7) when $\lambda = \lambda_k$, $k \geq 2$, i.e., we deal with the resonance problem at higher eigenvalues $\lambda = (p-1)k^p$.

Following [268], we introduce the following notation. In addition to the orthogonality condition (6.2.2) (with f instead of h), the following numbers (for a fixed $k \in \mathbb{N}$) will appear

$$\begin{aligned} A_m &:= \int_{m\pi_p/k}^{(m+1)\pi_p/k} f(t) \sin_p kt \, dt, \\ B_m &:= \int_{(m-1/2)\pi_p/k}^{(m+1/2)\pi_p/k} f(t) \cos_p kt \, dt, \end{aligned}$$

and the "higher order" orthogonality condition

$$F_k(f) := \int_0^{\pi_p} f(t) \sin_p kt \, dt = 0 \tag{6.2.35}$$

will also play an important role. Further, we denote

$$L_k(f) := \int_0^{\pi_p} f(t) \cos_p kt \left(\int_0^t f(s) \sin_p ks \, ds \right) dt$$

and its "approximation" by a Riemann-like sum

$$S_k(f) := \sum_{0 \leq n < m \leq k-1} A_n B_m.$$

Similarly as in the case of resonance at the first eigenvalue, we will need the following statement concerning the asymptotics of the first zero of the solution for the initial value problem

$$(\Phi(u'))' + \lambda_k \Phi(u) = f(t), \quad u(0) = 0, \ u'(0) = \alpha, \tag{6.2.36}$$

this time with a higher eigenvalue λ_k, $k \geq 2$ (compare the statement with Lemma 6.2.1). We skip the proof. Note only that it is based on a modified Prüfer transformation.

Lemma 6.2.5. *Assume that $p \neq 2$, $k \geq 2$, $f \in L^1(0, \pi_p)$ and $f \not\equiv 0$. Then, given $0 < \varepsilon < \pi_p/k$, there exists a constant $\alpha^* > 0$ such that for $|\alpha| > \alpha^*$ any solution of (6.2.36) has precisely one zero t^1_α in the interval $[\pi(1 - 1/k) + \varepsilon, \pi_p(1 + 1/k) - \varepsilon]$. As $|\alpha| \to \infty$, the dependence of t^1_α on α is*

$$t^1_\alpha = \pi_p + \frac{1}{k(p-1)\Phi(\alpha)} F_k(f) + o(|\alpha|^{1-p}). \tag{6.2.37}$$

If $F_k(f) = 0$, we have the formula

$$t_\alpha^1 = \pi_p + \frac{p-2}{k(p-1)^2|\alpha|^{2(1-p)}} L_k(f) + o(|\alpha|^{2(1-p)}). \tag{6.2.38}$$

If $L_k(f) \neq 0$, then

$$\operatorname{sgn}(t_\alpha^1 - \pi_p) = \operatorname{sgn}(p-2)\operatorname{sgn}(L_k(f)).$$

In particular, if both $F_k(f) = 0$ and $S_k(f) \geq 0$, then $\operatorname{sgn}(t_\alpha^1 - \pi_p) = \operatorname{sgn}(p-2)$.

Theorem 6.2.4. *Assume that $p \neq 2$, $k \geq 2$, $f \not\equiv 0$ in $(0, \pi_p)$ and $f \in L^1(0, \pi_p)$ satisfies (6.2.35) and $L_k(f) \neq 0$. Then BVP (6.1.7) with $\lambda = \lambda_k = (p-1)k^p$ has at least one solution. The set of all possible solutions is bounded in $C^1[0, \pi_p]$. The inequality $L_k(f) > 0$ is satisfied if f satisfies both (6.2.35) and $S_k(f) \geq 0$. The equation $S_k(f) = 0$ holds true if at least one of the following two sets of $k-1$ orthogonality conditions is satisfied:*

(a) $A_m = 0$ for $m = 0, 1, \ldots, k-2$;

(b) $B_m = 0$ for $m = 0, 1, \ldots, k-1$.

Proof. The proof is in principle the same as the proof of Theorem 6.2.1 and it is based on the topological degree argument. We define the operator $T_{\mu,f}$ in the same way as in the proof of that theorem.

Case 1 ($(p-2)L_k(f) < 0$). We claim that there exists a constant $R > 0$ such that for any $\mu \in [1, (1+1/k)^{1/p}]$ the BVP

$$(\Phi(u'))' + \mu\lambda_k\Phi(u) = \mu f(\mu^{1/p}t), \quad u(0) = 0 = u(\pi_p) \tag{6.2.39}$$

has no solution for which $\|u\|_X \geq R$, where $X = C_0^1[0, \pi_p]$. By contradiction, suppose that there are sequences

$$\{\mu_k\}_{n=1}^\infty \subset \left[1, (1+k^{-1})^{1/p}\right] \quad \text{and} \quad \{u_n\}_{n=1}^\infty \subset X$$

such that each pair (μ_n, u_n) satisfies equation (6.2.39) and $\|u_n\|_X \to \infty$ as $n \to \infty$. In analogy with the proof of Theorem 6.2.1 (for $p \in (1,2)$) we may assume also that $\mu_n \to \mu^* \in [1, (1+1/k)^{1/p}]$ as $n \to \infty$, together with $\alpha_n := u_n'(0) \to \infty$. Each u_n can be extended to a solution of the initial value problem

$$(\Phi(u_n'))' + \mu_n\lambda_k\Phi(u_n) = \mu_n f(\mu_n^{1/p}t), \quad u_n(0) = 0,\ u_n'(0) = \alpha_n,$$

on $\mathbb{R}_+ = [0, \infty)$, and hence $v_n(t) = u_n(\mu_n^{-1/p}t)$ solves the initial value problem

$$(\Phi(v_n'))' + \lambda_k\Phi(v_n) = f(t), \quad v_n(0) = 0,\ v_n'(0) = \tilde{\alpha}_n,$$

where $\tilde{\alpha}_n := \mu_n^{1/p}\alpha_n \to \infty$. Let $0 < \varepsilon < (\pi_p/k)$ be sufficiently small. By Lemma 6.2.5 there is precisely one zero point t_α^1 of v_n in the interval $[(1-1/k)\pi_p + \varepsilon, (1+1/k)\pi_p - \varepsilon]$ and $t_\alpha^1 \to \pi_p$ as $n \to \infty$, according to (6.2.37). Similarly, the next larger zero point t_α^2 of v_n approaches $(1+1/k)\pi_p$ as $n \to \infty$. Moreover, equation (6.2.38) yields $t_\alpha^1 < \pi_p$. On the other hand, for every $n \in \mathbb{N}$

$$v_n(\mu^{1/p}\pi_p) = u_n(\pi_p) = 0 \quad \text{with } 1 \le \mu_n^{1/p} \in [1, (1+1/k)^{1-(1/p)}]$$

forces $t_\alpha^1 < \mu_n^{1/p} < t_\alpha^2$ which contradicts the definition of t_α^1 and t_α^2 and proves boundedness of solutions of (6.2.39).

Case 2 ($(p-2)L_k(f) > 0$). Similarly as in the previous part of the proof we reach contradiction assuming that there is a solution of (6.2.39) with $\|u\|_X$ arbitrarily large for any $\mu[(1-1/k)^{p-1}, 1]$, only the inequality $t_\alpha^1 > \pi_p$ is to be reversed.

The remaining part of the proof of the statement up to "... bounded in $C^1[0,\pi_p]$." is similar to that of Theorem 6.2.1. The proof of the statement starting with "The inequality $L_k(f) > 0$..." is a matter of a direct computation, we refer to [268] for details. □

Now we will present a statement showing that the boundedness of solution space of (6.1.7) at resonance is not "stable" in the sense that a small perturbation of the right-hand side f by a term which is transversal to the hyperplane $\{f \in L^1(0,\pi_p) : \int_0^{\pi_p} f(t)\sin_p t\,dt = 0\}$. This statement is proved in [268] as a corollary of a more general statement, but we prefer to formulate here only that corollary, and we refer to [268] for more details.

Theorem 6.2.5. *Suppose that $p \ne 2$, $g : (0,\pi_p) \to \mathbb{R}$ is a continuous bounded function, $g \not\equiv 0$ and $\int_0^{\pi_p} g(t)\sin_p t\,dt = 0$. Let $\alpha, \varepsilon > 0$ be arbitrary. Then there exists a constant $\gamma \in \mathbb{R}$, $0 < |\gamma| < \varepsilon$, such that (6.1.7) with $\lambda = \lambda_1 = (p-1)$ and $f(t) = g(t) + \gamma$ has a solution satisfying $|u'(0)| \ge \alpha$. In particular, there exist a sequence $\gamma_n \ne 0$, $\gamma_n \to 0$, and a sequence of solutions u_n of (6.1.7) with $f(t) = g(t) + \gamma_n$ such that $|u_n'(0)| \to \infty$ as $n \to \infty$, i.e., $\|u_n\|_{C^1} \to \infty$.*

6.3 Boundary value problems at resonance

We start this section with a multiplicity result of Ambrosetti-Prodi type, i.e., a statement concerning the BVP of the form (6.2.3), where the right hand-side h is split as in the below formula (6.3.1). In the subsequent subsection we extend the classical Landesman-Lazer linear solvability condition to half-linear BVP's, and the last subsection is devoted to the Fučík spectrum and solvability of half-linear BVP's with asymmetric nonlinearities.

6.3.1 Ambrosetti-Prodi type result

To simplify the presentation of the results of this subsection, we will use the following notation. We set

$$L^p = L^p(0,\pi_p), \quad W_0^{1,p} = W_0^{1,p}(0,\pi_p), \quad C^1 = C^1[0,\pi_p],$$

and

$$\begin{aligned} C_0 &= \{u \in C[0,\pi_p],\ u(0) = 0 = u(\pi_p)\}, \\ C_0^1 &= \{u \in C^1[0,\pi_p],\ u(0) = 0 = u(\pi_p)\}. \end{aligned}$$

Let us split any $h \in L^\infty$ as follows

$$h(t) = \tilde{h}(t) + \bar{h}\sin_p t, \tag{6.3.1}$$

where $\bar{h} \in \mathbb{R}$ and

$$\int_0^{\pi_p} \tilde{h}(t)\sin_p t\,dt = 0. \tag{6.3.2}$$

The aim of this subsection is to prove the so-called Ambrosetti-Prodi type results for the BVP

(6.3.3)

$$(\Phi(u'))' + (p-1)\Phi(u) = h(t), \qquad u(0) = 0 = u(\pi_p), \quad h(t) = \tilde{h}(t) + \bar{h}\sin_p t.$$

We define the spaces $\tilde{L}^\infty$, $\tilde{W}_0^{1,p}$, $\tilde{L}^p$, $\tilde{C}^1$, $\tilde{C}_0^1$ as those formed respectively by elements of L^∞, $W_0^{1,p}$, L^p, C^1, C_0^1, which satisfy (6.3.2). We also define the energy functional associated with (6.3.3)

$$E_h(u) := \frac{1}{p}\int_0^{\pi_p} |u'|^p\,dt - \frac{1}{q}\int_0^{\pi_p} |u|^p\,dt + \int_0^{\pi_p} hu\,dt, \quad q = \frac{p}{p-1}. \tag{6.3.4}$$

The following two concepts concern the geometry of the functional E_h.

Definition 6.3.1. We say that a functional $E : W_0^{1,p} \to \mathbb{R}$ has a *local saddle point geometry*, if there exist $u, v \in W_0^{1,p}$ which are separated by $\tilde{W}_0^{1,p}$ in the sense that

$$E(u) < \inf_{w \in \tilde{W}_0^{1,p}} E(w), \quad E(v) < \inf_{w \in \tilde{W}_0^{1,p}} E(w),$$

and any continuous path from u to v has a nonempty intersection with $\tilde{W}_0^{1,p}$.

We say that E has a *local minimizer geometry* if there exists $R > 0$ so that

$$\inf_{\|u\|_{1,p} < R} E(u) < \inf_{\|u\|_{1,p} = R} E(u), \quad \|u\|_{1,p} = \left(\int_0^{\pi_p} |u'|^p\right)^{1/p}.$$

The following statement can be regarded, in a certain sense, as a complement of Theorem 6.2.2.

Theorem 6.3.1. *Assume $\tilde{h} \in \tilde{C}^1$, $\tilde{h} \neq 0$, then E_h has a local saddle point geometry for $1 < p < 2$ and a local minimizer geometry for $p > 2$.*

Proof. For $1 < p < 2$, let $u(t) = \tilde{u}(t) + \bar{u}\sin_p t$, $\tilde{u} \in \tilde{W}_0^{1,p}$. By the variational character of the first eigenvalue $\lambda_1 = p - 1$ it follows that there exists $\eta > 0$ such that

$$E_{\tilde{h}}(\tilde{u}) = \frac{1}{p}\int_0^{\pi_p} |\tilde{u}'|^p - \frac{1}{q}\int_0^{\pi_p} |\tilde{u}|^p + \int_0^{\pi_p} \tilde{h}\tilde{u} \geq \eta\int_0^{\pi_p} |\tilde{u}'|^p - \left|\int_0^{\pi_p} \tilde{h}\tilde{u}\right|$$

and $E_{\tilde{h}}(\tilde{u})$ is bounded from below on the subspace $\tilde{W}_0^{1,p}$. On the other hand, from the proof of Theorem 6.2.2 we find that there exist sequences $\{u_n\}, \{v_n\} \subset W_0^{1,p}$ such that $E_{\tilde{h}}(u_n) \to -\infty$, $E_{\tilde{h}}(v_n) \to -\infty$ and $\bar{u}_n \to \infty$, $\bar{v}_n \to -\infty$, respectively (note that $\bar{v}_n$ has the analogous meaning as $\bar{u}$ at the beginning of the proof). This means that for n large enough, u_n, v_n are separated by $\tilde{W}_0^{1,p}$ in the sense of the previous definition.

Next, for $p > 2$, from Theorem 6.2.2, there exists $R > 0$ such that

$$I := \inf_{v \in W_0^{1,p}} E_{\tilde{h}}(v) < E_{\tilde{h}}(u), \tag{6.3.5}$$

for any $u \in W_0^{1,p}$, $\|u\|_{1,p} = R$ and all global minimizers of $E_{\tilde{h}}$ belong to the ball of radius R in $W_0^{1,p}$. Assume now that there exists a sequence $u_n \in W_0^{1,p}$ with $\|u_n\|_{1,p} = R$ such that $E_{\tilde{h}}(u_n) \to I$. Since we can assume that this sequence is weakly convergent in $W_0^{1,p}$, say to u, then $u_n \to u$ in L^p. The weak lower semicontinuity of $E_{\tilde{h}}$ and (6.3.5) imply

$$E_{\tilde{h}}(u) \le \liminf_{n\to\infty} E_{\tilde{h}}(u_n) = I$$

and hence $\|u_n\|_{1,p} \to \|u\|_{1,p}$. Thus u_n strongly converges to u in $W_0^{1,p}$, which implies that $\|u\|_{1,p} = R$ and $E_{\tilde{h}}(u) = I$. But this is a contradiction, hence $E_{\tilde{h}}$ has a local minimizer geometry. □

In the proof of the main result of this section we will need the following statements. The first one concerns the Palais-Smale condition and it is a variant of Lemma 6.1.2.

Lemma 6.3.1. *Let h be written in the form (6.3.1) and $\bar{h} \neq 0$. Then E_h satisfies a Palais-Smale condition, i.e., if $E_h(u_n) \to c \in \mathbb{R}$, $E_h'(u_n) \to 0$, then u_n contains a subsequence which converges strongly in $W_0^{1,p}$.*

We will also need the concepts of the upper and lower solutions and associated statements.

We call a function $\underline{u} \in C^1$ with absolutely continuous $\Phi(\underline{u}')$ a *lower solution* of (6.3.3) if $\underline{u}(0) \le 0$, $\underline{u}(\pi_p) \le 0$ and

$$(\Phi(\underline{u}'))' + (p-1)\Phi(\underline{u}) \ge h$$

a.e. in $(0, \pi_p)$. If all inequalities are reversed, we have the definition of an *upper solution* $\bar{u}$. We write $u \prec v$ if $u(t) < v(t)$ on $(0, \pi_p)$ and either $u(0) < v(0)$ ($u(\pi_p) < v(\pi_p)$) or $u(0) = v(0)$ ($u(\pi_p) = v(\pi_p)$) and $u'(0) < v'(0)$ ($u'(\pi_p) > v'(\pi_p)$). A lower solution $\underline{u}$ is said to be *strict* if every solution u of (6.3.3) such that $\underline{u} \le u$ on $[0, \pi_p]$ satisfies $\underline{u} \prec u$. The *strict upper solution* is defined in an analogous way.

For $h \in L^\infty$ we define an operator $T_h : C_0^1 \to C_0^1$ by $T_h(v) = u$ if and only if

$$(\Phi(u'))' = h(t) - (p-1)\Phi(v), \ t \in (0, \pi_p), \quad u(0) = 0 = u(\pi_p).$$

Lemma 6.3.2. *Assume that $\underline{u}$ and $\overline{u}$ are respectively lower and upper solutions of* (6.3.3) *with $\underline{u} \le \overline{u}$. Then this problem has at least one solution u satisfying*

$$\underline{u}(t) \le u(t) \le \overline{u}(t), \quad t \in [0, \pi_p].$$

Moreover, if $\underline{u}$ and $\overline{u}$ are strict and satisfy $\underline{u} \prec \overline{u}$, then there exists $R_0 > 0$ such that for $R > R_0$

$$\deg[I - T_h; \Omega, 0] = 1,$$

where $\Omega = \{u \in C_0^1 : \ \underline{u} \prec u \prec \overline{u}\} \cap B_{C_0^1}(0, R)$.

The previous statement is used in the proof of the next results which play the fundamental role in the proof of the main statement of this subsection.

Lemma 6.3.3. *Let* (6.3.3) *be solvable for $h_i(t) = \tilde{h}(t) + \bar{h}_i \sin_p t$, $i = 1, 2$, $\bar{h}_1 < \bar{h}_2$. Then it is solvable for any $h(t) = \tilde{h}(t) + \bar{h} \sin_p t$ with $\bar{h} \in (\bar{h}_1, \bar{h}_2)$.*

Lemma 6.3.4. *Let $h \in L^\infty \setminus \tilde{L}^\infty$. Then either* (6.3.3) *has no solution or all solutions of* (6.3.3) *are a-priori bounded in C_0^1 by a constant depending on the value of*

$$\left| \int_0^{\pi_p} h(t) \sin_p t \, dt \right|,$$

and there exists R_0 such that for all $R > R_0$,

$$\deg[I - T_h; B_{C_0^1}(0, R), 0] = 0.$$

Lemma 6.3.5. *Let the assumptions of Lemma* 6.3.3 *be fulfilled. Then* (6.3.3) *has at least two solutions for any $h(t) = \tilde{h}(t) + \bar{h} \sin_p t$ with $\bar{h} \in (\bar{h}_1, \bar{h}_2)$, $\bar{h} \ne 0$.*

Now we are in a position to formulate and to prove the main statement of this subsection which, in addition, summarizes previous partial results. This statement can be regarded as a complement of Theorem 6.2.3.

Theorem 6.3.2. *Let $p \ne 2$. Then there exists an open dense set (with respect to L^∞ norm) $\mathcal{S} \subset \tilde{L}^\infty$ with $C^1[0, \pi_p] \cap \tilde{L}^\infty \setminus \{0\} \subset \mathcal{S}$. For every $\tilde{h} \in \mathcal{S}$ there exists a ball in $L^\infty(0, \pi_p)$ centered at $\tilde{h}$ such that* (6.3.3) *has solution for every h belonging to this ball. Moreover, for any $\tilde{h} \in \mathcal{S}$ there exist real numbers $H_\pm = H_\pm(\tilde{h})$, $H_- < 0 < H_+$, such that* (6.3.3) *with $h(t) = \tilde{h}(t) + \bar{h} \sin_p t$ has*

(i) no solution if $\bar{h} \notin [H_-, H_+]$;

(ii) at least two solutions if $\bar{h} \in (H_-, H_+) \setminus \{0\}$;

(iii) at least one solution if $\bar{h} \in \{H_-, 0, H_+\}$.

Proof. Let $\mathcal{S} \subset \tilde{L}^\infty$ be defined as follows: $\tilde{h} \in \mathcal{S}$ if and only if $E_{\tilde{h}}$ has a local saddle point geometry if $p \in (1, 2)$ or has a local minimizer geometry if $p > 2$. From Theorem 6.2.2 it follows that $C^1 \setminus \{0\} \subset \mathcal{S}$, hence $\mathcal{S}$ is dense in $\tilde{L}^\infty$. Since the local geometry of the functional $E_{\tilde{h}}$ is invariant with respect to small perturbations of $\tilde{h} \in L^\infty$, the set $\mathcal{S}$ is also open.

Let us fix for a moment some $\tilde{h}_0 \in \mathcal{S}$ and consider $\rho > 0$ so small that the local geometry of E_h is the same as that of $E_{\tilde{h}_0}$ for any $h \in B_{L^\infty}(\tilde{h}_0;\rho)$ (the ball of radius ρ around $\tilde{h}_0$ in L^∞). According to Lemma 6.3.1, for $h \in B_{L^\infty}(\tilde{h}_0,\rho)\setminus \tilde{L}^\infty$ the functional E_h satisfies the Palais-Smale condition. Hence for any $h \in B_{L^\infty}(\tilde{h}_0;\rho)\setminus \tilde{L}^\infty$ the problem (6.3.3) has at least one solution by a standard variational argument applied to E_h (Saddle Point Theorem for $p \in (1,2)$ and the minimization argument when $p > 2$, see [339]) But from Corollary 6.3.3 it follows that problem (6.3.3) has also a solution for any $h \in B_{L^\infty}(\tilde{h}_0;\rho) \cap \tilde{L}^\infty$, and hence, in particular, problem (6.3.3) has a solution for any $\tilde{h} \in \mathcal{S}$.

Let us fix now $\tilde{h} \in \mathcal{S}$ and consider $h(t) = \tilde{h}(t) + \bar{h}\sin_p t$. Define

$$H_- = H_-(\tilde{h}) := \inf \bar{h}, \quad H_+ = H_+(\tilde{h}) := \sup \bar{h},$$

where the infimum and supremum are taken over all $\bar{h}$ such that (6.3.3) (with above fixed $\tilde{h}$) has a solution. Then from the above formulated lemmata we have $H_- < H_+$. Let us prove that $H_\pm$ are finite. Suppose, by contradiction, that there exist sequences $\bar{h}_n \in \mathbb{R}$, $u_n \in C_0^1$ such that $\bar{h}_n \to \infty$ and u_n is a solution of (6.3.3) with $\bar{h} = \bar{h}_n$. Dividing the differential equation in (6.3.3) by $\bar{h}_n$, and setting $v_n = \bar{h}_n^{-\frac{1}{p-1}} u_n$, we find that $v_n \in C_0^1$ satisfies

$$(\Phi(v_n'))' + (p-1)\Phi(v_n) = \frac{\tilde{h}(t)}{\bar{h}_n} + \sin_p t,$$

or equivalently

(6.3.6)

$$v_n = \int_0^t \Phi^{-1}\left[\Phi(v_n'(0)) + \int_0^s \left(\frac{\tilde{h}(\tau)}{\bar{h}_n} + \sin_p \tau - (p-1)\Phi(v_n(\tau))\right) d\tau\right] ds.$$

It also follows from Lemma 6.3.4 that v_n is uniformly bounded in C_0^1. Now, the Ascoli-Arzelá Theorem implies that for a subsequence of v_n (denoted again v_n) we have $v_n \to v_0$ in C_0. The Lebesgue Dominated Convergence Theorem and (6.3.6) imply that

$$v_0(t) = \int_0^t \Phi^{-1}\left[\Phi(v_0'(0)) + \int_0^s (\sin_p \tau - (p-1)\Phi(v_0(\tau)))\, d\tau\right] ds,$$

i.e., v_0 is a solution of

$$(\Phi(v_0'))' + (p-1)\Phi(v_0) = \sin_p t, \quad v_0(0) = 0 = v_0(\pi_p),$$

which contradicts the later given Corollary 7.1.2 (see Subsection 7.1.3). This proves that $H_+ < \infty$ and similarly we prove that also $H_- > -\infty$.

Now, from the definition of the numbers $H_\pm$ it follows that (6.3.3) has no solution if $\bar{h} \notin [H_-, H_+]$. On the other hand, from Lemma 6.3.3, it follows that (6.3.3) has a solution for any $h \in (H_-, H_+)$. Then Corollary 6.3.5 implies that (6.3.3) has at least two solutions if $\bar{h} \in (H_-, H_+) \setminus \{0\}$.

It remains to prove that (6.3.3) has a solution if $\bar{h} = H_\pm$. To this end, let us assume that $\bar{h}_n \to H_+$ as $n \to \infty$ and let u_n be a solution of (6.3.3) with

$h(t) = \tilde{h}(t) + \bar{h}_n \sin_p t$. According to Lemma 6.3.4 the sequence u_n is bounded in C_0^1 and the Ascoli-Arczelá Theorem implies that for some subsequence (denoted again u_n) $u_n \to u_0$ in C_0. The same argument as above yields that u_0 is a solution of

$$(\Phi(u'))' + (p-1)\Phi(u) = \tilde{h}(t) + H_+ \sin_p t, \quad u(0) = 0 = u(\pi_p).$$

Similarly we can prove that (6.3.3) has a solution for $\bar{h} = H_-$. This completes the proof. □

6.3.2 Landesman-Lazer solvability condition

The paper of Landesman and Lazer [232] published in 1970 is the pioneering work concerning solvability of the linear BVP at resonance. Since that time, the conditions ensuring solvability of BVP's in this situation (the so-called Landesman-Lazer conditions) have been extended in many directions. The main statement of this subsection establishes the Landesman-Lazer solvability conditions for the the half-linear BVP

$$(\Phi(x'))' + \lambda_n \Phi(x) + g(x) = h(t), \quad x(0) = 0 = x(\pi_p). \tag{6.3.7}$$

It is supposed that $h \in L^q(0, \pi_p)$, g is a bounded continuous function such that there exist finite limits $\lim_{x\to\pm\infty} g(x) = g_\pm$. By φ_n we denote the normalized eigenfunction corresponding to the n-th eigenvalue, i.e., $\lambda_n = (p-1)n^p$, $\varphi_n(t) = \alpha_n \sin_p(nt)$, where $\alpha_n > 0$ is such that $\|\varphi_n\|_{L^p} = 1$.

Let the functionals A, B be given by (6.1.4), $G(t) = \int_0^t g(s)\,ds$,

$$C(x) := \int_0^{\pi_p} [G(x(t)) - h(t)x(t)]\,dt$$

and

$$J(u) := A(u) - \lambda_n B(u) - C(u), \quad E(u) = \frac{A(u)}{B(u)}. \tag{6.3.8}$$

The main result of this subsection reads as follows.

Theorem 6.3.3. *The boundary value problem* (6.3.7) *has a solution provided one of the following two conditions is satisfied*

$$\begin{aligned} g_+ \int_0^{\pi_p} \varphi_n^+(t)\,dt + g_- \int_0^{\pi_p} \varphi_n^-(t)\,dt \quad &> \int_0^{\pi_p} \varphi_n(t)h(t)\,dt \\ &> g_+ \int_0^{\pi_p} \varphi_n^-(t)\,dt + g_- \int_0^{\pi_p} \varphi_n^+(t)\,dt \end{aligned} \tag{6.3.9}$$

or

$$\begin{aligned} g_+ \int_0^{\pi_p} \varphi_n^+(t)\,dt + g_- \int_0^{\pi_p} \varphi_n^-(t)\,dt \quad &< \int_0^{\pi_p} \varphi_n(t)h(t)\,dt \\ &< g_+ \int_0^{\pi_p} \varphi_n^-(t)\,dt g_- \int_0^{\pi_p} \varphi_n^+(t)\,dt, \end{aligned} \tag{6.3.10}$$

where $\varphi_n^+ = \max\{0, \varphi_n\}$, $\varphi_n^- = \min\{0, \varphi_n\}$.

Before giving the proof which is based on the fact that critical points of the functional J are weak solutions of (6.3.7), we introduce some concepts and auxiliary statements which we will need in this proof. The proof relies on a saddle point-type theorem for *linked sets*, which in turn relies on a variational characterization of eigenvalues λ_n.

Let $\mathcal{E}$ be a closed subset of $W_0^{1,p}(0,\pi_p)$ and $\mathcal{Q}$ be a submanifold of $W_0^{1,p}(0,\pi_p)$ with relative boundary $\partial\mathcal{Q}$. Following [339, Definition 8.1] we say that $\mathcal{E}$ and $\partial\mathcal{Q}$ *link* if $\mathcal{E}\cap\partial\mathcal{Q}=\emptyset$, and for any continuous map $h: W_0^{1,p}(0,\pi_p)\to W_0^{1,p}(0,\pi_p)$ such that $h|_{\partial\mathcal{Q}}=\mathrm{id}$, there holds $h(\mathcal{Q})\cap\mathcal{E}\neq\emptyset$. The following lemma is proved in [339, Theorem 8.4].

Lemma 6.3.6. *Let $J\in C^1(W_0^{1,p}(0,\pi_p))$ satisfy the Palais-Smale condition. Consider a closed subset $\mathcal{E}\subset W_0^{1,p}(0,\pi_p)$ and a submanifold $\mathcal{Q}\subset W_0^{1,p}(0,\pi_p)$ with relative boundary $\partial\mathcal{Q}$, and let $\Gamma:=\{h\in C(W_0^{1,p}(0,\pi_p),W_0^{1,p}(0,\pi_p)):\ h|_{\partial\mathcal{Q}}=id\}$. Suppose that $\mathcal{E}$ and $\partial\mathcal{Q}$ link, and* $\inf_{\mathcal{E}}J(u)>\sup_{\partial\mathcal{Q}}J(u)$. *Then*

$$\beta=\inf_{h\in\Gamma}\sup_{\mathcal{Q}}J(h(u))$$

is a critical value of J.

Lemma 6.3.7. *If either* (6.3.9) *or* (6.3.10) *is satisfied, then J satisfies the Palais-Smale condition.*

Proof. Let $\{u_k\}\subset W_0^{1,p}(0,\pi_p)$ be a sequence such that $|J(u_k)|\le c$ and $J'(u_k)\to 0$ in $(W_0^{1,p}(0,\pi_p))^*$. We must show that $\{u_k\}$ has a subsequence which converges in $W_0^{1,p}(0,\pi_p)$. First we show that $\{u_k\}$ is bounded. By contradiction, suppose that $\|u_k\|\to\infty$ and consider the sequence $v_k=u_k/\|u_k\|$. Then $\{v_k\}$ is bounded and hence without lost of generality we can suppose that this sequence is weakly convergent to some v_0. We assume that

$$J'(u_k)=A'(u_k)-\lambda_n B'(u_k)-C'(u_k)\to 0,$$

hence, dividing this relation by $\|u_k\|^{p-1}$, we have

$$A'(v_k)-\lambda_n B'(v_k)-\frac{C'(u_k)}{\|u_k\|^{p-1}}\to 0.$$

By the boundedness of C' we know that $C'(u_k)/\|u_k\|^{p-1}\to 0$ and by the compactness of B' we know that $B'(v_k)\to B'(v_0)$. Thus $v_k\to v_0=(A')^{-1}(\lambda_n B'(v_0))$ in $W_0^{1,p}(0,\pi_p)$. It follows that $v_0=\pm\varphi_n$. We assume that $v_0=\varphi_n$, the case $v_0=-\varphi_n$ can be treated analogously.

Now we add the inequalities

$$-cp\le pJ(u_k)\le cp$$

and ($\|\cdot\|_*$ denotes the norm in $(W_0^{1,p}(0,\pi_p))^*$)

$$-\|J'(u_k)\|_*\|u_k\|\le -\langle J'(u_k),u_k\rangle\le \|J'(u_k)\|_*\|u_k\|$$

to get

$$\begin{aligned} -cp - \|J'(u_k)\|_* \|u_k\| &\le -p \int_0^{\pi_p} G(u_k) + \int_0^{\pi_p} g(u_k)u_k + (p-1) \int_0^{\pi_p} hu_k \\ &\le cp + \|J'(u_k)\|_* \|u_k\|. \end{aligned}$$

Dividing by $\|u_k\|$ and writing $G(u_k)/\|u_k\| = \tilde{g}(u_k)v_k$, where

$$\tilde{g}(s) := \begin{cases} G(s)/s & \text{for } s \neq 0, \\ 0 & \text{for } s = 0, \end{cases}$$

we get

$$\left| \int_0^{\pi_p} [g(u_k) - p\tilde{g}(u_k)]v_k + (p-1) \int_0^{\pi_p} hv_k \right| \le \frac{cp}{\|u_k\|} + \|J'(u_k)\|_*.$$

The right-hand side of the last inequality approaches 0 and $\int_0^{\pi_p} hv_k \to \int_0^{\pi_p} h\varphi_n$ as $k \to \infty$, so

$$\lim_{k\to\infty} \int_0^{\pi_p} [g(u_k) - p\tilde{g}(u_k)]v_k = (1-p) \int_0^{\pi_p} h\varphi_n.$$

Recall that $W_0^{1,p}(0,\pi_p)$ embeds compactly into $C[0,\pi_p]$, so without loss of of generality, $v_k = u_k/\|u_k\| \to \varphi_n$ uniformly, and hence $u_k(t) \to \infty$ for $\{t : \varphi_n(t) > 0\}$ and $u_k(t) \to -\infty$ for $\{t : \varphi_n(t) < 0\}$. But $u_k(t) \to \pm\infty$ implies $g(u_k(t)) \to g_\pm$ as well as $\tilde{g}(u_k(t)) \to g_\pm$, by the application of L'Hospital's rule to $G(s)/s$. Thus, by the Lebesgue Dominated Convergence Theorem,

$$\lim_{k\to\infty} \int_0^{\pi_p} [g(u_k) - p\tilde{g}(u_k)]v_k = (1-p) \left[g_+ \int_0^{\pi_p} \varphi_n^+ + g_- \int_0^{\pi_p} \varphi_n^- \right],$$

and so

$$g_+ \int_0^{\pi_p} \varphi_n^+ + g_- \int_0^{\pi_p} \varphi_n^- = \int_0^{\pi_p} h\varphi_n,$$

which contradicts (6.3.9) or (6.3.10). Hence, the sequence $\{u_k\}$ is bounded. By compactness there is a subsequence such that $B'(u_k)$ and $C'(u_k)$ converge in $(W_0^{1,p}(0,\pi_p))^*$. Since $J'(u_k) \to 0$, $A'(u_k)$ converges in $(W_0^{1,p}(0,\pi_p))^*$. Finally, $u_k = (A')^{-1}(A'(u_k))$ converges in $W_0^{1,p}(0,\pi_p)$. This completes the proof. □

Now we are ready to prove the Landesmann-Lazer solvability condition (Theorem 6.3.3) for half-linear BVP at resonance.

Proof. We show, using below given auxiliary results (a) – (g), that the assumptions of Lemma 6.3.6 are satisfied provided (6.3.9) and (6.3.10) hold. This implies then the statement of theorem.

Let $\mathcal{Q}_{n,T} := \{tu : 0 \le t \le T, u \in \Lambda_n\}$ for $T > 0$, where Λ_n is defined by (6.1.6), and let $\mathcal{E}_c := \{u \in W_0^{1,p}(0,\pi_p) : A(u) \ge cB(u)\}$. We will divide our arguments into several steps.

(a) If $h : \mathcal{Q}_{n,T} \to W_0^{1,p}(0,\pi_p)$ is a continuous map such that $h|_{\partial\mathcal{Q}_{n,T}}$ is odd, then we have $h(\mathcal{Q}_{n,T}) \cap \mathcal{E}_{\lambda_{n+1}} \neq \emptyset$. Indeed, suppose not, so $h(\mathcal{Q}_{n,T}) \subset (\mathcal{E}_{\lambda_{n+1}})^c$, where $(\cdot)^c$ denotes the complement of the set indicated. Since $0 \in \mathcal{E}_{\lambda_{n+1}}$, we have $(\mathcal{E}_{\lambda_{n+1}})^c \subset W_0^{1,p}(0,\pi_p)\setminus\{0\}$ and we can compose with the radial projection onto $\mathcal{S}$ to get, without loss of generality, $h(\mathcal{Q}_{n,T}) \subset \mathcal{S}\cap(\mathcal{E}_{\lambda_{n+1}})^c$. Since $E(h(u)) < \lambda_{n+1}$ for $u \in \mathcal{Q}_{n,T}$, where E is given by (6.3.8), is a compact set, we may assume that there exists an $\varepsilon > 0$ such that $E(h(u)) \le \lambda_{n+1} - \varepsilon$ for every $u \in \mathcal{Q}_{n,T}$. Now, Theorem 6.1.2 implies that the genus $\gamma(\{u \in \mathcal{S} : \ E(u) \le \lambda_{n+1} - \varepsilon\}) \le n$, so there is a continuous odd $\tilde{h} : \{u \in \mathcal{S} : \ E(u) \le \lambda_{n+1} - \varepsilon\} \to \mathbb{R}^n \setminus \{0\}$. Hence, the composed map $\tilde{h} \circ h : \mathcal{Q}_{n,T} \to \mathbb{R}^n \setminus \{0\}$ is continuous such that $\tilde{h} \circ h(-x) = -\tilde{h} \circ h(x)$ for $x \in \partial\mathcal{Q}_{n,T}$. But $\mathcal{Q}_{n,T}$ is homeomorphic to the closed unit ball in $\mathbb{R}^n$, so the previous statement contradicts the classical Borsuk-Ulam theorem (see [86, p. 21]).

(b) The next statement we present without proof, we refer to [131] for the details.
Given $\varepsilon < \min\{|\lambda_{n+1} - \lambda_n|, |\lambda_n - \lambda_{n-1}|\}$, there exists an $\tilde{\varepsilon} \in (0,\varepsilon)$ and a one-parameter family of homeomorphisms $\eta : [-1,1] \times \mathcal{S} \to \mathcal{S}$ such that (i) $\eta(t,u) = u$ if $E(u) \in (-\infty, \lambda_n - \varepsilon] \cup [\lambda_n + \varepsilon, \infty)$ or if $u \in \mathcal{K}_{\lambda_n}$, where $\mathcal{K}_{\lambda_n}$ is defined in the Subsection 6.1.2, (ii) $E(\eta(t,u))$ is strictly decreasing in t if $E(u) \in (\lambda_n - \tilde{\varepsilon}, \lambda_n + \tilde{\varepsilon})$ and $u \in \mathcal{K}_{\lambda_n}$, (iii) $\eta(t,-u) = -\eta(t,u)$.

(c) In this part of the proof we define

$$\begin{aligned} \tilde{\mathcal{E}}_{\lambda_n} &= \{tu : \ t \in \mathbb{R}, \ u \in \eta(-1, \mathcal{E}_{\lambda_n} \cap \mathcal{S}\}, \\ \tilde{\mathcal{Q}}_{n,T} &= \{tu : 0 \le t \le T, u \in \eta(1, \Lambda_n)\}, \end{aligned}$$

where Λ_n is defined by (6.1.6). By the part (b), $A(u) - \lambda_n B(u) \ge 0$ for $u \in \tilde{\mathcal{E}}_{\lambda_n}$, with equality if and only if $u = c\varphi_n$ for some $c \in \mathbb{R}$. Similarly, $A(u) - \lambda_n B(u) \le 0$ for $\tilde{\mathcal{Q}}_{n,T}$ with equality if and only if $u = c\varphi_n$ for some $c \in \mathbb{R}$. Using the part (a) and the fact that $\eta(t,\cdot)$ is an odd homeomorphism, we can prove the following statement.

(d) If $h : \tilde{\mathcal{Q}}_{n,T} \to W_0^{1,p}(0,\pi_p)$ is continuous such that $h|_{\partial\tilde{\mathcal{Q}}_{n,T}}$ is odd, then $h(\tilde{\mathcal{Q}}_{n,T}) \cap \mathcal{E}_{\lambda_{n+1}} \neq \emptyset$. To show this, suppose, by contradiction, that $h : \tilde{\mathcal{Q}}_{n,T} \to W_0^{1,p}(0,\pi_p)$ is continuous such that $h|_{\partial\tilde{\mathcal{Q}}_{n,T}}$ is odd with $h(\tilde{\mathcal{Q}}_{n,T}) \cap \mathcal{E}_{\lambda_{n+1}} = \emptyset$. Then the function $\tilde{h} : \mathcal{Q}_{n,T} \to W_0^{1,p}(0,\pi_p)$ defined by $\tilde{h}(tu) := h(t\eta(1,u))$ for $u \in \Lambda_n$ and $0 \le t \le T$ does not satisfy the conlusion (a), a contradiction.

(e) Let $h : \mathcal{Q}_{n-1,T} \to W_0^{1,p}(0,\pi_p)$ be continuous such that $h|_{\mathcal{Q}_{n-1,T}}$ is odd, then we have

$$h(\mathcal{Q}_{n-1,T}) \cap \tilde{\mathcal{E}}_{\lambda_n} = \emptyset.$$

To prove this claim, suppose that it does not hold. Then, as in the proof of the part (a), we may assume that $h(\mathcal{Q}_{n-1,T}) \subset (\tilde{\mathcal{E}}_{\lambda_{n+}})^c \cap \mathcal{S}$. Then $\tilde{h} : \mathcal{Q}_{n-1,T} \to W_0^{1,p}(0,\pi_p)$ given by $\tilde{h}(u) = \eta(1, h(u))$ is a continuous function which is odd on its boundary and with the image in $(\mathcal{E}_{\lambda_n})^c \cap \mathcal{S}$, a contradiction with the part (a).

In the remaining part of the proof we suppose that (6.3.9) holds and we prove the inf sup assumption of Lemma 6.3.6 with $\mathcal{E} = \mathcal{E}_{\lambda_{n+1}}$ and $\mathcal{Q} = \tilde{\mathcal{Q}}_{n,T}$. It follows that $\partial\mathcal{Q} \cap \mathcal{E} = \emptyset$. This fact and the statement of the part (c) imply that $\mathcal{E}$ and $\partial\mathcal{Q}$ link.

(f) If condition (6.3.9) is satisfied, then there exists $R > 0$ and $\delta > 0$ such that $\langle J'(tu), u\rangle \le -\delta$ for all t, u with $t \ge R$ and $u \in \eta(1, \Lambda_n)$. To prove this, suppose $t_k \to \infty$ and $u_k \in \eta(1, \Lambda_n)$ such that

$$\limsup_{k\to\infty} \langle J'(t_k u_k), u_k\rangle \ge 0.$$

Since $\eta(1, \Lambda_n)$ is compact, we may assume that $u_k \to u_0$ in $\eta(1, \Lambda_n)$. If $u_0 \ne \pm p^{1/p}\varphi_n$, then

$$\int_0^{\pi_p} |u_0'|^p - \lambda_n \int_0^{\pi_p} |u_0|^p \le -\varepsilon$$

for some $\varepsilon > 0$ (note that this is the stage of the proof where it is technically important to use $\eta(1, \Lambda_n)$ rather than Λ_n). Thus

$$\int_0^{\pi_p} |u_k'|^p - \lambda_n \int_0^{\pi_p} |u_k|^p \le -\frac{\varepsilon}{2}$$

for k large enough. Hence

$$\langle J'(t_k u_k), u_k\rangle \le -\frac{\varepsilon}{2} t_k^{p-1} - \int_0^{\pi_p} [g(t_k u_k) - h] u_k$$

for large k, which leads to a contradiction of the lim sup assumption. Hence, suppose that $u_0 = p^{1/p}\varphi_n$. We still have

$$\int_0^{\pi_p} |u_k'|^p - \lambda_n \int_0^{\pi_p} |u_k|^p \le 0,$$

so

$$\langle J'(t_k u_k), u_k\rangle \le - \int_0^{\pi_p} [g(t_k u_k) - h] u_k$$

for all $k \in \mathbb{N}$. The fact that g is bounded and that $u_k \to p^{1/p}\varphi_n$ allows to apply the Lebesgue Dominated Convergence Theorem to get

$$\lim_{k\to\infty} \langle J'(t_k u_k), u_k\rangle \le -p^{1/p} \left(g_+ \int_0^{\pi_p} \varphi_n^+ + g_- \int_0^{\pi_p} \varphi_n^- - \int_0^{\pi_p} h\varphi_n \right) < 0,$$

by (6.3.9). Once again a contradiction is reached. The case $u_0 = -p^{1/p}\varphi_n$ is similar.

(g) The proof is finished by the following statement. If (6.3.9) is satisfied, then there exists $T > 0$ such that

$$\inf_{\mathcal{E}_{\lambda_{n+1}}} J(u) > \sup_{\partial \tilde{Q}_{n,T}} J(u).$$

To prove this, notice that for $u \in \mathcal{E}_{\lambda_{n+1}}$ we have

$$J(u) \ge \frac{1}{p} (\lambda_{n+1} - \lambda_n) \|u\|_{L^p}^p - \int_0^{\pi_p} [G(u) - hu],$$

which is clearly bounded below by some value α. By the previous part of the proof we have

$$\begin{aligned} J(tu) &= J(Ru) + J(tu) - J(Ru) \\ &= J(Ru) + \int_0^t \langle J'(su), u\rangle\, ds \\ &\le c - \delta(t-R) \end{aligned}$$

for all $u \in \eta(1, \Lambda_n)$, all $t > R$, and some $c \in \mathbb{R}$. Thus there exists a $T > R$ such that $J(tu) \le c - \delta(T-R) < \alpha$ for all $t \ge T$ and $u \in \eta(1, \Lambda_n)$. □

6.3.3 Fučík spectrum

The concept of the Fučík spectrum has been introduced in connection with the boundary value problems associated with the equation

$$u'' + f(u) = 0 \tag{6.3.11}$$

involving the so-called *asymmetric nonlinearity* f (another terminology is the *jumping nonlinearity*), i.e., a function f such that there exist finite limits

$$f_- = \lim_{u\to-\infty} \frac{f(u)}{u}, \quad f_+ = \lim_{u\to+\infty} \frac{f(u)}{u},$$

but $f_- \neq f_+$.

If these limits are equal, solvability problem is closely connected with the relationship of this limit to the classical spectrum of the linear part $-u''$ (together with considered boundary conditions, Dirichlet, periodic, Neuman,...). If $f_- \neq f_+$, the crucial role is played by the pairs $[\mu, \nu]$, $\mu, \nu \in \mathbb{R}$, for which the equation

$$-y'' = \mu y_+ - \nu y_- \tag{6.3.12}$$

has a nontrivial solution satisfying boundary conditions under consideration. Here $y^+ = \max\{0, y\}$, $y_- = \max\{-y, 0\}$. In case of the Dirichlet boundary condition (other types of boundary conditions can be treated in a similar way), it is not difficult to verify that (6.3.12) together with the boundary condition $y(0) = 0 = y(\pi)$ has a nontrivial solution if and only if $\mu = 1$ ($\nu \in \mathbb{R}$ arbitrary), or $\nu = 1$ ($\mu \in \mathbb{R}$ arbitrary), or μ and ν are related by one of the following identities with $k \in \mathbb{N}$

$$\frac{k}{\sqrt{\mu}} + \frac{k}{\sqrt{\nu}} = 1, \quad \frac{k+1}{\sqrt{\mu}} + \frac{k}{\sqrt{\nu}} = 1, \quad \frac{k}{\sqrt{\mu}} + \frac{k+1}{\sqrt{\nu}} = 1.$$

In case of the scalar p-Laplacian $-\left(|y'|^{p-2}y'\right)'$ (instead of the operator $-u''$) the situation is very similar. Consider the BVP

$$(\Phi(x'))' + \mu\Phi(x^+) - \nu\Phi(x^-) = 0, \quad x(0) = 0 = x(\pi_p). \tag{6.3.13}$$

Using the generalized sine function $\sin_p$ (which satisfies the boundary condition in (6.3.13)), it easy to compute that the Fučík spectrum of (6.3.13) consists of the trivial part

$$[\mu, \nu] \in \Sigma_1 := \{[p-1, \lambda], \lambda \in \mathbb{R}\} \cup \{[\lambda, p-1], \lambda \in \mathbb{R}\}$$

and the hyperbola like curves Σ_{2k}, $\Sigma_{2k+1,1}$, and $\Sigma_{2k+1,2}$, where

$$\begin{aligned}\Sigma_{2k} &= \left\{[\mu,\nu]: \frac{k}{\mu^{1/p}}+\frac{k}{\nu^{1/p}}=1\right\},\\ \Sigma_{2k+1,1} &= \left\{[\mu,\nu]: \frac{k+1}{\mu^{1/p}}+\frac{k}{\nu^{1/p}}=1\right\},\\ \Sigma_{2k+1,2} &= \left\{[\mu,\nu]: \frac{k}{\mu^{1/p}}+\frac{k+1}{\nu^{1/p}}=1\right\}.\end{aligned}$$

A general treatment of the Fučík spectrum for the one-dimensional p-Laplacian can be found e.g. in [118, 332], see also [55] for some computational aspects of the problem. More precisely, the results of [332] concern the more general differential equation

$$t^{-\alpha}(t^{\alpha}\Phi(u'))' + c(t)\Phi(u) + w(t)[\mu(u^+)^{p-1} - \nu(u^-)^{p-1}] = 0, \tag{6.3.14}$$

where c, w are continuous functions on an interval $I = [a,b]$, $w(t) > 0$ for $t \in I$ and $\alpha \geq 0$ is a real constant. Problem (6.3.14) is considered together with the Sturm-Liouville boundary conditions

$$\gamma_1\Phi(u)(a) + \gamma_2(t^{\alpha}\Phi(u'))(a) = 0, \tag{6.3.15}$$

$$\gamma_3\Phi(u)(b) + \gamma_4(t^{\alpha}\Phi(u'))(b) = 0, \tag{6.3.16}$$

where $\gamma_1^2+\gamma_2^2>0$, $\gamma_3^2+\gamma_4^2>0$. If $a>0$ or $a=0$ and $0\leq\alpha<p-1$, the BVP is referred to as *regular*, denoted by (R), and is said to be *singular*, denoted by (S), in the opposite case. In the latter case, the boundary condition (6.3.15) at $t=0$ is always $u'(0)=0$.

In the next theorem, a function u is said to be *initially positive* or *initially negative* at a if $u(a+)>0$ or $u(a+)<0$, respectively. Also, we denote by λ_k the k-th eigenvalue of the "normal" BVP

$$t^{-\alpha}(t^{\alpha}\Phi(u'))' + [c(t)+\lambda w(t)]\Phi(u) = 0 \tag{6.3.17}$$

with boundary conditions (6.3.15), (6.3.16). The proof of the next theorem is based on a version of the Prüfer transformation applied to (6.3.14), we skip this proof because of its technical complexity.

Theorem 6.3.4. *The Fučík spectrum σ of* (6.3.14), (6.3.15), (6.3.16) *is closed in the $\mu\nu$-plane $\mathbb{R}^2$. It takes the form $\sigma=\sigma^+\cup\sigma^-$, where $(\mu,\nu)\in\sigma^+$, if and only if $(\nu,\mu)\in\sigma^-$. The sets σ^+, σ^- are closed and the eigenfunctions corresponding to points of these sets are initially positive and negative, respectively. Furthermore,*

$$\sigma\cap\{(\lambda,\lambda):\lambda\in\mathbb{R}\} = \{(\lambda_k,\lambda_k): k\in\mathbb{N}\}\subset\sigma^+\cap\sigma^-.$$

The eigenfunctions corresponding to the connected components σ_k^+, σ_k^-, $k\in\mathbb{N}$, of σ^+, σ^- have exactly $k-1$ zeros on (a,b). The first (trivial) component of σ_1 consists of $\{\lambda_1\}\times\mathbb{R}$, while for $k\geq 2$, σ_k^+ is a C^1 curve $(\mu,\nu(\mu))$ with $\nu'(\mu)<0$ and with the following asymptotics (the relations for σ_k^+ follow by symmetry):

(i) *Singular case $a = 0$ and $\alpha \geq p-1$:*

$$\begin{aligned} k = 2i: \quad & \nu(\infty) = \lambda_i, \quad \nu(\lambda_i^b+) = \infty, \\ k = 2i+1: \quad & \nu(\infty) = \lambda_i^b, \quad \nu(\lambda_{i+1}+) = \infty. \end{aligned}$$

(ii) *Regular case $a > 0$ or $0 \leq \alpha < p-1$:*

$$\begin{aligned} k = 2i: \quad & \nu(\infty) = \lambda_i^a, \quad \nu(\lambda_i^b+) = \infty, \\ k = 2i+1: \quad & \nu(\infty) = \lambda_i^{ab}, \quad \nu(\lambda_{i+1}+) = \infty, \end{aligned}$$

where λ_i^{ab}, λ_i^a, λ_i^b are eigenvalues of (6.3.16) *with the boundary conditions*

$$\begin{aligned} \lambda^{ab}: \quad & u(a) = 0, & & u(b) = 0, \\ \lambda^{a}: \quad & u(a) = 0, & & \gamma_3\Phi(u)(b) + \gamma_4(t^\alpha\Phi(u'))(b) = 0, \\ \lambda^{b}: \quad & \gamma_1\Phi(u)(a) + \gamma_2(t^\alpha\Phi(u'))(a) = 0, & & u(b) = 0, \end{aligned}$$

respectively.

Using this description of the Fučík spectrum, one can prove the following statement concerning solvability of the BVP's associated with the differential equation

$$t^{-\alpha}(t^\alpha\Phi(u'))' + f(t,u) = 0. \tag{6.3.18}$$

Note that the uniqueness condition on the initial value problem associated with (6.3.18) supposed in this theorem can be found e.g. in [332].

Theorem 6.3.5. *Let f be continuous in $[a,b] \times \mathbb{R}$ and the initial value problem* (6.3.18) *with the initial values $u(0) = u_0$, $u'(0) = 0$ in case* (S) *(i.e., $a = 0$), and $u(a) = u_0$, $(t^\alpha\Phi(u'))(a) = u_1$ in case* (R)*, has a unique solution. Suppose that there exists a function $h \in L^\infty(a,b)$, $0 \not\equiv h(t) \geq 0$, such that*

$$-\infty < -K \leq \liminf_{u\to\pm\infty} \frac{f(t,u)}{\Phi(u)} \leq \limsup_{u\to\infty} \frac{f(t,u)}{\Phi(u)} \leq (\lambda_1 w + c - h)(t)$$

or

$$(\mu_k w + c + h)(t) \leq \liminf_{u\to\infty} \frac{f(t,u)}{\Phi(u)} \leq \limsup_{u\to\infty} \frac{f(t,u)}{\Phi(u)} \leq (\mu_{k+1} w + c - h)(t)$$

and

$$(\nu_{k+1} w + c + h)(t) \leq \liminf_{u\to-\infty} \frac{f(t,u)}{\Phi(u)} \leq \limsup_{u\to-\infty} \frac{f(t,u)}{\Phi(u)} \leq (\nu_{k+1} w + c - h)(t),$$

where $(\lambda_1, 0) \in \sigma^+$, $(\mu_k, \nu_k) \in \sigma_k^+$, $(\mu_{k+1}, \nu_{k+1}) \in \sigma_{k+1}^+$ are elements of the Fučík spectrum of (6.3.14), (6.3.15), (6.3.16) *with no, k, $k+1$ zeros in (a,b), respectively. Then* (6.3.18), (6.3.15), (6.3.16) *has a solution. The same conclusion holds if $(0, \lambda_1) \in \sigma_1^-$, $(\mu_k, \nu_k) \in \sigma_k^-$ and $(\mu_{k+1}, \nu_{k+1}) \in \sigma_{k+1}^-$.*

We finish this subsection with a statement concerning the unique solvability of "Fučík type" initial value problem. The proof of this statement can be found in [332].

Theorem 6.3.6. *Let $c, d \in C(I)$. Then the initial value problem*

$$\begin{aligned} t^{-\alpha}(t^{\alpha}\Phi(u'))' + c(t)(u^+)^{p-1} - d(t)(u^-)^{p-1} &= 0, && t \in I, \\ u(0) = u_0, \quad u'(0) &= 0 && \textit{in case (S)}, \\ u(0) = u_0, \quad (t^{\alpha}\Phi(u'))(a) &= u_1 && \textit{in case (R)} \end{aligned}$$

has a unique solution.

6.4 Notes and references

The description of eigenvalues and eigenfunctions of the boundary value (6.1.2) can already be found in the paper of Elbert [139]. Let us mention very recent paper [41] by Binding, Boulton, Čepička, Drábek and Girg, where it is shown that for $p \geq 12/11$ (see also the detailed numerical analysis in that paper) the functions $f_n(t) = \sin_p(n\pi_p t)$ form a Riesz basis in $L^2(0,1)$ and a Schauder basis in $L^{\alpha}(0,1)$ for any $1 < \alpha < \infty$. The variational characterization of the eigenvalues of the BVP (6.1.2) presented in Subsection 6.1.2 is taken from the paper of Drábek and Robinson [131]. The results of Subsection 6.1.3, in particular, the construction of the forcing term f in (6.1.7) for which this BVP with $\lambda < \lambda_1$ has at least two solutions are originally presented in the papers of Del Pino, Elgueta and Manasevich [91] (case $p > 2$) and of Fleckinger, Hernández, Takáč and de Thélin [164] (the case $p \in (1,2)$). The statement given in Theorem 6.1.4 which extends the previous results to general λ is taken from Drábek and Takáč [134]. The results of Subsection 6.1.4 are presented in the paper of Del Pino, Elgueta and Manasevich [91]. Finally, statements given in Subsection 6.1.5 form the main part of the paper Del Pino, and Manasevich [92]. The method used in that paper follows the idea introduced in the paper of Guedda and Veron [170].

Concerning the statements presented in Section 6.2, the main results of this section, i.e., those in Subsection 6.2.1, come from Del Pino, Drábek and Manasevich [88], an abbreviated version of this paper is [87]. Note that the proof of the main auxiliary statement of this subsection, Lemma 6.2.1 follows essentially the proof of Lemma 6.1.3 and requires the assumption that the forcing term is of the class C^1. The second part of this section (Subsection 6.2.2) is a very brief extract of the paper of Manasevich and Takáč [268], unfortunately, we found no space to present other interesting results of that paper here. Note only that Lemma 6.2.5 extends the statement of the above mentioned Lemma 6.2.1, it requires the forcing term to be only in L^1 and its proof follows completely different idea than that of Lemma 6.2.1. Related results concerning both resonant and nonresonant BVP's can be found in the papers of Binding, Drábek, Huang [43], Drábek [118, 122], Lindqvist [257], Manasevich, Mawhin [266], and Zhang [381].

The first subsection of Section 6.3 is a substantial part of the paper of Drábek, Girg and Manasevich [124]. Note that the linear motivation of the results of that paper is the paper Ambrosetti and Prodi [18]. The extension of the linear Landesman-Lazer solvability conditions (originally published in [232]) is taken from the paper of Drábek and Robinson [131]. The concept of the Fučík spectrum of the linear second order differential operator is introduced in the papers of Fučík [167] and of Dancer [85]. The half-linear version of the basic results of that paper can be found e.g. in the book of Drábek [119]. Theorems 6.3.4 and 6.3.4 are taken from the paper of Reichel and Walter [332], the related results can be found e.g. in the paper Fabry, Manasevich [158]. Concerning an attempt to extend the results on the Fredholm alternative to the Fučík-type BVP we refer to the paper of X. Yang [375].

Finally note that we have presented in this chapter only the results which are relatively closely related to the equation (1.1.1). There are many papers devoted to BVP's associated with the equation $(\Phi(y'))' = f(t, y, y')$, these papers extend corresponding results for the equation $y'' = f(t, y, y')$. As an example let us mention at least the papers of J. Wang [357] and of Averna and Bonanno [27], see also the references given therein. Another possible line of the generalization of linear results is suggested in the paper of Manasevich and Sedziwy [267], where the Lienard type equation is investigated.

CHAPTER 7

PARTIAL DIFFERENTIAL EQUATIONS WITH p-LAPLACIAN

Similarly to the boundary value problems for half-linear ordinary differential equations, also partial differential equations with p-Laplacian are treated in many papers. Recall that the p-Laplacian is the partial differential operator

$$\Delta_p u(x) := \operatorname{div}(\|\nabla u(x)\|^{p-2} \nabla u(x)), \quad x = (x_1, \dots, x_N) \in \mathbb{R}^N, \tag{7.1.1}$$

where $p > 1$, $\operatorname{div} := \sum_{k=1}^{N} \frac{\partial}{\partial x_k}$ is the usual divergence operator,

$$\nabla u(x) = \left(\frac{\partial u}{\partial x_1}, \dots, \frac{\partial u}{\partial x_N} \right)$$

is the Hamilton nabla operator and $\| \cdot \|$ denotes the Euclidean norm in $\mathbb{R}^N$.

In the first section we deal with Dirichlet BVP for partial differential equations involving p-Laplacian (7.1.1). The second section is devoted to higher dimensional BVP at resonance, in particular, to higher dimensional analogue of the results given in Section 6.2. The last section of the chapter contains a brief treatment of the oscillation theory of PDE's with p-Laplacian.

7.1 Eigenvalues and comparison principle

In this section we deal with the properties of the first two eigenvalues of the Dirichlet BVP for p-Laplacian. Then we present maximum and comparison principles for this operator and we conclude this section with a brief description of the Fučík spectrum of the Dirichlet boundary value problem associated with p-Laplacian.

7.1.1 Dirichlet BVP with p-Laplacian

In this subsection we deal with the properties of the first eigenvalue and the associated eigenfunction of the Dirichlet boundary value problem

$$\begin{cases} \Delta_p u + \lambda \Phi(u) = 0, & x \in \Omega \subset \mathbb{R}^N, \\ u(x) = 0, & x \in \partial\Omega, \end{cases} \tag{7.1.2}$$

where Ω is a bounded domain in $\mathbb{R}^N$ and λ is an eigenvalue parameter.

The solution of problem (7.1.2) is understood in the weak sense; we say that λ is an *eigenvalue* if there exists a function $u \in W_0^{1,p}(\Omega)$, $u \not\equiv 0$, such that

$$\int_\Omega \|\nabla u\|^{p-2} \langle \nabla u, \nabla \eta \rangle \, dx = \lambda \int_\Omega \Phi(u) \eta \, dx, \tag{7.1.3}$$

for every $\eta \in W_0^{1,p}(\Omega)$, where $\langle \cdot, \cdot \rangle$ denotes the scalar product in $\mathbb{R}^N$. The function u is called an *eigenfunction.*

The first eigenvalue $\lambda_1 = \lambda_1(\Omega)$ is obtained as the minimum of the Rayleigh quotient

$$\lambda_1 = \inf_v \frac{\int_\Omega \|\nabla v\|^p \, dx}{\int_\Omega |v|^p \, dx}, \tag{7.1.4}$$

where the infimum is taken over all $v \in W_0^{1,p}(\Omega)$, $v \not\equiv 0$. If u realizes the infimum in (7.1.4), so does also $|u|$, and this leads immediately to the following statement.

Theorem 7.1.1. *The eigenfunction u associated with the first eigenvalue λ_1 does not change its sign in Ω. Moreover, if $u \geq 0$, then actually $u > 0$ in the interior of Ω.*

Proof. The statement concerning the positivity of u follows from the Harnack inequality [352, p. 724]. □

In the proof of the main result of this subsection we will need the following inequalities, for the proof see [258].

Lemma 7.1.1. *Let $w_1, w_2 \in \mathbb{R}^N$.*

(i) If $p \geq 2$, then

$$\|w_2\|^p \geq p\|w_1\|^p \langle w_1, (w_2 - w_1) \rangle + \frac{\|w_2 - w_1\|^p}{2^{p-1} - 1}. \tag{7.1.5}$$

(ii) If $1 < p < 2$, then

$$\|w_2\|^p \geq p\|w_1\|^p \langle w_1, (w_2 - w_1) \rangle + C(p) \frac{\|w_2 - w_1\|^p}{(\|w_1\| + \|w_2\|)^{2-p}}, \tag{7.1.6}$$

where $C(p)$ is a positive constant depending only on p.

The main statement of this subsection reads as follows.

Theorem 7.1.2. *The first eigenvalue of* (7.1.2) *is simple and isolated for any bounded domain* $\Omega \subset \mathbb{R}^N$.

Proof. Here we follow Lindqvist's modification [258] of the original proof of Anane [19] where it is supposed that the boundary $\partial\Omega$ is of the Hölder class $C^{2,\alpha}$. Recall that the space $C^{n,\alpha}$, $\alpha \in (0,1)$, is a subspace of C^n formed by the functions f such that

$$\sup_{x\neq y} \frac{\|D^\beta f(x) - D^\beta f(y)\|}{\|x-y\|^\alpha} < \infty,$$

for differentiation indices $|\beta| \leq n$, where $|\beta| = \beta_1 + \cdots + \beta_N$ and

$$D^\beta = \frac{\partial^{|\beta|}}{\partial x_1^{\beta_1} \cdots \partial x_N^{\beta_N}}.$$

This assumption on the boundary of Ω is removed in Lindqvist's proof by introducing the functions $u+\varepsilon$, $v+\varepsilon$ instead of u, v, respectively (used by Anane).

Suppose that u, v are eigenfunctions of (7.1.2) with $\lambda = \lambda_1$. Let $\varepsilon > 0$ and denote $v_\varepsilon = v+\varepsilon$, $u_\varepsilon = u+\varepsilon$. Further, let $\eta = u_\varepsilon - v_\varepsilon^p u_\varepsilon^{1-p}$, $\tilde\eta = v_\varepsilon - u_\varepsilon^p v_\varepsilon^{1-p}$. Then $\eta, \tilde\eta \in W_0^{1,p}(\Omega)$ and

$$\nabla\eta = \left\{1 + (p-1)\left(\frac{v_\varepsilon}{u_\varepsilon}\right)^p\right\}\nabla u - p\left(\frac{v_\varepsilon}{u_\varepsilon}\right)^{p-1}\nabla v.$$

A similar formula holds for $\nabla\tilde\eta$. Inserting the test functions η and $\tilde\eta$ into (7.1.3) and adding both equations, we get

(7.1.7)

$$\begin{aligned}
&\lambda_1 \int_\Omega \left[\frac{u^{p-1}}{u_\varepsilon^{p-1}} - \frac{v^{p-1}}{v_\varepsilon^{p-1}}\right](u_\varepsilon^p - v_\varepsilon^p)\,dx \\
&= \int_\Omega \left[\left\{1+(p-1)\left(\frac{v_\varepsilon}{u_\varepsilon}\right)^p\right\}\|\nabla u_\varepsilon\|^p + \left\{1+(p-1)\left(\frac{u_\varepsilon}{v_\varepsilon}\right)^p\right\}\|\nabla v_\varepsilon\|^p\right]dx \\
&\quad - \int_\Omega \left[p\left(\frac{v_\varepsilon}{u_\varepsilon}\right)^{p-1}\|\nabla u_\varepsilon\|^{p-2}\langle\nabla u_\varepsilon, \nabla v_\varepsilon\rangle \right.\\
&\quad \left. + p\left(\frac{u_\varepsilon}{v_\varepsilon}\right)^{p-1}\|\nabla v_\varepsilon\|^{p-2}\langle\nabla v_\varepsilon, \nabla u_\varepsilon\rangle\right]dx \\
&= \int_\Omega (u_\varepsilon^p - v_\varepsilon^p)(\|\nabla\log u_\varepsilon\|^p - \|\nabla\log v_\varepsilon\|^p)\,dx \\
&\quad - \int_\Omega p v_\varepsilon^p \|\nabla\log u_\varepsilon\|^{p-2}\langle\nabla\log u_\varepsilon, (\nabla\log v_\varepsilon - \nabla\log u_\varepsilon)\rangle\,dx \\
&\quad - \int_\Omega p u_\varepsilon^p \|\nabla\log v_\varepsilon\|^{p-2}\langle\nabla\log v_\varepsilon, (\nabla\log u_\varepsilon - \nabla\log v_\varepsilon)\rangle\,dx \geq 0
\end{aligned}$$

by the inequality given in Lemma 7.1.1. It is obvious that

$$\lim_{\varepsilon\to 0+}\int_\Omega \left[\frac{u^{p-1}}{u_\varepsilon^{p-1}}-\frac{v^{p-1}}{v_\varepsilon^{p-1}}\right](u_\varepsilon^p-v_\varepsilon^p)\,dx=0. \tag{7.1.8}$$

Let us first consider the case $p\geq 2$. According to inequality (7.1.5) we have

$$\begin{aligned}0 &\leq \frac{1}{2^{p-1}-1}\int_\Omega\left(\frac{1}{v_\varepsilon^p}+\frac{1}{u_\varepsilon^p}\right)\|v_\varepsilon\nabla u_\varepsilon-u_\varepsilon\nabla v_\varepsilon\|^p\,dx\\ &\leq -\lambda_1\int_\Omega\left[\left(\frac{u}{u_\varepsilon}\right)^{p-1}-\left(\frac{v}{v_\varepsilon}\right)^{p-1}\right](u_\varepsilon^p-v_\varepsilon^p)\,dx\end{aligned}$$

for every $\varepsilon>0$ (here we have used inequality (7.1.5) with $w_1=\nabla\log u_\varepsilon$, $w_2=\nabla\log v_\varepsilon$ and vice versa). In view of (7.1.8), taking a sequence $\varepsilon_k\to 0+$ as $k\to\infty$ and using Fatou's lemma in the previous computations we finally arrive at the conclusion that $v\nabla u=u\nabla v$ a.e. in Ω. Hence there is a constant κ such that $u=\kappa v$ a.e. in Ω and by continuity this equality holds everywhere in Ω.

Now we turn the attention to the case $1<p<2$ where the situation is similar as in the previous case. Applying inequality (7.1.6) in (7.1.7) we obtain

$$\begin{aligned}0 &\leq C(p)\int_\Omega (u_\varepsilon v_\varepsilon)^p(u_\varepsilon^p+v_\varepsilon^p)\frac{\|v_\varepsilon\nabla u_\varepsilon-u_\varepsilon\nabla v_\varepsilon\|^2}{(v_\varepsilon\|\nabla u_\varepsilon\|+u_\varepsilon\|\nabla v_\varepsilon\|)^{2-p}}\,dx\\ &\leq -\lambda_1\int_\Omega\left[\left(\frac{u}{u_\varepsilon}\right)^{p-1}-\left(\frac{v}{v_\varepsilon}\right)^{p-1}\right](u_\varepsilon^p-v_\varepsilon^p)\,dx\end{aligned}$$

for every $\varepsilon>0$. Using (7.1.8), we again arrive at the desired dependence $u=\kappa v$ for some constant κ.

As for the isolation of the first eigenvalue λ_1, we proceed as follows. Since λ_1 is defined as the minimum of the quotient (7.1.4), it is isolated from the left. If v is an eigenfunction associated with an eigenvalue $\lambda>\lambda_1$ then v changes its sign in Ω. Indeed, suppose that v does not change its sign in Ω. Then using the same method as in the previous part of the proof, we get (for details we refer to [19])

$$0\leq\int_\Omega(\lambda_1-\lambda)(u^p-v^p)\,dx=(\lambda_1-\lambda)\left(\frac{1}{\lambda_1}-\frac{1}{\lambda}\right)$$

which is a contradiction.

Now, suppose, by contradiction, that there exists a sequence of eigenvalues $\lambda_n\to\lambda_1+$ and let u_n be the sequence of associated eigenfunctions such that $\|u_n\|=1$. This sequence contains a weakly convergent subsequence in $W_0^{1,p}(\Omega)$, denoted again u_n, and hence strongly convergent in $L^p(\Omega)$. Since we have $u_n=-\lambda_n\Delta_p^{-1}(\Phi(u_n))$ (this is a usual argument in the theory of partial equations with p-Laplacian, we refer e.g. to the monograph [168]), the sequence u_n converges strongly in $W_0^{1,p}(\Omega)$ to a function of the $W_0^{1,p}$ norm equal to 1 associated with λ_1. However, by the Jegorov theorem, the sequence u_n converges uniformly to a function u except for a set of arbitrarily small Lebesgue measure. However, this is a contradiction with the fact that the eigenfunction associated with the first eigenvalue does not change its sign in Ω. □

7.1.2 Second eigenvalue of p-Laplacian

In this subsection we briefly deal with the variational description of the second eigenvalue of p-Laplacian and with a nodal domain property of the associated eigenfunction. We again suppose that Ω is a bounded domain in $\mathbb{R}^n$.

We consider the eigenvalue problem (7.1.2) and we introduce the functionals

$$A(u) = \frac{1}{p}\int_\Omega \|\nabla u\|^p\,dx, \quad B(u) = \frac{1}{p}\int_\Omega |u|^p\,dx, \quad F(u) = A^2(u) - B(u).$$

It is clear that the critical point u of F associated to a critical value c (i.e., $F(u) = c$ and $F'(u) = 0$) is an eigenfunction associated to the eigenvalue

$$\lambda = \frac{1}{2\sqrt{-c}}.$$

Conversely, if $u \neq 0$ is an eigenfunction associated to a positive eigenvalue λ, $v = (2\lambda A(u))^{-\frac{1}{p}} u$ will be also an eigenfunction associated to $\lambda = 1/[2A(v)]$ and v is a critical point of F associated to the critical value $c = -1/[4\lambda^2]$.

Let us consider the sequence $\{c_n\}_{n\in\mathbb{N}}$ defined by

$$c_n = \inf_{K\in\mathcal{A}_n}\ \sup_{v\in K} F(v), \tag{7.1.9}$$

where

$$\mathcal{A}_n = \{K \subset W_0^{1,p}(\Omega) : K \text{ is a symmetrical compact and } \gamma(K) \geq n\}$$

and $\gamma(K)$ denotes the Krasnoselskii genus of K, i.e., the minimal integer n such that there exists a continuous odd mapping of $K \to \mathbb{R}^n \setminus \{0\}$. It can be proved (using the Palais-Smale condition for F) that the sequence c_n consists of the critical values of F and $c_n \to 0-$. The sequence of eigenvalues λ_n defined by

$$\lambda_n = \frac{1}{2\sqrt{-c_n}} \tag{7.1.10}$$

is positive, nondecreasing and tends to ∞. Note that it is an open problem whether (7.1.10) describes *all* eigenvalues of (7.1.2) (in contrast to the scalar case $N = 1$, compare with Subsection 6.1.2).

We denote by $Z(u) = \{x \in \Omega : \ u(x) = 0\}$ the so-called *nodal contour* of the function u and let $N(u)$ denote the number of components (the so-called *nodal domains*) of $\Omega \setminus Z(u)$. For each eigenfunction u associated to λ we define

$$N(\lambda) = \max\{N(u) : u \text{ is a solution of (7.1.2)}\}.$$

At the end of this subsection, we present without proof the main result of [21]. The next statement shows, among others, that the second eigenvalue λ_2 can be characterized by (7.1.10).

Theorem 7.1.3. *For each eigenvalue λ of* (7.1.2) *it holds $\lambda_{N(\lambda)} \leq \lambda$. Moreover, the value λ_2 given by* (7.1.10) *satisfies*

$$\lambda_2 = \inf\{\lambda : \lambda \text{ is positive eigenvalue of (7.1.2)},\ \lambda > \lambda_1\}.$$

7.1.3 Comparison and antimaximum principle for p-Laplacian

The (strong) comparison principle and antimaximum principle are well known statements for the classical linear Laplace operator (see e.g. [78] for the antimaximum principle). Here we present some of their extensions to the p-Laplacian. We do not present details of all proofs, in some cases we give only a brief outline of ideas used there.

We start with the so-called *strong comparison principle* and *boundary point principle.* We will use the usual convention that for $f, g \in L^\infty(\Omega)$ the relation $f \not\equiv g$ in a domain $\Omega \subset \mathbb{R}^N$ means that $f(x) \neq g(x)$ for $x \in \Omega' \subset \Omega$ with Ω' having positive measure.

The strong comparison and boundary points principles concern a pair of nonautonomous Dirichlet BVP's with the p-Laplacian

$$\begin{aligned} (7.1.11) \qquad -\Delta_p u &= \lambda\Phi(u) + f, \ x \in \Omega, \quad u = 0, \ x \in \partial\Omega, \\ (7.1.12) \qquad -\Delta_p v &= \lambda\Phi(v) + g, \ x \in \Omega, \quad v = 0, \ x \in \partial\Omega. \end{aligned}$$

Theorem 7.1.4. *Let $\Omega \subset \mathbb{R}^N$ be a bounded domain whose boundary is a connected $C^{2,\alpha}$ manifold for some $\alpha \in (0,1)$ and let $0 \le \lambda < \lambda_1$, where λ_1 is the first eigenvalue of $-\Delta_p$ in Ω. Further, let $0 \le f \le g$ with $f \not\equiv g$ in Ω. If $u, v \in W_0^{1,p}(\Omega)$ are solutions of* (7.1.11) *and* (7.1.12)*, respectively, then the strong comparison principle holds:*

$$(7.1.13) \qquad 0 \le u < v \ \text{in}\ \Omega \quad \text{and} \quad \frac{\partial v}{\partial \nu} < \frac{\partial u}{\partial \nu} \le 0 \ \text{on}\ \partial\Omega,$$

where $\frac{\partial}{\partial \nu}$ denotes the derivative in the direction of the exterior normal.

For $f \equiv 0$ and $u \equiv 0$ the previous statement is known as the *strong maximum principle* and can be found e.g. in [355, Theorem 5]. The proof of Theorem 7.1.4 is based on the so-called *weak comparison principle* combined with certain regularity results for weak solutions of (7.1.11), (7.1.12). Recall that the weak comparison principle reads as follows.

Theorem 7.1.5. *Let $\Omega \subset \mathbb{R}^N$ be a bounded domain with $C^{1,\alpha}$ boundary $\partial\Omega$ for some $\alpha \in (0,1)$. Further let $f, g \in L^q(\Omega)$, $q = p/(p-1)$, $\tilde{f}, \tilde{g} \in W^{1-(1/p),p}(\partial\Omega)$ with $f \le g$ in Ω and $\tilde{f} \le \tilde{g}$ in $\partial\Omega$. Consider a pair of Dirichlet BVP*

$$\begin{aligned} (7.1.14) \qquad -\Delta_p u &= \lambda\Phi(u) + f, \quad x \in \Omega, \ u = \tilde{f} \ \text{on}\ \partial\Omega, \\ (7.1.15) \qquad -\Delta_p v &= \lambda\Phi(v) + g, \quad x \in \Omega, \ v = \tilde{g} \ \text{on}\ \partial\Omega. \end{aligned}$$

and let u, v be weak solutions of (7.1.14) *and* (7.1.15)*, respectively. Then*

(i) if $\lambda \le 0$, then $u \le v$ almost everywhere in Ω;

(ii) if $\lambda < \lambda_1$, $0 \le f \le g$ in Ω and $\tilde{f} \equiv \tilde{g} \equiv 0$ in $\partial\Omega$, then $0 \le u \le v$ almost everywhere in Ω with $u, v \in L^\infty(\Omega)$.

As we have shown in Subsection 6.1.3, if $0 \le \lambda < \lambda_1$, the uniqueness of a solution of Dirichlet BVP is generally violated. However, if the forcing term f in (7.1.11) is nonnegative, the uniqueness is preserved as the next statement shows.

Corollary 7.1.1. *Suppose that the assumptions of Theorem 7.1.5 concerning* (7.1.14) *are satisfied with* $\tilde{f} \equiv 0$ *and* $f \ge 0$, *but* $f \not\equiv 0$ *in* Ω. *Then* (7.1.11) *has a unique solution.*

Proof. Suppose that $u_1, u_2 \in W_0^{1,p}(\Omega)$ are two solutions of (7.1.11). Then these solutions satisfy

$$u_i > 0 \text{ in } \Omega, \quad \frac{\partial u_i}{\partial \nu} < 0 \text{ on } \partial\Omega, \ i = 1, 2,$$

and this enables to apply the inequality of Díaz and Saa [94, Lemma 2] which reads

$$\int_\Omega \left(\frac{-\Delta_p u_1}{u_1^{p-1}} - \frac{-\Delta_p u_2}{u_2^{p-1}} \right) (u_1^p - u_2^p)\, dx \ge 0, \tag{7.1.16}$$

with equality if and only if u_1, u_2 are proportional (see also the proof of Theorem 7.1.1). Substituting from (7.1.11) we obtain

$$\int_\Omega f(x) \left(\frac{1}{u_1^{p-1}} - \frac{1}{u_2^{p-2}} \right) (u_1^p - u_2^p)\, dx \ge 0.$$

However, the integrand in the last integral is nonpositive, and hence it must vanish a.e. in Ω and (7.1.16) implies that $u_2 = \mu u_1$ for some positive μ. Inserting this relation into (7.1.11) we conclude that $(\mu - 1) f \equiv 0$ in Ω, so $\mu = 1$ and hence $u_1 = u_2$. □

Another consequence of the maximum principle is the following nonexistence result for the resonance problem at the first eigenvalue of $-\Delta_p$.

Corollary 7.1.2. *Let* $f \in L^\infty(\Omega)$, $f \ge 0$, $f \not\equiv 0$ *and* $\lambda = \lambda_1$ *in* (7.1.11). *Then this BVP has no solution in* $W_0^{1,p}(\Omega)$.

Proof. Let $u \in W_0^{1,p}(\Omega)$ be a solution of (7.1.11) with $\lambda = \lambda_1$. First, similarly as in the proof of the maximum principle one can show that $u > 0$ in Ω and $\frac{\partial u}{\partial \nu} < 0$ on $\partial\Omega$. Let φ be the (positive) eigenfunction associated with λ_1. Next, we apply again (7.1.16) for u and $t\varphi$, $t > 0$. We have

$$\int_\Omega \left(\frac{-\Delta_p u}{u^{p-1}} - \frac{-\Delta_p (t\varphi)}{(t\varphi)^{p-1}} \right) (u^p - (t\varphi)^p)\, dx \ge 0,$$

and letting $t \to \infty$ we arrive at the inequality $\int_\Omega f\varphi / \Phi(u)\, dx \le 0$, but this is a contradiction since $f \ge 0$ and $f \not\equiv 0$. □

Now pass to the antimaximum principle for p-Laplacian. This principle concerns the situation when the spectral parameter λ is such that $\lambda > \lambda_1$ but it is close enough to λ_1.

Theorem 7.1.6. *Let $\Omega \subset \mathbb{R}^N$ be a bounded domain with a $C^{1,\alpha}$ boundary, for some $\alpha \in (0,1)$. Assume that $f \in L^\infty(\Omega)$ $f \geq 0$, $f \not\equiv 0$ in Ω. Then there exists a constant $\delta = \delta(f)$ with the following property: if $\lambda \in (\lambda_1, \lambda_1 + \delta)$ and $u \in W_0^{1,p}(\Omega)$ is a weak solution of (7.1.11), then $u \in C^{1,\beta}(\overline{\Omega})$ for some $\beta > 0$ and the antimaximum principle holds:*

$$u < 0 \text{ in } \Omega \quad \text{and} \quad \frac{\partial u}{\partial \nu} > 0 \text{ on } \partial\Omega. \tag{7.1.17}$$

Proof. Suppose, by contradiction, that there is no constant $\delta > 0$ with the claimed property. Then there exists a sequence $\{\alpha_k\} \subset (\lambda_1, \infty)$ with $\alpha_k \to \lambda_1$ such that (7.1.11) with $\lambda = \alpha_k$ has a (weak) solution $u_k \in W_0^{1,p}(\Omega) \cap L^\infty(\Omega)$ which does not satisfy inequalities (7.1.17). This means

$$-\Delta_p u_k = \alpha_k \Phi(u_k) + f(x), \ x \in \Omega, \quad u_k = 0, \ x \in \partial\Omega. \tag{7.1.18}$$

We claim that

$$\|u_k\|_\infty \to \infty \quad \text{as} \quad k \to \infty, \tag{7.1.19}$$

where $\|\cdot\|_\infty$ is L^∞ norm. Suppose not, then there is a subsequence of $\{u_k\}$, denoted again $\{u_k\}$, which is bounded in $L^\infty(\Omega)$. A regularity result of Lieberman [256, Theorem 1] implies that $\{u_k\}$ is bounded in $C^{1,\beta}(\overline{\Omega})$ for some $\beta \in (0,1)$. Moreover, by the Ascoli-Arzelá Theorem, $\{u_k\}$ is relatively compact in $C^{1,\beta^*}(\overline{\Omega})$ for some $\beta^* \in (0,\beta)$. Thus, we may extract a convergent subsequence $u_{n_k} \to u^*$ in $C^{1,\beta^*}(\overline{\Omega})$ as $k \to \infty$. Letting $k \to \infty$ in the weak formulation of (7.1.18), we arrive at

$$\int_\Omega \|\nabla u^*\|^{p-2} \langle \nabla u^*, \nabla w \rangle \, dx = \lambda_1 \int_\Omega \Phi(u^*) w \, dx + \int_\Omega f w \, dx$$

for all $w \in W_0^{1,p}(\Omega)$. So $u^* \in C^{1,\beta^*}(\overline{\Omega})$ is a weak solution of (7.1.11) with $\lambda = \lambda_1$, a contradiction to Corollary 7.1.2, which proves (7.1.19).

Now set $v_k = u_k / \|u_k\|_\infty$, i.e., $\|v_k\|_\infty = 1$. Thus, BVP (7.1.18) becomes

$$-\Delta_p v_k = \alpha_k \Phi(v_k) + \frac{f(x)}{\|u_k\|_\infty^{p-1}}, \ x \in \Omega, \quad v_k = 0, \ x \in \partial\Omega. \tag{7.1.20}$$

Since $\{v_k\}$ is relatively compact in $C^{1,\beta^*}(\overline{\Omega})$, let us extract a convergent subsequence $v_{n_k} \to v^*$ in $C^{1,\beta^*}(\overline{\Omega})$. Again, letting $k \to \infty$ in the weak formulation of (7.1.20), we arrive at

$$\int_\Omega \|\nabla v^*\|^{p-2} \langle \nabla v^*, \nabla w \rangle \, dx = \lambda_1 \int_\Omega \Phi(v^*) w \, dx + \int_\Omega f w \, dx$$

for all $w \in W_0^{1,p}(\Omega)$. We conclude that $v^* \in C^{1,\beta^*}(\overline{\Omega})$ is an eigenfunction of (7.1.2) with $\|v^*\|_\infty = 1$, hence $v^* = k\varphi$ for some nonzero $\gamma \in \mathbb{R}$. We distinguish the cases $\gamma > 0$ and $\gamma < 0$.

(I) Case $\gamma > 0$. Then there exists an integer k_0 such that each v_{n_k}, for $k \geq k_0$, satisfies the strong maximum principle of Theorem 7.1.4, i.e.,

$$v_{n_k} > 0, \ x \in \Omega \quad \text{and} \quad \frac{\partial v_{n_k}}{\partial \nu} < 0, \ x \in \partial\Omega. \tag{7.1.21}$$

We rewrite (7.1.20) as follows
(7.1.22)

$$-\Delta_p v_{n_k} = \lambda_1\Phi(v_{n_k}) + (\alpha_{n_k} - \lambda_1)\Phi(v_{n_k}) + \frac{h(x)}{\|u_{n_k}\|_\infty^{p-1}},\ x \in \Omega, \quad v_{n_k} = 0,\ x \in \partial\Omega,$$

for $k \geq k_0$. Since

$$f_{n_k}(x) := (\alpha_{n_k} - \lambda_1)\Phi(v_{n_k}) + \frac{h(x)}{\|u_{n_k}\|_\infty^{p-1}} \geq 0 \quad \text{in } \Omega,$$

the nonexistence result of Corollary 7.1.2 applies to (7.1.22) and gives the required contradiction.

(II) Case $\gamma < 0$. For large k, the function $-v_{n_k}$ satisfies (7.1.21), but this contradicts our assumption that $u_{n_k} = \|u_{n_k}\|_\infty v_{n_k}$ does not satisfy inequality (7.1.17). □

Remark 7.1.1. A similar statement as presented in the previous section can be formulated also for "Fučík type" equation

$$-\Delta_p = \alpha\Phi(u^+) - \beta(u^-) + f(x),\ x \in \Omega, \quad u = 0,\ x \in \partial\Omega,$$

we refer to [162] for details.

7.1.4 Fučík spectrum for p-Laplacian

The situation with the Fučík spectrum of the p-Laplacian in higher dimension is more complicated in comparison with the one-dimensional p-Laplacian. To show this difference, we briefly present the main results of the paper [81]. Recall that we look for pairs (α, β) for which there exists a nontrivial solution of the BVP

$$-\Delta_p u = \alpha(u^+)^{p-1} - \beta(u^-)^{p-1},\ x \in \Omega, \quad u = 0,\ x \in \partial\Omega. \tag{7.1.23}$$

Here, again, $u^+ = (|u| + u)/2$, $u^- = (|u| - u)/2$.

Similarly as in the one-dimensional case, the Fučík spectrum Σ_p of the operator $-\Delta_p$ in $\Omega \subset \mathbb{R}^N$ consists of the trivial part $\lambda_1 \times \mathbb{R}$ and $\mathbb{R} \times \lambda_1$, where λ_1 is the first eigenvalue of $-\Delta_p$ in Ω. In this subsection we show the variational description of the first nontrivial curve of the Fučík spectrum and we also show the application of this result in the investigation of solvability of the BVP

$$-\Delta_p u = f(x, u),\ x \in \Omega, \quad u = 0,\ x \in \partial\Omega. \tag{7.1.24}$$

We start with the variational construction of the first nontrivial curve of Σ_p. Let $s \geq 0$ and consider the functional

$$J_s(u) = \int_\Omega \|\nabla u\|^p\,dx - s\int_\Omega (u^+)^p\,dx \tag{7.1.25}$$

over $W_0^{1,p}(\Omega)$ and its restriction $\tilde{J}_s$ over the manifold

$$S = \left\{u \in W_0^{1,p}(\Omega) : \int_\Omega |u|^p\,dx = 1\right\}.$$

Using the method of Lagrange multipliers, one can verify that the points of Σ_p on the line parallel to the axis of the first quadrant passing through the point $(0, s)$ are exactly of the form $(s + \tilde{J}_s(u), \tilde{J}_s(u))$, where u is a critical point of $\tilde{J}$.

The first critical point of $\tilde{J}$ is obtained by the minimization; this is the eigenfunction φ_1 corresponding to the first eigenvalue λ_1 of $-\Delta_p$ in Ω and $\tilde{J}_s(\varphi_1) = \lambda_1 - s$. The corresponding point of Σ_p is $(\lambda_1, \lambda_1 - s)$ which is a point of the trivial part $\lambda_1 \times \mathbb{R}$. One can also show that $-\varphi_1$ is a strict local minimum of $\tilde{J}_s$ with $\tilde{J}_s(-\varphi_1) = \lambda_1$. The corresponding point of Σ_p is $(\lambda_1 + s, \lambda_1)$ which is a point of $\mathbb{R} \times \lambda_1$.

To find a third critical point, we use the Mountain Pass Theorem combined with the following result.

Theorem 7.1.7. *Let*

$$\Gamma := \{\gamma \in C([-1,1], S) : \gamma(-1) = -\varphi_1, \ \gamma(1) = \varphi_1\}$$

and let

$$c(s) = \inf_{\gamma \in \Gamma} \max_{u \in \gamma[-1,1]} \tilde{J}(u). \tag{7.1.26}$$

Then $c(s)$ is the critical value of $\tilde{J}_s$ with $c(s) > \lambda_1$.

The idea of the proof of the main results of this subsection is based on the statements of the following two auxiliary results (which we present without proofs, we refer to [81] for details).

Lemma 7.1.2. *The straight lines of the trivial part of Σ_p $\lambda_1 \times \mathbb{R}$ and $\mathbb{R} \times \lambda_1$ are isolated in Σ_p.*

Lemma 7.1.3. *Let $r \in \mathbb{R}$ and denote $\mathcal{O} := \{u \in S : \tilde{J}_s(u) < r\}$. Then every nonempty connected component of $\mathcal{O}$ contains a critical point of $\tilde{J}_s$.*

The point $(s + c(s), c(s))$ obviously does not belong to the trivial part of Σ_p. Continuing in this way for every $s \geq 0$, and then taking the points symmetric with respect to the diagonal of the first quadrant, we obtain *the first nontrivial part $\mathcal{C}$* of Σ_p. This fact is summarized in the next theorem.

Theorem 7.1.8. *Let $s \geq 0$. The point $(s + c(s), c(s))$ is the first nontrivial point of Σ_p on the line parallel to the diagonal passing through the point $(s, 0)$. In particular, if $s = 0$, we have*

$$\lambda_2 = \inf_{\gamma \in \Gamma} \max_{u \in \gamma[-1,1]} \int_\Omega \|\nabla u\|^p \, dx. \tag{7.1.27}$$

Proof. Suppose, by contradiction, the existence of a point of the form $(s + \mu, \mu) \in \Sigma_p$ such that $\lambda_1 < \mu < c(s)$. By Lemma 7.1.2 one can suppose that $\tilde{J}_s$ has no critical value in the interval $[\lambda_1, \mu]$. To reach a contradiction with the definition of $c(s)$, we construct a path $\gamma \in \Gamma$ on which $\tilde{J}_s$ attains values less than or equal to μ what yields the required contradiction. This construction can be briefly described as follows. Let $u \in S$ be a solution of (7.1.23) corresponding to $\alpha = s + \mu$, $\beta = \mu$. Necessarily, u changes its sign in Ω. Define the path

$$u_1(t) = \frac{(1-t)u + tu^+}{\|(1-t)u + tu^+\|_p}, \qquad \|\cdot\|_p := \|\cdot\|_{W^{1,p}},$$

which starts at u and goes to $u^+/\|u^+\|_p$. Then, the construction consists of the functions $u_2(t)$, $u_3(t)$ which are defined analogously; u_2 starts at $u^+/\|u^+\|_p$ and ends at $u^-/\|u^-\|_p$, u_3 goes from $-u^-/\|u^-\|_p$ to u. A simple calculation, using (7.1.23), then shows that $\tilde{J}_s$ attains values less than or equal to μ on these curves. We have $\tilde{J}_s(u^-/\|u^-\|_p) = \mu - s$. Using the fact that $u^-/\|u^-\|_p$ is not a critical point and applying Lemma 7.1.3 with $r = \mu - s$, one can construct a path $u_4(t)$ going from $u^-/\|u^-\|_p$ to φ_1 or $-\varphi_1$ and the value of $\tilde{J}_s$ on this curve is less than or equal to $\mu - s$. To explain the idea, suppose that the path goes to φ_1. Since $|\tilde{J}_s(v) - \tilde{J}_s(-v)| \le s$ for every $v \in S$, the value of $\tilde{J}_s$ over the path $-u_4$ remains less than or equal to $(\mu - s) + s = \mu$ and allows to continue from $-\varphi_1$ to $-u^-/\|u^-\|_p$. Drawing a diagram of these curves, one can see that they can be glued in such a way that the resulting path starts at $-\varphi_1$, ends at φ_1 and the value of $\tilde{J}$ along this curve remains less than or equal to μ. □

Note that the characterization of the second eigenvalue (7.1.27) is slightly different than that presented in Subsection 7.1.2 which is based on the Lusternik-Schnirelmann procedure.

Now we present a statement showing that the first nontrivial curve $\mathcal{C}$ is decreasing and asymptotically approaches the trivial part of Σ_p.

Theorem 7.1.9. *The curve $\mathcal{C}$ is decreasing in the sense that $0 \le s < s'$ implies $s + c(s) < s' + c(s')$ and $c(s) > c(s')$. Moreover, if $p > N$ and there exists a point of the boundary $\partial\Omega$ in whose neighborhood $\partial\Omega$ is regular, then $c(s) \to \lambda_1$ as $s \to \infty$.*

To apply the previous description of the curve $\mathcal{C}$, consider BVP (7.1.24) and denote

$$\begin{aligned} \gamma_\pm &:= \liminf_{s\to\pm\infty} \frac{f(x,s)}{\Phi(s)}, \quad \Gamma_\pm := \liminf_{s\to\pm\infty} \frac{f(x,s)}{\Phi(s)}, \\ \delta_\pm &:= \liminf_{s\to\pm\infty} \frac{pF(x,s)}{\Phi(s)}, \quad \Delta_\pm := \liminf_{s\to\pm\infty} \frac{pF(x,s)}{\Phi(s)}, \end{aligned}$$

where $F(x,s) = \int_0^s f(x,t)\,dt$ and the limits are supposed to be uniform with respect to x.

Theorem 7.1.10. *Let $(\alpha, \beta) \in \mathcal{C}$ and suppose*

(i) $\lambda_1 \le \gamma_+(x) \le \Gamma(x) \le \alpha$, $\lambda_1 \le \gamma_-(x) \le \Gamma_-(x) \le \beta$ a.e. in Ω;

(ii) $\delta_-(x) > \lambda_1$, $\delta_+ > \lambda_1$ on a subset of Ω of positive measure;

(iii) $\Delta_+(x) < \alpha$ or $\Delta_-(x) < \beta$ a.e. in Ω.

Then BVP (7.1.24) has at least one (weak) solution in $W_0^{1,p}(\Omega)$.

We skip the proof of the previous theorem, it follows a similar idea as that of Theorem 6.1.6 in the previous chapter. The significant role is played by the following Sturmian type statement.

Lemma 7.1.4. *Let $(\alpha, \beta) \in \mathcal{C}$ and let the functions $a, b \in L^\infty(\Omega)$ satisfy*

(i) $\lambda_1 \le a(x) \le \alpha$, $\lambda_1 \le b(x) \le \beta$ *a.e. in* Ω*;*

(ii) $a(x) > \lambda_1$ *and* $b(x) > \lambda_1$ *on a subset of* Ω *of positive measure;*

(iii) $a(x) < \alpha$ *or* $\beta(x) < \beta$ *a.e. in* Ω.

Then the BVP

$$-\Delta_p u = a(x)(u^+)^{p-1} - b(x)(u^-)^{p-1}, \ x \in \Omega, \quad u = 0, \ x \in \partial\Omega$$

has a solution.

Remark 7.1.2. Combining the idea used in the above described variational construction of the first nontrivial curve of the Fučík spectrum of $-\Delta_p$ with the construction of higher variational eigenvalues given in Subsection 7.2.3, one can construct "higher variational curves" of the Fučík spectrum, see [311], where also the application of that variational result is used to study solvability of (7.1.24) with nonlinearity g "near" the variational part of the Fučík spectrum.

7.2 Boundary value problems at resonance

This section is in a certain sense higher-dimensional analogue of some parts of Subsections 6.2 and 6.3. In particular, we show how the Fredholm alternative and Landesman-Lazer solvability conditions extend to partial differential equations with p-Laplacian.

7.2.1 Resonance at the first eigenvalue in higher dimension

Consider the boundary value problem at resonance

$$-\Delta_p u = \lambda_1 \Phi(u) + f(x), \ x \in \Omega, \quad u = 0, \ x \in \partial\Omega, \tag{7.2.1}$$

where $f \in L^\infty(\Omega)$, λ_1 is the first eigenvalue of $-\Delta_p$ in Ω. We denote by φ_1 the associated eigenfunction. Similarly as in the one-dimensional case, we will suppose the orthogonality condition

$$\langle f, \varphi_1 \rangle := \int_\Omega f(x)\varphi_1(x)\, dx = 0. \tag{7.2.2}$$

In this subsection, we will not give proofs of the statements, we rather present *ideas* used in these proofs. These ideas are in principle different in case $p > 2$ and in the case $p \in (1, 2)$. The former case is usually referred to as the *degenerate* case, while the case $p \in (1, 2)$ is called *singular* case. However, unifying point in both cases is the well-known fact that (weak) solutions of (7.2.2) are critical points of the energy functional

$$\mathcal{J}_\lambda(u) := \frac{1}{p}\int_\Omega \|\nabla u\|^p\,dx - \frac{\lambda}{p}\int_\Omega |u|^p\,dx - \int_\Omega f(x)u\,dx. \tag{7.2.3}$$

If $p > 2$, it is verified that the set of global minimizers u_λ of $\mathcal{J}_\lambda$, parametrized by $\lambda \in [0, \lambda_1)$, contains a sequence that converges to a solution of (7.2.1) in $C^1(\overline{\Omega})$ as $\lambda \to \lambda_1$. On the other hand, if $p \in (1, 2)$, the existence of a solution of (7.2.1) is obtained from a minimax principle performed in the orthogonal decomposition

$$W_0^{1,p}(\Omega) = \mathrm{Lin}\{\varphi_1\} + [\mathrm{Lin}\{\varphi_1\}]^\perp, \quad u = \tau\varphi_1 + u^\perp,$$

relative to the scalar product (7.2.2). Note that the case $p > 2$ is more difficult; the proof is based on the "quadratization" of the functional $\mathcal{J}_\lambda$ (i.e., a second-order Taylor formula) near the principal eigenfunction φ_1. Note also that this quadratization procedure, as pointed out in [342], can be regarded, in a certain sense, as a higher dimensional counterpart of the Prüfer transformation which plays the crucial role in the proofs of many results for one-dimensional p-Laplacian. Again, we refer to the above mentioned paper of Takáč [342] for details.

Now we turn our attention to the formulation the main results of this subsection.

(I) *Degenerate case* $p > 2$. First we formulate technical assumptions on the domain Ω. In the degenerate case, we suppose the following two hypotheses.

Hypothesis (H1). If $N \geq 2$, then Ω is a bounded domain whose boundary $\partial\Omega$ is a compact manifold of the class $C^{1,\alpha}$ for some $\alpha \in (0, 1)$ and Ω also satisfies the interior sphere condition at every point of $\partial\Omega$. If $N = 1$, then Ω is simply an open interval in $\mathbb{R}$.

Note that for $N \geq 2$ the hypothesis (H1) is satisfied provided $\Omega \subset \mathbb{R}^N$ is a bounded domain with C^2 boundary. In addition to properties of the first eigenvalue λ_1 and the associated eigenfunction φ_1 established in Subsection 7.1.1, let us mention that under the hypothesis (H1) there exists a constant $\beta \in (0, \alpha)$ such that $\varphi_1 \in C^{1,\beta}(\bar{\Omega})$, see [351] and [256]. Throughout this and the next subsections the constants α, β have just this meaning.

To introduce the second (rather technical) hypothesis, we use the following notation. By $\mathcal{D}_{\varphi_1}$ we denote the completion of $W_0^{1,p}(\Omega)$ with respect to the norm

$$\|v\|_{\varphi_1} := \left(\int_\Omega \|\nabla\varphi_1\|^{p-2}\|\nabla v\|^2\,dx\right)^{1/2}$$

and by $\mathcal{Q}_0$ we denote the quadratic form

$$\begin{aligned} 2\mathcal{Q}_0(\phi) \;&:=\; \int_\Omega \|\nabla\varphi_1\|^{p-2}\left\{\|\phi\|^2 + (p-2)\left|\frac{\langle\nabla\varphi_1, \nabla\phi\rangle}{\|\nabla\phi\|}\right|^2\right\} dx \\ &\quad -\lambda_1(p-1)\int_\Omega \varphi_1^{p-2}\phi^2\,dx, \end{aligned}$$

and finally let

$$U := \{x \in \Omega : \nabla\varphi_1(x) \neq 0\}. \tag{7.2.4}$$

Hypothesis (H2). If $N \geq 2$ and $\partial\Omega$ is not connected, there is no function $v \in \mathcal{D}_{\varphi_1}$, $\mathcal{Q}_0(v) = 0$, with the following four properties:

(i) $v = \varphi_1\chi_S$ a.e. in Ω, where $S \subset \Omega$ is a set with $0 < m_N(S) < m_N(\Omega)$, where $m_N(\cdot)$ and χ are the N-dimensional Lebesgue measure and the characteristic function of the set indicated, respectively;

(ii) $\bar{S}$ is connected and $\bar{S} \cap \partial\Omega \neq \emptyset$;

(iii) every connected component of the set U is entirely contained either in S or else in $\Omega \setminus S$;

(vi) $(\partial S) \cap \Omega \subset \Omega \setminus U$.

Note that it is conjectured in [342] that (H2) holds provided (H1) does.

The first result concerns *a-priori* uniform boundedness of solutions of (7.2.1) for $\lambda \in [0, \lambda_1]$.

Theorem 7.2.1. *Let $\mathcal{K}$ be a nonempty, $*$-weakly compact set in $L^\infty(\Omega)$ such that $0 \neq \mathcal{K}$ and (7.2.2) holds for all $f \in \mathcal{K}$. Then there exists a constant $C = C(\mathcal{K})$ with the following property: If $\lambda \in [0, \lambda_1]$, $f \in \mathcal{K}$, and if $u \in W_0^{1,p}(\Omega)$ is a critical point of $\mathcal{J}_\lambda$, i.e., a weak solution of (7.2.1), then $\|u\|_{C^{1,\beta}} \leq C$.*

The previous statement is used in the main result of this subsection concerning the degenerate case.

Theorem 7.2.2. *Let $f \in L^\infty(\Omega)$ satisfies the orthogonality condition (7.2.2). Then (7.2.1) possesses a weak solution $u \in W_0^{1,p}(\Omega)$. Moreover, if $f \not\equiv 0$ in Ω, the set of weak solutions of (7.2.1) is bounded in $C^{1,\beta}(\overline{\Omega})$.*

Note that the boundedness result of the previous theorem is uniform for $f \in \mathcal{K}$ (compare with the analogical statement for the one-dimensional p-Laplacian). The following result complements Theorem 7.2.1 for $\mathcal{K} \cap [\text{Lin}\{\varphi_1\}]^\perp = \emptyset$.

Theorem 7.2.3. *Let $\mathcal{K}$ be a nonempty $*$-weakly compact set in $L^\infty(\Omega)$ that satisfies $\langle f, \varphi_1\rangle \neq 0$ for every $f \in \mathcal{K}$. Then there exists a constant $C = C(\mathcal{K}) > 0$ with the following property: If $f \in \mathcal{K}$ and if $u \in W_0^{1,p}(\Omega)$ is a weak solution of (7.2.1), then $\|u\|_{C^{1,\beta}} \leq C$.*

The next corollary to Theorem 7.2.3 shows that (7.2.1) may have no solution if the orthogonality condition (7.2.2) is violated.

Corollary 7.2.1. *Given an arbitrary $g \in L^\infty(\Omega)$ with $0 \leq g \not\equiv 0$ in Ω, there exists a constant $\gamma = \gamma(g) > 0$ with the following property: If $f \in L^\infty(\Omega)$, $f \not\equiv 0$, is such that*

$$f = \mu g + \tilde{f}, \quad \text{with some } \mu \in \mathbb{R} \text{ and } \tilde{f} \in L^\infty(\Omega),$$

and $\|\tilde{f}\|_{L^\infty} \leq \gamma|\mu|$, then (7.2.1) has no weak solution $u \in W_0^{1,p}(\Omega)$.

(II) *Singular case* $1 < p < 2$. In the statements of this part of subsection we suppose only the hypothesis (H1). We need to redefine the space $\mathcal{D}_{\varphi_1}$ as follows. We define $v \in \mathcal{D}_{\varphi_1}$ if and only if $v \in W_0^{1,p}(\Omega)$, $\nabla v(x) = 0$ a.e. in $\Omega \setminus U$, where U is defined in (7.2.4), and

$$\|v\|_{\varphi_1} := \left(\int_U \|\nabla \varphi_1\|^{p-2} \|\nabla v\|^2 \, dx \right)^{1/2} < \infty.$$

We refer to [342] for some comments concerning this definition of $\mathcal{D}_{\varphi_1}$.

Theorem 7.2.4. *Let $f \in L^\infty(\Omega)$ satisfy $f \notin [\mathcal{D}_{\varphi_1}]^\perp$ and $\langle f, \varphi_1 \rangle = 0$. Then problem (7.2.1) possesses a weak solution $u \in W_0^{1,p}(\Omega)$. Moreover, if $\mathcal{K}$ is a nonempty $*$-weakly compact set in $L^\infty(\Omega)$ such that $\mathcal{K} \cap [\mathcal{D}_{\varphi_1}]^\perp = \emptyset$ and $\langle f, \varphi_1 \rangle = 0$ for all $f \in \mathcal{K}$, then there exists a constant $C = C(\mathcal{K}) > 0$ with the property: If $f \in \mathcal{K}$ and $u \in W_0^{1,p}(\Omega)$ is a weak solution of (7.2.1), then $\|u\|_{C^{1,\beta}} \leq C$.*

Theorem 7.2.5. *Let $f \in L^\infty(\Omega)$ satisfy $f \notin [\mathcal{D}_{\varphi_1}]^\perp$ and $\langle f, \varphi_1 \rangle = 0$. Then there exists a constant $\lambda_0 = \lambda_0(\mathcal{K}) \in (0, \lambda_1)$ such that for $\lambda_0 \leq \lambda < \lambda_1$ and $f \in \mathcal{K}$, BVP (7.2.1) with λ instead of λ_1 has at least three (pairwise distinct) weak solutions u_1, u_2, u_3 specified as follows:*

(i) *The energy functional (7.2.3) possesses a critical point $u_1 \in W_0^{1,p}(\Omega)$ (hence a weak solution of the BVP under consideration) such that u_1 is not a strict local minimizer and satisfies $\|u_1\|_{C^{1,\beta}} \leq C$, where $C = C(\mathcal{K})$ is a constant independent of both $\lambda \in [\lambda_0, \lambda_1)$ and $f \in \mathcal{K}$.*

(ii) *The functional (7.2.3) possesses two (distinct) local minimizers $u_2, u_3 \in W_0^{1,p}(\Omega)$, at least one of which is global.*

In analogy with Theorem 7.2.3 for $p > 2$, if f stays away from the subspace $[\mathrm{Lin}\{\varphi_1\}]^\perp$, one can show the following uniform boundedness result and its corollary.

Theorem 7.2.6. *Let $\mathcal{K}$ be a nonempty $*$-weakly compact set in $L^\infty(\Omega)$ that satisfies $\langle f, \varphi_1 \rangle \neq 0$ for every $f \in \mathcal{K}$. Then the conclusion of Theorem 7.2.3 is valid also for $p \in (1, 2)$.*

Corollary 7.2.2. *Let $g \in L^\infty(\Omega)$ be an arbitrary function with $0 \leq g \not\equiv 0$ in Ω. Then the conclusion of Corollary 7.2.1 remains valid.*

7.2.2 Resonance at the first eigenvalue – multiplicity results

The results of this subsection can be understood, in a certain sense, as a multidimensional analogue of results given in Subsection 6.3.1. We adopt the same notation and hypotheses (H1), (H2) as in the previous subsection. Also, the principal idea of the proof is similar and consists in quadratization of the energy functional $\mathcal{J}$ given by (7.2.3). In contrast to the previous section, we formulate the results simultaneously for the degenerate case $p > 2$ and the singular case $p \in (1, 2)$.

Let $f \in W_0^{1,p}(\Omega)$, we express f as the orthogonal sum

$$f = \zeta \varphi_1 + f^T, \quad \text{where } \zeta := \frac{\langle f, \varphi_1 \rangle}{\langle \varphi_1, \varphi_1 \rangle} \quad \text{and } \langle f^T, \varphi_1 \rangle = 0.$$

In the singular case, instead of (H2) (which holds in this case automatically), we suppose the following restriction on the function f.

Hypothesis (H3): $f^T \in L^\infty(\Omega)$ satisfies $f^T \notin [\mathcal{D}_{\varphi_1}]^\perp$.

For the formulation convenience, we will consider our BVP in the form

$$-\Delta_p u = \lambda_1 \Phi(u) + f^T(x) + \zeta\varphi_1(x), \ x \in \Omega, \quad u = 0, \ x \in \partial\Omega. \tag{7.2.5}$$

Theorem 7.2.7. *Let $f^T \in L^\infty(\Omega)$ satisfy $f^T \not\equiv 0$ and $\langle f^T, \varphi_1\rangle = 0$. Then there exist two constants $\xi_* < 0 < \xi^*$ such that BVP (7.2.5) has at least one weak solution $u \in W_0^{1,p}(\Omega)$ if and only if $\zeta_* \le \zeta \le \zeta^*$. Moreover, there are two additional constants $\zeta_\sharp, \zeta^\sharp \in \mathbb{R}$, such that (7.2.5) possesses at least two distinct weak solutions provided $\zeta_\sharp < \zeta < \zeta^\sharp$.*

Remark 7.2.1. In contrast to the one-dimensional case treated in Subsection 6.3.1, it is an open problem whether $\zeta_\sharp = \zeta_*$ and $\zeta^* = \zeta^\sharp$.

Theorem 7.2.8. *Let f^T be the same as in the previous theorem. If $\zeta = 0$ in (7.2.5), then the set of all weak solutions to this BVP is bounded in $C^{1,\beta}(\Omega)$. If $\delta > 0$ is given, then the set of all weak solutions of (7.2.5) is bounded in $C^{1,\beta}(\bar{\Omega})$ uniformly for $|\zeta| \ge \delta$.*

7.2.3 Landesman-Lazer result in higher dimension

In this subsection we present the extension of the results given in Subsection 6.3.2. We consider the boundary value problem

$$-\Delta_p u - \lambda\Phi(u) + g(x,u) = f(x), \ x \in \Omega, \quad u = 0, \ x \in \partial\Omega, \tag{7.2.6}$$

where $\Omega \subset \mathbb{R}^N$ is a bounded domain whose boundary $\partial\Omega$ (in case $N \ge 2$) is a compact connected manifold of the class C^2.

As we have already mentioned before, in contrast to the one-dimensional case $N = 1$, the structure of all eigenvalues of (7.1.2) is not known. On the other hand, there are several possibilities how to find a sequence of the so-called *variational eigenvalues* which tend to infinity. Here we present the method used in [133] which enables to find one such sequence and also to give a variational proof of the existence of at least one solution of (7.2.6) if g satisfies some additional conditions (of Landesman-Lazer type) and λ is any (even non-variational) eigenvalue of (7.1.2).

Consider the functional

$$I(u) = \frac{\int_\Omega \|\nabla u\|^p}{\int_\Omega |u|^p}$$

for $u \in W_0^{1,p}(\Omega) \setminus \{0\}$, and the manifold

$$\mathcal{S} := \{u \in W_0^{1,p}(\Omega) \colon \|u\|_{L^p} = 1\}.$$

It is a matter of straightforward computation to verify that the eigenvalues and eigenfunctions of $-\Delta_p$ correspond to the critical values and critical points of $I|_{\mathcal{S}}$, respectively.

For any $k \in \mathbb{N}$, denote

$$\mathcal{F}_k\colon = \{\mathcal{A} \subset \mathcal{S}\colon \exists \text{ a continuous odd surjection } h\colon \mathcal{S}^{k-1} \to \mathcal{A}\},$$

where $\mathcal{S}^{k-1}$ represents the unit sphere in $\mathbb{R}^k$. Next define

$$\lambda_k\colon = \inf_{\mathcal{A}\in\mathcal{F}_k} \sup_{u\in\mathcal{A}} I(u).$$

It is proved in [133] that $\lambda_k, k = 1, 2, \dots,$ are the critical values of $I|_{\mathcal{S}}$, and hence the eigenvalues of $-\Delta_p$.

Let $\{\mu_k\}$ be the eigenvalues defined by the Lusternik–Schnirelmann characterization (see e.g. [168]) involving a minimax over sets of genus greater than k. Then $\lambda_1 = \mu_1, \lambda_2 = \mu_2$ and $\lambda_k \geq \mu_k, k = 3, 4, \dots$. In particular, $\lambda_k \to \infty$ as $k \to \infty$. It is worth mentioning that whether equalities $\lambda_k = \mu_k, k = 3, 4, \dots,$ hold true is an open problem if $N \geq 2$. Also, as already pointed out, it is not clear if $\{\lambda_k\}_{k=1}^{\infty}$ form the complete set of eigenvalues of $-\Delta_p$ if $N \geq 2$. On the other hand both problems are solved positively for $N = 1$ as we have shown in Subsection 6.1.2.

In our further considerations we will use the standard spaces $W_0^{1,p}(\Omega)$, $L^p(\Omega)$, $C(\bar{\Omega})$ and $C^1(\bar{\Omega})$ (or $C_0^1(\bar{\Omega})$, respectively), with the corresponding norms

$$\begin{aligned} \|u\| &= \left(\int_\Omega \|\nabla u\|^p\, dx\right)^{\frac{1}{p}}, \quad \|u\|_{L^p} = \left(\int_\Omega |u|^p\, dx\right)^{1/p}, \\ \|u\|_C &= \max_{x\in\bar{\Omega}} |u(x)|, \quad \|u\|_{C^1} = \|u\|_C + \max_{x\in\bar{\Omega}} \|\nabla u(x)\|, \end{aligned}$$

respectively. The subscript 0 indicates that the traces (or values) of functions are equal zero on $\partial\Omega$. Moreover, for the element $h \in C(\bar{\Omega})$ we use the following (L^2-non-orthogonal) decomposition

$$h(x) = \tilde{h}(x) + \hat{h},$$

where $\hat{h} \in \mathbb{R}$ and

$$\int_\Omega \tilde{h}(x)\varphi_1(x)dx = 0.$$

The particular subspace formed by $\tilde{h}(x)$ will be denoted by $\tilde{C}(\bar{\Omega})$. By $B_C(\tilde{f}, \rho)$ we denote the open ball in the space $C(\bar{\Omega})$ with the center $\tilde{f}$ and radius ρ.

In this subsection we will assume that $g = g(x, s)$ is a continuous function in both variables, which is bounded and the limits

$$g^{\pm}(x)\colon = \lim_{s\to\pm\infty} g(x, s)$$

exist finite for all $x \in \bar{\Omega}$. The reader can have in mind e.g. the function

$$g(x, s) = \arctan s, \quad x \in \bar{\Omega}, s \in \mathbb{R},$$

for which $g^-(x) \equiv -\pi/2, \;\; g^+(x) \equiv \pi/2$.

Our main results concern the solvability of (7.2.6) and read as follows, the proofs can be found in [133] and [121].

Theorem 7.2.9. *Assume that λ is an eigenvalue of $-\Delta_p$ (variational or nonvariational) and either*

$$\int_{v(x)>0} g^+(x)v(x)\,dx + \int_{v(x)<0} g^-(x)v(x)\,dx > \int_\Omega f(x)v(x)\,dx, \tag{7.2.7}$$

or

$$\int_{v(x)>0} g^+(x)v(x)\,dx + \int_{v(x)<0} g^-(x)v(x)\,dx < \int_\Omega f(x)v(x)\,dx \tag{7.2.8}$$

hold for any nonzero eigenfunction v associated with the eigenvalue λ. Then (7.2.6) has at least one solution.

Note that according to the previous theorem the boundary value problem

$$-\Delta_p u - \lambda_1 \Phi(u) + \varepsilon \arctan u = f,\ x \in \Omega, \quad u = 0,\ x \in \partial\Omega, \tag{7.2.9}$$

has at least one solution if

$$-\varepsilon\frac{\pi}{2} < \frac{1}{\|\varphi_1\|_{L^1}} \int_\Omega f(x)\varphi_1(x)\,dx < \varepsilon\frac{\pi}{2}$$

The above inequalities do not make any sense if $\varepsilon = 0$, i.e., Theorem 7.2.9 does not cover the solvability of (7.2.6) with $\lambda = \lambda_1$ and $g \equiv 0$, i.e., of

$$-\Delta_p u - \lambda_1 \Phi(u) = f,\ x \in \Omega, \quad u = 0,\ x \in \partial\Omega. \tag{7.2.10}$$

However, we can apply the following result, which is, in a certain sense, similar to Theorem 7.2.7

Theorem 7.2.10. *Let $p \neq 2$ and $\tilde{f} \in \tilde{C}(\bar{\Omega})$. Then the problem (7.2.10) has at least one solution if $f = \tilde{f}$. For $0 \neq \tilde{f} \in \tilde{C}(\bar{\Omega})$ there exists $\rho = \rho(\tilde{f}) > 0$ such that (7.2.10) has at least one solution for any $f \in B_C(\tilde{f}, \rho)$. Moreover, there exist real numbers $F_- < 0 < F_+$ such that the problem (7.2.10) with $f = \tilde{f} + \hat{f}$ has*

(i) no solution for $\hat{f} \notin [F_-, F_+]$;

(ii) at least two distinct solutions for $\hat{f} \in (F_-, 0) \cup (0, F_+)$;

(iii) at least one solution for $\hat{f} \in \{F_-, 0, F_+\}$.

Remark 7.2.2. (i) Let us emphasize that Theorem 7.2.9 generalizes the classical result of Landesman and Lazer [232] and its proof can be found in [133]. However, we have to admit that both (7.2.7), (7.2.8) are sufficient conditions only and their necessity is an open problem even in the special case $g(x, s) = \arctan s$. One might get the impression that letting $\varepsilon \to 0$ in (7.2.9) we obtain that

$$\int_\Omega f(x)\varphi_1(x)\,dx = 0$$

is sufficient condition for the solvability of (7.2.10). Even if this is the case, it cannot be proved passing to the limit for $\varepsilon \to 0$ in (7.2.9) because of the lack of a-priori estimates of corresponding solutions.

(ii) Note that Theorem 7.2.9 provides necessary and sufficient condition for the solvability of the problem (7.2.10). This condition is in fact also of Landesman-Lazer type ([232], [133]). Indeed, given $\tilde{f} \in \tilde{C}(\bar{\Omega})$, $\tilde{f} \neq 0$, the problem (7.2.10) with the right-hand side $f(x) = \tilde{f}(x) + \hat{f}$ has a solution if and only if

$$F_-(\tilde{f}) \le \frac{1}{\|\varphi_1\|_{L^1}} \int_\Omega f(x)\varphi_1(x)\,dx \le F_+(\tilde{f}).$$

However, it should be pointed out that this condition differs from the original condition of Landesman and Lazer due to the fact that F_- and F_+ depend on the component $\tilde{f}$ of the right-hand side f (and not on the perturbation term $g = g(x,u)$). By homogeneity we have that for any $t > 0$,

$$F_\pm(t\tilde{f}) = tF_\pm(\tilde{f}).$$

(iii) The proof of Theorem 7.2.9 relies on the combination of the variational approach and the method of lower and upper solutions. One of the principal troubles is connected with the fact that usual apriori estimates and Palais-Smale condition fail. Let us also point out that the proof of Theorem 7.2.10 essentially uses the results obtained in [127], [342] and [165]. Finally note that the approach used in this proof is very different from that used to prove Theorem 7.2.9.

7.3 Oscillation theory of PDE's with p-Laplacian

In this section we briefly present main ideas of the oscillation theory of the partial differential equation

$$(7.3.1) \qquad \Delta_p u + c(x)\Phi(u) = 0, \quad \Delta_p u = \operatorname{div}(\|\nabla u\|^{p-2}\nabla u), \quad x \in \mathbb{R}^N,\ p > 1.$$

First, using the Picone identity, we show that Sturmian comparison theory extends to (7.3.1). Then we present criteria which guarantee nonexistence of positive solutions of (7.3.1) in the whole $\mathbb{R}^N$, we also give some oscillation criteria for this equation. The last subsection deals with partial differential equations with the so-called pseudolaplacian which is another partial differential operator which reduces to the operator of the form $(r(t)\Phi(x'))'$ in the scalar case $N = 1$.

Throughout this section we use the following notation:

$$\begin{aligned} \Omega_r &= \{x \in \mathbb{R}^N : \|x\| \ge r\}, \\ S_r &= \partial\Omega_r = \{x \in \mathbb{R}^N : \|x\| = r\}, \end{aligned}$$

ω_N is the area of the unit sphere in $\mathbb{R}^N$.

7.3.1 Picone's identity for equations with p-Laplacian

Consider a pair of partial differential operators with p-Laplacian

$$l[u] := \operatorname{div}(r(x)\|\nabla u\|^{p-2}\nabla u) + c(x)\Phi(u)$$

and

$$L[u] := \operatorname{div}(R(x)\|\nabla u\|^{p-2}\nabla u) + C(x)\Phi(u).$$

It is assumed that r, c, R, C are defined in some bounded domain $G \subset \mathbb{R}^N$ with piecewise smooth boundary ∂G and that $r, R \in C^1(\bar{G})$ are positive functions in $\bar{G}$, and $c, C \in C(\bar{G})$. The domain $\mathcal{D}_l(G)$ of l is defined to be the set of all functions of the class $C^1(\bar{G})$ with the property that $r\|\nabla u\|^{p-2}\nabla u \in C^1(G) \cap C(\bar{G})$. The domain $\mathcal{D}_L(G)$ of L is defined analogously.

The proof of the below given N-dimensional extension of Picone's identity is similar as in the scalar case, see [191].

Theorem 7.3.1. *Let $u \in \mathcal{D}_l(G)$, $v \in \mathcal{D}_L(G)$ and $v(x) \neq 0$ for $x \in G$. Then*

$$\begin{aligned}\operatorname{div}\left(\frac{u}{\Phi(v)}\left[\Phi(v)r(x)\|\nabla u\|^{p-2}\nabla u - \Phi(u)R(x)\|\nabla v\|^{p-2}v\right]\right)\\ = [r(x) - R(x)]\|\nabla u\|^p + [C(x) - c(x)]|u|^p\\ + R(x)\left[\|\nabla u\|^p + (p-1)\left\|\frac{u}{v}\nabla v\right\|^p - p\left\|\frac{u}{v}\nabla v\right\|^{p-2}(\nabla u)\left(\frac{u}{v}\nabla v\right)\right]\\ + \frac{u}{\Phi(v)}\left[\Phi(v)l[u] - \Phi(u)L[v]\right].\end{aligned}$$

Taking $r = R$, $c = C$ in the previous theorem, and using the fact that if v is a solution of $l[v] = 0$ for which $v(x) \neq 0$ in G, then the function $w = [r(x)\|\nabla v\|^{p-2}/\Phi(v)]\nabla v$ is a solution of the Riccati type partial differential equation

$$\operatorname{div} w + c(x) + (p-1)r^{1-q}(x)\|w\|^q = 0, \quad q = \frac{p}{p-1}, \tag{7.3.2}$$

we have Picone's identity in the special form

$$r(x)\|\nabla u\|^p - c(x)|u|^p = \operatorname{div}(w(x)|u|^p) + pr^{1-q}(x)\tilde{P}(r^{q-1}(x)\nabla u, w(x)\Phi(u)),$$

where

$$\tilde{P}(x, y) = \frac{\|x\|^p}{p} - \langle x, y\rangle + \frac{\|y\|^q}{q}.$$

As a consequence of Theorem 7.3.1 we have the following extension of the Leighton comparison theorem.The proof of this statement is again similar to the "ordinary" case, compare Subsection 2.3.2.

Theorem 7.3.2. *Suppose that the boundary ∂G is of the class C^1. If there exists a nontrivial solution $u \in \mathcal{D}_l(G)$ of $l[u] = 0$ such that $u = 0$ on ∂G and*

$$\int_G \{[R(x) - r(x)]\|\nabla u\|^p - [C(x) - c(x)]|u|^p\}\, dx \leq 0,$$

then every solution $v \in \mathcal{D}_L(G)$ of $L[v] = 0$ must vanish at some point of G, unless v is a constant multiple of u.

Another consequence of Picone's identity is the following Sturmian separation theorem.

Theorem 7.3.3. *Suppose that G is the same as in the previous theorem and there exists a nontrivial solution $u \in \mathcal{D}_l(G)$ of $l[u] = 0$ with $u = 0$ on ∂G. Then every solution v of $l[u] = 0$ must vanish at some point of G, unless v is a constant multiple of u.*

Remark 7.3.1. Recall that if we study properties of solutions of PDE's with p-Laplacian (7.3.1) in a radially symmetric domain $G = B_R = \{x \in \mathbb{R}^N : \|x\| \le R\}$ with a radially symmetric potential c, i.e., $c(x) = b(\|x\|)$ for some $b : [0, \infty) \to \mathbb{R}$, then one can look for solutions in the radial form $u(x) = v(r) = v(\|x\|)$ and v solves the ODE of the form (1.1.1)

$$\frac{d}{dr}\left[r^{N-1}\Phi\left(\frac{d}{dr}v\right)\right] + r^{N-1}b(r) = 0.$$

This method of the investigation of oscillatory properties of (7.3.1) has been used e.g. in [113, 219], see also the references given therein.

7.3.2 Nonexistence of positive solutions in $\mathbb{R}^N$

Concerning the linear differential equation

$$\Delta u + c(x)u = 0, \tag{7.3.3}$$

there is a voluminous literature dealing with oscillatory properties of (7.3.3). As for the classical results concerning oscillation of this equation, we refer to [341, Chapter 4], the paper [337], and the references given therein. The oscillation theory of (7.3.3) recognizes two types of oscillation. Equation (7.3.3) is said to be *weakly oscillatory*, if every solution has a zero outside of every ball in $\mathbb{R}^N$ and it is said to be *strongly* or *nodally oscillatory* if every solution has a nodal domain outside of any ball in $\mathbb{R}^N$. Recall that a bounded domain $D \subset \mathbb{R}^N$ is the *nodal domain* of a function u, if $u(x) = 0$ for $x \in \partial D$ and $u(x) \not\equiv 0$ in D. Moss and Piepenbrick [296] showed that both definitions are equivalent in the linear case if the function c is locally Hölder continuous. Equivalence of these definitions for (7.3.1) is an open problem.

We investigate here properties of (7.3.1) using essentially the following two methods:

(i) *Variational principle* – consisting in the relationship between the existence of a positive solution of (7.3.1) in a domain $\Omega \subset \mathbb{R}^N$ and positivity of the "p-degree" functional

$$\mathcal{F}_p(u; \Omega) := \int_\Omega \{\|\nabla u(x)\|^p - c(x)|u(x)|^p\}\, dx \tag{7.3.4}$$

over the class of functions satisfying $u|_{\partial\Omega} = 0$, i.e., over $W_0^{1,p}(\Omega)$.

(ii) *Riccati technique* – this method is based on the fact that if u is a nonzero solution to (7.3.1) then the vector function

$$w = \frac{\|\nabla u\|^{p-2}\nabla u}{\Phi(u)} \tag{7.3.5}$$

satisfies the Riccati type equation

$$\operatorname{div} w + c(x) + (p-1)\|w\|^q = 0, \tag{7.3.6}$$

where q is the conjugate number of p, i.e., $1/p + 1/q = 1$.

We start with two auxiliary statements based on the Riccati technique. Denote

$$Q(r) = \frac{1}{\omega_N}\int_{\|x\|<r} c(x)\, dx, \tag{7.3.7}$$

and for some $r_0 \geq 0$ denote

$$W(r) = \frac{p-1}{\omega_N}\int_{r_0<\|x\|<r} \|w\|^q\, dx, \tag{7.3.8}$$

where w is the solution of Riccati equation (7.3.6), defined on Ω_{r_0}.

Lemma 7.3.1. *Suppose that u is a solution of equation (7.3.1) and there exists a number r_0 such that $u(x) > 0$ on Ω_{r_0}. Let w be the corresponding solution of Riccati equation defined by (7.3.5). Denote by*

$$c_0 = \begin{cases} Q(r_0) + \frac{1}{\omega_N}\int_{S_{r_0}} \langle w, \nu\rangle dS & \text{if } r_0 > 0, \\ 0 & \text{if } r_0 = 0, \end{cases}$$

where ν in the surface integral is the unit outside normal vector to the sphere S_{r_0}. The following inequality holds for every $r \geq r_0$

$$Q(r) + W(r) \leq c_0 + r^{\frac{N-1}{p}}\left(\frac{1}{p-1}W'(r)\right)^{\frac{1}{q}}.$$

Proof. Let us compute $Q(r) + W(r)$. By (7.3.7) and (7.3.8) we have

$$Q(r) + W(r) = Q(r_0) + \frac{1}{\omega_N}\int_{r_0<\|x\|<r}\left(c(x) + (p-1)\|w\|^q\right) dx.$$

Using this, Riccati equation (7.3.2) and the Gauss theorem, we get

$$\begin{aligned} Q(r) + W(r) &= Q(r_0) + \frac{1}{\omega_N}\int_{S_{r_0}} \langle w, \nu\rangle dS - \frac{1}{\omega_N}\int_{S_r} \langle w, \nu\rangle dS \\ &= c_0 - \frac{1}{\omega_N}\int_{S_r} \langle w, \nu\rangle dS. \end{aligned}$$

The Schwarz and Hölder inequalities imply

$$\begin{aligned}
Q(r)+W(r) &\le c_0+\frac{1}{\omega_N}\int_{S_r}\|w\|\,\|\nu\|dS\\
&\le c_0+\frac{1}{\omega_N}\Big[\int_{S_r}\|w\|^q dS\Big]^{\frac{1}{q}}\Big[\int_{S_r}dS\Big]^{\frac{1}{p}}\\
&= c_0+\Big(\frac{1}{p-1}\Big)^{\frac{1}{q}}\Big[\frac{p-1}{\omega_N}\int_{S_r}\|w\|^q dS\Big]^{\frac{1}{q}}r^{\frac{N-1}{p}}\\
&= c_0+\Big(\frac{1}{p-1}W'(r)\Big)^{\frac{1}{q}}r^{\frac{N-1}{p}}
\end{aligned}$$

which completes the proof. □

Lemma 7.3.2. *Suppose that u is a solution of (7.3.1) which is positive in $\mathbb{R}^N$ and $Q(r)\ge 0$ for all $r\ge 0$. Further, suppose that there exist continuous nonnegative functions m, $\tilde m$, M, $\tilde M$ such that the inequalities*

$$\begin{aligned}
\tilde m(r) &\le (p-1)\int_0^r [Q(t)+m(t)]^q t^{(1-N)\frac{q}{p}}dt,\\
\tilde M(r) &\le \int_r^\infty [Q(t)M^{p-1}(t)+1]^q t^{(1-N)\frac{q}{p}}dt,\\
m(r) &\le W(r) \quad \text{and} \quad W(r)M^{p-1}(r)\le 1
\end{aligned}$$

hold for all $r>0$. Then

$$\tilde m(r)\le W(r) \quad \text{and} \quad W(r)\tilde M^{p-1}(r)\le 1 \quad \text{for all } r>0. \tag{7.3.9}$$

Proof. From the assumptions and from Lemma 7.3.1 with $r_0=0$ it follows that

$$\begin{aligned}
\tilde m(r) &\le (p-1)\int_0^r [Q(t)+m(t)]^q t^{(1-N)\frac{q}{p}}dt\\
&\le (p-1)\int_0^r [Q(t)+W(t)]^q t^{(1-N)\frac{q}{p}}dt\\
&\le \int_0^r W'(t)dt=W(r)
\end{aligned}$$

and the first inequality in (7.3.9) holds. Moreover, for every $R>r$

$$\begin{aligned}
&\int_r^R [Q(t)M^{p-1}(t)+1]^q t^{(1-N)\frac{q}{p}}dt\\
&\quad= \int_r^R [Q(t)M^{p-1}(t)W(t)+W(t)]^q W^{-q}(t)t^{(1-N)\frac{q}{p}}dt\\
&\quad\le (p-1)(q-1)\int_r^R [Q(t)+W(t)]^q W^{-q}(t)t^{(1-N)\frac{q}{p}}dt\\
&\quad\le (q-1)\int_r^R W'(t)W^{-q}(t)dt\le [W^{-q+1}]_R^r\le W^{-q+1}(r)
\end{aligned}$$

holds and the limit process $R \to \infty$ implies $\tilde{M}(t)W^{q-1}(t) \le 1$ which is equivalent to the second inequality in (7.3.9). □

Now we use the previous statements to prove results concerning nonexistence of positive solutions of (7.3.1) in the whole space $\mathbb{R}^N$. In the linear case ($p = 2$) statements of this kind are formulated either in terms of spectral properties of the Schrödinger operator defined by the left-hand side of (7.3.3) (more precisely, by the existence of negative eigenvalues of the associated differential operator, see [337]) or in terms of the existence of a nodal domain of a solution of (7.3.3) (like in [166, 297, 298]).

However, in the general case $p > 1$ we have no complete analogue of the spectral theory of linear differential operators (since the additivity of the solution space of (7.3.1) is lost and remains only homogeneity). Also, we miss a "systematic" oscillation theory of (7.3.1) leaned on the (modified) Courant-Hilbert variational principle consisting essentially in *equivalence* between oscillation of linear equations and positivity of associated quadratic functionals.

Theorem 7.3.4. *Let $c(x) \not\equiv 0$ and $p \ge N$. If*

$$\liminf_{r\to\infty} \int_{\|x\|<r} c(x)\,dx \ge 0, \tag{7.3.10}$$

then (7.3.1) *possesses no positive solution in* $\mathbb{R}^N$.

Proof. Suppose, by contradiction, that u is a solution of (7.3.1) positive on $\mathbb{R}^N$. We will use the functions $Q(r)$ and $W(r)$ defined by (7.3.7) and (7.3.8), where $r_0 = 0$. Due to the fact that $c \not\equiv 0$ we have $u \not\equiv$ const, hence $w \not\equiv 0$ and there exists a point $r_1 > 0$ such that $W(r_1) > 0$. Then, in view of (7.3.10), there exists r_2, $r_2 > r_1$, such that

$$Q(r) \ge -\frac{1}{2}W(r_1) \quad \text{for } r \ge r_2.$$

From here and from the fact that $W(r)$ is a nondecreasing function we obtain

$$\frac{1}{2}W(r) \le W(r) - \frac{1}{2}W(r_1) \le W(r) + Q(r)$$

for all $r \ge r_2$. Then using Lemma 7.3.1 we get

$$\left[\frac{1}{2}W(r)\right]^q \le \Big[Q(r) + W(r)\Big]^q \le r^{(N-1)\frac{q}{p}}\frac{W'(r)}{p-1}$$

and equivalently

$$\frac{p-1}{2^q} r^{(1-N)\frac{q}{p}} \le \frac{W'(r)}{W^q(r)}$$

for $r \ge r_2$. Integrating this inequality we obtain

$$\frac{p-1}{2^q}\int_{r_2}^{r} t^{(1-N)\frac{q}{p}}\,dt \le \int_{r_2}^{r} \frac{W'(t)}{W^q(t)}\,dt = \int_{W(r_2)}^{W(r)} \frac{dW}{W^q} \le \int_{W(r_2)}^{\infty} \frac{dW}{W^q}$$

for every $r \geq r_2$. If r tends to infinity, then the integral on the left-hand side diverges, since $p \geq N$ implies $q(1-N)/p \geq -1$ and the integral on the right-hand side converges. This contradiction completes the proof. □

In the next theorem we treat the case without the restriction $p \geq N$.

Theorem 7.3.5. *Let $Q(t) \geq 0$, $\{m_i\}_{i=1}^\infty$, $\{M_i\}_{i=1}^\infty$ be the sequences of nonnegative functions, continuous on $(0,\infty)$, satisfying*

$$(7.3.11) \qquad m_1 \equiv M_1 \equiv 0,$$

$$(7.3.12) \qquad m_{i+1}(r) \leq (p-1)\int_0^r [Q(t)+m_i(t)]^q t^{(1-N)\frac{q}{p}}\,dt,$$

$$(7.3.13) \qquad M_{i+1}(r) \leq \int_r^\infty [Q(t)M_i^{p-1}(t)+1]^q t^{(1-N)\frac{q}{p}}\,dt,$$

for every $r \geq 0$. If

$$(7.3.14) \qquad \sup_{i,j\in\mathbb{N}}\sup_{r>0} m_i(r)M_j^{p-1}(r) > 1,$$

then (7.3.1) possesses no positive solution in $\mathbb{R}^N$.

Proof. Suppose, by contradiction, that u is a solution of (7.3.1) which is positive in $\mathbb{R}^N$. By Lemma 7.3.2, it holds for every $i \in \mathbb{N}$

$$m_i(r) \leq W(r), \quad W(r)M_i^{p-1}(r) \leq 1.$$

Combining these results we have

$$m_i(r)M_j^{p-1}(r) \leq 1$$

for every $i, j \in \mathbb{N}$ and $r > 0$ which contradicts (7.3.14). □

Corollary 7.3.1. *Let $Q \geq 0$ and $p < N$. If*

$$\sup_{r>0} r^{p-N}\int_0^r Q^q(t)t^{\frac{1-N}{p-1}}\,dt > \frac{(N-p)^{p-1}}{(p-1)^p}$$

then (7.3.1) possesses no positive solution on $\mathbb{R}^N$.

Proof. Define the set of functions as follows:

$$m_i \equiv M_i \equiv 0 \quad \text{for } i \in \mathbb{N}\setminus\{2\},$$
$$m_2(r) = (p-1)\int_0^r Q^q(t)t^{(1-N)\frac{q}{p}}\,dt = (p-1)\int_0^r Q^q(t)t^{\frac{1-N}{p-1}}\,dt$$

and

$$M_2(r) = \int_r^\infty t^{(1-N)\frac{q}{p}}\,dt = -\frac{p-1}{p-N}r^{\frac{p-N}{p-1}}.$$

From here it follows

$$m_2(r)M_2^{p-1}(r) = \frac{(p-1)^p}{(N-p)^{p-1}}r^{p-N}\int_0^r Q^q(t)t^{\frac{1-N}{p-1}}\,dt.$$

Now Theorem 7.3.5 implies the conclusion. □

7.3.3 Oscillation criteria

In this subsection we turn our attention to criteria for nonexistence of positive solutions to (7.3.1) in the exterior domain Ω_{r_0} for $r_0 > 0$ arbitrarily large. The following theorem shows that the (deeply) developed oscillation theory of ordinary differential equation (1.1.1) can be used to study oscillations of (7.3.1) in the sense of weak oscillation.

Theorem 7.3.6. *Suppose that the half-linear ODE*

$$\left(r^{n-1}\Phi(z')\right)' + \frac{1}{\omega_N} C(r)\Phi(z) = 0, \quad ' = \frac{d}{dr} \tag{7.3.15}$$

is oscillatory, where

$$C(r) := \int_{S_r} c(x) dS.$$

Then equation (7.3.1) *has no positive solution in the exterior domain* Ω_r *for* $r > 0$ *arbitrarily large.*

Proof. Let $z = z(r)$ be an oscillatory solution of (7.3.15) and $r_n \to \infty$ be its zeros. Then integration by parts yields

$$\int_{r_{n-1}}^{r_n} \left[r^{n-1}|z'(r)|^p - \frac{1}{\omega_N} C(r)|z(r)|^p \right] dr = 0. \tag{7.3.16}$$

Now, for the function $y : \mathbb{R}^N \to \mathbb{R}$ defined by $y(x) = z(\|x\|)$ and $D_n = \{x : r_n \le \|x\| \le r_{n+1}\}$, we have by a direct computation

$$\begin{aligned}
\mathcal{F}_p(y, D_n) \quad &:= \quad \int_{D_n} \left(\|\nabla y(x)\|^p - c(x)|y(x)|^p \right) dx \\
&= \int_{r_n}^{r_{n+1}} \left[|z'(r)|^p \left(\int_{S_r} dS_r \right) - \left(\int_{S_r} c(x) dS_r \right) |z(r)|^p \right] dr \\
&= \int_{r_n}^{r_{n+1}} \left[\omega_N r^{N-1} |z'(r)|^p - C(r)|z(r)|^p \right] dr \\
&= \omega_N \int_{r_n}^{r_{n+1}} \left[r^{N-1}|z'(r)|^p - \frac{1}{\omega_N} C(r)|z(r)|^p \right] dr = 0.
\end{aligned}$$

Now, if a positive solution $u = u(x)$ exists in an exterior domain Ω_R for some R, then integrating the Picone identity given in Theorem 7.3.1 over D_n with n sufficiently large, and with w given by (7.3.5), we have

$$\int_{D_n} \left(\|\nabla y(x)\|^p - c(x)|y(x)|^p \right) dx \ge 0.$$

Moreover, since $u(x) > 0$ in $\overline{\Omega}$ and $y\,|_{\partial\Omega} = 0$, the functions u, y are not proportional, which means that last inequality is strict, a contradiction. □

As a consequence of Theorems 2.3.5 and 2.2.3 applied to (7.3.15) and using the fact that

$$\int_{\|x\|\leq r} c(x)\,dx = \int_0^r \left(\int_{S_r} c(x)\,dS_r\right)\,dr$$

we have the following statement.

Corollary 7.3.2. *Equation* (7.3.1) *has no solution positive in the exterior domain* Ω_{r_0} *for every* $r_0 \geq 0$ *provided one of the following conditions holds:*

(i) Let $p \geq N$ *hold and*

$$\lim_{r\to\infty} \int_{\|x\|<r} c(x)\,dx = \infty.$$

(ii) Let $p-1 \geq N$, $\alpha \in (-\frac{N}{p}, p-N-1]$ *and*

$$\lim_{t\to\infty} \frac{1}{t^{\alpha+1}} \int_0^t r^\alpha \left(\int_{\|x\|\leq r} c(x)\,dx\right)\,dr = \infty.$$

The disadvantage of Theorem 7.3.6 lies in the fact that integrating the function over the sphere S_r, we loose the information about the distribution of the potential c over this sphere, which may be important in some cases. This disadvantage is improved in the next theorem which introduces the H-functions averaging technique (compare with Subsection 3.2.3) in the oscillation theory of PDE's with p-Laplacian. We present the statement without proof, this proof can be found in [275].

Theorem 7.3.7. *Let* $r_0 \geq 0$ *be fixed and denote* $D := \{(r,x) \in \mathbb{R}\times\mathbb{R}^N : r_0 \leq \|x\| \leq r\}$, $D_0 := \{(r,x) \in \mathbb{R}\times\mathbb{R}^N : r_0 \leq \|x\| < r\}$. *Let* $H(r,x) \in C(D, [0,\infty))$, $\alpha(x)$ *is a nonnegative continuous function for* $\|x\| \geq r_0$ *such that* H *has a continuous partial derivative with respect to* x_i, $i = 1,\dots,N$, *on* D_0 *and the following conditions hold:*

(i) $H(r,x) = 0$ *if and only if* $r = \|x\|$.

(ii) There exists a positive function $k \in C[r_0,\infty)$ *such that the function* $f(r,\rho) := k(\rho)\int_{S_\rho} H(r,x)\,dS$ *is nonincreasing with respect to* ρ *for every* $r \geq \rho \geq r_0$.

(iii) The vector-valued function $\mathbf{h}(r,x)$ *defined on* D_0 *by*

$$\mathbf{h}(r,x) := \nabla_x H(r,x) + \frac{H(r,x)}{\alpha(x)}\nabla\alpha(x)$$

satisfies

$$\int_{r_0\leq\|x\|\leq r} H^{1-p}(r,x)\|\mathbf{h}(r,x)\|^p\alpha(x)\,dx < \infty.$$

If

$$\limsup_{r\to\infty}\left(\int_{S_{r_0}} H(r,x)\,dS\right)^{-1} \times \int_{r_0\le\|x\|\le r}\left[H(r,x)\alpha(x)c(c) - \frac{\|\mathbf{h}(r,x)\|^p\alpha(x)}{p^pH^{p-1}(r,x)}\right]dx = \infty,$$

then (7.3.1) *has no positive solution in the exterior domain* Ω_r *for arbitrary* $r > 0$.

7.3.4 Equations involving pseudolaplacian

Another partial differential equation which reduces to half-linear equation (1.3.2) in the "ordinary" case is the partial differential equation with the so-called *pseudolaplacian*

$$\tilde{\Delta}_p u := \sum_{i=1}^N \frac{\partial}{\partial x_i}\Phi\left(\frac{\partial u}{\partial x_i}\right).$$

We consider the partial differential equation

$$\tilde{\Delta}_p u + c(x)\Phi(u) = 0 \tag{7.3.17}$$

and the associated energy functional

$$\begin{aligned}\mathcal{F}_p(u;\Omega) : &= \int_\Omega\left\{\sum_{i=1}^N\left|\frac{\partial u}{\partial x_i}\right|^p - c(x)|u|^p\right\}dx\\ &= \int_\Omega\left\{\|\nabla u\|_p^p - c(x)|u|^p\right\}dx,\end{aligned}$$

where $\|x\|_p = \left(\sum_{i=1}^N |x_i|^p\right)^{\frac{1}{p}}$ denotes the p-norm in $\mathbb{R}^N$. This functional plays an important role in variational principle for equations with pseudolaplacian. Another important object associated with (7.3.17) is a Riccati type equation which we obtain as follows. Let u be a solution of (7.3.17) which is nonzero in Ω and denote

$$v := \left(\Phi\left(\frac{\partial u}{\partial x_1}\right),\dots,\Phi\left(\frac{\partial u}{\partial x_n}\right)\right), \qquad w := \frac{v}{\Phi(u)}.$$

Then, using the fact that (7.3.17) can be written in the form $\operatorname{div} v = -c(x)\Phi(u)$, we have

$$\begin{aligned}\operatorname{div} w &= \frac{1}{\Phi^2(u)}\left\{\Phi(u)\operatorname{div} v - \Phi'(u)\langle\nabla u, v\rangle\right\}\\ &= -c(x) - (p-1)\frac{|u|^{p-2}}{|u|^{2p-2}}\sum_{i=1}^N\left|\frac{\partial u}{\partial x_i}\right|^p\\ &= -c(x) - (p-1)\sum_{i=1}^N\left|\Phi\left(\frac{\partial u/\partial x_i}{u}\right)\right|^q\\ &= -c(x) - (p-1)\|w\|_q^q,\end{aligned}$$

where $\langle \cdot, \cdot \rangle$ denotes the usual scalar product in $\mathbb{R}^N$, $q = p/(p-1)$ is the conjugate exponent of p and $\|x\|_q = \left(\sum_{i=1}^N |x_i|^q\right)^{\frac{1}{q}}$ denotes the q-norm in $\mathbb{R}^N$. Consequently, the vector variable w satisfies the Riccati type equation

$$\operatorname{div} w + c(x) + (p-1)\|w\|_q^q = 0. \tag{7.3.18}$$

For equation (7.3.17) we can establish oscillation theory and theory for eigenvalue problems similar to that for classical p-Laplacian. An important role is played in this theory by the following Picone type identity.

Theorem 7.3.8. *Let w be a solution of (7.3.18) which is defined in $\bar{\Omega}$ and $u \in W^{1,p}(\Omega)$. Then*

$$\begin{aligned} \mathcal{F}_p(u;\Omega) &= \int_{\partial\Omega} |u(x)|^p w(x)\, dS \\ &\quad + p\int_\Omega \left\{ \frac{\|\nabla u(x)\|_p^p}{p} - \langle \nabla u(x), \Phi(u(x)) w(x)\rangle + \frac{\|w(x)\|_q^q |\Phi(u(x))|^q}{q} \right\} dx. \end{aligned}$$

Moreover, the last integral in this formula is always nonnegative, it equals zero only if $u \neq 0$ in $\bar{\Omega}$ and

$$w = \frac{1}{\Phi(u)}\left(\Phi\left(\frac{\partial u}{\partial x_1}\right), \dots, \Phi\left(\frac{\partial u}{\partial x_n}\right)\right).$$

Having at disposal the previous theorem, one can extend many results of previous subsections to equations with pseudolaplacian. We refer to [48, 104] for details.

7.4 Notes and references

The results concerning the properties of the first eigenvalue and of the associated eigenfunction of (7.1.2), as presented here, are taken from the paper of Lindqvist [258]. Under various restrictions on the domain Ω, statements of this kind are proved in several preceeding papers, let us mention at least the paper of Anane [19]. The properties of the second eigenvalue of (7.1.2) are established in Anane, Tsouli [21]. The main statements of Subsection 7.1.3 can be found e.g. in the paper of Cuesta and Takáč [83], see also the paper of the same authors [84]. The proof of Theorem 7.1.6 as presented here can be found in the book of Drábek, Krejčí and Takáč [128, Theorem 7.3, p. 156]. Recent papers dealing with (anti)maximum principle for p-Laplacian are the papers of Godoy, Gossez and Paczka [169] and of Fleckinger, Gossez, Hernández, de Thélin and Takáč [162, 163, 164], see also references given in those papers. The presentation of Subsection 7.1.4 follows the paper Cuesta, de Figueiredo and Gossez [82] which is essentially abbreviated version of the paper [81] of the same authors. Related recent papers are Alif [12], Arias,[24], Drábek, Robinson [132], Micheletti and Pistoia [287] and Perera [311, 312].

The results of first two subsections of Section 7.2 are taken from the papers of Takáč [342, 343]. Note that several additional results concerning the structure of the solutions of (7.2.5) are hidden in the proofs of statements of Subsection 7.2.1, 7.2.2. However, these proofs are technically rather complicated and long to be presented here (each of the above mentioned two Takáč's papers has more than 30 pages), so we refer to [343, 342] for details. Related results can also be found in the papers of Alziary, Drábek, Fleckinger, Girg, Takáč and Ulm [17, 121, 126]. The results of Subsection 7.2.3 can be found in the papers of Drábek and Robinson [123, 133], for related results and references we also refer to the papers of Arcoya and Orsina [23], Drábek [120] and Drábek and Holubová [127]. Let us also mention the recent papers [68], [69], and [125] by Čepička, Drábek, Girg, and Takáč dealing with various aspects of the boundary value problems associated with p-Laplacian, where, in particular, topics like bifurcation from infinity near eigenvalues and solvability of BVP's under Landesman-Lazer type conditions are discussed.

Picone's identity, as presented in Subsection 7.3.1, was proved by Jaroš, Kusano and Yoshida [191]. However, this identity can be found in various modifications (sometimes implicitly) also in other papers, e.g. in the papers of Allegretto [13, 14] and Allegretto, Huang [15, 16] and in the paper of Dunninger [135]. The criteria for the nonexistence of positive solutions of (7.3.1) given in Subsection 7.3.2 are presented in Došlý and Mařík [113], this paper also contains some criteria given in Subsection 7.3.3. The linear version of these results can be found in the paper of Schminke [337]. The concluding part of this subsection is taken from Mařík's paper [275], this statement is a direct extension of [360, Theorem 1] of Q. R. Wang to (7.3.1). Related results and references can be found also in another Mařík's paper [279].

There exist many papers dealing with various oscillation and spectral properties of PDE's with p-Laplacian, recall here at least the papers Anane, Charkone and Gossez [20] Bennewitz and Saitö [34], Fiedler [161], and the papers of Jaroš, Kusano, Mařík, M. Naito,, Y. Naito, Ogata, Usami and Yoshida [193, 218, 219, 224, 277, 278, 301], but this is really only a very limited sample of papers where equations of the form (7.3.1) are treated.

The basic properties of solutions of PDE's with pseudolaplacian (7.3.17) can be found in the papers of Bognár [45, 46, 47]. Oscillation and nonoscillation criteria for this equation as well as the properties of the first eigenvalue of the Dirichlet BVP associated with pseudolaplacian are presented in the papers of Bognár and Došlý [48] and Došlý [104].

Finally note that throughout the whole book, the exponent p in the p-Laplacian is *a constant*. Recently, several papers appeared (see, e.g. X. L. Fan, Q. Zhang, D. Zhao [160] and the references given therein), where the exponent p may depend on the independent variable $x \in \mathbb{R}^N$, i.e., the investigated operator is of the form $\operatorname{div}(\|\nabla u(x)\|^{p(x)-2}\nabla u(x))$. However, this more general situation is not treated in our book.

CHAPTER 8

HALF-LINEAR DIFFERENCE EQUATIONS

The aim of this chapter is to present discrete versions of some of the results for (1.1.1) given in the previous parts of the book. First we recall some basic facts on *linear* difference equations, then we mention difficulties related to the discrete case. The largest part is devoted to the oscillation theory of half-linear difference equations. For comparison purposes, some of the statements will be proved in details. The chapter is concluded with the theory of half-linear dynamic equations on time scales, which unifies and extends the continuous and the discrete theory. We focus our attention to those types of results which explain the discrepancies between the continuous and the discrete cases. We also show some phenomena that are not usual in the differential/difference equations case.

If it will not be said otherwise, by an interval we mean the discrete interval in this chapter, e.g., $[0, N] = \{0, 1, 2, \dots, N\} \subseteq \mathbb{Z}$, etc. We also introduce a usual convention, namely for any sequence $\{a_k\}$ and any $m \in \mathbb{Z}$ we put $\sum_{k=m}^{m-1} a_k = 0$ and $\prod_{k=m}^{m-1} a_k = 1$.

8.1 Basic information

We start with pointing out basic differences and similarities between discrete and continuous oscillation theories, we also briefly discuss the discretization procedure which leads from (half-linear) differential equations to difference equations.

8.1.1 Linear difference equations

In the last two decades, a considerable attention has been devoted to the oscillation theory (as well as to some other aspects of the qualitative theory) of the Sturm-

Liouville difference equation

$$\Delta(r_k \Delta x_k) + c_k x_{k+1} = 0, \tag{8.1.1}$$

where $\Delta x_k = x_{k+1} - x_k$ is the usual forward difference operator, r, c are real-valued sequences and $r_k \neq 0$. Oscillation theory parallel to that for the Sturm-Liouville differential equation (1.1.2) has been established and many oscillation and nonoscillation criteria have now their discrete counterparts for (8.1.1). We refer to monographs [1, 11, 138, 175, 199] for general background. Basic tools of the linear discrete oscillation theory are the discrete quadratic functional

$$\mathcal{F}_d(x; 0, N) = \sum_{k=0}^{N} [r_k (\Delta x_k)^2 - c_k x_{k+1}^2],$$

the Riccati difference equation (related to (8.1.1) by the substitution $w = r\Delta x / x$)

$$\Delta w_k + c_k + \frac{w_k^2}{r_k + w_k} = 0 \tag{8.1.2}$$

and the link between them, the (reduced) discrete Picone identity

$$\mathcal{F}_d(x; 0, N) = w_k y_k^2 \Big|_{k=0}^{N+1} + \sum_{k=0}^{N} \frac{1}{r_k + w_k} (r_k \Delta x_k - w_k x_k)^2, \tag{8.1.3}$$

w being a solution of the Riccati equation, which is defined for $k = 0, \ldots, N+1$.

A natural idea, suggested by similarity of oscillation theories for linear equation (1.1.2) and half-linear equation (1.1.1), is to look for half-linear extension of these results and to establish a discrete half-linear oscillation theory parallel to that for (1.1.1). Therefore, the subject of this chapter is the theory of the half-linear difference equation

$$\Delta(r_k \Phi(\Delta x_k)) + c_k \Phi(x_{k+1}) = 0, \tag{8.1.4}$$

where r, c are real-valued sequences and $r_k \neq 0$. At a first glance we see the difference from the continuous case, namely the presence of the shift in the second term, i.e., x_{k+1} instead of x_k. This is due to the discretization, see the next subsection. One can use a different discretization scheme, but what we have used here is the most usual one. Moreover, any index different from the index $k+1$ in the second term in (8.1.4) would destroy the below discrete Sturmian theory.

8.1.2 Discretization, difficulties versus eases

As we have said above, (8.1.4) can be understood either as a generalization of (8.1.1) or as a discrete counterpart of (1.1.1). Before starting qualitative investigation of (8.1.4), let us recall the process of discretization of (1.1.1). Thus consider differential equation of the form (1.1.1) with the coefficients $\tilde{r}$, $\tilde{c}$, i.e.,

$(\tilde{r}(t)\Phi(z'))' + \tilde{c}(t)\Phi(z) = 0$, where $\tilde{r}(t) > 0$ and $\tilde{c}(t)$ are continuous functions on (a real interval) $[a, b]$. For small $h = (b-a)/N$, $N \in \mathbb{N}$, we have

$$z'(t) \approx \frac{z(t) - z(t-h)}{h}$$

and

$$(\tilde{r}(t)\Phi(z'(t)))' \approx \frac{1}{h}\left\{\frac{\tilde{r}(t+h)\Phi[z(t+h) - z(t)]}{h} - \frac{\tilde{r}(t)\Phi[z(t) - z(t-h)]}{h}\right\}.$$

Let $t = a + kh$, where k is a discrete variable taking on the integer values $0 \le k \le N$. If $z(t)$ is a solution of the above half-linear differential equation on $[a, b]$, then we have

$$\begin{aligned}&\tilde{r}(a + (k+1)h)\Phi[z(a + (k+1)h) - z(a + kh)] - \\ &\quad - \tilde{r}(a + kh)\Phi[z(a + kh) - z(a + (k-1)h)] + h^2\tilde{c}(a + kh)\Phi(z(a + kh)) \approx 0.\end{aligned}$$

Now we set $y_{k+1} = z(a + kh)$, $r_k = \tilde{r}(a + kh)$ and $c_k = h^2\tilde{c}(a + kh)$. Hence we get

$$r_{k+1}\Phi(y_{k+2} - y_{k+1}) - r_k\Phi(y_{k+1} - y_k) + c_k\Phi(y_{k+1}) \approx 0$$

and thus

$$\Delta(r_k\Phi(\Delta y_k)) + c_k\Phi(y_{k+1}) \approx 0$$

for $0 \le k \le N-2$. Note that y_k is defined for $0 \le k \le N$. Observe that the resulting r_k is positive. But since we want to establish the theory in a full generality, we allow r_k to attain also negative values. This enables to consider equations like the Fibonacci recurrence relation $x_{k+2} = x_{k+1} + x_k$ for which, when rewritten into the self-adjoint form (8.1.1), one gets $r_k = (-1)^k$, see e.g. [11]. We will see that the principal results of our theory apply also in such cases. Possible negativity of r is the first interesting example of differences between the discrete and the continuous cases.

In contrast to the continuous case, there is no problem with the existence and uniqueness for solutions of (8.1.4). Expanding the forward differences, this equation can be written as

$$r_{k+1}\Phi(x_{k+2} - x_{k+1}) - r_k\Phi(x_{k+1} - x_k) + c_k\Phi(x_{k+1}) = 0$$

and hence

$$x_{k+2} = x_{k+1} + \Phi^{-1}\left(\frac{1}{r_{k+1}}[r_k\Phi(x_{k+1} - x_k) - c_k\Phi(x_{k+1})]\right).$$

This means that given the initial conditions $x_0 = A, x_1 = B$, we can compute explicitly all other x_k. Moreover, given any $N \in \mathbb{N}$, the values $x_2, \dots, x_N$ depend continuously (in the norm of $\mathbb{R}^{N-1}$) on x_0, x_1.

Clearly, all of the problems, which are due to the lack of additivity of the solution space (see Section 1.3 for the continuous case), are transferred into the discrete case. However, in the discrete case one has to overcome another difficulties. Indeed, we will see that the results for (8.1.4) are similar to those for (1.1.1), but the proofs are often more difficult. The reason is that the calculus of finite differences and sums is sometimes more cumbersome than the differential and integral calculus. For example, we have no discrete analogue of the chain rule for the differentiation of the composite function, or no discrete analogue of the method of substitution in integration. On the other hand, there are some points where the discrete calculus is "easier", for example, if an infinite series $\sum^{\infty} a_k$ is convergent, we have $\lim_{n\to\infty} a_n = 0$, while the convergence of the integral $\int^{\infty} f(t)\, dt$ gives generally no information about $\lim_{t\to\infty} f(t)$.

In the continuous case we have seen how the transformation of independent variable enables to rewrite equation into the equation of the same form but with $r(t) \equiv 1$, see Section 1.2.7. Some of the results obtained for this easier equation then can be easily rewritten for original equation. Since there is no discrete analogy of this transformation, it is convenient to investigate half-linear difference equations with general r_k, if possible.

8.2 Half-linear discrete oscillation theory

This is the main section of the chapter devoted to half-linear difference equations. First we establish the basic facts of the discrete half-linear oscillation theory. In particular, we show that the variational principle and the Riccati technique, properly modified, are the fundamental methods of this theory similarly as in the continuous case. Then we use these methods to illustrate the main difficulties in the "discretization" of (non)oscillation criteria and some other results presented in the previous parts of this book.

8.2.1 Discrete roundabout theorem and Sturmian theory

Let us again emphasize that general discrete oscillation theory can be established under the mere assumption $r_k \neq 0$, while we have to suppose that $r(t) > 0$ in the continuous case. This fact affects the following definition of the basic concepts of oscillation theory.

Definition 8.2.1. We say that an interval $(m, m+1]$ contains a *generalized zero* of a solution x of (8.1.4) if $x_m \neq 0$ and $x_m x_{m+1} r_m \leq 0$. Equation (8.1.4) is said to be *disconjugate* on $[0, N]$ provided the solution x of (8.1.4) given by the initial conditions $x_0 = 0$, $r_0 \Phi(x_1) = 1$ has no generalized zero in $(0, N+1]$.

If $r_m > 0$, a generalized zero of x is just the zero of x at $m+1$ or the sign change $x_m x_{m+1} < 0$.

In order to present the central statement of half-linear discrete oscillation theory, namely the discrete Roundabout theorem, we have to introduce another important concepts. Along with (8.1.4) consider the *generalized Riccati difference equation*

$$\Delta w_k + c_k + w_k \left(1 - \frac{r_k}{\Phi(\Phi^{-1}(r_k) + \Phi^{-1}(w_k))}\right) = 0. \tag{8.2.1}$$

Define the class $\mathcal{D} = \mathcal{D}(0, N)$ of the so-called *admissible sequences* by

$$\mathcal{D}(0, N) = \{x : [0, N+1] \to \mathbb{R};\ x_0 = 0 = x_{N+1}\}$$

and the *discrete p-degree functional* $\mathcal{F}_d$ on $\mathcal{D}(0, N)$ by

$$\mathcal{F}_d(y; 0, N) = \sum_{k=0}^{N} \left[r_k |\Delta y_k|^p - c_k |y_{k+1}|^p \right].$$

Often we will write just $\mathcal{F}_d(y)$. Note that (8.1.4) can be viewed as the Euler-Lagrange equation associated to certain discrete variational problem involving $\mathcal{F}_d$. Very important role in the proof of the Roundabout theorem is played by Picone's identity, that is the discrete counterpart to Theorem 1.2.1, which was proved in [323].

Theorem 8.2.1. *Consider a pair of the second order half-linear difference operators of the form*

$$l[y_k] \equiv \Delta(r_k \Phi(\Delta y_k)) + c_k \Phi(y_{k+1})$$

and

$$L[z_k] \equiv \Delta(R_k \Phi(\Delta z_k)) + C_k \Phi(z_{k+1})$$

on $[0, N]$, *with* $r_k \neq 0 \neq R_k$. *Let* y_k, z_k *be defined on* $[0, N+2]$ *and let* $z_k \neq 0$ *for* $k \in [0, N+1]$. *Then for* $k \in [0, N]$,

$$\begin{aligned}(8.2.2)\quad \Delta\left\{ \frac{y_k}{\Phi(z_k)} [\Phi(z_k) r_k \Phi(\Delta y_k) - \Phi(y_k) R_k \Phi(\Delta z_k)] \right\} &= \\ = (C_k - c_k)|y_{k+1}|^p + (r_k - R_k)|\Delta y_k|^p & \\ + \frac{y_{k+1}}{\Phi(z_{k+1})} \{ l[y_k]\Phi(z_{k+1}) - L[z_k]\Phi(y_{k+1}) \} + \frac{R_k z_k}{z_{k+1}} G(y_k, z_k),&\end{aligned}$$

where

$$(8.2.3)\quad G(y_k, z_k) := \frac{z_{k+1}}{z_k} |\Delta y_k|^p - \frac{z_{k+1}\Phi(\Delta z_k)}{z_k \Phi(z_{k+1})} |y_{k+1}|^p + \frac{z_{k+1}\Phi(\Delta z_k)}{z_k \Phi(z_k)} |y_k|^p \geq 0$$

with equality if and only if $\Delta y_k = y_k(\Delta z_k / z_k)$.

Now we are ready to formulate the main statement of this chapter, a discrete Roundabout theorem.

Theorem 8.2.2. *The following statements are equivalent:*

(i) Equation (8.1.4) *is disconjugate on* $[0, N]$.

(ii) There exists a solution of (8.1.4) *having no generalized zero in* $[0, N+1]$.

(iii) There exists a solution w *of the generalized Riccati difference equation* (8.2.1) *(related to* (8.1.4) *by the substitution* $w_k = r_k \Phi(\Delta x_k / x_k)$*), which is defined for every* $k \in [0, N+1]$ *and satisfies* $r_k + w_k > 0$ *for* $k \in [0, N]$.

(iv) *The discrete p-degree functional* $\mathcal{F}_d(y;0,N)$ *is positive for every nontrivial* $y \in \mathcal{D}(0,N)$.

Proof. (i) $\Rightarrow$ (ii): First note that for the solution x of (8.1.4) given by $x_0 = 0$, $x_1 = \Phi^{-1}(1/r_0)$, we have $r_k x_k x_{k+1} > 0$ for $k \in [1, N]$. Now consider the solution $x^{[\varepsilon]}$ of (8.1.4) satisfying the initial conditions $x_0^{[\varepsilon]} = \varepsilon > 0$, $x_1^{[\varepsilon]} = \Phi^{-1}(1/r_0)$. Then, according to the above mentioned continuous dependence on initial values, $x^{[\varepsilon]} \to x$ on $[0, N+1]$ as $\varepsilon \to 0$. Hence, if we choose ε sufficiently small, then $\tilde{x} \equiv x^{[\varepsilon]}$ satisfies $r_k \tilde{x}_k \tilde{x}_{k+1} > 0$ for $k \in [0, N]$.

(ii) $\Rightarrow$ (iii): Let x be a solution of (8.1.4) having no generalized zeros in $[0, N+1]$, and let $w_k = r_k \Phi(\Delta x_k / x_k)$. Then

$$\begin{aligned}
\Delta w_k &= \frac{\Delta(r_k\Phi(\Delta x_k))\Phi(x_k) - r_k\Phi(\Delta x_k)(\Phi(x_{k+1}) - \Phi(x_k))}{\Phi(x_{k+1})\Phi(x_k)} \\
&= -c_k - w_k + \frac{r_k\Phi(\Delta x_k)}{\Phi(x_k + \Delta x_k)} = -c_k - w_k + \frac{r_k\Phi(\Delta x_k)}{\Phi(x_k)\Phi\left(1 + \Delta x_k/x_k\right)} \\
&= -c_k - w_k\left(1 - \frac{1}{\Phi\left(1 + \Phi^{-1}\left(w_k/r_k\right)\right)}\right) \\
&= -c_k - w_k\left(1 - \frac{r_k}{\Phi\left(\Phi^{-1}(r_k) + \Phi^{-1}(w_k)\right)}\right).
\end{aligned}$$

Moreover, $r_k x_k x_{k+1} > 0$ if and only if $r_k \Phi(x_k)\Phi(x_{k+1}) > 0$ and

$$\begin{aligned}
r_k\Phi(x_k)\Phi(x_{k+1}) &= r_k\Phi(x_k)\Phi(x_k + \Delta x_k) \\
&= \Phi^2(x_k)\Phi\left(\Phi^{-1}(r_k) + \Phi^{-1}(r_k)\frac{\Delta x_k}{x_k}\right) \\
&= \Phi^2(x_k)\Phi(\Phi^{-1}(r_k) + \Phi^{-1}(w_k))
\end{aligned}$$

for $k \in [0, N]$. Hence $r_k x_k x_{k+1} > 0$ if and only if $\Phi^{-1}(r_k) + \Phi^{-1}(w_k) > 0$, i.e., if and only if $r_k + w_k > 0$.

(iii) $\Rightarrow$ (iv): Assume that w_k is a solution of (8.2.1) with $r_k + w_k > 0$. Note that then z_k given by $w_k = r_k\Phi(\Delta z_k)/\Phi(z_k)$, i.e., $\Delta z_k = \Phi^{-1}(w_k/r_k) z_k$, is a solution of (8.1.4). From the Picone identity (8.2.2) applied to the case $c_k \equiv C_k$, $r_k \equiv R_k$ and $w_k = r_k\Phi(\Delta z_k)/\Phi(z_k)$ we obtain

$$\Delta[y_k r_k \Phi(\Delta y_k)] - \Delta[|y_k|^p w_k] = y_{k+1}\Delta(r_k\Phi(\Delta y_k)) + p_k|y_{k+1}|^p + \tilde{G}(y_k, w_k),$$

where

$$\tilde{G}(y_k, w_k) = r_k|\Delta y_k|^p - \frac{w_k r_k}{\Phi(\Phi^{-1}(r_k) + \Phi^{-1}(w_k))}|y_{k+1}|^p + w_k|y_k|^p.$$

Hence

$$r_k|\Delta y_k|^p - p_k|y_{k+1}|^p = \Delta[w_k|y_k|^p] + \tilde{G}(y_k, w_k).$$

The summation of the above given equality from 0 to N yields

$$\mathcal{F}_d(y) = [w_k|y_k|^p]_{k=0}^{N+1} + \sum_{k=0}^{N}\tilde{G}(y_k, w_k),$$

compare with the quadratic case (8.1.3). Then $\mathcal{F}_d(y) \geq 0$ by inequality (8.2.3), since $r_k z_{k+1}/z_k > 0$. In addition, if $\mathcal{F}_d(y) = 0$, then, again by (8.2.3), $\Delta y_k = y_k \Delta z_k / z_k$. Further, we have $y_0 = 0$ and therefore $y \equiv 0$. Consequently, $\mathcal{F}_d(y) > 0$ for every nontrivial admissible sequence y.

(iv) $\Rightarrow$ (i): Suppose that $\mathcal{F}_d$ is positive for all nontrivial admissible sequences and (8.1.4) is not disconjugate in $[0, N+1]$, i.e., the solution x given by the initial condition $x_0 = 0$, $x_1 = \Phi^{-1}(1/r_0)$ has a generalized zero in the interval $[0, N+1]$, i.e., $r_m x_m x_{m+1} < 0$ or $x_{m+1} = 0$ for some $m \in \{1, \dots, N\}$. Define $y = \{y_k\}_{k=0}^{N+1}$ as follows

$$y_k = \begin{cases} x_k & k = 0, \dots, m, \\ 0 & k = m+1, \dots, N+1. \end{cases}$$

Then we have (using summation by parts applied to $\mathcal{F}_d(x; 0, m-1)$)

$$\begin{aligned} \mathcal{F}_d(y; 0, N) &= \mathcal{F}_d(x; 0, m-1) + [r_m |\Delta y_m|^p] = r_k \Phi(\Delta x_k) x_k \big|_0^m + r_m |x_m|^p \\ &= |x_m|^p \left[r_m \frac{\Phi(\Delta x_m)}{\Phi(x_m)} + r_m \right] = |x_m|^p \left[w_m + r_m \right] \leq 0 \end{aligned}$$

since $w_m + r_m \leq 0$ if and only if $r_m x_m x_{m+1} \leq 0$ as we have shown in the previous part of this proof. □

Remark 8.2.1. (i) Similarly as in the continuous case, it can be shown that the disconjugacy of (8.1.4) on $[0, N]$ may be equivalently defined as the property of (8.1.4) when any its nontrivial solution has at most one generalized zero in the interval $(0, N+1]$, and the solution $\tilde{y}$ satisfying $\tilde{y}_0 = 0$ has no generalized zero in $(0, N+1]$.

(ii) Picone's identity could be used to prove (ii) $\Rightarrow$ (iv) directly.

The Roundabout theorem may serve to provide very easy proofs of Sturm type theorems. Indeed, the proof of the following comparison theorem is based on the equivalence (i) $\Leftrightarrow$ (iv), while the proof of the subsequent separation theorem is based on the implication (ii) $\Rightarrow$ (i) and the homogeneity of the solution space of (8.1.4).

Theorem 8.2.3. *Let the operators l and L be as in Theorem 8.2.1. Suppose that $R_k \geq r_k$ and $c_k \geq C_k$ for $k \in [0, N]$. If $l[y_k] = 0$ is disconjugate on $[0, N]$, then so is the equation $L[z_k] = 0$.*

Theorem 8.2.4. *Two nontrivial solutions $y^{[1]}$ and $y^{[2]}$ of (8.1.4), which are not proportional, cannot have a common zero. Let $m < n$. If $y^{[1]}$ satisfying $y_m^{[1]} = 0$ has a generalized zero in $(n, n+1]$, then $y^{[2]}$ has a generalized zero in $(m, n+1]$. If $y^{[1]}$ has generalized zeros in $(m, m+1]$ and $(n, n+1]$, then $y^{[2]}$ has a generalized zero in $(m, n+1]$.*

Remark that the last theorem does not exclude the situation where two linearly independent solutions have common generalized zero. Taking e.g. the equation $y_{k+2} + 2y_{k+1} + 2y_k = 0$ (every three-term linear recurrence relation $A_k y_{k+2} + B_k y_{k+1} + C_k y_k = 0$ can be written in the form (8.1.1)), one gets the example, where such a case may really happen. Indeed, its two linearly independent solutions are $y_k = 2^{k/2}\sin(3\pi k/4)$ and $z_k = 2^{k/2}\cos(3\pi k/4)$. Both these solutions have a generalized zero in $(1, 2]$.

In the next sections, the following concepts will be widely used.

Definition 8.2.2. Equation (8.1.4) is said to be *nonoscillatory* if there exists $M \in \mathbb{N}$ such that this equation is disconjugate on $[M, N]$ for every $N > M$. In the opposite case, (8.1.4) is said to be *oscillatory*. Oscillation of (8.1.4) may be equivalently defined as follows. A nontrivial solution of (8.1.4) is called *oscillatory* if it has infinitely many generalized zeros. In view of the fact that the Sturm type separation theorem extends to (8.1.4), we have the following equivalence: One solution of (8.1.4) is oscillatory if and only if every solution of (8.1.4) is oscillatory. Hence we can classify *equation* (8.1.4) as *oscillatory* or *nonoscillatory.*

8.2.2 Methods of half-linear discrete oscillation theory

We start with what is usually refereed to as the *(discrete) Riccati technique.* This method is based on the equivalence (i) $\Leftrightarrow$ (iii) of Theorem 8.2.2. Note that this technique can be refined in various ways, as shown in the next subsection. Observe that in contrast to the continuous theory the condition $r_k + w_k > 0$ is involved here.

Theorem 8.2.5. *Equation* (8.1.4) *is nonoscillatory if and only if there exists a sequence w_k, with $r_k + w_k > 0$ for large k, satisfying generalized Riccati equation* (8.2.1).

(Discrete) variational principle extends to (8.1.4) as follows, and it is based on the equivalence (i) $\Leftrightarrow$ (iv) of Theorem 8.2.2.

Theorem 8.2.6. *(i) Equation* (8.1.4) *is nonoscillatory if and only if there exists $N \in \mathbb{N}$ such that*

$$\mathcal{F}_d(y; N, \infty) = \sum_{k=N}^{\infty} [r_k|\Delta y_k|^p - c_k|y_{k+1}|^p] > 0$$

for every nontrivial $y \in \mathcal{D}(N)$, where

$$\mathcal{D}(N) := \{x : \mathbb{N} \to \mathbb{R}\,;\ \exists M > N \text{ with } x_k = 0 \text{ if } k \notin (N, M)\}.$$

(ii) Equation (8.1.4) *is oscillatory if and only if for any $N \in \mathbb{N}$ there exists a (nontrivial) $y \in \mathcal{D}(N)$ such that $\mathcal{F}_d(y; N, \infty) \le 0$.*

Another well-known method, which is available in the half-linear discrete oscillation theory is the *(discrete) reciprocity principle.* Here we suppose that $r_k > 0$ and $c_k > 0$ (even if for the transformation itself, it suffices to assume $r_k \neq 0$ and $c_k \neq 0$). If we denote $u_k = r_k\Phi(\Delta y_k)$, where y is a solution of (8.1.4), then u satisfies the reciprocal equation

$$\Delta(c_k^{1-q}\Phi^{-1}(\Delta u_k)) + r_{k+1}^{1-q}\Phi^{-1}(u_{k+1}) = 0, \tag{8.2.4}$$

where Φ^{-1} is the inverse function of Φ, i.e., $\Phi^{-1}(x) = |x|^{q-1} \operatorname{sgn} x$ and q is the conjugate number of p, i.e., $1/p + 1/q = 1$. Conversely, if $y_k = c_{k-1}^{1-q}\Phi^{-1}(\Delta u_{k-1})$, where u is a solution of (8.2.4), then y_k solves the original equation (8.1.4). Since the discrete version of the Rolle mean value theorem holds, see e.g. [1], we have the following equivalence: (8.1.4) is oscillatory [nonoscillatory] if and only if (8.2.4) is oscillatory [nonoscillatory].

Finally note that till now we have no reasonable discrete version of the generalized Prüfer transformation. This is caused, in particular, by the absence of a discrete chain rule.

8.2.3 Refinements of Riccati technique

Define the operator $\mathcal{R}$ by

$$\mathcal{R}[w_k] := \Delta w_k + c_k + w_k\left(1 - \frac{r_k}{\Phi\left(\Phi^{-1}(w_k) + \Phi^{-1}(r_k)\right)}\right).$$

Before presenting variants and improvements of the Riccati technique, let us give one technical result. One can see that the third term in the operator $\mathcal{R}$, i.e., the third term in the generalized Riccati difference equation, is of quite complicated form in contrast to its continuous counterpart and this sometimes causes difficulties when handling with it. Nevertheless, the next lemma shows what could be expected, namely that the function

$$S(x,y) = S(x,y,p) = x\left(1 - \frac{y}{\Phi(\Phi^{-1}(x) + \Phi^{-1}(y))}\right)$$

exhibits behavior similar to that of the function $x^2/(x+y)$, which appears in Riccati difference equation (8.1.2) associated to linear difference equation (8.1.1).

Lemma 8.2.1. *The function $S(x,y,p)$ has the following properties:*

(i) $S(x,y,p)$ is continuously differentiable on

$$D := \{(x,y,p) \in \mathbb{R} \times \mathbb{R} \times (1,\infty), x \neq -y\}.$$

(ii) Let $y > 0$. Then $x\frac{\partial S}{\partial x}(x,y,p) \geq 0$ for $x + y > 0$, where $\frac{\partial S}{\partial x}(x,y,p) = 0$ if and only if $x = 0$.

(iii) Let $x + y > 0$. Then $\frac{\partial S}{\partial y}(x,y,p) \geq 0$, where the equality holds if and only if $x = 0$.

(iv) $S(x,y,p) \geq 0$ for $x + y > 0$, where the equality holds if and only if $x = 0$.

(v) Suppose that the sequence (x_k, y_k), $k = 1, 2, \ldots$, is such that $x_k + y_k > 0$ and there exists a constant $M > 0$ such that $y_k \leq M$ for $k = 1, 2, \ldots$. Then $S(x_k, y_k, p) \to 0$ implies $x_k \to 0$. Moreover, $\liminf_{k\to\infty} y_k \geq 0$.

(vi) Let $\bar{S}(x,y,p) = x - S(x,y,p)$ hold. Then $\bar{S}(x,y,p) = \bar{S}(y,x,p)$ on D and $\frac{\partial \bar{S}}{\partial x}(x,y,p) \geq 0$ for $x + y > 0$, where the equality holds if and only if $y = 0$. If $y \leq 1$, then $\bar{S}(x,y,p) < 1$ for all $x + y > 0$.

(vii) *Let* $x, y > 0$. *Then* $\frac{\partial S}{\partial p}(x, y, p) \geq 0$.

(viii) *Suppose that* $\operatorname{sgn} y = \operatorname{sgn}(x+y)$ *and* $y \neq 0$. *Then*

$$S(x, y, p) = \frac{(p-1)|x|^q|\xi|^{p-2}}{\Phi(\Phi^{-1}(x) + \Phi^{-1}(y))},$$

where ξ *is between* $\Phi^{-1}(y)$ *and* $\Phi^{-1}(x) + \Phi^{-1}(y)$.

The equivalence of disconjugacy of (8.1.4) and solvability of (8.2.1) (satisfying $r_k + w_k > 0$), coupled with the Sturmian comparison theorem for (8.1.4) (Theorem 8.2.3), lead to the following refinement of the Riccati equivalence from the previous subsection.

Theorem 8.2.7. *The following statements are equivalent:*

(i) *Equation* (8.1.4) *is nonoscillatory.*

(ii) *There is* $N \in \mathbb{N}$ *and a sequence* w *such that* $\mathcal{R}[w_k] = 0$ *and* $r_k + w_k > 0$ *for* $k \in [N, \infty)$.

(iii) *There is* $N \in \mathbb{N}$, *a constant* $A \in \mathbb{R}$ *and a sequence* w *such that*

$$w_k = A - \sum_{j=N}^{k-1} [c_j + S(w_j, r_j)]$$

and $r_k + w_k > 0$ *for* $k \in [N, \infty)$.

(iv) *There is* $N \in \mathbb{N}$ *and a sequence* w *such that* $\mathcal{R}[w_k] \leq 0$ *and* $r_k + w_k > 0$ *for* $k \in [N, \infty)$.

(v) *There is* $N \in \mathbb{N}$ *and a sequence* y *such that*

$$r_k y_k y_{k+1} > 0 \quad \text{and} \quad y_{k+1} l[y_k] \leq 0 \tag{8.2.5}$$

for $k \in [N, \infty)$, *where* l *is defined in Theorem 8.2.1.*

Proof. We show the following implications:

(i) $\Rightarrow$ (ii): This implication is in fact a part of Theorem 8.2.5.

(ii) $\Rightarrow$ (iii): Trivial.

(iii) $\Rightarrow$ (iv): Trivial.

(iv) $\Rightarrow$ (v): Let w satisfy $\mathcal{R}[w_k] \leq 0$ with $r_k + w_k > 0$ on $[N, \infty)$ and let

$$u_k = \prod_{j=N}^{k-1} \left(1 + \Phi^{-1}(w_j/r_j)\right), \quad k \geq N,$$

be a solution of the first order difference equation

$$\Delta u_k = \Phi^{-1}(w_k/r_k) u_k, \quad u_N = 1,$$

Then $u_k \neq 0$ since

$$1 + \Phi^{-1}(w_k/r_k) = \frac{1}{\Phi^{-1}(r_k)} \left[\Phi^{-1}(r_k) + \Phi^{-1}(w_k)\right] \neq 0.$$

Recall that $\Phi^{-1}(r_k) + \Phi^{-1}(w_k) > 0$ if and only if $w_k + r_k > 0$. Further,

$$\begin{aligned}
u_{k+1}l[u_k] &= u_{k+1}\left[\Delta(r_k\Phi(\Delta u_k)) + c_k\Phi(u_{k+1})\right] - \frac{|u_{k+1}|^p r_k\Phi(\Delta u_k)\Delta\Phi(u_k)}{\Phi(u_k)\Phi(u_{k+1})} \\
&\quad + \frac{|u_{k+1}|^p r_k\Phi(\Delta u_k)\Delta\Phi(u_k)}{\Phi(u_k)\Phi(u_{k+1})} \\
&= u_{k+1}\Phi(u_{k+1})\frac{\Delta(r_k\Phi(\Delta u_k))\Phi(u_k) - r_k\Phi(\Delta u_k)\Delta\Phi(u_k)}{\Phi(u_k)\Phi(u_{k+1})} \\
&\quad + |u_{k+1}|^p c_k + |u_{k+1}|^p \frac{r_k\Phi(\Delta u_k)}{\Phi(u_k)}\left(1 - \frac{\Phi(u_k)}{\Phi(u_{k+1})}\right) \\
&= |u_{k+1}|^p R[w_k] \leq 0,
\end{aligned}$$

for $k \in [N, \infty)$, since $w_k = r_k\Phi(\Delta u_k/u_k)$ and

$$\frac{\Phi(u_k)}{\Phi(u_{k+1})} = \frac{1}{\Phi\left(1 + \Delta u_k/u_k\right)} = \frac{r_k}{\Phi\left(\Phi^{-1}(r_k) + \Phi^{-1}(w_k)\right)}.$$

Hence (v) holds.

(v) $\Rightarrow$ (i): Suppose that a sequence u_k satisfying (8.2.5) on $[N, \infty)$ exists. Then $\varphi_k := -u_{k+1}l[u_k]$ is a nonnegative sequence on this discrete interval. Further, set $\tilde{r}_k = r_k$ and $\tilde{c}_k = c_k - \varphi_k/|u_{k+1}|^p$. Hence $\tilde{c}_k \geq c_k$ and

$$\Delta(\tilde{r}_k\Phi(\Delta u_k)) + \tilde{c}_k\Phi(u_{k+1}) = \Delta(r_k\Phi(\Delta u_k)) + \left(c_k - \frac{\varphi_k}{|u_{k+1}|^p}\right)\Phi(u_{k+1}) = 0.$$

Thus equation $\Delta(\tilde{r}_k\Phi(\Delta u_k)) + \tilde{c}_k\Phi(u_{k+1}) = 0$ is disconjugate on $[N, \infty)$ and therefore (8.1.4) is also disconjugate on $[N, \infty)$ by the Sturm comparison theorem (Theorem 8.2.3) and hence nonoscillatory. □

The following two statements can be understood as the discrete counterparts of the Hartman-Wintner theorem. In fact, they describe the asymptotic behavior of $\sum^k S(w_j, r_j)$ in the dependence on the behavior of $\sum^k c_j$. We have already seen that the results (the techniques of proofs) in the discrete case are mostly more complicated than those in the corresponding continuous case. However, the next result shows that this is not always true. Thanks to the necessary condition for the convergence of infinite series and the special form of the third term of the generalized Riccati difference equation together with the condition $r_k + w_k > 0$, the next two statements and their proofs (and also some of their applications) are simpler (and stronger) in a certain sense than those for differential equations. Note that these statements have no continuous analogies. It is worthy to emphasize that in the following theorems we do not require the sequence r to be positive. On the other hand, it is still an open problem to prove them for a more general class of positive sequences r than those which are bounded above (the boundedness substantially simplifies the problem). Note that even in the linear case we have no "full discrete version" for (8.1.1) of the Hartman-Wintner theorem for equation (1.1.2) with the condition $\int^\infty r^{-1}(s)\,ds = \infty$. At present, the "corresponding" discrete version is known only under the restriction $\limsup_{k\to\infty} k^{-3/2}\sum^k r_j < \infty$. In fact, this can be slightly extended using the "weighted averaging" technique, see [73, 156]. Finally note that by the "full discrete version" we mean that there is no additional restriction on c. Actually, subsequent theorems in the second part of this subsection show that it is possible to obtain similar kind of results under the condition $\sum^\infty r_k^{1-q} = \infty$, but with additional condition on c.

Theorem 8.2.8. *Assume that*

$$\text{there exists } M > 0 \text{ such that } r_k \le M \text{ for } k \in \mathbb{N}. \tag{8.2.6}$$

Further suppose that (8.1.4) *is nonoscillatory. Then the following statements are equivalent:*

(i) It holds

$$\liminf_{k\to\infty} \sum_{j=1}^{k} c_j > -\infty. \tag{8.2.7}$$

(ii) For any nonoscillatory solution y with $r_k y_k y_{k+1} > 0$, $k \ge N$, for some $N \in \mathbb{N}$, the sequence $w_k = r_k \Phi(\Delta y_k)/\Phi(y_k)$, $k \ge N$, satisfies

$$\sum_{j=N}^{\infty} S(w_j, r_j) < \infty. \tag{8.2.8}$$

Moreover, this implies $\liminf_{k\to\infty} r_k \ge 0$.

(iii) The limit

$$\lim_{k\to\infty} \sum_{j=1}^{k} c_j \quad \textit{exists (as a finite number).} \tag{8.2.9}$$

Proof. (i) $\Rightarrow$ (ii): In view of the nonnegativity of the function S, see Lemma 8.2.1 (iv), the sequence $\sum_{j=N}^k S(w_j, r_j)$ is nondecreasing for $k \ge N$. Therefore there exists the limit of this sequence equal either to a finite (positive) number or to ∞. Suppose, for the contrary, that there is a nonoscillatory solution y of (8.1.4) such that

$$w_k = \frac{r_k \Phi(\Delta y_k)}{\Phi(y_k)} > -r_k \tag{8.2.10}$$

and

$$\sum_{j=N}^{\infty} S(w_j, r_j) = \infty. \tag{8.2.11}$$

From (8.2.1) we have

$$w_{k+1} = w_N - \sum_{j=N}^{k} c_j - \sum_{j=N}^{k} S(w_j, r_j), \ k \geq N. \tag{8.2.12}$$

Now, from (8.2.7), (8.2.11) and the last equation we obtain $\lim_{k\to\infty} w_k = -\infty$. But this contradicts (8.2.10) since (8.2.6) holds. Therefore we must have (8.2.8). Moreover, $S(w_k, r_k) \to 0$ is a necessary condition for $\sum_{j=N}^{\infty} S(w_j, r_j) < \infty$, and this implies $w_k \to 0$ by Lemma 8.2.1 [(v)]. The condition $r_k + w_k > 0$ now implies $\liminf_{k\to\infty} r_k \geq 0$.

(ii) $\Rightarrow$ (iii): Let w be a sequence as in (ii). According to the above observation, we know that $w_k \to 0$. Now, by letting $k \to \infty$ in equation (8.2.12) we obtain the statement (iii).

(iii) $\Rightarrow$ (i): This implication is obvious. □

The following result is a counterpart to the previous theorem.

Theorem 8.2.9. *Assume that* (8.2.6) *holds and* (8.1.4) *is nonoscillatory. Then the following statements are equivalent:*

(i) It holds

$$\liminf_{k\to\infty} \sum_{j=1}^{k} c_j = -\infty. \tag{8.2.13}$$

(ii) There exists a nonoscillatory solution y of (8.1.4) *with $r_k y_k y_{k+1} > 0$, $k \geq N$ for some $N \in \mathbb{N}$, such that* (8.2.11) *holds, where $w_k = r_k \Phi(\Delta y_k)/\Phi(y_k) > -r_k$ for $k \geq N$.*

(iii) It holds

$$\lim_{k\to\infty} \sum_{j=1}^{k} c_j = -\infty. \tag{8.2.14}$$

Proof. (i) $\Rightarrow$ (ii): This follows from Theorem 8.2.8.

(ii) $\Rightarrow$ (iii): From (8.2.1) using $r_k + w_k > 0$, $k \geq N$, and (8.2.6) we get

$$\sum_{j=N}^{k} c_j = -w_{k+1} + w_m - \sum_{j=N}^{k} S(w_j, r_j) \leq M + w_N - \sum_{j=N}^{k} S(w_j, r_j) \to -\infty.$$

(iii) $\Rightarrow$ (i): Trivial. □

The following theorem gives a necessary condition for nonoscillation of (8.1.4) in terms of the existence of solution of generalized Riccati difference equation in a summation form. Compare with Theorem 2.2.4.

Theorem 8.2.10. *Let* (8.2.6) *and* (8.2.9) *hold. If* (8.1.4) *is nonoscillatory, then there exists a sequence* w_k *such that* $r_k + w_k > 0$, $k \geq N$ *for some* $N \in \mathbb{N}$, *and*

$$w_k = \sum_{j=k}^{\infty} c_j + \sum_{j=k}^{\infty} S(w_j, r_j). \tag{8.2.15}$$

Proof. In view of the assumptions, (8.2.9) holds. From (8.2.1) we get (8.2.12) and letting $k \to \infty$ in this equation, we obtain

$$0 = w_N - \sum_{j=N}^{\infty} c_j - \sum_{j=N}^{\infty} S(w_j, r_j)$$

by Theorem 8.2.8. Replacing N by k we obtain (8.2.15). □

It is clear that the necessary condition in the above theorem is also sufficient for nonoscillation of (8.1.4). One can easily verify it by applying the difference operator to the both sides of (8.2.15) and making use the fact that $r_k + w_k > 0$, $k \geq N$. Then the statement follows from Theorem 8.2.7. But the next theorem shows that such a type of condition guaranteeing nonoscillation can be somewhat relaxed.

Theorem 8.2.11. *Let* (8.2.6) *and* (8.2.9) *hold. Suppose that* $r_k > 0$ *for large* k. *If there exists a sequence* z_k *such that* $r_k + z_k > 0$, $k \geq N$ *for some* $N \in \mathbb{N}$, *satisfying*

$$z_k \geq \sum_{j=k}^{\infty} c_j + \sum_{j=k}^{\infty} S(z_j, r_j) \geq 0, \tag{8.2.16}$$

or

$$z_k \leq \sum_{j=k}^{\infty} c_j + \sum_{j=k}^{\infty} S(z_j, r_j) \leq 0, \tag{8.2.17}$$

then (8.1.4) *is nonoscillatory.*

Proof. Suppose that (8.2.6) and (8.2.7) hold and either (8.2.16) or (8.2.17) is fulfilled. Let

$$w_k = \sum_{j=k}^{\infty} c_j + \sum_{j=k}^{\infty} S(z_j, r_j).$$

Then $\Delta w_k = -c_k - S(z_k, r_k)$. We have $z_k \geq w_k \geq 0$ or $z_k \leq w_k \leq 0$ and hence $S(z_k, r_k) \geq S(w_k, r_k)$, $k \geq N$, according to Lemma 8.2.1 (ii). Obviously, $r_k + w_k > 0$ and $\Delta w_k + p_k + S(w_k, r_k) \leq 0$ for $k \geq N$. Now, equation (8.1.4) is nonoscillatory by Theorem 8.2.7. □

In what follows we will show that the conditions for nonoscillation similar to those in the last two theorems can be obtained also in a different way, but with different assumptions. We present the statements without proofs. In fact, they are similar to the continuous case, see Subsection 2.2.5. We start with an auxiliary statement showing that under certain assumptions one can find a positive solution of (8.2.1). This fact plays an important role in the proof of the next two theorems.

Lemma 8.2.2. *Assume* $r_k > 0$,

$$\liminf_{k\to\infty} \sum_{j=N}^{k} c_j \geq 0 \quad \textit{and} \quad \not\equiv 0$$

for all large N, *and*

$$\sum_{k=1}^{\infty} r_k^{1-q} = \infty. \tag{8.2.18}$$

If (8.1.4) *is nonoscillatory, then* (8.2.1) *possesses an eventually positive solution.*

Compare the next statement with Theorems 8.2.10 and 8.2.11. Note that if $c_k \geq 0$ in these statements, then it is easy to prove the existence of a (positive) sequence satisfying (8.2.15). See also the continuous case (Theorem 2.2.4).

Theorem 8.2.12. *Let the assumptions of the previous lemma hold and let* $\sum_{j=1}^{\infty} c_j$ *be convergent. Then* (8.1.4) *is nonoscillatory if and only if there exists a positive sequence* w *satisfying*

$$w_k \geq \sum_{j=k}^{\infty} c_j + \sum_{j=k}^{\infty} S(w_j, r_j) \tag{8.2.19}$$

for large k. *In fact,* w *is given by* $w_k = r_k \Phi(\Delta y_k / y_k) > 0$, *where* y *is an eventually positive solution of* (8.1.4).

We conclude this section with the statement which claims that under slightly stronger assumptions than those of the previous theorem, a positive solution of the generalized Riccati difference equation can be estimated from above by a known sequence. This will be important in some of the subsequent applications of the Riccati technique.

Theorem 8.2.13. *Let the assumptions of the previous theorem hold. Assume further that* $c_k \geq 0$ *(and eventually nontrivial) for all large* k, *say* $k \geq N$. *If* (8.1.4) *is nonoscillatory with an eventually positive solution* y, *then* $w_k = r_k \Phi(\Delta y_k / y_k) > 0$ *for* $k \geq N$ *and satisfies* $w_k \to 0$ *as* $k \to \infty$. *Moreover, the inequality*

$$w_k \leq \left(\sum_{j=N}^{k-1} r_j^{1-q} \right)^{1-p}$$

holds for $k \geq N$.

8.2.4 Discrete oscillation criteria

In a discrete Leighton-Wintner criterion, similarly as in the continuous case, equation (8.1.4) is viewed as a perturbation of the one-term equation

$$\Delta(r_k \Phi(\Delta x_k)) = 0. \tag{8.2.20}$$

In accordance with the continuous case, we need (8.2.20) to be nonoscillatory in this approach, so we suppose that $r_k > 0$ for large k, otherwise this equation is oscillatory – each sign change of r_k is a generalized zero of the constant solution $x_k \equiv 1$.

Theorem 8.2.14. *Suppose that $r_k > 0$ for large k,*

$$\sum^{\infty} r_k^{1-q} = \infty \quad \text{and} \quad \sum^{\infty} c_k = \infty. \tag{8.2.21}$$

Then (8.1.4) *is oscillatory.*

Proof. We will see that the idea of the proof is exactly the same as in the continuous case. Let $N \in \mathbb{N}$ be arbitrary. For $N < n < m < M$ (which will be determined later) define a sequence $y \in \mathcal{D}(N)$, $\mathcal{D}$ being defined in Theorem 8.2.6, as follows

$$y_k = \begin{cases} 0 & k = N, \\ \left(\sum_{j=N}^{k-1} r_j^{1-q}\right)\left(\sum_{j=N}^{n-1} r_j^{1-q}\right)^{-1} & N+1 \le k \le n, \\ 1 & n+1 \le k \le m-1, \\ \left(\sum_{j=k}^{M-1} r_j^{1-q}\right)\left(\sum_{j=m}^{M-1} r_j^{1-q}\right)^{-1} & m \le k \le M-1, \\ 0 & k \ge M. \end{cases}$$

Then we have

$$\begin{aligned} \mathcal{F}_d(y; N, \infty) &= \sum_{k=N}^{\infty} \left[r_k |\Delta y_k|^p - c_k |y_{k+1}|^p\right] = \sum_{k=N}^{M-1} \left[r_k |\Delta y_k|^p - c_k |y_{k+1}|^p\right] \\ &= \left(\sum_{k=N}^{n-1} + \sum_{k=n}^{m-1} + \sum_{k=m}^{M-1}\right) \left[r_k |\Delta y_k|^p - c_k |y_{k+1}|^p\right] \\ &= \left(\sum_{k=N}^{n-1} r_k^{1-q}\right)^{1-p} - \sum_{k=N}^{n-1} c_k |y_{k+1}|^p - \sum_{k=n}^{m-1} c_k \\ &\quad - \sum_{k=m}^{M-1} c_k |y_{k+1}|^p + \left(\sum_{k=m}^{M-1} r_k^{1-q}\right)^{1-p}. \end{aligned}$$

Now, using the discrete version of the second mean value theorem of the summation calculus (see e.g. [99]), there exists $\tilde{m} \in [m-1, M-1]$ such that

$$\sum_{k=m}^{M-1} c_k |y_{k+1}|^p \ge \sum_{k=m}^{\tilde{m}} c_k.$$

Let $n > N$ be fixed. Since (8.2.21) holds, for every $\varepsilon > 0$ there exist $M > m > n$ such that

$$\sum_{k=n}^{\tilde{m}} c_k > \mathcal{F}_d(y; N, n-1) + \varepsilon \quad \text{whenever } \tilde{m} > m \text{ and } \left(\sum_{k=m}^{M-1} r_k^{1-q}\right)^{1-p} < \varepsilon.$$

Consequently, we have

$$\mathcal{F}_d(y;N,\infty) \le \mathcal{F}_d(y;N,n-1) - \sum_{k=n}^{\tilde{m}} c_k + \left(\sum_{k=m}^{M-1} r_k^{1-q}\right)^{1-p} < 0$$

what we needed to prove. □

In Subsection 1.2.10 we have presented an alternative proof of the continuous Leighton-Wintner criterion – based on the Riccati technique. Next we show the difficulties in an attempt to follow this idea in the discrete case. The "Riccati proof" goes by contradiction. Suppose that (8.2.21) holds and (8.1.4) is nonoscillatory, i.e., there exists a solution of (8.2.1) satisfying $r_k + w_k > 0$ for large k. The summation of (8.2.1) from N to $k-1$, where N, k are sufficiently large, yields

$$w_k = w_N - \sum_{j=N}^{k-1} c_j - \sum_{j=N}^{k-1} S(w_j, r_j) \le -\sum_{j=N}^{k-1} S(w_j, r_j) =: G_k.$$

In the continuous case we obtained the analogous inequality

$$w(t) \le -(p-1)\int_T^t r^{1-q}(s)|w(s)|^q\, ds =: G(t)$$

which leads to the inequality

$$\text{(8.2.22)} \qquad \frac{G'(t)}{G^q(t)} \le \frac{r^{1-q}(t)}{\int_T^t r^{1-q}(s)\, ds}$$

and integrating it we get $\int^\infty r^{1-q}(t)\, dt < \infty$, a contradiction.

The inequality $w_k \le G_k$ is the discrete analogue of (8.2.22), and to get a contradiction from this inequality is a difficult problem even in the linear case $p = 2$.

On the other hand, if the first condition in (8.2.21) is changed to (8.2.6), then oscillation of (8.1.4) follows from a very simple argument, which is in fact based on the Riccati technique. Indeed, if (8.1.4) is nonoscillatory, then there is w satisfying $r_k + w_k > 0$ and (8.2.12) for large $k \ge N$. Since S is nonnegative, we get $w_k \le w_N - \sum_{j=N}^{k-1} c_j$, consequently $w_k \to -\infty$ as $k \to \infty$. From $r_k + w_k > 0$ we obtain $w_k \ge -M$, a contradiction. Note that in contrast to the above Leighton-Wintner type criterion, r does not need to be positive. But if it is positive, then the previous criterion is better. Also observe, that the statement which was just proved can be easily obtained from Theorems 8.2.8 and 8.2.9 as well. Another criterion which immediately follows from these two theorems is that a sufficient condition for oscillation is

$$\text{(8.2.23)} \qquad \liminf_{k\to\infty} \sum_{j=1}^{k} c_j < \limsup_{k\to\infty} \sum_{j=1}^{k} c_j$$

provided (8.2.6) holds. Notice that this criterion has no continuous analogue. Also it is interesting to see what role is played by the condition $r_k + w_k > 0$. In a certain sense, (8.2.23) can be understood as a discrete counterpart to Theorem 2.2.10.

Note that the Leighton-Wintner type criterion is completed by the Hille-Nehari type criterion, when $\sum^\infty c_j$ converges, see Theorem 3.1.1 for the continuous version. In [3], the Hille-Nehari type criterion was proved in a more general setting, for half-linear dynamic equations. Its difference equations version is given in the next statement. In the proof, the equivalence (i) $\Leftrightarrow$ (iv) from Theorem 8.2.7 is used. Additional assumptions are required in comparison with the continuous case. The criterion was for the first time proved under condition (8.2.6), but later it was shown that (8.2.6) can be replaced by a weaker condition, namely (8.2.24). A nonoscillatory counterpart to the following theorem is the discrete Hille-Nehari criterion presented in the next section.

Theorem 8.2.15. *Let $r_k > 0$ and $c_k \geq 0$ for large k with* (8.2.9) *and*

$$\lim_{k\to\infty} \frac{r_k^{1-q}}{\sum^{k-1} r_j^{1-q}} = 0. \tag{8.2.24}$$

If

$$\liminf_{k\to\infty} \left(\sum^{k-1} r_j^{1-q}\right)^{p-1} \left(\sum_{j=k}^{\infty} c_j\right) > \frac{1}{p}\left(\frac{p-1}{p}\right)^{p-1}, \tag{8.2.25}$$

then (8.1.4) *is oscillatory.*

In [323], the same technique as that in Theorem 8.2.14 was used to prove the Hille-Nehari type criterion which differs from the last theorem as follows: c_k does not need to be eventually positive, r_k does not need to satisfy (8.2.24), but (8.2.18) holds, the constant on the right-hand side of (8.2.25) being replaced by (the larger one) 1. Similarly as in the continuous case, the complement to that criterion, in the sense of the convergence of $\sum^\infty r_k^{1-q}$, can be proved easily by means of the discrete reciprocity principle mentioned in Subsection 8.2.2.

From the above criteria it can be seen that if (8.2.6) holds, then the nonexistence of the limit in (8.2.9) as a finite number (except in the case (8.2.14)) implies oscillation. From this point of view, the cases when (8.2.9) or (8.2.14) holds seem to be interesting for the examination. More precisely, we look for additional conditions, which guarantee oscillation of (8.1.4) in these cases. In the case when (8.2.9) holds, such conditions already exist. See e.g. the above Hille-Nehari type criterion. Many other criteria in this case are presented in the following two papers: In [324], oscillation criteria are proved making use Theorem 8.2.12 and Theorem 8.2.13. For example, under the assumptions of Theorem 8.2.13, equation (8.1.4) is oscillatory provided there is a $\nu > 0$ such that $p - \nu > 1$ and

$$\lim_{k\to\infty} \sum_{j=1}^{k} \left(\sum_{i=1}^{j} r_i^{1-q}\right)^{\nu} c_j = \infty.$$

Note that the Hille-Nehari type criterion with $r_k \equiv 1$ is given there as a consequence of a more general theorem, which is based on the discrete function sequence technique. In [325], discrete counterparts to some of the criteria from Section 3.3 are proved. The main tool in that paper is Theorem 8.2.10. Also, the Hille-Nehari type oscillation criterion with $r_k \equiv 1$ can be found there, as a corollary of the following theorem. The sequence c_k does not need to be nonnegative.

Theorem 8.2.16. *Suppose that* (8.2.9) *holds,* $r_k \equiv 1$, *and*

$$\limsup_{k\to\infty} \frac{\sum_{j=1}^{k}(j+1)^{p-2}\sum_{i=j+1}^{\infty} c_i}{\sum_{j=1}^{k} 1/(j+1)} > \frac{1}{p}\left(\frac{p-1}{p}\right)^{p-1}.$$

Then (8.1.4) *is oscillatory.*

Concerning the case when (8.2.14) holds, the situation becomes more difficult and we still have no reasonable conditions of such type. But it really makes a sense to look for these conditions since both oscillation and nonoscillation are possible in this case. Indeed, for example, if $c_k \equiv a < 0$ and $r_k \equiv 1$, then equation (8.1.4) is nonoscillatory by comparison theorem (since $\Delta(\Phi(\Delta y_k)) = 0$ is nonoscillatory) and (8.2.14) holds. The next criterion, which is in fact corollary of [322, Theorem 5], proved by the Riccati technique, enables to give an example of oscillatory equation (8.1.4) with c_k satisfying (8.2.14).

Theorem 8.2.17. *If there exist two sequences of integers* m_k *and* n_k, $n_k \geq m_k+1$, *such that* $m_k \to \infty$ *for* $k \to \infty$, *and*

$$\sum_{j=m_k}^{n_k-1} c_j \geq r_{m_k} + r_{n_k},$$

then equation (8.1.4) *is oscillatory.*

Example 8.2.1. Let $m_k = 4k$, $k \in \mathbb{N}$. Put $r_k \equiv 1$ and $c_{m_k} = 1$, $c_{m_k+1} = 1$, $c_{m_k+2} = 1$, $c_{m_k+3} = -4$ for $k \in \mathbb{N}$. Then

$$\sum_{j=m_k}^{m_k+2} c_j = 3 > 2 = r_{m_k} + r_{m_k+3}$$

for all $k \in \mathbb{N}$. Equation (8.1.4) is oscillatory by Theorem 8.2.17. It is clear that $\sum_{j=1}^{\infty} c_j = -\infty$.

8.2.5 Hille-Nehari discrete nonoscillation criteria

The following theorem is a discrete version of Theorem 2.2.9.

Theorem 8.2.18. *Suppose that* $r_k > 0$ *for large* k, (8.2.9) *and* (8.2.24) *hold. If*

$$\limsup_{k\to\infty} \left(\sum^{k-1} r_j^{1-q}\right)^{p-1} \left(\sum_{j=k}^{\infty} c_j\right) < \frac{1}{p}\left(\frac{p-1}{p}\right)^{p-1} \tag{8.2.26}$$

and

$$\liminf_{k\to\infty}\left(\sum^{k-1} r_j^{1-q}\right)^{p-1}\left(\sum_{j=k}^{\infty} c_j\right) > -\frac{2p-1}{p}\left(\frac{p-1}{p}\right)^{p-1}, \tag{8.2.27}$$

then (8.1.4) *is nonoscillatory.*

Proof. It is sufficient to show that the generalized Riccati inequality $\mathcal{R}[w_k] \le 0$ has a solution w with $r_k + w_k > 0$ in a neighborhood of infinity. We recommend the reader to compare this proof with that of Theorem 2.2.9 to see difference between the discrete and the continuous case.

Set

$$w_k = C\left(\sum^{k-1} r_j^{1-q}\right)^{1-p} + \sum_{j=k}^{\infty} c_j, \tag{8.2.28}$$

where C is a suitable constant, which will be specified later. The following equalities hold by the Lagrange Mean Value Theorem,

$$\Delta\left(\sum^{k-1} r_j^{1-q}\right)^{1-p} = (1-p)r_k^{1-q}\eta_k^{-p},$$

where $\sum^{k-1} r_j^{1-q} \le \eta_k \le \sum^{k} r_j^{1-q}$. Similarly,

$$\begin{aligned} 1 &- \frac{r_k}{\Phi(\Phi^{-1}(r_k)+\Phi^{-1}(w_k))} \\ &= \frac{1}{\Phi(\Phi^{-1}(r_k)+\Phi^{-1}(w_k))}\left\{\Phi(\Phi^{-1}(r_k)+\Phi^{-1}(w_k)) - \Phi(\Phi^{-1}(r_k))\right\} \\ &= \frac{p-1}{\Phi(\Phi^{-1}(r_k)+\Phi^{-1}(w_k))}|\xi|^{p-2}\Phi^{-1}(w_k), \end{aligned}$$

where ξ_k is between $\Phi^{-1}(r_k)$ and $\Phi^{-1}(r_k)+\Phi^{-1}(w_k)$. Hence

$$\Phi^{-1}(r_k) - |\Phi^{-1}(w_k)| \le \xi_k \le \Phi^{-1}(r_k) + |\Phi^{-1}(w_k)|$$

and

$$\frac{|w_k|}{r_k} = \left(\frac{r_k^{1-q}}{\sum^{k-1} r_j^{1-q}}\right)^{p-1}\left|C + \left(\sum^{k-1} r_j^{1-q}\right)^{p-1}\left(\sum_{j=k}^{\infty} c_j\right)\right|$$

Therefore $w_k/r_k \to 0$ for $k \to \infty$ according to (8.2.24), (8.2.26) and (8.2.27). Further, we have

$$\Delta w_k + c_k - w_k\left(1 - \frac{r_k}{\Phi(\Phi^{-1}(r_k)+\Phi^{-1}(w_k))}\right)$$

$$
\begin{aligned}
&= (1-p)Cr_k^{1-q}\eta_k^{-p} + \frac{(p-1)|w_k|^q\xi_k^{p-2}}{\Phi(\Phi^{-1}(r_k)+\Phi^{-1}(w_k))} \\
&= (p-1)r_k^{1-q}\left\{\frac{|w_k|^q|\xi_k|^{p-2}r_k^{q-1}}{\Phi(\Phi^{-1}(r_k)+\Phi^{-1}(w_k))} - \frac{C}{\eta_k^p}\right\} \\
&= (p-1)\frac{r_k^{1-q}}{\left(\sum^{k-1} r_j^{1-q}\right)^p} \\
&\quad \times \left\{\frac{\left|C + \left(\sum^{k-1} r_j^{1-q}\right)^{p-1} \sum_{j=k}^{\infty} c_j\right|^q |\xi_k|^{p-2} r_k^{q-1}}{\Phi(\Phi^{-1}(r_k)+\Phi^{-1}(w_k))} - \frac{C\left(\sum^{k-1} r_j^{1-q}\right)^p}{\eta_k^p}\right\} \\
&\le \frac{(p-1)r_k^{1-q}}{\left(\sum^{k-1} r_j^{1-q}\right)^p}\left\{\left|C + \left(\sum^{k-1} r_j^{1-q}\right)^{p-1}\left(\sum_{j=k}^{\infty} c_j\right)\right|^q \gamma_k - \frac{C\left(\sum^{k-1} r_j^{1-q}\right)^p}{\left(\sum^{k} r_k^{1-q}\right)^p}\right\},
\end{aligned}
$$

where

$$
\gamma_k = \frac{|\xi_k|^{p-2} r_k^{1-q}}{\Phi(\Phi^{-1}(r_k)+\Phi^{-1}(w_k))}.
$$

Concerning the asymptotic behavior of this sequence and of

$$
\left(\sum^{k-1} r_j^{1-q}\right)\left(\sum^{k} r_k^{1-q}\right)^{-1},
$$

we have

$$
\lim_{k\to\infty} \frac{\sum^{k} r_j^{1-q}}{\sum^{k-1} r_j^{1-q}} = \lim_{k\to\infty} \frac{r_k^{1-q} + \sum^{k-1} r_j^{1-q}}{\sum^{k-1} r_j^{1-q}} = 1
$$

since (8.2.24) holds. Further,

$$
\frac{|\xi_k|^{p-2} r_k^{1-q}}{\Phi(\Phi^{-1}(r_k)+\Phi^{-1}(w_k))} \le \frac{r_k^{1-q}(\Phi^{-1}(r_k)+\Phi^{-1}(|w_k|))^{p-2}}{\Phi(\Phi^{-1}(r_k)-\Phi^{-1}(|w_k|))} \to 1
$$

as $k \to \infty$ since $|w_k|/r_k \to 0$ as $k \to \infty$. Consequently,

$$
\limsup_{k\to\infty} \gamma_k \le 1. \tag{8.2.29}
$$

Now, inequalities (8.2.26) (8.2.27) imply the existence of $\varepsilon > 0$ such that

(8.2.30)

$$-\frac{2p-1}{p}\left(\frac{p-1}{p}\right)^{p-1}+\varepsilon<\left(\sum^{k-1} r_j^{1-q}\right)^{p-1}\left(\sum_{j=k}^{\infty} c_j\right)<\frac{1}{p}\left(\frac{p-1}{p}\right)^{p-1}-\varepsilon$$

for k sufficiently large. Let $\tilde{\gamma}_k=\gamma_k^{\frac{1}{q}}$, $\tilde{\varepsilon}=\varepsilon\left(\frac{p}{p-1}\right)^{p-1}$ and let $C=\left(\frac{p-1}{p}\right)^{p}$ in (8.2.28). According to (8.2.29), $\tilde{\gamma}_k<1/(1-\tilde{\varepsilon})$ for large k. Further,

$$\begin{aligned}
\tilde{\gamma}_k<\frac{1}{1-\tilde{\varepsilon}} \quad &\Longleftrightarrow \quad 1>\left[1-\left(\frac{p}{p-1}\right)^{p-1}\varepsilon\right]\tilde{\gamma}_k\\
&\Longleftrightarrow \quad \left(\frac{p-1}{p}\right)^{p-1}>\left(\frac{p-1}{p}\right)^{p-1}\left[\frac{p-1}{p}+\frac{1}{p}-\left(\frac{p}{p-1}\right)^{p-1}\varepsilon\right]\tilde{\gamma}_k\\
&\Longleftrightarrow \quad \left(\frac{p-1}{p}\right)^{\frac{p}{q}}>\left[\left(\frac{p-1}{p}\right)^{p}+\frac{1}{p}\left(\frac{p-1}{p}\right)^{p-1}-\varepsilon\right]\tilde{\gamma}_k\\
&\Longleftrightarrow \quad \frac{C^{\frac{1}{q}}-C\tilde{\gamma}_k}{\tilde{\gamma}_k}>\frac{1}{p}\left(\frac{p-1}{p}\right)^{p-1}-\varepsilon.
\end{aligned}$$

Therefore, the second inequality in (8.2.30) implies

$$C^{\frac{1}{q}}>\left[C+\left(\sum^{k-1} r_j^{1-q}\right)^{p-1}\left(\sum_{j=k}^{\infty} c_j\right)\right]\tilde{\gamma}_k.$$

By a similar computation (using the first inequality in (8.2.30)) we get

$$-C^{\frac{1}{q}}<\left[C+\left(\sum^{k-1} r_j^{1-q}\right)^{p-1}\left(\sum_{j=k}^{\infty} c_j\right)\right]\tilde{\gamma}_k.$$

Consequently,

$$\left|C+\left(\sum^{k-1} r_j^{1-q}\right)^{p-1}\left(\sum_{j=k}^{\infty} c_j\right)\right|^{q}<C$$

for large k and hence $R[w_k]\leq 0$. Finally, since $r_k>0$ and $w_k/r_k\to 0$ as $k\to\infty$, we have $r_k+w_k>0$ for large k and the proof is complete. □

If we compare the previous statement with Theorem 2.2.9 (which is a continuous counterpart of this theorem), we see that assumption (8.2.24) has no continuous analogue. This is a consequence of the fact that we have no "reasonable" version of the differentiation chain rule in the discrete case, and its partial discrete substitution – the Lagrange Mean Value Theorem – needs additional assumptions. On the other hand, there is the discrete Hille-Nehari type nonoscillation criterion, given also in [115], where (8.2.24) is not needed. However, another price has to be paid: the constant is not as good as that at the right-hand side of (8.2.26), and only a positive part of c is involved. The proof is based on the variational method (Theorem 8.2.6). The crucial role is played by the half-linear discrete version of the Wirtinger type inequality. In the proof of this inequality we need the following technical result.

Lemma 8.2.3. *Let*

$$\mu := \begin{cases} \sup_{t>s>0} \frac{1}{t-s}\left[\Phi^{-1}\left(\frac{t^p-s^p}{p(t-s)}\right) - s\right] & p \geq 2, \\ \sup_{t>s>0} \frac{1}{t-s}\left[t - \Phi^{-1}\left(\frac{t^p-s^p}{p(t-s)}\right)\right] & p \leq 2. \end{cases} \tag{8.2.31}$$

Then for given $\beta > \alpha > 0$ and for $\xi = \lambda\beta + (1-\lambda)\alpha$ given by the Lagrange mean value theorem $\beta^p - \alpha^p = p\Phi(\xi)(\beta - \alpha)$ we have $\max\{\lambda, (1-\lambda)\} \leq \mu$.

Proof. If $p \geq 2$, then $\lambda \geq 1/2$, i.e., $\max\{\lambda, 1-\lambda\} = \lambda$ and for $p \leq 2$ we have $\lambda \leq 1/2$. The conclusion now can be easily verified by a direct computation via the Lagrange Mean Value Theorem applied to the function $t \to t^p$, $t \geq 0$. □

Lemma 8.2.4. *Let M_k be a positive sequence such that ΔM_k is of one sign for $k \geq N \in \mathbb{N}$. Then for every $y \in \mathcal{D}(N)$ we have*

$$\sum_{k=N}^{\infty} |\Delta M_k||y_{k+1}|^p \leq p^p[\mu(1+\psi_N)]^{p-1} \sum_{k=N}^{\infty} \frac{M_k^p}{|\Delta M_k|^{p-1}}|\Delta y_k|^p, \tag{8.2.32}$$

where

$$\psi_N = \sup_{k \geq N} \frac{|\Delta M_k|}{|\Delta M_{k-1}|} \tag{8.2.33}$$

and μ is given in the previous lemma.

Proof. Suppose that $\Delta M_k > 0$ for $k \geq N$; in case $\Delta M_k < 0$ we would proceed in the same way. Using summation by part, the Hölder inequality, the Lagrange Mean Value Theorem and the Jensen inequality for the convex function $\xi \to |\xi|^p$, we have

$$\begin{aligned}
&\sum_{k=N}^{\infty} |\Delta M_k||y_{k+1}|^p \\
&\quad = \sum_{k=N}^{\infty} M_k \Delta(|y_k|^p) = p\sum_{k=N}^{\infty} M_k|\Phi(\xi_k)||\Delta y_k| \\
&\quad \leq p\left(\sum_{k=N}^{\infty} \frac{M_k^p}{|\Delta M_k|^{p-1}}|\Delta y_k|^p\right)^{1/p}\left(\sum_{k=N}^{\infty} |\Delta M_k||\Phi(\xi_k)|^q\right)^{1/q} \\
&\quad = p\left(\sum_{k=N}^{\infty} \frac{M_k^p}{|\Delta M_k|^{p-1}}|\Delta y_k|^p\right)^{1/p}\left(\sum_{k=N}^{\infty} |\Delta M_k|\left[\lambda_k|y_k|^p + (1-\lambda_k)|y_{k+1}|^p\right]\right)^{1/q} \\
&\quad \leq p\left(\sum_{k=N}^{\infty} \frac{M_k^p}{|\Delta M_k|^{p-1}}|\Delta y_k|^p\right)^{1/p}\left[\left(\sum_{k=N}^{\infty} \frac{|\Delta M_k|}{|\Delta M_{k-1}|}|\Delta M_{k-1}||y_k|^p\right.\right.
\end{aligned}$$

$$+ \sum_{k=N}^{\infty} |\Delta M_k||y_{k+1}|^p \Bigg) \max\{\lambda_k, (1-\lambda_k)\}\Bigg]^{1/q},$$

where $\xi_k = \lambda_k y_k + (1-\lambda_k) y_{k+1}$ is a number between y_k, y_{k+1}, i.e., $\lambda_k \in [0,1]$. Now, by Lemma 8.2.3, $\max\{\lambda_k, 1-\lambda_k\} \le \mu$ and since $y_k = 0$ for $k \le N$, we have

$$\sum_{k=N}^{\infty} \frac{|\Delta M_k|}{|\Delta M_{k-1}|} |\Delta M_{k-1}||y_k|^p \le \left(\sup_{k\ge N} \frac{|\Delta M_k|}{|\Delta M_{k-1}|}\right) \sum_{k=N}^{\infty} |\Delta M_k||y_{k+1}|^p.$$

Consequently, we have

$$\sum_{k=N}^{\infty} |\Delta M_k||y_{k+1}|^p \le p\left[\mu\left(1+\sup_{k\ge N}\frac{|\Delta M_k|}{|\Delta M_{k-1}|}\right)\right]^{1/q}$$
$$\times \left(\sum_{k=N}^{\infty} \frac{M_k^p}{|\Delta M_k|^{p-1}} |\Delta y_k|^p\right)^{1/p} \left(\sum_{k=N}^{\infty} |\Delta M_k||y_{k+1}|^p\right)^{1/q},$$

and hence

$$\sum_{k=N}^{\infty} |\Delta M_k||y_{k+1}|^p \le p^p \left[\mu(1+\psi_N)\right]^{p-1} \sum_{k=N}^{\infty} \frac{M_k^p}{|\Delta M_k|^{p-1}} |\Delta y_k|^p,$$

what we needed to prove. □

Theorem 8.2.19. *Suppose that $r_k > 0$ for large k, (8.2.18) holds, $\sum^{\infty} c_k^+ < \infty$, $c^+ := \max\{0, c\}$,*

(8.2.34)
$$\varphi_N := \left[\sup_{k\ge N} \frac{\sum^k r_j^{1-q}}{\sum^{k-1} r_j^{1-q}}\right]^{p(p-1)} < \infty, \quad \psi_N := \left[\sup_{k\ge N} \frac{r_k}{r_{k-1}}\right]^{1-q} < \infty.$$

Further suppose that

(8.2.35)
$$0 < \limsup_{N\to\infty} (1+\psi_N)^{p-1} \varphi_N =: \Psi < \infty.$$

If

(8.2.36)
$$\limsup_{k\to\infty} \left(\sum^{k-1} r_j^{1-q}\right)^{p-1} \sum_{j=k}^{\infty} c_j^+ < \frac{1}{p\mu^{p-1}} \left(\frac{p-1}{p}\right)^{p-1} \frac{1}{\Psi},$$

then (8.1.4) is nonoscillatory.

Proof. According to Theorem 8.2.6, it suffices to find $N \in \mathbb{N}$ such that we have

$$\sum_{k=N}^{\infty} \left[r_k |\Delta y_k|^p - c_k |y_{k+1}|^p\right] > 0$$

for any nontrivial $y \in \mathcal{D}(N)$. To this end, let

$$M_k = \left(\sum^{k-1} r_j^{1-q}\right)^{1-p}.$$

Then, using the Lagrange Mean Value Theorem,

$$|\Delta M_k| = \frac{p-1}{\xi_k^p}\Delta\left(\sum^{k-1} r_j^{1-q}\right) = \frac{p-1}{\xi_k^p} r_k^{1-q},$$

where $\xi_k \in \left(\sum^{k-1} r_j^{1-q}, \sum^{k} r_j^{1-q}\right)$. Hence

$$\frac{p-1}{\left(\sum^k r^{1-q}\right)^p} r_k^{1-q} \le |\Delta M_k| \le \frac{p-1}{\left(\sum^{k-1} r^{1-q}\right)^p} r_k^{1-q}, \tag{8.2.37}$$

thus

$$\frac{|\Delta M_k|}{|\Delta M_{k-1}|} \le \left(\frac{r_k}{r_{k-1}}\right)^{1-q}$$

and

$$\frac{M_k^p}{|\Delta M_k|^{p-1}} \le r_k \left(\frac{1}{p-1}\right)^{p-1} \left[\frac{\sum^k r^{1-q}}{\sum^{k-1} r^{1-q}}\right]^{p(p-1)}.$$

Now, according to (8.2.36), there exists $\varepsilon > 0$ such that

$$\limsup_{k\to\infty} \left(\sum^{k-1} r_j^{1-q}\right)^{p-1} \sum_{j=k}^{\infty} c_j^+ < \frac{1}{p\mu^{p-1}}\left(\frac{p-1}{p}\right)^{p-1} \frac{1}{\Psi+\varepsilon},$$

and (8.2.35) implies the existence of $N_0 \in \mathbb{N}$ such that $(1+\psi_N)^{p-1}\varphi_N < \Psi + \varepsilon$ for $N \ge N_0$. Now, using the summation by parts and applying the same idea as in the proof of Lemma 8.2.4, we have for any nontrivial $y \in \mathcal{D}(N)$

$$\begin{aligned}
\sum_{k=N}^{\infty} c_k |y_{k+1}|^p &\le \sum_{k=N}^{\infty} c_k^+ |y_{k+1}|^p = \sum_{k=N}^{\infty}\left(\sum_{j=N}^{\infty} c_j^+\right)\Delta\left(|y_k|^p\right) \\
&= \sum_{k=N}^{\infty} \frac{1}{M_k}\left(\sum_{j=N}^{\infty} c_j^+\right) M_k \Delta(|y_k|^p) \\
&< \left(\frac{p-1}{p}\right)^{p-1} \frac{1}{\mu^{p-1}(\Psi+\varepsilon)} \sum_{k=N}^{\infty} M_k |\Phi(\xi_k)||\Delta y_k| \\
&\le \frac{(p-1)^{p-1}}{p^{p-1}\mu^{p-1}(\Psi+\varepsilon)}\left(\sum_{k=N}^{\infty} \frac{M_k^p}{|\Delta M_k|^{p-1}}|\Delta y_k|^p\right)^{1/p}
\end{aligned}$$

$$
\begin{aligned}
&\times \left[\mu(1+\psi_N)\sum_{k=N}^{\infty}|\Delta M_k||y_{k+1}|^p\right]^{1/q}\\
\le\ & \frac{(p-1)^{p-1}}{p^{p-1}\mu^{p-1}(\Psi+\varepsilon)}\left[\mu(1+\psi_N)\right]^{1/q}p^{p/q}\\
&\times[\mu(1+\psi_N)]^{(p-1)/q}\left(\sum_{k=N}^{\infty}\frac{M_k}{|\Delta M_k|^{p-1}}|\Delta y_k|^p\right)^{\frac{1}{p}+\frac{1}{q}}\\
\le\ & (1+\psi_N)^{p-1}\varphi_N\frac{1}{\Psi+\varepsilon}\sum_{k=N}^{\infty}r_k|\Delta y_k|^p \le \sum_{k=N}^{\infty}r_k|\Delta y_k|^p.
\end{aligned}
$$

Hence

$$\sum_{k=N}^{\infty}\left[r_k|\Delta y_k|^p - c_k|y_{k+1}|^p\right] > 0$$

for every nontrivial $y \in \mathcal{D}(N)$, what we needed to prove. □

8.2.6 Some discrete comparison theorems

We start with a discrete version of Theorem 2.3.12, where at the same time we make the comparison in a Hille-Wintner sense (for differential equations case see Theorem 2.3.1). Observe that in contrast to the continuous counterpart, i.e., Theorem 2.3.12, we do not need here a condition of type (2.3.34) or (2.3.35). This is owing to the fact that for S to be nondecreasing with respect to p, the expression w/r does not need to be small, as it is in the continuous case. On the other hand, we have to use more complicated technique in the proof: the Riccati one combined with the Schauder fixed point theorem. The reason is that the inequality (8.2.19) is involved instead of the discrete counterpart of equation from Theorem 2.2.4. However, if $c_k \ge 0$ or r_k is bounded, then we can simply use just the Riccati technique in view of Theorem 8.2.10 and the note before Theorem 8.2.12. Along with (8.1.4) consider the equation of the same form

(8.2.38) $$\Delta(R_k\Phi_\alpha(\Delta x_k)) + C_k\Phi_\alpha(x_{k+1}) = 0,$$

where $\Phi_\alpha(x) = |x|^{\alpha-1}\operatorname{sgn} x$, $\alpha > 1$. Let β stand for the conjugate number of α.

Theorem 8.2.20. *Suppose that* $0 < R_k \le r_k$ *and*

(8.2.39) $$\sum_{j=k}^{\infty} C_j \ge \sum_{j=k}^{\infty} c_j \ge 0 \quad (\not\equiv 0)$$

for large k *(in particular, we assume that these series are convergent). Further assume that* $\sum_{k=1}^{\infty} R^{1-\beta} = \infty$. *If* $\alpha \ge p$ *and equation* (8.2.38) *is nonoscillatory, then* (8.1.4) *is also nonoscillatory.*

Proof. By Theorem 8.2.12, the nonoscillation of (8.2.38) implies the existence of $m_1 \in \mathbb{N}$ such that

$$z_k \geq \sum_{j=k}^{\infty} C_j + \sum_{j=k}^{\infty} S(z_j, R_j, \alpha) =: Z_k \tag{8.2.40}$$

for $k \geq m_1$ (clearly, with $z_k + R_k > 0$). Let $m_2 \in \mathbb{N}$ be such that (8.2.39) holds and $\sum_{j=k}^{\infty} c_j \geq 0$ for $k \geq m_2$. Set $m = \max\{m_1, m_2\}$ and define the set Ω and the mapping $\mathcal{T}$ by

$$\Omega = \{w \in \ell^\infty : 0 \leq w_k \leq Z_k \text{ for } k \geq m\}$$

and

$$(\mathcal{T}w)_k = \sum_{j=k}^{\infty} c_j + \sum_{j=k}^{\infty} S(w_j, R_j, p), \quad k \geq m, \ w \in \Omega,$$

respectively. We show that $\mathcal{T}$ has a fixed point in Ω. We must verify that

1) Ω is bounded, closed and convex subset of ℓ^∞,
2) $\mathcal{T}$ maps Ω into itself,
3) $\mathcal{T}\Omega$ is relatively compact,
4) $\mathcal{T}$ is continuous.

ad 1) Clearly, Ω is bounded and convex. Let $x^n = \{x_k^n\}$, $n = 1, 2, \dots$, be any sequence in Ω such that x^n approaches x (in the sup norm) as $n \to \infty$. From our assumptions, for any $\varepsilon > 0$ there exists $N \in \mathbb{N}$ such that $\sup_{k \geq m} |x_k^n - x_k| < \varepsilon$ for all $n \geq N$. Thus, for any fixed k, we have $\lim_{n\to\infty} x_k^n = x_k$. Since $0 \leq x_k^n \leq z_k$ for all n, then $0 \leq x_k \leq z_k$. We have $k \geq m$ arbitrary and hence x belongs to Ω

ad 2) Suppose that $w \in \Omega$ and define $x_k = (\mathcal{T}w)_k$, $k \geq m$. Obviously, $x_k \geq 0$ for $k \geq m$. We must show that $x_k \leq Z_k$, $k \geq m$. We have

$$\begin{aligned} x_k &= \sum_{j=k}^{\infty} c_j + \sum_{j=k}^{\infty} S(w_j, R_j, p) \leq \sum_{j=k}^{\infty} C_j + \sum_{j=k}^{\infty} S(w_j, R_j, p) \\ &\leq \sum_{j=k}^{\infty} C_j + \sum_{j=k}^{\infty} S(w_j, R_j, \alpha) \leq \sum_{j=k}^{\infty} C_j + \sum_{j=k}^{\infty} S(z_j, R_j, \alpha) \end{aligned}$$

by the assumptions of theorem and by Lemma 8.2.1 [(i),(iv)]. Hence, $\mathcal{T}\Omega \subset \Omega$.

ad 3) According to [77, Theorem 3.3], it suffices to show that $\mathcal{T}\Omega$ is uniformly Cauchy. Let $\varepsilon > 0$ be given. We show that there exists $N \in \mathbb{N}$ such that for any $k, l > N$ $|(\mathcal{T}x)_k - (\mathcal{T}x)_l| < \varepsilon$ for any $x \in \Omega$. Without loss of generality, suppose $k < l$. Then we have

$$\begin{aligned} |(\mathcal{T}x)_k - (\mathcal{T}x)_l| &= \left| \sum_{j=k}^{l-1} c_j + \sum_{j=k}^{l-1} S(x_j, R_j, p) \right| \\ &= \sum_{j=k}^{l-1} c_j + \sum_{j=k}^{l-1} S(x_j, R_j, p) \end{aligned} \tag{8.2.41}$$

for large k. Taking into account the properties of c_k and $S(x_k, R_k, p)$, for any $\varepsilon > 0$ one can find $N \in \mathbb{N}$ such that

$$\sum_{j=k}^{l-1} c_j < \frac{\varepsilon}{2} \quad \text{and} \quad \sum_{j=k}^{l-1} S(x_j, R_j, p) < \frac{\varepsilon}{2} \quad \text{for } l > k > N.$$

From here and (8.2.41), $|(\mathcal{T}x)_k - (\mathcal{T}x)_l| < \varepsilon$, hence $\mathcal{T}\Omega$ is relatively compact.

ad 4) Let $x^n = \{x_k^n\}$, $k \geq m$, be a sequence in Ω converging to x. The fact that $\mathcal{T}x^n$ converges to $\mathcal{T}x$ can be easily shown by means of the discrete version of the Lebesgue Dominated Convergence Theorem.

Therefore, it follows from the Schauder fixed point theorem that there exists an element $w \in \Omega$ such that $w = \mathcal{T}w$. In view of the definition of $\mathcal{T}$, this (positive) sequence w satisfies the equation

$$w_k = \sum_{j=k}^{\infty} c_j + \sum_{j=k}^{\infty} S(w_j, R_j, p), \quad k \geq m,$$

and hence also the equation (8.2.1) with R instead of r. Consequently, the sequence y given by

$$y_m = m_0 \neq 0 \quad \text{and} \quad y_{k+1} = \left(1 + (w_k/R_k)^{q-1}\right) y_k, \quad k \geq m,$$

is a nonoscillatory solution of

$$\Delta(R_k \Phi(\Delta y_k)) + c_k \Phi(y_{k+1}) = 0$$

and hence this equation is nonoscillatory. The statement now follows from Theorem 8.2.3. □

The next result is an extension of the so-called *(discrete) telescoping principle*, which was introduced in [208] for the second order linear difference equation (8.1.1). Note that in [208] the authors consider equation (8.1.1) only under the assumption $r_k > 0$ and hereby the following result with $r_k \neq 0$ is new even in the linear case (in spite of the fact that the idea of the proof remains essentially the same). Compare with the continuous case, see Subsection 2.3.4.

Before presenting the main result, let us introduce some concepts and assumptions. Denote by $\mathcal{S}$ the set of all real sequences $y = \{y_k : k \in \mathbb{N}\}$. Assume

$$I = \bigcup_{i=1}^{j} I_i, \quad I_i = (m_i, n_i], \ i = 1, \dots, j, \ j \leq \infty, \tag{8.2.42}$$

where $m_i, n_i \in \mathbb{N}$, $i = 1, \dots, j$, are such that $m_i < n_i < m_{i+1}$ and $\operatorname{card}(\mathbb{N} \setminus I) = \infty$. Based on the set I, we define an interval shrinking transformation $\tau = \tau_I : \mathbb{N} \to \mathbb{N}$ as follows:

$$K = \tau(k) = \operatorname{card}([1, k] \cap I^C),$$

where $I^C = \mathbb{N} \setminus I$. Let $M_i = \tau(m_i)$. Then $M_i = \tau(k)$ for $k \in [m_i, n_i]$, $i = 1, \ldots, j$. This transformation τ induces a transformation $T = T_I : \mathcal{S} \to \mathcal{S}$ defined as follows: For $y \in \mathcal{S}$

$$Ty = Y = \{Y_K : K \in \mathbb{N}\} \text{ with } Y_K = y_k \text{ when } \tau(k) = K.$$

Theorem 8.2.21. *Let $r_k \neq 0$, $k \in \mathbb{N}$, and assume that (8.2.42) holds. Let $R = Tr$ and $C = Tc$ for $T = T_I$. Assume*

$$\sum_{k=m_i+1}^{n_i} c_k \geq 0, \quad i = 1, \ldots, j. \tag{8.2.43}$$

Suppose that $X = \{X_K : K \in \mathbb{N}\}$ is a solution of the equation

$$\Delta(R_K \Phi(\Delta X_K)) + C_K \Phi(X_{K+1}) = 0, \tag{8.2.44}$$

such that $R_K X_K X_{K+1} > 0$ for $K < N$ and $R_N X_N X_{N+1} \leq 0$. If the sequence y is a solution of equation (8.1.4) such that $y_1 \neq 0$ and $r_1 \Phi(\Delta y_1)/\Phi(y_1) \leq R_1 \Phi(\Delta X_1)/\Phi(X_1)$, then there exists $l \leq n$ such that $r_l y_l y_{l+1} \leq 0$, where $N = \tau(n)$. More precisely, if $N \leq M_i$, then there exists $l \leq m_i$ such that $r_l y_l y_{l+1} \leq 0$, $i = 1, 2, \ldots, j$.

Proof. In this proof, by $y \not< X$ we mean either $y \geq X$ or y does not exist. The proof is by induction. Assume that the conclusion is not true. Then $w_k = -r_k \Phi(\Delta y_k)/\Phi(y_k)$ satisfies

$$\Delta w_k = c_k + w_k \left(\frac{r_k}{(\Phi^{-1}(r_k) - \Phi^{-1}(w_k))^{p-1}} - 1 \right) \tag{8.2.45}$$

or, equivalently,

$$w_{k+1} = c_k + \tilde{S}(w_k, r_k), \quad k = 1, \ldots, n, \tag{8.2.46}$$

and $w_k < r_k$, $k = 1, \ldots, n$, where

$$\tilde{S}(w_k, r_k) = \frac{w_k r_k}{(\Phi^{-1}(r_k) - \Phi^{-1}(w_k))^{p-1}}.$$

Observe that the behavior of the function $\tilde{S}$ is similar to the behavior of the function $\bar{S}$ from Lemma 8.2.1. In particular, $\tilde{S}(w_k, r_k)$ is nondecreasing with respect to the first variable for $r_k > w_k$. Let $V_K = -R_K \Phi(\Delta X_K)/\Phi(X_K)$. Then

$$V_{K+1} = C_K + \tilde{S}(V_K, R_K), \quad K = 1, \ldots, N-1, \tag{8.2.47}$$

$V_K < R_K$, $K = 1, \ldots, N-1$ and $V_N \not< R_N$.

If $N \leq M_1 = m_1$, then for $k = 1, \ldots, N$, $K = k$, and hence $R_K = r_k$, $P_K = p_k$ and equation (8.2.47) is the same as (8.2.46). By the hypothesis $w_1 \geq V_1$, comparing (8.2.46) and (8.2.47) step by step (using the above property of $\tilde{S}$), we find that $w_{k+1} \geq V_{k+1}$, $k = 1, \ldots, N-1$. In particular,

$$w_n = w_N \geq V_N \not< R_N = r_n.$$

This implies that $w_n \not< r_n$, contradicting the assumption.

If $M_1 < N \le M_2$, then arguing as above we find that $w_{m_1+1} = w_{M_1+1} \ge V_{M_1+1}$. Adding (8.2.45) for k from m_1+1 to n_1 and using (8.2.43), we obtain

$$w_{n_1+1} - w_{m_1+1} = \sum_{k=m_1+1}^{n_1} c_k + \sum_{k=m_1+1}^{n_1} \tilde{S}(w_k, r_k) \ge 0,$$

hence $w_{n_1+1} \ge w_{m_1+1} \ge V_{m_1+1}$. Noting that $\tau(n_1+1) = N_1$, we see that w_k, V_K satisfy the same generalized Riccati equation for $n_1+1 \le k \le n$ and $M_1+1 \le K \le N$, respectively. As before, we see that $w_n \ge V_N \not< R_N = r_n$ and, again, this implies that $w_n \not< r_n$, contradicting the assumption. The proof of inductive step from i to $i+1$ is similar and hence is omitted. □

Theorem 8.2.22 (Telescoping principle). *Under the conditions and with the notation of Theorem 8.2.21, if* (8.2.44) *is oscillatory, then* (8.1.4) *is oscillatory.*

Proof. Let X_k be a solution of (8.2.44) with $X_1 \neq 0$. Let y_k be a solution of (8.1.4) satisfying $y_1 \neq 0$, $r_1\Phi(\Delta y_1)/\Phi(y_1) \le R_1\Phi(\Delta X_1)/\Phi(X_1)$. By Theorem 8.2.21, there exists $l_1 > 0$ such that $r_{l_1} y_{l_1} y_{l_1+1} \le 0$. Now, working on the solution for $k \ge l_1+1$ instead of $k \ge 1$ and proceeding as before, we show that there exists $l_2 \ge l_1+1$ such that $r_{l_2} y_{l_2} y_{l_2+1} \le 0$. Continuing this process leads to the conclusion that y is oscillatory, hence (8.1.4) is oscillatory. □

8.3 Half-linear dynamic equations on time scales

In this section we develop the theory of half-linear dynamic equations on time scales, which unifies and extends the continuous and the discrete theory. In addition, such a theory explains some discrepancies between them. The understanding of these discrepancies is important, for example, for numerical approximations. The statements are presented without proofs. Our attention is focused mainly to those types of the results which explain the discrepancies, or show some phenomena that are not usual in the differential/difference equations case.

8.3.1 Essentials on time scales, basic properties

In 1988, Stefan Hilger [178] introduced the calculus on time scales in order to unify continuous and discrete analysis. By a *time scale* $\mathbb{T}$ (an alternative terminology is *measure chain*) we understand any closed subset of the real numbers $\mathbb{R}$ with the usual topology inherited from $\mathbb{R}$. Typical examples of time scales are $\mathbb{T} = \mathbb{R}$ and $\mathbb{T} = \mathbb{Z}$ – the set of integers (or $h\mathbb{Z}$ defined as $\{hk : k \in \mathbb{Z}\}$ with a positive h). The operators $\rho, \sigma : \mathbb{T} \to \mathbb{T}$ are defined by

$$\sigma(t) = \inf\{s \in \mathbb{T} : s > t\}, \quad \rho(t) = \sup\{s \in \mathbb{T} : s < t\}$$

and are called the *right jump operator* and *left jump operator*, respectively. The quantity $\mu(t) = \sigma(t) - t$ is called the *graininess* of $\mathbb{T}$. A point $t \in \mathbb{T}$ is said to

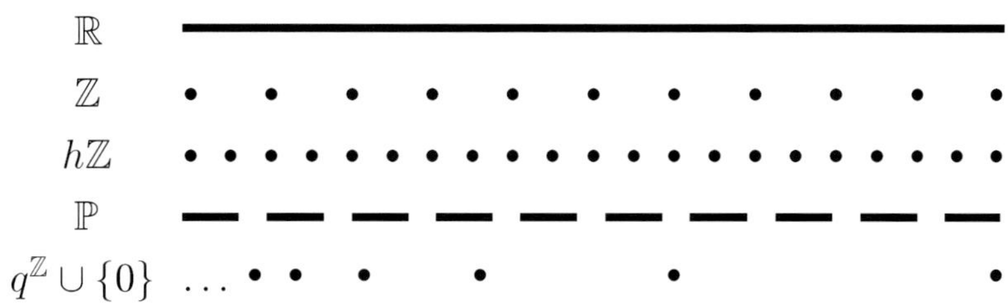

Figure 8.3.1: Examples of time scales

be *right-dense*, *right-scattered*, if $\sigma(t) = t$, $\sigma(t) > t$, respectively. Their "left-counterparts" are defined similarly via $\rho(t)$. If $f : \mathbb{T} \to \mathbb{R}$, the *delta-derivative* is defined by

$$f^{\Delta}(t) = \lim_{s \to t, \sigma(s) \neq t} \frac{f(\sigma(s)) - f(t)}{\sigma(s) - t}.$$

A function $f : \mathbb{T} \to \mathbb{R}$ is called *rd-continuous* provided it is continuous at all right-dense points in $\mathbb{T}$ and its left-sided limits exist (finite) at all left-dense points in $\mathbb{T}$. For $a, b \in \mathbb{T}$ and a delta differentiable function f, the *"Newton integral"* is defined by $\int_a^b f^{\Delta}(t)\,\Delta t = f(b) - f(a)$. For the concept of the *Riemann delta integral* and the *Lebesgue delta integral* see [50, Chapter 5]. Note that we have

$$\sigma(t) = t, \ \mu(t) \equiv 0, \ f^{\Delta} = f', \ \int_a^b f(t)\,\Delta t = \int_a^b f(t)\,dt, \ \text{when } \mathbb{T} = \mathbb{R},$$

while

$$\sigma(t) = t + 1, \ \mu(t) \equiv 1, \ f^{\Delta} = \Delta f, \ \int_a^b f(t)\,\Delta t = \sum_{t=a}^{b-1} f(t), \ \text{when } \mathbb{T} = \mathbb{Z}.$$

These are the most typical time scales, but there exist much more examples, which may bring quite surprising unusual (and unpleasant, sometimes) phenomena in some aspects of the theory. Let us mention at least $\mathbb{T} = q^{\mathbb{N}_0} = \{q^k : k \in \mathbb{N}_0\}$ (or $\mathbb{T} = q^{\mathbb{Z}} \cup \{0\}$), where $q > 1$ is a real number. Then $\sigma(t) = qt$ and $\mu(t) = (q-1)t$, and a dynamic equation considered on such a time scale is called q-difference equation. The last example of the time scale which we present here is the set $\mathbb{P}$ defined as the union of closed mutually disjoint intervals. The basic facts of the time scale calculus can be found in [178] and the general theory of dynamic equations on time scales along with an excellent introduction into the subject is presented in [49, 50].

Now consider the linear dynamic equation on a time scale $\mathbb{T}$

$$(r(t)x^{\Delta})^{\Delta} + c(t)x^{\sigma} = 0, \tag{8.3.1}$$

where $x^{\sigma} = x \circ \sigma$ and $r, c : \mathbb{T} \to \mathbb{R}$ with $r(t) \neq 0$, and its half-linear extension

$$(r(t)\Phi(x^{\Delta}))^{\Delta} + c(t)\Phi(x^{\sigma}) = 0. \tag{8.3.2}$$

Obviously, (8.3.2) reduces to (1.1.1) if $\mathbb{T} = \mathbb{R}$ and to (8.1.4) if $\mathbb{T} = \mathbb{Z}$, respectively.

By means of the approach, which is an extension of that in Subsection 1.1.6, it can be shown that the initial value problem involving equation (8.3.2) is globally uniquely solvable provided the coefficients r, c are rd-continuous. However, it has to mentioned, that the part with a reciprocal equation cannot be extended reasonably here, since it requires Δ and σ to be commutative (i.e., $f^{\Delta\sigma} = f^{\sigma\Delta}$), which is not true in general. Also the approach based on the Prüfer transformation has not been developed yet. The reason is that there is no "real" chain rule for differentiation on time scales.

8.3.2 Oscillation theory of half-linear dynamic equations

Concerning (8.3.1), the foundation of oscillation theory of this dynamic equation was established in [154], and then elaborated in many works. Here we offer the half-linear extension of this theory.

Similarly as in the above cases, the central role is played by the Roundabout theorem. In that theorem, the Riccati dynamic equation

$$w^{\Delta} + c(t) + S[w, r](t) = 0, \tag{8.3.3}$$

where

$$S[w, r](t) = \lim_{\lambda \to \mu(t)} \frac{w(t)}{\lambda} \left(1 - \frac{r(t)}{\Phi(\Phi^{-1}(r(t)) + \lambda \Phi^{-1}(w(t)))}\right),$$

and the p-degree functional

$$\mathcal{F}(y; a, b) = \int_a^b \left[r(t)|y^{\Delta}|^p - c(t)|y^{\sigma}|^p\right] \Delta t$$

play the same role as their continuous and discrete counterparts. Observe how the function S looks like when $p = 2$ or $\mathbb{T} = \mathbb{R}$ or $\mathbb{T} = \mathbb{Z}$. Here are another important concepts: We say that a solution y of (8.3.2) has a *generalized zero* at t in case $y(t) = 0$. We say y has a *generalized zero* in $(t, \sigma(t))$ in case $r(t)y(t)y(\sigma(t)) < 0$. We say that (8.3.2) is *disconjugate* on the interval I, if there is no nontrivial solution of (8.3.2) with two (or more) generalized zeros in I. The definition of (non)oscillation of (8.3.2) is obvious. Note that in the generalized Roundabout theorem, which formally looks the same as its special cases $\mathbb{T} = \mathbb{R}$ or $\mathbb{T} = \mathbb{Z}$. the solution w of (8.3.3) has to satisfy the additional condition $\Phi^{-1}(r(t)) + \mu(t)\Phi^{-1}(w(t)) > 0$.

Thanks to this theorem, an extension of the Sturmian theory can be developed, and the generalized Riccati technique and variational principle are at disposal. Consequently, many statements (in particular, (non)oscillation criteria and comparison theorems) for (8.3.2) can be stated. Our aim here is not to present all those which have already been established. We just want to stress some interesting points where, in particular, the role of graininess (i.e., the role of what time scale is just chosen) can be seen.

For instance, the unification and extension of Theorems 3.1.1 and 8.2.15 is the following Hille-Nehari type criterion.

Theorem 8.3.1. *Suppose that* $\int^\infty r^{1-q}(t)\,\Delta t = \infty$ *and* $\int^\infty c(t)\,\Delta t$*, with* $c(t) \ge 0$*, converges. Let*

$$\lim_{t\to\infty} \frac{\mu(t) r^{1-q}(t)}{\int_a^t r^{1-q}(s)\,\Delta s} = 0. \tag{8.3.4}$$

If

$$\liminf_{t\to\infty} \left(\int^t r^{1-q}(s)\,\Delta s \right)^{p-1} \int_t^\infty c(s)\,\Delta s > \frac{1}{p}\left(\frac{p-1}{p}\right)^{p-1},$$

then (8.3.2) *is oscillatory.*

Observe that for $\mathbb{T} = \mathbb{R}$, condition (8.3.4) is trivially satisfied, while it reduces to (8.2.24) when $\mathbb{T} = \mathbb{Z}$. It is easy to see that if $r(t) \equiv 1$, then (8.3.4) holds when $\mathbb{T} = \mathbb{Z}$. However, this fails to hold when $\mathbb{T} = q^{\mathbb{N}_0}$. In general, a "larger" graininess is "worse" for this condition to be satisfied.

On the other hand, a "larger" graininess may make some conditions to be satisfied "more easily". Indeed, a closer examination of the proof of the comparison theorem with respect to p (Theorem 2.3.12) shows that a crucial role is played the monotonicity of the function $S = S(x, y, p)$ with respect to p. For that, in the continuous case, we need $z - z\log z \ge 0$, where $z = (x/y)^{q-1}$. As said above, in the corresponding discrete case, there is no such a condition. In the general time scale case, this condition reads as

$$\lim_{\lambda\to\mu} \frac{(1+\lambda z)\log(1+\lambda z) - \lambda z \log z}{\lambda} \ge 0.$$

Observe that this condition really reduces to $z - z\log z \ge 0$ if $\mu = 0$ ($\mathbb{T} = \mathbb{R}$) and it is trivially satisfied if $\mu = 1$ ($\mathbb{T} = \mathbb{Z}$).

We conclude this section by showing another phenomenon, concerning half-linear dynamic equations, where it can be seen how the graininess affects oscillatory properties in such a manner, that is not known from the differential/difference equations case. Consider the generalized Euler type dynamic equation

$$\left[\Phi\left(y^\Delta\right)\right]^\Delta + \frac{\gamma}{(\sigma(t))^p}\Phi(y^\sigma) = 0. \tag{8.3.5}$$

By means of the Hardy inequality on time scales combined with the variational principle and the Sturm type comparison theorem, it can be shown that this equation is nonoscillatory provided $\gamma \le q^{-p}$. Now consider the generalized Euler type dynamic equation in a slightly different form

$$[\Phi(y^\Delta)]^\Delta + \frac{\gamma}{t^p}\Phi(y^\sigma) = 0. \tag{8.3.6}$$

Let $0 < \gamma < q^{-p}$. Assume that $\mathbb{T}$ is such that $\mu(t)/t \to 0$ as $t \to \infty$. Then (8.3.6) is nonoscillatory by the time scale extension of Theorem 3.1.3. Note that this extension requires condition (8.3.4), which in our case reduces to $\mu(t)/t \to 0$. Now pick a time scale such that $\int_a^\infty t^{-p}\Delta t = \infty$, e.g., let $\mathbb{T} = \left\{2^{p^k} : k \in \mathbb{N}_0\right\}$.

Let γ be the same as before. Equation (8.3.6) is then oscillatory by the criterion corresponding to the Hille-Wintner theorem. Thus we have an example showing that oscillatory properties of equation (8.3.6) may be completely changed when one replaces a time scale by a different one, leaving the form of the equation the same. In particular, there is no "important" (time scale-invariant) critical constant q^{-p} in (8.3.6) as it seems to be in (8.3.5) (till now we know that (8.3.5) is oscillatory provided $\gamma > q^{-p}$ and $\mu(t)/t \to 0$ as $t \to \infty$ by the above Hille-Nehari type criterion; unfortunately, for other time scales this question remains open). On the other hand, interesting questions arise like what a "critical" graininess of the time scale is, which is a "border" between oscillation and nonoscillation of (8.3.6). In other words, whether there exists a time scale whose graininess is bounded by certain critical function, and (8.3.6) is nonoscillatory on such a time scale, while if we take a time scale which has the graininess greater than this critical function at infinitely many points, then (8.3.6) becomes oscillatory.

8.4 Notes and references

The results concerning half-linear difference equation are taken from Řehák [322, 323, 321, 324, 325, 326] and Došlý, Řehák [115]. The paper [326] by Řehák deals with strong (non)oscillation of (8.1.4) and also contains the examination of generalized discrete Euler equation, which cannot be solved explicitly. The concept of recessive solution for (8.1.4) (i.e., the discrete counterpart of principal solution) is introduced in Došlý, Řehák [116] via the minimal solution of generalized Riccati difference equation. Some of its basic properties and applications are given there as well. Another characterization of recessive solution can be found in Cecchi, Došlá, Marini [62]. The papers [273, 280] by Mařík deals with the discrete p-degree functional considered for sequences satisfying another type of boundary conditions than those mentioned above. Forced oscillation is investigated in Došlý, Graef, Jaroš [109], while oscillation and nonoscillation of half-linear difference equations generated by deviating arguments is studied in Wong, Agarwal [365]. Various aspects of qualitative theory of (8.1.4), like existence/nonexistence of positive nondecreasing solutions or comparison theorems can be found [75, 76, 77, 238, 261] by Cheng, Li, Lu, Patula, Yeh. Very recent monograph [2] by Agarwal, Bohner, Grace and O'Regan deals specially with the discrete oscillation theory. The literature concerning the results on qualitative theory of difference equations, which are quasilinear or involving similar types of nonlinearities is very extensive. The monograph [1] by Agarwal as well as above mentioned [2] can serve as a good source for searching such references.

The part devoted to half-linear dynamic equations is based on the papers [3] by Agarwal, Bohner, Řehák, and [327, 328, 330] by Řehák. Further related results can be found in Řehák's paper [329].

CHAPTER 9

RELATED DIFFERENTIAL EQUATIONS AND INEQUALITIES

The aim of this chapter is to study the equations which are in various relationships to half-linear second order equations. We start with a natural generalization of half-linear equations – the so-called quasilinear equations, i.e., the equations where the exponents in nonlinearities of the first and second term are generally different. In two subsequent sections we discuss how a forcing term and the presence of deviating arguments, respectively, affect oscillatory properties. In the fourth section we mention few words about half-linear equations of higher order. The chapter is concluded by the section devoted to the classical inequalities that are related to half-linear equations.

9.1 Quasilinear differential equations

In this section we change the notation which we have used throughout the whole book. Till now, q was the conjugate number of p, i.e., $q = p/(p-1)$. In this section, *q is any real number satisfying $q > 1$* and the conjugate number of p will be denoted by p^*. We will mainly deal with the *quasilinear equation*

$$(9.1.1) \qquad (r(t)\Phi_p(x'))' + c(t)\Phi_q(x) = 0, \quad \Phi_p(s) = |s|^{p-2}s, \ \Phi_q(s) := |s|^{q-2}s;$$

we will briefly treat also some more general equations. Note that in some literature, if $p = 2$ and q is general (with $q > 1$), then (9.1.1) is called *semilinear equation*, while for p and q general (with $p, q > 1$), it is called to be quasilinear. The functions r, c satisfy the same assumptions as in (1.1.1). Observe that in general, the solution space to (9.1.1) is neither additive nor homogeneous. If $p = q$, then (9.1.1) reduces to (1.1.1). This means that the investigation of (9.1.1) is more complicated than that of (1.1.1). We will see that the loss of homogeneity brings some phenomena which are not usual in the (half-)linear case.

In the literature, under the quasilinear equations are often understood also equations in slightly more general forms. In fact, many of the results, in particular, asymptotic and oscillatory properties, can be easily extended to such equations. For instance, the second term may read as $\pm c(t)f(x)$ (with $c(t)>0$) or $\pm F(t,x)$, f and F being continuous, where the sign condition

$$\operatorname{sgn} f(x) = \operatorname{sgn} x, \quad \text{resp. } \operatorname{sgn} F(t,x) = \operatorname{sgn} x \ (\text{for each fixed } t) \tag{9.1.2}$$

is satisfied, and some monotonicity assumption, as well as (strong) sub/superlinearity, are usually required. Sometimes, the monotonicity assumption can be relaxed to the existence of suitable monotone functions which estimate the nonlinearity. Another generalization lies in replacing the first term in (9.1.1) by $(r(t)\varphi(x'))'$, where $\varphi : \mathbb{R} \to \mathbb{R}$ is continuous, strictly increasing and such that

$$\operatorname{sgn} \varphi(u) = \operatorname{sgn} u \quad \text{and} \quad \varphi(\mathbb{R}) = \mathbb{R}. \tag{9.1.3}$$

Recall that very important role is played by the fact whether the integral

$$\int^{\infty} r^{1/(1-p)}(s)\, ds$$

is convergent or divergent – asymptotic behavior of solutions may be then substantially changed. Similarly as for half-linear equations, in some cases equation (9.1.1) can be studied with $r \equiv 1$, owing to the transformation of independent variable. Since the structure of solution spaces of quasilinear equations is much more complicated, in the most of cases we have to assume that the coefficient c is eventually nonnegative or nonpositive – clearly, the character of results may be then completely different.

Note that using the substitution $r\Phi_p(x') =: u$, equation (9.1.1) can be written as the first order system of the form

$$x' = a_1(t)|u|^{\lambda_1} \operatorname{sgn} u, \quad u' = a_2(t)|x|^{\lambda_2} \operatorname{sgn} x \tag{9.1.4}$$

with suitable functions a_1, a_2 and real constants λ_1, λ_2. The last system has been investigated in several papers of Mirzov and the results are summarized in his book [292]. In some works, more general systems appear, e.g. of the form $x' = a(t)f_1(x)g_1(y)$, $y' = b(t)f_2(x)g_2(y)$, where the functions a, b, f_1, f_2, g_1, g_2 are subject to suitable conditions. There is very extensive literature on quasilinear equations. As a sample of papers dealing with (9.1.1) and related equations we refer to [28, 58, 155, 212, 263, 334] and the references given therein. Let us mention also the monographs [6, 202]. We do not want to repeat that "quasilinear" theory which has already been processed in various works. The principal aim of this section is to point at relationships with the "half-linear" theory.

9.1.1 Quasilinear equations with constant coefficients

First we focus our attention to the initial value problem

$$(\Phi_p(x'))' + \lambda\Phi_q(x) = 0, \quad x(0) = a, \ x'(0) = b. \tag{9.1.5}$$

We will modify the method used in the definition of the half-linear sine function $\sin_p$ and of other half-linear trigonometric functions.

Theorem 9.1.1. *For any $\lambda \geq 0$, the initial value problem (9.1.5) has a unique solution defined on the whole real line $\mathbb{R}$.*

Proof. The crucial fact used in the proof is that

$$\frac{|x'(t)|^p}{p^*} + \lambda \frac{|x(t)|^q}{q} = \frac{|b|^p}{p^*} + \lambda \frac{|a|^q}{q}, \tag{9.1.6}$$

as can be verified by differentiation. Clearly, if $a = 0 = b$, the last identity implies that the trivial solution is the unique solution. If $a = 0$ or $b = 0$, supposing that there are two different solutions x_1, x_2 satisfying the same initial conditions, we find that the absolute value of their difference $z = |x_1 - x_2|$ satisfies a Gronwall type inequality and hence $z \equiv 0$. This idea, slightly modified, applies also to the case when both $a \neq 0$ and $b \neq 0$. □

The remaining part of this subsection will be devoted to the initial value problem

$$(\Phi_p(x'))' + \lambda \Phi_q(x) = 0, \quad x(0) = 0, \ x'(0) = \alpha > 0. \tag{9.1.7}$$

Denote by t_α the first positive zero of the derivative x', i.e., $x(t) > 0$, $x'(t) > 0$ for $t \in (0, t_\alpha)$. Further denote by $R := x(t_\alpha)$. Then using the same idea as above we have the identity

$$\frac{(x'(t))^p}{p^*} + \lambda \frac{x^q(t)}{q} = \lambda \frac{R^q}{q}. \tag{9.1.8}$$

Solving this equality for x' and integrating, we find

$$\left(\frac{q}{\lambda p^*}\right)^{\frac{1}{p}} \int_0^t \frac{x'(s)\, ds}{(R^q - x^q(s))^{\frac{1}{p}}} = t, \tag{9.1.9}$$

which after a change of variables can be written as

$$t = \left(\frac{q}{\lambda p^*}\right)^{\frac{1}{p}} \frac{1}{R^{\frac{q-p}{p}}} \int_0^{\frac{x}{R}} \frac{ds}{(1 - s^q)^{\frac{1}{p}}}. \tag{9.1.10}$$

For $t \in [0, q/2]$, let us set

$$\operatorname{arcsin}_{pq} t := \frac{q}{2} \int_0^{\frac{2t}{q}} \frac{ds}{(1 - s^q)^{\frac{1}{p}}}, \tag{9.1.11}$$

and note that this integral converges for $t \in [0, q/2]$. Substituting $t = \tau^{\frac{1}{q}}$ in (9.1.11), we obtain

$$\operatorname{arcsin}_{pq} t = \frac{1}{2} \tilde{B}\left(\frac{1}{q}, \frac{1}{p^*}, \left(\frac{2t}{q}\right)^q\right), \tag{9.1.12}$$

where

$$\tilde{B}\left(\frac{1}{q}, \frac{1}{p^*}, y\right) = \int_0^y \tau^{\frac{1}{q}-1}(1-\tau)^{-\frac{1}{p}}\, d\tau$$

denotes the incomplete beta function. Next, substituting $t = q/2$ in (9.1.12), we define

$$\pi_{pq} := 2\arcsin_{pq}\left(\frac{q}{2}\right) = B\left(\frac{1}{q}, \frac{1}{p^*}\right),$$

where B denoted the classical Euler beta function.

The function $\arcsin_{pq}$ is a bijection of $[0, q/2]$ onto $[0, \pi_{pq}/2]$, so we can define first $\sin_{pq} : [0, \pi_{pq}/2] \to [0, q/2]$ as the inverse function of $\arcsin_{pq}$ and then to define this function for $t \in \mathbb{R}$ in the obvious way: $\sin_{pq} t = \sin_{pq}(\pi_{pq} - t)$ for $t \in [\pi_{pq}/2, \pi_{pq}]$ and then extend this function over $\mathbb{R}$ as odd and $2\pi_{pq}$ periodic function. It is a simple matter to verify that $\sin_{pq}$ is the unique (global) solution of the initial value problem

$$(\Phi_p(x'))' + \frac{2^q}{p^* q^{q-1}}\Phi_q(x) = 0, \quad x(0) = 0,\ x'(0) = 1. \tag{9.1.13}$$

Similarly as in case $p = q$ we denote $\cos_{pq} t = \frac{d}{dt}\sin_{pq} t$. Then from (9.1.6) and (9.1.13) we have

$$|\cos_{pq} t|^p + \left(\frac{2}{q}\right)^q |\sin_{pq} t|^q \equiv 1. \tag{9.1.14}$$

From (9.1.10) and (9.1.11) we find that

$$t = \frac{2}{(\lambda p^*)^{\frac{1}{p}} q^{\frac{1}{p^*}}} R^{\frac{p-q}{p}} \arcsin_{pq}\left(\frac{qx}{2R}\right), \tag{9.1.15}$$

and hence

$$x(t) = \frac{2R}{q}\sin_{pq}\left(\frac{(\lambda p^*)^{\frac{1}{p}} q^{\frac{1}{p^*}}}{2} R^{\frac{q-p}{p}} t\right), \tag{9.1.16}$$

for all $t \in \mathbb{R}$. From (9.1.8) we can express R in terms of α to obtain

$$R^{\frac{q-p}{p}} = \left(\frac{q}{\lambda p^*}\right)^{\frac{q-p}{pq}} \alpha^{\frac{q-p}{q}}.$$

Substituting this expression into (9.1.16), and setting

$$A_{pq}(\alpha, \lambda) := \frac{(\lambda p^*)^{\frac{1}{q}} q^{\frac{1}{q^*}}}{2}\alpha^{\frac{q-p}{q}}, \quad q^* = \frac{q}{q-1},$$

we find that the solution of (9.1.7) is

$$x(t) = \frac{\alpha}{A_{pq}(|\alpha|, \lambda)}\sin_{pq}(A_{pq}(|\alpha|, \lambda)t), \tag{9.1.17}$$

and this solution is $\tau(\alpha)$-periodic function with

$$\tau(\alpha) = \frac{2\pi_{pq}}{A_{pq}(\alpha, \lambda)} = 4t_\alpha.$$

Theorem 9.1.2. *For any given $\alpha \neq 0$, the set of eigenvalues of the problem*

$$(\Phi_p(x'))' + \lambda\Phi_q(x) = 0, \quad x(0) = 0 = x(T), \tag{9.1.18}$$

whose (nontrivial) solution x satisfies $x'(0) = \alpha$, is given by

$$\lambda(\alpha) = \left(\frac{2n\pi_{pq}}{T}\right)^q \frac{|\alpha|^{p-q}}{p^* q^{q-1}}, \quad n \in \mathbb{N}, \tag{9.1.19}$$

with the corresponding eigenfunctions

$$x_{n,\alpha}(t) = \frac{\alpha T}{n\pi_{pq}} \sin_{pq}\left(\frac{n\pi_{pq}}{T}t\right), \quad n \in \mathbb{N}.$$

Proof. For a given $\alpha \in \mathbb{R}$, by imposing that x in (9.1.17) satisfies the boundary conditions in (9.1.18), we obtain that λ is an eigenvalue of this problem if and only if

$$\frac{1}{2}(p^*)^{\frac{1}{q}} q^{\frac{1}{q^*}} \lambda^{\frac{1}{q}} |\alpha|^{\frac{q-p}{q}} T = n\pi_{pq}, \quad n \in \mathbb{N},$$

and hence (9.1.19) follows. The expression for eigenfunctions follows then directly from (9.1.17). □

9.1.2 Existence, uniqueness and singular solutions

Equation (9.1.1) and system (9.1.4) are sometimes called of *Emden-Fowler type* (or *generalized Emden-Fowler*) since if $p = 2$ in (9.1.1), this equation reduces to Emden-Fowler equation. It is known that this equation may possess the so-called singular solution, i.e., the global existence or uniqueness may be violated.

Recall that a solution x of (9.1.1) is called the *singular solution of the first kind* if x becomes eventually trivial, i.e., there exists $T \in \mathbb{R}$ such that $x \not\equiv 0$ for $t < T$ and $x(t) = 0$ for $t \geq T$, and a solution x is *singular solution of the second kind* if there exists a finite time T such $\lim_{t\to T-} |x(t)| = \infty$. The set of singular solutions of the first and second kind will be denoted by $\mathbb{S}_1$ and $\mathbb{S}_2$, respectively. A solution which is not singular is called *proper*. There is also an alternative terminology concerning singular solutions. A solution x of (9.1.1) is called the *extinct solution (of the first kind)* if there exists $T \in \mathbb{R}$ such that $x \not\equiv 0$ for $t < T$ and $\lim_{t\to T-} y(t) = 0 = \lim_{t\to T-} y'(t)$, i.e., it has the same meaning as the singular solution of the first kind. For the meaning of the *extinct solution of the second kind* see the end of Subsection 9.1.7. Finally note that the singular solution of the second kind is sometimes called *blowing-up solution*.

Theorem 9.1.3. *Suppose that $r(t) > 0$, $c(t) < 0$ for large t.*

(i) If $p = q$, i.e., (9.1.1) reduces to (1.1.1), then $\mathbb{S}_1 = \emptyset$, $\mathbb{S}_2 = \emptyset$.

(ii) If $p < q$, then $\mathbb{S}_1 = \emptyset$ and $\mathbb{S}_2 \neq \emptyset$.

(iii) If $p > q$, then $\mathbb{S}_1 \neq \emptyset$ and $\mathbb{S}_2 = \emptyset$.

Proof. Equation (9.1.1) can be written as a system of the Emden-Fowler type for the vector $(x,u) = (x, r\Phi_p(x'))$

$$x' = a_1(t)|u|^{\lambda_1} \operatorname{sgn} u, \quad u' = a_2(t)|x|^{\lambda_2} \operatorname{sgn} x, \tag{9.1.20}$$

where $\lambda_1 = 1/(p-1)$, $\lambda_2 = q-1$, $a_1(t) = 1/\Phi_{p^*}(r(t))$, $a_2(t) = -c(t)$. Using results of Mirzov [292, Theorems 9.1, 9.2] on the nonexistence of singular solutions of system (9.1.20), we obtain for (9.1.1) that $\mathbb{S}_1 = \emptyset$ when $p \le q$ and $\mathbb{S}_2 = \emptyset$ when $p \ge q$. The existence of singular solutions of (9.1.1) was proved by Chanturia [71, Theorems 1, 3]. From there it follows for (9.1.1) that $\mathbb{S}_2 \ne \emptyset$ if $p < q$ and $\mathbb{S}_1 \ne \emptyset$ if $p > q$. □

Observe that $c(t)$ is assumed to be negative in the theorem. The different situation happens when $c(t)$ is positive and satisfies additional conditions, as the following theorem shows.

Theorem 9.1.4. *Suppose that $r(t) \equiv 1$ and $c(t)$ is continuously differentiable and positive on $[a,\infty)$. Then for any A and B, the solution of the initial value problem (9.1.1), $x(t_0) = A$, $x'(t_0) = B$, $t_0 \in [a,\infty)$, exists on $[a,\infty)$ and is unique.*

Proof. We give only the sketch of the proof. Similarly as in Subsection 1.1.6 we distinguish the cases where the right-hand side of system (9.1.20) satisfies or not the Lipschitz condition with respect to the initial conditions and the values p and q. Then we apply Peano theorem, Picard theorem and Gronwall inequality, similarly as in Lemmas 1.1.3–1.1.5. A different approach needs to be used to prove the uniqueness for the only remaining case where $q < 2$ and $A = B = 0$. This approach requires the continuous differentiability of c, it is based on certain energy functions related to (9.1.1), and serves to show the global existence in the general case as well. □

In fact, the last theorem can be established under less restrictive assumptions on $c(t)$, namely $c(t) > 0$ is locally of bounded variation on $[a,\infty)$.

9.1.3 Asymptotic of nonoscillatory solutions

The classification of nonoscillatory solutions of (1.1.1) can be extended under the assumption that $c(t) \ne 0$ for large t also to (9.1.1):

$$\mathbb{M}^+ = \{x : \exists t_x \ge 0 : \ x(t)x'(t) > 0 \text{ for } t > t_x\},$$
$$\mathbb{M}^- = \{x : \exists t_x \ge 0 : \ x(t)x'(t) < 0 \text{ for } t > t_x\}.$$

The following theorem deals with the existence of solutions in these classes. It is closely related to Theorem 9.1.3.

Theorem 9.1.5. *Suppose that $r(t) > 0$, $c(t) < 0$ for large t.*

(i) If $p = q$, i.e., (9.1.1) reduces to (1.1.1), then $\mathbb{M}^- \ne \emptyset$ and $\mathbb{M}^+ \ne \emptyset$.

(ii) If $p < q$, then $\mathbb{M}^- \ne \emptyset$.

(iii) If $p > q$, then $\mathbb{M}^+ \neq \emptyset$.

Proof. We will use Theorem 9.1.3. Assume $p \leq q$. Using a result of Chanturia [72, Theorem 1], we get that there exists a solution x of (9.1.1) such that $|x(0)| > 0$, $x(t)x'(t) \leq 0$ for $t > 0$. Since $\mathbb{S}_1 = \emptyset$, we obtain $x \in \mathbb{M}^-$. Now assume $p \geq q$. It remains to show that $\mathbb{M}^+ \neq \emptyset$. Let x be a solution of (9.1.1) satisfying the initial condition $x(0)x'(0) > 0$. Then we can assume that x is defined on $(0, \infty)$ since $\mathbb{S}_2 = \emptyset$. Consider the function F_x given by $F_x(t) = r(t)\Phi_p(x'(t))x(t)$. Then $F_x(0) > 0$ and

$$\begin{aligned}\frac{d}{dt}F_x(t) &= [r(t)\Phi_p(x'(t))]'x(t) + r(t)\Phi_p(x'(t))x'(t)\\ &= -c(t)\Phi_q(x(t))x(t) + r(t)\Phi_p(x'(t))x'(t) \geq 0.\end{aligned}$$

Thus F_x is a nondecreasing function, which implies $x(t)x'(t) > 0$. Taking into account that, in this case, $\mathbb{S}_2 = \emptyset$, the assertion follows. □

In Section 4.1 we have seen that certain integrals of functions r, c play an important role in the asymptotic classification of nonoscillatory solutions of (1.1.1). As an illustration of the extension of these results to (9.1.1) we give two statements. The first one deals with the existence in $\mathbb{M}^-$ and the Schauder-Tychonov fixed point theorem plays a crucial role in its proof.

Theorem 9.1.6. *Suppose that $r(t) > 0$, $c(t) < 0$ for large t, $\int^\infty r^{1-p^*}(t)\,dt < \infty$ and*

$$\int^\infty |c(t)|\Phi_q\left(\int_t^\infty r^{1-p^*}(s)\,ds\right)dt < \infty,$$

where $\Phi_q(s) = |s|^{q-1}\operatorname{sgn} s$. Then there exists at least one solution x of (9.1.1) in the class $\mathbb{M}^-$ such that $\lim_{t\to\infty} x(t) = 0$ and

$$\lim_{t\to\infty}\frac{x(t)}{\int_t^\infty r^{1-p^*}(s)\,ds} = \ell_x, \quad 0 < \ell_x < \infty. \tag{9.1.21}$$

Proof. Let t_0 be so large that

$$\int_{t_0}^\infty |c(t)|\Phi_q\left(\int_t^\infty r^{1-p^*}(s)\,ds\right)dt < 1 - \Phi_p(1/2)$$

and denote with $C[t_0, \infty)$ the Fréchet space of all continuous functions on $[t_0, \infty)$ endowed with the topology of uniform convergence on compact subintervals of $[t_0, \infty)$. Let Ω be the nonempty subset of $C[t_0, \infty)$ given by

$$\Omega = \left\{u \in C[t_0, \infty) : \frac{1}{2}\int_t^\infty \Phi_{p^*}\left(\frac{1}{r(s)}\right)ds \leq u(t) \leq \int_t^\infty \Phi_{p^*}\left(\frac{1}{r(s)}\right)ds\right\}.$$

Clearly Ω is bounded, closed and convex. Consider the operator $\mathcal{T} : \Omega \to C[t_0, \infty)$ defined by

$$\mathcal{T}(u)(t) = \int_t^\infty \Phi_{p^*}\left[\frac{1}{r(s)}\left(1 + \int_{t_0}^s c(\tau)\Phi_q(u(\tau))\,d\tau\right)\right]ds.$$

Now it is not difficult to verify that $\mathcal{T}$ is continuous in Ω and $\mathcal{T}(\Omega)$ is relatively compact in $C[t_0,\infty)$. Therefore, by the Schauder-Tychonov fixed point theorem, there exists a fixed element $x \in \Omega$ which is a desired solution. The assertion on the asymptotic estimation follows from the fact that the argument of the limit in (9.1.21) is bounded and $x'(t)/r^{1-p^*}(t)$ is monotone. Indeed, we apply the L'Hospital rule to the limit in (9.1.21). □

It is known (see [200, Theorem 4] and [202, Corollary 17.4]) that if $p = 2$, $r(t) \equiv 1$, $q > 2$ and $\liminf_{t\to\infty} t^q c(t) > 0$, then $\mathbb{M}^+ = \emptyset$. Consequently, when $p < q$ the class $\mathbb{M}^+$ may be empty. The next theorem gives the conditions ensuring that $\mathbb{M}^+$ is nonempty. Following [203], an eventually positive solution $x \in \mathbb{M}^+$ is said to be *strongly increasing* if $\lim_{t\to\infty} x(t) = \infty$, $\lim_{t\to\infty} r(t)\Phi_p(x'(t)) = \infty$, and an eventually positive solution $x \in \mathbb{M}^+$ is said to be *weakly increasing* if at least one of the limits $\lim_{t\to\infty} x(t)$ and $\lim_{t\to\infty} r(t)\Phi_p(x'(t))$ exists finitely.

Theorem 9.1.7. *Suppose that $r(t) > 0$, $c(t) < 0$ for large t, $p < q$ and*

$$\int^{\infty} r^{1-p^*}(t)\Phi_{p^*}\left(\int^t |c(s)|\,ds\right) dt < \infty$$

or

$$\int^{\infty} |c(t)|\Phi_q\left(\int^t r^{1-p^*}(s)\,ds\right) dt < \infty.$$

Then $\mathbb{M}^+$ contains a one parametric family of strongly increasing solutions and a one parametric family of weakly increasing solutions.

Proof. This result was shown in [229, Theorem 1 and its corollary] for system (9.1.20). Putting $\lambda_1 = 1/(p-1)$, $\lambda_2 = q-1$, $a_1(t) = 1/\Phi_{p^*}(r(t))$ and $a_2(t) = -c(t)$, the assertion follows. □

9.1.4 Sufficient and necessary conditions for oscillation

Since the Sturmian theory does not extend to (9.1.1) in general, the concept of *oscillation of equation* cannot be defined by means of the existence of one (arbitrary) oscillatory solution, and so we simply require *oscillation of all nontrivial (proper) solutions* (recall that a solution is said to be oscillatory if it has arbitrarily large zeros). Nevertheless, there are quite many oscillation criteria in the (half-)linear case which can be more or less extended to the quasilinear case. For instance, if $r(t) \equiv 1$ and $\int^\infty c(s)\,ds = \lim_{t\to\infty}\int^t c(s)\,ds = \infty$, then all proper solutions of (9.1.1) are oscillatory, see e.g. [206]. Of course, some of the refined criteria for (1.1.1), which need very sophisticated methods, are impossible to extend. On the other hand, in the quasilinear case, one can find equations (in particular, the so-called strongly sub/super-linear ones, which do not include (half-)linear ones), for which some phenomena occur that are not known in the (half-)linear case. By this we mean, for instance, the below mentioned effective conditions for oscillation which are necessary and sufficient at the same time.

For illustration, we consider a more general equation of the form

$$(r(t)\varphi(y'))' + F(t,y) = 0, \tag{9.1.22}$$

where r, φ and F satisfy the conditions presented at the beginning of this section (in particular, (9.1.2) and (9.1.3)). In the course of the progress, we will show how the situation looks like for the special case (9.1.1), i.e., $\varphi(\cdot) = \Phi_p(\cdot)$ and $F(t,\cdot) = c(t)\Phi_q(\cdot)$ with $c(t) > 0$.

In the sequel, we will assume that

(9.1.23) $$\int^{\infty} \left|\varphi^{-1}\left(\frac{C}{r(t)}\right)\right| dt = \infty \quad \text{for every constant } C \neq 0$$

and

(9.1.24) $$\varphi^{-1}(uv) \geq \varphi^{-1}(u)\varphi^{-1}(v) \quad \text{for all } u, v \text{ with } uv > 0.$$

Further, we will use the condition: for each fixed $B > 0$

(9.1.25) $$\lim_{A \to 0, AB > 0} \frac{\int_a^t \varphi^{-1}(A/r(s))}{\int_a^t \varphi^{-1}(B/r(s))} = 0$$

uniformly on any interval of the form $[t_0, \infty)$, $t_0 > a$. Clearly, in the case of equation (9.1.1), condition (9.1.23) reduces to $\int^{\infty} r^{1-p^*}(s) = \infty$ and conditions (9.1.24), (9.1.25) are trivially satisfied.

The crucial role is played by the following concepts. We say that equation (9.1.22) is *strongly superlinear* if there exists a constant $\gamma > 0$ such that the function $|v|^{-\gamma}|F(t,v)|$ is nondecreasing in $|v|$ for each fixed t and

(9.1.26) $$\int_M^{\infty} \frac{dv}{\varphi^{-1}(v^{\gamma})} < \infty \quad \text{and} \quad \int_{-\infty}^{-M} \frac{dv}{\varphi^{-1}(|v|^{\gamma})} < \infty \quad \text{for any } M > 0;$$

equation (9.1.22) is *strongly sublinear* if there exists a constant $\delta > 0$ such that $|v|^{-\delta}|F(t,v)|$ is nonincreasing in $|v|$ for each fixed t,

$$\int_0^N \frac{dv}{[\varphi^{-1}(v)]^{\delta}} < \infty \quad \text{and} \quad \int_{-N}^0 \frac{dv}{|\varphi^{-1}(v)|^{\delta}} < \infty \quad \text{for any } N > 0.$$

According to this definition, (9.1.1) is strongly superlinear if $p < q$ and strongly sublinear if $p > q$. Observe that the (half-)linear case is not included.

The proofs of the "only if" parts of the following theorems are based on the Schauder-Tychonov fixed point theorem: we show that there exist certain nonoscillatory solutions provided that the integral in below given conditions (9.1.27) or (9.1.28) is convergent. The idea is similar to that of the proof of Theorem 9.1.6; strong super/sub-linearity is not needed there. The "if" parts are proved by contradiction where a proper analysis of nonoscillatory solutions of (9.1.22) is made. For illustration, in order to see the role of the strong superlinearity, let us mention at least that the existence of an eventually positive solution y of (9.1.22) leads to the inequality

$$A_1 \int_{t_0}^{t} \varphi^{-1}\left(\frac{1}{r(s)} \int_s^{\infty} F(\tau, A_2)\, d\tau\right) ds \leq \int_{y(t_0)}^{y(t)} \frac{du}{\varphi^{-1}(u^{\gamma})},$$

where A_1, A_2 are positive constants, provided the assumptions of the following theorem hold. This however contradicts (9.1.27) because of (9.1.26).

Theorem 9.1.8. *Let* (9.1.23) *and* (9.1.24) *hold. Suppose that equation* (9.1.22) *is strongly superlinear. Then all proper solutions of* (9.1.22) *are oscillatory if and only if*

$$\int_a^\infty \left|\varphi^{-1}\left(\frac{1}{r(t)}\int_t^\infty F(s,C)\,ds\right)\right| dt = \infty \tag{9.1.27}$$

for every constant $C \neq 0$.

Theorem 9.1.9. *Let* (9.1.23), (9.1.24) *and* (9.1.25) *hold. Suppose that equation* (9.1.22) *is strongly sublinear. Then all proper solutions of* (9.1.22) *are oscillatory if and only if*

$$\int_a^\infty \left|F\left(t, C\int_a^t \varphi^{-1}\left(\frac{B}{r(s)}\right) ds\right)\right| dt = \infty \tag{9.1.28}$$

for all constants $B \neq 0$, $C > 0$.

It is easy to see that in the case of (9.1.1) (with $c(t) > 0$), conditions (9.1.27) and (9.1.28) reduce to

$$\int_a^t \left(\frac{1}{r(t)}\int_t^\infty c(s)\,ds\right)^{p^*-1} dt = \infty \quad \text{and} \quad \int_a^\infty c(t)\left(\int_a^t r^{1-p^*}(s)\,ds\right)^{q-1} dt = \infty,$$

respectively.

Similar kind of problems and further extensions are discussed e.g. in [67] for a slightly more special equation of the form

$$(r(t)x')' + c(t)f(x) = 0.$$

In particular, both cases when $\int^\infty \frac{1}{r(t)}\,dt$ is either convergent or divergent are considered. A comparison technique is employed there. Recall also that when considering certain special case, namely the equation $x'' + c(t)\Phi_q(x) = 0$, above theorems go back to the classical results of Atkinson [25] and of Belohorec [33].

9.1.5 Generalized Riccati transformation and applications

In this subsection we show that the Riccati type substitution may be helpful in the theory of quasilinear equations, even if the resulting first order generalized Riccati equation is not "pure" in the sense that besides the independent variable w it contains a solution x of the original second order equation. Here we prove oscillation criteria for quasilinear equation. Other application will be shown in the next subsection. Observe that in the two applications given below, the forms of Riccati substitutions are different.

For illustration, we consider the more general equation

$$(\Phi(x'))' + F(t,x) = 0 \tag{9.1.29}$$

under the assumptions that the continuous function F satisfies $\operatorname{sgn} F(t,x) = \operatorname{sgn} x$ for $t \in [t_0, \infty)$.

Theorem 9.1.10. *All proper solutions of* (9.1.29) *are oscillatory if one of the following three conditions is satisfied:*

(i) *for all* $\delta > 0$

$$\int^{\infty} \inf_{\delta \le |y|} |F(t,y)|\, dt = \infty,$$

(ii) *for some* $0 < \lambda < p-1$ *and all* $\delta > 0$

$$\int^{\infty} t^{\lambda} \inf_{\delta \le |y|} \frac{F(t,y)}{|y|^{p-1}}\, dt = \infty,$$

(iii) *for all* δ, δ' *with* $\delta' > \delta > 0$

$$\int^{\infty} \inf_{\delta \le |y| \le \delta'} |F(t,y)|\, dt = \infty,$$

and there exists a positive continuous function φ *satisfying* $\int^{\infty} \varphi(y)\, dy = \infty$, *such that* $|F(t,y)| \ge \varphi(|y|)$ *for large* t *and large* $|y|$.

Proof. To illustrate ideas used in the proof, we prove the part (ii), the proof of the remaining two statements is analogical. Suppose, by contradiction, that (9.1.29) has a proper solution x which is positive for large t (if x is negative, we proceed analogously). Then from (9.1.29) we have that also $x'(t) > 0$ and we put $w(t) = \Phi(x'/x)$. Then w satisfies the Riccati type equation

$$w' + (p-1)|w|^{p^*} + \frac{F(t,x(t))}{x^{p-1}(t)} = 0, \tag{9.1.30}$$

recall that p^* denotes the conjugate number of p. Multiplying (9.1.30) by t^{λ} and integrating over $[t_0, t]$, t_0 sufficiently large, we have

(9.1.31)
$$t^{\lambda} w(t) - \lambda \int_{t_0}^{t} s^{\lambda-1} w(s)\, ds + (p-1) \int_{t_0}^{t} s^{\lambda} (w(s))^{p^*}\, ds + \int_{t_0}^{t} s^{\lambda} \frac{F(s,x(s))}{x^{p-1}(s)}\, ds \le c,$$

where $c > 0$ is a real constant.

Suppose first that $\int^{\infty} s^{\lambda-1} w(s)\, ds < \infty$. Then it follows from (9.1.31) that

$$\int_{t_0}^{t} s^{\lambda} \frac{F(s,x(s))}{x^{p-1}(s)}\, ds \le c + \lambda \int_{t_0}^{t} s^{\lambda-1} w(s)\, ds,$$

and taking the limit as $t \to \infty$, we get

$$\int_{t_0}^{\infty} s^{\lambda} \frac{F(s,x(s))}{x^{p-1}(s)}\, ds < \infty.$$

However, this is impossible since assumptions of our theorem imply that for t_0 sufficiently large

$$\int_{t_0}^{\infty} s^{\lambda} \frac{F(s,x(s))}{x^{p-1}(s)}\, ds \ge \int_{t_0}^{\infty} s^{\lambda} \inf_{\delta \le x} \frac{F(s,x)}{x^{p-1}}\, ds = \infty, \tag{9.1.32}$$

where $\delta = x(t_0) > 0$.

Suppose next that

$$\int^{\infty} s^{\lambda-1} w(s)\, ds = \infty. \tag{9.1.33}$$

Then, by (9.1.31),

$$\int_{t_0}^{t} s^{\lambda} \frac{F(s, x(s))}{x^{p-1}(s)}\, ds \le c + \lambda \int_{t_0}^{t} s^{\lambda-1} w(s)\, ds - (p-1) \int_{t_0}^{t} s^{\lambda} |w(s)|^{p^*}\, ds. \tag{9.1.34}$$

Note that the second integral in equation (9.1.34) is estimated by means of the Hölder inequality as follows

$$\begin{aligned} \int_{t_0}^{t} s^{\lambda-1} w(s)\, ds &= \int_{t_0}^{t} s^{(\lambda-p)/p} s^{\lambda(p-1)/p} w(s)\, ds \\ &\le \left(\int_{t_0}^{t} s^{\lambda-p}\, ds \right)^{\frac{1}{p}} \left(\int_{t_0}^{t} s^{\lambda} w^{p*}(s)\, ds \right)^{\frac{1}{p^*}} \\ &\le \left(\frac{t_0^{\lambda-p+1}}{p-1-\lambda} \right)^{\frac{1}{p}} \left(\int_{t_0}^{t} s^{\lambda} w^{p^*}(s)\, ds \right)^{\frac{1}{p^*}} \\ &= \frac{(t_0^{\lambda-p+1}/(p-1-\lambda))^{\frac{1}{p}}}{\left(\int_{t_0}^{t} s^{\lambda} w^{p^*}(s)\, ds \right)^{\frac{1}{p}}} \int_{t_0}^{t} s^{\lambda} w^{p^*}(s)\, ds. \end{aligned} \tag{9.1.35}$$

Since (9.1.33) implies that

$$\int_{t_0}^{t} s^{\lambda} w^{p^*}(s)\, ds \to \infty \quad \text{as } t \to \infty,$$

we see from (9.1.35) that there exists $t_1 \ge t_0$ such that

$$\int_{t_0}^{t} s^{\lambda-1} w(s)\, ds \le \frac{p-1}{\lambda} \int_{t_0}^{t} s^{\lambda} w^{p^*}(s)\, ds, \quad t \ge t_1.$$

Using this inequality in (9.1.34) we conclude that

$$\int_{t_0}^{\infty} s^{\lambda} \frac{F(s, x(s))}{x^{p-1}(s)}\, ds \le c.$$

This however contradicts (9.1.32) which holds also in this case. □

9.1.6 (Half-)linearization technique

The linearization method is a typical method of the qualitative investigation of various nonlinear differential equations. The reason is obvious: linear equations are generally easier to investigate than the nonlinear ones. Naturally, a quasilinear equation can be compared also with the half-linear one. In Subsection 2.3.5 we have presented comparison theorem, where two half-linear equations with the same coefficients but different power functions were compared. Here we first compare the linear equation

$$(r(t)x')' + c(t)x = 0 \tag{9.1.36}$$

with the nonlinear one

$$(r(t)x')' + c(t)f(x) = 0, \tag{9.1.37}$$

where $r(t) > 0, c(t) > 0$ and f is a continuous function satisfying $uf(u) > 0$ for $u \neq 0$. Further, oscillatory properties of quasilinear equation (9.1.1) are investigated using the half-linear oscillation theory.

Oscillation of nonlinear equation is defined as in Subsection 9.1.4. To recall the concept of strong and conditional oscillation of (half-)linear equations, see Section 5.4. The proof of the following theorem is omitted. Note only that several cases are distinguished, according to whether $\int^\infty q(s)\, ds$ is convergent or divergent, and the value of

$$\limsup_{t\to\infty} \left(\int^t r^{-1}(s)\, ds\right)\left(\int_t^\infty q(s)\, ds\right) \quad \text{and} \quad \limsup_{t\to\infty} \left(\int^t q(s)\, ds\right)\left(\int_t^\infty r^{-1}(s)\, ds\right)$$

is zero or a positive number or infinity. Further, among others, Sturmian comparison theorem and strongly (non)oscillation criteria play important roles.

Theorem 9.1.11. (i) *Assume*

$$\int_a^\infty \frac{1}{r(t)}\, dt = \infty \quad \textit{and} \quad \lim_{|u|\to\infty} \frac{f(u)}{u} = \infty.$$

If (9.1.36) *is either strongly oscillatory or conditionally oscillatory, then* (9.1.37) *is oscillatory.*

(ii) *Assume*

$$\int_a^\infty \frac{1}{r(t)}\, dt < \infty \quad \textit{and} \quad \lim_{|u|\to 0} \frac{f(u)}{u} = \infty.$$

If (9.1.36) *is either strongly oscillatory or conditionally oscillatory, then* (9.1.37) *is oscillatory.*

Next we present the half-linearization method, which is a partial generalization of (i) of the previous theorem. A Riccati-type substitution is employed there.

Theorem 9.1.12. *Let $p > q$. Denote $\eta(t) = \int^t r^{1-q}(s)\, ds$ and assume that*

$$\int^\infty r^{1-p^*}(t)\, dt = \infty \tag{9.1.38}$$

and

$$\liminf_{t\to\infty} \int^t c(s)\, ds > -\infty. \tag{9.1.39}$$

If the half-linear differential equation

$$(r(t)\eta^{p-1}(t)\Phi(y'))' + \alpha c(t)\Phi(y) = 0 \tag{9.1.40}$$

is oscillatory for every positive constant α *(i.e., it is strongly oscillatory), then* (9.1.1) *is oscillatory, i.e., it has no nonoscillatory proper solution.*

Proof. Let x be a nonoscillatory proper solution of (9.1.1), say $x(t) > 0$ for $t \geq t_0$. Define

$$w(t) = \frac{r(t)\Phi(x'(t))}{x^{q-1}(t)}.$$

Then we have

$$w' = -c(t) - (q-1)r(t)\frac{|x'(t)|^p}{x^q(t)} = -c(t) - (q-1)r^{1-p^*}|w|^{p^*} x^{\frac{q-p}{p-1}}(t), \tag{9.1.41}$$

and hence

$$w(t) = w(t_0) - \int_{t_0}^t c(s)\,ds - (q-1)\int_{t_0}^t r(s)\frac{|x'(s)|^p}{x^q(t)}\,ds. \tag{9.1.42}$$

Now, with respect to the integral

$$\mathcal{L}(t_0, t) := \int_{t_0}^t r(s)\frac{|x'(s)|^p}{x^q(s)}\,ds,$$

we need to consider the following cases.

I) $\mathcal{L}(t_0, \infty) < \infty$. In this case, there exists a positive constant R such that $\mathcal{L}(t_0, t) \leq R$ for $t \geq t_0$. Using the Hölder inequality we obtain, with $\gamma = q/p > 1$,

$$\begin{aligned} x^{1-\gamma}(t) - x^{1-\gamma}(t_0) &\leq (\gamma - 1)\int_{t_0}^t \left|\frac{x'(s)}{x^\gamma(s)}\right| ds \\ &\leq (\gamma-1)(\mathcal{L}(t_0,t))^{1/p}\left(\int_{t_0}^t r^{1-p^*}(s)\,ds\right)^{1/p^*} \\ &\leq (\gamma-1)R^{1/p}\eta^{1/p^*}(t). \end{aligned}$$

By (9.1.38) there exists a constant $M > 0$ and $T \geq t_0$ such that $x^{1-\gamma}(t) \leq M\eta^{1/p^*}(t)$ for $t \geq T$, or $x(t) \geq ((1/M)\eta^{-\frac{1}{p^*}}(t))^{1/(\gamma-1)}$ for $t \geq T$ and

$$x^{\frac{q-p}{p-1}}(t) \geq \frac{d_1}{\eta(t)}, \quad \text{for } t \geq T, \tag{9.1.43}$$

where $d_1 = (1/M)^{\frac{q-p}{(p-1)(\gamma-1)}}$. Using (9.1.43) in equation (9.1.41) we obtain

$$w'(t) \leq -c(t) - \frac{d_1(q-1)}{p-1}\left(\frac{1}{r(t)\eta^{p-1}}\right)^{p^*-1}|w(t)|^{p^*}, \quad t \geq T.$$

Hence, by Theorem 2.2.1 we find that equation (9.1.40) is nonoscillatory, a contradiction.

II) $\mathcal{L}(t_0, \infty) = \infty$. In view of condition (9.1.39), for some constant L, equation (9.1.42) gives

$$-w(t) \geq L + (q-1)\mathcal{L}(t_0, t), \quad t \geq t_0. \tag{9.1.44}$$

Let $T \geq t_0$ be such that $\Lambda = L + (q-1)\mathcal{L}(t_0, t) > 0$ for $t \in [T, \infty)$. Then (9.1.44) ensures that w is negative on $[T, \infty)$. Now, (9.1.44) gives

$$\frac{(q-1)r(t)|x'(t)|^p}{[L + (q-1)\mathcal{L}(t_0, t)]x^q(t)} \geq -(q-1)\frac{x'(t)}{x(t)}, \quad t \geq T,$$

and consequently, for all $t \geq T$

$$\log\left[\frac{1}{\Lambda}\left(L + (q-1)\mathcal{L}(t_0, t)\right)\right] \geq \log\left(\frac{x(T)}{x(t)}\right)^{q-1}.$$

Hence, $L+(q-1)\mathcal{L}(t_0, t) \geq \Lambda(X(T)/x(t))^{q-1}$. So, inequality (9.1.44) yields $x'(t) \leq -\Lambda_1 r^{1-p^*}(t)$, for $t \geq T$, where $\Lambda_1(\Lambda x^{q-1}(T))^{1/(p-1)}$. Thus, we have

$$x(t) \leq x(T) - \Lambda_1 \int_T^t r^{1-p^*}(s)\, ds, \quad t \geq T,$$

and this, taking into account (9.1.38), implies $x(t) \to -\infty$ as $t \to \infty$, a contradiction. □

9.1.7 Singular solutions of black hole and white hole type

Looking at the form of equation (9.1.1), a natural question arises, namely what we can say about equation of the type (9.1.1) (or of a similar type), where p and/or q are less than 1. Such a question has already been extensively discussed in the literature. The qualitative behavior of the *singular Emden-Fowler equation* $(r(t)y')' + c(t)y^{q-1} = 0$ and its generalization $(r(t)\Phi(y'))' + c(t)y^{q-1} = 0$, where $p > 1$, $q < 1$ and r, c are positive continuous functions, has been investigated in details by many authors, see e.g [204, 225, 345, 353]. We emphasize that here by singularity we mean the singularity of equations at dependent variables (sometimes called singularity at the phase variables) in contrast to the equations with singular coefficients. Since the theory of the above mentioned singular Emden-Fowler equations has been well processed in the literature we rather focus our attention to other types of singular equations where some new phenomena may occur. Equations of the form $(|y'|^{p-1})' + c(t)|y|^{q-1} = 0$, or of similar forms, where $p, q \in \mathbb{R}$ with $p \neq 1$ and c is either always positive or always negative continuous function, have appeared very recently, see e.g. [186, 187, 188, 225, 346]. Note that such forms allow us to consider *equations with singular nonlinearity in the differential operator* (when $p < 1$) or even *doubly singular equations* (when $p < 1$ and $q < 1$). Some new interesting phenomena may occur for the equations of the latter forms. To be more more precise, first let us consider the equation

$$(|y'|^{p-1})' + c(t)y^{q-1} = 0, \tag{9.1.45}$$

where $p < 1$, $q \in \mathbb{R}$ and c is a positive continuous function defined on $[a, \infty)$. We want to show that equation of the form (9.1.45) may have, along with proper and "usual" blowing-up and extinct solutions, also singular solutions of a new type that we call *black hole solution.* It is defined as a solution y of (9.1.45) which exists on some interval $[t_0, T)$, $T < \infty$, and has the properties $\lim_{t\to T-} y(t) \in (0, \infty)$ and $\lim_{t\to T-} |y'(t)| = \infty$. A simple example of a singular equation of the form (9.1.45) (with $q = 1$) having this type of solution is

$$(|y'|^{p-1})' + \left(\frac{p}{p-1}\right)^{p-1} = 0, \tag{9.1.46}$$

where $p < 1$. For any given $T > a$ and $M > 0$, the function $y(t) = M+(T-t)^{p/(p-1)}$ defined and positive on $[a, T)$ is a decreasing solution of (9.1.46) with a singularity of black hole type at T. Similarly, for any $M > 0$ the function $y(t) = M - (T - t)^{p/(p-1)}$ provides an example of a "local" increasing black hole solution of (9.1.46) which is defined and positive in some sufficiently small left neighborhood of T.

Next we show that there does exist a wide class of equations of type (9.1.45) which admit black hole solutions.

Theorem 9.1.13. *For any $T > a$ and any $M > 0$, equation* (9.1.45) *possesses an increasing black hole solution defined on some interval $[t_0, T)$, $a \le t_0 < T$, if and only if $p < 0$.*

Proof. "$\Rightarrow$": Assuming that y is a positive increasing black hole solution of (9.1.45) defined on $[t_0, T)$, after some easy computation, we obtain

$$\int_{t_0}^{T} \left(\int_s^T c(\tau)\, d\tau\right)^{\frac{1}{p-1}} ds < \infty,$$

which is possible only if $p < 0$.

"$\Leftarrow$": To prove this part we use a standard argument based on the Schauder fixed point theorem. We show that the operator $\mathcal{T} : \Omega \to C[t_0, \infty]$ given by

$$\mathcal{T}(u)(t) = M - \int_t^T \left(\int_s^T c(\tau)u^{q-1}(\tau)\, d\tau\right)^{\frac{1}{p-1}} ds,$$

$t \in [t_0, T]$, with $T > a$ and $M > 0$ being given arbitrarily, has a fixed point in the set Ω which is defined by

$$\Omega = \left\{u \in C[t_0, T] \,:\, \frac{M}{2} \le u(t) \le M, t \in [t_0, T]\right\}.$$

Details are omitted. □

A closer examination of the previous proof shows that if there is made a special choice of M, then the desired black hole solution is guaranteed to exist on the entire given interval $[a, T)$ and the above theorem can be restated as the following "global" existence result.

Theorem 9.1.14. *Suppose that $p \neq q$. Then, for any $T > a$, equation (9.1.45) possesses an increasing black hole solution defined on $[a, T)$ if and only if $p < 0$.*

Corresponding theorems dealing with decreasing black hole solutions can be proved in a very similar way. We stress that black hole solutions exist regardless q is greater than or less than 1.

Introducing certain energy functions, under the assumption $c \in C^1[a, \infty)$, it can be shown that the only decreasing solutions of (9.1.45) [resp. all increasing singular solutions of (9.1.45)] are black hole solutions provided $p < 0$ and $q < 0$ [resp. $p < 0 < q$].

Similarly as for nonsingular quasilinear equations, it can be shown the existence of "classical" singular solutions and proper solutions with some prescribed asymptotic behavior.

By analogy with the concept of black hole solutions we introduce the so-called white hole solutions that appear to be singular solutions of a new type. Consider the equation

$$(9.1.47) \qquad (|y'|^{p-1})' + c(t)|y|^{q-1} = 0,$$

where $p, q \in \mathbb{R}$ are constants with $p > 1$, and c is a positive continuous function defined on $[a, \infty)$. A solution y of (9.1.47) is said to be *white hole solution* of (9.1.47) if it is defined on some interval $[t_0, T)$, $T < \infty$, $\lim_{t \to T-} y(t) \in \mathbb{R} \setminus \{0\}$ and $\lim_{t \to T} y'(t) = 0$. An example of a nonlinear equation of the form (9.1.47) (with $q = 1$) which possesses a singular solution of this new type is (9.1.46), where $p > 1$. The desired solutions are defined exactly as those ones right after (9.1.46). However, now we assume $p > 1$, and so they are of white hole type. Another simple example is the "almost linear" equation $(|y'|)' + |y| = 0$. As easily seen, for any real T and any $M > 0$, the function $y(t) = M \cos(T - t)$ defined and positive on $[t_0, T)$, $t_0 \geq T - \pi/2$, is its increasing singular solution which is of the white hole type. In a very similar way as in the above theorems on black hole solutions, it can be shown that, regardless of q is greater or less than one, equation (9.1.47) always has singular solutions of white hole type if and only if $p > 1$. It concerns both solutions, decreasing and increasing, and the "global" existence result can be stated as well provided $p \neq q$. Note that in the results concerning both cases, i.e., black hole and white solutions, the assumption on the positivity of c may be somewhat relaxed, which however then has to be compensated by some additional requirements.

We finish this section by examining the (singular) equation

$$(9.1.48) \qquad (|y'|^{p-1})' + \lambda|y|^{q-1} = 0,$$

$t \in [a, \infty)$, where $\lambda > 0$, $p \in \mathbb{R} \setminus \{0, 1\}$, $q \in \mathbb{R} \setminus \{0\}$ and $p \neq q$. Equation (9.1.48) has for any given $T > a$ the solution $y(t) = M(T - t)^{p/(p-q)}$, $t \in [a, T)$, where

$$M = (\lambda K^{p-1})^{1/(p-q)}, \quad K = \left|\frac{p-q}{p}\right| \left|\frac{p-q}{(p-1)q}\right|^{1/(p-1)}.$$

Now we distinguish three cases:

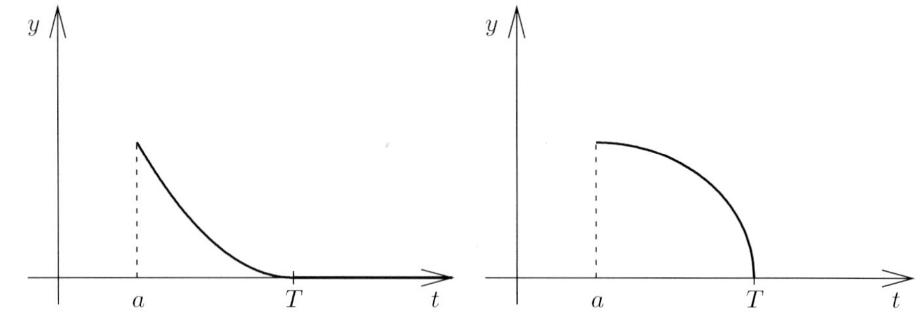

Figure 9.1.1: Extinct solutions of the first kind and of the second kind, respectively

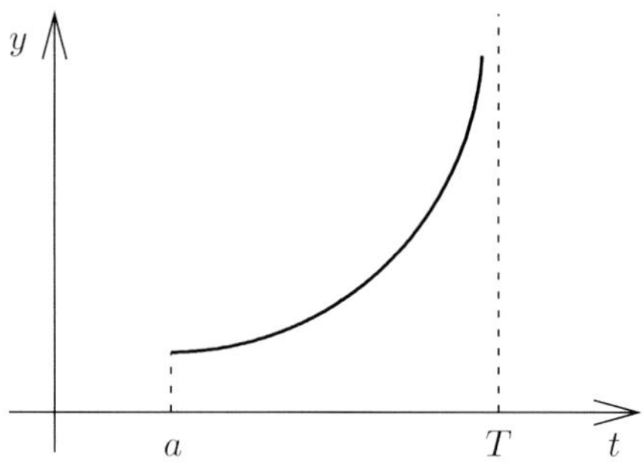

Figure 9.1.2: Blowing-up solution

(i) If $p/(p-q) > 1$, then (9.1.48) has an extinct solution of the first kind, i.e., a solution y such that $\lim_{t\to T-} y(t) = 0 = \lim_{t\to T-} y'(t)$.

(ii) If $0 < p/(p-q) < 1$, then (9.1.48) has an extinct solution of the second kind, i.e., a solution y such that $\lim_{t\to T-} y(t) = 0$ and $\lim_{t\to T-} y'(t) = -\infty$. Sometimes we call it a solution with a "zero black hole" at T.

(iii) If $p/(p-q) < 0$, then (9.1.48) has a blowing-up solution, i.e., a solution y such that $\lim_{t\to T-} y(t) = \infty = \lim_{t\to T-} y'(t)$.

9.1.8 Curious doubly singular equations

The so-called doubly singular equations are also of interest. Some of them have already been discussed in the previous subsection. Besides this, note that, for instance in [186], equations of the form $(|y'|^{p-1})' - c(t)|y|^{q-1} = 0$, where $p < 1$, $q < 1$ are constants, and c is a positive continuous function, are studied. The results presented there are more or less "standard" (i.e., conditions for the existence of singular solutions or of solutions with prescribed asymptotic behavior). However, a closer look at doubly singular equations reveals that they may offer very interesting possibilities. A very natural question comes up: What about the doubly singular case where the exponents are equal (i.e., the *half-linear singular case*)?

What we present next comes from Jaroslav Jaroš who started the study which could answer the above question. In 1999, in [184], he gave an interesting example of doubly singular equation of the second order, namely,

$$\left(r(t)\frac{1}{y'}\right)' + c(t)\frac{1}{y} = 0, \tag{9.1.49}$$

where $r > 0$ and c are continuous functions. This equation may play a very significant role in the theory of the class of quasilinear equations where the exponents in nonlinearities are less than 0 (or, more generally, where the exponents may be arbitrary real numbers). Its importance is compared with the classical Sturm-Liouville equation. Even, it seems to be more important from a certain point of view. Indeed, as easily seen, equation (9.1.49) can be *explicitly solved* – the corresponding Riccati equation is a *linear equation of the first order.*

In the particular case $r(t) \equiv 1$ and $c(t) = \lambda$, $\lambda \in (0, \infty)$, we obtain the equation

$$\left(\frac{1}{y'}\right)' + \lambda\frac{1}{y} = 0. \tag{9.1.50}$$

The solution of (9.1.50) is

$$y(t) = \begin{cases} |Kt + M|^{1/(1-\lambda)} & \text{if } \lambda \neq 1 \quad (K \neq 0), \\ K\mathrm{e}^{Mt} & \text{if } \lambda = 1 \quad (K, M \neq 0). \end{cases}$$

Now, if $\lambda > 1$, then y is blowing-up, if $0 < \lambda < 1$, then y is extinct of the first kind, and if $\lambda < 0$, then y is extinct of the second kind.

9.1.9 Coupled quasilinear systems

Quite recently, it has been started the investigation of the *coupled quasilinear systems*, i.e., the systems of the form

$$\begin{aligned} (r(t)\Phi_\alpha(x'))' &= \sigma\varphi(t)\Phi_\gamma(y), \\ (q(t)\Phi_\beta(y'))' &= \psi(t)\Phi_\delta(x), \end{aligned} \tag{9.1.51}$$

where $\sigma \in \{-1, 1\}$, r, q, φ, ψ are positive continuous functions defined on $[a, \infty)$, and $\alpha, \beta, \gamma, \delta$ are constants greater than 1. Various extensions can also be considered; for instance, the expression $\varphi(t)\Phi_\gamma(y)$ is replaced by $F(t, y)$ (with F being the same as at the beginning of this section) or some of the constants $\alpha, \beta, \gamma, \delta$ are allowed to be less than 1 which then covers a singular case. It is not difficult to see that system (9.1.51) includes a wide class of fourth order nonlinear equations, for instance those of the form $y^{(4)} + \varphi(t)\Phi_\gamma(y) = 0$. An advantage of the investigation of system (9.1.51) is that its form may enable better understanding the structure of solution space of (9.1.51) than in the case of fourth order equations (there indeed exist results that nicely illustrate this fact – the situations which seem to be quite different for the fourth order equations are shown to be "symmetric" when rewritten into the system). Moreover, since (9.1.51) consists of two second order equations in a formally "self-adjoint" form, theory of such systems can be understood as an extension of the approach known from the extensive theory of second order scalar nonlinear differential equations of the form (9.1.1). Motivated ourselves in such a way, some kind of results can be expected, even though the raise of the order makes usually problems much more difficult.

We choose here some of the typical results – problems of the (non)existence of singular solutions, and some aspects of oscillation and asymptotic theory. We present them without proofs. Note only that the analysis of nonoscillatory solutions and the Schauder or Schauder-Tychonov fixed point theorem play an important role there. Usually we look for fixed points of certain vector operators in the subsets of Cartesian products of the Banach spaces $C[a,b]$ or of the Fréchet spaces $C[a,\infty)$. Compare the next statement with Theorem 9.1.3.

Theorem 9.1.15. *Let $\sigma = 1$ and $\alpha,\beta,\gamma,\delta > 1$.*

(i) If $(\alpha-1)(\beta-1) > (\gamma-1)(\delta-1)$, then, for any $T > a$, system (9.1.51) *possesses an extinct singular solution, i.e., the solution (x,y) such that $x(t) > 0$, $y(t) > 0$ on $[a,T)$ and $x(t) = y(t) = 0$ on $[T,\infty)$.*

(ii) If $(\alpha-1)(\beta-1) < (\gamma-1)(\delta-1)$, then system (9.1.51) *has no extinct singular solution.*

For comparison purposes, the next statement is presented for a more general system of the form

$$\begin{aligned} (r(t)\Phi_\alpha(x'))' &= -F(t,y), \\ (q(t)\Phi_\beta(y'))' &= G(t,x), \end{aligned} \tag{9.1.52}$$

where $F, G : [a,\infty)\times\mathbb{R} \to \mathbb{R}$ are continuous functions, nondecreasing with respect to the second variable, and such that $uF(k,u) > 0$, $uG(k,u) > 0$ for every $u \neq 0$ and $k \in [a,\infty)$. Further we assume here that $\alpha > 1$, $\beta > 1$,

$$\int_a^\infty \Phi_{\alpha^*}\left(\frac{1}{r(t)}\right) dt = \infty = \int_a^\infty \Phi_{\beta^*}\left(\frac{1}{q(t)}\right)$$

and

$$\int_a^\infty |F(t,C)|\, dt < \infty, \quad \int_a^\infty |G(t,C)|\, dt < \infty,$$

where α^* and β^* are the conjugate numbers of α and β, respectively, and C is any nonzero constant. We will need the next definition: We call system (9.1.52) *strongly superlinear* (resp. *strongly sublinear*) if $\lambda > 0$ and $\mu > 0$ exist such that $|F(t,u)|/|u|^\lambda$ and $|G(t,u)|/|u|^\mu$ are nondecreasing (resp. nonincreasing) in $|u|$ for each fixed $t \in [a,\infty)$, and $\lambda\mu > (\alpha-1)(\beta-1)$ (resp. $\lambda\mu < (\alpha-1)(\beta-1)$). The following notation will be useful:

$$\begin{aligned} S^\alpha_{rF}(C) &= \int_a^\infty \Phi_{\alpha^*}\left(\frac{1}{r(t)}\int_t^\infty |F(s,C)|\, ds\right) dt, \\ S^\beta_{qG}(C) &= \int_a^\infty \Phi_{\beta^*}\left(\frac{1}{q(t)}\int_t^\infty |G(s,C)|\, ds\right) dt, \end{aligned}$$

$$\mathcal{P}^{\alpha}_{Gr}(C) = \int_a^\infty \left| G\left(t, C\int_a^t \Phi_{\alpha^*}\left(\frac{1}{r(s)}\right) ds\right)\right| dt,$$
$$\mathcal{P}^{\beta}_{Fq}(C) = \int_a^\infty \left| F\left(t, C\int_a^t \Phi_{\beta^*}\left(\frac{1}{q(s)}\right) ds\right)\right| dt,$$

where C is a nonzero real constant. The next result deals with oscillation of (9.1.52), which means oscillation of all its solutions. An oscillatory solution of (9.1.52) is defined as the solution whose both components oscillate. Note that in view of the sign conditions, one component of solution of (9.1.52) is oscillatory if and only if another component is so. This can be easily seen by using the Rolle Theorem. Compare the following statement with the results of Subsection 9.1.4.

Theorem 9.1.16. *(i) Assume $\mathcal{S}^{\beta}_{qG}(C) = \mathcal{P}^{\alpha}_{Gr}(C) = \mathcal{P}^{\beta}_{Fq}(C) = \infty$, $\mathcal{S}^{\alpha}_{rF}(M) < \infty$, for all $C \neq 0$ and a constant $M \neq 0$. Let (9.1.52) be strongly superlinear. Then (9.1.52) is oscillatory if and only if*

$$\int_a^\infty \left| \Phi_{\beta^*}\left\{\frac{1}{q(t)}\int_t^\infty G\left[s, \int_s^\infty \Phi_{\alpha^*}\left(\frac{1}{r(\tau)}\int_\tau^\infty F(\eta, C)\, d\eta\right) d\tau\right] ds\right\}\right| dt = \infty$$

for every $C \neq 0$.

(ii) Assume $\mathcal{S}^{\beta}_{qG}(C) = \mathcal{P}^{\alpha}_{Gr}(C) = \mathcal{S}^{\alpha}_{rF}(C) = \infty$, $\mathcal{P}^{\beta}_{Fq}(M) < \infty$ for all $C \neq 0$ and a constant $M \neq 0$. Let (9.1.52) be strongly sublinear. Then (9.1.52) is oscillatory if and only if

$$\int_a^\infty \left| F\left(t, \int_a^t \Phi_{\beta^*}\left[\int_a^s G\left(\tau, \int_a^\tau \Phi_{\alpha^*}\left(\frac{C}{r(\eta)}\right) d\eta\right) d\tau\right] ds\right)\right| dt = \infty$$

for every $C \neq 0$.

The last result which we choose to present here deals with the so-called regularly decaying solutions, and allows to consider singular systems.

Theorem 9.1.17. *Let $\sigma = -1$, $\alpha > 1$, $\beta > 1$, $\gamma \neq -1$ and $\delta \neq -1$. System (9.1.51) has a regularly decaying solution, i.e., a solution (x, y) such that $x(\infty) = 0 = y(\infty)$, $r(t)\Phi_\alpha(x')$ is bounded and $q(t)\Phi(y')$ tends to a nonzero limit, if and only if*

$$\int^\infty r^{1/(1-\alpha)}(s)\, ds < \infty, \quad \int^\infty q^{1/(1-\beta)}(s)\, ds < \infty,$$

$$\int^\infty \varphi(t)\left(\int_t^\infty q^{1/(1-\beta)}(s)\, ds\right)^{\gamma-1} dt < \infty$$

and

$$\int^\infty \psi(t)\left(\int_t^\infty r^{1/(1-\alpha)}(s)\, ds\right)^{\delta-1} dt < \infty.$$

Note that many other results concerning asymptotic and oscillatory behavior of (9.1.51) have been established, even for the systems with "worse" nonlinearities, with forcing terms, and singularities. On the other hand, the theory of half-linear coupled systems (i.e., when $\alpha = \beta = \gamma = \delta$ in (9.1.51)) has not been developed yet. Of course, the results for the general system may be applicable here, but what we have in mind is the theory which is more refined, and uses the homogeneity of the solution space. In other words, an extension of the theory described in the first two chapters of this book to the half-linear systems of the form (9.1.51) is still missing. Finally observe that instead of two equations one can consider n second order equations, e.g., of the form $(r_1(t)\Phi_{\alpha_1}(x_1'))' = \varphi_1(t)\Phi_{\gamma_1}(x_2)$, $(r_2(t)\Phi_{\alpha_2}(x_2'))' = \varphi_2(t)\Phi_{\gamma_2}(x_3)$, ..., $(r_n(t)\Phi_{\alpha_n}(x_n'))' = \varphi_n(t)\Phi_{\gamma_n}(x_1)$, $n \in \mathbb{N}$. Such a system then may be reduced to $2n$-th order (scalar) quasi- or half-linear equation. There exist some isolated results for scalar equations of this form, see Subsection 9.4.2.

9.2 Forced half-linear differential equations

The main concern of this section is to study the influence of the forcing term on oscillatory properties of investigated equations. First we deal with forced half-linear equations and then we turn our attention to forced quasilinear equation (9.1.1).

9.2.1 Two oscillatory criteria

Consider the forced half-linear equation

$$(r(t)\Phi(x'))' + c(t)\Phi(x) = f(t), \tag{9.2.1}$$

where r, c and f are continuous functions on $[a, \infty)$ with $r(t) > 0$. We start with a weighted oscillatory criterion where an important role is played by the sign of the forced term in subintervals. Throughout this subsection, q has its usual meaning, i.e., it is a conjugate number to p.

Theorem 9.2.1. *Suppose that for any $T \geq 0$ there exist $T \leq s_1 < t_1 \leq s_2 < t_2$ such that $f(t) \leq 0$ for $t \in [s_1, t_1]$ and $f(t) \geq 0$ $t \in [s_2, t_2]$. Denote*

$$D(s_i, t_i) = \{u \in C^1[s_i, t_i] : u(t) \not\equiv 0, \; u(s_i) = 0 = u(t_i)\}, \quad i = 1, 2.$$

If there exists a function $h \in D(s_i, t_i)$ and a positive nondecreasing function $\varphi \in C^1[T, \infty)$ such that

$$\int_{s_i}^{t_i} h^2(t)\varphi(t)c(t)\, dt > \frac{1}{p^p} \int_{s_i}^{t_i} \frac{r(t)\varphi(t)}{|h(t)|^{p-2}} \left(2|h'(t)| + h(t)\frac{\varphi'(t)}{\varphi(t)}\right)^p dt, \tag{9.2.2}$$

for $i = 1, 2$, then every solution of (9.2.1) is oscillatory.

Proof. Suppose that x is a nonoscillatory solution of (9.2.1) which is eventually of one sign, say $x(t) > 0$ for $t \geq T_0$, and let the function w be defined by the modified Riccati substitution

$$w(t) = \varphi(t)\frac{r(t)\Phi(x'(t))}{\Phi(x(t))}.$$

Then w solves the generalized Riccati equation

$$w' = -\varphi(t)c(t) + \frac{\varphi'(t)}{\varphi(t)}w - (p-1)\frac{|w|^q}{(r(t)\varphi(t))^{q-1}} + \frac{\varphi(t)f(t)}{\Phi(x(t))}. \tag{9.2.3}$$

By the assumptions of theorem, one can choose $s_1, t_1 \geq T_0$ so that $f(t) \leq 0$ on $I = [s_1, t_1]$ with $s_1 < t_1$. From (9.2.3) we have for $t \in I$

$$\varphi(t)c(t) \leq -w'(t) + \frac{\varphi'(t)}{\varphi(t)}w(t) - (p-1)\frac{|w(t)|^q}{(r(t)\varphi(t))^{q-1}}. \tag{9.2.4}$$

Multiplying (9.2.4) by h^2 and integrating over I we obtain

$$\begin{aligned}\int_{s_1}^{t_1} h^2(t)\varphi(t)c(t)\,dt \quad &\leq \quad -\int_{s_1}^{t_1} h^2(t)w'(t)\,dt + \int_{s_1}^{t_1} h^2(t)\frac{\varphi'(t)}{\varphi(t)}w(t)\,dt \\ &\quad -(p-1)\int_{s_1}^{t_1} h^2(t)\frac{|w(t)|^q}{(r(t)\varphi(t))^{q-1}}\,dt.\end{aligned}$$

Integrating the last inequality by parts and using the fact that $h(s_1) = 0 = h(t_1)$, we get

$$\begin{aligned}\int_{s_1}^{t_1} h^2(t)\varphi(t)c(t)\,dt \quad &\leq \quad \int_{s_1}^{t_1} 2|h(t)||h'(t)||w(t)|\,dt + \int_{s_1}^{t_1} h^2(t)\frac{\varphi'(t)}{\varphi(t)}w(t)\,dt \\ &\quad -(p-1)\int_{s_1}^{t_1} h^2(t)\frac{|w(t)|^q}{(r(t)\varphi(t))^{q-1}}\,dt \\ &\leq \quad \int_{s_1}^{t_1}\left(2|h(t)||h'(t)| + \frac{\varphi'(t)}{\varphi(t)}h^2(t)\right)|w(t)|\,dt \\ &\quad -(p-1)\int_{s_1}^{t_1} h^2(t)\frac{|w(t)|^q}{(r(t)\varphi(t))^{q-1}}\,dt.\end{aligned}$$

Now, the application of the Young inequality yields for $t \in [s_1, t_1]$

$$\begin{aligned}&\left(2|h(t)||h'(t)| + \frac{\varphi'(t)}{\varphi(t)}h^2(t)\right)|w(t)| - (p-1)\frac{|w(t)|^q}{(r(t)\varphi(t))^{q-1}}h^2(t) \\ &\qquad\qquad \leq \frac{1}{p^p}\frac{r(t)\varphi(t)}{|h(t)|^{p-2}}\left(2|h'(t)| + |h(t)|\frac{\varphi'(t)}{\varphi(t)}\right)^p,\end{aligned}$$

thus

$$\int_{s_1}^{t_1} h^2(t)\varphi(t)c(t)\,dt \leq \frac{1}{p^p}\int_{s_1}^{t_1}\frac{r(t)\varphi(t)}{|h(t)|^{p-2}}\left(2|h'(t)| + |h(t)|\frac{\varphi'(t)}{\varphi(t)}\right)^p\,dt,$$

which contradicts our assumption. When x is eventually negative, we may employ the fact that $f(t) \geq 0$ on some interval in any neighborhood of ∞ to reach a similar contradiction. □

Remark 9.2.1. If $p = 2$ and $\varphi(t) \equiv 1$, then the above theorem reduces to the result of Wong [363]. In particular, condition (9.2.2) reads as $\int_{s_i}^{t_i} [c(t)h^2(t) - r(t)(h'(t))^2]\, dt > 0$.

Now we will see that a "large amplitude" of the forcing term may generate oscillation of all solutions of (9.2.1).

Theorem 9.2.2. *Let* $c(t) \geq 0$ *for* $t \geq a$. *If*

$$\liminf_{t\to\infty} \int_a^t f(s)\, ds = -\infty, \qquad \limsup_{t\to\infty} \int_a^t f(s)\, ds = \infty, \tag{9.2.5}$$

and

$$\begin{aligned} \liminf_{t\to\infty} \int_a^t r^{1-q}(s) \left(\int_a^s f(u)\, du \right)^{q-1} ds &= -\infty, \\ \limsup_{t\to\infty} \int_a^t r^{1-q}(s) \left(\int_a^s f(u)\, du \right)^{q-1} ds &= \infty, \end{aligned}$$

then every solution of (9.2.1) *is oscillatory.*

Proof. Assume the contrary. Then without loss of generality, we can assume that there is a nonoscillatory solution x of (9.2.1), say, $x(t) > 0$ for $t \geq T \geq a$. From (9.2.1), we have $(r(t)\Phi(x'))' \leq f(t)$ for $t \geq T$. Thus, it follows that

$$r(t)\Phi(x'(t)) - r(T)\Phi(x'(T)) \leq \int_T^t f(s)\, ds. \tag{9.2.6}$$

By (9.2.5) there exists $t_0 \geq T$ sufficiently large so that $x'(t_0) < 0$ and $x'(t) < 0$ for $t \geq t_0$. Replacing T by t_0 in (9.2.6), we get

$$x'(t) \leq r^{1-q}(t) \left(\int_{t_0}^t f(s)\, ds \right)^{q-1}$$

and

$$x(t) \leq x(t_0) + \int_{t_0}^t r^{1-q}(s) \left(\int_{t_0}^s f(u)\, du \right)^{q-1} ds.$$

Therefore, $\liminf_{t\to\infty} x(t) = -\infty$, which contradicts the fact that $x(t) > 0$ eventually. □

9.2.2 Forced super-half-linear oscillation

Along with (unforced) half-linear equation (1.1.1) consider the forced super-half-linear equation

$$\mathcal{M}[y] \equiv (R(t)\Phi_p(y'))' + C(t)\Phi_q(y) = f(t). \tag{9.2.7}$$

The functions Φ_p, Φ_q have the same meaning as for quasilinear equation (9.1.1), here with $1 < p \leq q$. Further, R, C, f are continuous functions with $R(t) > 0$. Now we give without proof a nonlinear variant of Picone's identity which will be needed to prove the main results of this subsection.

Theorem 9.2.3. *Let x and y be continuously differentiable functions such that $R\Phi_p(y')$ is also continuously differentiable and $y(t) \neq 0$ in an interval $I \subset \mathbb{R}$. Then in this interval*

$$\left\{\frac{|x|^p}{\Phi_p(y)} R(t)\Phi_p(y')\right\}' = R(t)|x'|^p - \left\{C(t)|y|^{q-p} - \frac{f(t)}{\Phi_p(y)}\right\}|x|^p \\ - R(t)Q(x', xy'/y) + \frac{|x|^p}{\Phi_p(y)}\{\mathcal{M}[y] - f(t)\}, \tag{9.2.8}$$

where Q denotes the form defined by $Q(u,v) := |u|^p - pu\Phi_p(y) + (p-1)|v|^p$ which satisfies $Q(u,v) \geq 0$ for all $u, v \in \mathbb{R}$ with the equality if and only if $u = v$.

To obtain the first result concerning equation (9.2.7), assume that $C(t) \geq 0$ on some interval $[a,b]$. Further, consider the functional

$$\mathcal{G}(y; a, b) = \int_a^b \left\{R(t)|y'|^p - (p-1)^{\frac{1-p}{q-1}}(q-1)(q-p)^{\frac{p-q}{q-1}}[C(t)]^{\frac{p-1}{q-1}}|f(t)|^{\frac{q-p}{q-1}}|y|^p\right\} dt,$$

over $W_0^{1,p}(a,b)$, with the convention that $0^0 = 1$.

Theorem 9.2.4. *If there exists a nontrivial $\eta \in W_0^{1,p}(a,b)$ such that*

$$\mathcal{G}(\eta, a, b) \leq 0, \tag{9.2.9}$$

then every solution y of (9.2.7) defined on $[a,b]$ and satisfying $y(t)f(t) \leq 0$ in this interval must have a zero in $[a,b]$.

Proof. Assume by a contradiction that (9.2.7) has a solution y satisfying $y(t)f(t) \leq 0$ and $y(t) \neq 0$ on $[a,b]$. Then identity (9.2.8) with $x(t)$ replaced by $\eta(t)$ reduces to

$$\left\{\frac{|\eta|^p}{\Phi_p(y)} R(t)\Phi_p(y')\right\}' = R(t)|\eta'|^p - \left\{C(t)|y|^{q-p} - \frac{f(t)}{\Phi_p(y)}\right\}|\eta|^p - R(t)Q(\eta', \eta y'/y). \tag{9.2.10}$$

Denote by $F(y)$ the expression in the brackets on the right-hand side of (9.2.10) considered as the function of y and observe that

$$\min_{y\neq 0} F(y) = \min_{y \neq 0}\left\{C(t)|y|^{q-p} + \frac{|f|}{|y|^{p-1}}\right\} = (p-1)^{-\frac{p-1}{q-1}}(q-1)(q-p)^{\frac{p-q}{q-1}}C^{\frac{p-1}{q-1}}|f|^{\frac{q-p}{p-1}} \tag{9.2.11}$$

if $p < q$, and $F(y) \geq C(t)$ if $p = q$. Thus, in both cases (9.2.10) reduces to

$$\left\{\frac{|\eta|^p}{\Phi_p(y)} R(t)\Phi_p(y')\right\}' \leq R(t)|\eta'|^p \\ - (p-1)^{-\frac{p-1}{q-1}}(q-1)(q-p)^{\frac{p-q}{q-1}}C^{\frac{p-1}{q-1}}|f|^{\frac{q-p}{p-1}}|\eta|^p - R(t)Q(\eta', \eta y'/y). \tag{9.2.12}$$

Integrating inequality (9.2.12) from a to b we obtain

$$0 \leq \mathcal{G}(\eta; a, b) - \int_a^b R(t)Q\left(\eta', \frac{\eta y'}{y}\right) dt,$$

which is a contradiction unless $\mathcal{G}(\eta; a, b) \equiv 0$ in $[a, b]$. The last relation implies that y must be a constant multiple of η, and so we get, in particular, $y(a) = y(b) = 0$. □

Corollary 9.2.1. *Let there exist two sequences of disjoint intervals* (a_n^-, b_n^-), (a_n^+, b_n^+), $t_0 \le a_n^- < b_n^- \le a_n^+ < b_n^+$, $a_n^- \to \infty$ *as* $n \to \infty$ *such that* $C(t) \ge 0$ *on* $[a_n^-, b_n^-] \cup [a_n^+, b_n^+]$, $f(t) \le 0$ *on* $[a_n^-, b_n^-]$, $f(t) \ge 0$ *on* $[a_n^+, b_n^+]$, $n = 1, 2, \ldots$, *and two sequences of nontrivial continuously differentiable functions* $\eta_n^-(t)$ *and* $\eta_n^+(t)$ *defined on* $[a_n^-, b_n^-]$ *and* $[a_n^+, b_n^+]$, *respectively, such that*

$$\eta_n^-(a_n^-) = \eta_n^-(b_n^-) = \eta_n^+(a_n^+) = \eta_n^+(b_n^+) = 0,$$

for $n = 1, 2, \ldots$, *and* $\mathcal{G}(\eta_n^{\pm}; a_n^{\pm}, b_n^{\pm}) \le 0$ *for* $n \in \mathbb{N}$. *Then all solutions of* (9.2.7) *are oscillatory.*

The next result is a Leighton type comparison theorem where equation (9.2.7) is compared with (1.1.1). Theorem 9.2.3 plays an important role there.

Theorem 9.2.5. *If there exists a nontrivial solution* x *of* (1.1.1) *in* $[a, b]$ *such that* $x(a) = x(b) = 0$ *and*

$$\begin{aligned}\mathcal{H}(x; a, b) = \int_a^b \Big\{&(r(t) - R(t))|x'|^p \\ &+ \Big[(p-1)^{\frac{1-p}{q-1}}(q-1)(q-p)^{\frac{p-q}{q-1}}[C(t)]^{\frac{p-1}{q-1}}|f(t)|^{\frac{q-p}{q-1}} - c(t)\Big]|x|^p\Big\}\, dt \ge 0,\end{aligned}$$

then every solution y *of* (9.2.7) *satisfying* $y(t)f(t) \le 0$ *in* (a, b) *has a zero in* $[a, b]$.

Proof. If x is a nontrivial solution of (1.1.1) satisfying $x(a) = x(b) = 0$, then integration by parts yields

$$\int_a^b \{r(t)|x'|^p - c(t)|x|^p\}\, dt = 0. \tag{9.2.13}$$

Thus, combining (9.2.9) with (9.2.13) we obtain $\mathcal{H}(x; a, b) = -\mathcal{G}(x; a, b) \ge 0$ and the conclusion follows from Theorem 9.2.3. □

Now we give a Sturm-Picone type comparison theorem involving equations (9.2.7) and (1.1.1).

Corollary 9.2.2. *Let* $C(t) \ge 0$ *in* $[a, b]$. *If* $r(t) \ge R(t)$,

$$(p-1)^{\frac{1-p}{q-1}}(q-1)(q-p)^{\frac{p-q}{q-1}}[C(t)]^{\frac{p-1}{q-1}}|f(t)|^{\frac{q-p}{q-1}} \ge c(t)$$

in $[a, b]$ *and there exists a nontrivial solution* x *of* (1.1.1) *such that* $x(a) = x(b) = 0$, *then any solution of* (9.2.7) *satisfying* $y(t)f(t) \le 0$ *in* (a, b) *has a zero in* $[a, b]$.

As a consequence of Theorem 9.2.5, we have the following general comparison result which relates oscillation of the forced super-half-linear equation (9.2.7) to that of conjugacy of two sequences of associated "minorant" half-linear equations (9.2.14) and (9.2.15) below considered on the sequences of corresponding disjoint intervals $[a_n^-, b_n^-]$ and $[a_n^+, b_n^+]$, respectively.

Corollary 9.2.3. *Let there exist two sequences of disjoint intervals* (a_n^-, b_n^-) *and* (a_n^+, b_n^+), $t_0 \le a_n^- < b_n^- \le a_n^+ < b_n^+$, $a_n^- \to \infty$ *as* $n \to \infty$ *such that* $C(t) \ge 0$ *on* $[a_n^-, b_n^-] \cup [a_n^+, b_n^+]$, $f(t) \le 0$ *on* $[a_n^-, b_n^-]$, $f(t) \ge 0$ *on* $[a_n^+, b_n^+]$, $n = 1, 2, \dots,$ *and two sequences of half-linear equations*

$$(r_n^-(t)\Phi(x'))' + c_n^-(t)\Phi(x) = 0, \tag{9.2.14}$$

$$(r_n^+(t)\Phi(x'))' + c_n^+(t)\Phi(x) = 0, \tag{9.2.15}$$

where $r_n^-, c_n^- : [a_n^-, b_n^-] \to \mathbb{R}$ *and* $r_n^+, c_n^+ : [a_n^+, b_n^+] \to \mathbb{R}$ *are continuous functions with* $r_n^-(t) > 0$ *and* $r_n^+(t) > 0$, *with respective nontrivial solutions* x_n^- *and* x_n^+ *satisfying* $x_n^-(a_n^-) = x_n^-(b_n^-) = x_n^+(a_n^+) = x_n^+(b_n^+) = 0$, $n = 1, 2, \dots,$ *and*

$$\int_{a_n^\pm}^{b_n^\pm} \Big\{ \big[r_n^\pm(t) - R(t)\big] |(x_n^\pm)'|^p \\ + \Big[(p-1)^{\frac{1-p}{q-1}} (q-1)(q-p)^{\frac{p-q}{q-1}} [C(t)]^{\frac{p-1}{q-1}} |f(t)|^{\frac{q-p}{q-1}} - c_n^\pm(t)\Big] |x_n^\pm|^p \Big\}\, dt \ge 0$$

for every $n \in \mathbb{N}$. *Then all solutions of* (9.2.7) *are oscillatory.*

In the next result, by *consecutive sign change points* of the oscillatory forcing function f we understand points $t_1, t_2 \in [t_0, \infty)$, $t_1 < t_2$, such that $f(t) \ge 0$ (resp. $f(t) \le 0$) on $[t_1, t_2]$ and $f(t) < 0$ (resp. $f(t) > 0$) on $(t_1 - \varepsilon, t_1) \cup (t_2, t_2 + \varepsilon)$ for some $\varepsilon > 0$.

Corollary 9.2.4. *Assume that* $C(t) \ge 0$ *on* $[t_0, \infty)$

$$r(t) \ge R(t), \tag{9.2.16}$$

$$(p-1)^{\frac{1-p}{q-1}} (q-1)(q-p)^{\frac{p-q}{q-1}} [C(t)]^{\frac{p-1}{q-1}} |f(t)|^{\frac{q-p}{q-1}} \ge c(t) \tag{9.2.17}$$

for $t \ge t_0$, *and either* (9.2.16) *or* (9.2.17) *do not become an identity on any open interval where* $f(t) \equiv 0$. *Moreover, suppose that* (1.1.1) *is oscillatory and the distance between consecutive zeros of any solution of* (1.1.1) *is less than the distance between consecutive sign change points of the forcing function* f. *Then every nontrivial solution of* (9.2.7) *is oscillatory, too.*

Before presenting the last result of this section, recall that the concept of quick oscillation has been introduced in Subsection 5.6.2.

Corollary 9.2.5. *Let* $C(t) \ge 0$ *for* $t \ge t_0$. *If* (9.2.16) *and* (9.2.17) *hold and every solution of* (1.1.1) *is quickly oscillatory, then every nontrivial solution of the forced equation* (9.2.7) *is oscillatory, too, provided that the forcing function* f *is moderately oscillatory, i.e., it changes sign on* $[T, \infty)$ *for each* $T \ge t_0$ *and the distance between consecutive sign change points of* f *is bounded from below.*

9.3 Half-linear differential equations with deviating arguments

The aim of this section is to study half-linear equations where the argument (arguments) in the non-differential term (terms) are deviating. The presence of such deviation may cause phenomena which are not usual in the "ordinary case". For instance, the equation $(\Phi(x'))' = c(t)\Phi(x)$ with $c(t) \geq 0$ is nonoscillatory by the Sturmian comparison theorem since its Sturmian majorant $(\Phi(x'))' = 0$ is nonoscillatory. However, the presence of a deviating argument may generate oscillation of some or all of its solutions. Indeed, for example the function $\sin_p t$ is an oscillatory solution of the functional differential equations $(\Phi(x'(t)))' = (p-1)\Phi(x(t-\pi_p))$ and $(\Phi(x'(t)))' = (p-1)\Phi(x(t+\pi_p))$. This is obvious if one realizes that $\sin_p(t-\pi_p) = \sin_p(t+\pi_p) = -\sin_p t$. On the other hand, we will see that some of the aspects of behavior of solutions of the functional differential equations are shared with the corresponding ordinary differential equations, in particular, when a deviation is small or when one deals with the existence of some nonoscillatory solutions.

Since the Sturmian theory in general does not extend to the deviating case (in particular, oscillatory and nonoscillatory solutions may coexist), oscillation of a given functional equation is defined here as the oscillation of all its nontrivial solutions. Recall that a solution is said to be oscillatory if it has arbitrarily large zeros. A solution which is of eventually one sign is said to be nonoscillatory.

Some of the parts of proofs, which are rather technical, are omitted; we refer to the sources where they can be found.

9.3.1 Oscillation of equation with nonnegative second coefficient

Let us consider the half-linear functional equation

$$(r(t)\Phi(x'(t)))' + c(t)\Phi(x(\tau(t))) = 0 \tag{9.3.1}$$

on the interval $[a, \infty)$. In addition to the usual assumptions on $c(t)$ and $r(t)$, throughout this subsection we suppose that $\int^\infty r^{1-q}(s)\, ds = \infty$, $c(t) \geq 0$ (and eventually nontrivial), $\tau(t)$ is a continuously differentiable function with $\tau'(t) \geq 0$ and $\lim_{t\to\infty} \tau(t) = \infty$.

In the first theorem we compare oscillation of (9.3.1) with oscillation of certain first order equation.

Theorem 9.3.1. *Let $\tau(t) \leq t$. If for all large $T \geq b \geq a$ so that $\tau(t) \geq b$, $t \geq T$, the first order delay equation*

$$y'(t) + c(t)\left(\int_b^{\tau(t)} r^{1-q}(s)\, ds\right)^{p-1} y(\tau(t)) = 0 \tag{9.3.2}$$

is oscillatory, then (9.3.1) *is oscillatory.*

Proof. Let x be a nonoscillatory solution of (9.3.1). Then there is $t_1 \geq a$ such that $x'(t) > 0$ for $t \geq t_1$. It can be shown, see [6, Lemma 3.13.1], that there is $t_2 \geq t_1$ such that $\tau(t) \geq t_1$ and

$$x(\tau(t)) \geq r^{q-1}(\tau(t))x'(\tau(t)) \int_{t_1}^{\tau(t)} r^{1-q}(s)\,ds \tag{9.3.3}$$

for $t \geq t_2$. Using (9.3.3) in equation (9.3.1), we obtain

$$\begin{aligned}(r(t)(x'(t))^{p-1})' + c(t)\left(r^{q-1}(\tau(t))x'(\tau(t)) \int_{t_1}^{\tau(t)} r^{1-q}(s)\,ds\right)^{p-1} \\ \leq (r(t)(x'(t))^{p-1})' + c(t)x^{p-1}(\tau(t)) = 0\end{aligned}$$

for $t \geq t_2$. Setting $w(t) = r(t)(x'(t))^{p-1}$ in the above inequality, we get

$$w'(t) + c(t)\left(\int_{t_1}^{\tau(t)} r^{1-q}(s)\,ds\right)^{p-1} w(\tau(t)) \leq 0 \tag{9.3.4}$$

for $t \geq t_2$. Integrating (9.3.4) from $t \geq t_2$ to ξ and letting $\xi \to \infty$, we find

$$w(t) \geq \int_t^\infty c(s)\left(\int_{t_1}^{\tau(s)} r^{1-\eta}(\eta)\,d\eta\right)^{p-1} w(\tau(s))\,ds.$$

Clearly, the function $w(t)$ is strictly decreasing for $t \geq t_2$. Hence, by [6, Lemma 3.13.6], there exists a positive solution y of (9.3.2) with $\lim_{t\to\infty} y(t) = 0$. But this contradicts the assumption that (9.3.2) is oscillatory. □

Remark 9.3.1. The statement of Theorem 9.3.1 can easily be extended to the quasilinear equation

$$(\Phi_\alpha(x'(t)))' + c(t)\Phi_p(x(\tau(t))) = 0 \tag{9.3.5}$$

with $\alpha > 1$. Equation (9.3.2) then reads as

$$y'(t) + c(t)\left(\int_b^{\tau(t)} r^{1-q}(s)\,ds\right)^{p-1} |y(\tau(t))|^{(p-1)/(\alpha-1)} \operatorname{sgn} y(\tau(t)) = 0. \tag{9.3.6}$$

The next lemma is a classical result, see e.g. [6]. A similar result can be stated also when c is nonpositive and $\tau(t) \geq t$. Such a modification will find an application in the next subsection.

Lemma 9.3.1. *Let $\tau(t) \leq t$ for $t \geq a$. If*

$$\liminf_{t\to\infty} \int_{\tau(t)}^t c(s)\,ds > \frac{1}{e},$$

then

(a) *the inequality* $x'(t) + c(t)x(\tau(t)) \le 0$ *has no eventually positive solution,*

(b) *the inequality* $x'(t) + c(t)x(\tau(t)) \ge 0$ *has no eventually negative solution,*

(c) *the equation* $x'(t) + c(t)x(\tau(t)) = 0$ *is oscillatory.*

The above lemma when applied to (9.3.2) leads to the following statement.

Theorem 9.3.2. *If for every* $T \ge b \ge a$ *so that* $t \ge \tau(t) \ge b$, $t \ge T$,

$$\liminf_{t\to\infty} \int_{\tau(t)}^{t} c(s)\left(\int_b^{\tau(s)} r^{1-q}(\eta)\,d\eta\right)^{p-1} ds > \frac{1}{e}, \tag{9.3.7}$$

then equation (9.3.1) *is oscillatory.*

Remark 9.3.2. Utilizing (9.3.6), the above theorem can be extended to (9.3.5) provided (9.3.7) is replaced by

$$\int^{\infty} c(s)\left(\int_b^{\tau(s)} r^{1-q}(\eta)\,d\eta\right)^{\alpha-1} ds = \infty$$

when $\alpha < p$.

In the proof of the next theorem, a generalized Riccati type transformation plays a crucial role.

Theorem 9.3.3. *Let* $\tau(t) \le t$ *for* $t \ge a$. *If there exists a positive function* $\rho(t) \in C^1[a,\infty)$ *such that*

$$\limsup_{t\to\infty} \int_a^t \left(\rho(s)c(s) - p^{-p} r(\tau(s)) \frac{(\rho'(s))^p}{(\rho(s)\tau'(s))^{p-1}}\right) ds = \infty, \tag{9.3.8}$$

then equation (9.3.1) *is oscillatory.*

Proof. Let x be a nonoscillatory solution of (9.3.1), say $x(t) > 0$ for $t \ge a$. As in Theorem 9.3.1 there exists $t_1 \ge a$ such that for $t \ge t_1$,

$$x'(t) > 0 \quad \text{and} \quad r(t)(x'(t))^{p-1} \le r(\tau(t))(x'(\tau(t)))^{p-1}. \tag{9.3.9}$$

Define $w(t) = \rho(t)r(t)(x'(t)/x(\tau(t)))^{p-1}$ for $t \ge t_1$. Then for $t \ge t_1$, w satisfies the Riccati type equation

$$w'(t) = -\rho(t)c(t) + \frac{\rho'(t)}{\rho(t)}w(t) - (p-1)\rho(t)\tau'(t)\frac{r(t)(x'(t))^{p-1}x'(\tau(t))}{x^p(\tau(t))}. \tag{9.3.10}$$

Using (9.3.9) in (9.3.10), we get

$$w'(t) \le -\rho(t)c(t) + \frac{\rho'(t)}{\rho(t)}w(t) - (p-1)\rho(t)\tau'(t)r(t)\left(\frac{r(t)}{r(\tau(t))}\right)^{q-1}\left(\frac{x'(t)}{x(\tau(t))}\right)^p,$$

or

$$w'(t) \le -\rho(t)c(t) + \frac{\rho'(t)}{\rho(t)} w(t) - (p-1)\tau'(t)(r(\tau(t))\rho(t))^{1-q} w^q(t). \tag{9.3.11}$$

From the Young inequality we get

$$\frac{\rho'(t)}{\rho(t)} w(t) - (p-1)\tau'(t)(r(\tau(t))\rho(t))^{1-q} w^q \le p^{-p} \frac{r(\tau(t))(\rho'(t))^p}{(\rho(t)\tau'(t))^{p-1}}.$$

Thus, (9.3.11) gives

$$w'(t) \le -\rho(t)c(t) + p^{-p} r(\tau(t)) \frac{(\rho'(t))^p}{(\rho(t)\tau'(t))^{p-1}}.$$

Integrating the above inequality from t_1 to t, we get

$$0 < w(t) \le w(t_1) - \int_{t_1}^{t} \left(\rho(s)c(s) - p^{-p} r(\tau(s)) \frac{(\rho'(s))^p}{(\rho(s)\tau'(s))^{p-1}} \right) ds.$$

Taking lim sup of both sides as $t \to \infty$, we find a contradiction to (9.3.8). □

The following corollary is immediate. Note that it can be proved also by a slightly different approach based on the Riccati type transformation and Hölder's inequality, see [6, Corollary 3.13.3].

Corollary 9.3.1. *Let condition* (9.3.8) *of Theorem 9.3.3 be replaced by*

$$\limsup_{t\to\infty} \int_a^t \rho(s)c(s)\, ds = \infty \quad \text{and} \quad \lim_{t\to\infty} \int_a^t r(\tau(s)) \frac{(\rho'(s))^p}{(\rho(s)\tau'(t))^{p-1}}\, ds < \infty.$$

The the conclusion of Theorem 9.3.3 holds.

For simplicity, the next criterion is proved for equation (9.3.1) where $r(t) \equiv 1$, i.e., for

$$(\Phi(x'))' + c(t)\Phi(x(\tau(t))) = 0. \tag{9.3.12}$$

Its extension to the general case is not difficult; the formulation is given in the subsequent remark. It is shown that if the delay $\tau(t)$ is sufficiently close to t, in a certain sense, then some oscillation criteria for (1.3.2) can be extended to (9.3.12). Oscillation criteria presented here are half-linear extensions of some results for the linear second order retarded equations (the case $p = 2$ in (9.3.12)) given in [153, 306].

First we present without proof a technical auxiliary statement, the proof can be found in [8].

Lemma 9.3.2. *Suppose that the following conditions hold:*

(i) $x(t) \in C^2[T, \infty)$ *for some* $T > 0$,

(ii) $x(t) > 0$, $x'(t) > 0$, *and* $x''(t) \le 0$ *for* $t \ge T$.

Then for each $k_1 \in (0,1)$ *there exists a constant* $T_{k_1} \ge T$ *such that*

$$x(\tau(t)) \ge \frac{k_1\tau(t)}{t}x(t), \quad \text{for } t \ge T_{k_1},$$

and for every $k_2 \in (0,1)$ *there exists a constant* $T_{k_2} \ge T$ *such that*

$$x(t) \ge k_2 t x'(t), \quad \text{for } t \ge T_{k_2}.$$

Theorem 9.3.4. *Let* $\tau(t) \le t$ *for* $t \ge a$. *Denote for* $t \ge a$

$$\gamma(t) := \sup\{s \ge t_0 : \tau(s) \le t\}.$$

Equation (9.3.12) *is oscillatory if either of the following holds:*

$$\limsup_{t\to\infty} t^{p-1}\int_t^\infty c(s)\left(\frac{\tau(s)}{s}\right)^{p-1} ds > 1, \tag{9.3.13}$$

or

$$\limsup_{t\to\infty} t^{p-1}\int_{\gamma(t)}^\infty c(s)\, ds > 1. \tag{9.3.14}$$

Proof. Suppose to the contrary that (9.3.12) has a nonoscillatory solution $x(t)$. Without loss of generality we may suppose that $x(t) > 0$ for large t, say $t \ge t_1$. Then also $x(\tau(t)) > 0$ on $[t_1,\infty)$ for t_1 large enough. Since $c(t) \ge 0$ on $[t_1,\infty)$,

$$(\Phi(x'(t)))' = -c(t)(\Phi(x(\tau(t)))) \le 0. \tag{9.3.15}$$

Hence, the function $\Phi(x')$ is decreasing. Since $\sup\{c(t); t \ge T\} > 0$ for any $T \ge 0$, we see that either

(a) $x'(t) > 0$ for all $t \ge t_1$, or

(b) there exists $t_2 \ge t_1$ such that $x'(t) < 0$ on $[t_2,\infty)$.

If (b) holds, then it follows from (9.3.15) that

$$0 \ge [|x'(t)|^{p-2}x'(t)]' = (p-1)|x'(t)|^{p-2}x''(t), \qquad \text{for } t \ge t_2.$$

Thus, $x''(t) \le 0$ for $t \in [t_2,\infty)$. This and $x'(t) < 0$ on $[t_2,\infty)$ imply that there exists $t_3 > t_2$ such that $x(t) \le 0$ for $t \ge t_3$. This contradicts $x(t) > 0$. Thus, (a) holds.

Integrating (9.3.15) from $t \ge t_1$ to ∞, we obtain

$$\begin{aligned}
-\int_t^\infty c(s)\Phi(x(\tau(s)))ds &= \int_t^\infty \Phi(x'(s))'\, ds \\
&= \int_t^\infty ([x'(s)]^{p-1})' ds = \lim_{s\to\infty}[x'(s)]^{p-1} - [x'(t)]^{p-1}.
\end{aligned}$$

Since $x'(t) > 0$ for $t \geq t_1$, we find
(9.3.16)

$$[x'(t)]^{p-1} = \lim_{s\to\infty}[x'(s)]^{p-1} + \int_t^\infty c(s)\Phi(x(\tau(s)))\,ds \geq \int_t^\infty c(s)[x(\tau(s))]^{p-1}ds.$$

It follows from (ii) of Lemma 9.3.2 that, for each $k_2 \in (0,1)$, there exists a $T_{k_2} \geq t_1$ such that

$$[x(t)]^{p-1} \geq k_2^{p-1}t^{p-1}[x'(t)]^{p-1} \geq k_2^{p-1}t^{p-1}\int_t^\infty c(s)[x(\tau(s))]^{p-1}ds, \tag{9.3.17}$$

for $t \geq T_{k_2}$. By (i) of Lemma 9.3.2, for each $k_1 \in (0,1)$, there exists a T_{k_1}, such that

$$[x(\tau(s))]^{p-1} \geq k_1^{p-1}\left(\frac{\tau(t)}{t}\right)^{p-1}(x(t))^{p-1}, \tag{9.3.18}$$

for $t \geq T_{k_1}$. Then, by (9.3.17) and (9.3.18), for $t \geq t_4 := \max\{T_{k_1}, T_{k_2}\}$,

$$\begin{aligned}[x(t)]^{p-1} &\geq k_2^{p-1}t^{p-1}\int_t^\infty c(s)[x(\tau(s))]^{p-1}ds \\ &\geq k_1^{p-1}k_2^{p-1}t^{p-1}\int_t^\infty c(s)\left(\frac{\tau(s)}{s}\right)^{p-1}[x(s)]^{p-1}ds \\ &\geq k^{2(p-1)}t^{p-1}\int_t^\infty c(s)\left(\frac{\tau(s)}{s}\right)^{p-1}[x(s)]^{p-1}ds,\end{aligned} \tag{9.3.19}$$

where $k := \min\{k_1, k_2\}$. Since $x'(t) > 0$, it follows that

$$\begin{aligned}1 &\geq \frac{k^{2p-2}t^{p-1}}{[x(t)]^{p-1}}\int_t^\infty c(s)\left(\frac{\tau(s)}{s}\right)^{p-1}[x(s)]^{p-1}ds \\ &\geq k^{2p-2}t^{p-1}\int_t^\infty c(s)\left(\frac{\tau(s)}{s}\right)^{p-1}ds, \qquad \text{for } t \geq t_4.\end{aligned} \tag{9.3.20}$$

Hence,

$$\limsup_{t\to\infty} t^{p-1}\int_t^\infty c(s)\left(\frac{\tau(s)}{s}\right)^{p-1}ds := a < \infty.$$

Suppose that (9.3.13) holds, then there exists a sequence $\{s_n\}$ with the properties $\lim_{n\to\infty} s_n = \infty$ and

$$\lim_{n\to\infty} s_n^{p-1}\int_{s_n}^\infty c(s)\left(\frac{\tau(s)}{s}\right)^{p-1}ds = a > 1.$$

For $\varepsilon_1 := (a-1)/2 > 0$, there exists an integer $N_1 > 0$ such that

$$\frac{a+1}{2} = a - \varepsilon_1 < s_n^{p-1}\int_{s_n}^\infty c(s)\left(\frac{\tau(s)}{s}\right)^{p-1}ds, \tag{9.3.21}$$

for $n > N_1$. Choose k such that

$$\left(\frac{2}{a+1}\right)^{1/2(p-1)} < k < 1.$$

By (9.3.20) and (9.3.21),

$$1 \geq k^{2(p-1)} s_n^{p-1} \int_{s_n}^{\infty} c(s) \left(\frac{\tau(s)}{s}\right)^{p-1} ds > \left(\frac{2}{a+1}\right)\left(\frac{a+1}{2}\right) = 1,$$

for s_n large enough. This contradiction shows that (9.3.13) does not hold.

Now, by $\gamma(t) \geq t$ and (9.3.17), we have

$$[x(t)]^{p-1} \geq k_2^{p-1} t^{p-1} \int_{\gamma(t)}^{\infty} c(s)[x(\tau(s))]^{p-1} ds,$$

for $t \geq T_{k_2}$. Since $x(t)$ is increasing and $\tau(s) \geq t$ for $s \geq \gamma(t)$, it follows that

$$[x(t)]^{p-1} \geq k_2^{p-1} t^{p-1} \int_{\gamma(t)}^{\infty} c(s)[x(\tau(s))]^{p-1} ds \geq k_2^{p-1} t^{p-1} [x(t)]^{p-1} \int_{\gamma(t)}^{\infty} c(s) ds.$$

Dividing $[x(t)]^{p-1}$ in both sides of the above inequality, we get

$$k_2^{p-1} t^{p-1} \int_{\gamma(t)}^{\infty} c(s) ds \leq 1, \tag{9.3.22}$$

for $t \geq T_{k_2}$. Thus,

$$\limsup_{t\to\infty} t^{p-1} \int_{\gamma(t)}^{\infty} c(s) ds := b < \infty.$$

Suppose that (9.3.14) holds. Then there exists a sequence $\{t_n\}$ with $\lim_{n\to\infty} t_n = \infty$ such that

$$\lim_{n\to\infty} t_n^{p-1} \int_{\gamma(t_n)}^{\infty} c(s) ds = b > 1.$$

Thus, for $\varepsilon_2 := (b-1)/2 > 0$, there exists an integer $N_2 > 0$ such that

$$\frac{b+1}{2} = b - \varepsilon_2 < t_n^{p-1} \int_{\gamma(t_n)}^{\infty} c(s) ds, \tag{9.3.23}$$

for $n > N_2$. Choose $k_2 \in \left(2(b+1)^{1-q}, 1\right)$. By (9.3.22) and (9.3.23),

$$1 \geq k_2^{p-1} t_q^{p-1} \int_{\gamma(t_n)}^{\infty} c(s) ds > \frac{2}{b+1} \cdot \frac{b+1}{2} = 1,$$

for t_n large enough. This contradiction proves that (9.3.14) does not hold. □

Remark 9.3.3. In an extension of the above theorem to equation (9.3.1), condition (9.3.13) is split into the following two conditions provided r is differentiable: If $r'(t) \leq 0$, then for $t \geq T \geq a$,

$$\limsup_{t\to\infty} \left(\int_T^t r^{1-q}(s)\, ds \right)^{p-1} \int_t^\infty \frac{r(s)}{r(T)} \left(\frac{\tau(s)}{s} \right)^{p-1} c(s)\, ds > 1,$$

if $r'(t) \geq 0$, then for $t \geq T \geq a$,

$$\limsup_{t\to\infty} \left(\int_T^t r^{1-q}(s)\, ds \right)^{p-1} \int_t^\infty \left(\frac{\tau(s)}{s} \right)^{p-1} c(s)\, ds > 1.$$

Condition (9.3.14) takes the form

$$\limsup_{t\to\infty} \left(\int_T^t r^{1-q}(s)\, ds \right)^{p-1} \int_{\gamma(t)}^\infty c(s)\, ds > 1.$$

Example 9.3.1. Consider the functional differential equation

$$[\Phi(x')]' + \frac{2^p(p-1)}{t^p} \Phi(x(t/2)) = 0. \tag{9.3.24}$$

Since

$$\begin{aligned} \limsup_{t\to\infty} t^{p-1} \int_t^\infty \frac{2^p(p-1)}{s^p} \left(\frac{(s/2)}{s} \right)^{p-1} ds &= 2(p-1) \limsup_{t\to\infty} t^{p-1} \int_t^\infty \frac{1}{s^p} ds \\ &= 2(p-1) \limsup_{t\to\infty} t^{p-1} \left(\frac{1}{(p-1)t^{p-1}} \right) = 2 > 1, \end{aligned}$$

it follows from (9.3.13) of Theorem 9.3.4 that (9.3.24) is oscillatory. In fact, if the coefficient $2^p(p-1)t^{-p}$ of (9.3.24) is replaced by kt^{-p} with $k > 2^{p-1}(p-1)$, (9.3.24) will again be oscillatory.

In the next theorem we assume that $\tau(t) \leq t$ and we denote

$$\mu(t) = \left(\frac{\tau(t)}{t} \right)^{p-1}.$$

Theorem 9.3.5. *Equation* (9.3.12) *is oscillatory if the differential equation*

$$(\Phi(x'))' + \lambda\mu(t)c(t)\Phi(x) = 0 \tag{9.3.25}$$

is oscillatory for some $\lambda \in (0,1)$.

Proof. By contradiction, suppose that there exists an eventually positive solution x of (9.3.12) and we may also assume that $x(\tau(t)) > 0$ on $[t_1, \infty)$ for some $t_1 \geq t_0$. Then $x''(t) \leq 0$, $x'(t) > 0$ on $[t_2, \infty)$ for some $t_2 \geq t_1$. Since $\lambda \in (0,1)$, it follows from Lemma 9.3.2 that

$$x(\tau(t)) \geq \lambda^{1/(p-1)} x(t) \frac{\tau(t)}{t},$$

for t large enough. Thus,

$$(9.3.26) \qquad [x(\tau(t))]^{p-1} \geq \lambda [x(t)]^{p-1} \left(\frac{\tau(t)}{t}\right)^{p-1},$$

for t large enough. Let

$$w(t) = \frac{\Phi(x'(t))}{\Phi(x(t))}.$$

Then, by (9.3.26),

$$\begin{aligned}
w'(t) &+ \frac{\lambda[\tau(t)]^{p-1}}{t^{p-1}} c(t) + (p-1)|w(t)|^q \\
&= \frac{[(x'(t))^{p-1}]'[x(t)]^{p-1} - [x'(t)]^{p-1}[(x(t))^{p-1}]'}{[(x(t))^{p-1}]^2} \\
&\quad + \frac{\lambda(\tau(t))^{p-1}}{t^{p-1}} c(t) + (p-1)\left(\frac{[x'(t)]^{p-1}}{[x(t)]^{p-1}}\right)^q \\
&= \frac{[(x'(t))^{p-1}]'[x(t)]^{p-1} - (p-1)[x'(t)]^{p-1}[x(t)]^{p-2}x'(t)}{[(x(t))^{p-1}]^2} \\
&\quad + \frac{\lambda(\tau(t))^{p-1}}{t^{p-1}} c(t) + \frac{(p-1)[x'(t)]^p}{[x(t)]^p} \\
&= \frac{[(x'(t))^{p-1}]'}{[x(t)]^{p-1}} - \frac{(p-1)[x'(t)]^p}{[x(t)]^p} + \frac{\lambda(\tau(t))^{p-1}}{t^{p-1}} c(t) + \frac{(p-1)[x'(t)]^p}{[x(t)]^p} \\
&= \frac{-c(t)|x(\tau(t))|^{p-2}x(\tau(t))}{[x(t)]^{p-1}} + \frac{\lambda(\tau(t))^{p-1}c(t)}{t^{p-1}} \\
&= \frac{c(t)}{[x(t)]^{p-1}} \left(\frac{\lambda[\tau(t)]^{p-1}}{t^{p-1}}[x(t)]^{p-1} - [x(\tau(t))]^{p-1}\right) \\
&\leq 0.
\end{aligned}$$

This and Theorem 2.2.1 imply that (9.3.25) is nonoscillatory, but this is a contradiction. Hence, (9.3.12) is oscillatory. □

Remark 9.3.4. Theorem 9.3.5 is an extension of Theorem 2.2 of [153].

Theorem 9.3.6. *Let $\tau(t) \leq t$ for $t \geq a$. If*

$$(9.3.27) \qquad \limsup_{t\to\infty} \int^t c(s) \left(\frac{\tau(s)}{s}\right)^{p-1} ds = \infty,$$

then equation (9.3.12) *is oscillatory.*

Proof. Suppose to the contrary that (9.3.12) has a nonoscillatory solution $x(t)$ which may be assumed to be eventually positive. As in the proof of the Theorem 9.3.4, there exists $t_1 > t_0$ such that $x(\tau(t)) > 0$, $x'(t) > 0$, and $x''(t) < 0$ for $t > t_1$. By (i) of Lemma 9.3.2, there exists $t_2 \geq t_1$ such that

$$x(\tau(t)) \geq \left(\frac{1}{2}\right)^{1/(p-1)} \frac{\tau(t)}{t} x(t),$$

or

$$(x(\tau(t)))^{p-1} \geq \frac{1}{2}\left(\frac{\tau(t)}{t}\right)^{p-1}[x(t)]^{p-1}$$

for $t \geq t_2$. Since $x'(t) > 0$ and $x(\tau(t)) > 0$ for $t \geq t_1$,

$$-((x'(t))^{p-1})' = c(t)[x(\tau(t))]^{p-1} \geq \frac{1}{2}c(t)\left(\frac{\tau(t)}{t}\right)^{p-1}[x(t)]^{p-1}$$

for $t \geq t_2$. Integrating the above inequality from t_2 to t and using the increasing property of $x(t)$, we get

$$\begin{aligned}
[x'(t)]^{p-1} - [x'(t_2)]^{p-1} &\leq -\frac{1}{2}\int_{t_2}^{t} c(s)\left(\frac{\tau(s)}{s}\right)^{p-1}[x(s)]^{p-1}ds \\
&\leq -\frac{1}{2}[x(t_2)]^{p-1}\int_{t_2}^{t} c(s)\left(\frac{\tau(s)}{s}\right)^{p-1}ds,
\end{aligned}$$

or

$$[x'(t)]^{p-1} \leq [x'(t_2)]^{p-1} - \frac{1}{2}[x(t_2)]^{p-1}\int_{t_2}^{t} c(s)\left(\frac{\tau(t)}{t}\right)^{p-1}ds$$

for $t \geq t_2$. This and (9.3.27) imply $[x'(t)]^{p-1} < 0$ for t large enough. This is a contradiction. Thus, (9.3.12) is oscillatory. □

9.3.2 Oscillation of equation with nonpositive second coefficient

Let us consider half-linear functional equations in the form

$$(r(t)\Phi(x'(t)))' - c(t)\Phi(x(\tau(t))) = 0 \tag{9.3.28}$$

and

$$(r(t)\Phi(x'(t)))' - \sum_{i=1}^{n} c_i(t)\Phi(x(\tau_i(t))) = 0, \tag{9.3.29}$$

on the interval $[a, \infty)$, where c_i and τ_i, $i = 1, \dots, n$, satisfy the same conditions as c and τ, respectively, and the conditions imposed on r, c and τ are the same as in the previous subsection. We will see that (9.3.28) has all bounded [resp. unbounded] solutions oscillatory provided the deviation $|\tau(t) - t|$ is large enough in some sense and $\tau(t)$ is retarded [resp. advanced]. Even though from such results oscillation of (9.3.28) does not follow, for equations of mixed type, i.e., involving both retarded and advanced arguments, it is possible to establish oscillation of all solutions. This may happen for equation (9.3.29).

For simplicity, similarly as in some of the above results, we will deal with details only the case when $r(t) \equiv 1$. The extension to the general case is not difficult. Nevertheless, the main statement will be formulated for general r.

We begin by considering the functional differential inequality

$$\{(\Phi(x'(t)))' - c(t)\Phi(x(\tau(t)))\} \operatorname{sgn} x(\tau(t)) \geq 0. \tag{9.3.30}$$

Let x be a nonoscillatory solution of (9.3.30). It is easy to see that x' is eventually of constant sign, so that either $x(t)x'(t) < 0$ or $x(t)x'(t) > 0$ for large t. Thus x is bounded of unbounded according to whether the first or the second inequality holds. Note that if $x(t) > 0$, say for $t \geq b$, then (9.3.30) implies that $x'(t)$ is increasing for $t \geq T$, where $T \geq a$ is chosen so large that $\lim_{t \geq T} \tau(t) \geq b$, and hence x is a convex function on $[T, \infty)$. Our first result shows that in the case when $\tau(t)$ is a retarded argument, it may happen that (9.3.28) admits no bounded nonoscillatory solution.

Theorem 9.3.7. *Suppose that $\tau(t) < t$ for $t \geq a$ and either*

$$\limsup_{t\to\infty} \int_{\tau(t)}^{t} c(s)(\tau(t) - \tau(s))^{p-1} ds > 1 \tag{9.3.31}$$

or

$$\limsup_{t\to\infty} \int_{\tau(t)}^{t} \left(\int_{s}^{t} c(\eta)\, d\eta \right)^{q-1} ds > 1 \tag{9.3.32}$$

or

$$\liminf_{t\to\infty} \int_{(t+\tau(t))/2}^{t} c(s) \left(\frac{s - \tau(s)}{2} \right)^{p-1} ds > \frac{1}{e}. \tag{9.3.33}$$

Then every bounded solution of (9.3.30) *is oscillatory.*

Proof. Let x be a bounded nonoscillatory solution of (9.3.30). Without loss of generality we may assume that $x(t) > 0$ and $x'(t) < 0$ for $t \geq b \geq a$.

Suppose first that (9.3.31) holds. Let $T > b$ be such that $\inf_{t \geq T} \tau(t) \geq b$. Since x is convex on $[T, \infty)$, we have $x(\tau(s)) \geq -x'(\tau(s))(\tau(t) - \tau(s))$, $t \geq s \geq T$. Multiplying this inequality by $c(t)$, substituting for the left-hand side by $(\Phi(x'(s)))' = -((-x'(s))^{p-1})'$ and integrating from $\tau(t)$ to t, we have

$$(-x'(\tau(t)))^{p-1} - (-x'(t))^{p-1} \geq (-x'(\tau(t)))^{p-1} \int_{\tau(t)}^{t} c(s)(\tau(t) - \tau(s))^{p-1} ds,$$

whence it follows that

$$(-x'(\tau(t)))^{p-1} \left(\int_{\tau(t)}^{t} c(s)(\tau(t) - \tau(s))^{p-1} ds - 1 \right) \leq 0,$$

$t \geq T$. But this is inconsistent with (9.3.31).

Suppose next that (9.3.32) holds. Integration of (9.3.30) over $[\sigma, t]$ gives

$$(-x'(\sigma))^{p-1} = (-x'(t))^{p-1} + \int_{\sigma}^{t} c(\eta)(x(\tau(\eta)))^{p-1} d\eta \geq \int_{\sigma}^{t} c(\eta)(x(\tau(\eta)))^{p-1},$$

which implies

$$-x'(\sigma) \geq \left(\int_{\sigma}^{t} c(\eta)(x(\tau(\eta)))^{p-1} d\eta \right)^{q-1}, \tag{9.3.34}$$

$t \geq \sigma \geq T$. Substituting (9.3.34) into

$$x(s) = x(t) + \int_{s}^{t} (-x'(\eta))\, d\eta, \tag{9.3.35}$$

we have

$$x(s) \geq \int_{s}^{t} \left(\int_{\eta}^{t} c(z)(x(\tau(z)))^{p-1} dz \right)^{q-1} d\eta,$$

$t \geq s \geq T$. Putting $s = \tau(t)$ in (9.3.35) and using the fact that $x(\tau(t))$ is decreasing, we conclude that

$$x(\tau(t)) \left(\int_{\tau(t)}^{t} \left(\int_{\eta}^{t} c(z)\, dz \right)^{q-1} d\eta - 1 \right) \leq 0,$$

$t \geq T$, which contradicts (9.3.32).

Concerning the part with condition (9.3.33), we proceed similarly as in the proof of Theorem 9.3.1 and application of Lemma 9.3.1, and so we omit details. Note only that condition (9.3.33) ensures oscillation of the delayed equation

$$y'(t) + c(t) \left(\frac{t - \tau(t)}{2} \right)^{p-1} y \left(\frac{t + \tau(t)}{2} \right) = 0.$$

This completes the proof. □

A dual statement to Theorem 9.3.7 holds in the case where $\tau(t)$ is an advanced argument.

Theorem 9.3.8. *Suppose that $\tau(t) > t$ for $t \geq a$ and either*

$$\limsup_{t\to\infty} \int_{t}^{\tau(t)} c(s)(\tau(s) - \tau(t))^{p-1} ds > 1$$

or

$$\limsup_{t\to\infty} \int_{t}^{\tau(t)} \left(\int_{t}^{s} c(\eta)\, d\eta \right)^{q-1} ds > 1$$

or

$$\liminf_{t\to\infty} \int_{t}^{(t+\tau(t))/2} c(s) \left(\frac{\tau(s) - s}{2} \right)^{p-1} ds > \frac{1}{e}.$$

Then every unbounded solution of (9.3.30) *is oscillatory.*

Proof. The proof is similar to that of Theorem 9.3.7, and hence is omitted. □

The following theorem follows from the above results, now extended to the case of general r. Recall that throughout this section we assume $\int^{\infty} r^{1-q}(t)\, dt = \infty$.

Theorem 9.3.9. *(i) All bounded solutions of (9.3.29) are oscillatory if there exists $i \in \{1, 2, \ldots, n\}$ such that $\tau_i(t) < t$ for $t \geq a$ and either*

$$\limsup_{t\to\infty} \int_{\tau_i(t)}^{t} c_i(s) \left(\int_{\tau_i(s)}^{\tau_i(t)} r^{1-q}(\eta)\, d\eta \right)^{p-1} ds > 1$$

or

$$\limsup_{t\to\infty} \int_{\tau_i(t)}^{t} r^{1-q}(s) \left(\int_{s}^{t} c_i(\eta)\, d\eta \right)^{q-1} ds > 1$$

or

$$\liminf_{t\to\infty} \int_{(t+\tau_i(t))/2}^{t} c_i(s) \left(\int_{\tau_i(s)}^{(s+\tau_i(s))/2} r^{1-q}(\eta)\, d\eta \right)^{p-1} ds > \frac{1}{e}.$$

(ii) All unbounded solutions of (9.3.29) are oscillatory if there is $j \in \{1, 2, \ldots, n\}$ such that $\tau_j(t) > t$ for $t \geq a$ and either

$$\limsup_{t\to\infty} \int_{t}^{\tau_j(t)} c_j(s) \left(\int_{\tau_j(t)}^{\tau_j(s)} r^{1-q}(\eta)\, d\eta \right)^{p-1} ds > 1$$

or

$$\limsup_{t\to\infty} \int_{t}^{\tau_j(t)} r^{1-q}(s) \left(\int_{s}^{t} c_j(\eta)\, d\eta \right)^{q-1} ds > 1$$

or

$$\liminf_{t\to\infty} \int_{t}^{(t+\tau_j(t))/2} c_j(s) \left(\int_{(s+\tau_j(s))/2}^{\tau_j(s)} r^{1-q}(\eta)\, d\eta \right)^{p-1} ds > \frac{1}{e}.$$

(iii) All solutions of (9.3.29) are oscillatory if there are $i, j \in \{1, 2, \ldots, n\}$ such that $\tau_i(t)$ and $\tau_j(t)$ satisfy the conditions of (i) and (ii), respectively.

Proof. In view of the assumption $\int^{\infty} r^{1-q}(s)\, ds = \infty$, we can prove the statement only for the case where $r(t) \equiv 1$, and an extension is immediate.

(i) Suppose to the contrary that (9.3.29) has a bounded nonoscillatory solution x. Then, from (9.3.29) we see that x satisfies the differential inequality

$$\{(\Phi(x'(t)))' - c_i(t)\Phi(x(\tau_i(t)))\} \operatorname{sgn} x(\tau_i(t)) \geq 0 \tag{9.3.36}$$

for all sufficiently large t. This, however, is impossible since the possibility of the existence of bounded nonoscillatory solutions to (9.3.36) is excluded by Theorem 9.3.7.

(ii) Argumentation is similar to that of (i), with making use of Theorem 9.3.8.

(iii) This is an immediate consequence of (i) and (ii). □

9.3.3 Existence and asymptotic behavior of nonoscillatory solutions

We are now interested in the existence and asymptotic behavior of nonoscillatory solutions of equation (9.3.29). If x is a nonoscillatory solution of (9.3.29), then there is $t_0 \geq a$ such that either

(9.3.37) $$x(t)x'(t) > 0 \quad \text{for } t \geq t_0,$$

or

(9.3.38) $$x(t)x'(t) < 0 \quad \text{for } t \geq t_0.$$

If (9.3.37) holds, then it is not difficult to show that x is unbounded, and the limit $M_x = \lim_{t\to\infty} r(t)\Phi(x'(t))$ exists, either infinite or finite. If (9.3.38) holds, then x is bounded and the finite limit $x(\infty) = \lim_{t\to\infty} x(t)$ exists. Compare with Section 4.1 where asymptotic behavior of solutions to "ordinary" half-linear equations is studied.

In what follows we only need to consider eventually positive solutions of equation (9.3.29), since if x satisfies (9.3.29), then so does $-x$. Let x be an eventually positive solution of (9.3.29) satisfying (9.3.37) whose quasiderivative $r\Phi(x')$ has a finite limit M_x. The twice integration of (9.3.29) for $t \geq T$ yields

(9.3.39) $$x(t) = x(T) + \int_T^t r^{1-q}(s)\left(M_x - \int_s^\infty \sum_{i=1}^n c_i(u)x^{p-1}(\tau_i(u))\,du\right)^{q-1} ds,$$

where $T > t_1$ is chosen so that $\inf_{t\geq T} \tau_i(t) \geq t_1$, $i = 1, \dots, n$.

Similarly, if x is an eventually positive solution of (9.3.29) satisfying (9.3.38), then we obtain

(9.3.40) $$x(t) = x(\infty) + \int_t^\infty r^{1-q}(s)\left(\int_s^\infty \sum_{i=1}^n c_i(u)x^{p-1}(\tau_i(u))\,du\right)^{q-1} ds,$$

$t \geq T$.

Based on these integral representations of (9.3.29), we can prove the following existence theorems.

Theorem 9.3.10. *Equation* (9.3.29) *has a nonoscillatory solution x such that*

(9.3.41) $$\lim_{t\to\infty} \frac{x(t)}{\int_a^t r^{1-q}(s)\,ds} \in \mathbb{R} \setminus \{0\}$$

if and only if

(9.3.42) $$\int^\infty c_i(s)\left(\int^{\tau_i(s)} r^{1-q}(u)\,du\right)^{p-1} ds < \infty,$$

$i = 1, 2, \dots, n$.

Proof. The necessity follows from (9.3.39). To prove the sufficiency, let $k > 0$ be arbitrarily fixed, and let $T > a$ be so large that

$$T^* = \min\left\{\inf_{t \geq T} \tau_i(t)\right\} \geq t_0$$

and

$$\sum_{i=1}^{n} \int_T^\infty c_i(s) \left(\int_T^{\tau_i(s)} r^{1-q}(u)\, du\right)^{p-1} ds \leq \frac{2^{p-1}-1}{2^{p-1}},$$

which is possible thanks to (9.3.42). Consider the subset Ω of the Fréchet space $C[T^*, \infty)$ and the mapping $\mathcal{T} : \Omega \to C[T^*, \infty)$ defined by

$$\Omega = \Big\{x \in C[T^*, \infty) : \frac{k}{2}\int_T^t r^{1-q}(s)\, ds \leq x(t) \leq k\int_T^t r^{1-q}(s)\, ds,\ t \geq T,\ x(t) = 0,\ T^* \leq t \leq T\Big\}$$

and

$$(\mathcal{T}x)(t) = \begin{cases} \displaystyle\int_T^t r^{1-q}(s)\left(k^{p-1} - \int_s^\infty \sum_{i=1}^{n} c_i(u) x^{p-1}(\tau_i(u))\, du\right)^{q-1} ds & t \geq T, \\ 0 \quad T^* \leq t \leq T. \end{cases}$$

Now, by means of the Schauder-Tychonov fixed point theorem, it is easy to show that $\mathcal{T}$ has a fixed element in Ω, which satisfies equation (9.3.29) and condition (9.3.41). □

The proof of the next theorem is similar to that of the previous one, and hence it is omitted.

Theorem 9.3.11. *Equation* (9.3.29) *has a nonoscillatory solution* x *such that*

$$\lim_{t \to \infty} x(t) \in \mathbb{R} \setminus \{0\}$$

if and only if

$$\int^\infty \left(\frac{1}{r(s)}\int_s^\infty c_i(u)\, du\right)^{q-1} ds < \infty, \tag{9.3.43}$$

$i = 1, 2, \ldots, n.$

It remains to discuss the existence of an unbounded nonoscillatory solution x of (9.3.29) which has the property that the limit in (9.3.41) tends to $\pm\infty$, and of bounded solution x of (9.3.29) having the property that $\lim_{t\to\infty} x(t) = 0$. However, this is a difficult problem and there are no general criteria available for the existence of such solutions. Therefore, we confine ourselves to the case where at least one of the $\tau_i(t)$ is retarded and show that some sufficient conditions can be derived under which (9.3.29) has a nonoscillatory solution tending to zero as $t \to \infty$. Such a solution is often referred to as a *decaying nonoscillatory solution.* Our derivation is based on the following theorem.

Theorem 9.3.12. *Suppose that there exists an* $i_0 \in \{1, 2, \dots, n\}$ *such that*

$$\tau_{i_0}(t) < t, \quad c_{i_0}(t) > 0 \tag{9.3.44}$$

for $t \geq a$. *Further, assume that there exists a positive decreasing function* $\varphi(t)$ *on* $[T, \infty)$ *satisfying*

$$\varphi(t) \geq \int_t^\infty r^{1-q}(s) \left(\int_s^\infty c_i(u)(\varphi(\tau_i(u)))^{p-1} du \right)^{q-1} ds, \tag{9.3.45}$$

$t \geq T$, *where* T *is chosen so that* $\inf_{t \geq T} \tau_i(t) \geq a$, $i = 1, \dots, n$. *Then equation* (9.3.29) *has a decaying nonoscillatory solution.*

Proof. Let $\Lambda = \{z \in C[T, \infty) : 0 < z(\varphi(t))\}$. With each $z \in \Lambda$ we associate the function $\tilde{z} \in C[a, \infty)$ defined by

$$\tilde{z}(t) = \begin{cases} z(t) & \text{for } t \geq T, \\ z(T) + [\varphi(t) - \varphi(T)] & \text{for } a \leq t \leq T. \end{cases}$$

Define the mapping $\mathcal{H} : \Lambda \to C[T, \infty)$ as follows

$$(\mathcal{H}z)(t) = \int_t^\infty r^{1-q}(s) \left(\int_s^\infty \sum_{i=1}^n c_i(u)(\tilde{z}(\tau_i(u)))^{p-1} du \right)^{q-1} ds,$$

$t \geq T$. It now follows from the Schauder-Tychonov fixed point theorem that $\mathcal{H}$ has a fixed element $z(t)$ in Λ which is a solution to (9.3.29) tending to zero as $t \to \infty$. The fact that $z(t) > 0$ for $t \geq T$ can be verified exactly as in [313, p. 170]. □

To apply the last theorem, it is convenient to distinguish the following three cases:

$$\int^\infty \sum_{i=1}^n c_i(s)\, ds < \infty \quad \text{and} \quad \int^\infty r^{1-q}(s) \left(\int_s^\infty \sum_{i=1}^n c_i(u)\, du \right)^{q-1} ds < \infty, \tag{9.3.46}$$

$$\int^\infty \sum_{i=1}^n c_i(s)\, ds < \infty \quad \text{and} \quad \int^\infty r^{1-q}(s) \left(\int_s^\infty \sum_{i=1}^n c_i(u)\, du \right)^{q-1} ds = \infty, \tag{9.3.47}$$

and

$$\int^\infty \sum_{i=1}^n c_i(s)\, ds = \infty. \tag{9.3.48}$$

Condition (9.3.46), which is nothing else but (9.3.43), always guarantees the existence of a decaying nonoscillatory solution of (9.3.29).

Theorem 9.3.13. *Suppose that* (9.3.44) *holds for some* $i_0 \in \{1, 2, \dots, n\}$. *If* (9.3.43) *is satisfied, then* (9.3.29) *has a nonoscillatory decaying solution.*

Proof. Let T be large enough so that $\min_i\{\inf_{t\geq T}\tau_i(t)\}\geq\max\{1,a\}$ and

$$\int_T^\infty r^{1-q}(s)\left(\int_s^\infty\sum_{i=1}^n c_i(u)\,du\right)^{q-1}ds\leq\frac{1}{2}.$$

We shall show that $\varphi(t)=1+(1/t)$ satisfies (9.3.45). Indeed, we have

$$\begin{aligned}&\int_t^\infty r^{1-q}(s)\left(\int_s^\infty\sum_{i=1}^n c_i(u)(\varphi(\tau_i(u)))^{p-1}\,du\right)^{q-1}ds\\ &\quad=\int_t^\infty r^{1-q}(s)\left(\int_s^\infty\sum_{i=1}^n c_i(u)\left(1+\frac{1}{\tau_i(u)}\right)^{p-1}du\right)^{q-1}ds\\ &\qquad\leq 2\int_t^\infty r^{1-q}(s)\left(\int_s^\infty\sum_{i=1}^n c_i(u)\,du\right)^{q-1}ds\leq 1<\varphi(t),\end{aligned}$$

$t\geq T$. The conclusion now follows from Theorem 9.3.12. □

We now state the existence theorems which are applicable to the cases (9.3.47) and (9.3.48).

Theorem 9.3.14. *Suppose that* (9.3.44) *holds for some* $i_0\in\{1,2,\dots,n\}$, *and*

$$\limsup_{t\to\infty}\int_{\sigma(t)}^\infty r^{1-q}(s)\left(\int_s^\infty\sum_{i=1}^n c_i(u)\,du\right)^{q-1}ds<\frac{1}{e},$$

where $\sigma(t)=\min_i\tau_i(t)$. *Then equation* (9.3.29) *has a nonoscillatory decaying solution.*

Proof. Let

$$Q(t)=\left(\frac{1}{r(t)}\int_t^\infty\sum_{i=1}^n c_i(s)\,ds\right)^{q-1}$$

and choose $T>a$ so that $\inf_{t\geq T}\sigma(t)\geq a$, and

$$Q_T=\sup_{t\geq T}\int_{\sigma(t)}^t Q(s)\,ds\leq\frac{1}{e}. \tag{9.3.49}$$

Define $\varphi(t)=\exp\left(-(1/Q_T)\int_a^t Q(s)\,ds\right)$. Since for $i=1,2,\dots,n$,

$$\begin{aligned}\varphi(\tau_i(t))&=\exp\left(\frac{1}{Q_T}\int_{\tau_i(t)}^t Q(s)\,ds\right)\exp\left(-\frac{1}{Q_T}\int_a^t Q(s)\,ds\right)\\ &\leq e\exp\left(-\frac{1}{Q_T}\int_a^t Q(s)\,ds\right)=e\varphi(t),\end{aligned}$$

$t \geq T$, in view of (9.3.49), we have

$$\begin{aligned}
&\int_t^\infty r^{1-q}(s) \left(\int_s^\infty \sum_{i=1}^n c_i(u)(\varphi(\tau_i(u)))^{p-1}\, du \right)^{q-1} ds \\
&\quad \leq e \int_t^\infty Q(s)\varphi(s)\, ds \leq e \int_t^\infty Q(s) \exp\left(-\frac{1}{Q_T} \int_a^s Q(u)\, du \right) ds \\
&\quad \leq eQ_T \exp\left(-\frac{1}{Q_T} \int_a^t Q(s)\, ds \right) = eQ_T\varphi(t) \leq \varphi(t),
\end{aligned}$$

$t \geq T$. The conclusion now follows from Theorem 9.3.12. □

Theorem 9.3.15. *Suppose that* (9.3.44) *holds for some* $i_0 \in \{1, 2, \ldots, n\}$. *Further, assume that there exists* $T > a$ *such that* $\inf_{t \geq T} \sigma(t) \geq a$,

$$Q_T^* = \inf_{t \geq T} Q(t) > 0 \quad \textit{and} \quad \sup_{t \geq T} \int_{\sigma(t)}^t \sum_{i=1}^n c_i(u)\, du < \frac{p}{e} \left(\frac{Q_T^*}{p-1} \right)^{1/q},$$

where the functions $Q(t)$ *and* $\sigma(t)$ *are as in the previous theorem. Then equation* (9.3.29) *has a nonoscillatory decaying solution.*

Proof. Let

$$P_T = \sup_{t \geq T} \int_{\sigma(t)}^t \sum_{i=1}^n c_i(s)\, ds$$

and

$$\varphi(t) = \exp\left(-\frac{q}{P_T} \int_a^t \sum_{i=1}^n c_i(s)\, ds \right).$$

Then we have $\varphi(\tau_i(t)) \leq \exp q\varphi(t)$, $t \geq T$, $i = 1, 2, \ldots, n$, and hence

$$\begin{aligned}
\int_t^\infty \sum_{i=1}^n c_i(s)(\varphi(\tau_i(s)))^{p-1} ds &\leq e^p \int_t^\infty \left(\sum_{i=1}^n c_i(s) \right) (\varphi(s))^{p-1} ds \\
&= e^p \int_t^\infty \left(\sum_{i=1}^n c_i(s) \right) \exp\left(-\frac{p}{P_T} \int_a^s c_i(u)\, du \right) ds \\
&\leq \frac{P_T e^p}{p} \exp\left(-\frac{p}{P_T} \int_a^t \sum_{i=1}^n c_i(s)\, ds \right),
\end{aligned}$$

$t \geq T$. Consequently, we obtain

$$\begin{aligned}
&\int_t^\infty r^{1-q}(s) \left(\int_s^\infty \sum_{i=1}^n c_i(u)(\varphi(\tau_i(u)))^{p-1}\, du \right)^{q-1} ds \\
&\quad \leq \left(\frac{P_T}{p} \right)^{q-1} e^q \int_t^\infty \exp\left(-\frac{q}{P_T} \int_a^s \sum_{i=1}^n c_i(u)\, du \right) ds
\end{aligned}$$

$$\begin{aligned}
&\le \frac{1}{Q_T^*}\left(\frac{P_T}{p}\right)^{q-1} e^q \int_t^\infty c_i(s)\exp\left(-\frac{q}{P_T}\int_a^s \sum_{i=1}^n c_i(u)\,du\right) ds \\
&\le \frac{P_T}{qQ_T^*}\left(\frac{P_T}{p}\right)^{q-1} e^q \exp\left(-\frac{q}{P_T}\int_a^t \sum_{i=1}^n c_i(s)\,ds\right) \\
&\le \varphi(t),
\end{aligned}$$

This establishes the existence of a strictly decreasing function $\varphi(t) > 0$ satisfying (9.3.45). The conclusion now follows from Theorem 9.3.12. □

9.4 Higher order half-linear differential equations

This section is devoted to a brief discussion of oscillatory properties and solvability of boundary value problems associated with higher order half-linear differential equations, i.e., with equations whose solution spaces are homogeneous but generally not additive.

9.4.1 p-biharmonic operator

In this short subsection we consider the eigenvalue problem for the so-called *p-biharmonic operator*

$$\Delta(|\Delta u|^{p-2}\Delta u) = \lambda\Phi(u),\ x \in \Omega, \quad u(x) = 0 = \Delta u(x),\ x \in \partial\Omega, \tag{9.4.1}$$

where $\Omega \subset \mathbb{R}^N$ is a bounded domain with the smooth boundary $\partial\Omega$ and Δ is the classical Laplace operator. As a special case, we consider the one-dimensional case $N = 1$

$$(\Phi(u''))'' = \lambda\Phi(u),\ t \in (0,1), \quad u(0) = u''(0) = 0 = u(1) = u''(1), \tag{9.4.2}$$

where the general results for (9.4.1) can be considerably improved.

We start with (9.4.2) where we have a complete picture about eigenvalues and eigenfunctions, as the next statement shows.

Theorem 9.4.1. *The set of eigenvalues of* (9.4.2) *is formed by a sequence*

$$0 < \lambda_1(p) < \lambda_2(p) < \cdots < \lambda_n(p) \to \infty.$$

For any $n \in \mathbb{N}$, the function $p \mapsto \lambda_n(p)$ is continuous. Every eigenvalue $\lambda_n(p) = n^{2p}\lambda_1$ is simple and the corresponding one-dimensional space of solutions of problem (9.4.2) *with $\lambda = \lambda_n(p)$ is spanned by a function having exactly n bumps (i.e., the points at which u and u'' equal zero) in* $(0,1)$. *Each n-bump solution is constructed by the reflection and compression of the eigenfunction u_1 associated with the first eigenvalue λ_1.*

More precisely, one can verify that $u_1(t) = u_1(1-t)$ for $t \in [0,1]$, i.e., u_1 is symmetric with respect to $t = 1/2$. Now, the eigenfunction u_n corresponding to the eigenvalue $\lambda_n = n^{2p}\lambda_1$ is of the form

$$u_n(t) = \begin{cases} u_1(nt) & t \in \left[0, \frac{1}{n}\right], \\ -u_1(nt-1) & t \in \left[\frac{1}{n}, \frac{2}{n}\right], \\ \vdots & \\ (-1)^n u_1(nt-n+1) & t \in \left[\frac{n-1}{n}, 1\right]. \end{cases}$$

Now we turn our attention to the general case (9.4.1). The solution of this problem is defined as follows. Consider the Dirichlet problem for the Poisson equation

$$-\Delta w = f, \ x \in \Omega, \quad w = 0, \ x \in \partial\Omega. \tag{9.4.3}$$

This problem has a unique solution for every $f \in L^p(\Omega)$, $p > 1$, so the linear operator $T : L^p(\Omega) \to W^{2,p} \cap W_0^{1,p}(\Omega)$ which assigns to the right-hand side f of (9.4.3) its solution w, i.e., $w = T(f)$, is well defined. Using this operator and denoting $v = -\Delta u$ in (9.4.1), this differential equation can be rewritten as the operator equation

$$\Phi(v) = \lambda T(\Phi(T(v))), \quad x \in \Omega. \tag{9.4.4}$$

Now, a function $u \in W^{2,p}(\Omega) \cap W_0^{1,p}(\Omega)$ is called a *solution of* (9.4.1) if $v = -\Delta u$ solves (9.4.4) in $L^q(\Omega)$, $q = p/(p-1)$ being the conjugate exponent. The parameter λ is called an *eigenvalue* if the solution u is nontrivial.

The next result can be seen as an extension of the results of Section 7.1 concerning the first eigenvalue of the p-Laplacian.

Theorem 9.4.2. *The problem* (9.4.1) *has the least positive egenvalue* $\lambda_1(p)$ *which is simple and isolated in the sense that the set of solutions of* (9.4.1) *with* $\lambda = \lambda_1(p)$ *forms an one-dimensional linear space spanned by a positive eigenfunction* u_1 *such that* $\Delta u_1 < 0$ *in* Ω *and* $\frac{\partial u_1}{\partial \nu} < 0$ *on* $\partial\Omega$, *where* ν *is the exterior normal to* Ω. *Moreover, there exists* $\delta > 0$ *such that* (9.4.1) *possesses no eigenvalue on the interval* $(\lambda_1(p), \lambda_1(p) + \delta)$. *Problem* (9.4.1) *has a positive solution if and only if* $\lambda = \lambda_1$. *The function* $p \mapsto \lambda_1(p)$ *is continuous.*

9.4.2 Higher order half-linear eigenvalue problem

We consider the eigenvalue problem

(9.4.5)

$$\left(r(t)\Phi(u^{(n)})\right)^{(n)} + (-1)^n \lambda c(t)\Phi(u) = 0, \quad u^{(i)}(a) = 0 = u^{(i)}(b), \ i = 0, \dots, n-1,$$

where $r \in C^n[a,b]$, $c \in C[a,b]$ are positive functions and λ is the eigenvalue parameter. There are only a few papers dealing with higher order half-linear differential equations. The reason is that the lack of additivity of the solution space makes here much more "damage" than in the second order case.

Theorem 9.4.3. *The eigenvalue problem* (9.4.5) *has an infinite sequence of real eigenvalues*

$$0 < \lambda_1 < \lambda_2 < \cdots < \lambda_k \to \infty.$$

To each λ_k *there corresponds an essentially unique eigenfunction* u_k *that has exactly* $k-1$ *zeros in* (a,b)*, all simple. The zeros of* u_k *interlace with those of* u_{k+1}.

Remark 9.4.1. The problem (9.4.5) is a special case of the general n-th order BVP which using the Elbert's notation (see Subsection 1.1.1) $y^{p*} := |y|^p \operatorname{sgn} y$ can be written as

$$(a_{n-1}(t)(\dots(a_1(t)((a_0(t)u^{p_0*})')^{p_1*})'\dots)^{p_{n-1}*})' = \lambda b(t)u^{p*}, \tag{9.4.6}$$

where a_i, b are positive functions, $p_i > 0$, $i = 0, \dots, n-1$ and

$$p_0 p_1 \cdots p_{n-1} = p.$$

The last condition assures that the solution space of (9.4.6) is homogeneous. We have presented here the main result of [150] in a simplified form. The reason is that the main general statement of [150] requires several technical restriction on the boundary conditions associated with (9.4.6) which are automatically satisfied in the simplified setting of Theorem 9.4.3.

Another problem associated with the equation

$$(-1)^n (r(t)\Phi(u^{(n)}))^{(n)} + c(t)\Phi(u) = 0, \tag{9.4.7}$$

or with the more general equation

$$\sum_{k=0}^{n} (-1)^k \left(a_k(t)\Phi(u^{(k)})\right)^{(k)} = 0, \quad a_n(t) > 0, \tag{9.4.8}$$

where $a_i(t)$, $i = 0, \dots, n$, are continuous functions, is their oscillation theory, analogous to the linear case ($p = 2$ in (9.4.8)). Following the linear case, two point t_1, t_2 are said to be *conjugate* relative to (9.4.8) if there exists a nontrivial solution of (9.4.8) satisfying $u^{(i)}(t_1) = 0 = u^{(i)}(t_2)$, $i = 0, \dots, n-1$. Equation (9.4.8) is said to be *nonoscillatory*, if there exists $T \in \mathbb{R}$ such that the interval $[T, \infty)$ contains no pair of points conjugate relative to (9.4.8). In the opposite case, this equation is said to be *oscillatory*. The main tool of the higher order *linear* oscillation theory is the fact that (9.4.8) with $p = 2$ is nonoscillatory if and only if there exists $T \in \mathbb{R}$ such that

$$\int_T^\infty \left[\sum_{k=0}^{n} a_k(t)(y^{(k)})^2\right] dt > 0 \tag{9.4.9}$$

for every nontrivial $y \in W_0^{n,2}(T, \infty)$. The reason for the fact that half-linear higher order oscillation theory is almost "terra incognita" is that we have only the following implication at disposal.

Theorem 9.4.4. *If there exists $T \in \mathbb{R}$ such that the (energy) functional*

$$\mathcal{F}_p^{[n]}(y;T,\infty) := \int_T^\infty \left[\sum_{k=0}^n a_k(t)|y^{(k)}|^p\right] dt > 0 \tag{9.4.10}$$

for every nontrivial $y \in W_0^{n,p}(T,\infty)$, then (9.4.8) is nonoscillatory.

Proof. Suppose that (9.4.8) is oscillatory, i.e., for any $T \in \mathbb{R}$ there exists a pair of conjugate points $t_1, t_2 \in [T,\infty)$ relative to this equation and let u be the solution having zero points of multiplicity n at t_1 and t_2. Define the function

$$y(t) := \begin{cases} 0 & t \in [T, t_1], \\ u(t) & t \in [t_1, t_2], \\ 0 & t \in [t_2, \infty). \end{cases}$$

Then $y \in W_0^{n,p}(T,\infty)$. Now, multiplying (9.4.8) by y, integrating the obtained equation from T to ∞, using by parts integration (similarly as in Subsection 1.2.2) we find that $\mathcal{F}_p^{[n]}(y) = 0$, a contradiction. □

The previous statement, coupled with the Wirtinger inequality (Lemma 2.1.1), gives the following result concerning the higher order Euler type differential equation. This result can be understood as a partial extension of the statement which claims that the differential equation

$$(-1)^n y^{(2n)} + \frac{\gamma}{t^{2n}} y = 0 \tag{9.4.11}$$

is nonoscillatory if and only if $\gamma \le [(2n-1)!!]^2/4^n$.

Theorem 9.4.5. *The differential equation*

$$(-1)^n \left(\Phi(u^{(n)})\right)^{(n)} + \frac{\gamma}{t^{np}}\Phi(u) = 0 \tag{9.4.12}$$

is nonoscillatory provided

$$\gamma < p^{-np} \prod_{k=1}^n (kp-1)^p.$$

Proof. The Wirtinger inequality (2.1.1) with $M(t) = (p-1-\alpha)^{p-1} t^{\alpha-p+1}$, $\alpha \ne p-1$, gives for any $y \in (T,\infty)$

$$\int_T^\infty t^\alpha |y'|^p \, dt \ge \left(\frac{|p-1-\alpha|}{p}\right)^p \int_t^\infty t^{\alpha-p}|y|^p \, dt.$$

Applying this inequality, respectively, with $\alpha = 0, -p, \dots, -(n-1)p$, we find that the functional

$$\int_T^\infty \left[|y^{(n)}|^p - \frac{\gamma}{t^{np}}|y|^p\right] dt$$

is positive for every $y \in W_0^{n,p}(T,\infty)$, which implies the required result. □

9.5 Inequalities related to half-linear differential equations

Here we want to present integral inequalities involving a function and its derivative (or its integral) which are related to (1.1.1). There are two main connections: we give inequalities which can be viewed as a necessary (and sometimes sufficient) condition for the existence of certain solutions of (1.1.1). Further there are inequalities which may serve as a tool for proving some qualitative results for (1.1.1). In fact, there are inequalities like that of Lyapunov or of Vallée-Poussin (see Section 5.1) which belong to the class satisfying these connections but we decided to present them in the framework of a different topic since they do not belong to the wide class of inequalities involving a function and its derivative, in contrast to the inequalities of Wirtinger type, of Hardy type and of Opial type.

9.5.1 Inequalities of Wirtinger and Hardy type

We start with Wirtinger type inequalities. These are studied in the literature in various modifications, let us mention at least [29]. We recall here the inequality, proved in Section 2.1, which will be later shown to be related to other inequalities: Let $p > 1$, M be a positive continuously differentiable function for which $M'(t) \neq 0$ in $[a, b]$ and let $u \in W_0^{1,p}(a, b)$. Then

$$\int_a^b |M'(t)||u|^p \, dt \leq p^p \int_a^b \frac{M^p(t)}{|M'(t)|^{p-1}} |u'|^p \, dt. \tag{9.5.1}$$

We call this inequality the "half-linear" version of the Wirtinger inequality because it fits to the variational principle and it is used to prove nonoscillatory criteria for (1.1.1), see Section 2.1.

Next we will discuss well-known Hardy inequality. First recall its classical version from [173]: Let $p > 1$, a function f be nonnegative and such that $\int_a^\infty f^p(t)\, dt$ exists. Denote $F(t) := \int_a^t f(s)\, ds$. Then

$$\int_a^\infty \left(\frac{F(t)}{t-a}\right)^p dt < \left(\frac{p}{p-1}\right)^p \int_a^\infty f^p(t)\, dt, \tag{9.5.2}$$

unless $f \equiv 0$. The constant $[p/(p-1)]^p$ is the best possible.

There exist several proofs of this statement. In one of them, (9.5.2) may be viewed as a necessary condition for the existence of a positive increasing solution of generalized Euler equation (1.4.20), where $\gamma = \gamma_p = ((p-1)/p)^p$, which indeed exists (see Section 1.4.2). In particular, this equation is nonoscillatory. On the other hand, the following estimation shows that (9.5.2) is a sufficient condition for nonoscillation of this equation. Indeed, let a function f be such that $\xi \in W_0^{1,p}(a, \infty)$, where $\xi(t) = \int_a^t f(s)\, ds$, $a > 0$. Then we have

$$\mathcal{F}(\xi; a, \infty) = \int_a^\infty \left\{ |\xi'|^p - \frac{\gamma_p}{t^p} |\xi|^p \right\}(t)\, dt = \int_a^\infty \left\{ |f(t)|^p - \frac{\gamma_p}{t^p} \left| \int_a^t f(s)\, ds \right|^p \right\} dt$$

$$\geq \int_a^\infty \left\{ |f(t)|^p - \frac{\gamma_p}{(t-a)^p} \left(\int_a^t |f(s)|\, ds \right)^p \right\} dt.$$

Now the last expression is positive by (9.5.2) provided f is nontrivial, which is our case, if we assume that $\xi \not\equiv 0$. Hence, (1.4.20) with $\gamma = \gamma_p$ is nonoscillatory by Theorem 2.1.1.

One of the possible connections between Hardy and Wirtinger inequality is the following one. If $b = \infty$ and $M(t) = t$ in (9.5.1), we get the inequality

$$\int_a^\infty |y'|^p\, dt \geq \gamma_p \int_a^\infty \frac{|y|^p}{t^p}\, dt,$$

which is the same type of inequality as (9.5.2), but with arguments satisfying different boundary conditions.

There is a large amount of various extensions of the Hardy inequality. We mention here those ones which are related to our topic. Beesack [30] (see also his nice survey [31]) showed that if $f \geq 0$, r' and c are continuous functions in (a,b), $r > 0$, $p > 1$, $\int_a^b r(t) f^p(t)\, dt < \infty$, $r(t) = \mathcal{O}[(t-a)^{p-1}]$ or $r^{q-1}(t) \int_a^t r^{1-q}(s)\, ds = \mathcal{O}(t-a)$ as $t \to a+$, q being the conjugate number of p, and (1.1.1) has a solution y with $y(t) > 0$ and $y'(t) > 0$ in (a,b), then

$$\int_a^b c(t) F^p(t)\, dt \leq \int_a^b r(t) f^p(t)\, dt, \tag{9.5.3}$$

where F is defined as above. A similar theorem holds when $p < 0$. If $0 < p < 1$, then inequality in (9.5.3) is reversed. Analogous theorems are obtained with $G(t) := \int_t^b f(s)\, ds$ instead of F.

As a consequence of more general statements, Li and Yeh [241] obtained the result similar to the one of Beesack: Let r' and c be continuous functions in (a,b) with $r(t) > 0$. If (1.1.1) has a positive increasing [decreasing] solution y on (a,b) and a nonnegative $u \in AC(a,b)$ satisfies

$$\liminf_{t\to b-} u^p(t) r(t) \frac{\Phi(y'(t))}{\Phi(y(t))} \geq \limsup_{t\to a+} u^p(t) r(t) \frac{\Phi(y'(t))}{\Phi(y(t))}$$

[in case of a decreasing y, lim sup and lim inf are interchanged], then

$$\liminf_{A\to a+, B\to b-} \int_A^B \{r|u'|^p - cu^p\}(t)\, dt \geq 0, \tag{9.5.4}$$

where equality holds if and only if u is a constant multiple of y. If $0 < p < 1$, then "$\geq$" should be replaced by "$\leq$". Li and Yeh used these inequalities to prove the variational principle differently from the proof of (i)$\Rightarrow$(ii) of the Roundabout theorem (Theorem 1.2.2).

Last but not least it should be mentioned the work of Kufner and his colleagues [210, 211]. They substantially generalized the results of Beesack and others, e.g., in the following manner, which is interesting from our point of view. Let c and r

be positive, measurable and finite almost everywhere on (a, b), where $-\infty \leq a < b \leq \infty$. If $1 < p \leq \alpha < \infty$, $r \in AC(a, b)$ and $r \in L^{1-q}(a, t)$ with $1/p + 1/q = 1$ for every $t \in (a, b)$, then

$$\left(\int_a^b c(t)|u(t)|^\alpha dt\right)^{\frac{1}{\alpha}} \leq K \left(\int_a^b r(t)|u'(t)|^p dt\right)^{\frac{1}{p}} \tag{9.5.5}$$

holds for every $u \in AC(a, b)$ satisfying $u(t) \to 0$ as $t \to a+$, where K is some positive constant independent of u, if and only if there exists $\lambda > 0$ such that the differential equation

$$\lambda(r^{\alpha/p}(t)(y')^{\alpha/q})' + c(t)(y)^{\alpha/q} = 0 \tag{9.5.6}$$

has a solution satisfying $y' \in AC(a, b)$, $y(t) > 0$, $y'(t) > 0$ for every $t \in (a, b)$. These extensions could be perhaps used to establish new sophisticated nonoscillation criteria. Note that it is known several other sufficient and necessary conditions in terms of r and c for the validity of (9.5.5).

9.5.2 Inequalities of Opial type

Opial inequality is another inequality involving a function and its derivative, which has appeared in many various modifications. Here we present the version of Boyd and Wong [53]. For this we consider half-linear equation of the form

$$[r(t)\Phi(y')]' - \lambda s'(t)\Phi(y) = 0, \tag{9.5.7}$$

where r is a positive continuous function on $[0, a]$, s is a nonnegative continuously differentiable function on $[0, a]$ and $\lambda > 0$ is a constant. Suppose that the boundary value problem (9.5.7), $y(0) = 0$, $r(a)\Phi(y'(a)) = \lambda s(a)\Phi(y(a))$ has a nonnegative solution y. Then for an absolutely continuous u with $u(0) = 0$,

$$\frac{1}{\lambda_0}\int_0^a r(t)|u'(t)|^p dt \geq p \int_0^a s(t)|u'(t)u^{p-1}(t)|\, dt,$$

where λ_0 is the largest eigenvalue. Equality is attained if and only if u is a constant multiple of y. Let us mention, for example, one special case. If $r \equiv 1$, then (9.5.7) reduces to the explicitly solvable equation $[r(t)\Phi(y')]' = 0$, and the eigenfunction corresponding to the eigenvalue $\lambda_0 = \left(\int_0^a r^{1-q}(s)\, ds\right)^{1-p}$ is $y(t) = \int_0^t r^{1-q}(s)\, ds$. Opial inequality then reads as

$$\left(\int_0^a r^{1-q}(s)\, ds\right)^{p-1} \int_0^a r(t)|u'(t)|^p dt \geq p \int_0^a |u'(t)u^{p-1}(t)|\, dt,$$

with equality if and only if $u(t) = K\int_0^t r^{1-q}(s)\, ds$, $K \in \mathbb{R}$. From many other works with variants of Opial inequality which hold for the functions with various boundary conditions let us again mention the survey by Beesack [31] of integral inequalities modeled on the classical inequalities of Hardy, Opial and Wirtinger. The inequalities therein are mainly of the "generalized Opial type", namely

$$\int_a^b \left\{ r|u'|^{p+\alpha} - c|u|^p|u'|^\alpha \right\}(t)\, dt \ge 0,$$

where (a,b) is finite or infinite interval, r and c are positive functions. They hold for all functions u in $C^1(a,b)$ that satisfy certain boundary conditions and they are again viewed as necessary conditions for the existence of certain solutions of some differential equations. If $\alpha = 0$, $u(a) = 0$ or $u(b) = 0$ or both, then we get an inequality of Hardy type, while for $\alpha = 1$, $u(a) = 0$ or $u(b) = 0$ or both, we have an inequality of Opial type. Some authors call the case when $\alpha = 0$ and $p = 2$ as the inequality of Wirtinger type but it does not seem to match the terminology in the most of other literature. Another variant of Opial inequality similar to that of Boyd and Wong, which is related to half-linear equation, can be found in Li and Yeh [241], and reads as follows. Let $r > 0$ and s be continuously differentiable functions on (a,b). If the equation

$$[r(t)\Phi(y')]' - \frac{1}{p}s'(t)\Phi(y) = 0$$

has a positive increasing solution y on (a,b) and a nonnegative absolutely continuous function u satisfies

$$\liminf_{t\to b-} \left\{ r(t)\frac{\Phi(y'(t))}{\Phi(y(t))} - \frac{s(t)}{p} \right\} u^p(t) \ge \limsup_{t\to a+} \left\{ r(t)\frac{\Phi(y'(t))}{\Phi(y(t))} - \frac{s(t)}{p} \right\} u^p(t),$$

then

$$\liminf_{A\to a+, B\to b-} \int_A^B \left\{ r|u'|^p - su^{p-1}u' \right\}(t)\, dt \ge 0$$

with equality only if u is a constant multiple of y. A similar statement holds when y is a positive decreasing solution.

Finally it should be mentioned that there exists a monograph devoted to Opial inequalities and their applications, namely [9].

9.6 Notes and references

There is a huge amount of papers dealing with quasilinear equations; a sample of them is mentioned at the beginning of this chapter. Quasilinear equations with constant coefficients (see Subsection 9.1.1) were studied by Drábek and Manasevich [129] and Otani [308, 309], see also the paper of Talenti [344] which gives another point of view on the problem. The results of Subsections 9.1.2 and 9.1.3 are modeled on Cecchi, Došlá, Marini [57], except of Theorem 9.1.4 proved in Kitano, Kusano [206]. The statements of Subsection 9.1.4 are taken from Elbert, Kusano [144]. Another results concerning asymptotic properties of nonoscillatory solutions of (9.1.1) and of more general equations of this type can be found e.g.

in [4, 58, 59, 157, 205, 222, 227, 248, 253, 299, 319, 334, 347, 364, 372] by Agarwal, Cecchi, Došlá, Evtuchov, Fan, Grace, Kiomura, Kusano, W. T. Li, Marini, M. Naito, Ogata, Rabtsevich, Rogovchenko, Tanigawa, Yang and Yoshida. The results of Subsection 9.1.5 are due to J. Wang [358]. Theorem 9.1.11 was proved by Cecchi, Marini, Villari [67]. Theorem 9.1.12 is taken from Agarwal, Grace, O'Regan [6]. Black resp. white hole solutions were studied by Jaroš, Kusano in [187] resp. [188]. Subsection 9.1.8 is based on the lecture by Jaroš [184]. Theorem 9.1.15 was proved by Tanigawa [348]. The statements of Theorem 9.1.16 are the continuous versions of the results due to Marini, Matucci, Řehák [284]. The same authors proved Theorem 9.1.17 in [282]. Further results on coupled nonlinear differential systems (in particular, existence of singular solutions and asymptotic theory) can be found in some papers of the four last quoted authors, as well as of Jaroš and Kusano, e.g. in [189, 226, 283]. First order systems of four nonlinear equations, which cover coupled systems, were studied by Kusano, Naito and Wu [221].

The results of Subsection 9.2.1 are taken from Li and Cheng [252], related results are given in the paper of Yang [371]. Oscillation of a more general half-linear forced equation than (9.2.1) is investigated in Kusano, Ogata [223] and Wong, Agarwal [366], but for simplicity we present here the results of [252]. Damped half-linear equations were studied in Agarwal, Grace, O'Regan [6]. Subsection 9.2.2 is based on Jaroš, Kusano, Yoshida [192].

Subsection 9.3.1 contains the result taken from Agarwal, Grace, O'Regan [6]. The remaining subsections of Section 9.3 follow Agarwal, Grace, O'Regan [6] and Kusano, Lalli [213], se also the papers of Agarwal, Grace and O'Regan [5, 7].

The results of Subsection 9.4.1 are taken from the paper of Drábek and Otani [130] and the result of Subsection 9.4.2 is a simplified version of the main result of the paper of Elias and Pinkus [150]. See also the paper of Pinkus [316]. Recent results and references concerning higher order half-linear and quasilinear equations can be found e.g. in the paper of Naito and Wu [302], we also refer to the older paper of Kratochvíl and Nečas [209].

The last section, which is devoted to inequalities, presents classical results or their extensions. A detailed discussion concerning a relevant literature is given within the text.

Bibliography

[1] R. P. Agarwal, Difference Equations and Inequalities, Theory, Methods and Applications, Second Edition, M. Dekker, New York-Basel, 2000.

[2] R. P. Agarwal, M. Bohner, S. R. Grace, D. O'Regan, Discrete Oscillation Theory, in preparation.

[3] R. P. Agarwal, M. Bohner, P. Řehák, *Half-linear dynamic equations on time scales: IVP and oscillatory properties*, In Nonlinear Analysis and Applications: To V. Lakshmikantham on his 80th Birthday, Ravi P. Agarwal, Donal O'Regan editors, Dordrecht: Kluwer Academic Publishers, 2003, Vol. I, 1–56.

[4] R. P. Agarwal, S. R. Grace, *On the oscillation of certain second order differential equations*, Georgian Math. J. **7** (2000), 201–213.

[5] R. P. Agarwal, S. R. Grace, *Oscillation criteria for second order half-linear differential equations with deviating arguments*, Dyn. Contin. Discrete Impuls. Syst., Ser. A, Math. Anal. **9** (2002), 217–224.

[6] R. P. Agarwal, S. R. Grace, D. O'Regan, Oscillation Theory of Second Order Linear, Half-Linear, Superlinear and Sublinear Dynamic Equations, Kluwer Academic Publishers, Dordrecht/Boston/London, 2002.

[7] R. P. Agarwal, S. R. Grace, D. O'Regan, *Linearization of the second-order superlinear oscillation theorems*, Math. Comput. Modelling **39** (2004), 1165–1183.

[8] R. P. Agarwal, S. L. Shieh, C. C. Yeh, *Oscillation criteria for second order retarded differential equations*, Math. Comput. Modelling **26** (1997), 1–11.

[9] R. P. Agarwal, P. Y. H. Pang, Opial inequalities with applications in differential and difference equations, Kluwer Academic Publishers, Dordrecht, 1995.

[10] C. D. Ahlbrandt, D. B. Hinton, R. T. Lewis, *The effect of variable change on oscillation and disconjugacy criteria with applications to spectral theory and asymptotic theory*, J. Math. Anal. Appl. **81** (1981), 234–277.

[11] C. D. Ahlbrandt, A. C. Peterson, Discrete Hamiltonian Systems: Difference Equations, Continued Fractions, and Riccati Equations, Kluwer Academic Publishers, Boston, 1996.

[12] M. Alif, *Sur le spectre de Fučík de p-Laplacien avec poids indefinis*, C. R. Acad. Sci. Paris, Ser. I **334** (2002), 1061-1066.

[13] W. Allegretto, *Sturmian type theorems for solutions of elliptic nonlinear problems*, Nonlin. Differ. Equ. Appl. **7** (2000), 309–321.

[14] W. Allegretto, *Sturmian theorems for degenerate elliptic equations*, Proc. Amer. Math. Soc. **129** (2001), 3031–3035.

[15] W. Allegretto, Y. X. Huang, *A Picone's identity for the p-Laplacian and applications*, Nonlin. Anal. **32** (1998), 819–830.

[16] W. Allegretto, Y. X. Huang, *Principal eigenvalues and Sturm comparison via Picone's identity*, J. Differential Equations **156** (1999), 427–438.

[17] B. Alziary, J. Fleckinger, P. Takáč, *Variational methods for a resonant problem with the p-Laplacian in $\mathbb{R}^N$*, Electron. J. Differential Equations **2004** (2004), No. 76, 1–32.

[18] A. Ambrosetti, G. Prodi *On the inversion of some mapping with singularities between Banach spaces*, Annali Mat. Pura Appl. **93** (1973), 321–347.

[19] A. Anane, *Simplicite et isolation de la premiére valuer propre du p-laplacien avec poids*, C. R. Acad. Sci. Paris, Ser. I **305** (1987), 725–728.

[20] A. Anane, O. Charkone, J. P. Gossez, *Spectre d'ordre supérieur et probléms de non-résonance*, C. R. Acad. Sci. Paris, Ser. I **325** (1997), 33–36.

[21] A. Anane, N. Tsouli, *On the second eigenvalue of the p-laplacian*, in "Nonlinear Notes in Mathematics" (A. Benkirane, J. P. Gosez, Eds.) Pitman Research Notes in Mathematics, vol. 343, Longman, Harlow, 1996, pp. 1–9.

[22] J. Andres, *On the criterion of de la Vallée Poussin*, Publ. Math. (Debrecen) **45** (1994), 145–152.

[23] D. Arcoya, L. Orsina, *Landesman-Lazer conditions for quasilinear elliptic equations*, Nonlin. Anal. **28** (1997), 1623–1632.

[24] M. ARIAS, *Some remarks on the antimaximum principle and the Fučík spectrum for the p-laplacian*, Partial differential equations, Lecture Notes in Pure and Appl. Math. 229, Dekker, New York, 2002, pp. 103-109.

[25] F. V. ATKINSON, *On second-order non-linear oscillations*, Pacific J. Math. **5** (1955), 643–647.

[26] F. V. ATKINSON, Á. ELBERT, *An extension of Milloux theorem to half-linear differential equations*, EJQTDE, Proc. 6th Coll. QTDE, **2000** (2000), No. 8, 1–10.

[27] D. AVERNA, G. BONANNO, *Three solutions for a quasilinear two-point boundary value problem involving the one-dimensional p-Laplacian*, Proc. Edinburgh Math. Soc. **47** (2004), 257–270.

[28] B. AYNLAR, A. TIRYAKI, *Oscillation theorems for nonlinear second order differential equations*, Comput. Math. Appl. **44** (2002), 529–538.

[29] P. R. BEESACK, *Integral inequalities of the Wirtinger type*, Duke Math. J. **25** (1958), 477–498.

[30] P. R. BEESACK, *Hardy's inequality and its extensions*, Pacific J. Math. **11** (1961), 39–61.

[31] P. R. BEESACK, *Integral inequalities involving function and its derivative*, Amer. Math. Monthly **78** (1971), 705–741.

[32] P. R. BEESACK, K. M. DAS, *Extensions of Opial's inequality*, Pacific J. Math. **26** (1968), 215–232.

[33] Š. BELOHOREC, *On some properties of the equation* $y'' + f(x)y^\alpha = 0$, $0 < \alpha < 1$, Mat. Časopis Sloven. Akad. Vied. **17** (1967), 10–19.

[34] CH. BENNEWITZ, Y. SAITÖ, *An embedding norm and the Lindqvist trigonometric functions*, Electron. J. Differential Equations **2002** (2002), No. 86, pp. 1–6.

[35] A. BESICOVITCH, Almost Periodic Functions, Dover, New York, 1954.

[36] I. BIHARI, *Ausdehung der Sturmschen Oszillations and vergleichungsärte auf die Lösungen gewisser nicht-linearen Differenzialgleichungen zweiter Ordnung*, Publ. Math. Inst. Hungar. Acad. Sci. **2** (1957), 159–173.

[37] I. BIHARI, *Oscillation and monotonicity theorems concerning non-linear differential equations of the second order*, Acta Math. Acad Sci. Hungar. **9** (1958), 83–104.

[38] I. BIHARI, *An oscillation theorem concerning the half-linear differential equation of the second order*. Publ. Math. Inst. Hungar. Acad. Sci. Ser. A **8** (1964), 275–280.

[39] I. Bihari, *On the second order half-linear differential equation*, Studia Sci. Math. Hungar. **3** (1968), 411–437.

[40] I. Bihari, *Asymptotic results concerning the equation* $x''|x'|^{n-1} + a(t)x^{\overset{*}{n}} = 0$, Studia Sci. Math. Hungar. **19** (1984), 151–157.

[41] P. Binding, L. Boulton, J. Čepička, P. Drábek, P. Girg, *Basic properties of eigenfunctions of the p-Laplacian*, submitted (2005).

[42] P. Binding, P. Drábek, *Sturm-Liouville theory for p-Laplacian*, Acta Math. Hungar. **40** (2003), 375–396.

[43] P. A. Binding, P. Drábek, Y. X. Huang, *The range of the p-Laplacian*, Appl. Math. Lett. **10** (1997), 77–82.

[44] N. H. Bingham, C. M. Goldie, J. L. Teugels, Regular Variation, Encyclopedia of Mathematics and its Applications 27, Cambridge Univ. Press, 1987.

[45] G. Bognár, *On the radial symmetric solution of a nonlinear partial differential equation*, Publ. Univ. Miskolc, Ser. D, Nat. Sci. Math. **36** (1995), 13–22.

[46] G. Bognár, *Existence theorems for eigenvalues of a nonlinear eigenvalue problem*, Commun. Appl. Nonlinear Anal. **4** (1997), 93–102.

[47] G. Bognár, *On the roots of the solution of generalized Bessel equation*, Proceedings of CDDE 2002, M. Bartušek, O. Došlý ed., Masaryk Univesity Press, Brno 2003, 33–43,.

[48] G. Bognár, O. Došlý, *The application of Picone's identity for some nonlinear elliptic differential equations*, Acta. Math. Univ. Comenianae **72** (2003), 45–57.

[49] M. Bohner, A. Peterson, Dynamic Equations on Time Scales: An Introduction with Applications, Birkhäuser, Boston, 2001.

[50] M. Bohner, A. Peterson ed., Advaces in Dynamic Equations on Time Scales, Birkhäuser, Boston, 2003.

[51] H. Bohr, Almost-periodic Functions, Chelsa, New York, 1951.

[52] O. Borůvka, Lineare Differentialtransformationen 2. Ordnung, Deutscher Verlag der Wissenschaften, Berlin, 1961.

[53] D. W. Boyd, J. S. W. Wong, *An extension of Opial's inequality*, J. Math. Anal. Appl. **19** (1967), 100–102.

[54] P. J. Browne, *A Prüfer approach to half-linear Sturm-Liouville problems*, Proc. Edinburgh Math. Soc. **41** (1998), 573–583.

[55] B. M. Brown, W. Reichel, *Computing eigenvalues and Fučík-spectrum of radially symmetric p-Laplacian*, J. Comput. Appl. Math. **148** (2002), 183–211.

[56] A. Cañada, A. Fonda, P. Drábek ed., Handbook of Differential Equations, Ordinary Differential Equations, Volume I, Elsevier, North Holland, Amsterdam, 2004.

[57] M. Cecchi, Z. Došlá, M. Marini, *On the dynamics of the generalized Emden-Fowler equations*, Georgian Math. J. **7** (2000), 269–282.

[58] M. Cecchi, Z. Došlá, M. Marini, *On decaying solutions for functional differential equations with p-Laplacian*, Nonlin. Anal. **47** (2001), 4387–4398.

[59] M. Cecchi, Z. Došlá, M. Marini, *On the nonoscillatory solutions of differential equations with p-Laplacian*, Adv. Math. Sci. Appl. **11** (2001), 419–436.

[60] M. Cecchi, Z. Došlá, M. Marini, *Principal solutions and minimal sets of quasilinear differential equations*, Dynam. Systems Appl. **13** (2004), 223–234.

[61] M. Cecchi, Z. Došlá, M. Marini, *Half-linear equations and characteristic properties of principal solutions*, J. Differential Equations **208** (2005), 494–507.

[62] M. Cecchi, Z. Došlá, M. Marini, *On recessive and dominant solutions for half-linear difference equations*, J. Difference Equ. Appl. **10** (2004), 797–808.

[63] M. Cecchi, M. Marini, G. Villari, *On the monotonicity property for a certain class of second order differential equations*, J. Differential Equations **82** (1989), 15–27.

[64] M. Cecchi, M. Marini, G. Villari, *Integral criteria for a classification of solutions of linear differential equations*, J. Differential Equations **9** (1992), 381–397.

[65] M. Cecchi, M. Marini, G. Villari, *On a nonlinear second order differential equation with oscillating coefficient*, Differential Integral Equations **6** (1993), 47–61.

[66] M. Cecchi, M. Marini, G. Villari, *On some classes of continuable solutions of a nonlinear differential equation*, J. Differential Equations **118** (1995), 403–419.

[67] M. Cecchi, M. Marini, G. Villari, *Comparison results for oscillation of nonlinear differential equations*, NoDEA Nonlinear Differential Equations Appl. **6** (1999), 173–190.

[68] J. ČEPIČKA, P. DRÁBEK, P. GIRG, *Quasilinear boundary value problems: existence and multiplicity results*, in Variational methods: open problems, recent progress, and numerical algorithms, 111–139, Contemp. Math., 357, Amer. Math. Soc., Providence, RI, 2004.

[69] J. ČEPIČKA, P. DRÁBEK, P. GIRG, *Open problems related to p-Laplacian*, to appear in Bol. Soc. Esp. Mat. Apl. (2005).

[70] T. CHANTLADZE, N. KANDELAKI, A. LOMTATIDZE, *On zeros of solutions of a second order singular half-linear equation*, Mem. Diff. Equ. Math. Phys. **17** (1999), 127–154.

[71] T. A. CHANTURIA, *On singular solutions of nonlinear systems of ordinary differential equations*, Colloq. Math. Soc. J. Bolyai **15** (1975), 107–119.

[72] T. A. CHANTURIA, *On monotonic solutions of systems of nonlinear differential equations*, Ann. Polon. Math. **37** (1980) 59–70 (in Russian).

[73] S. CHEN, L. H. ERBE, *Riccati techniques and discrete oscillations*, J. Math. Anal. Appl. **142** (1989), 468–487.

[74] S. CHEN, Q. HUANG, L. H. ERBE, *Bounded and zero-convergent solutions of a class of Stieltjes integro-differential equations*, Proc. Amer. Math. Soc. **113** (1991), 999–1008.

[75] S. S. CHENG, *Hille-Wintner comparison theorems for nonlinear difference equations*, Funkcial. Ekvac. **37** (1994), 531–535.

[76] S. S. CHENG, R. F. LU, *A generalization of the discrete Hardy's inequality*, Tamkang J. Math. **24** (1993), 469–475.

[77] S. S. CHENG, W. T. PATULA, *An existence theorem for a nonlinear difference equation*, Nonlinear Anal. **20** (1993), 193–203.

[78] P. CLEMENT, L. PELETIER, *An anti-maximum principle for second order elliptic equations*, J. Differential Equations **34** (1079), 218–229.

[79] W. J. COLES, *An oscillation criterion for second order differential equations*, Proc. Amer. Math. Soc. **19** (1969), 775–779.

[80] W. A. COPPEL, Disconjugacy, Lecture Notes in Math. 220 (1971).

[81] M. CUESTA, D. DE FIGUEIREDO, J. P. GOSSEZ, *The beginning of the Fučík spectrum for the p-Laplacian*, J. Differential Equations **159** (1999), 212–238.

[82] M. CUESTA, D. DE FIGUEIREDO, J. P. GOSSEZ, *A nodal domain property for the p-Laplacian*, C. R. Acad. Sci. Paris, Sér. I **330** (2000), 669–673.

[83] M. CUESTA, P. TAKÁČ, *A strong comparison principle for the Dirichlet p-Laplacian*, Reaction diffusion systems (Trieste, 1995), Lecture Notes in Pure and Appl. Math. 194, Dekker, New York, 1998, 79–87.

[84] M. CUESTA, P. TAKÁČ, *A strong comparison principle for positive solutions of degenerate elliptic equations*, Differential Integral Equations **13** (2000), 721–746.

[85] E. N. DANCER, *On the Dirichlet for weakly non-linear elliptic partial differential equations*, Proc. Roy Soc. Edinburgh, Sect. A, **89** (1976), 283–300.

[86] K. DEIMLING, Nonlinear Functional Analysis, Springer-Verlag, Berlin/Heidelberg/New York, 1985.

[87] M. DEL PINO, P. DRÁBEK, R. MANASEVICH, *The Fredholm alternative at the first eigenvalue for the one-dimensional p-Laplacian*, C. R. Acad. Sci. Paris, Ser. I, **227** (1998), 461–465.

[88] M. DEL PINO, P. DRÁBEK, R. MANASEVICH, *The Fredholm alternative at the first eigenvalue for the one dimensional p-Laplacian*, J. Differential Equations **151** (1999), 386–419.

[89] M. DEL PINO, R. MANASEVICH, *Oscillation and nonoscillation for* $(|u'|^{p-2}u')' + a(t)|u|^{p-2}u = 0$*,* $p > 1$, Houston J. Math. **14** (1988), 173–177.

[90] M. DEL PINO, M. ELGUETA, R. MANASEVICH, *Generalizing Hartman's oscillation result for* $(|x'|^{p-2}x')' + c(t)|x|^{p-2}x = 0$*,* $p > 1$, Houston J. Math. **17** (1991), 63–70.

[91] M. DEL PINO, M. ELGUETA, R. MANASEVICH, *A homotopic deformation along p of a Leray-Schauder degree result and existence for* $(|u'|^{p-2}u')' + f(t,u) = 0$*,* $u(0) = 0 = u(T)$*,* $p > 1$, J. Differential Equations **80** (1989), 1–13.

[92] M. DEL PINO, R. MANASEVICH, *Multiple solutions for the p-Laplacian under global nonresonance*, Proc. Amer. Math. Soc. **112** (1991), 131–138.

[93] J. I. DÍAZ, Nonlinear Partial Differential Equations and Free Boundaries. Vol I. Elliptic Equations, Pitman, London 1985.

[94] J. I. DÍAZ, J. E. SAA, *Existence et unicite de solutions positives pour certaines équations elliptiques quasiliéaries*, C. R. Acad. Sci. Paris, Ser. I, **305** (1987), 521–524.

[95] Z. DOŠLÁ, O. DOŠLÝ, *Quadratic functionals with general boundary conditions*, Appl. Math. Optim. **36** (1997), 243–262.

[96] Z. DOŠLÁ, I. VRKOČ, *On an extension of the Fubini theorem and its application in ODEs*, Nonlin. Anal. **57** (2004), 531–548.

[97] O. DOŠLÝ, *The multiplicity criteria for zero points of second order differential equations*, Math. Slovaca **42** (1992), 183–192.

[98] O. DOŠLÝ, *Conjugacy criteria for second order differential equations*, Rocky Mountain J. Math. **23** (1993), 849–861.

[99] O. DOŠLÝ, *Oscillation criteria for higher order Sturm-Liouville difference equations*, J. Difference Equ. Appl. **4** (1998), 425–450.

[100] O. DOŠLÝ, *Oscillation criteria for half-linear second order differential equations*, Hiroshima Math. J. **28** (1998), 507–521.

[101] O. DOŠLÝ, *A remark on conjugacy of half-linear second order differential equations*, Math. Slovaca **50** (2000), 67–79.

[102] O. DOŠLÝ, *Methods of oscillation theory of half-linear second order differential equations*, Czechoslovak Math. J. **50 (125)** (2000), 657–671.

[103] O. DOŠLÝ, *Half-linear oscillation theory*, Studies Univ. Žilina Math.-Phys. Ser. **13** (2001), 65–73.

[104] O. DOŠLÝ, *The Picone identity for a class of partial differential equations*, Math. Bohem. **127** (2002), 581–589.

[105] O. DOŠLÝ, *Qualitative theory of half-linear second order differential equations*, Math. Bohem. **127** (2002), 181–195.

[106] O. DOŠLÝ, *Half-linear second order differential equations: oscillation theory and principal solution*, Studies Univ. Žilina Math.-Phys. Ser. **17** (2003), 69-78.

[107] O. DOŠLÝ, Á. ELBERT, *Conjugacy of half-linear second-order differential equations*, Proc. Roy. Soc. Edinburgh **130A** (2000), 517–525.

[108] O. DOŠLÝ, Á. ELBERT, *Integral characterization of principal solution of half-linear differential equations*, Studia Sci. Math. Hungar. **36** (2000), 455–469.

[109] O. DOŠLÝ, J. R. GRAEF, J. JAROŠ, *Forced oscillation of second order linear and half-linear difference equations*, Proc. Amer. Math. Soc. **131** (2003), 2859–2867.

[110] O. DOŠLÝ, J. JAROŠ, *Singular version of Leighton's comparison theorem for forced quasilinear second order differential equations*, Arch. Math. (Brno) **39** (2003), 335–349.

[111] O. DOŠLÝ, A. LOMTATIDZE, *Disconjugacy and disfocality criteria for singular half-linear second order differential equations*, Ann. Polon. Math. **72** (1999), 273–284.

[112] O. DOŠLÝ, A. LOMTATIDZE *Oscillation and nonoscillation criteria for half-linear second order differential equations*, submitted.

[113] O. DOŠLÝ, R. MAŘÍK, *Nonexistence of positive solutions of PDE's with p-Laplacian*, Acta Math. Hungar. **90** (2001), 89–107,

[114] O. DOŠLÝ, S. PEÑA, *A linearization method in oscillation theory of half-linear second order differential equations*, to appear in J. Inequal. Appl.

[115] O. DOŠLÝ, P. ŘEHÁK, *Nonoscillation criteria for half-linear second-order difference equations*, Comput. Math. Appl. **42** (2001), 453–464.

[116] O. DOŠLÝ, P. ŘEHÁK, *Recessive solution of half-linear second order difference equations*, J. Difference Equ. Appl. **9** (2003), 49–61.

[117] O. DOŠLÝ, J. ŘEZNÍČKOVÁ, *Regular half-linear second order differential equations*, Arch. Math. (Brno) **39** (2003), 233–245.

[118] P. DRÁBEK, *Fredholm alternative for the p-Laplacian: yes or no?*, Mustonen, Vesa (ed.) et al, Function spaces, differential operators and nonlinear analysis: Proceedings of the conference FSDONA-99, Syöte, Finland, June 10-16, 1999, Prague, Mathematical Institute of the Academy of Sciences of the Czech Republic, 2000, 57–64,.

[119] P. DRÁBEK, Solvability and Bifurcations of Nonlinear Equations, Research Notes in Mathematics 264, Longman, Harlow, 1992.

[120] P. DRÁBEK, *Nonlinear eigenvalue problem for p-Laplacian in* $\mathbb{R}^N$, Math. Nachr. **173** (1995), 131–139.

[121] P. DRÁBEK, *Geometry of energy functional and the Fredholm alternative for the p-Laplacian*, Electron. J. Differential Equations, Conference 08, 2002, 103–120.

[122] P. DRÁBEK, *On Fredholm alternative for certain quasilinear boundary value problems*, Math. Bohem. **127** (2002), 197–202.

[123] P. DRÁBEK, *Landesman-Lazer type like results for the p-Laplacian*, Function spaces, differential operators and nonlinear analysis (Teistungen, 2001), Birkhäuser, Basel, 2003, 289–295.

[124] P. DRÁBEK, P. GIRG, R. MANASEVICH, *Generic Fredholm results for the one-dimensional p-Laplacian*, Nonlinear Differential Equations Appl. **8** (2001), 285–298.

[125] P. DRÁBEK, P. GIRG, P. TAKÁČ, *Bounded perturbations of homogeneous quasilinear operators using bifurcations from infinity*, J. Differential Equations **204** (2004), 265–291.

[126] P. DRÁBEK, P. GIRG, P. TAKÁČ, M. ULM, *The Fredholm alternative for p-Laplacian: bifurcation from infinity, existence and multiplicity*, Indiana Univ. Math. J. **53** (2004), 433–482.

[127] P. DRÁBEK, G. HOLUBOVÁ, *Fredholm alternative for the p-Laplacian in higher dimension*, J. Math. Anal. Appl. **263** (2001), 182–194.

[128] P. DRÁBEK, P. KREJČÍ, P. TAKÁČ, eds., Nonlinear Differential Equations, Chapman & Hall/CRC Research Notes in Math. Series, Vol **404**, CRC Press LLC. Boca Raton, FL, U.S.A., 1999.

[129] P. Drábek, R. Manasevich, *On the closed solution to some nonhomogeneous eigenvalue problems with p-Laplacian*, Differential Integral Equations **12** (1999), 773–788.

[130] P. Drábek, M. Otani, *Global bifurcation result for the p-biharmonic operator*, Electron. J. Differential Equations Vol **2001** (2001), 1–19.

[131] P. Drábek, S. B. Robinson, *Resonance problems for the one-dimensional p-Laplacian*, Proc. Amer. Math. Soc. **128** (1999), 755–765.

[132] P. Drábek, S. B. Robinson, *On the generalization of the Courant nodal domain theorem*, J. Differential Equations **181** (2002), 58–71.

[133] P. Drábek, S. B. Robinson, *Resonance problem for p-Laplacian*, J. Funct. Anal. **169** (1999), 189–200.

[134] P. Drábek, P. Takáč, *A counterexample to the Fredholm alternative for the p-Laplacian*, Proc. Amer. Math. Soc. **127** (1999), 1079–1087.

[135] D. R. Dunninger, *A Sturm comparison theorem for some degenerate quasilinear elliptic operators*, Boll. U. M. I. **9-A** (1995), 117–121.

[136] A. Dzurnak, A. B. Mingarelli, *Sturm-Liouville equations with Besicovitch almost-periodicity*, Proc. Amer. Math. Soc. **106** (1989), 647–653.

[137] W. Eberhard, Á. Elbert, *On the eigenvalues of half-linear boundary value problems*, Math. Nachr. **183** (1997), 55–72.

[138] S. Elaydi, An Introduction to Difference Equations, Academic Press, 2nd edition, San Diego, 1999.

[139] Á. Elbert, *A half-linear second order differential equation*, Colloq. Math. Soc. János Bolyai **30** (1979), 158–180.

[140] Á. Elbert, *The Wronskian and the half-linear differential equations*, Studia Sci. Math. Hungar. **15** (1980), 101–105.

[141] Á. Elbert, *Oscillation and nonoscillation theorems for some non-linear ordinary differential equations*, Lectures Notes in Math. No. 964 (1982), 187–212.

[142] Á. Elbert, *Asymptotic behaviour of autonomous half-linear differential systems on the plane*, Studia Sci. Math. Hungar. **19** (1984), 447–464.

[143] Á. Elbert, *On the half-linear second order differential equations*, Acta Math. Hungar. **49** (1987), 487–508.

[144] Á. Elbert, T. Kusano, *Oscillation and non-oscillation theorems for a class of second order quasilinear differential equations*, Acta Math. Hungar. **56** (1990), 325–336.

[145] Á. Elbert, T. Kusano, *Principal solutions of nonoscillatory half-linear differential equations*, Adv. Math. Sci. Appl. **18** (1998), 745–759.

[146] Á. Elbert, T. Kusano, M. Naito, *On the number of zeros of nonoscillatory solutions to second order half-linear differential equations*, Ann. Univ. Sci. Budapest **42** (1999), 101–131.

[147] Á. Elbert, T. Kusano, M. Naito, *Counting the number of zeros of nonoscillatory solutions of second order half-linear differential equations*, P. Marušiak (ed) et al., Proceedings of the international scientific conference of mathematics, Žilina, Slovakia, June 30-July 3, 1998, Žilina University Publisher (1999), 133–140.

[148] Á. Elbert, T. Kusano, T. Tanigawa, *An oscillatory half-linear differential equation*, Arch. Math. (Brno) **33** (1997), 355–361.

[149] Á. Elbert, A. Schneider, *Perturbations of the half-linear Euler differential equation*, Result. Math. **37** (2000), 56–83.

[150] U. Elias, A. Pinkus, *Nonlinear eigenvalue problem for a class of ordinary differential equations*, Proc. Roy. Soc. Edinburgh **132A** (2002), 1333–1359.

[151] R. Emden, Gaskugeln, Anwendungen der mechanischen Warmentheorie auf Kosmologie und metheorologische Probleme, Leipzig, 1907.

[152] L. H. Erbe, *Oscillation theorems for second order linear differential equations*, Pacific J. Math. **35** (1970), 337–343.

[153] L. H. Erbe, *Oscillation criteria for second order nonlinear delay equations*, Canad. Math. Bull. **16** (1973), 49–56.

[154] L. H. Erbe, S. Hilger, *Sturmian theory on measure chains*, Differential Equations Dynam. Systems **1** (1993), 223–244.

[155] L. H. Erbe, Z. Liang, *Qualitative behavior of a generalized Emden-Fowler differential system*, Czechoslovak Math. J. **41** (1991), 454–466.

[156] L. H. Erbe, P. Yan, *Weighted averaging techniques in oscillation theory for second order difference equations*, Canad. Math. Bull. **35** (1992), 61–69.

[157] V. M. Evtuchov, *On conditions for nonoscillation of a nonlinear second order differential equation*, Mat. Zametki **67** (2000), 201–210.

[158] Ch. Fabry, R. Manasevich, *Equations with p-Laplacian and an asymmetric nonlinear term*, Discr. Cont. Dynam. Syst. **7** (2001), 545–547.

[159] X. Fan, W. T. Li, C. Zhong, *A Classification scheme for positive solutions of second order iterative differential equations*, Electron. J. Diff. Equations. **25** (2000), 1–14.

[160] X. Fan, Q. Zhang, D. Zhao, *Eigenvalues of $p(x)$-Laplacian Dirichlet problem*, J. Math. Anal. Appl. **302** (2005), 306-317.

[161] F. FIEDLER, *Oscillation criteria of Nehari-type for Sturm-Liouville operators and elliptic differential operators of the second order and the lower spectrum*, Proc. Roy. Soc. Edinburgh **109A** (1988), 127-144.

[162] J. FLECKINGER, J. P. GOSSEZ, P. TAKÁČ, *Existence, nonnexistence et principe de l'antimaximum pour le p-laplacian*, C. R. Acad. Sci. Paris, Sér. I **321** (1995), 731–734.

[163] J. FLECKINGER-PELLÉ, J. P. GOSSEZ, F. DE THELIN, *Antimaximum principle in* $\mathbb{R}^N$*: local versus global*, J. Differential Equations **196** (2004), 119–133.

[164] J. FLECKINGER, J. HERNÁNDEZ, P. TAKÁČ, F. DETHÉLIN, *Uniqueness and positivity for solutions of equations with p-Laplacian*, Proc. Conference on Reaction-Diffusion Equations, Trieste, Italy, October 1995, Marcel Dekker, Inc. New York-Basel, 1998, 141–151.

[165] J. FLECKINGER-PELLÉ, P. TAKÁČ, *An improved Poincaré inequality and the p-Laplacian at resonance for* $p > 2$, Adv. Differential Equations **7** (2002), 951–971.

[166] R. H. FOWLER, *The solutions of Emden's and similar differential equations*, Monthly Notices Roy. Astronom. Soc. **91** (1930), 63–91.

[167] S. FUČÍK, *Boundary value problems with jumping nonlinearities*, Čas. Pěst. Mat. **101** (1976), 69–87.

[168] S. FUČÍK, J. NEČAS, J. SOUČEK, V. SOUČEK, Spectral Analysis of Nonlinerar Operators, Lectures Notes in Mathematics No. 346, Springer Verlag, New York-Berlin-Heidelberg, 1973.

[169] T. GODOY, J. P. GOSSEZ, S. PACZKA, *On the antimaximum principle for the p-Laplacian with indefinite weight*, Nonlin. Anal. **51** (2002), 449–467.

[170] M. GUEDDA, L. VERON, *Bifurcation phenomena associated to the p-Laplace operator*, Trans. Amer. Math. Soc. **310** (1988), 419–431.

[171] J. K. HALE, Ordinary Differential Equations, John Wiley, New York 1969.

[172] S. G. HALVORSEN, A. B. MINGARELLI, *On the oscillation of almost-periodic Sturm-Liouville operators with an arbitrary coupling constant*, Proc. Amer. Math. Soc. **97** (1986), 269–272.

[173] G. H. HARDY, J. E. LITTLEWOOD, G. PÓLYA, Inequalitites, Cambridge, 1934.

[174] P. HARTMAN, Ordinary Differential Equations, John Wiley & Sons, Inc., New York-London-Sydney, 1964.

[175] P. HARTMAN, *Difference equations: disconjugacy, principal solutions, Green's function, complete monotonicity*, Trans. Amer. Math. Soc. **246** (1978), 1–30.

[176] L. HATVANI, *On stability properties of solutions of second order differential equations*, Proc. Colloq. Qual. Theory. Differ. Equ., Electron. J. Qual. Theory Differ. Equ., Szeged **2000**, Supl. 6 p.

[177] L. HATVANI, *On the existence of small solutions to linear second order differential equations with step function coefficients*, Dynam. Contin. Discrete Impuls. Systems **4** (1998), 321–330.

[178] S. HILGER, *Analysis on measure chains – a unified approach to continuous and discrete calculus*, Results Math. **18** (1990), 18–56.

[179] H. L. HONG, W. CH. LIAN, C. C. YEH, *The oscillation of half-linear differential equations with oscillatory coefficient*, Math. Comput. Modelling **24** (1996), 77–86.

[180] H. HOSHINO, R. IMABAYASHI, T. KUSANO, T. TANIGAWA, *On second order half-linear oscillations*, Adv. Math. Sci. Appl. **8** (1998), 199–216.

[181] H. C. HOWARD, V. MARIĆ, *Regularity and nonoscillation of solutions of second order linear differential equations*, Bull. Cl. Sci. Math. Nat. Sci. Math. **22** (1997), 85–98.

[182] H. B. HSU, C. C. YEH, *Oscillation theorems for second-order half-linear differential equations*, Appl. Math. Lett. **9** (1996), 71–77.

[183] Y. X. HUANG, *Eigenvalues of the p-Laplacian with indefinite weight*, Comment. Math. Univ. Carolinae **36** (1995), 519–527.

[184] J. JAROŠ, *On a curious doubly singular equation of the second order*, talk presented at the International Scientific Conference on Mathematics, Herl' any, Slovak Republic, October 21-23, 1999.

[185] J. JAROŠ, T. KUSANO, *A Picone type identity for half-linear differential equations*, Acta Math. Univ. Comenianea **68** (1999), 137–151.

[186] J. JAROŠ, T. KUSANO, *On a class of doubly singular differential equations of second order*, Fukuoka Univ. Sci. Rep. **29** (1999), 7–12.

[187] J. JAROŠ, T. KUSANO, *On black hole solutions of second order differential equations with a singular nonlinearity in the differential operator*, Funkcial. Ekvac. **43** (2000), 491–509.

[188] J. JAROŠ, T. KUSANO, *On white hole solutions of a class of nonlinear ordinary differential equations of the second order*, Funkcial. Ekvac. **45** (2002), 319–339.

[189] J. JAROŠ, T. KUSANO, T. TANIGAWA, *Existence of singular solutions for a class of systems of singular differential equations* Fukuoka Univ. Sci. Rep. **30** (2000), 169–177.

[190] J. Jaroš, T. Kusano, T. Tanigawa, *Nonoscillation theory for second order half-linear differential equations in the framework of regular variation*, Result. Math. **43** (2003), 129–149.

[191] J. Jaroš, T. Kusano, N. Yoshida, *A Picone-type identity and Sturmian comparison and oscillation theorems for a class of half-linear partial differential equations of the second order*, Nonlin. Anal. TMA **40** (2000), 381–395.

[192] J. Jaroš, T. Kusano, N. Yoshida, *Forced superlinear oscillations via Picone's identity*, Acta Math. Univ. Comenianae **69** (2000), 107–113.

[193] J. Jaroš, T. Kusano, N. Yoshida, *Generalized Picone's formula and forced oscillations in quasilinear differential equations of the second order*, Arch. Math. (Brno) **38** (2002), 53–59.

[194] J. Jiang, *Oscillation and nonoscillation for second order quasilinear differential equations*, Math. Sci. Res. **4** (2000), 39–47.

[195] R. Johnson, J. Moser, *The rotation number for almost periodic potentials*, Comm. Math. Phys. **84** (1982), 403–438.

[196] I. V. Kamenev, *An integral criterion for oscillation of linear differential equations of second order*, (Russian) Mat. Zametki **23** (1978), 249–251.

[197] N. Kandelaki, A. Lomtatidze, D. Ugulava, *On the oscillation and nonoscillation of a second order half-linear differential equation*, Georgian Math. J. **7** (2000), 329–346.

[198] J. Karsai, *On the existence of a solution tending to zero of nonlinear differential equations (a nonlinear extension of a theorem of Milloux, Prodi, Trevisian and Hartman)*, Dynam. Systems Appl. **6** (1997), 429–440.

[199] W. G. Kelley, A. Peterson: Difference Equations: An Introduction with Applications, Acad. Press, San Diego, 1991.

[200] I. Kiguradze, *Asymptotic properties of solutions of a nonlinear differential equation of Emden-Fowler type,* (Russian), Izvestiya Akad. Nauk SSSR. Ser. Mat. **29** (1965), 965–986.

[201] I. Kiguradze, Some singular boundary value problems for ordinary differential equations, (Russian) Tbilisi University Press, Tbilisi, 1975.

[202] I. Kiguradze, T. A. Chanturia, Asymptotic Properties of Solutions of Nonautonomous Ordinary Differential Equations, Kluwer Acad. Publishers, Dordrecht, 1993.

[203] I. Kiguradze, G. Kvinikadze, *On strongly increasing solutions of nonlinear ordinary differential equations*, Ann. Mat. Pura Appl. **130** (1982), 67–87.

[204] I. T. KIGURADZE, B. L. SHEKHTER, *Singular boundary value problems for second-order ordinary differential equations*, (Russian) Translated in J. Soviet Math. **43** (1988), Itogi Nauki i Tekhniki, Current problems in mathematics. Newest results, Vol. 30 (Russian), 105–201, 204, Akad. Nauk SSSR, Vsesoyuz. Inst. Nauchn. i Tekhn. Inform., Moscow, 1987, 2340–2417.

[205] J. KIOMURA, T. KUSANO, M. NAITO, *Positive solutions of second order quasilinear differential equations with general nonlinearities*, Studia Sci. Math. Hungar. **35** (1999), 39–51.

[206] M. KITANO, T. KUSANO, *On a class of second order quasilinear ordinary differential equations*, Hiroshima Math. J. **25** (1995), 321–355.

[207] Q. KONG, *Oscillation criteria for second order half-linear differential equations*, Ruan, Shigui (ed) et al., Differential equations with applications in biology. Proceedings of the international conference, Halifax, Canada, July 25-29, 1997, Providence, RI: Amer. Math. Soc., Fiedlds Inst. Commun. **21** (1999), 317–323.

[208] Q. KONG, A. ZETTL, *Interval oscillation conditions for difference equations*, SIAM J. Math. Anal. **26** (1995), 1047–1060.

[209] A. KRATOCHVÍL, J. NEČAS, *On the discreteness of the spectrum on the fourth-order nonlinear Sturm-Liouville problem* (in Russian), Comment. Math. Univ. Carolinae **12** (1971), 639-653.

[210] A. KUFNER, B. OPIC, Hardy-type Inequalities, Pitman Research Notes in Mathematics Series, 219, Longman Scientific & Technical, Harlow, 1990.

[211] A. KUFNER, L. E. PERSSON, Weighted Inequalities of Hardy Type, World Scientific Publishing Co., Inc., River Edge, NJ, 2003.

[212] T. KUSANO, M. KITANO, *On a class of second order quasilinear ordinary differential equations*, Hiroshima Math. J. **25** (1995), 321–355.

[213] T. KUSANO, B. S. LALLI, *On oscillation of half-linear functional differential equations with deviating arguments*, Hiroshima Math. J. **24** (1994), 549–563.

[214] T. KUSANO, V. MARIĆ, *Nonoscillatory linear and half-linear differential equations having regularly varying solutions*, Adv. Math. Sci. Appl. **14** (2004), 351–5357.

[215] T. KUSANO, V. MARIĆ, T. TANIGAWA, *Asymptotics of some classes of nonoscillatory solutions of second-order half-linear differential equations*, Bull. Cl. Sci. Math. Nat. Sci. Math. **28** (2003), 61–74.

[216] T. KUSANO, M. NAITO, *Sturm-Liouville eigenvalue problems for half-linear ordinary differential equations*, Rocky Mountain J. Math. **31** (2001), 1039–1054.

[217] T. KUSANO, M. NAITO, *On the number of zeros of nonoscillatory solutions of half-linear ordinary differential equations involving a parameter*, Trans. Amer. Math. Soc. **324** (2002), 4751–4767.

[218] T. KUSANO, Y. NAITO, *Oscillation and nonoscillation criteria for second order quasilinear differential equations*, Acta. Math. Hungar. **76** (1997), 81–99.

[219] T. KUSANO, Y. NAITO, A. OGATA, *Strong oscillation and nonoscillation of quasilinear differential equations of second order*, Differential Equations Dynam. Systems **2** (1994), 1–10.

[220] T. KUSANO, M. NAITO, T. TANIGAWA, *Second order half-linear eigenvalue problem*, Fukuoka Univ. Sci. Rep. **27** (2999), 1–7.

[221] T. KUSANO, M. NAITO, F. WU, *On the oscillation of solutions of 4-dimensional Emden-Fowler differential systems*, Adv. Math Sci. Appl. **11** (2001), 685–719.

[222] T. KUSANO, A. OGATA, *Existence and asymptotic behavior of positive solutions of second order quasilinear differential equations*, Funkcial. Ekvac. **37** (1994), 345–361.

[223] T. KUSANO, A. OGATA, *On oscillation of functional differential equations with forcing term*, Fukuoka Univ. Sci. Rep. **29** (1999), 225–229.

[224] T. KUSANO, A. OGATA, H. USAMI, *Oscillation theory for a class of second order quasilinear ordinary differential equations with application to partial differential equations*, Japan. J. Math. **19** (1993), 131–147.

[225] T. KUSANO, T. TANIGAWA *Positive solutions to a class of second order differential equations with singular nonlinearities*, Appl. Anal. **69** (1998), 315–331.

[226] T. KUSANO, T. TANIGAWA, *Positive decreasing solutions of systems of second order singular differential equations*, J. Inequal. Appl. **5** (2000), 581–602.

[227] T. KUSANO, N. YOSHIDA, *Nonoscillation theorems for a class of quasilinear differential equations of second order*, J. Math. Anal. Appl. **189** (1995), 115–127.

[228] T. KUSANO, J. WANG, *Oscillation properties of half-linear functional differential equations of the second order*, Hiroshima Math. J. **25** (1995), 371–385.

[229] G. KVINIKADZE, *On strongly increasing solutions of system of nonlinear ordinary differential equations*, Trudy Inst. Prikl. Mat. I. Vekua **37** (1990), 41–49.

[230] M. K. KWONG, A. ZETTL, *Integral inequalities and second order linear oscillation*, J. Differential Equations **45** (1982), 16–33.

[231] M. K. Kwong, A. Zettl, *A new approach to second order linear oscillation theory*, Ordinary differential equations and operators, Lect. Notes Math. 1032 (1983), 328–345.

[232] E. M. Landesman, A. C. Lazer, *Nonlinear perturbations of linear elliptic boundary value problems at resonance*, J. Math. Mech. **19** (1970), 609–623.

[233] W. Leighton, *The detection of oscillation of solutions of a second order linear differential equations*, Duke J. Math. **17** (1950), 57–62.

[234] W. Leighton, *Comparison theorems for linear differential equations of second order*, Proc. Amer. Math. Soc. **13** (1962), 603–610.

[235] W. Leighton, *Some oscillation theory*, Z. Angew. Math. Mech. **63** (1983), 303–315.

[236] W. Leighton, M. Morse, *Singular quadratic functionals*, Trans. Amer. Math. Soc. **40** (1936), 252–286.

[237] H. J. Li, *Oscillation criteria for second order linear differential equations*, J. Math. Anal. Appl. **194** (1995), 217–234.

[238] H. J. Li, C. C. Yeh, *Existence of positive nondecreasing solutions of nonlinear difference equations*, Nonlin. Anal. **22** (1994), 1271–1284.

[239] H. J. Li, C. C. Yeh, An integral criterion for oscillation of nonlinear differential equations, Math. Japonica **41** (1995), 185–188.

[240] H. J. Li, C. C. Yeh, *An oscillation criterion of almost-periodic Sturm-Liouville equations*, Rocky Mountain J. Math. **25** (1995), 1417–1429.

[241] H. J. Li, C. C. Yeh, *Inequalities for a function involving its integral and derivative*, Proc. Royal Soc. Edinburgh **125A** (1995), 133–151.

[242] H. J. Li, C. C. Yeh, *Nonoscillation criteria for second-order half-linear differential equations*, Appl. Math. Lett. **8** (1995), 63–70.

[243] H. J. Li, C. C. Yeh, *Oscillation criteria for half-linear second order differential equations*, Hiroshima Math. J. **25** (1995), 571–583.

[244] H. J. Li, C. C. Yeh, *Oscillations of half-linear second order differential equations*, Hiroshima Math. J. **25** (1995), 585–594.

[245] H. J. Li, C. C. Yeh, *Sturmian comparison theorem for half-linear second order differential equations*, Proc. Roy. Soc. Edinburgh **125A** (1995), 1193–1204.

[246] H. J. Li, C. C. Yeh, *Nonoscillation of half-linear differential equations*, Publ. Math. (Debrecen) **49** (1996), 327–334.

[247] W. T. Li, *Positive solutions of second order nonlinear differential equations*, J. Math. Anal. Appl. **221** (1998), 326–337.

[248] W. T. Li, *Classification schemes for positive solutions of nonlinear differential systems*, Math. Comput. Modelling **36** (2002) 411–418.

[249] W. T. Li, *Interval oscillation criteria for second order damped half-linear differential equations with forcing term*, Math. Inequal. Appl. **6** (2003), 487–495.

[250] W. T. Li, *Interval oscillation criteria for second-order half-linear functional differential equations*, Appl. Math. Comput. **155** (2004), 451–468.

[251] W. T. Li, S. S. Cheng, *Limiting behavior of nonoscillatory solutions of a pair of coupled nonlinear differential equations*, Proc. Edingburgh Math. Soc., **43**, (2000), 457–473.

[252] W. T. Li, S. S. Cheng, *An oscillation criterion for nonhomogeneous half-linear differential equations*, Appl. Math. Lett. **15** (2002), 259–263.

[253] W. T. Li, X. L. Fan, *Monotone solutions of second-order nonlinear differential equations*, Appl. Math. Lett. **13** (2000), 65–70.

[254] W. T. Li, C. Zhong, X. L. Fan, *Oscillation criteria for second-order half-linear ordinary differential equations with damping*, Rocky Mountain J. Math. **33** (2003), 927–951.

[255] W. C. Lian, C. C. Yeh, H. J. Li, *The distance between zeros of an oscillatory solution to a half-linear differential equation*, Comput. Math. Appl. **29** (1995), 39–43.

[256] G. Lieberman, *Boundary regularity for solutions of degenerate elliptic equations*, Nonlin. Anal. **12** (1988), 1203–1219.

[257] P. Lindqvist, *Note on a nonlinear eigenvalue problem*, Rocky Mountain J. Math. **23** (1993), 281–288.

[258] P. Lindqvist, *On the equation* $\mathrm{div}(|\nabla u|^{p-2}\nabla u)+\lambda|u|^{p-2}u=0$, Proc. Amer. Math. Soc. **109** (1990), 157–164.

[259] P. Lindqvist, J. Peetre, *Two remarkable identities, called twos, for inverses to some Abelian integrals*, Amer. Math. Monthly **108** (2001), 403–410.

[260] P. Lindqvist, *Some remarkable sine and cosine functions*, Ric. Mat. **44** (1995), 269–290.

[261] B. Liu, S. S. Cheng, *Positive solutions of second order nonlinear difference equations*, J. Math. Anal. Appl. **204** (1996), 482–493.

[262] A. Lomtatidze, *Oscillation and nonoscillation of Emden-Fowler type equation of second order*, Arch. Math. (Brno) **32** (1996), 181–193.

[263] J. Luo, L. Debnath, *Oscillation criteria for second order functional differential equations*, Comput. Math. Appl. **44** (2002), 731–739.

[264] W. Magnus, S. Winkler, Hille's Equations, John Wiley, New York, 1966.

[265] J. V. Manojlović, *Oscillation criteria for second-order half-linear differential equations*, Math. Comput. Modelling **30** (1999), 109–119.

[266] R. Manásevich, J. Mawhin, *The spectrum of p-Laplacian systems with various boundary conditions*, Adv. Diff. Equ. **5** (2000), 1289–1318.

[267] R. Manásevich, S. Sedziwy, *p-Laplacian and Lienard-type equation*, Rocky Mountain J. Math. **27** (1997), 611–617.

[268] R. Manásevich, P. Takáč, *On the Fredholm alternative for the p-Laplacian*, Proc. London Math. Soc., III Ser. **84** (2002), 324–342.

[269] V. Marić, Regular Variation and Differential Equations, Lecture Notes in Math. 1726, Springer-Verlag, Berlin 2000.

[270] L. Markus, R. A. Moore, *Oscillation and disconjugacy for linear differential equations with almost periodic coefficients*, Acta Math. **96** (1956), 99–123.

[271] R. Mařík, *On p-degree functionals with one free end point*, Proc. Int. Sci. Conf. Math. Žilina 1998, 183–188.

[272] R. Mařík, *Nonnegativity of functionals corresponding to the second order half-linear differential equation*, Arch. Math. (Brno) **35** (1999), 155–164.

[273] R. Mařík, *Comparison theorems for half-linear second order difference equations*, Arch. Math (Brno) **36** (2000), Proceedings of CDDE 2000 issue, 513–518.

[274] R. Mařík, *Focal points of half-linear differential equations*, Differential Equations Dynam. Systems **8** (2000), 111–124.

[275] R. Mařík, *Oscillation criteria for PDE with p-Laplacian via the Riccati technique*, J. Math. Anal. Appl. **248** (2000), 290–308.

[276] R. Mařík, *A half-linear differential equation corresponding to variational problem*, Nonlin. Anal. **45A** (2001), 203–211.

[277] R. Mařík, *Positive solutions of inequality with p-Laplacian in exterior domains*, Math. Bohem. **127** (2002), 597–604.

[278] R. Mařík, *Integral averages and oscillation criteria for half-linear partial differential equation*, Appl. Math. Comput. **150** (2004), 69–87.

[279] R. Mařík, *Riccati-type inequality and oscillation criteria for a half-linear PDE with damping*, Electron. J. Differential Equations 2004, No. 11, 17 pp. (electronic).

[280] R. Mařík, *Discrete singular functionals*, to appear in Arch. Math. (Brno).

[281] M. Marini, *On nonoscillatory solutions of a second order nonlinear differential equation*, Boll. U.M.I., Ser. VI **3-C** (1984), 189–202.

[282] M. Marini, S. Matucci, P. Řehák, *On decay of coupled nonlinear differential systems*, Adv. Math. Sci. Appl. **12** (2002), 521–533.

[283] M. Marini, S. Matucci, P. Řehák, *Zero convergent solutions for a class of p-laplacian systems*, In Proceedings of Equadiff 10 (CD ROM), Praha: Masaryk University Publishing House, 2002, 279–287.

[284] M. Marini, S. Matucci, P. Řehák, *Oscillation of coupled nonlinear discrete systems*, J. Math. Anal. Appl. **295** (2004), 459–472.

[285] M. Marini, P. Zezza, *On the asymptotic behavior of the solutions of second order linear differential equations*, J. Differential Equations **28** (1978), 1–17.

[286] S. Matucci, *On asymptotic decaying solutions for a class of second order differential equations,* Arch. Math. (Brno) **35** (1999), 275–284.

[287] C. Micheletti, A. Pistoia, *On the Fučík spectrum for the p-Laplacian*, Differential Integral Equations, **14** (2001), 867–882.

[288] H. Milloux, *Sur l'équation différentielle* $x'' + A(t)x = 0$, Prace Mat. **41** (1934), 39–53.

[289] A. B. Mingarelli, S. G. Halvorsen, Non-oscillation domains of differential equations with two parameters, Lecture Notes in Math. 1338 (1988).

[290] J. D. Mirzov, *On some analogs of Sturm's and Kneser's theorems for nonlinear systems*, J. Math. Anal. Appl. **53** (1976), 418–425.

[291] J. D. Mirzov, *Principal and nonprincipal solutions of a nonoscillatory system*, Tbiliss. Gos. Univ. Inst. Prikl. Mat. Trudy **31** (1988), 100–117.

[292] J. D. Mirzov, Asymptotic Properties of Nonlinear Systems of Nonautonomous Ordinary Differential Equations, Majkop, 1993 (in Russian).

[293] J. D. Mirzov, Asymptotic Properties of Solutions of Systems of Nonlinear Nonautonomous Ordinary Differential Equations, Folia Facultatis Scientiarium Naturalium Universitatis Masarykianae Brunensis. Mathematica, 14. Masaryk University, Brno, 2004.

[294] R. A. Moore, *The least eigenvalue of Hill's equation*, J. Anal. Math. **5** (1956-7), 183–196.

[295] J. Moser, Integrable Hamiltonian Systems and Spectral Theory, Leziomi Fermiane, Accademia Nazionale dei Lincei, Roma, 1983.

[296] W. F. Moss, J. Piepenbrick, *Positive solutions of elliptic equations*, Pacific J. Math. **75** (1978), 219–226.

[297] E. Müller-Pfeiffer, *On the existence of nodal domain for elliptic differential operators*, Proc. Roy. Soc. Edinburgh **94A** (1983), 287–299.

[298] E. Müller-Pfeiffer, *An extension of the Sturm-Picone theorem to elliptic differential equations*, Proc. Roy. Soc. Edinburgh **97A** (1984), 209–215.

[299] M. Mizukami, M. Naito, H. Usami, *Asymptotic behavior of solutions of a class of second order quasilinear ordinary differential equations*, Hiroshima Math. J. **32** (2002), 51–78.

[300] Y. Nagabuchi, M. Yamamoto, *On the monotonicity of solutions for a class of nonlinear differential equations of second order*, Math. Japonica **37** (1992), 665–673.

[301] M. Naito, Y. Naito, H. Usami, *Oscillation theory for semilinear elliptic equations with arbitrary nonlinearities*, Funkcial. Ekvac. **40** (1997), 41–55.

[302] M. Naito, F. Wu, *On the existence of eventually positive solutions of fourth-order quasilinear differential equations*, Nonlin. Anal. **57** (2004), 253–263.

[303] R. Navarro, *Focal oscillation for second order systems*, Dynam. Systems Appl. **6** (1997), 535–539.

[304] Z. Nehari, *Oscillation criteria for second-order linear differential equations*, Trans. Amer. Math. Soc. **85** (1957), 428–445.

[305] F. Neuman, Global Properties of Linear Differential Equations, Kluwer Academic Publishers, Dordrecht, 1991.

[306] J. Ohriska, *Oscillation of second order delay and ordinary differential equation*, Czech. Math. J. **34** (1984), 107–112.

[307] C. Olech, Z. Opial, T. Wazewski, *Sur le probleme d'oscillation des integrales de l'equation $y''+g(t)y=0$*, Bull. Acad. Polon. Sci. **5** (1957), 621–626.

[308] M. Otani, *A remark on certain nonlinear elliptic equations*, Proc. Fac. Sci. Tokai Univ. **19** (1984), 23–28.

[309] M. Otani, *On certain second order ordinary differential equations associated with Sobolev-Poincare-type inequalities*, Nonlin. Anal. Appl. **8** (1984), 1255–1270.

[310] S. Peña, *Conjugacy criteria for half-linear differential equations*, Arch. Math. (Brno) **35** (1999), 1–11.

[311] K. Perera, *Resonance problem with respect to the Fučík spectrum of the p-Laplacian*, Electron. J. Differential Equations **2002** (2002), No. 36, 1–10.

[312] K. Perera, *On the Fučík spectrum for p-Laplacian*, to appear.

[313] Ch. Philos, *On the existence of nonoscillatory solutions tending to zero at ∞ for differential equations with positive delays*, Arch. Math. (Basel) **36** (1981), 168–178.

[314] Ch. Philos, *Oscillation theorems for linear differential equations of second order*, Arch. Math. (Basel) **53** (1989), 493–492.

[315] M. Picone, *Sui valori eccezionalle di un parametro da cui dipende un' equatione differenzialle lineare del secondo ordine*, Ann. Scuola Norm. Sup. Pisa **11** (1910), 1–141.

[316] A. Pinkus, *n-widths of Sobolev spaces in L^p*, Constr. Approx. **1** (1985), 15–62.

[317] M. Piros, *On the solutions of a half-linear differential equation*, Studia Sci. Math. Hungar. **19** (1984), 193–211.

[318] M. Ráb, *Kriterien für oscillation der Lösungen der Differentialgleichung* $[p(x)y']' + q(x)y = 0$, Čas. Pěst. Mat. **84** (1957), 335–370.

[319] V. A. Rabtsevich, *On strongly increasing solutions of arbitrary order Emden-Fowler system*, Differ. Uravn. **34** (1998), 222–227 (in Russian).

[320] W. T. Reid, Ordinary Differential Equations, Wiley, New York, 1971.

[321] P. Řehák, *Half-linear discrete oscillation theory*, Proceedings of the 6th Coll. Qual. Theory DE (EJQTDE), Szeged 1999 (2000), 1–14.

[322] P. Řehák, *Hartman-Wintner lemma, oscillation and conjugacy criteria for half-linear difference equations*, J. Math. Anal. Appl. **252** (2000), 813–827.

[323] P. Řehák, *Oscillatory properties of second order half-linear difference equations,* Czechoslovak Math. J. **51** (2001), 303–321.

[324] P. Řehák, *Generalized discrete Riccati equation and oscillation of half-linear difference equations*, Math. Comput. Modelling **34** (2001), 257–269.

[325] P. Řehák, *Oscillation criteria for second order half-linear difference equations*, J. Difference Equ. Appl. **7** (2001), 483–505.

[326] P. Řehák, *Comparison theorems and strong oscillation in the half-linear discrete oscillation theory*, Rocky Mountain J. Math. **33** (2003), 333–352.

[327] P. Řehák, *Half-linear dynamic equations on time scales: IVP and oscillatory properties*, J. Nonlinear Funct. Anal. Appl. **7** (2002), 361–403.

[328] P. Řehák, *On certain comparison theorems for half-linear dynamic equations on time scales*, Abstr. Appl. Anal. **7** (2004), 551–565.

[329] P. Řehák, *Opial and Lyapunov type inequalities for half-linear dynamic equations*, Dynam. Systems Appl. **13** (2004), 443–456.

[330] P. Řehák, *Hardy inequality on time scales and its application to half-linear dynamic equations*, Arch. Ineq. Appl., to appear.

[331] W. Reichel, W. Walter, *Radial solutions of equations and inequalities involving the p-Laplacian*, J. Inequal. Appl. **1** (1997), 47–91.

[332] W. Reichel, W. Walter, *Sturm-Liouville type problems for the p-Laplacian under asymptotic non-resonance conditions*, J. Differential Equations **156** (1999), 50–70.

[333] J. Řezníčková, *An oscillation criterion for half-linear second order differential equations*, Miskolc Math. Notes **5** (2004), 203–212.

[334] S. P. Rogovchenko, Y. V. Rogovchenko, *Asymptotic behavior of certain second order differential equations*, Dynam. Systems Appl. **10** (2001), 185–200.

[335] K. Rostás, *Comparison theorems for pseudoconjugate points of half-linear ordinary differential equations of the second order*, Acta Math. Univ. Comenianae, **72** (2003), 223–228.

[336] G. Sansone, Equazioni diferenziali nel campo reale I, II, Zanichelli, Bologna, 1949.

[337] U. W. Schminke: *The lower spectrum of Schrödinger operators*, Arch. Rat. Mech. Anal. **75** (1981), 147-155.

[338] E. Seneta, Regularly Varying Functions, Lecture Notes in Mathematics 508, Springer-Verlag, Berlin-Heidelberg-New York, 1976.

[339] M. Struwe, Variational Methods: Applications to Nonlinear Partial Differential Equations and Hamiltonian Systems, Springer-Verlag, New York, 1990.

[340] J. Sugie, N. Yamaoka, *Growth conditions and oscillation of nonlinear differential equations with p-Laplacian*, J. Math. Anal. Appl. (2005), to appear.

[341] C. A. Swanson, Comparison and Oscillation Theory of Linear Differential Equations, Acad. Press, New York-London, 1968

[342] P. Takáč, *On the Fredholm alternative for the p-Laplacian at the first eigenvalue*, Indiana Univ. Math. J. **51** (2002), 187–237.

[343] P. Takáč, *On the number and structure of solutions for a Fredholm alternative with the p-Laplacian*, J. Differential Equations **185** (2002), 306–347.

[344] G. Talenti, *Best constant in Sobolev inequality*, Ann. Mat. Pura Appl. **110** (1976), 353–272.

[345] S. Taliaferro, *On the positive solutions of* $y'' + \varphi(t)y^{-\lambda} = 0$, Nonlinear Anal. **2** (1978), 437–446.

[346] T. Tanigawa *Asymptotic behavior of positive solutions to nonlinear singular differential equations of second order*, Studia Sci. Math. Hungar. **35** (1999), 427–444.

[347] T. Tanigawa, *Existence and asymptotic behavior of positive solutions of second order quasilinear differential equations*, Adv. Math. Sci. Appl. **9** (1999), 907–938.

[348] T. Tanigawa, *Positive decreasing solutions of systems of second order quasilinear differential equations*, Funkcial. Ekvac. **43** (2000) 361–380.

[349] L. H. Thomas, *The calculation of atomic fields*, Proc. Cambridge Phil. Soc. **23** (1927), 542–548.

[350] F. J. Tipler, *General relativity and conjugate differential equations*, J. Differential Equations **30** (1978), 165–174.

[351] P. Tolksdorf, *On the Dirichlet problem for quasilinear equations in domains with conical boundary points*, Comm. P.D.E. **8** (1983), 773–817.

[352] N. Trudinger, *On Harnack type inequalities and their application to quasilinear elliptic equations*, Comm. Pure Appl. Math. **20** (1967), 721–747.

[353] H. Usami, *On positive decaying solutions of singular Emden-Fowler-type equations*, Nonlinear Anal. **16** (1991), 795–803.

[354] Ch. de Valleé Poussin, *Sur l'équation différentielle linéqire du second order. Détermination d'une intégrale par deux valuers assignés. Extension aux équasions d'ordre n*, J. Math. Pures Appl. **8** (1929), 125–144.

[355] J. L. Vasques, *A strong maximum principle for some quasilinear elliptic equations*, Appl. Math. Optim. **12** (1884), 191–202.

[356] W. Walter, *Sturm-Liouville for the radial Δ_p-operator*, Math. Z. **227** (1998), 175–195.

[357] J. Wang, *The existence of positive solutions for the one-dimensional p-Laplacian*, Proc. Amer. Math. Soc. **125** (1997), 2275–2283.

[358] J. Wang, *Oscillation and nonoscillation theorems for a class of second order quasilinear functional differential equations*, Hiroshima Math. J. **27** (1997), 449–466.

[359] J. Wang, *On second order quasilinear oscillations*, Funkc. Ekvac. **41**, (1998), 25–54.

[360] Q. R. Wang, *Oscillation and asymptotics for second-order half-linear differential equations*, Appl. Math. Comput. **122** (2001), 253–266.

[361] Q. R. Wang, Q. G. Yang, *Interval oscillation criteria for second order half-linear differential equations*, J. Math. Anal. Appl. **291** (2004), 224–236.

[362] D. Willet, *On the oscillatory behavior of the solutions of second order differential equations*, Ann. Polon. Math. **21** (1969), 175–194.

[363] J. S. Wong, *Oscillation criteria for forced second order linear differential equation*, J. Math. Anal. Appl. **231** (1999), 235–240.

[364] P. J. Y. Wong, R. P. Agarwal, *The oscillation and asymptotically monotone solutions of second-order quasilinear differential equations*, Appl. Math. Comput. **79** (1996), 207–237.

[365] P. J. Y. Wong, R. P. Agarwal, *Oscillations and nonoscillations of half-linear difference equations generated by deviating arguments*, Comput. Math. Appl. **36** (1998), 11–26.

[366] P. J. Y. Wong, R. P. Agarwal, *Oscillation criteria for half-linear differential equations*, Adv. Math. Sci. Appl. **9**, (1999), 649–663.

[367] J. Yan, *A note on an oscillation criterion for an equation with damping term*, Proc. Amer. Math. Soc. **90** (1984), 277–280.

[368] J. Yan, *Oscillation theorems for second order linear differential equations with damping*, Proc. Amer. Math. Soc. **98** (1986), 276–282.

[369] J. Yan, *On the distance between zeros and the limit-point problem*, Proc. Amer. Math. Soc. **107** (1989), 971–975.

[370] Q. Yang, S. S. Cheng, *Oscillation of second-order half-linear differential equations with damping*, Georgian Math. J. **10** (2003), 785–797.

[371] X. Yang, *Oscillation results for second order half-linear differential equations*, Math. Comput. Modelling **36** (2002), 503–507.

[372] X. Yang, *Positive solutions for the one-dimensional p-Laplacian*, Math. Comput. Modelling **35** (2002), 129–135.

[373] X. Yang, *On inequalities of Lyapunov type*, Appl. Math. Comput. **134** (2003), 293–300.

[374] X. Yang, *Nonoscillation criteria for second-order nonlinear differential equations*, Appl. Math. Comput. **131** (2002), 125–131.

[375] X. Yang, *On the Fredholm alternative for the p-Laplacian*, Appl. Math. Comput. **153** (2004), 537–556.

[376] X. Yang, *Oscillation criteria for a class of quasilinear differential equations*, Appl. Math. Comput. **153** (2004), 225–229.

[377] C. C. Yeh, *Levin's comparison theorems for second order nonlinear differential equations and inequalities*, Math. Japon. **36** (1991), 703–710.

[378] A. Yu, A. Levin, *A comparison principle for second order differential equations*, Sov. Math. Dokl. **1** (1960), 1313–1316.

[379] V. Zeidan, P. Zezza, *Variable end points in the calculus of variations: Coupled points*, Anal. and Opt. of Syst., Lecture notes in control and information, Sci. II, Springer Verlag (1988), 372–380.

[380] V. Zeidan, P. Zezza, *Coupled points in the calculus of variations and applications to periodic solutions*, Trans. Amer. Math. Soc. **315** (1989), 323–325.

[381] M. Zhang, *Nonuniform nonresonance at the first eigenvalue of the p-Laplacian*, Nonlin. Anal. **29** (1999), 41–51.

[382] M. Zhang, *The rotation number approach to eigenvalues of the one-dimensional p-Laplacian with periodic potentials*, Proc. London Math. Soc. **64** (2001), 125–143.

Index

H

Q

R

S

W

Y

Z

Author Index